Models, Logics, and Higher-Dimensional Categories

A Tribute to the Work of Mihály Makkai

Volume 53

CRM PROCEEDINGS & LECTURE NOTES

Centre de Recherches Mathématiques
Montréal

Models, Logics, and Higher-Dimensional Categories

A Tribute to the Work of Mihály Makkai

Bradd Hart
Thomas G. Kucera
Anand Pillay
Philip J. Scott
Robert A. G. Seely
Editors

The Centre de Recherches Mathématiques (CRM) of the Université de Montréal was created in 1968 to promote research in pure and applied mathematics and related disciplines. Among its activities are special theme years, summer schools, workshops, postdoctoral programs, and publishing. The CRM is supported by the Université de Montréal, the Province of Québec (FQRNT), and the Natural Sciences and Engineering Research Council of Canada. It is affiliated with the Institut des Sciences Mathématiques (ISM) of Montréal. The CRM may be reached on the Web at www.crm.math.ca.

American Mathematical Society
Providence, Rhode Island USA

The production of this volume was supported in part by the Fonds Québécois de la Recherche sur la Nature et les Technologies (FQRNT) and the Natural Sciences and Engineering Research Council of Canada (NSERC).

2000 *Mathematics Subject Classification.* Primary 03–06, 03C45, 03C52, 03C90, 03G30, 18C10, 18D05, 18D30.

Library of Congress Cataloging-in-Publication Data

Models, logics, and higher-dimensional categories : a tribute to the work of Mihály Makkai / Bradd Hart... [et al.], editors.
p. cm. — (CRM proceedings & lecture notes ; v. 53)
Proceedings of a conference held at Centre de recherches mathématiques of the Université de Montréal, June 18–20, 2009.
Includes bibliographical references.
ISBN 978-0-8218-7281-9 (alk. paper)
1. Model theory—Congresses. 2. Categories (Mathematics)—Congresses. I. Makkai, Mihály, 1939– II. Hart, Bradd T.

QA9.7.M64 2011
512′.62—dc23

2011024611

∞ The paper used in this book is acid-free and falls within the guidelines
established to ensure permanence and durability.
This volume was submitted to the American Mathematical Society
in camera-ready form by the Centre de Recherches Mathématiques.
Visit the AMS home page at `http://www.ams.org/`

10 9 8 7 6 5 4 3 2 1 16 15 14 13 12 11

Contents

Introduction

On June 18th – 20th, 2009, a meeting was held at the Centre de recherches mathématiques (at l'Université de Montréal) as a tribute to the work of Mihály Makkai. This was supplemented by a one-day workshop on July 18th at McGill University. Such a meeting was timely for several reasons: not only the occasion of Makkai's 70th birthday year, but also because of the importance and influence his work continues to hold in a wide range of areas. For example, there is his work in infinitary logic, admissible sets and traditional model theory; his papers in the early 80's introduced a whole generation of model theorists to stability theory. The 1960s and 70s saw the development of a considerable body of work creating a remarkably successful marriage of the ideas and techniques of category theory on the one hand and logic and model theory on the other hand. Makkai was and has continued to be a leading and original contributor to this endeavour. The monograph *First Order Categorical Logic* (Makkai & Reyes, 1977) was the first to successfully combine traditional model theory and (infinitary) logic with the work of the Grothendieck school in algebraic geometry; it was a milestone in the subject and set the stage for subsequent work. In succeeding years, Makkai has built up an impressive body of work in several related fields: categorical model theory, first order logic with dependent sorts, higher dimensional category theory, and most importantly, a coherent and far reaching view of categorical logic in mathematical foundations. We look forward to a greater integration of this categorical point of view into contemporary model theory, some examples of which are already present in this volume.

The meeting focussed on the main themes in Makkai's research career: traditional model theory, categorical model theory and logics, and higher-dimensional category theory. This volume is a partial proceedings of this event, and displays the vibrant nature of current research reflecting the influence of Makkai's ideas and his work. One notable sentiment often expressed at the meeting was the affection and regard all the participants felt for Mihály. Most of us are long-time colleagues, many are former students or postdocs of his (a blend reflected in the editorship of this volume), and have long benefited from his advice and example. It is with great pleasure we offer this tribute of our respect and friendship to him.

The Editors

Mihály Makkai: A Biographical Note, with Reminiscences

Mihály Makkai was born in Budapest on June 24, 1939, the son of László Makkai, a noted Hungarian historian, and the grandson of Sándor Makkai, a well-known writer and a bishop of the Hungarian Reformed Church in Transylvania. At the conference, former student Djordje Čubrić treated us to a brief account of Mihály's very early career as a mathematician: not the first professional publications from 1964 and 1965, but rather his frequent appearances in the early 1950s as a problem-solver in the famous Hungarian high-school mathematics magazine KőMaL.

Mihály was a student of Rósza Péter at the Eőtvős Loránd University in Budapest, and was profoundly influenced by studies with Andrzej Mostowksi in Warsaw in 1964 – 1965. He obtained his Ph.D. in 1966 and was awarded the Candidate's Degree in Mathematics from the Hungarian Academy of Sciences in 1969.

A significant early event for Mihály was a year (1966 – 1967) spent as a guest lecturer at University of California, Berkeley, where he taught two terms of model theory: Ehrenfeucht – Mostowski models, saturation and homogeneity, and Morley's Categoricity Theorem in the first term, and infinitary logic in the second. Julia Knight spoke at the conference of the impact that this course had on her and many other logic students; this was also the beginning of a long friendship. Already one of Mihály's best habits as a teacher was apparent: he gave many long written assignments and returned them to his students covered with detailed comments not only on errors, but on other ideas suggested by the solutions, or on alternate solutions. This is a practice still followed by many of his students in their teaching of logic, including the author of this note. As well, it was here that Mihály first met his long-time friend and collaborator Gonzalo Reyes.

In 1971, Mihály and his family were able to leave Hungary, and at the invitation of another well-known Hungarian mathematician George Grätzer, joined the Department of Mathematics and Astronomy at the University of Manitoba. It was here that the author of this note first encountered him: I was just at the end of the second year of my undergraduate program, when the Head of the Department tracked me down and told me that "We have to find some students for this new fellow from Hungary; you will be taking his fourth-year Honours logic course next year." I was thoroughly intimidated at first, but it was only four or five weeks into the course the following autumn I knew what I was going to be doing for the rest of my life. Mihály soon moved to Montréal, and after a year as a Research Associate at the Centre de recherches mathématiques, joined the faculty at McGill University, where he has been ever since. He was appointed Peter Redpath Chair of

---

Compiled from information provided by Mihály Makkai, Djordje Čubrić, Victor Harnik, Julia Knight, and from the personal recollections of these, the author, and others.

Pure Mathematics in 2000, and received the David Thompson Award for Graduate Teaching and Supervision in 2002.

At the conference, many others also spoke of the electric effect of Mihály's teaching on their mathematical careers. Hal Kierstead, while a student of Alfred Manaster's at University of California, San Diego, first met Mihály as a visiting student at the special year in Mathematical Logic at UCLA during 1975 – 1976. Hal followed Mihály to Montréal and McGill University, and (under the joint supervision of Manaster and Makkai) was the first student to receive a Ph.D. under Mihály's direction. Mihály's teaching style at all times reveals his passion for the subject matter, and whether it is in the classroom, in discussions with students, or at work with a collaborator, his obvious joy at the beauty of mathematics is contagious.

Mihály's published works (well over 80 as of this writing) show a very wide range of interests in foundational matters. His early reputation was founded on strong contributions to infinitary logic, and this interest has remained throughout his career. His reconnection with Reyes at the Centre de recherches mathématiques in 1973 – 1974 began the long series of papers in categorical logic, topos theory, and, more recently, higher dimensional category theory. At the same time he became deeply interested in Shelah's Classification Theory, with his 1984 paper *A survey of basic stability theory* serving as a guide for young researchers for years. Several people at the conference commented on the influence of this paper. It is a prime exemplar of Mihály's mathematical writing style: clear, well-organized, and direct, and always with the "right" way of motivating and explaining difficult ideas. The foundations of the work in categorical logic were laid with the Springer Lecture notes book with Reyes, *First order categorical logic*, followed 11 years later with the AMS Monograph *Accessible categories*, with R. Paré. Mihály's intuition about the importance of analogues of Stone Duality in first order logic were expressed in several papers, culminating in *Strong conceptual completeness for first order logic*; a paper presenting an entirely new kind of completeness theorem for first-order logic which could only be expressed using the language and tools of categorical logic. This body of work makes significant contributions towards understanding what it means to "interpret one theory in another." In recent years, Mihály (with students and collaborators) has studied higher-dimensional categories. This volume contains work by some of his collaborators in this domain.

Throughout his career, in research and in teaching, Mihály's work has been motivated by a desire to understand the foundations of mathematics in the broadest sense. This volume amply illustrates this theme.

Thomas G. Kucera

Centre de Recherches Mathématiques
CRM Proceedings and Lecture Notes
Volume **53**, 2011

The Number of Countable Models in Categories of Sheaves

Nathanael Leedom Ackerman

ABSTRACT. The goal of this paper is to extend Morley's results in [10] to categories of sheaves. We show that in this context there are four natural notions of "a countable model." We then show $\mathbf{\Pi}^1_3$ determinacy implies that for any of these four notions and for any sentence $T \in \mathcal{L}_{\omega_1,\omega}(L)$ either there is a perfect set of nonisomorphic models of T in our category of sheaves or else there are at most $\aleph_1$ many nonisomorphic models of T in our category of sheaves. We also show that for one of the four notions of countable we can remove the determinacy assumption.

1. Introduction

1.1. Summary. At the end of his seminal paper [12], Vaught asked the question: "Can it be proved, without the use of the continuum hypothesis, that there exists a complete theory having exactly $\aleph_1$ nonisomorphic denumerable models?". This question has stood to this day as one of the oldest open problems in model theory. The statement that it has a negative answer, i.e., that no countable first-order theory has exactly $\aleph_1$ many countable models (assuming $\neg$CH), has become known as "Vaught's Conjecture".

Over the years much work has been done on Vaught's conjecture and there have been several special situations where it has been shown to hold (such as for ω-stable theories in [11] by Shelah, Harrington, and Makkai). However the general case remains elusive. In his paper [10], Morley took one of the most significant steps towards resolving Vaught's conjecture. He proved that any sentence of $\mathcal{L}_{\omega_1,\omega}(L)$ which does not have a perfect set of countable models must have either countably many or $\aleph_1$ many countable models.

Among the many reasons why Morley's paper represented significant progress towards a resolution of Vaught's conjecture was that it extended the scope of the conjecture from first order theories to sentences of $\mathcal{L}_{\omega_1,\omega}(L)$. By extending the scope of Vaught's conjecture, Morley brought the conjecture into the realm of infinitary logic as well as descriptive set theory and thereby opened up the use of

2010 *Mathematics Subject Classification.* Primary 03C15, 03C90, 03G30; Secondary 03E15, 18F20.

Key words and phrases. Countable model, Vaught's conjecture, category theory, sheaf, site.

This is the final form of the paper.

techniques from these areas to its study. This has led to several important discoveries about the nature of counterexamples to Vaught's conjecture. For example it led to the discovery by Harnik and Makkai (in [3]) that if there is a counterexample to Vaught's conjecture then there must be a minimal one.

One of the most significant discoveries of categorical logic is that the operations of $\mathcal{L}_{\omega_1,\omega}(L)$ can be described categorically. This allows us, as is done by Makkai and Reyes in [9], to consider models of sentences of $\mathcal{L}_{\omega_1,\omega}(L)$ in categories other than the category of sets and functions.

The goal of this paper is to extend Morley's result to the context where the underlying category is a category of sheaves on a weak site. In the course of this paper we will consider four reasonable notions of "a countable model." We call these notions purely countable, countably generated, monic countable and epi countable.

The main results of this paper show that for any of our four notions of a countable model, any countable language L, any sentence $T \in \mathcal{L}_{\omega_1,\omega}(L)$ and any countable weak site (C, J_C), if there isn't a perfect set of countable models of T in $\mathrm{Sh}(C, J_C)$[1] then, under $\mathbf{\Pi}^1_3$ determinacy, there are no more than $\aleph_1$ many countable models of T in $\mathrm{Sh}(C, J_C)$. Further, in the case of purely countable models, we can remove the determinacy assumption.

1.2. Outline. We prove the main results of this paper by showing there is an equivalence (of the appropriate type) between the collections of countable models we care about in our category of sheaves and specific $\mathbf{\Sigma}^1_2$ subsets of the reals. This reduces the question of counting the number of countable models in a category of sheaves to counting the number of isomorphism classes under an appropriate isomorphism relation on our $\mathbf{\Sigma}^1_2$ set. These isomorphism relations turn out to be either $\mathbf{\Sigma}^1_1$ or $\mathbf{\Sigma}^1_2$ depending on the notion of countability.

We start this paper in Section 2 by reviewing some set theory and category theory which we will need in order to count the number of these isomorphism classes of models.

Next, in Section 3, we move on to a discussion of models in a category of sheaves. One of the most important ideas in this section is that when counting the number of models in a category of sheaves, it suffices to restrict ourselves to a particular collection of models in the category of separated presheaves. We further show that these separated presheaf models can be thought of as ordinary set models of a sentence of $\mathcal{L}_{\infty,\omega}(L')$ (for some specific language L').

In Section 4 we introduce the concept of Σ_1 definable and Δ_1 definable classes of models. Σ_1 definable classes of models are particularly well-behaved and most collections of models we consider will be Σ_1 definable classes. Specifically, in Section 4.2, we show that the collection of models of a sentence of $\mathcal{L}_{\infty,\omega}(L)$ in a category of sheaves forms a Δ_1 definable class of models.

In Section 5, we introduce our four notions of countability and show that the models of each notion form a Σ_1 definable class of models. We then prove, in Section 6, our various results concerning the number of countable models in any Σ_1 definable class (and in particular the number of countable models which satisfy a sentence of $\mathcal{L}_{\omega_1,\omega}(L)$).

[1] $\mathrm{Sh}(C, J_C)$ is defined in Section 2.2. It is equivalent to the category of sheaves on (C, J_C).

We end in Section 7 with some conjectures, including a generalization of Vaught's conjecture and a simple example of a category of sheaves in which Vaught's conjecture holds.

1.3. Acknowledgement. The author would like to thank Gerald Sacks for introducing him to Vaught's conjecture and Morley's paper [10] as well as for many interesting conversations. The author would also like to thank Leo Harrington, Jan Reimann, and Dana Scott for useful conversations during the development of this paper.

2. Background

We begin by recalling the definition of the formulas we are interested in.

Definition 2.1. If L is a language we define $\mathcal{L}_{\infty,\omega}(L)$ to be the smallest collection of formulas containing all atomic formulas of L and closed under: finite quantification, negation, implication and set sized conjunctions and disjunctions with the condition that every subformula of a formula of $\mathcal{L}_{\infty,\omega}(L)$ can have at most finitely many free variables. We let $\mathcal{L}_{\omega_1,\omega}(L)$ be the subset of $\mathcal{L}_{\infty,\omega}(L)$ consisting of those formulas all of whose conjunctions and disjunctions are countable.

2.1. Set theory. We begin this section with a brief discussion of the set theoretic notations and concepts used in this paper. We refer the reader to such standard texts as [6] for any set theoretic results or definitions not explicitly discussed.

In this paper we will assume Zermelo – Fraenkel Set Theory (ZF) as our ambient theory and we will assume all results take place in a fixed model of ZF which we refer to as SET. By a *standard model* (of set theory) we mean a transitive (not necessarily proper) class M. Notice that if (M, E) is such that M is a (not necessarily proper) class and E is a well-founded, extensional relation on M then by Mostowski's Collapsing Theorem ([6, Theorem 6.15]) (M, E) is isomorphic to a $(\text{mc}(M, E), \in)$ where $\text{mc}(M, E)$ is a standard model and $\in$ is the "element of" relation in SET.

Unless otherwise stated all standard models of set theory will satisfy ZF. While we will not necessarily assume this of all standard models of set theory, we will assume that SET $\models [\bigwedge_{\alpha<\omega_1} |X_\alpha| \leq \aleph_1] \rightarrow |\bigcup_{\alpha<\omega_1} X_\alpha| \leq \aleph_1$.[2]

There is one important situation when we will be interested in standard models of set theory which may not satisfy ZF. There are times when we will need to prove the existence of certain standard (countable) models of set theory. However, as ZF doesn't prove such models exist, we will instead use models of a finite fragment which we call ZF*. We want to think of ZF* as a fragment of ZF "large enough to prove every result which came before it in this paper." In particular this means that while in each result ZF* is a fixed finite fragment of ZF, ZF* is not necessarily fixed between results.

Many of our results depend on results from descriptive set theory which we collect here. As the reader considers these results, it is worth keeping in mind how they will be used. We will show that for our purposes it suffices to consider collections of models where the collection is encoded by a $\mathbf{\Sigma}^1_2$ set of reals. The question of counting the number of models then becomes a question of counting

[2]This follows from the axiom of choice, but it is consistent with ZF that it holds and the axiom of choice doesn't.

the number of equivalence classes under model isomorphism (in the appropriate category). The notion of model isomorphism will in general be a $\mathbf{\Sigma}^1_2$ relation, except in special situations when we can reduce it to a $\mathbf{\Sigma}^1_1$ relation.

If the reader wishes to take the descriptive set theory results on faith, he can browse Propositions 2.4, 2.9, 2.10, and 2.13 which are the fundamental results in this section and move on to Section 2.2.

We now give an important connection between $\mathbf{\Sigma}^1_2$ sets of reals and Σ_1 classes.

Definition 2.2. Suppose L is a first-order language. We say a formula $\varphi(x)$ of set theory (possibly with parameters) is an *L-formula* if

- φ is only satisfied by models of L in SET.
- φ is closed under isomorphism (i.e., if $\varphi(x)$ holds and x is isomorphic to y (as models in SET) then $\varphi(y)$ holds).

Definition 2.3. Suppose L is a countable language and let $\mathrm{Mod}_L(\omega) \subseteq 2^\omega$ be the collection of reals which encode (via fixed encoding[3]) models of L. We give $\mathrm{Mod}_L(\omega)$ the standard topology (see [2]). For each element $x \in \mathrm{Mod}_L(\omega)$ let x_M be the model of L on ω which x encodes.

Proposition 2.4. *Suppose L is a countable language and φ is an L formula which is Σ_1 and has only a hereditarily countable parameter P. Also suppose P^* is a real encoding the directed graph[4] of P. Then there is a $\Sigma^1_2(P^*)$ formula, $\varphi^*(x)$, such that $\varphi^*(x)$ holds if and only if $\varphi(x_M)$ holds. Further the choice of φ^* is independent of the standard model of set theory we are working in.*

Proof. Let $\varphi(x) = (\exists y)\psi(x, y, P)$ where ψ is a Δ_1 formula. We define $\varphi^*(x)$ to be the formula saying:

- $x \in \mathrm{Mod}_L(\omega)$.
- There exists $E \subseteq \omega \times \omega$ such that
  - E is well-founded and extensional.
  - Transitive Closure of $\{x, P\} \subseteq \mathrm{mc}(\omega, E)$
  - $\mathrm{mc}(\omega, E) \models \mathrm{ZF}^* \wedge (\exists y)\psi(x_M, y, P)$.

It is clear that φ^* is $\Sigma^1_2(P^*)$. Further $\varphi^*(x)$ holds if and only if there is a countable standard model of ZF^* containing x where $\varphi(x_M)$ holds. So by the absoluteness of Δ_1 formulas, if $\varphi^*(x)$ holds in SET then so does $\varphi(x_M)$. Now suppose $\mathrm{SET} \models \psi(x_M, y, P)$ for some set y. Let $(K, \in)$ be a transitive model of ZF^* which contains $\{x, y, P\}$ and satisfies $\psi(x, y, P)$. Let $(H, \in)$ be a countable, elementary substructure of $(K, \in)$ containing the transitive closure of $\{x, P\}$ (one exists as the transitive closure of $\{x, P\}$ is countable). So $H \models (\exists y)\psi(x, y, P)$ and satisfies ZF^*. But, as H is countable there is a bijection $f\colon \omega \to H$. So if $aEb \leftrightarrow f(a) \in f(b)$ then because the the transitive closure of $\{x, P\}$ is contained in $\mathrm{mc}(E, \omega)$, E witnesses that $\varphi^*(x)$ holds. □

---

[3]By a "fixed encoding" we mean any Δ_1 formula d (without parameters) such that for any standard models V of ZF^*

- $V \models (\forall x, y)d(x, y) \to x \in \mathrm{Mod}_L(\omega) \wedge |y| = \aleph_0 \wedge y$ is a model of the language L.
- For any model M of L in V, $V \models |M| = \aleph_0 \to [(\exists x \in \mathrm{Mod}_L(\omega))(\forall N \in V)d(x, N) \leftrightarrow N = M]$.
- $V \models (\forall x \in \mathrm{Mod}_L(\omega))(\exists y)d(x, y)$.

[4]A set P is hereditarily countable if its transitive closure, $tc(P)$, is countable. The directed graph of a set P is the pair $((\mathrm{tc}(P), \in), P)$ treated as a model of the language of directed graphs with a distinguished predicate.

We will need the following theorem due to Harrington and Shelah ([5]).

Definition 2.5. A set $X \subseteq \omega^\omega$ is *κ-Suslin* if there is a tree $T \subseteq (\omega \times \kappa)^{<\omega}$ such that $x \in X \leftrightarrow (\exists y \in \kappa^\omega)(x, y) \in [T]$.[5] We say $X = p[T]$. We say X is *co-κ-Suslin* if $(\omega^\omega - X)$ is κ-Suslin.

Definition 2.6. Suppose E is a co-κ-Suslin equivalence relation on ω^ω with $\omega^\omega \times \omega^\omega - E = p[T]$ where $T \subseteq (\omega \times \omega \times \kappa)^{<\omega}$. E is *strongly thick* if

($*_t$) There is a perfect set P and $t \subseteq T$ with $|t| = \aleph_0$ such that $\{(x, y) \in P \times P : x \neq y\} \subseteq p[t]$.

Notice that if an equivalence relation is strongly thick then it has a perfect set of inequivalent reals.

Proposition 2.7 ([5]). *Suppose*

- *E is a co-κ-Suslin equivalence relation on ω^ω with $\omega^\omega \times \omega^\omega - E = p[T]$ where $T \subseteq (\omega \times \omega \times \kappa)^{<\omega}$.*
- *In some Cohen forcing extension of $L[T]$, $\omega^\omega \times \omega^\omega - p[T]$ is an equivalence relation.*
- *E is not strongly thick.*

Then there are at most κ many E-inequivalent reals.

There are two conditions on our ambient set theoretic universe which we would like to consider.

Definition 2.8. We define the following statements

($*_\#$) Let ($*_\#$) be the statement:
  - $(\forall a \in \omega^\omega)$ $a^\#$ exists.
  - For every $T \subseteq \omega_2$ there is a real which is Cohen generic over $L[T]$.

($*_B$) Let B be the closure of the $\mathbf{\Sigma}^1_2$ sets under finite Boolean combinations and continuous preimage. Let ($*_B$) be the statement
  - All sets in B are determined.
  - Dependent Choice holds.

Proposition 2.9. *Suppose X is a $\mathbf{\Sigma}^1_2$ set of reals and E is a $\mathbf{\Sigma}^1_2$-equivalence relation on ω^ω which does not contain a perfect set of E-inequivalent reals in X. Further suppose ($*_\#$) holds. Then there are at most $\aleph_2$ many E-inequivalent reals in X.*

Proof. Let $E'(x, y) := \big(E(x, y) \wedge X(x) \wedge X(y)\big) \vee \big(\neg X(x) \wedge \neg X(y)\big)$. E' is then $\mathbf{\Pi}^1_3$ and so E' is co-ω_2-Suslin by a theorem of Martin (see [6, Theorem 32.15]). If there does not exists a perfect set of E'-inequivalent reals, then E' is not strongly thick. Hence, by Proposition 2.7, there are at most $\aleph_2$ many E'-inequivalent reals and so at most $\aleph_2$ many E-inequivalent reals in X. □

The following uses, in a fundamental way, a theorem of Harrington and Sami (from [4]).

Proposition 2.10. *Suppose X is a $\mathbf{\Sigma}^1_2$ set of reals and E is a $\mathbf{\Sigma}^1_2$-equivalence relation on ω^ω. If ($*_B$) holds and there is not a perfect set of E-inequivalent reals in X then there are at most $\aleph_1$ many E-inequivalent reals in X.*

---

[5]For a tree T on $(\omega \times \kappa)^{<\omega}$, $[T]$ is the collection $(x, y) \in (\omega \times \kappa)^\omega \cong \omega^\omega \times \kappa^\omega$ such that $(x|_n, y|_n) \in T$ for all $n \in \omega$.

PROOF. Let's assume there is not a perfect set of E-inequivalent reals in X. We know by a result of Sierpinski (see [6, Theorem 25.19]) that $X = \bigcup_{\alpha<\omega_1} X_\alpha$ where each X_α is Borel and disjoint. Define $xE^\alpha y \Leftrightarrow [xEy \wedge X_\alpha(x) \wedge X_\alpha(y)] \vee [\neg X_\alpha(x) \wedge \neg X_\alpha(y)]$.

Claim 2.11. *For each $\alpha < \omega_1$, E^α is a $\mathbf{\Delta}^1_2$ equivalence relation such that there does not exists a perfect set of E^α-inequivalent reals.*

PROOF. For each $\alpha < \omega_1$ it is immediate that E^α is a $\mathbf{\Sigma}^1_2$ equivalence relation. If there existed a perfect set of E^α-inequivalent reals, then there would be a perfect set of E-inequivalent reals in X (and we have assumed there isn't). So there must not be a perfect set of E^α-inequivalent reals. Further, by [4, Corollary 3], E^α is a $\mathbf{\Delta}^1_2$ set. $\square$

Claim 2.12. *For each $\alpha < \omega_1$, there are at most $\aleph_1$ many E^α-inequivalent reals.*

PROOF. First note E^α is a $\Delta^1_2(e_\alpha)$ equivalence relation for some $e_\alpha \in \omega^\omega$. Because $(*_B)$ holds all $\Sigma^1_1(e_\alpha)$ sets are determined and so, by [6, Theorem 33.19], $e_\alpha^\#$ exists. In particular this means that $|(\aleph_1)^{L[e_\alpha]}| = \aleph_0$ and so, by a (relativized) Corollary 3 of [5] there are at most $\aleph_1$ many E^α-inequivalent reals. $\square$

For each α, E and E^α agree on X_α. So in each X_α there are at most $\aleph_1$ many E-inequivalent reals. Hence in $X = \bigcup_{\alpha<\omega_1} X_\alpha$ there are at most $\aleph_1$ many E-inequivalent reals. $\square$

The results above concerning $\mathbf{\Sigma}^1_2$ equivalence relations use large cardinal and determinacy assumptions in a fundamental way. However, if we restrict ourselves to $\mathbf{\Sigma}^1_1$ relations instead of $\mathbf{\Sigma}^1_2$ relations, then we are able to remove these assumptions.

Proposition 2.13. *Suppose X is a $\mathbf{\Sigma}^1_2$ set of reals and E is a $\mathbf{\Sigma}^1_1$-equivalence relation on ω^ω. If there is not a perfect set of E-inequivalent reals in X, then there are at most $\aleph_1$ many E-inequivalent reals in X.*

PROOF. Lets assume there is not a perfect set of E-inequivalent reals in X. We know, by [6, Theorem 25.19], that $X = \bigcup_{\alpha<\omega_1} X_\alpha$ where each X_α is Borel and the X_α are disjoint. Define $xE^\alpha y \Leftrightarrow [xEy \wedge X_\alpha(x) \wedge X_\alpha(y)] \vee [\neg X_\alpha(x) \wedge \neg X_\alpha(y)]$.

Each E^α is then a $\mathbf{\Sigma}^1_1$ equivalence relation on ω^ω such that there is not a perfect set of E^α-inequivalent reals (because if there were then there would be a perfect set of E-inequivalent reals on X). Hence for each $\alpha < \omega_1$ there are at most $\aleph_1$ many E^α-inequivalent reals. But for each α, E and E^α agree on X_α. So in each X_α there are at most $\aleph_1$ many E-inequivalent reals. Hence in $X = \bigcup_{\alpha<\omega_1} X_\alpha$ there are at most $\aleph_1$ many E-inequivalent reals. $\square$

2.2. Category theory. In this section we review some of the categorical notions which we will need. For more information on the category theory in this paper the reader is referred to such standard texts as [7]. For more information on the sheaf theoretic ideas used in this paper the reader is referred to such standard texts as [8]. In this section we also mention several results concerning the category $\mathrm{Sh}(C, J_C)$ and the notion of a weak site. The proofs of these results we leave to the reader (as they are not difficult). However, for a precise treatment of these results we refer the reader to [1].

All categories in this paper will be locally small and we will use the convention that if C is a category with objects A and B, $C[A, B]$ is the set of morphisms whose domain is A and whose codomain is B. If C is a small category we also use $C[-, B]$ for the image of B under the Yoneda embedding. We will abuse notion (when no confusion can arise) and let SET be the category of sets and functions in our ambient model of ZF.

Definition 2.14. If $F\colon C^{\text{op}} \to \text{SET}$ is a functor we say $\langle b, U\rangle$ *is an element of* F ($\langle b, U\rangle \in F$), if $U \in \text{obj}(C)$ and $b \in F(U)$. We define the *size of* F ($|F|$) to be the size of the collection of elements, i.e., $|\{\langle b, U\rangle : b \in F(U)\}|$. If $F, G\colon C^{\text{op}} \to \text{SET}$ then we say $F \subseteq G$ if $(\forall U \in \text{obj}(C))F(U) \subseteq G(U)$ and $(\forall a \in F(U))(\forall f\colon X \to U)F(f)(a) = G(f)(a)$.

If F happens to be a sieve on Y, and $f \in C[X, Y]$ then we will abuse notation and say $f \in F$ when $\langle f, X\rangle \in F$. Similarly if S is a set that is the graph of a sieve S' on A we will use $S \in J_C(A)$ to mean $S' \in J_C(A)$.

Definition 2.15. A *weak site* is a pair (C, J_C) where C is a small category and J_C a function from the objects of C to collections of sieves such that for any $A \in \text{obj}(C)$:

- $J_C(A)$ consists of sieves on A.
- (Identity) $C[-, A] \in J_C(A)$.
- (Base Change) If $S \in J_C(A)$ and $f \in C[B, A]$ then $f^*S = \{g \in C[-, B] : f \circ g \in S\} \in J_C(B)$.

If in addition (C, J_C) satisfies:

- (Local Character) For all sieves T on A, if $S \in J_C(A)$ and $(\forall f \in S)f^*T \in J_C\big(\text{dom}(f)\big)$ then $T \in J_C(A)$.

then we say (C, J_C) is a *site*.

The notion of a weak site is the absolute analog of a site and for every weak site there is a smallest site containing it.

Definition 2.16. Let (C, J_C) be a weak site. Define $J_C^\alpha(A)$ on $A \in \text{obj}(C)$ as follows

- $J_C^0(A) = J_C(A)$.
- $J_C^{\alpha+1}(A) = \{T \text{ a sieve on } A : \big(\exists S \in J_C(A)\big)\big(\forall f \in S(B)\big)f^*T \in J_C^\alpha(B)\}$.
- $J_C^{\omega\cdot\gamma}(A) = \bigcup_{\beta<\omega\cdot\gamma} J_C^\beta(A)$.

Define $J_C^{\text{ORD}} = \bigcup_{\alpha\in\text{ORD}} J_C^\alpha$.

We can think of the structure (C, J_C^{ORD}) as the site obtained by closing the weak site (C, J_C) under local character.

Lemma 2.17. *For any weak site (C, J_C), (C, J_C^{ORD}) is a site.*

Lemma 2.18. *For any weak site (C, J_C), the set $\{(S, A) : S \in J_C^{\text{ORD}}(A)\}$ is Σ_1 definable with parameter (C, J_C).*

Definition 2.19. Suppose (C, J_C) is a weak site and $F\colon C^{\text{op}} \to \text{SET}$ is a presheaf.

- We say F is *separated for* (C, J_C) if for every compatible collection of elements $\langle (a_i, i) : i \in S\rangle$ with $S \in J_\alpha(A)$, there is at most one $a \in F(A)$ covered[6] by $\langle (a_i, i) : i \in S\rangle$.
- We say F is a *sheaf for* (C, J_C) if for every compatible collection of elements $\langle (a_i, i) : i \in S\rangle$ with $S \in J_\alpha(A)$, there is exactly one $a \in F(A)$ covered by $\langle (a_i, i) : i \in S\rangle$.

We denote by $\mathrm{Sep}(C, J_C)$ and $\mathrm{Sheaf}(C, J_C)$ the full subcategories of presheaves on C consisting of the separated presheaves and the sheaves for (C, J_C) respectively.

Suppose $F\colon C^{\mathrm{op}} \to \mathrm{SET}$ is a presheaf. Then F is separated if and only if

$$(\forall A \in \mathrm{obj}(C))(\forall S \in J_C(A))(\forall a, b \in F(A))\Big[\bigwedge_{f\in S} F(f)(a) = F(f)(b)\Big] \to a = b.$$

As such the property of being separated for the weak site (C, J_C) is Δ_0 and hence absolute between standard models of ZF containing (C, J_C).

On the other hand the statement that F is a sheaf is equivalent to the statement

$$(\forall A \in \mathrm{obj}(C))(\forall S \in J_C(A))(\forall B \in \mathrm{obj}(C))(\forall \alpha\colon S \Rightarrow F)$$
$$(\exists a \in F(A))(\forall f \in S(B))\alpha(f) = F(f)(a).$$

Hence, because there is an unbound universal quantifier over natural transformations from sieves to our functor, this is a Π_1 statement and hence only downward absolute.

Lemma 2.20. *We have the following equalities*:
- $\mathrm{Sheaf}(C, J_C) = \mathrm{Sheaf}(C, J_C^{\mathrm{ORD}})$.
- $\mathrm{Sep}(C, J_C) = \mathrm{Sep}(C, J_C^{\mathrm{ORD}})$.

This lemma says, for the purpose of considering sheaves and separated presheaves, we can restrict our attention to weak sites. This is important for two reasons. First, being a site is not an absolute property, however being a weak site is an absolute property. Second, as the proof of our main result will rely on facts about definable subsets of reals, it is important that all parameters under consideration be countable. However, there are countable weak sites (C, J_C) such that in any standard model of set theory any site containing (C, J_C) is uncountable (and in fact has the size of the continuum).

Lemma 2.21. *Suppose $V_0 \subseteq V_1$ are standard models of set theory, A and B are separated presheaves (in V_0) for a weak site $(C, J_C) \in V_0$ and $V_0 \models A \subseteq B$. For $b \in B(U)$ let $\mathrm{Cov}_A(b) \Leftrightarrow \{f \in C[-, U] : B(f)(b) \in A(\mathrm{dom}(f))\} \in J_C^{\mathrm{ORD}}(U)$. Then for any $b \in B$*

$$V_0 \models \mathrm{Cov}_A(b) \iff V_1 \models \mathrm{Cov}_A(b).$$

PROOF. This is by induction on the least α such that $\{f \in C[-, U] : B(f)(b) \in A(\mathrm{dom}(f))\} \in J_C^\alpha(U)$. □

In other words, if we are given a separated presheaf B, a subpresheaf A, and an element b of B then whether b is covered by elements of A is independent of the standard model of set theory we are in. This will be important later on.

[6]For $i \in J_C(A)$ and $a \in F$, $\langle (a_i, i) : i \in S\rangle$ covers a if and only if $F(i)(a) = a_i$ for all $i \in S$.

Definition 2.22. Suppose (C, J_C) is a weak site. Let $\mathrm{Sh}^*(C, J_C)$ be the category such that:

(a) The objects of $\mathrm{Sh}^*(C, J_C)$ are the separated presheaves for (C, J_C).
(b) The morphisms of $\mathrm{Sh}^*(C, J_C)[D, R]$ are the tuples $\langle d, r, f\rangle$ where
- d, r are separated presheaves for (C, J_C).
- $f\colon d \Rightarrow r$ is a natural transformation.
- $d \subseteq D$, $r \subseteq R$ and d, r cover D, R respectively in (C, J_C^{ORD}).[7]

(c) $\langle d, r, f\rangle \circ \langle d', r', g\rangle = \langle g^{-1}[r' \cap d], r, f \circ g\rangle$.
(d) If $X \in \mathrm{obj}(\mathrm{Sh}^*(C, J_C))$ then the identity on X is $\langle X, X, \mathrm{id}_X\rangle$.

Definition 2.23. For all $\langle d_f, r_f, f\rangle, \langle d_g, r_g, g\rangle \in \mathrm{Sh}^*(C, J_C)[D, R]$ we define $\langle d_f, r_f, f\rangle \equiv \langle d_g, r_g, g\rangle$ if and only if $(\forall x \in d_f \cap d_g) f(x) = g(x)$. We then define $\mathrm{Sh}(C, J_C) = \mathrm{Sh}^*(C, J_C)/\equiv$.

Lemma 2.24. *If $V_0 \subseteq V_1$ are standard models then $\mathrm{Sh}^*(C, J_C)^{V_0} \subseteq \mathrm{Sh}^*(C, J_C)^{V_1}$ and for all $f, g \in \mathrm{Sh}^*(C, J_C)^{V_0}[A, B]$, $V_0 \models f \equiv g$ if and only if $V_1 \models f \equiv g$.*

Lemma 2.25. *$\mathrm{Sh}(C, J_C)$ is definable by a Σ_1 formula whose only parameter is (C, J_C).*

PROOF. First notice that $\equiv$ and the property of being a separated presheaf for (C, J_C) are Δ_1 with parameter (C, J_C). So it suffices to show that there is a Σ_1 formulas $S(F, x, y, C, J_C)$ such that if x and y are separated presheaves for (C, J_C) then $S(F, x, y, C, J_C)$ holds if and only if $F = \langle d, r, f\rangle \in \mathrm{Sh}^*(C, J_C)[x, y]$ (i.e., F is a map from x to y in $\mathrm{Sh}^*(C, J_C)$). However, to show that S is Σ_1 it suffices to show that the statement "d is a cover of D", i.e., $(\forall a \in D) \operatorname{Cov}_d(a)$, is Σ_1. But this follows from Lemma 2.18. □

Notice that $\mathrm{Sh}(C, J_C)$ is a category and

Lemma 2.26. *There is an equivalence of categories:*

$$\mathbf{j}\colon \mathrm{Sh}(C, J_C) \to \mathrm{Sheaf}(C, J_C).$$

Further, if A is a separated presheaf then $A \subseteq \mathbf{j}(A)$.

Theorem 2.27. *Let $\mathbf{b}\colon \mathrm{Sep}(C, J_C) \to \mathrm{Sh}(C, J_C)$ be the identity on objects with $\mathbf{b}(f) = \langle A, B, f\rangle$ if $f\colon \mathrm{Sep}(C, J_C)[A, B]$. Then $\mathbf{j} \circ \mathbf{b}$ is left adjoint to the inclusion map $\mathbf{i}\colon \mathrm{Sheaf}(C, J_C) \to \mathrm{Sep}(C, J_C)$. In particular $\mathbf{j} \circ \mathbf{b}$ is isomorphic to the sheafification functor $\mathbf{a}\colon \mathrm{Sep}(C, J_C) \to \mathrm{Sheaf}(C, J_C)$.*

These results tell us that $\mathrm{Sh}(C, J_C)$ is a version of the category of sheaves which is, in some sense, absolute (as the category of sheaves is not absolute, given that the property of a functor being a sheaf is not absolute).

Lemma 2.28. *Suppose $A \in \mathrm{obj}(\mathrm{Sh}(C, J_C))$ and $S \in \mathrm{Sub}_{\mathrm{Sh}(C, J_C)}(A)$ is a subobject of A in $\mathrm{Sh}(C, J_C)$.[8] Then there is a (unique) subpresheaf $\lfloor S\rfloor \subseteq A$ such that $\langle \lfloor S\rfloor, A, \mathrm{in}\rangle$[9] $\in S$ and whenever $\langle d, r, f\rangle \in S$ then $f[d]$[10] $\subseteq \lfloor S\rfloor$ as separated*

---

[7] We say d covers D if for every $U \in \mathrm{obj}(C)$ and every $a \in D(U)$, $\operatorname{Cov}_d(a)$ holds.

[8] Recall that a subobject S of an object A is a maximal collection of monomorphisms with codomain A such that if $s, s' \in S$ then there is an isomorphism $\alpha\colon \mathrm{dom}(s) \to \mathrm{dom}(s')$ with $s' \circ \alpha = s$.

[9] in is the inclusion map from $\lfloor S\rfloor$ to A

[10] $f[d]$ is the image of d under f.

presheaves. Further $\lfloor S \rfloor$ is independent of the standard model of set theory we are working in.

PROOF. Suppose $\langle d, r, f\rangle \in S$. Let $\lfloor S\rfloor_f(U) = \{a \in A(U) : \mathrm{Cov}_{f[d]}(a)\} = \mathbf{j}(f[d]) \cap A$. It is immediate that $\lfloor S\rfloor_f \subseteq A$ and hence $\langle \lfloor S\rfloor_f, A, \mathrm{in}\rangle$ is a subobject of A. Further, $\langle \lfloor S\rfloor_f, A, \mathrm{in}\rangle \circ \langle d, r \cap \lfloor S\rfloor_f, f\rangle \equiv \langle d, r, f\rangle$ and so $\langle \lfloor S\rfloor_f, A, \mathrm{in}\rangle \in S$ (as $\langle d, r \cap \lfloor S\rfloor_f, f\rangle$ is an isomorphism).

Finally as $f[d] \subseteq \lfloor S\rfloor_f$ it suffices to show that if $\langle d, r, f\rangle, \langle d', r', f'\rangle \in S$ then $\lfloor S\rfloor_f = \lfloor S\rfloor_{f'}$. Now $\mathbf{j}(\langle d, r, f\rangle)$ and $\mathbf{j}(\langle d', r', f'\rangle)$ are elements of the same subobject of $\mathbf{j}(A)$ so $\mathrm{image}\big(\mathbf{j}(\langle d', r', f'\rangle)\big) = \mathrm{image}\big(\mathbf{j}(\langle d, r, f\rangle)\big)$ and hence $\lfloor S\rfloor_{f'} = \mathrm{image}\big(\mathbf{j}(\langle d', r', f'\rangle)\big) \cap A = \mathrm{image}\big(\mathbf{j}(\langle d, r, f\rangle)\big) \cap A = \lfloor S\rfloor_f$.

It is also immediate from Lemma 2.21 that $\lfloor S \rfloor$ is absolute between standard models of set theory. □

We know that in the category of separated presheaves or in the category of sheaves a subobject is determined by the image of one of its elements. Lemma 2.28 tells us is that the same holds for $\mathrm{Sh}(C, J_C)$, i.e., that a subobject is determined by a maximal separated subpresheaf covered by the range of one of its elements.

3. Model theory in a category of sheaves

In this section we begin our discussion of models in a category of sheaves. For more on the basics of categorical model theory in categories of sheaves we refer the reader to [8] and [9].

For simplicity of notation we will fix a first order language $L = \langle \mathcal{S}_L, \mathcal{R}_L, \mathcal{F}_L\rangle$ with sorts $\mathcal{S}_L$, relations $\mathcal{R}_L$ and functions $\mathcal{F}_L$. We will also fix a weak site (C, J_C).

3.1. Models and languages. In this section we discuss the relationships between models in $\mathrm{Sep}(C, J_C)$ and $\mathrm{Sh}(C, J_C)$.

Definition 3.1. Suppose D is an ∞-logos. A *model M of L in D* consists of:

- For every sort $S \in \mathcal{S}_L$ an object S^M of D.
- For every function $f \in \mathcal{F}_L$ of sort $\big((S_1, \dots, S_n), S\big)$ a map in D
$$f^M : S_1^M \times \cdots \times S_n^M \to S^M.$$
- For every relation $R \in \mathcal{R}_L$ of sort $(S_1, \dots, S_n)$ an object R_*^M and a monic in D
$$R^M : R_*^M \to S_1^M \times \cdots \times S_n^M.$$

We let $\mathrm{Mod}_L(D)$ be the collection of all models of the language L in the category D.

An important observation is that for any language L we can find an expansion L' and a first order theory T' in L' such that:

- T' says every finite product of sorts in L' is isomorphic to an actual sort via a map in L'.[11]
- T' says there are identity morphism for each sort in L'.
- Every model of L in any category with finite products has a unique expansion to L' making it a model of T'.

[11]This includes the empty product. In particular we have not mentioned constants because they can be thought of as functions whose domain is the empty product.

This observation allows us to assume, without loss of generality, that our models are such that each finite product of sorts is isomorphic to an actual sort and that all sorts have identity maps. While this will not add anything substantively it will make the presentation simpler.

Definition 3.2. Suppose that $M, N \in \mathrm{Mod}_L(D)$. A map of models $f\colon M \to N$ is a set of maps $\langle \alpha_S\colon S \in \mathcal{S}_L \cup \mathcal{R}_L\rangle$ such that:

- $\alpha_S \in D[S^M, S^N]$ if $S \in \mathcal{S}_L$ and $\alpha_R \in D[R^M_*, R^N_*]$ if $R \in \mathcal{R}_L$.
- For any function $f \in \mathcal{F}_L$ with $f\colon A \to B$, $\alpha_B \circ f^M = f^N \circ \alpha_A$.
- For any relation $R \in \mathcal{R}_L$ of sort S, $\alpha_S \circ R^M = R^N \circ \alpha_R$.

This makes $\mathrm{Mod}_L(D)$ into a category.

Lemma 3.3. *If D is Σ_1 definable with L as a parameter then $\mathrm{Mod}_L(D)$ is Σ_1 definable with L as a parameter.*

Lemma 3.4. *If $F\colon D \to E$ is a functor which preserves monomorphisms and finite products, then F induces a functor from $\mathrm{Mod}_L(D)$ to $\mathrm{Mod}_L(E)$ obtained by applying F to every sort, function and relation.*

When no confusion can arise we will use the same symbol for a functor between categories and for the functor between categories of models that it induces.

Corollary 3.5. $\mathbf{b}\colon \mathrm{Sep}(C, J_C) \to \mathrm{Sh}(C, J_C)$ *extends to a functor*

$$\mathbf{b}\colon \mathrm{Mod}_L\big(\mathrm{Sep}(C, J_C)\big) \to \mathrm{Mod}_L\big(\mathrm{Sh}(C, J_C)\big).$$

PROOF. Immediate. □

Corollary 3.5 is important because models in $\mathrm{Sep}(C, J_C)$ are much easier to handle from a set theoretic point of view than models in $\mathrm{Sh}(C, J_C)$. Further, as we will see, every model in $\mathrm{Sh}(C, J_C)$ is isomorphic (in $\mathrm{Sh}(C, J_C)$) to a model which is in the image of $\mathbf{b}$. So, when counting the number of isomorphism classes of models, it will suffice to restrict our attention to models in $\mathrm{Sh}(C, J_C)$ which are in the image of $\mathbf{b}$.

There is a special class of separated presheaf models which are of particular interest.

Definition 3.6. A model M in $\mathrm{Mod}_L\big(\mathrm{Sep}(C, J_C)\big)$ is *sheaflike* if for every relation R in $\mathcal{R}_L$, $R^M_* = \lfloor R^M \rfloor$ and R^M is an inclusion.

Lemma 3.7. *The collection of sheaflike models of a language L in* $\mathrm{Sep}(C, J_C)$ *is a Δ_1 class (with parameters (C, J_C) and L).*

PROOF. Immediate. □

Lemma 3.8. *For every model M in $\mathrm{Mod}_L\big(\mathrm{Sh}(C, J_C)\big)$ there is a sheaflike model M' in $\mathrm{Mod}_L\big(\mathrm{Sep}(C, J_C)\big)$ such that $M \cong \mathbf{b}(M')$ in $\mathrm{Mod}_L\big(\mathrm{Sh}(C, J_C)\big)$ and*

$$\Big|\omega \times L \times \coprod_{S \in \mathcal{S}_L} |S^M|\Big| = \Big|\omega \times L \times \coprod_{S \in \mathcal{S}_L} |S^{M'}|\Big|.$$

PROOF. By Lemma 2.26 every map $g \in \mathrm{Sh}(C, J_C)[A, B]$ is taken via $\mathbf{j}$ to a map $\mathbf{j}(g) \in \mathrm{Sheaf}[\mathbf{a}(A), \mathbf{a}(B)]$. For each sort S and $n \leq \omega$ we define S_n as follows:

- $S_0 = S^M$.
- $S_{n+1} = S_n \cup \bigcup_{g : P \to S \in \mathcal{F}_L} \mathbf{j}(g^M)[P_n]$.

- $S_\omega = \bigcup_{n\in\omega} S_n$.

Notice that $|\omega \times L \times \coprod_{S\in\mathcal{S}_L} |S^M|| = |\omega \times L \times \coprod_{S\in\mathcal{S}_L} |S_\omega||$.

It therefore suffices to create the desired model with sorts S_ω. In order to do this we first observe that for each sort S, $S^M \subseteq S_\omega \subseteq \mathbf{a}(S^M)$. For each sort S let $S^{M'} = S_\omega$ and for each function $f\colon S \to P$ let $f^{M'} = \mathbf{j}(f^M)|_{S^{M'}}$. Note that by construction $f^{M'}[S^{M'}] \subseteq P^{M'}$. Let $i^S_{M,M'}\colon S^M \to S^{M'}$ be the inclusion map from S^M into S_ω. For each relation $R \subseteq S$ suppose $R^M = \langle a^M, b^M, r^M\rangle$ and let $r^* = r[a^M]$. Then let $R^{M'}_* = \lfloor i^S_{M,M'}[r^*]\rfloor$ and $R^{M'}$ be the identity. It is then immediate from the definition that $M' \in \mathrm{Mod}_L(\mathrm{Sep}(C, J_C))$ and is sheaflike. Further the map of models $\{\langle S^M, S^{M'}, i^S_{M,M'}\rangle : S \in \mathcal{S}_L\} \cup \{\langle a^M, R^{M'}_*, r^M\rangle : R \in \mathcal{R}_L\}$ is an isomorphism between M and $\mathbf{b}(M')$ in $\mathrm{Mod}_L(\mathrm{Sh}(C, J_C))$ with inverse $\{\langle S^M, S^M, \mathrm{id}\rangle : S \in \mathcal{S}_L\} \cup \{\langle r^M[a^M], a^M, (r^M)^{-1}\rangle : R \in \mathcal{R}_L\}$. □

Lemma 3.8 shows that, for the purposes of counting isomorphism classes of models in $\mathrm{Mod}_L(\mathrm{Sh}(C, J_C))$, it suffices to restrict our attention to models of the form $\mathbf{b}(M)$ where M is sheaflike.

Lemma 3.9. *There is a Σ_1 formula $\mathrm{Iso}_{(C,J_C),L}(x, y)$ (with parameters (C, J_C) and L) which holds if and only if*

- *$x, y \in \mathrm{Mod}_L(\mathrm{Sep}(C, J_C))$ are sheaflike.*
- *$\mathbf{b}(x)$ is isomorphic to $\mathbf{b}(y)$ in $\mathrm{Mod}_L(\mathrm{Sh}(C, J_C))$.*

Proof. The formula $\varphi(f, g, x, y)$ which says

- $x, y \in \mathrm{Mod}_L(\mathrm{Sep}(C, J_C))$ and are sheaflike.
- f is a map from x to y in $\mathrm{Mod}_L(\mathrm{Sh}(C, J_C))$.
- g is a map from y to x in $\mathrm{Mod}_L(\mathrm{Sh}(C, J_C))$.

is Σ_1 by Lemma 2.25. So $\mathrm{Iso}_{(C,J_C),L}(x, y) := [(\exists f, g)\varphi(f, g, x, y) \wedge f \circ g \equiv \mathrm{id}_y \wedge g \circ f \equiv \mathrm{id}_x]$ is Σ_1 as $\equiv$ is Σ_1 definable. □

3.2. Separated presheaves as models. In this section we show the collection of models in $\mathrm{Sep}(C, J_C)$ which are sheaflike is easily described from a descriptive set theory point of view. For $T \in \mathcal{L}_{\infty,\omega}(L)$ we let $\mathrm{Mod}_L(T, \mathrm{SET})$ be the full subcategory of $\mathrm{Mod}_L(\mathrm{SET})$ consisting of those models which satisfy T.

Proposition 3.10. *There is a language L^S_L and a sentence T^S_L of $\mathcal{L}_{\infty,\omega}(L^S_L)$ such that:*

1. *$|L^S_L| = |L|$.*
2. *$\mathrm{Mod}_{L^S_L}(T^S_L, \mathrm{SET})$ is equivalent to the category of sheaflike models in $\mathrm{Mod}_L(\mathrm{Sep}(C, J_C))$.*
3. *If $k_0\colon \mathrm{Mod}_{L^S_L}(T^S_L, \mathrm{SET}) \to \mathrm{Mod}_L(\mathrm{Sep}(C, J_C))$ is the equivalence of categories from (2) then k_0 is Δ_1 with parameters (C, J_C) and L.*
4. *For every model M of T^S_L in SET, $|\omega \times \coprod_{S\in\mathcal{S}_{L^S_L}} S^M| = |\omega \times \coprod_{S\in\mathcal{S}_L} S^{k_0(M)}|$ (i.e., a model of T^S_L in SET is the same size as the corresponding model in $\mathrm{Sep}(C, J_C)$).*

Proof. Let L^S_L be the language consisting of the following:

- A sort S^X for each $X \in \mathrm{obj}(C)$ and each sort S of L.
- A function symbol $S^f : S^X \to S^Y$ for each $f \in C[Y, X]$ and each sort S.

- Function symbols $g_X : S^X \to P^X$ for each function symbol $g : S \to P$ in L and $X \in \mathrm{obj}(C)$.
- Relations R^X of sort S^X for each relation R of sort S and $X \in \mathrm{obj}(C)$.

The theory T^S_L then consists of axioms which say:

(1) (a) For each sort $S \in \mathcal{S}_L$, if $S^*(X) = S^X$ for each $X \in \mathrm{obj}(C)$ and $S^*(f) = S^f$ for each $f \in C[Y, X]$ then S^* is a presheaf on C.
(b) For each relation $R \in \mathcal{R}_L$, if R^* is defined similarly to S^* then R^* is a presheaf and $R^* \subseteq S^*$.

(2) (a) If $I \in J_C(A)$ and $S \in \mathcal{S}_L$, then $(\forall x, y \in S^A) x = y \leftrightarrow \bigwedge_{f \in I} S^f(x) = S^f(y)$.
(b) For each relation $R \in \mathcal{R}_L$ and each $I \in J_C(A)$, $(\forall x \in S^A) R^A(x) \leftrightarrow \bigwedge_{f \in I} R^{\mathrm{dom}(f)}(S^f(x))$.

(3) Each collection $\langle g_X : X \in \mathrm{obj}(C)\rangle$ is a natural transformation with respect to the presheaf structure given in (1).

Conditions (1)(a) and (1)(b) guarantee that models of T^S_L in SET have presheaves for sorts and relations. Condition (2)(a) guarantees that these presheaves are separated. Condition (3) further guarantees that every model in $\mathrm{Mod}_{L^S_L}(T^S_L, \mathrm{SET})$ gives rise to a model in $\mathrm{Mod}_L(\mathrm{Sep}(C, J_C))$. Lastly (2)(b) ensures that any such model is sheaflike.

It is also immediate that any sheaflike model in $\mathrm{Mod}_L(\mathrm{Sep}(C, J_C))$ gives rise to a model in $\mathrm{Mod}_{L^S_L}(T^S_L, \mathrm{SET})$, that this operation of transforming a model in $\mathrm{Mod}_L(\mathrm{Sep}(C, J_C))$ into models in $\mathrm{Mod}_{L^S_L}(T^S_L, \mathrm{SET})$ preserves maps, and this operation induces the required equivalence of categories. □

Proposition 3.10 allows us to translate results about models of $\mathrm{Mod}_{L^S_L}(T^S_L, \mathrm{SET})$ into results about sheaflike models in $\mathrm{Mod}_L(\mathrm{Sep}(C, J_C))$. But by Lemma 3.8 we can therefore also translate results about models of $\mathrm{Mod}_{L^S_L}(T^S_L, \mathrm{SET})$ into results about models in $\mathrm{Mod}_L(\mathrm{Sh}(C, J_C))$.

Corollary 3.11. *If (C, J_C) is countable and L is a countable language then $T^S_L \in \mathcal{L}_{\omega_1,\omega}(L^S_L)$.*

From now on we will let L be a countable language and (C, J_C) be a countable weak site.

4. Definable classes of models

4.1. Σ_1 definable classes of models.

Definition 4.1. φ_X is a *definable class of models* if it is a formula (possibly with parameters) such that whenever V is a standard model of ZF* containing (C, J_C) as well as the parameters of φ then

- $V \models$ "$\varphi_X(M)$ implies M is a sheaflike model in $\mathrm{Mod}_L(\mathrm{Sep}(C, J_C))$."
- For all sheaflike models $M, N \in \mathrm{Mod}_L(\mathrm{Sep}(C, J_C)) \cap V$, $V \models$ "If $\mathbf{b}(M) \cong \mathbf{b}(N)$ then $\varphi_X(M)$ if and only if $\varphi_X(N)$."

We say a definable class is Σ_1 (or Δ_1) if φ_X is Σ_1 (or Δ_1).

Lemma 4.2. *Suppose φ_X is a Σ_1 definable class of models. Then the collection of $x \in \mathrm{Mod}_{L^S_L}(\omega)$ such that*

- $x_M \in \mathrm{Mod}_{L^S_L}(T^S_L, \mathrm{SET})$.
- $\varphi\big(k_0(x_M)\big)$ *holds.*

is a $\mathbf{\Sigma}^1_2$ *set of reals.*

PROOF. By Propositions 2.4 and 3.10. □

Definition 4.3. If φ_X is a Σ_1 definable class of models let $\widehat{\varphi_X}$ be the $\mathbf{\Sigma}^1_2$ formula defined in Lemma 4.2.

There are a few simple results about Σ_1 definable classes of models which are worth stating explicitly.

Lemma 4.4. Σ_1 *definable classes of models are closed under finite intersection.*

Definition 4.5. Suppose $\varphi(x)$ is a formula which is only satisfied by sheaflike objects of $\mathrm{Mod}_L\big(\mathrm{Sep}(C, J_C)\big)$. Let $\overline{\varphi}(x)$ be the formula

$$\Big(\exists M \in \mathrm{Mod}_L\big(\mathrm{Sep}(C, J_C)\big)\Big)\, \mathrm{Iso}_{(C,J_C),L}(x, M) \wedge \varphi(M).$$

Lemma 4.6. *If* $\varphi(x)$ *is a* Σ_1 *formula which is only satisfied by sheaflike models then so is* $\overline{\varphi}(x)$. *Further* $\overline{\varphi}(x)$ *is a* Σ_1 *definable class of models.*

PROOF. It is immediate by Lemma 3.9. □

This tells us if we have a Σ_1 collection of sheaflike models, the class of sheaflike models whose image under **b** is isomorphic (in $\mathrm{Sh}(C, J_C)$) to the image under **b** of some element of our collection is a Σ_1 definable class of models.

Definition 4.7. Suppose L_1 is the language with no function symbols or relations and where $\mathcal{S}_{L_1} = \{S_1\}$ has only one sort. A *definable class of sheaves* is a definable class of models in L_1, i.e., a class of separated presheaves closed under isomorphism in $\mathrm{Sh}(C, J_C)$. We similarly define Σ_1 definable classes of sheaves and Δ_1 definable classes of sheaves.

Lemma 4.8. *If* X *is a* Σ_1 (Δ_1)*-definable class of sheaves and* $\mathrm{Mod}_L(X)$ *is the formula describing those sheaflike models of* L *all of whose sorts are in* X*, then* $\mathrm{Mod}_L(X)$ *is a* Σ_1 (Δ_1)*-definable class of models.*

PROOF. Immediate. □

4.2. Models of $\mathcal{L}_{\infty,\omega}(L)$. In this section we show that for any sentence $T \in \mathcal{L}_{\infty,\omega}(L)$ the collection of models which satisfy T in $\mathrm{Mod}_L\big(\mathrm{Sh}(C, J_C)\big)$ forms a Δ_1 definable class of models.

Definition 4.9. If $\varphi(\mathbf{x}) \in \mathcal{L}_{\infty,\omega}(L)$ with a free variable of sort S and if $M \in \mathrm{Mod}_L(D)$ then we define $\{\mathbf{x} : D \models \varphi(\mathbf{x})\}^M \subseteq S^M$ by induction in the standard way (see [8]). If φ is a sentence in $\mathcal{L}_{\infty,\omega}(L)$ then we say that $M \models \varphi$ if $\{\mathbf{x} : D \models \varphi\}^M \cong 1_D$.[12]

Definition 4.10. If $T \in \mathcal{L}_{\infty,\omega}(L)$ we define $\mathrm{Mod}_L(T, D)$ to be the full subcategory of $\mathrm{Mod}_L(D)$ consisting of those models which satisfy T.

---

[12]Here 1_D is terminal object of D.

It is worth stressing that for a formula $\varphi \in \mathcal{L}_{\infty,\omega}(L)$ and a sheaflike model M in $\mathrm{Mod}_L(\mathrm{Sep}(C, J_C))$, it is not necessarily the case that $\mathbf{b}(\{\mathbf{x} : \mathrm{Sep}(C, J_C) \models \varphi(\mathbf{x})\}^M)$ and $\{\mathbf{x} : \mathrm{Sh}(C, J_C) \models \varphi(\mathbf{x})\}^{\mathbf{b}(M)}$ are the same subobject. In particular it is not necessarily the case that M satisfies the same sentences of $\mathcal{L}_{\infty,\omega}(L)$ as $\mathbf{b}(M)$ does.

However, as we will see, for any model $M \in \mathrm{Mod}_L(\mathrm{Sh}(C, J_C))$, the satisfaction relation between M and sentences of $\mathcal{L}_{\infty,\omega}(L)$ is absolute.

Proposition 4.11. *Suppose $V_0 \subseteq V_1$ are standard models of* ZF*, $\varphi(\mathbf{x}) \in \mathcal{L}_{\infty,\omega}(L)^{V_0}$ *is a formula of sort* S, *and* $M \in \mathrm{Mod}_L(\mathrm{Sh}(C, J_C))^{V_0}$. *Further suppose that* $M = \mathbf{b}(M')$ *where* $M' \in \mathrm{Mod}_L(\mathrm{Sep}(C, J_C))$ *and is sheaflike. Then* $\lfloor\{\mathbf{x} : \mathrm{Sh}(C, J_C) \models \varphi(\mathbf{x})\}^M\rfloor^{V_0} = \lfloor\{\mathbf{x} : \mathrm{Sh}(C, J_C) \models \varphi(\mathbf{x})\}^M\rfloor^{V_1}$, *i.e., the maximal subpresheaf in the subobject* $\{\mathbf{x} : \mathrm{Sh}(C, J_C) \models \varphi(\mathbf{x})\}^M$ *is independent of the model of set theory we are working in.*

Proof. We will prove this by induction on the complexity of the formula.

Base Case: If $\varphi(\mathbf{x})$ is a relation, $R(\mathbf{x})$, then this follows by Lemma 2.28 and because M is the image of a sheaflike model. If $\varphi(\mathbf{x}) := $ "$f(\mathbf{x}) = g(\mathbf{x})$" for functions $f, g \in \mathcal{F}_L$, then $\lfloor\{\mathbf{x} : \mathrm{Sh}(C, J_C) \models \varphi(\mathbf{x})\}\rfloor^M$ is the equalizer of f^M and g^M in $\mathrm{Sh}(C, J_C)$. However as f^M, g^M are in the image of $\mathbf{b}$, $\lfloor\{\mathbf{x} : \mathrm{Sh}(C, J_C) \models \varphi(\mathbf{x})\}\rfloor^M$ is also $\lfloor\{\mathbf{x} : \mathrm{Sep}(C, J_C) \models \varphi(\mathbf{x})\}\rfloor^M$ which is absolute.

Inductive Case:
- $\varphi(\mathbf{x}) = \bigvee_{i\in I} \psi_i(\mathbf{x})$. Let $X_i = \lfloor\{\mathbf{x} : \mathrm{Sh}(C, J_C) \models \psi_i(\mathbf{x})\}\rfloor$ (which is independent of the model of set theory by the inductive hypothesis) and let $X = \{\mathbf{x} : \mathrm{Sep}(C, J_C) \models \bigvee_{i\in I} X_i\}$. For any $A \in \mathrm{obj}(C)$, $X(A) = \bigcup_{i\in I} X_i(A)$ (which is also independent of the model of set theory). We then have $\lfloor\{\mathbf{x} : \mathrm{Sh}(C, J_C) \models \bigvee_{i\in I} X_i\}\rfloor = \mathbf{j} \circ \mathbf{b}(X) \cap S^M = \lfloor X\rfloor$ (in either V_0 or V_1). But $\lfloor X\rfloor^{V_0} = \lfloor X\rfloor^{V_1}$ by Lemma 2.28 and so $\lfloor\{\mathbf{x} : \mathrm{Sh}(C, J_C) \models \bigvee_{i\in I} X_i\}\rfloor^{V_0} = \lfloor\{\mathbf{x} : \mathrm{Sh}(C, J_C) \models \bigvee_{i\in I} X_i\}\rfloor^{V_1}$.
- $\varphi(\mathbf{x}) = \bigwedge_{i\in I} \psi_i(\mathbf{x})$. Identical to the case where $\varphi(\mathbf{x}) = \bigvee_{i\in I} \psi_i(\mathbf{x})$.
- $\varphi(\mathbf{x}) = (\exists_f)\psi(\mathbf{y})$. Because M is the image of a sheaflike model (from $\mathbf{b}$), f is a map in $\mathrm{Sep}(C, J_C)$. Let $Y = \lfloor\{\mathbf{y} : \mathrm{Sh}(C, J_C) \models \psi(\mathbf{y})\}\rfloor$ and let $X = f[Y]$ in $\mathrm{Sep}(C, J_C)$. Then $\lfloor\{\mathbf{x} : \mathrm{Sh}(C, J_C) \models (\exists_f)\psi(\mathbf{x})\}\rfloor = \mathbf{a}(X) \cap S^M = \lfloor X\rfloor$. But $\lfloor X\rfloor^{V_0} = \lfloor X\rfloor^{V_1}$ by Lemma 2.28.
- $\varphi(\mathbf{x}) = (\forall_f)\psi(\mathbf{y})$. Suppose $f\colon S_0 \to S_1$ and $A \in \mathrm{obj}(C)$. Further suppose $a \in S_1^M(A)$. Let $X_a = f^{-1}(a)$ in $\mathrm{Sep}(C, J_C)$. Then $a \in \lfloor\{\mathbf{x} : \mathrm{Sh}(C, J_C) \models (\forall_f)\psi(\mathbf{x})\}\rfloor$ if and only if $\lfloor X_a\rfloor \subseteq \lfloor\{\mathbf{y} : \mathrm{Sh}(C, J_C) \models \psi(\mathbf{y})\}\rfloor$. But this holds if and only if $X_a \subseteq \lfloor\{\mathbf{y} : \mathrm{Sh}(C, J_C) \models \psi(\mathbf{y})\}\rfloor$. Notice X_a is absolute and $\lfloor\{\mathbf{y} : \mathrm{Sh}(C, J_C) \models \psi(\mathbf{y})\}\rfloor$ is absolute by the inductive hypothesis. Hence the statement $X_a \subseteq \lfloor\{\mathbf{y} : \mathrm{Sh}(C, J_C) \models \psi(\mathbf{y})\}\rfloor$ is absolute and $\lfloor\{\mathbf{x} : \mathrm{Sh}(C, J_C) \models (\forall_f)\psi(\mathbf{x})\}\rfloor$ is also absolute.
- $\varphi(\mathbf{x}) = \psi(\mathbf{x}) \to \phi(\mathbf{x})$. Because for any subobjects A, B of E we have that $(A \Rightarrow B) = \forall_\alpha(A \wedge B)$ where $\alpha\colon A \to E$ is a monic and $A \wedge B$ is considered a subobject of A.
- $\varphi(\mathbf{x}) = \neg\psi(\mathbf{x})$. Because for any subobjects A of E, $\neg A = A \to \bot$ and $\bot$ is absolute.
- $\varphi(\mathbf{x}) = \psi(t_1(\mathbf{x}), \ldots, t_n(\mathbf{x}))$ for terms $t_i(\mathbf{x})$. Because $\lfloor\{\mathbf{x} : \mathrm{Sh}(C, J_C) \models \varphi(\mathbf{x})\}\rfloor^M$ is the same subobject as

$$\left\lfloor\left\{\mathbf{x} : \mathrm{Sh}(C, J_C) \models (\exists y_1, \ldots, y_n)\psi(y_1, \ldots, y_n) \wedge \bigwedge_{i\leq n} t_i(\mathbf{x}) = y_i\right\}\right\rfloor^M. \qquad \square$$

Corollary 4.12. *For any model* $M \in \mathrm{Mod}_L(\mathrm{Sh}(C, J_C))^{V_0}$ *and any sentence* $\varphi \in \mathcal{L}_{\infty,\omega}(L)^{V_0}$*,* $V_0 \models$ "$M \models \varphi$" *if and only if* $V_1 \models$ "$M \models \varphi$."

PROOF. Because the satisfaction relation of $\mathcal{L}_{\infty,\omega}(L)$ is closed under isomorphism, by Lemma 3.8 it suffices to restrict our attention to models of the form $M = \mathbf{b}(M')$ where M' is sheaflike. But by Proposition 4.11, the satisfaction relation is absolute for those models. □

Hence if $\mathrm{Th}(M) = \{\varphi \in \mathcal{L}_{\infty,\omega}(L) : M \models \varphi\}$ then $\mathrm{Th}(M)^{V_0} = \mathrm{Th}(M)^{V_1} \cap V_0$.

Definition 4.13. Suppose $T \in \mathcal{L}_{\infty,\omega}(L)$. Let $\mathrm{Mod}_L^{\mathrm{SL}}(C, J_C)$ be the full subcategory of $\mathrm{Mod}_L(\mathrm{Sep}(C, J_C))$ consisting of sheaflike models. Further let $\mathrm{Mod}_L^{\mathrm{SL}}(T, (C, J_C))$ be the full subcategory of $\mathrm{Mod}_L^{\mathrm{SL}}(C, J_C)$ consisting of models M such that $\mathbf{b}(M) \models T$ in $\mathrm{Sh}(C, J_C)$.

Lemma 4.14. $\mathrm{Mod}_L^{\mathrm{SL}}(T, (C, J_C))$ *is a* Δ_1 *definable class of models.*

PROOF. As $\mathrm{Mod}_L^{\mathrm{SL}}(T, (C, J_C))$ is closed under isomorphism in $\mathrm{Sh}(C, J_C)$ (for sheaflike models) it suffices to prove that it is a Δ_1 class. We know, by Proposition 4.11, that for $M \in \mathrm{Mod}_L^{\mathrm{SL}}(T, (C, J_C))$ the following are equivalent

- SET $\models$ "$\mathbf{b}(M) \models T$" in $\mathrm{Mod}_L(\mathrm{Sh}(C, J_C))$.
- $(\exists W$ a standard model$)(W, \in) \models \mathrm{ZF}^*$, $M \in W$ and $(W, \in) \models$ "$\mathbf{b}(M) \models T$" in $\mathrm{Mod}_L(\mathrm{Sh}(C, J_C))$.
- $(\forall W$ a standard model$)[(W, \in) \models \mathrm{ZF}^*$ and $M \in W]$ implies $[(W, \in) \models$ "$\mathbf{b}(M) \models T$" in $\mathrm{Mod}_L(\mathrm{Sh}(C, J_C))]$. □

5. Notions of countable sheaves

We are now ready to say what we mean by a countable model. We will do this as follows.

Definition 5.1. Suppose $M \in \mathrm{Mod}_L(\mathrm{Sh}(C, J_C))$ is a model of L (a countable language). We say M is *countable* if for each $S \in \mathcal{S}_L$, S^M is a countable sheaf.

The astute reader will notice that this definition isn't quite complete as we still haven't defined what we mean by a countable sheaf. Unfortunately it turns out that there are four notions of countability which we want to consider[13]: purely countable, countably generated, monic countable and epi countable. We will discuss each of these in this section. We will also refer to a model as being purely countable, countably generated, etc. if it satisfies Definition 5.1 with the corresponding notion of countable.

5.1. Purely countable sheaves.

Definition 5.2. We say that $A \in \mathrm{obj}(\mathrm{Sh}(C, J_C))$ is *purely countable* if $|\mathbf{j}(A)| = \aleph_0$ (i.e., the sheafification of A is countable as a functor).

Lemma 5.3. $A \in \mathrm{obj}(\mathrm{Sh}(C, J_C))$ *is purely countable if and only if every separated presheaf which is isomorphic to it (in* $\mathrm{Sh}(C, J_C)$*) is countable as a functor.*

[13]There are of course other reasonable notions of countable. One such, suggested by a helpful referee, is A is *internally epi countable* if it is valid in the internal language of the topos that there is a surjection from a natural number object to A. However, due to a limitation of space, we will restrict our attention to the four mentioned here.

PROOF. This is because every presheaf which is isomorphic to A in $\mathrm{Sh}(C, J_C)$ is isomorphic (in $\mathrm{Sep}(C, J_C)$) to a subpresheaf of $\mathbf{a}(A)$. □

From a set theoretic point of view being a purely countable sheaf is a natural notion and is absolute between standard models of set theory.

Proposition 5.4. *There is a* $\mathbf{\Pi}^1_1$ *statement* $\mathrm{PC}(x)$ *which holds of* x *if and only if*

(1) $x \in \mathrm{Mod}_{L^S_{L_1}}(\omega)$.
(2) $x_M \models T^S_{L_1}$.
(3) $S_1^{k_0(x_M)}$ *is a sheaf.*

PROOF. (1) and (2) are Borel properties so it suffices to show that (3) is a $\mathbf{\Pi}^1_1$ property. However (3) is equivalent to:

$$[(\forall I \in J_C(A))(\forall \langle y_i : i \in I\rangle) \text{ with } (\forall i \in I) y_i \in S^{k_0(x_M)}]$$
$$[\exists \text{ a unique } y \in S^{k_0(x_M)}] \text{ such that } y \text{ is covered by } \langle y_i : i \in S\rangle.$$

which is $\mathbf{\Pi}^1_1$ because (C, J_C) is countable and hence any compatible collection $\langle y_i : i \in I\rangle$ can be encoded by a real. □

Corollary 5.5. *For any countable language* L *there is a* $\mathbf{\Pi}^1_1$ *statement* $\mathrm{PC}_L(x)$ *which holds if and only if*

(1) $x \in \mathrm{Mod}_{L^S_L}(\omega)$.
(2) $x_M \models T^S_L$.
(3) *Every sort of* $k_0(x_M)$ *is a sheaf.*

Corollary 5.6. *The collection of purely countable models is a* Δ_1 *definable collection of models.*

PROOF. As all $\mathbf{\Pi}^1_1$ formulas are absolute between standard models of ZF* being purely countable is absolute between standard models of ZF*. So a model M is purely countable if either of the following equivalent statements holds:

- $(\exists W \text{ transitive}) M \in W$ and $(W, \in) \models$ "ZF* $\wedge\, M$ is purely countable."
- $(\forall W \text{ transitive}) M \in W$ and $(W, \in) \models$ "ZF* $\wedge\, M$ is purely countable." □

Corollary 5.7. *If* (A *is a purely countable sheaf*)^{SET} *then* A *is a sheaf in all standard models of* ZF* *in which* A *is countable.*

PROOF. Because $\mathbf{\Pi}^1_1$ sentences are absolute between standard models of ZF*, if x is a real encoding A (i.e., $x_M = A$) then $\mathrm{PC}(x)$ holds in any standard model of ZF* in which x exists. But in any standard model of ZF* in which A is countable there is some real encoding A. Hence in any standard model of ZF* where A is countable there is some real x encoding A satisfying $\mathrm{PC}(x)$. Therefore in any standard model of set theory where A is countable, A is a sheaf. □

While the notion of being purely countable is absolute from a set theoretic point of view, it is unfortunately very sensitive to the underlying weak site (C, J_C). In particular it is not preserved by passing to an equivalent category of sheaves.

Lemma 5.8. *If* $A \in \mathrm{Sh}(C, J_C)$ *and* A *is not a subobject of* 1 *then there exists a countable* (D, J_D) *such that*

- $C \subseteq D$, $|\text{obj}(D) - \text{obj}(C)| = 1$, $J_C(A) = J_D(A)$ *if* $A \in \text{obj}(C)$, *and* $|J_D(*_x)| = 2$ *if* $*_x \notin \text{obj}(C)$.
- *If* $i \colon C \to D$ *is the inclusion map, then* $i \circ - \colon \text{Sh}(D, J_D) \to \text{Sh}(C, J_C)$ *is an equivalence of categories (with inverse* i'*).*
- $|i'(A)|$ *has (at least) continuum many elements (and hence isn't purely countable).*

PROOF. By assumption there is some $x \in \text{obj}(C)$ where $|A(x)| > 1$. Let D be the free category generated by

- $\text{obj}(D) = \text{obj}(C) \cup \{*_x\}$.
- $D[y, z] = C[y, z]$ if $y, z \in \text{obj}(C)$.
- $D[x, *_x] = \{f_i^x : i \in \omega\}$ (where the f_i^x are new morphisms).
- $D[y, *_x] = \{f_i^x \circ g : g \in D[y, x]\}$.
- $D[*_x, y] = \varnothing$ if $y \in \text{obj}(C)$ and $D[*_x, *_x] = \text{id}_{*_x}$.

We then let $J_D(y) = J_C(y)$ if $y \in \text{obj}(C)$ and $J_D(*_x) = \{D[-, *_x], S_x\}$ where S_x is the sieve generated by $D[x, *_x]$. It is then immediate that $i \circ -$ is an equivalence of categories as there is a unique way to extend any sheaf on (C, J_C) to a sheaf on (D, J_D) which preserves all maps. Further, it is clear that the unique (nontotal) cover of $*_x$ induces a bijection (for any sheaf B on (D, J_D)) between $B(*_x)$ and $B(x)^\omega$. Hence, as $|A(x)| \geq 2$, we have $i'(A)(*_x)$ has at least 2^ω elements. □

A similar argument also shows

Lemma 5.9. *For all* (C, J_C) *there is a* (D, J_D) *with* $C \subseteq D$, $|\text{morph}(D)| = |\omega \times \text{morph}(C)|$, $\bigcup |\bigcup_{A \in \text{obj}(D)} J_D(A) - J_C(A)| = |\omega \times \text{obj}(C)|$, *and* $\text{Sh}(C, J_C)$ *is equivalent to* $\text{Sh}(D, J_D)$ *but all purely countable objects in* $\text{Sh}(D, J_D)$ *are subobjects of* 1.

Not only is the notion of being purely countable very sensitive to the underlying site, but we also have that, unlike in the case of a purely countable set, there are sheaves which are not purely countable in any extension of the universe.

Corollary 5.10. *There is a countable weak site* (D, J_D) *such that the natural number object in* $\text{Sh}(D, J_D)$ *is not purely countable in any standard model of set theory.*

5.2. Countably generated. While the notion of being purely countable is very dependent on the underlying weak site, there is a notion of countability which is much less so. This is what we call being countably generated. While a sheaf is purely countable if every object to which it is isomorphic (in $\text{Sh}(C, J_C)$) is countable (as a functor), a sheaf is countably generated if there is some object in $\text{Sh}(C, J_C)$ to which it is isomorphic and which is countable as a functor.

Definition 5.11. We say that $A \in \text{obj}(\text{Sh}(C, J_C))$ is *countably generated* if there is an $A^* \in \text{obj}(\text{Sh}(C, J_C))$ such that $\langle A^*, A, \text{in}\rangle \equiv \langle A, A, \text{id}\rangle$ in $\text{Sh}(C, J_C)$ and $|A^*| = \aleph_0$ (i.e., there is a countable A^* which covers A).

Lemma 5.12. *Suppose* $V_0 \subseteq V_1$ *are standard models of set theory. If* A *is countably generated in* $\text{Sh}(C, J_C)^{V_0}$ *then* A *is countably generated in* $\text{Sh}(C, J_C)^{V_1}$.

Lemma 5.13. *The collection of countably generated sheaves forms a* Σ_1 *definable class.*

PROOF. Because $\mathrm{Sh}(C, J_C)$ is Σ_1 definable, so is the property $\langle A^*, A, \mathrm{in}\rangle \in \mathrm{Sh}(C, J_C)$. Further the collection of $\langle A^*, A, \mathrm{in}\rangle$ such that $|A^*| \leq \aleph_0$ is Σ_1 definable. $\square$

Corollary 5.14. *The collection of countably generated models in $\mathrm{Mod}_L^{\mathrm{SL}}(C, J_C)$ is a Σ_1 definable class of models.*

PROOF. By Lemma 5.13 and because being countably generated is closed under isomorphism (in $\mathrm{Mod}_L(\mathrm{Sh}(C, J_C))$). $\square$

However, unlike in the purely countable case, being countably generated is preserved under equivalences of categories, at least as long as we restrict ourselves to nice countable weak sites.

Lemma 5.15. *Suppose $A \in \mathrm{obj}(\mathrm{Sh}(C, J_C))$, $C \subseteq D$, $|\mathrm{morph}(D) - \mathrm{morph}(C)| \leq \omega$ and $i \circ -: \mathrm{Sh}(D, J_D) \cong \mathrm{Sh}(C, J_C)$ is an equivalence of categories. Then for any $A \in \mathrm{obj}(\mathrm{Sh}(C, J_C))$, $i \circ A$ is countably generated if and only if A is.*

PROOF. If $\langle A^*, A, \mathrm{in}\rangle \equiv \langle A, A, \mathrm{id}\rangle$ then $\langle i \circ A^*, i \circ A, \mathrm{in}\rangle \equiv \langle i \circ A, i \circ A, \mathrm{id}\rangle$ and $|i \circ A^*| \leq |A^*|$. So if A is countably generated, so is $i \circ A$.

In the other direction, if $\langle A_i, i\circ A, \mathrm{in}\rangle \equiv \langle i\circ A, i\circ A, \mathrm{in}\rangle$ let $A^*(U) = \{x \in A(U) : (\exists V \in \mathrm{obj}(C))(\exists f \in D[U, V])(\exists y \in A_i(V))A(f)(y) = x\}$. Then $|A^*| = |\omega \times A_i|$ and $\langle A^*, A, \mathrm{in}\rangle \equiv \langle A, A, \mathrm{id}\rangle$. So if $i \circ A$ is countably generated so is A. $\square$

We also have that every sheaf is countably generated in some model of set theory.

Lemma 5.16. *If A is an object of $\mathrm{Sh}(C, J_C)$ then there is a forcing extension where A is countably generated.*

PROOF. We always have $\langle A, A, \mathrm{id}\rangle \equiv \langle A, A, \mathrm{id}\rangle$ so if $|A| \leq \aleph_0$ then A is countably generated. But we can always find a generic extension where $|A|$ is countable (as $|A| = |\{\langle a, U\rangle : a \in A(U)\}|$ and $\{\langle a, U\rangle : a \in A(U)\}$ is absolute). $\square$

While the notion of being countably generated is independent of countable changes to the site, it is not in general preserved under arbitrary equivalences of categories.

Lemma 5.17. *There is a (D, J_D) such that $C \subseteq D$, (C, J_C) is equivalent to (D, J_D) (as sites [14]), but no element of $\mathrm{Sh}(D, J_D)$ (other than the initial element $0^{\mathrm{Sh}(D,J_D)}$) is countably generated.*

PROOF. Let D be a category equivalent to C but where each object of C is isomorphic to $\kappa > \omega$ many objects. If A is any separated presheaf for (D, J_D) (such that $|A| \neq 0$) then $|A| \geq \kappa > \omega$. So if $A \neq 0^{\mathrm{Sh}(D,J_D)}$ is an object of $\mathrm{Sh}(D, J_D)$ then A is not countably generated. $\square$

5.3. Monic countable. In this section we introduce the notion of monic countability. This is the first of two notions of countability which are determined solely by the existence of certain maps to/from the natural number object (and hence are preserved by equivalence of categories).

Definition 5.18. Let $\mathbb{N} = (\coprod_\omega 1)^{\mathrm{Sep}(C,J_C)}$, i.e., $\mathbb{N}$ = the coproduct of ω many copies of the terminal object (in $\mathrm{Sep}(C, J_C)$).

[14]See [6] for a definition of equivalence of sites.

Notice that $\mathbf{b}(N)$ is a natural number object in $\mathrm{Sh}(C, J_C)$.

Definition 5.19. A sheaf A is *monic countable* if there is a monomorphism (in $\mathrm{Sh}(C, J_C)$) from A into $\mathbf{b}(\mathbb{N})$.

Lemma 5.20. *The monic countable sheaves form a Σ_1 definable class.*

PROOF. Notice that being monic countable is preserved by isomorphism. Further, $\langle d, r, f\rangle \colon A \to \mathbb{N}$ is a monomorphism if and only if f is injective. So, as $\mathrm{Sh}(C, J_C)$ is Σ_1 definable, so are the monic countable sheaves. $\square$

Lemma 5.21. *Every monic countable sheaf is countably generated.*

PROOF. Suppose A is monic countable and $\langle d, r, f\rangle \colon A \to N$ is a monomorphism. Then d covers A and so $\langle d, A, \mathrm{in}\rangle \equiv \langle A, A, \mathrm{id}\rangle$. But $|d| = |f[d]|$ as f is injective and $|f[d]| \leq |r| \leq |\mathbb{N}| = \aleph_0$ (as $|\mathrm{obj}(C)| = \aleph_0$). Hence A is countably generated. $\square$

Corollary 5.22. *The monic countable models form a Σ_1 definable class of countably generated models.*

Lemma 5.23. *Being monic countable is preserved under equivalence of categories.*

5.4. Epi countable.

Definition 5.24. We say a sheaf A is *epi countable* if there is an epimorphism in $\mathrm{Sh}(C, J_C)$ from $\mathbb{N}$ onto A.

Lemma 5.25. *The epi countable sheaves form a Σ_1 definable class.*

PROOF. Notice that being epi countable is preserved by isomorphism. Further, $\langle d, r, f\rangle \colon \mathbf{b}(N) \to A$ is an epimorphism in $\mathrm{Sh}(C, J_C)$ if and only if $f[d]$ covers A (i.e., if and only if $\mathbf{j}(\langle d, r, f\rangle)_U \colon \mathbf{j}\circ\mathbf{b}(\mathbb{N})(U) \to \mathbf{j}\circ\mathbf{b}(A)(U)$ is surjective for all $U \in \mathrm{obj}(C)$). If we let $\varphi_{\text{e.c.}}(A) := \exists\langle d, f[d], f\rangle \in \mathrm{Sh}(C, J_C)[\mathbf{b}(\mathbb{N}), A]$ then $\varphi_{\text{e.c.}}(A) \leftrightarrow A$ is epi countable. Hence, as $\mathrm{Sh}(C, J_C)[\mathbf{b}(\mathbb{N}), A]$ is Σ_1 definable we have $\varphi_{\text{e.c.}}(A)$ is Σ_1. $\square$

Lemma 5.26. *Epi countable sheaves are countably generated.*

PROOF. Suppose A is epi countable and $\langle d, r, f\rangle \in \mathrm{Sh}(C, J_C)[\mathbb{N}, A]$ is an epimorphism. Then $f[d]$ covers A. Hence $\langle f[d], A, \mathrm{in}\rangle \equiv \langle A, A, \mathrm{id}\rangle$. But $d \subseteq \mathbb{N}$ and $|\mathbb{N}| = \aleph_0$ (as $|\mathrm{obj}(C)| = \aleph_0$) so $|f[d]| \leq \aleph_0$. Hence A is countably generated. $\square$

Corollary 5.27. *The epi countable models of L in $\mathrm{Sh}(C, J_C)$ form a Σ_1 definable class of countably generated models.*

Lemma 5.28. *Being epi countable is preserved under equivalence of categories.*

6. Number of countable models

We are now ready to prove our bounds on the number of countable models of a sentence of $\mathcal{L}_{\infty,\omega}(L)$. The following definitions will be useful.

Definition 6.1. For $x, y \in \mathrm{Mod}_{L^S_L}(\omega)$ with $x_M, y_M \models T^S_L$ let $x \cong_{\mathrm{Sh}} y$ if and only if $k_0(x_M) \cong k_0(x_M)$ in $\mathrm{Mod}_L(\mathrm{Sh}(C, J_C))$.

Definition 6.2. For $x, y \in \mathrm{Mod}_{L^S_L}(\omega)$ with $x_M, y_M \models T^S_L$ let $x \cong_{\mathrm{Sep}} y$ if and only if $k_0(x_M) \cong k_0(x_M)$ in $\mathrm{Mod}_L(\mathrm{Sep}(C, J_C))$.

Definition 6.3. Suppose φ_X is a Σ_1 definable class of models. We say φ_X has the *Morley Property* if either there is a perfect set of $\cong_{\mathrm{Sh}}$-inequivalent reals in $\widehat{\varphi_X}$ or there are at most $\aleph_1$ many $\cong_{\mathrm{Sh}}$-inequivalent reals in $\widehat{\varphi_X}$.

We say φ_X has the *Weak Morley Property* either there is a perfect set of $\cong_{\mathrm{Sh}}$-inequivalent reals in $\widehat{\varphi_X}$ or there are at most $\aleph_2$ many $\cong_{\mathrm{Sh}}$-inequivalent reals in $\widehat{\varphi_X}$.

Proposition 6.4. *Suppose φ_X is a Σ_1 definable class of countably generated models. Then there are at most 2^ω many isomorphism classes (for isomorphism in $\mathrm{Mod}_L(\mathrm{Sh}(C, J_C))$) in φ_X.*

PROOF. This follows immediately from Proposition 3.10 and the fact that there are at most 2^ω many countable models in $\mathrm{Mod}_{L_L^S}(T, \mathrm{SET})$. □

6.1. Purely countable.

Lemma 6.5. *For every model $M \in \mathrm{Mod}_L(\mathrm{Sh}(C, J_C))$ there is a unique (up to isomorphism in $\mathrm{Mod}_L(\mathrm{Sep}(C, J_C))$) model $M' \in \mathrm{Mod}_L(\mathrm{Sep}(C, J_C))$ such that*

- *M' is sheaflike.*
- *For all $S \in \mathcal{S}_L$, $S^{M'}$ is a sheaf for (C, J_C).*
- *$M \cong \mathbf{b}(M')$ (in $\mathrm{Mod}_L(\mathrm{Sh}(C, J_C))$).*

PROOF. By Lemma 3.8 there is a sheaflike model M^* such that $M \cong \mathbf{b}(M^*)$. If we let $M' = \mathbf{i}(\mathbf{a}(M^*))$, i.e the model obtained by applying the sheafification functor $\mathbf{a}$ to M^*, then M' has the desired properties. Further it is unique up to isomorphism in $\mathrm{Mod}_L(\mathrm{Sep}(C, J_C))$ as $\mathbf{i}$ is fully faithful. □

Corollary 6.6. *Two models $M_0, M_1 \in \mathrm{Mod}_L(\mathrm{Sh}(C, J_C))$ are isomorphic if and only if $\mathbf{i}(\mathbf{j}(M_0)), \mathbf{i}(\mathbf{j}(M_1)) \in \mathrm{Mod}_L(\mathrm{Sep}(C, J_C))$ are isomorphic.*

The following corollary collects much of what we have shown about purely countable models.

Corollary 6.7. *Suppose $\mathrm{Mod}_L^{\mathrm{PCS}}(C, J_C) = \{\mathbf{i}(\mathbf{j}(M)) : M \in \mathrm{Mod}_L(\mathrm{Sh}(C, J_C))$ and M is purely countable$\} \subseteq \mathrm{Mod}_L^{\mathrm{SL}}(C, J_C)$. Then*

- *$(\forall M \in \mathrm{Mod}_L(\mathrm{Sh}(C, J_C)), M$ purely countable$)(\exists M' \in \mathrm{Mod}_L^{\mathrm{PCS}}(C, J_C))$ $\mathbf{b}(M') \cong M$ (in $\mathrm{Mod}_L(\mathrm{Sh}(C, J_C))$).*
- *$(\forall M_0, M_1 \in \mathrm{Mod}_L^{\mathrm{PCS}}(C, J_C))\mathbf{b}(M_0) \cong \mathbf{b}(M_1)$ in $\mathrm{Mod}_L(\mathrm{Sh}(C, J_C))$ if and only if $M_0 \cong M_1$ in $\mathrm{Mod}_L(\mathrm{Sep}(C, J_C))$.*
- *$(\forall M \in \mathrm{Mod}_L^{\mathrm{PCS}}(C, J_C))(\exists x \in \mathrm{Mod}_L(\omega))\, \mathrm{PC}_L(x)$ and x_M is isomorphic to M in $\mathrm{Mod}_L(\mathrm{Sep}(C, J_C))$.*

Corollary 6.7 tells us that for the purposes of counting isomorphism classes in $\mathrm{Mod}_L(\mathrm{Sh}(C, J_C))$ of purely countable models, it suffices to consider models which are encoded by reals satisfying $\mathrm{PC}_L(x)$. However, when considering these models, it also suffices to only consider isomorphism in $\mathrm{Mod}_L(\mathrm{Sep}(C, J_C))$. This is important as isomorphism between separated presheaves is, from a set theoretic point of view, much simpler than isomorphism between sheaves.

Theorem 6.8. *Let φ_D be any Σ_1 definable class of purely countable models with only hereditarily countable parameters. Then D has the Morley property.*

PROOF. Let $D^* = \widehat{\varphi_D} \cap \mathrm{Mod}_L^{\mathrm{PCS}}(C, J_C)$. By Corollary 6.7 we know two things. First, $D^* = \{x \in \widehat{\varphi_D} : \mathrm{PC}_L(x)\}$ and second there is a bijection between $\cong_{\mathrm{Sep}}$-isomorphism classes of reals in D^* and $\cong_{\mathrm{Sh}}$-isomorphism classes of reals in $\widehat{\varphi_D}$.

Claim 6.9. *$x \cong_{\mathrm{Sep}} y$ is a $\mathbf{\Sigma}_1^1$ relation on* $\mathrm{Mod}_{L_L^S}(\omega)$.

PROOF. Because $x \cong_{\mathrm{Sep}} y$ if and only if $x_M \cong y_M$ as models of L_L^S (which is a $\mathbf{\Sigma}_1^1$ property). □

Claim 6.10. *D^* is a $\mathbf{\Sigma}_2^1$ set of reals.*

PROOF. By Proposition 2.4 and the fact that $\mathrm{PC}_L(x)$ is $\mathbf{\Pi}_1^1$. □

But by Proposition 2.13 these claims imply that D^* contains either a perfect set of $\cong_{\mathrm{Sep}}$-inequivalent reals or at most $\aleph_1$ many $\cong_{\mathrm{Sep}}$-inequivalent reals. So by Corollary 6.7 $\widehat{\varphi_D}$ contains either a perfect set of $\cong_{\mathrm{Sh}}$-inequivalent reals or at most $\aleph_1$ many $\cong_{\mathrm{Sh}}$-inequivalent reals and hence φ_D has the Morley property. □

Corollary 6.11. *For any theory $T \in \mathcal{L}_{\omega_1,\omega}(L)$ the collection of purely countable models of T has the Morley property.*

PROOF. This follows from Theorem 6.8 because the purely countable models satisfying T form a Σ_1 definable class of models. □

6.2. Countably generated.

Lemma 6.12. *$\cong_{\mathrm{Sh}}$ is a $\mathbf{\Sigma}_2^1$ equivalence relation.*

PROOF. This follows Lemma 3.9 and Proposition 2.4. □

Now recall the definition of $(*_{\#})$ and $(*_B)$ from Definition 2.8.

Theorem 6.13. *Let φ_D be any Σ_1 definable class of countably generated models with only hereditarily countable parameters. We then have the two results:*

(1) *If $(*_{\#})$ holds then φ_D has the weak Morley property.*
(2) *If $(*_B)$ holds then φ_D has the Morley property.*

PROOF. We know that $\widehat{\varphi_D}$ is a $\mathbf{\Sigma}_2^1$ set by Lemma 4.2 and that $\cong_{\mathrm{Sh}}$ is $\mathbf{\Sigma}_2^1$ by Lemma 6.12. (1) then follows from Proposition 2.9 and (2) then follows from Proposition 2.10. □

Corollary 6.14. *For any theory $T \in \mathcal{L}_{\omega_1,\omega}(L)$ the following hold.*

(1) *If $(*_{\#})$ holds then the collection of countably generated models which satisfy T has the weak Morley property.*
(2) *If $(*_B)$ holds then the collection of countably generated models which satisfy T has the Morley property.*

6.3. Monic and epi countable. We know by Corollary 5.27 and Corollary 5.22 that the collection of monic and epi countable models are Σ_1 definable classes of countably generated models. Hence as an immediate consequence of Theorem 6.13 we have:

Corollary 6.15. *For any theory $T \in \mathcal{L}_{\omega_1,\omega}(L)$ the following hold.*

(1) *If $(*_{\#})$ holds then*
   (a) *The collection of monic countable models which satisfy T has the weak Morley property.*

(b) *The collection of epi countable models which satisfy T has the weak Morley property.*

(2) *If $(*_B)$ holds then*

(a) *The collection of monic countable models which satisfy T has the Morley property.*

(b) *The collection of epi countable models which satisfy T has the Morley property.*

7. Vaught's (and other) conjectures

We end with a few conjectures.

7.1. Absoluteness of isomorphism. In Lemma 6.12 we showed that the isomorphism relation between reals encoding countably generated models was $\mathbf{\Sigma}^1_2$. As a consequence of this we see, by Shoenfield's Absoluteness Theorem, that the isomorphism relation between two countably generated models is absolute between standard models of set theory which contain ω_1^{SET}. However it is the author's belief that we can remove the condition that ω_1^{SET} be in our standard model of set theory when talking about absoluteness.

Conjecture 7.1. *If M and N in $\mathrm{Mod}_L(\mathrm{Sh}(C, J_C))$ are countably generated models then the statement "M is isomorphic to N in $\mathrm{Mod}_L(\mathrm{Sh}(C, J_C))$" is absolute between all standard models of ZF^*.*

If this conjecture holds then it is possible to remove the large cardinal and determinacy assumptions from Theorem 6.13.

Theorem 7.2. *Let φ_D be a Σ_1 definable class of countably generated models with only a hereditarily countable parameter P. If Conjecture 7.1 holds and there is a real Cohen generic over $L[P]$ then φ_D has the Morley property.*

PROOF. We define $\mathrm{Iso}^*(x, y)$ as the formula saying

- $x, y \in \mathrm{Mod}_{L^S_L}(\omega)$ and $x_M, y_M \models T^S_L$.
- For all $E \subseteq \omega \times \omega$, if
  - (ω, E) is a well-founded model of ZF^*.
  - Transitive Closure of $\{x, y, P\} \subseteq \mathrm{mc}(\omega, E)$

  then $\mathrm{mc}(\omega, E) \models x \cong_{\mathrm{Sh}} y$.

Claim 7.3. $\mathrm{Iso}^*(x, y)$ *holds if and only if* $x \cong_{\mathrm{Sh}} y$.

PROOF. The implication from left to right holds because $\cong_{\mathrm{Sh}}$ is upwards absolute. The implication from right to left is immediate from Conjecture 7.1. □

$\mathrm{Iso}^*(x, y)$ is clearly a $\mathbf{\Pi}^1_2$ relation (and by Claim 7.3 an equivalence relation). Let $\mathrm{Iso}^*_D(x, y) \leftrightarrow \mathrm{Iso}^*(x, y) \vee [x \notin \widehat{\varphi_D} \wedge y \notin \widehat{\varphi_D}]$. $\mathrm{Iso}^*_D(x, y)$ is then $\mathbf{\Pi}^1_2$ as $\widehat{\varphi_D}$ is a $\mathbf{\Sigma}^1_2$ set and is an equivalence relation as $Iso^*(x, y) \wedge x \in \widehat{\varphi_D} \rightarrow y \in \widehat{\varphi_D}$. So $Iso^*_D(x, y)$ is a co-ω_1-Suslin equivalence relation. By Proposition 2.7 it then follows that there are either a perfect set of Iso^*_D-inequivalent reals or there are at most $\aleph_1$ many Iso^*_D-inequivalent reals.

But as $\cong_{Sh}$ and Iso^*_D agree on $\widehat{\varphi_D}$ (by Claim 7.3) we therefore have φ_D has the Morley property. □

Corollary 7.4. *Suppose $T \in \mathcal{L}_{\omega_1,\omega}(L)$, there is a real Cohen generic over $L[(C,J_C),L,T]$*[15] *and Conjecture 7.1 holds. Then*

(a) *The collection of countably generated models which satisfy T has the Morley property.*
(b) *The collection of monic countable models which satisfy T has the Morley property.*
(c) *The collection of epi countable models which satisfy T has the Morley property.*

7.2. Absoluteness conjecture. In [10] Morley proved more than just that there is a trichotomy of options for the number of countable models of a sentence of $\mathcal{L}_{\omega_1,\omega}(L)$. He showed that, for any particular sentence $T \in \mathcal{L}_{\omega_1,\omega}(L)$, which of the trichotomy was realized was independent of the standard model of set theory where it was considered.

It seems reasonable that this would hold for models in a category of sheaves as well. So we make the following conjecture:

Conjecture 7.5. *For any sentence $T \in \mathcal{L}_{\omega_1,\omega}(L)$ and any countable weak site (C,J_C) the following are each independent of the standard model of set theory where we consider them:*

(1) (a) *Whether there is a perfect set of isomorphism classes of purely countable models of T in $\mathrm{Sh}(C,J_C)$.*
    (b) *The number of isomorphism classes of purely countable models of T in $\mathrm{Sh}(C,J_C)$ (if there is not a perfect set of isomorphism classes).*
(2) (a) *Whether there is a perfect set of isomorphism classes of countably generated models of T in $\mathrm{Sh}(C,J_C)$.*
    (b) *The number of isomorphism classes of countably generated models of T in $\mathrm{Sh}(C,J_C)$ (if there is not a perfect set of isomorphism classes).*
(3) (a) *Whether there is a perfect set of isomorphism classes of monic countable models of T in $\mathrm{Sh}(C,J_C)$.*
    (b) *The number of isomorphism classes of monic countable models of T in $\mathrm{Sh}(C,J_C)$ (if there is not a perfect set of isomorphism classes).*
(4) (a) *Whether there is a perfect set of isomorphism classes of epi countable models of T in $\mathrm{Sh}(C,J_C)$.*
    (b) *The number of isomorphism classes of epi countable models of T in $\mathrm{Sh}(C,J_C)$ (if there is not a perfect set of isomorphism classes).*

We now prove a partial absoluteness result for the number of purely countable models. For the purpose of this result let $(\neg\mathrm{CH})^*$ be the statement:

- There exists an $i\colon \omega_2 \to 2^\omega$ where i is an injection.[16]

Lemma 7.6. *Suppose $E \subseteq \omega^\omega \times \omega^\omega$ is a $\mathbf{\Sigma}^1_1$ equivalence relation on ω^ω and that $X \subseteq \omega^\omega$ is a $\mathbf{\Sigma}^1_2$ set. Then the statement:*

There exists a perfect set of E-inequivalent reals in X.

is absolute between standard models of set theory which satisfy $(\neg\mathrm{CH})^*$ and contain ω_1^{SET}.

[15] The first L represents the constructible universe while the second L is the language we are working in.

[16] This is equivalent to $\neg$CH in the presence of the axiom of choice.

PROOF. Let $L_{\leq}$ be the language of linear orders and $\mathrm{LO} \subseteq \mathrm{Mod}_{L_{\leq}}(\omega)$ be the collection of linear orders. Let $T(x, y)$ be a tree on $(\omega \times \omega) \times \omega$ such that for all $x, y \in \omega^\omega$

$$E(x, y) \iff T(x, y) \text{ is ill-founded.}$$

There is a $\mathbf{\Sigma}^1_1$ set $E' \subseteq \omega^\omega \times \omega^\omega \times \mathrm{LO}$ such that $E'(x, y, \alpha) = \neg(\|T(x, y)\| < \alpha)$ if α is well-founded.

We also know (by [6, Theorem 25.19]) that there is a $\mathbf{\Delta}^1_1$ subset $D_X \subseteq \omega^\omega \times \omega^\omega \times \mathrm{LO}$ such that if $D'(x, a) = (\forall y \in \omega^\omega) D_X(x, y, a)$ then

- $X(x) \leftrightarrow (\exists \alpha < \omega_1) D'(x, \alpha)$
- For all well-founded $\alpha \in \mathrm{LO}$, $D'(x, \alpha)$ is a Borel relation.

Further the Borel code of D_X is dependent only on the formula describing X (i.e. is independent of the standard model of set theory we are working in).

Notice the following statement is $\mathbf{\Sigma}^1_2$:

$$(\star) \quad (\exists \beta, \gamma \in \mathrm{LO})(\exists \text{ perfect set of reals } S)$$
$$(\beta, \gamma \text{ are well-founded}) \wedge (\forall x \in S) D'(x, \gamma) \wedge (\forall x, y \in S) \neg E'(x, y, \beta).$$

Hence, by Shoenfield's Absoluteness Theorem (see [6, Theorem 25.20]), $(\star)$ is absolute between standard models of set theory which contain the ω_1^{SET}. Hence it suffices to prove the following claim:

Claim 7.7. *Suppose V is a standard model of set theory satisfying $(\neg\mathrm{CH})^*$. Then $(\star)$ is equivalent to the statement "There are a perfect set of E-inequivalent reals in X".*

PROOF. The implication from left to right is immediate.

Assume that there are a perfect set of E-inequivalent reals in X in V. Let G be a generic such that $V[G] \models |\aleph_1^V| = \aleph_0$ and $|(2^\omega)^V| \geq |\aleph_2^V| \geq \aleph_1$. We then have $E^{V[G]} \cap V = E^V$. Similarly we have $X^{V[G]} \cap V = X^V$ and $[D'(-, \beta)]^{V[G]} \cap V = [D'(-, \beta)]^V$ for any $\beta \in \omega_1^V$. But in $V[G]$ we have that $E^{\omega_1^V}$ is a Borel equivalence relation (by [6, Lemma 30.10]) and has more than countably many equivalence classes on $D'(-, \omega_1^V)^{V[G]} = \bigcap_{\alpha < \omega_1^V} D'(-, \alpha)^{V[G]}$ (which is also a Borel set). Hence, by a theorem of Silver (see [6, Theorem 32.1]), there is a perfect set of $E^{\omega_1^V}$-inequivalent reals on $D'(-, \omega_1^V)$ in $V[G]$. So $V[G] \models (\star)$ and by absoluteness we then also have that $V \models (\star)$. □

□

The above proof follows closely the proof of Theorem 32.9 of [6] (which is a theorem of Burgess).

In general we cannot remove $(\neg\mathrm{CH})^*$ from the assumption. To see this notice that if L is the constructible universe then $\mathbb{R}^L$ is a Σ^1_2 set of reals. However in L, $\mathbb{R}^L$ contains a perfect set (and hence a perfect set of $=$-inequivalent reals) while there are generics G such that $L[G] \models |\mathbb{R}^L| = \aleph_0$.

Corollary 7.8. *Let φ_D be any Σ_1 definable class of purely countable models with only hereditarily countable parameters. Then whether or not $\widehat{\varphi_D}$ has a perfect set of $\cong_{\mathrm{Sh}}$-inequivalent reals is absolute between standard models of set theory which satisfy $(\neg\mathrm{CH})^*$ and contain ω_1^{SET}.*

PROOF. This follows immediately from the proof of Theorem 6.8 and Lemma 7.6. □

7.3. Topos Vaught's conjecture. Over the years Vaught's conjecture has been studied in ever widening contexts, such as "Vaught's conjecture for $\mathcal{L}_{\omega_1,\omega}(L)$" and the "Topological Vaught's Conjecture". Each time Vaught's conjecture is extended to a new area, we obtain access to a new collection of techniques for its study.

In this section we present a new generalization of Vaught's conjecture. It is our hope that the study of Vaught's conjecture in this context will shed light on the original conjecture as well as its variants.

Definition 7.9. Let (C, J_C) be a countable weak site and let L be a countable language. Suppose φ_D is a Σ_1 definable class of models in $\mathrm{Mod}_L(C, J_C)$. We then say φ_D has the *Vaught Property* if either:

- There is a perfect set of $\cong_{\mathrm{Sh}}$-inequivalent reals in $\widehat{\varphi_X}$.
- There are at most $\aleph_0$ many $\cong_{\mathrm{Sh}}$-inequivalent reals in $\widehat{\varphi_X}$.

Definition 7.10. We say a countable weak site (C, J_C) has the *Vaught property* if for every countable language L and every $T \in \mathcal{L}_{\omega_1,\omega}(L)$ the collections of purely countable, countably generated, monic countable and epi countable models of T all have the Vaught property.

In particular, a weak site has the Vaught property if, inside the category of sheaves on the weak site, Vaught's conjecture holds for any of the four notions of countable presented in this paper. It isn't hard to see that there are weak sites which have the Vaught property.

Proposition 7.11. *There is a countable weak site (C, J_C) which satisfies:*

- *J_C has only identity sieves (so $\mathrm{Sh}(C, J_C) = \mathrm{SET}^{C^{\mathrm{op}}}$).*
- *For every sheaf A the following are equivalent:*
  - *A is purely countable.*
  - *A is countably generated.*
  - *A is monic countable.*
  - *A is epi countable.*
- *(C, J_C) has the Vaught property.*

PROOF. Let C be the set ω treated as a discrete category. Fix $T \in \mathcal{L}_{\omega_1,\omega}(L)$, for some countable language L, and let φ_T be the collection of countably generated models of T in $\mathrm{SET}^{C^{\mathrm{op}}}$. (Note φ_T is also the collection of purely countable, epi countable and monic countable models of T in $\mathrm{SET}^{C^{\mathrm{op}}}$.)

A model of T in $\mathrm{SET}^{C^{\mathrm{op}}}$ consists of ω many models of T in SET. In particular a model of T in $\mathrm{SET}^{C^{\mathrm{op}}}$ is in φ_T if and only if each of the corresponding models in SET are countable.

We now break into two cases. First suppose that T has at least two countable models in SET. Then T must have at least 2^{ω} many countably generated models in $\mathrm{SET}^{C^{\mathrm{op}}}$. And in particular $\widehat{\varphi_T}$ must have a perfect set of $\cong_{\mathrm{Sh}}$-inequivalent reals. On the other hand suppose T has at most one countable model in SET. Then T has at most one countably generated model in $\mathrm{SET}^{C^{\mathrm{op}}}$ and so $\widehat{\varphi_T}$ is either empty or has no $\cong_{\mathrm{Sh}}$-inequivalent reals.

Hence, (C, J_C) has the Vaught property. □

In the other direction we then also have:

Proposition 7.12. *The following are equivalent.*

- *Vaught's Conjecture.*
- *The terminal site*[17] *has the Vaught property.*

This suggests the following generalization of Vaught's conjecture (which implies the actual Vaught's conjecture).

Conjecture 7.13 (Topos Vaught's Conjecture). *Every countable weak site* (C, J_C) *has the Vaught property.*

References

1. N. Ackerman, *Relativized Grothendieck topoi*, Ann. Pure Appl. Logic **161** (2010), no. 10, 1299 – 1312.
2. H. Becker and A. S. Kechris, *The descriptive set theory of Polish group actions*, London Math. Soc. Lecture Note Ser., vol. 232, Cambridge Univ. Press, Cambridge, 1996.
3. V. Harnik and M. Makkai, *A tree argument in infinitary model theory*, Proc. Amer. Math. Soc. **67** (1977), no. 2, 309 – 314.
4. L. Harrington and R. Sami, *Equivalence relations, projective and beyond*, Logic Colloquium '78 (Mons, 1978), Stud. Logic Foundations Math., vol. 97, North-Holland, Amsterdam, 1979, pp. 247 – 264.
5. L. Harrington and S. Shelah, *Counting equivalence classes for co-κ-Souslin equivalence relations*, Logic Colloquium '80 (Prague, 1980), Stud. Logic Foundations Math., vol. 108, North-Holland, Amsterdam, 1982, pp. 147 – 152.
6. T. Jech, *Set theory: The third millennium edition, revised and expanded*, Springer Monogr. Math., Springer, Berlin, 2003.
7. S. Mac Lane, *Categories for the working mathematician*, 2nd ed., Grad. Texts in Math., vol. 5, Springer-Verlag, New York, 1998.
8. S. Mac Lane and I. Moerdijk, *Sheaves in geometry and logic: A first introduction to topos theory*, Universitext, Springer, New York, 1992.
9. M. Makkai and G. E. Reyes, *First order categorical logic: Model-theoretical methods in the theory of topoi and related categories*, Lecture Notes in Math., vol. 611, Springer, Berlin, 1977.
10. M. Morley, *The number of countable models*, J. Symbolic Logic **35** (1970), 14 – 18.
11. S. Shelah, L. Harrington, and M. Makkai, *A proof of Vaught's conjecture for ω-stable theories*, Israel J. Math. **49** (1984), no. 1-3, 259 – 280.
12. R. L. Vaught, *Denumerable models of complete theories*, Infinitistic Methods (Warsaw, 1959), Pergamon, Oxford, 1961, pp. 303 – 321.

Department of Mathematics, University of California, Berkeley, 970 Evans Hall, Berkeley, CA 94720-3840, USA

E-mail address: nate@math.berkeley.edu

[17]The terminal site is the unique site on the category with one object and one morphism.

Centre de Recherches Mathématiques
CRM Proceedings and Lecture Notes
Volume **53**, 2011

Pseudomonadicity and 2-Stack Completions

Marta Bunge and Claudio Hermida

Dedicated to Michael Makkai on his 70th birthday

ABSTRACT. We consider the notion of stack, originally due to Grothendieck and Giraud [15], in the intrinsic sense (of Lawvere [26, 27]), that is, relative to the class of all epimorphisms in a topos **S**. We extend to dimension 2 the result of [7] on the fibrational (1-)stack completion of (the externalization of) a category object $\mathbb{C}$ in a Grothendieck topos **S**. In dimension 1, the monadicity and descent theorems of [2] and [4] are employed in [11] to show that **S** (or 0-**Stack**) is a 1-stack over itself, which yields Diaconescu's theorem [22] on the classification of G-torsors for a groupoid G as an application. Likewise, in dimension 2, we resort to the pseudomonadicity and 2-descent theorems of [20] in order to prove that **Stack** (or 1-**Stack**) is a 2-stack over **S**. We then derive a classification theorem for $\mathbb{G}$-2-torsors for a 2-gerbe $\mathbb{G}$. An axiom of stack completions (ASC) states, for a topos **S**, that the fibrational stack completion of any category object in **S** is representable, and similarly for 2-stack completions. ASC holds for any Grothendieck topos **S**, by a general argument in [22]. Alternatively, we could appeal to the Quillen model structure on **Cat**(**S**) of [23] (for **S** a Grothendieck topos) to obtain (strong) stack completions internal to **S**. While we leave the question of giving a similar construction (via Quillen model structure) in the 2-dimensional case open, we indicate how the passages of stack theory from dimension 1 to dimension 2 pave the way for similar results in higher dimensions.

Introduction

Section 1 discusses the beautiful theory of stacks and nonabelian topos cohomology. This material is considered by many to be one of the pinnacles of 20th century mathematics. The theory of stacks (or *champs*) was developed first by Grothendieck [16] and Giraud [15] in terms of sites, and then recast in an intrinsic fashion and clarified considerably in a second wave by the topos theory community based mostly in North America.

2010 *Mathematics Subject Classification.* 18D05, 18D30, 18F20, 18G60, 14A20.

Key words and phrases. Toposes, 2-categories, 2-descent, 2-stacks, 2-gerbes, 2-torsors, Quillen model structures, classifying structures, Morita equivalence theorems, higher dimensional stacks.

This is the final form of the paper.

Section 1, which is largely expository, is based on [11] and [7], though formulated in terms of fibrations over a given topos $\mathbf{S}$, as opposed to $\mathbf{S}$-indexed categories as in those sources. We use the formal definition-theorem-proof environment just in order to pave the way for an extension to the 2-dimensional case, which is the purpose of this paper. However, an informal description of the contents of these two sections that should be sufficient as a motivation for our program, is given next. The notion of an intrinsic stack was proposed by Lawvere [26, 27]. The intrinsic (or canonical) topology on $\mathbf{S}$ is the large site consisting of the class of all epimorphisms in $\mathbf{S}$ [1]. Denote by $\mathrm{Fib}(\mathbf{S})$ the 2-category of fibrations over $\mathbf{S}$, which is Cartesian closed as a bicategory.

Fundamental to the theory of intrinsic (1-)stacks are two key results, both from [11]. These are recalled in Section 1. The first of these, Theorem 1.9, is the characterization of stacks in terms of the weak equivalences recalled in Definition 1.1. This result says that $\mathscr{A}$ in $\mathrm{Fib}(\mathbf{S})$ is a stack iff $\mathscr{A}^{(-)}$ sends weak equivalences to equivalences. Whether one takes this characterization as the definition of a stack (a suggestion of Joyal) or not, is irrelevant, as the explicit definition of a stack (Definition 1.5) cast in terms of (1-)descent, is also needed for the deep applications of stacks theory.

From this characterization of stacks, it is immediate that any weak equivalence from a fibration to a stack enjoys the universal property of a stack completion. The facts that stacks are closed in $\mathrm{Fib}(\mathbf{S})$ under exponentiation with arbitrary objects, that weakly equivalent fibrations have equivalent stack completions (when such exist), and other results mentioned, all follow pretty immediately from the universal properties.

It is common to regard $\mathbf{S}$ as a fibration over itself via the codomain fibration

$$\mathrm{cod}\colon \mathbf{S}^{\to} \to \mathbf{S},$$

and to denote this fibration also as $\mathbf{S}$. The second key result, Corollary 1.12, is that $\mathbf{S}$ is a stack. This result is easily obtained from the monadicity and descent theorems of [2] and [4].

This stack $\mathbf{S}$ plays the role of the 'object of sets' giving rise to a Yoneda structure on $\mathrm{Fib}(\mathbf{S})$ (see [33, 35, 38] for a discussion of this Yoneda structure), and so all the presheaf objects $\mathbf{S}^{\mathscr{A}^{\mathrm{op}}}$ are stacks too. Using the above results it is established in [7] that, for any locally internal (= admissible for the Yoneda structure) fibration $\mathscr{A}$, one has a canonical embedding of it into a stack

$$y_{\mathscr{A}}\colon \mathscr{A} \to \mathbf{S}^{\mathscr{A}^{\mathrm{op}}}$$

namely the Yoneda embedding. Factoring this through its 'weakly-essential image' gives the stack completion of $\mathscr{A}$ of Theorem 1.13.

As nontrivial consequences of the results just mentioned, one obtains, for a nonempty and connected groupoid G in $\mathbf{S}$, both the fact that the fibration $\mathrm{Tors}^1(G)$ over $\mathbf{S}$ is the stack completion of G, and that the latter is classified by the topos $\mathscr{B}(G) = \mathbf{S}^{G^{\mathrm{op}}}$ (Diaconescu's theorem [22]). This application of stacks [7] shows how nonabelian cohomology includes Galois theory, thus making stacks such a terrific unifying perspective. By restricting to the case of discrete (localic) groupoids, we also recover a classification and Morita theorems given in [8], with applications in [9].

In Section 2 we present, in a unified manner, examples of stack completions in Commutative Algebra, all related to the Zariski topos and motivated by an example of [27]. In particular we obtain, as applications of stack completions, conceptual proofs of Kaplansky's theorem and Swan's theorem in addition to those given by Mulvey [29]. Other examples had already been given at the end of Section 2 of [7].

It is well known [13,22] that, if **S** is a Grothendieck topos, then the fibrational stack completion of a category object $\mathbb{C}$ in **S** is representable on account of the existence of a generating set for **S**. A direct proof of this result is given by Joyal and Tierney in [23] by means of a Quillen model structure on **Cat**(**S**) whose fibrant objects are the (so-called) strong stacks. In particular, this gives an internal (strong) stack completion of any category object $\mathbb{C}$ in **S**. In Section 3, we interpret this to say that any Grothendieck topos **S** satisfies (ASC), an 'axiom of stack completions.' Also in Section 3, we discuss another Quillen model structure on **Cat**(**S**) given by Lack in [24,25]. In **Cat**($\mathrm{Set}^{\mathbf{2}}$), and as shown independently in [11] and [24,25], the iso-fibrations, that is, those internal functors with the isomorphism lifting property, are precisely the stacks. We argue then that, for a general Grothendieck topos **S**, the data for the constructions of [23] and [24,25] need not agree. We expand further on this issue.

Sections 4 and 5 then develop the 2-dimensional analogue of the theory of (1-)stacks. 2-stacks are in particular 2-fibrations, and defined in terms of 2-descent —involving the three-(rather than two)-truncated simplicial nerve of the kernel groupoid of an epi, discussed in [20]. The two main results of the paper are the 2-dimensional analogues of the two fundamental results on stacks isolated above. The analogue of Theorem 1.9 is Theorem 4.9, and that of Corollary 1.12 is Theorem 5.7.

With these results established, the theory of 2-stack completions then unfolds as in dimension 1. The notion of an intrinsic 2-stack is introduced here via 2-descent and shown to be equivalent to a condition of inverting weak 2-equivalence functors. The characterization of intrinsic 2-stacks is given directly in Section 4 after stating the notion of a weak 2-equivalence. In Section 5 we adapt the pseudomonadicity and 2-descent theorems of [20] in order to prove that the 2-fibration cod: $\mathbf{Stack}^{\to} \to \mathbf{S}$ is a 2-stack. The representability of stack completions in **S** is a crucial ingredient in the proof.

A notable application is then given in Section 6, in which we construct the 2-stack completion of any 'generalized 2-gerbe,' a name we give to any 2-category object **C** in **S** which is furthermore 'hom-by-hom' a stack. Restricting to 2-groupoids, we obtain, for a 2-gerbe **G**, the 'classification' of **G**-torsors by the '2-topos' [34,36] $\mathscr{B}^2(\mathbf{G}) = \mathbf{Stack}^{\mathbf{G}^{\mathrm{op}}}$. To make this precise we need to specify a notion of morphism of 2-toposes, this task being left at the moment as an open question.

In Section 7, on equiv-fibrations, we exhibit an example of a 2-stack completion in connection with the Quillen model structure on 2-**Cat**(**S**) for **S** a Grothendieck topos as given by Lack in [24,25], applied to the Grothendieck topos $\mathbf{S} = \mathrm{Set}^{\mathbf{2}}$. This Quillen model structure is then not what we are looking for in the general case. Explicitly, by an argument involving the existence of a generating set, it can be shown in a manner analogous to that of the 1-dimensional case that every 2-gerbe in a Grothendieck topos S admits a representable 2-stack completion when regarded as a 2-fibration. The question of giving a suitable Quillen model structure on 2-**Cat**(S) whose fibrant objects are the '(strong) 2-stacks' is still open.

Finally, Section 8 gives an outline of how one would, in the same spirit as this paper, proceed to higher dimensions for a theory of n-stacks and their completions. We argue that, in principle, the results obtained in dimensions 1 and 2 pave the way for similar theorems about intrinsic n-stacks in higher dimensions $n > 2$, to be shown by induction by analogy with the passage from dimension 1 to dimension 2. Alternative outlines have been given in [14] and [37].

1. Intrinsic stacks

In this largely expository section we review the definition of a stack, state the two fundamental results of the theory of stacks (described informally in the Introduction), and recall an application of stack completions to nonabelian cohomology [11] and [7]. Throughout $\mathbf{S}$ is an elementary topos in the sense of Lawvere and Tierney [22], We assume familiarity with fibrations over a topos [26] and [31].

Definition 1.1. Let $F\colon \mathscr{D} \to \mathscr{C}$ be a functor between fibrations over $\mathbf{S}$. It is said to be a *weak equivalence* if the following conditions hold:

(1) (essentially surjective) For each $I \in \mathbf{S}$, $F^I : |\mathscr{D}^I| \to |\mathscr{C}^I|$, and $c \in |\mathscr{C}^I|$, there exists an epimorphism $e\colon J \to I$ in $\mathbf{S}$, $b \in |\mathscr{D}^I|$, and an isomorphism $\theta\colon F^J(b) \to e^*(c)$.

(2) (fully faithful) $\forall I \in \mathbf{S}\ \forall x, x' \in |\mathscr{D}^I|$, the function

$$\mathrm{Hom}_{\mathscr{D}^I}(x, x') \xrightarrow{F_{x,x'}} \mathrm{Hom}_{\mathscr{C}^I}(Fx, Fx')$$

is an isomorphism.

An explicit construction in $\mathbf{Cat}(\mathbf{S})$ of the 2-*pullback* (or *iso-comma-object*) of two functors $F\colon \mathbb{A} \to \mathbb{B}$ and $G\colon \mathbb{B}' \to \mathbb{B}$ is given [11] by the data

$$\begin{array}{ccc} F\|G & \xrightarrow{\ F'\ } & \mathbb{B}' \\ {\scriptstyle G'}\downarrow & \Rightarrow t & \downarrow{\scriptstyle G} \\ \mathbb{A} & \xrightarrow[\ F\ ]{} & \mathbb{B} \end{array}$$

where $F\|G$ is explicitly described as the category whose objects are triples (a, b', θ), $\theta\colon Fa \to Gb'$ an isomorphism, and whose morphisms are pairs of morphisms making the obvious diagrams commute. The functors F' and G' are given, on objects by $F'(a, b', \theta) = b'$ and $G'(a, b', \theta) = a$, and similarly for morphisms. The natural isomorphism $t\colon GF' \to GF'$ is defined by $t(a, b', \theta) = \theta$.

Remark 1.2. This construction, which is easily extended to fibrations over $\mathbf{S}$, satisfies the required universal property. If the codomain $\mathbb{B}$ is discrete, the square is simply a pullback.

The following results are taken from [11].

Proposition 1.3. *Weak equivalences between fibrations over* $\mathbf{S}$ *are*

(1) *stable under* 2*-pullbacks, and*

(2) *if, in a commutative triangle*

$$\begin{array}{ccc} \mathscr{A} & \xrightarrow{\ F\ } & \mathscr{B} \\ & {\scriptstyle H}\searrow \quad \swarrow{\scriptstyle G} & \\ & \mathscr{C} & \end{array}$$

any two of F, G, H *are wef, then so is the third.*

Given a morphism $e\colon J \to I$ in $\mathbf{S}$, its 1-kernel pair $\mathbb{J}_e$ is the 2-truncated simplicial complex

$$J \times_I J \times_I J \mathrel{\substack{\longrightarrow\\ \longrightarrow\\ \longrightarrow}} J \times_I J \mathrel{\substack{\longrightarrow\\ \longleftarrow\\ \longrightarrow}} J$$

where the arrows are labelled by π for the projections and δ for the diagonal.

Definition 1.4. For a morphism $e\colon J \to I$ in $\mathbf{S}$, and any fibration $\mathscr{A}$ over $\mathbf{S}$, *the category* $\mathrm{Des}_{\mathscr{A}}(e)$ *of descent objects and morphisms in* $\mathscr{A}$ *with respect to e*, has, as *objects*, pairs (a, θ) with $a \in \mathscr{A}^J$, and $\theta : \pi_0{}^*(a) \cong \pi_1{}^*(a)$, satisfying the

(1) (normalization condition)

$$\delta^*(\theta) = 1_a$$

(2) (cocycle condition)

$$\pi_{12}{}^*(\theta).\pi_{01}{}^*(\theta) :=: \pi_{02}{}^*(\theta),$$

and whose *morphisms* $f\colon (a,\theta) \to (b,\sigma)$ are morphisms $f\colon a \to b$ in $\mathbf{S}$ such that

$$\pi_1{}^*(f).\theta = \sigma.\pi_1{}^*(f).$$

It is understood that canonical isomorphisms are to be inserted wherever necessary.

Definition 1.5 ([11]). A fibration $\mathscr{A} \to \mathbf{S}$ is a *stack* (for the intrinsic topology on $\mathbf{S}$) if for every epimorphism $e\colon J \to I$ in $\mathbf{S}$, the functor $e^*\colon \mathscr{A}^I \to \mathscr{A}^J$ is of effective descent. This means that the canonical functor Φ_e in the diagram below, is an equivalence.

$$\begin{array}{ccc} \mathscr{A}^I & \xrightarrow{\Phi_e} & \mathrm{Des}_{\mathscr{A}}(e) \\ & {\scriptstyle e^*}\searrow \quad \swarrow{\scriptstyle U} & \\ & \mathscr{A}^J & \end{array}$$

Remark 1.6. A category object $\mathbb{A}$ in $\mathbf{S}$ is a stack if Definition 1.5 holds for the fibration $[\mathbb{A}] \to \mathbf{S}$, which is the *externalization* of $\mathbb{A}$.

Any epimorphism $e\colon J \twoheadrightarrow I$ in $\mathbf{S}$ gives rise to an internal functor $F_e\colon \mathbb{J}_e \to \mathbb{I}_{\mathrm{dis}}$ where $\mathbb{J}_e$ is the 1-kernel of e and $\mathbb{I}_{\mathrm{dis}}$ is the discrete category on I.

Proposition 1.7. *Let* $\mathscr{A}$ *be a fibration over* $\mathbf{S}$. *Then,* $\mathscr{A}$ *is a stack if and only if the internal functor* $F_e\colon \mathbb{J}_e \to \mathbb{I}_{\mathrm{dis}}$ *induces an equivalence of categories*

$$\mathscr{A}^{F_e}\colon \mathscr{A}^{\mathbb{I}_{\mathrm{dis}}} \to \mathscr{A}^{\mathbb{J}_e}.$$

PROOF. The diagram

$$\begin{array}{ccc} \mathscr{A}^I & \xrightarrow{\Phi_e} & \mathrm{Des}_{\mathscr{A}}(e) \\ {\scriptstyle \kappa_I}\Big\downarrow & & \Big\downarrow{\scriptstyle \Psi_e} \\ \mathscr{A}^{\mathbb{I}_{\mathrm{dis}}} & \xrightarrow[\mathscr{A}^{F_e}]{} & \mathscr{A}^{\mathbb{J}_e} \end{array}$$

is commutative, with κ_I an equivalence [31]. That Ψ_e is an equivalence is shown in [11]. Therefore, for any epimorphism $e\colon J \to I$ in $\mathbf{S}$, Φ_e is an equivalence (hence $\mathscr{A}$ is a stack), if and only of $\mathscr{A}^{F_e}$ is an equivalence, as claimed. $\square$

Lemma 1.8. *For any epimorphism e in* $\mathbf{S}$, *the internal functor* $F_e\colon \mathbb{J}_e \to \mathbb{I}_{\mathrm{dis}}$ *is a weak equivalence functor.*

Theorem 1.9. *A fibration $p\colon \mathscr{A} \to \mathbf{S}$ is a stack if and only if for every weak equivalence functor $F\colon \mathscr{B} \to \mathscr{C}$, the induced functor*

$$\mathscr{A}^F : \mathscr{A}^{\mathscr{C}} \to \mathscr{A}^{\mathscr{B}}$$

is an equivalence of fibrations over $\mathbf{S}$.

PROOF. It follows from Proposition 1.7 and Lemma 1.8 that the condition of the theorem is sufficient. The converse is shown by localization from the fact that, for $\mathscr{A}$ a stack and $F\colon \mathbb{B} \to \mathbb{C}$ a weak equivalence functor, the $\mathbf{S}$-indexed functor $\mathscr{A}^F$ is an equivalence 'at 1.' This is done by a standard descent argument. □

Corollary 1.10. *Let $\mathscr{A}$ be a fibration over $\mathbf{S}$, and let $F\colon \mathscr{A} \to \mathscr{B}$ be a weak equivalence functor, with $\mathscr{B}$ a stack over $\mathbf{S}$. Then, the pair $(\mathscr{B}, F)$ is the* stack completion *of $\mathscr{A}$ in the sense of satisfying the obvious universal property. Stack completions of a given $\mathscr{A}$ are unique up to equivalence.*

We next recall the construction [7] of the fibrational stack completion of any category object $\mathbb{C}$ in an elementary topos $\mathbf{S}$. The crucial result is Theorem 1.12 below, proved in [11]. We need to recall first the following two basic theorems relating monadicity and descent:

Beck Monadicity Criterion ([2, 12]). *If a fibration $\mathscr{A} \to \mathbf{S}$ has Σ subject to the Beck – Chevalley condition* (BCC) *then, for each morphism $e\colon J \to I$ in $\mathbf{S}$ satisfying the conditions that e^* creates e^*-contractible coequalizers and e^* reflects isomorphisms, the canonical functor*

$$\begin{array}{ccc} \mathscr{A}^I & \xrightarrow{\quad\Phi\quad} & (\mathscr{A}^J)^{\mathbb{T}_e} \\ & {\scriptstyle e^*}\searrow \quad \swarrow{\scriptstyle U_{\mathbb{T}_e}} & \\ & \mathscr{A}^J & \end{array}$$

where $\mathbb{T}_e$ is the monad on $\mathscr{A}^J$ induced by the adjoint pair $\Sigma_e \dashv e^*$, is an equivalence of categories.

Bénabou – Roubaud Monadicity and Descent Theorem ([4]). *If a fibration $\mathscr{A} \to \mathbf{S}$ has Σ subject to the Beck – Chevalley condition* (BCC) *then, for each morphism $e\colon J \to I$ in $\mathbf{S}$, the canonical functor*

$$\begin{array}{ccc} \mathrm{Des}_{\mathscr{A}}(e) & \xrightarrow{\quad\Psi\quad} & (\mathscr{A}^J)^{\mathbb{T}_e} \\ & {\scriptstyle U}\searrow \quad \swarrow{\scriptstyle U_{\mathbb{T}_e}} & \\ & \mathscr{A}^J & \end{array}$$

is an equivalence of categories.

Remark 1.11. Let $\mathscr{A}$ be a fibration over $\mathbf{S}$, with Σ subject to the Beck – Chevalley condition (BCC).

(1) The Bénabou – Roubaud Monadicity and Descent Theorem [4] has the following important consequence in our context: $\mathscr{A}$ is a stack if and only if for each epimorphism $e\colon J \twoheadrightarrow I$ in $\mathbf{S}$, the adjoint pair

$$\Sigma_e \dashv e^* : \mathscr{A}^I \to \mathscr{A}^J$$

is monadic.

(2) In turn, the Beck Monadicity Criterion [2, 12] gives a sufficient condition for the stack condition, to wit: If for each epimorphism $e\colon J \twoheadrightarrow I$ in $\mathbf{S}$, e^* creates e^*-contractible coequalizers and e^* reflects isomorphisms, then $\mathscr{A}$ is a stack.

Corollary 1.12. *The fibration*

$$\mathrm{cod}\colon \mathbf{S}^{\to} \to \mathbf{S}$$

is a stack.

PROOF. An exact category is a regular category in which equivalence relations are kernel pairs. Any topos $\mathbf{S}$ is exact and so are all of its slices. For $e\colon J \to I$ a morphism in $\mathbf{S}$, $e^*\colon \mathbf{S}/I \to \mathbf{S}/J$ is exact [1]. Since there is a left adjoint $\Sigma_e \dashv e^*$ that satisfies the (BCC), in order to show that the fibration in question is a stack, we need only verify monadicity, according to Remark 1.11(1). The first condition for monadicity for any morphism $e : J \to I$ follows from the exactness of e^*. Furthermore, for e an epimorphism, e^* is faithful, so it reflects isomorphisms. Therefore also the second condition for monadicity holds. The adjoint pair $\Sigma_e \dashv e^*$ is then monadic, hence the result. □

Theorem 1.13. *The (fibrational) stack completion of a category $\mathbb{C}$ in $\mathbf{S}$ is identified with the first factor in the factorization of* $\mathrm{yon}\colon \mathbb{C} \to \mathbf{S}^{\mathbb{C}^{\mathrm{op}}}$ *given by*

$$[\mathbb{C}] \xrightarrow{\mathrm{yon}} \mathbf{LR}(\mathbf{S}^{\mathbb{C}^{\mathrm{op}}}) \hookrightarrow \mathbf{S}^{\mathbb{C}^{\mathrm{op}}}$$

where $[\mathbb{C}]$ denotes the externalization of $\mathbb{C}$, and $\mathbf{LR}(\mathbf{S}^{\mathbb{C}^{\mathrm{op}}})$ denotes the full subcategory of $\mathbf{S}^{\mathbb{C}^{\mathrm{op}}}$ whose objects are the locally representable presheaves.

PROOF. The argument given in [7] uses Theorem 1.12 and the characterization of stacks in Proposition 1.9. It follows from these two that the fibration $\mathbf{S}^{\mathbb{C}^{\mathrm{op}}}$ is a stack for any category object $\mathbb{C}$ in $\mathbf{S}$. Further, the Yoneda embedding

$$\mathrm{yon}\colon \mathbb{C} \hookrightarrow \mathbf{S}^{\mathbb{C}^{\mathrm{op}}}$$

is already fully faithful, but it need not be essentially surjective. By cutting down to the full subcategory of locally representable objects, the first factor in the factorization

$$[\mathbb{C}] \xrightarrow{\mathrm{yon}} \mathbf{LR}(\mathbf{S}^{\mathbb{C}^{\mathrm{op}}}) \hookrightarrow \mathbf{S}^{\mathbb{C}^{\mathrm{op}}}$$

is not only fully faithful but also essentially surjective, hence a weak equivalence. It is easy to see that $\mathbf{LR}(\mathbf{S}^{\mathbb{C}^{\mathrm{op}}})$ is a stack, hence the stack completion of the category $\mathbb{C}$ in $\mathbf{S}$ (equivalently, of the representable fibration $[\mathbb{C}] \to \mathbf{S}$). □

Denote by $\mathbf{Top}_{\mathbf{S}}$ the 2-category whose objects are $\mathbf{S}$-bounded toposes, its 1-cells are geometric morphisms over $\mathbf{S}$, and where the 2-cells are natural isomorphisms between (inverse image parts) of geometric morphisms. For any pair of $\mathbf{S}$-bounded toposes $\mathscr{E}$ and $\mathscr{F}$ there is then a category denoted $\mathbf{Top}_{\mathbf{S}}[\mathscr{E}, \mathscr{F}]$, whose 1-cells and 2-cells are as described above.

This structure is to be distinguished from that of the bicategory of $\mathbf{S}$-bounded toposes, whose 2-cells are arbitrary natural transformations subject to coherence conditions. We work with the former and not the latter since it is what is suitable in dealing with nonabelian topos cohomology, in particular, taking into account the meaning of a (locally trivial) covering [9, 10].

We end this review section with some important connections between stacks theory and Galois theory. We recall some applications of stack completions to first degree nonabelian cohomology of a topos from [8] and [9].

In [16], Grothendieck defines the *fundamental groupoid* $\Pi_1(\mathscr{G})$ of a Galois topos $\mathscr{E}$ to be the fibration $\mathbf{Points}(\mathscr{E})$ over Set. We have argued coincidentally elsewhere [9] that this is the correct choice over an arbitrary base topos $\mathbf{S}$ as well, whereas a natural candidate for the *Galois groupoid* of $\mathscr{E}$ is the groupoid $G = \mathrm{Aut}(A)$ where A is a universal cover in $\mathscr{E}$. The Galois topos $\mathscr{E}$ is the classifying topos $\mathscr{B}(G)$ of G, and the fundamental groupoid of $\mathscr{E}$ is then identified with the stack completion of the Galois groupoid G. We refer to [9] for a proof of the following result:

Theorem 1.14. *Let $\mathscr{E}, \mathscr{F}$ be Galois toposes bounded over $\mathbf{S}$, considered as classifying toposes of the (etale complete) discrete groupoids G, K.*

- *There is an equivalence of categories*

$$\mathrm{Hom}(G, K) \simeq \mathbf{Top_S}[\mathscr{E}, \mathscr{F}]_+$$

where the symbol $+$ indicates commutation with the canonical (bags of) points.

- *There is an equivalence of categories*

$$\mathrm{Hom}(G^*, K^*) \simeq \mathbf{Top_S}[\mathscr{E}, \mathscr{F}]$$

where G^* and K^* are the (representable) stack completions of G and K, respectively.

We see then that by adding multiplicity (in the stack completions) one eliminates the dependence on the chosen points.

Proposition 1.15. *Let $\mathscr{E}$ be an $\mathbf{S}$-bounded topos. There is an $\mathbf{S}$-fibration* $\mathbf{Points}(\mathscr{E})$ *of points of $\mathscr{E}$. For G an* etale complete *groupoid in $\mathbf{S}$, the stack completion of G may be identified with*

$$[G] \xrightarrow{\Phi} \mathbf{Points}(\mathbf{S}^{G^{\mathrm{op}}}).$$

PROOF. This follows by restricting Proposition 3.1 and Theorem 3.2 of [8] to the discrete case. $\square$

Recall that 1-*dimensional cohomology* of $\mathbf{S}$ with coefficients in an etale complete groupoid $\mathbb{G}$ is given by the formula

$$\mathrm{H}^1(\mathbf{S}; G) = \Pi_0(\mathrm{Tors}^1(G))$$

where Π_0 denotes 'isomorphism classes'.

Proposition 1.16. *Let G be any nonempty and connected groupoid in $\mathbf{S}$. Then there is an equivalence*

$$\mathbf{LR}(\mathbf{S}^{G^{\mathrm{op}}}) \cong \mathrm{Tors}^1(G)$$

as $\mathbf{S}$-fibrations.

PROOF. The proof can be easily adapted from that of [8, Proposition 4.9] in the discrete case. $\square$

Theorem 1.17. *Let G be any etale complete groupoid in $\mathbf{S}$.*

(1) *The canonical morphism*

$$G \xrightarrow{\mathrm{triv}} \mathrm{Tors}^1(G)$$

exhibits $\mathrm{Tors}^1(G)$ as the stack completion of G.

(2) (Diaconescu's theorem [22]) *The topos $\mathbf{S}^{G^{\mathrm{op}}}$ classifies G-torsors, in the sense that there is an equivalence*

$$\mathrm{Tors}^1(G) \cong \mathbf{Points}(\mathbf{S}^{G^{\mathrm{op}}}).$$

PROOF. Theorem 1.13 applies to G. That is, we have that

$$[G] \xrightarrow{\text{yon}} \mathbf{LR}(\mathbf{S}^{G^{\mathrm{op}}})$$

is a stack completion of G. We now apply Proposition 1.16 to prove the first assertion. The second assertion follows from the first and Proposition 1.15. □

2. Stack completions in Algebra and Analysis

In this section we apply the criterion from Corollary 1.10 directly in order to find examples of stack completions in mathematics in addition to those given at the end of the second section of [7], as well as the main examples in nonabelian cohomology.

We illustrate this first by deriving a result of [26] involving the Zariski topos **Zar** from a more basic theorem in Commutative Algebra, and we then use the fact that **Zar** classifies local rings in order to derive further examples, such as Kaplansky's theorem and Swan's theorem. Different conceptual proofs of these theorems were given by Mulvey [29].

Theorem 2.1 ([5]). *For a ring A and a finitely generated A-module M, M is projective if and only if there exists a finite family $(a_i \mid i \in I)$, with $a_i \in A$, $1 = \sum c_i a_1$ for some $c_i \in A$, such that for each $i \in I$, the A_{a_i}-module M_{a_i} is free of finite rank. (By A_a and M_a it is meant localization at the multiplicative subset $\{1, a, a^2, \dots\} \subseteq A$.)*

Corollary 2.2 ([26]). *In the* Zariski topos **Zar**, *for U the generic local ring, the canonical functor*

$$\alpha_U : \mathbb{F}_U \to \mathbb{P}_U,$$

where $\mathbb{F}_U$ is the internal category of free U-modules of finite rank, and $\mathbb{P}_U$ is the internal category of finitely generated projective U-modules, is a weak equivalence.

PROOF. It is a simple exercise to show that Theorem 2.1 in Commutative Algebra can be interpreted as claimed in the Zariski topos in terms of the given site for it. □

Corollary 2.3 (Kaplansky's theorem). *For a local ring L in* Set, *the canonical functor*

$$\alpha_L : \mathbb{F}_L \to \mathbb{P}_L$$

is an equivalence.

PROOF. This follows from the fact that $L = \varphi^*(U)$ for a (unique) geometric morphism φ: Set $\to$ Zar, that any inverse image part of a geometric morphism preserves weak equivalence functors, and that in any topos satisfying the axiom of choice, every weak equivalence is an equivalence. □

Corollary 2.4 (Swan's theorem). *Let X be a paracompact topological space. Then, there is an equivalence between the category of* real vector bundles over X *and that of* finitely generated projective $\mathrm{Cont}(X, \mathbb{R})$-modules.

PROOF. In $\mathrm{Sh}(X)$, with X paracompact, and $\mathscr{C}_\mathbb{R}$ the sheaf of germs of $\mathbb{R}$-valued continuous functions, we have the following commutative diagram:

$$\begin{array}{ccc} \mathbb{F}_{\mathscr{C}_\mathbb{R}} & \xrightarrow{\alpha_{\mathscr{C}_\mathbb{R}}} & \mathbb{P}_{\mathscr{C}_\mathbb{R}} \\ {\scriptstyle F}\downarrow & & \downarrow{\scriptstyle P} \\ \widetilde{\mathbb{F}_{\mathscr{C}_\mathbb{R}}} & \xrightarrow[\widetilde{\alpha_{\mathscr{C}_\mathbb{R}}}]{} & \widetilde{\mathbb{P}_{\mathscr{C}_\mathbb{R}}} \end{array}$$

where $\alpha_{\mathscr{C}_\mathbb{R}}$ is a wef (same argument as for Kaplansky's theorem), and where P and F are weak equivalence functors into the stack completions, so that the induced $\widetilde{\alpha_{\mathscr{C}_\mathbb{R}}}$ is also one by properties of weak equivalence functors but, between stacks, any weak equivalence is an equivalence. In view of classical theorems from Analysis, this equivalence translates in turn into the statement that there is an equivalence between the categories of *real vector bundles over* X and that of *finitely generated projective* $\mathrm{Cont}(X, \mathbb{R})$*-modules*. □

Remark 2.5. A different proof of Swan's theorem is given in [29]. The strategy is to obtain first a constructive proof of Kaplansky's theorem and then interpreting it in the topos $\mathrm{Sh}(X)$, with X paracompact, and with $L = \mathscr{C}_\mathbb{R}$ the sheaf of germs of $\mathbb{R}$-valued continuous functions.

3. The isomorphism lifting property

Joyal and Tierney [23] have shown that, for $\mathbf{S}$ a Grothendieck topos, there is a Quillen model structure [21, 32] on $\mathbf{Cat}(\mathbf{S})$ whose weak equivalences are the weak equivalence functors, whose cofibrations are the internal functors injective on objects, and whose fibrations are those functors with the right lifting property with respect to all trivial cofibrations. The fibrant objects are called *strong stacks* and are shown to be stacks via a cone construction applied to a weak equivalence functor. In particular, this development provides a second justification for the assertion in Proposition 3.2.

It follows that any category object $\mathbb{C}$ in a Grothendieck topos $\mathbf{S}$ has an internal stack completion: factor the unique functor $\mathbb{C} \to \mathbf{1}$ into

$$\mathbb{C} \to \mathbb{C}^* \to \mathbf{1}$$

where $\mathbb{C} \to \mathbb{C}^*$ is a trivial cofibration and $\mathbb{C}^* \to \mathbf{1}$ is a fibration. Then,

$$\mathbb{C} \to \mathbb{C}^*$$

is the strong stack completion of $\mathbb{C}$, in particular, the stack completion of $\mathbb{C}$ in $\mathbf{Cat}(\mathbf{S})$.

The stack completion of a category object in an elementary topos $\mathbf{S}$ always exists as a fibration, but it need not be representable if $\mathbf{S}$ is not a Grothendieck topos. Notice also that the Grothendieck topos assumption is crucial for the Quillen model structure on $\mathbf{Cat}(\mathbf{S})$ to exist because of the *small object argument.*

Definition 3.1. Let $\mathbf{S}$ be an elementary topos. We say that an elementary topos $\mathbf{S}$ satisfies the *axiom of stack completions* (or (ASC) for short) if, for any a category object $\mathbb{C}$ in $\mathbf{S}$, there there is a representable stack completion of the fibration $[\mathbb{C}] \to \mathbf{S}$ over $\mathbf{S}$.

Proposition 3.2. *Any Grothendieck topos* **S** *satisfies the axiom of stack completions.*

PROOF. That the fibrational stack completion of the externalization of any category object $\mathbb{C}$ in **S** is representable can be shown in the style of [22, Lemma 8.35] using the existence of a generating set (see also [13]). $\square$

Another Quillen model structure on **Cat**(**S**) is that of Lack [24] whose fibrations are those morphisms with the isomorphism lifting propety. It has nothing to do with stacks in **Cat**(**S**), but it provides examples of stacks in $\mathbf{Cat}(\mathbf{S})^2$, as constructed already in [11].

Definition 3.3. A functor $F\colon \mathbb{A} \to \mathbb{B}$ in **Cat**(**S**) is said to satisfy the *isomorphism lifting property* ((ILP) for short), if the following holds:

- For every object A in $\mathbb{A}$, and every isomorphism $B \cong FA$ in $\mathbb{B}$ consisting of $b\colon B \to FA$ and $b'\colon FA \to B$ such that $bb' = 1$ and $b'b = 1$, there is given an isomorphism $A' \cong A$ in $\mathbb{A}$ consisting of $a\colon A' \to A$ and $a'\colon A \to A'$ such that $aa' = 1$ and $a'a = 1$, with $F(A') = B$, $Fa = b$, and $F(a') = b'$. (The indicated liftings are unique up to iso, hence Cartesian.)

Remark 3.4. The stability of wef in **Cat**(**S**) under 2-pullbacks (or iso-comma-objects), shown directly in [11], can alternatively be seen as a consequence a more basic fact, to wit, stability under pullback along any functor with the (ILP). In the diagram below, the projection $\widetilde{G}\colon \mathbb{B}\|G \to \mathbb{B}$ is the *free* (ILP) *functor* on G. The left square is the usual pullback of $\tilde{G}$ along F.

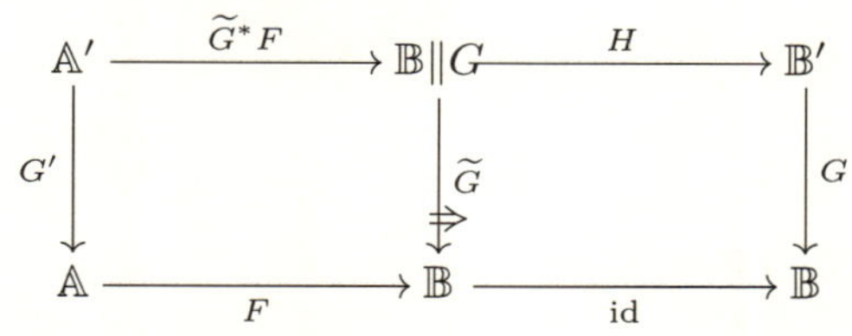

We know from [34] that the other projection $H\colon \mathbb{B}\|G \to \mathbb{B}'$ has a left adjoint left inverse, in fact, a pseudoinverse K, so that H is an adjoint equivalence. If F is a wef, then, since $\widetilde{G}$ has the (ILP), $\widetilde{G}^*F$ is a wef, and therefore the composite $F' = G^*(F) = H \cdot \widetilde{G}^*F$ is a wef.

Lemma 3.5 ([24]). *A functor* $F\colon \mathbb{A} \to \mathbb{B}$ *in* **Cat**(**S**) *satisfies the* (ILP) *iff it has the right lifting property with respect to the trivial cofibration* $j\colon \mathbf{1} \to \mathbf{I}$, *with* **I** *the 'free-living isomorphism.'*

Remark 3.6. As shown in [11] as a consequence of Corollary 1.10, a functor $F\colon \mathbb{A} \to \mathbb{B}$ in Set satisfying the (ILP) is a stack as a category object in the topos $\mathrm{Set}^{\mathbf{2}}$. Furthermore, for an arbitrary category object $F\colon \mathbb{A} \to \mathbb{B}$ in $\mathrm{Set}^{\mathbf{2}}$, its stack completion exists internally and is given by the *free* (ILP) functor on F. Indeed, in the commutative diagram

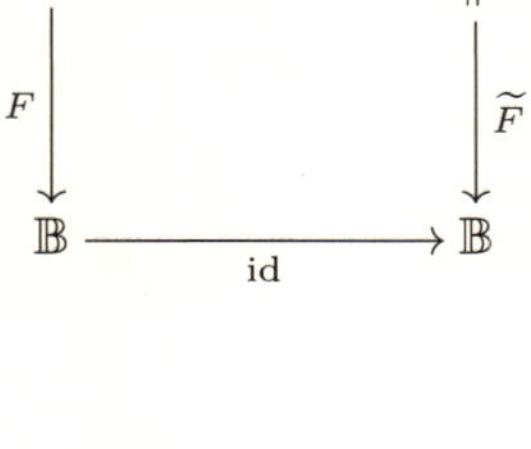

we have $\widetilde{F}\colon \mathbb{B}\|F \to \mathbb{B}$ an isofibration, and both K and id are equivalence functors. The pair $(K, \mathrm{id})\colon F \to \widetilde{F}$ is a wef in $\mathbf{Cat^2}$ and so, by Corollary 1.10, it exhibits $\widetilde{F}$ as the stack completion of F.

We now give an example of a stack completion that shows in particular that for an arbitrary topos $\mathbf{S}$ not every internal category in it is a stack.

Example 3.7. This example is an instance of Remark 3.6. Let $\mathbf{S} = \mathrm{Set}^{\mathbf{2}}$. Let G be a group (in Set). Consider the category object in $\mathbf{S}$ given by the functor

$$H\colon 1 \to G$$

which picks up the unit element e of G, that is, $H(0) = e$.

We claim that $\mathbb{C}$ is not a stack in $\mathbf{S}$. The best way to illustrate this is to exhibit its stack completion $\widetilde{\mathbb{C}}$ given by

$$\widetilde{H}\colon \mathrm{Ind}(G) \to G$$

where $\mathrm{Ind}(G)$ has as objects all elements $g \in G$, identified with triplets $(0, 0, g)$, and as morphisms $k\colon (0, 0, g) \to (0, 0, h)$ any element $k \in G$ such that $g = kh$ in G, that is $k = g.h^{-1}$. The functor $\widetilde{H}$ is defined by $\widetilde{H}(0, 0, g) = 0$ and $\widetilde{H}(k) = k$. We have a commutative square

$$\begin{array}{ccc}
1 & \xrightarrow{\Phi_0} & \mathrm{Ind}(G) \\
{\scriptstyle H}\downarrow & & \downarrow{\scriptstyle \widetilde{H}} \\
G & \xrightarrow[\Phi_1]{} & G
\end{array}$$

where Φ_1 is the identity functor, and $\Phi_1(0) = (0, 0, e)$. Notice that both Φ_0 and Φ_1 are equivalences in Set, yet the pair (Φ_0, Φ_1) is not an equivalence.

Remark 3.8. The fibrant objects of the Quillen Model structure of [23] are those functors of the form $\mathbb{C} \to \mathbf{1}$ with the right lifting property with respect to all trivial cofibrations. These are in turn identified with the category objects $\mathbb{C}$ which are stacks.

Contrast this with the notion of a category object $\mathbb{C}$ such that $\mathbb{C} \to \mathbf{1}$ has the right lifting property with respect to the trivial cofibration $j\colon \mathbf{1} \to \mathbb{I}$, where $\mathbb{I}$ is the free-living isomorphism, that is, such that $\mathbb{C} \to \mathbf{1}$ satisfies (ILP).

For $\mathbf{S}$ an arbitrary Grothendieck topos $\mathbf{S}$, although every internal functor $\mathbb{C} \to \mathbf{1}$ satisfies the (ILP), not every such functor is a fibration for the Quillen model structure on $\mathbf{Cat}(\mathbf{S})$ of [23]—equivalently, not every category object $\mathbb{C}$ in $\mathbf{S}$ is a stack. As argued in [11], a topos $\mathbf{S}$ satisfies the axiom of choice if and only if every internal category in it is a stack.

One may ask then what is the precise connection between the Quillen model structures of [23] and of [24]. It is clear from the above discussion that every fibration of [23] is also one for [24], that is, it satisfies the (ILP). This can also be inferred from the fact that orthogonality with respect to every trivial cofibration of [23] is in particular orthogonal to the trivial cofibration $j\colon \mathbf{1} \to \mathbb{I}$ and from the possibility in principle to find trivial cofibrations other than $j\colon \mathbf{1} \to \mathbb{I}$ that would force an additional injectivity requirement that is nontrivial without Choice. For instance, for any monomorphism $m\colon X \rightarrowtail Y$ with $X \twoheadrightarrow 1$, the codiscrete groupoid functor $G(m)\colon G(X) \to G(Y)$ is one such trivial cofibration.

It is claimed in [23] that the given Quillen model structure on $\mathbf{Cat}(\mathbf{S})$, for $\mathbf{S}$ a Grothendieck topos, is cofibrantly generated. The above shows that it is not the case that in general $j : \mathbf{1} \to \mathbb{I}$, where $\mathbb{I}$ is the free-living isomorphism, can be the single generating trivial cofibration. An explicit description of a generating set of trivial cofibrations is not, however, given therein. A general argument seems to be the only way to assert this fact.

4. Intrinsic 2-stacks

Let $\mathbf{S}$ be an elementary topos. When passing from the 0-level to level 1, we should note that the trivial identifications $\mathbf{Cat}^0(\mathbf{S}) = \mathbf{S} = \mathbf{Stack}^0(\mathbf{S})$ get differentiated. We shall denote by $\mathbf{Stack}(\mathbf{S})$ the full sub 2-category $\mathbf{Stack} \hookrightarrow \mathbf{Cat}$ whose objects are category objects in $\mathbf{S}$ which are intrinsic 1-stacks (we omit the superscript 1 in this case). More generally we want an inclusion of 2-fibrations, so we recall the basic definition from [18, 20]:

Definition 4.1. A 2-functor $P\colon F \to E$ is a *2-fibration* if:

- for any object X in F and any 1-cell $u\colon I \to PX$ in E, there is a 2-Cartesian 1-cell $\bar{u}\colon u^*(X) \to X$ with $P\bar{u} = u$.
- For any pair of objects X, Y in F, the induced functor $P_{X,Y}\colon F(X,Y) \to E(PX, PY)$ is a fibration, stable under precomposition: for every 1-cell $h\colon Z \to X$ in F, the functor $F(h,Y)\colon F(X,Y) \to F(Z,Y)$ preserves 2-Cartesian morphisms (from $P_{X,Y}$ to $P_{Z,Y}$).

Definition 4.2. Let $\mathrm{cod}\colon \mathbf{Stack}^{\to} \to \mathbf{S}$ be given by the following data.

(1) The fiber above an object I of $\mathbf{S}$ is given by the 2-category $\mathbf{Stack}/\mathbb{I}_{\mathrm{dis}}$,

- whose *objects* are pairs $\langle \mathbb{C}, P\rangle$, where $\mathbb{C}$ is a category object that is a stack in $\mathbf{S}$ and $P : \mathbb{C} \to \mathbb{I}_{\mathrm{dis}}$ a functor (necessarily both a fibration and a cofibration since $\mathbb{I}_{\mathrm{dis}}$ is discrete),
- whose *morphisms* are functors F which fit into a commutative diagram

$$\begin{array}{ccc} \mathbb{C} & \xrightarrow{\quad F \quad} & \mathbb{D} \\ & {\scriptstyle P}\searrow \quad \swarrow{\scriptstyle Q} & \\ & \mathbb{I}_{\mathrm{dis}} & \end{array}$$

- and whose *2-cells* are natural transformations $\alpha\colon F \Rightarrow F'\colon \mathbb{C} \to \mathbb{D}$ over $\mathbb{I}_{\mathrm{dis}}$.

(2) Change of base along a morphism $u\colon J \to I$ in $\mathbf{S}$ is given by the 2-functor

$$u^* : \mathbf{Stack}/\mathbb{I}_{\mathrm{dis}} \to \mathbf{Stack}/\mathbb{J}_{\mathrm{dis}},$$

pullback in $\mathbf{Cat}$ along the functor $u_{\mathrm{dis}}\colon \mathbb{J}_{\mathrm{dis}} \to \mathbb{I}_{\mathrm{dis}}$.

Proposition 4.3. *With the data as in Definition* 4.2, $\mathrm{cod}\colon \mathbf{Stack}^{\to} \to \mathbf{S}$ *is a 2-fibration.*

PROOF. The proof is straightforward. □

Given a morphism $e\colon J \to I$ in $\mathbf{S}$, its 2-kernel is the 2-groupoid given by the 3-truncated simplicial complex partially depicted by the following diagram whose horizontal right arrows are labelled d with subscripts, and whose horizontal left arrows are labelled s with subscripts.

$$J \times_I J \times_I J \times_I J \;\substack{\longrightarrow\\ \longrightarrow\\ \longrightarrow\\ \longrightarrow}\; J \times_I J \times_I J \;\substack{\longrightarrow\\ \longrightarrow\\ \longrightarrow}\; J \times_I J \;\substack{\longrightarrow\\ \longleftarrow\\ \longrightarrow}\; J.$$

Definition 4.4 ([20]). Let $\mathscr{A}$ be a 2-fibration. We denote by 2-$\mathrm{Des}_{\mathscr{A}}(e)$ the 2-category of 2-*descent data* on $\mathscr{A}$ with respect to $e\colon J \to I$, a morphism in $\mathbf{S}$:

- An *object* of 2-$\mathrm{Des}_{\mathscr{A}}(e)$ is a 4-tuple

$$(a, f, \gamma, \alpha)$$

with a an object of $\mathscr{A}^J$, together with an equivalence $f\colon {d_0}^*(a) \simeq {d_1}^*(a)$, and isomorphisms $\gamma\colon \mathrm{id}_a \cong {s_0}^*(f)$ and $\alpha\colon {d_1}^*(f) \cong {d_0}^*(f){d_2}^*(f)$ satisfying the coherence conditions ${d_0}^*(\gamma).f = {s_1}^*(\alpha)$, $f.{d_1}^*(\gamma) = {s_0}^*(\alpha)$.

- A 1-*cell*

$$(a, f, \gamma, \alpha) \to (b, g, \eta, \beta)$$

consists of a 1-cell $h : a \to b$, and a 2-cell $\theta\colon {d_0}^*(a).f \Rightarrow g.{d_1}^*(h)$, satisfying ${s_0}^*(\theta).\gamma = \eta.h$ and a commutative tetrahedron condition.

- A 2-*cell*

$$(h, \theta) \Rightarrow (k, \kappa)$$

is given by a 2-cell $\phi\colon h \Rightarrow k$ such that ${d_0}^*(\phi).\kappa = \theta.{d_1}^*(\phi)$.

Definition 4.5. • Let $\mathbf{S}$ be an elementary topos. A 2-fibration $\mathscr{A} \to \mathbf{S}$ is said to be an intrinsic 2-*stack* if for every epimorphism $e\colon J \to I$ in $\mathbf{S}$, the 2-functor $e^*\colon \mathscr{A}^I \to \mathscr{A}^J$ is of effective 2-descent, meaning that the canonical 2-functor $\Phi_e\colon \mathscr{A}^I \to$ 2-$\mathrm{Des}_{\mathscr{A}}(e)$ is an equivalence of 2-categories.

• Let $\mathscr{A} \to \mathbf{S}$ be a 2-fibration. Then, a 2-*stack completion* of $\mathscr{A}$ is given a pair $(\tilde{\mathscr{A}}, F)$, where $\tilde{\mathscr{A}} \to \mathbf{S}$ is a 2-stack, and $F\colon \mathscr{A} \to \tilde{\mathscr{A}}$ is a morphism of 2-fibrations, satisfying the obvious universal property among such pairs.

Definition 4.6. Let $F\colon \mathscr{B} \to \mathscr{C}$ be a 2-functor between 2-fibrations over $\mathbf{S}$. It is said to be a *weak* 2-*equivalence* if the following conditions hold:

(1) For each $I \in \mathbf{S}$, $F^I\colon \mathscr{B}^I \to \mathscr{C}^I$, and $c \in \mathscr{C}^I$, there exists an epimorphism $e\colon J \to I$ in $\mathbf{S}$, $b \in \mathscr{B}^J$, and an equivalence $\theta\colon F^J(b) \to e^*(c)$.

(2) $\forall I \in \mathbf{S}\ \forall b, b' \in \mathscr{B}^I$, the morphism

$$\mathrm{Hom}_{\mathscr{B}^I}(b, b') \xrightarrow{{F^I}_{b,b'}} \mathrm{Hom}_{\mathscr{C}^I}(F^I b, F^I b')$$

of 1-fibrations is a weak 1-equivalence functor.

Proposition 4.7. *Let $\mathscr{A}$ be a 2-fibration over $\mathbf{S}$. Then, $\mathscr{A}$ is a 2-stack if and only if the internal 2-functor $F_e\colon {\mathbb{J}^{(2)}}_e \to {\mathbb{I}^{(2)}}_{\mathrm{dis}}$, from the 2-kernel groupoid of e to the discrete 2-category on I, induces an equivalence of 2-fibrations*

$$\mathscr{A}^{F_e}\colon \mathscr{A}^{{\mathbb{I}^{(2)}}_{\mathrm{dis}}} \to \mathscr{A}^{{\mathbb{J}^{(2)}}_e}.$$

PROOF. The diagram

$$\begin{array}{ccc}
\mathscr{A}^I & \xrightarrow{\Phi_e} & 2\text{-}\mathrm{Des}_{\mathscr{A}}(e) \\
{\scriptstyle \kappa_I}\downarrow & & \downarrow{\scriptstyle \Psi_e} \\
\mathscr{A}^{{\mathbb{I}^{(2)}}_{\mathrm{dis}}} & \xrightarrow[\mathscr{A}^{F_e}]{} & \mathscr{A}^{{\mathbb{J}^{(2)}}_e}
\end{array}$$

is commutative, with κ_I a 2-equivalence. That Ψ_e is a 2-equivalence is shown by an argument analogous to that employed in the 1-dimensional case (exercise). Therefore, for any epimorphism $e\colon J \to I$ in $\mathbf{S}$, Φ_e is a 2-equivalence (hence $\mathscr{A}$ is a 2-stack), if and only of $\mathscr{A}^{F_e}$ is a 2-equivalence, as claimed. □

Lemma 4.8. *For any epimorphism e in $\mathbf{S}$, the internal 2-functor*

$$\mathscr{F}_e \colon \mathbb{J}^{(2)}{}_e \to \mathbb{I}^{(2)}{}_{\text{dis}}$$

from the 2-kernel groupoid of e to the discrete 2-category on I, is a weak 2-equivalence functor.

PROOF. The first condition in Definition 4.6 is obviously satisfied since $e\colon J \to I$ is an epimorphism. The second condition requires that

$$\mathrm{HOM}_{\mathbb{J}^{(2)}{}_e}(x, x') \xrightarrow{(F_e)x,x'} \mathrm{HOM}_{\mathbb{I}^{(2)}{}_{\text{dis}}}(Fx, Fx')$$

be a weak 1-equivalence of categories: fully faithfulness follows from the pullback construction of $\mathbb{J}^{(2)}{}_e$ while essential surjectivity follows once again from e being an epimorphism. □

Theorem 4.9. *A 2-fibration $\mathscr{A}$ over $\mathbf{S}$ is a 2-stack iff for every weak 2-equivalence $\Gamma\colon \mathscr{B} \to \mathscr{C}$, the induced $\mathscr{A}^\Gamma\colon \mathscr{A}^\mathscr{C} \to \mathscr{A}^\mathscr{B}$ is an equivalence of 2-fibrations.*

PROOF. It follows from Proposition 4.7 and Lemma 4.8 that the condition of the theorem is sufficient. We now prove necessity. Let $\mathscr{A}$ be an intrinsic 2-stack (Definition 4.5), and $\Gamma\colon \mathscr{B} \to \mathscr{C}$ a weak 2-equivalence (Definition 4.6).

Our goal is to define a 2-functor

$$H\colon \mathscr{A}^\mathscr{B} \to \mathscr{A}^\mathscr{C}$$

that is 'inverse' to $G = \mathscr{A}^\Gamma\colon \mathscr{A}^\mathscr{C} \to \mathscr{A}^\mathscr{B}$. Since all concepts involved are stable under localization, it will be enough to argue with the fibers over $1 \in \mathbf{S}$.

(1) Let $\Phi\colon \mathscr{B} \to \mathscr{A}$ be a 2-functor. In what follows we define the data for a 2-functor $\Psi\colon \mathscr{C} \to \mathscr{A}$, and let $H(\Phi) = \Psi$. We then extend the definition to 1-cells and 2-cells.

- Let $I \in \mathbf{S}$ and $c \in \mathscr{C}^I$. By the first condition on the weak 2-equivalence $\Gamma\colon \mathscr{B} \to \mathscr{C}$, there is an epimorphism $e\colon J \to I$ in $\mathbf{S}$, $b \in \mathscr{B}^J$, and an equivalence $\theta\colon \Gamma^J(b) \cong e^\star(c)$ in $\mathscr{C}^J$. Since $e^\star(c)$ has 2-descent data for e (see Definition 4.4), then so does $\Gamma^J(b)$.

Simplify the notation for the 2-kernel of e as the following 3-truncated simplicial complex:

$$J'''' \overrightarrow{\overrightarrow{\overrightarrow{\to}}} J'' \overrightarrow{\overrightarrow{\to}} J' \underset{\longleftarrow}{\overset{\longrightarrow}{\longleftarrow}} J.$$

Since $\Gamma\colon \mathscr{B} \to \mathscr{C}$ is a weak 2-equivalence, all three of the $\Gamma^{J'}$, $\Gamma^{J''}$ and $\Gamma^{J''''}$ are fully faithful by condition 2 in Definition 4.6. Since $\Gamma^J(b)$ has 2-descent data for e, so does b by the above observation. Therefore, also $\Phi^J(b)$ has 2-descent data for e. Since $\mathscr{A}$ is a 2-stack, there exists $a \in \mathscr{A}^I$ such that $e^\star(a) \cong \Phi^J(b)$. The choice of a is unique up to equivalence compatible with 2-descent data. If we choose some other e and b, by taking a common refinement, we see that there will be a unique equivalence between the new a and the old one, compatible with the 2-descent data. So, choose one such a and let $\Psi^I(c) = a$.

- If $h\colon c \to c'$ is a 1-cell in $\mathscr{C}^I$ then, by choosing a common refinement, we may assume that the same epimorphism $e\colon J \to I$ works for both c and c'. Then there exist $b, b' \in \mathscr{B}^J$ and equivalences $\theta\colon \Gamma^J(b) \cong e^\star(c)$ and $\theta'\colon \Gamma^J(b') \cong e^\star(c')$. Using that

$$\mathrm{Hom}_{\mathscr{B}^J}(b, b') \xrightarrow{\Gamma^J{}_{b,b'}} \mathrm{Hom}_{\mathscr{C}^J}\big(\Gamma^J(b), \Gamma^J(b')\big)$$

is a (1-)weak equivalence functor, and in particular full and faithful, there exists a unique $g\colon b \to b'$ such that the diagram

$$\begin{array}{ccc} \Gamma^J(b) & \xrightarrow{\theta} & e^\star(c) \\ {\scriptstyle \Gamma^J(g)}\downarrow & & \downarrow{\scriptstyle e^\star(h)} \\ \Gamma^J(b') & \xrightarrow[\theta']{} & e^\star(c') \end{array}$$

commutes (up to equivalence) with the horizontal arrows the given equivalences in the 2-category $\mathscr{C}^J$. Further, g is compatible with the 2-descent data on b and b', so that

$$\Phi^J(g)\colon \Phi^J(b) \to \Phi^J(b')$$

is also compatible with the corresponding 2-descent data. Therefore there exists a unique $f\colon a \to a'$ such that

$$\begin{array}{ccc} \Phi^J(b) & \longrightarrow & e^\star(a) \\ {\scriptstyle \Phi^J(g)}\downarrow & & \downarrow{\scriptstyle e^\star(f)} \\ \Phi^J(b') & \longrightarrow & e^\star(a') \end{array}$$

commutes (up to equivalence). Define $\Psi^I(h) = f$.

• For a 2-cell $\gamma\colon h \Rightarrow h'$, get $\beta\colon g \Rightarrow g'$ and then $\alpha\colon f \Rightarrow f'$, by arguments of the sort employed above. Define $\Psi^I(\gamma) = \alpha$. It is a routine verification to show that Ψ is a 2-functor.

• In a similar manner we define $H(t)$ for a 1-cell $t : \Phi \to \Phi'$, and $H(\delta)$ for a 2-cell $\delta\colon t \Rightarrow t'$. Verify that with these definitions H is a morphism of 2-fibrations.

(2) **Claim.** $(G \cdot H) \cong 1_{\mathscr{A}\mathscr{B}}$. By construction,

$$(G \cdot H)(\Phi) = \mathscr{A}^\Gamma\big(H(\Phi)\big) = \mathscr{A}^\Gamma(\Psi) = \Psi\Gamma.$$

Let $I \in \mathbf{S}$ and $b \in \mathscr{B}^I$. To calculate $\Psi^I\Gamma^I(b)$ as above, we may chose $e = 1_I\colon I \to I$ since $\Gamma^I(b) = (1_I)^\star\Gamma^I(b)$. Then, $\Psi^I\gamma^I(b) = a$ for some $a \in \mathscr{A}^I$ such that $1_I^\star(a) \cong \Phi^I(b)$, thus $\Psi^I\Gamma^I(b) \cong \Phi^I(b)$. We argue in a similar manner for 1-cells $g\colon b \to b'$ and for 2-cells $t\colon g \Rightarrow g'$. The conclusion reached says that

$$\Psi\Gamma \cong \Phi$$

hence (part of) the claim. One has to complete this with arguments involving 1-cells $\alpha\colon \Phi \to \Phi'$ and 2-cells $s\colon \alpha \Rightarrow \alpha'\colon \Phi \to \Phi'$. This is lengthy but entirely routine. □

(3) **Claim.** $(H \cdot G) \cong 1_{\mathscr{A}\mathscr{C}}$. Consider any 2-functor $\Psi' : \mathscr{C} \dashrightarrow \mathscr{A}$ such that $H(\mathscr{A}^\Gamma(\Psi')) = \Psi$. This time, to calculate $\Psi^I(c)$ we choose, as above, an epimorphism $e : J \to I$ and $b \in \mathscr{B}^I$ such that $\Gamma^J(b) \cong e^\star(c)$, and then find the 'unique' $a \in \mathscr{A}^J$ such that ${\Phi'}^J\Gamma^J(b) \cong e^\star(a)$. But

$${\Psi'}^J\Gamma^J(b) \cong {\Psi'}^J(e^\star(c)) \cong e^\star{\Psi'}^I(c)$$

so $\Psi^I(c) \cong {\Psi'}^I(c)$. This equivalence is natural, so $\Psi \cong \Psi'$. This proves the claim. □

Hence, $\mathscr{A}^\Gamma$ is an equivalence. Modulo the many routine details left to the patient reader, this finishes the proof. □

Corollary 4.10. *If $F\colon \mathscr{A} \to \mathscr{B}$ is a morphism of 2-fibrations that is a weak 2-equivalence, and $\mathscr{B}$ is a 2-stack, then the pair $(\mathscr{B}, F)$ is the 2-stack completion of $\mathscr{A}$.*

5. Stack is a 2-stack

In the construction [7] of the fibrational 1-stack completion of a category object in an elementary topos $\mathbf{S}$, a crucial step was to prove that the fibration $\mathrm{cod}\colon \mathbf{S}^{\to} \to \mathbf{S}$ is a 1-stack [11]. Recall that this was shown using the monadicity and descent theorems of [2] and [4].

It is our aim to construct the 2-stack completion of a 2-category object $\mathbf{C}$ in $\mathbf{S}$, possibly with certain assumptions on $\mathbf{C}$. By analogy with the case of dimension 1, we first show that $\mathrm{cod}\colon \mathbf{Stack}^{\to} \to \mathbf{S}$ is a 2-stack.

In the case of *groupoid stacks*, a direct proof of this fact is given in [30], but only for the fibers over 1. We shall resort instead to 2-dimensional analogues of the monadicity and descent theorems as given in [20], and prove the full 2-fibrational result. We assume, just as in [30], that $\mathbf{S}$ is a Grothendieck topos or, perhaps more generally, that $\mathbf{S}$ is an elementary topos satisfying the axiom of 1-stack completions.

We begin by recalling some definitions and results from [20] instantiated with $\mathbf{Cat}$ as base 2-category:

Definition 5.1. A 2-fibration $P\colon \mathscr{A} \to \mathbf{Cat}$ is said *to have Σ with the BCC for comma objects* if for every $u\colon \mathbb{B} \to \mathbb{A}$ in $\mathbf{Cat}$, the change of base $u^*\colon \mathscr{A}^{\mathbb{A}} \to \mathscr{A}^{\mathbb{B}}$ admits a left 2-adjoint $\Sigma_u \dashv u^*$ such that, for every comma square

$$\begin{array}{ccc} (v \downarrow u) & \xrightarrow{\;q\;} & \mathbb{B} \\ {\scriptstyle p}\downarrow & \Rightarrow \lambda & \downarrow{\scriptstyle u} \\ \mathbb{K} & \xrightarrow[\;v\;]{} & \mathbb{A} \end{array}$$

in $\mathbf{Cat}$, the induced

$$\tilde{\lambda}\colon \Sigma_p \cdot q^* \Rightarrow v^* \cdot \Sigma_u$$

is an equivalence.

Theorem 5.2 ([20]). *If a 2-fibration $P\colon \mathscr{A} \to \mathbf{Cat}$ has Σ subject to the BCC for comma objects in $\mathbf{Cat}$ along a cofibration, then, given $e\colon \mathbb{T} \to \mathbb{Q}$ in $\mathbf{Cat}$, there is a canonical biequivalence*

$$\text{2-Des}_{\mathscr{A}}(e) \to \mathrm{Ps}(e^*\Sigma_e)\mathrm{Alg}$$

where $\mathrm{Ps}(e^*\Sigma_e)\mathrm{Alg}$ is the 2-category of pseudo-algebras of the pseudo-monad $e^*\Sigma_e$, induced by e.

Definition 5.3. A functor $q\colon \mathbb{O} \to \mathbb{Q}$ in Cat is said to be a *regular 2-epi* if the comma object

$$\begin{array}{ccc} (q \downarrow q) & \xrightarrow{\;d\;} & \mathbb{O} \\ {\scriptstyle c}\downarrow & \Rightarrow \lambda & \downarrow{\scriptstyle q} \\ \mathbb{O} & \xrightarrow[\;q\;]{} & \mathbb{Q} \end{array}$$

exhibits (q, λ) as the representable Kleisi object[1] of the bimodule monad $(d, c) \colon \mathbb{O} \rightharpoonup \mathbb{O}$.

Remark 5.4. The standard factorization of a functor in **Cat** is the one into a regular 2-epi followed by a fully faithful functor. Using an explicit construction of the representable Kleisli object of a monad in bimodules in **Cat**(**S**) [19], one can identify the ingredients of this factorization for any (regular) epi $e \colon J \to I$ in **S**, regarded as $e_{\text{dis}} \colon \mathbb{J}_{\text{dis}} \to \mathbb{I}_{\text{dis}}$ in **Cat**, into a regular 2-epi $\bar{e}$ followed by a fully faithful functor E

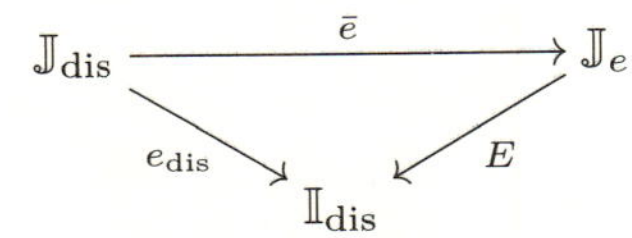

where $E \colon \mathbb{J}_e \to \mathbb{I}_{\text{dis}}$ is the 1-kernel of e. The regular 2-epi $\bar{e} \colon \mathbb{J}_{\text{dis}} \to \mathbb{J}_e$ is the left square in the diagram

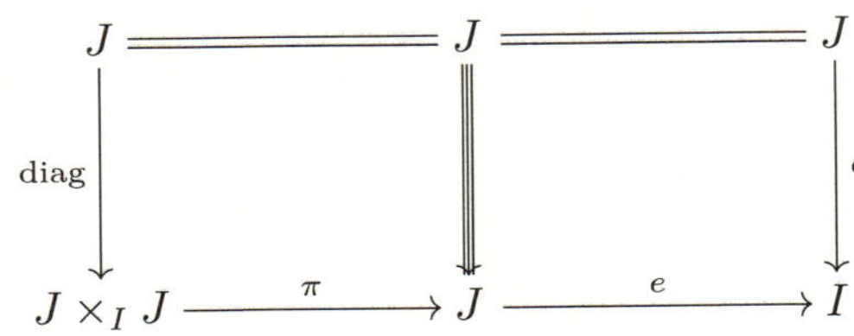

so that it is the identity at the level of objects and the diagonal at the level of morphisms.

Just as a regular epi (i.e., the coequalizer if its own kernel pair) in any regular category **R** is of effective 1-descent, a regular 2-epi in any strongly 2-regular 2-category $\mathscr{R}$ (in the sense of [20]) is of effective 2-descent. We state a particular case of this result, using that **Cat** is a strongly regular 2-category.

Theorem 5.5 ([20]). *Any regular 2-epi in* **Cat** *is of effective 2-descent for the basic 2-fibration*

$$\mathrm{cod} \colon \mathbf{Fib} \to \mathbf{Cat}\,.$$

Proposition 5.6. *Let* **S** *be a Grothendieck topos. Then* **Stack** *is a reflective subcategory of* **Cat**. *In particular, it has all colimits that exist in* **Cat**.

Proof. This is a consequence of the existence of representable stack completions of category objects in any Grothendieck topos (Proposition 3.2). The colimit of a diagram in **Stack** is the stack completion of the colimit of the diagram regarded in **Cat**. □

Theorem 5.7. *Let* **S** *be as above. Then, the 2-fibration* $\mathrm{cod} \colon \mathbf{Stack}^{\to} \to \mathbf{S}$ *is an intrinsic 2-stack. Equivalently, any regular epi in* **S** *is of effective 2-descent for* $\mathrm{cod} \colon \mathbf{Stack}^{\to} \to \mathbf{S}$.

Proof. Since **S** is assumed to be a Grothendieck topos, the 2-category **Stack** has all bicolimits that exist in **Cat**(**S**), by Proposition 5.6. The same is true of the fiber of **Stack** at any I (not just at 1) and, for any $u \colon J \to I$,

$$u^* \colon \mathbf{Stack}/\mathbb{I}_{\text{dis}} \to \mathbf{Stack}/\mathbb{J}_{\text{dis}}$$

[1] Such a Kleisli object for bimodules is a *lax colimit* of the corresponding diagram of functors and 2-cells in **Cat**.

preserves such bicolimits (in particular, pseudo-coequalizers). Moreover, since the 2-fibration cod : /*colon***Fib** $\to$ **Cat** has Σ with the BCC for comma objects, the same is true for cod: $(\mathbf{Stack}^{\to}) \to \mathbf{S}$.

By Theorem 5.2, in order to prove that **Stack**is a 2-stack, it is enough (by the pseudomonadicity theorem of [20]) to prove that, for any epimorphism $e\colon J \to I$ in $\mathbf{S}$, $e^*\colon \mathbf{Stack}/\mathbb{I}_{\mathrm{dis}} \to \mathbf{Stack}/\mathbb{J}_{\mathrm{dis}}$ reflects equivalences.

Recall from Remark 5.4 that the regular 2-epi/fully faithful factorization of $e_{\mathrm{dis}}\colon \mathbb{J}_{\mathrm{dis}} \to \mathbb{I}_{\mathrm{dis}}$ is given by

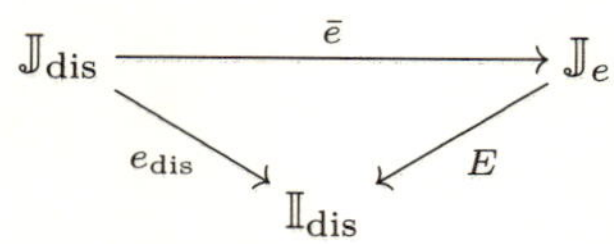

where $E\colon \mathbb{J}_e \to \mathbb{I}_{\mathrm{dis}}$ is the 1-kernel of e. Let x be such that $e^*x = (\bar{e}^*E^*)x$ is an equivalence. Since $\bar{e}$ is a regular 2-epi, it is of effective 2-descent for cod: **Fib** $\to$ **Cat** by Theorem 5.5, hence, it reflects equivalences. It follows then that E^*x is an equivalence.

It remains to show that x is an equivalence. For this, we use the fact that $E\colon \mathbb{J}_e \to \mathbb{I}_{dis}$ is a wef as follows. Consider the prism diagram

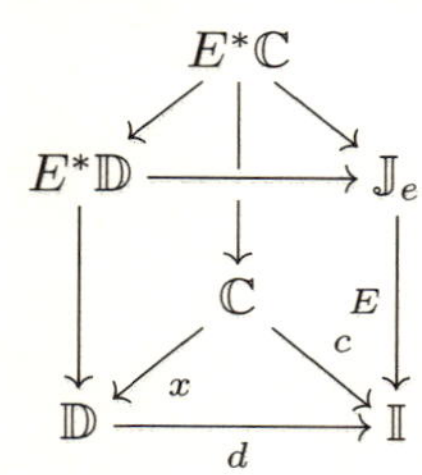

where the front and back square faces are pullbacks, and where $E^*x\colon E^*\mathbb{C} \to E^*\mathbb{D}$ is an equivalence. The morphisms $c^*E\colon E^*\mathbb{C} \to \mathbb{C}$ and $d^*E\colon E^*\mathbb{D} \to \mathbb{D}$ are weak equivalence functors, since E is one and the codomains of the pullbacks are discrete, so that they are 2-pullbacks in the sense of [11]. (Recall that weak equivalence functors are stable under 2-pullbacks, by Proposition 1.3.) Therefore, $d^*E \cdot E^*x = x \cdot c^*E$ (the lhs commuting square) is a weak equivalence functor since wef are composable and any equivalence is a wef. Now, since both the composite $x \cdot c^*E$ and c^*E are wef, so is x. But $\mathbb{C}$ is a stack, hence the wef x is an equivalence. This concludes the proof. □

Remark 5.8. The argument that, since E is a wef so are c^*E and d^*E, may be given two alternative justifications using that c and d are cofibrations (as their codomain is discrete) and wef are pullback stable along cofibrations, or else that E is a swef (surjective weak equivalence, i.e., with the object part an epi), and those are are pullback stable, so that c^*E and d^*E are in fact swef.

6. Applications to second degree nonabelian cohomology of a topos

The notion of a 2-gerbe [6, 13] is normally reserved for 2-groupoids. We are interested in a generalization of this notion which includes, as instances, both the 2-gerbes and the 2-stacks. In particular, the 2-stack completion of a generalized 2-gerbe, which we will show exists, is a generalized 2-gerbe.

Definition 6.1. A 2-category $\mathbf{C}$ in $\mathbf{S}$ is said to be a *generalized 2-gerbe* if it is locally a 1-stack, in the sense that all the Hom-categories $\mathbf{C}(x, y)$ are 1-stacks in $\mathbf{Cat}(\mathbf{S})$.

Denote by 2-**Gerb** the full subcategory of 2-$\mathbf{Cat}(\mathbf{S})$ whose objects are the generalized 2-gerbes. For any generalized 2-gerbe $\mathbf{C}$, we have a Yoneda 2-functor

$$\mathbf{C} \xrightarrow{Y} \mathbf{Stack}^{\mathbf{C}^{\mathrm{op}}}.$$

Remark 6.2. (1) Any 2-gerbe is a generalized 2-gerbe.
(2) Any 2-stack in $\mathbf{S}$ is a generalized 2-gerbe.

Definition 6.3. Let $F\colon A \to B$ be a 2-functor between 2-fibrations over $\mathbf{S}$. It is said to be a *special weak 2-equivalence* if the following conditions hold.

(1) For each $I \in \mathbf{S}$, $F^I\colon |A^I| \to |B^I|$, and $b \in |B^I|$, there exists an epimorphism $e : J \to I$ in $\mathbf{S}$, $a \in |A^I|$, and a (strong) equivalence $\theta\colon F^J(e) \to e^*(b)$.

(2) $\forall I \in \mathbf{S}\ \forall x, x' \in |A^I|$, the morphism

$$\mathrm{Hom}_{AA^I}(x, x') \xrightarrow{F_{x,x'}} \mathrm{Hom}_{B^I}(Fx, Fx')$$

of fibrations is a (strong) equivalence functor.

Remark 6.4. Any weak 2-equivalence in 2-**Gerb** is special since, at the level of the homs, it is a strong (not just weak) equivalence. In fact, since the fiber functors (at $I \in \mathbf{S}$)

$$\mathrm{Hom}(x, x') \xrightarrow{\mathrm{Y}_{x,x'}} \mathrm{Hom}\big(Y(x), Y(x')\big)$$

of fibrations is a weak equivalence functor with domain a stack, it is necessarily a strong equivalence.

Definition 6.5. Let $\mathbb{C}$ be a generalized 2-gerbe. An I-indexed family X of the 2-category $(\mathbf{Stack}^{\mathbf{C}^{\mathrm{op}}})^I$ is said to be *locally representable* if here exists an epimorphism $k\colon K \twoheadrightarrow I$ in $\mathbf{S}$, as well as an object $a \in \mathscr{A}^K$ and an equivalence $\mathrm{Y}(a) \cong k^*(X)$.

Theorem 6.6. *Let $\mathbf{S}$ be a Grothendieck topos. Let $\mathbf{C}$ be a generalized 2-gerbe in $\mathbf{S}$. Then the 2-stack completion of $\mathbf{C}$ can be identified with the weak equivalence 2-functor*

$$\mathbf{C} \xrightarrow{\mathrm{Y}} \mathbf{LR}(\mathbf{Stack}^{\mathbf{C}^{\mathrm{op}}})$$

where $\mathbf{LR}(\mathbf{Stack}^{\mathbf{C}^{\mathrm{op}}})$ is the full sub 2-fibration of $\mathbf{Stack}^{\mathbf{C}^{\mathrm{op}}}$ determined by the locally representable objects in each fiber.

PROOF. Since **Stack** is a 2-stack (over $\mathbf{S}$), so is $\mathbf{Stack}^{\mathbf{C}^{\mathrm{op}}}$ by Theorem 4.9. Factor the Yoneda embedding as

$$\mathbf{C} \xrightarrow{Y} \mathbf{LR}(\mathbf{Stack}^{\mathbf{C}^{\mathrm{op}}}) \hookrightarrow \mathbf{Stack}^{\mathbf{C}^{\mathrm{op}}}$$

and prove (just as in the case $n = 1$ from [11] that, since $\mathbf{Stack}^{\mathbf{C}^{\mathrm{op}}}$ is a 2-stack, so is $\mathbf{LR}(\mathbf{Stack}^{\mathbf{C}^{\mathrm{op}}})$. Furthermore, the first factor is a (special) weak 2-equivalence. □

We shall now restrict our attention from categories to groupoids and from 2-categories to 2-groupoids. By a 2-*groupoid* $\mathbf{G}$ we understand here a 2-category whose 1-cells are *equivalences* and whose 2-cells are *isomorphisms*. There is a different notion of 2-groupoid with isos at both levels, that we shall not consider here. What we call a 2-groupoid has been called a pseudogroupoid in [24].

Definition 6.7. Denote by cod: $\text{stack}^{\rightarrow} \to \mathbf{S}$ the corresponding data of Definition 4.2 restricted to *groupoids* that are stacks.

Corollary 6.8 ([30]). *Let* $\mathbf{S}$ *be a Grothendieck topos. Then, the* 2*-fibration* cod: $\text{stack}^{\rightarrow} \to \mathbf{S}$ *is a* 2*-stack. Equivalently, any regular epi in* $\mathbf{S}$ *is of effective* 2*-descent for* cod: $\mathbf{Stack} \to \mathbf{S}$.

PROOF. The proof of Theorem 5.7 is valid also if restricted to groupoids, in fact, with some simplifications in the construction of Σ from [20], as there is then no need to invert certain morphisms. $\square$

Remark 6.9. A particular case of the notion of a generalized 2-gerbe in $\mathbf{Gpd}$ is that of a 2-gerbe [6, 12]. A 2-groupoid $\mathbf{G}$ in $\mathbf{S}$ is said to be a 2-*gerbe* if for some ('nonempty' and 'connected') 1-groupoid 1-stack $\mathbb{A}$ (a *bouquet*), there is a 2-equivalence

$$\mathbf{G} \simeq \text{Equ}(\mathbb{A}).$$

We shall define a 2-fibration $\text{Tors}^2(\mathbf{G}) \to \mathbf{S}$ for any 2-gerbe $\mathbf{G}$ in 2-$\mathbf{Gpd}$. For each $I \in \mathbf{S}$ the 2-category $\left(\text{Tors}^2(\mathbf{G})\right)^I$ has:

(1) as *objects* the $\mathbf{G}$-2-torsors over I, where a 2-torsor over I is a groupoid stack $\mathbf{T}$ equipped with a structure $\mathbf{T} \times \mathbf{G} \xrightarrow{a} \mathbf{T}$ of a right $\mathbf{G}$-object, and with an epi $p\colon T_0 \to I$, such that

$$\mathbf{T} \times \mathbf{G} \xrightarrow{\langle \pi_1, a\rangle} \mathbf{T} \times_I \mathbf{T}$$

is an equivalence.

(2) as 1-*cells* the $\mathbf{G}$-equivariant 2-functors $h\colon \mathbf{T} \to \mathbf{R}$ (over I), in the sense that the diagram

$$\begin{array}{ccc}
\mathbf{T} \times \mathbf{G} & \xrightarrow{h \times \text{id}} & \mathbf{R} \times \mathbf{G} \\
{\scriptstyle \langle \pi_1, a\rangle}\big\downarrow & & \big\downarrow{\scriptstyle \langle \pi_1, b\rangle} \\
\mathbf{T} \times_I \mathbf{T} & \xrightarrow[h \times h]{} & \mathbf{R} \times_I \mathbf{R}
\end{array}$$

is commutative,

(3) and as 2-*cells*, natural isomorphisms $h \Rightarrow k\colon \mathbf{T} \to \mathbf{R}$ over I.

Proposition 6.10. $\text{Tors}^2(\mathbf{G})$ *is a* 2*-groupoid.*

PROOF. This is shown in basically the same way as in [1] (or [8] in the discrete case), using the properties of $\mathbf{S}$ as a regular (exact) category. $\square$

Theorem 6.11. *Let* $\mathbf{S}$ *be a Grothendieck topos, and* $\mathbf{G}$ *a* 2*-gerbe in* 2-$\mathbf{Gpd}$. *Then there is a biequivalence of* 2*-fibrations over* $\mathbf{S}$

$$\mathbf{LR}(\text{stack}^{\mathbf{G}^{\text{op}}}) \cong \text{Tors}^2(\mathbf{G}).$$

PROOF. The proof is entirely analogous to the same statement in [8] but for 2-groupoids and discrete spaces. That any 2-torsor is locally representable is inmediate. The converse uses properties of regular epis, and that pulling back along an epi reflects equivalences. $\square$

Corollary 6.12. *Let* $\mathbf{G}$ *be a 2-gerbe in* $\mathbf{S}$. *The canonical morphism*

$$[\mathbf{G}] \xrightarrow{\text{triv}} \mathrm{Tors}^2(\mathbf{G})$$

exhibits $\mathrm{Tors}^2(\mathbf{G})$ *as the 2-stack completion of* $[\mathbf{G}]$.

PROOF. Apply Theorem 6.6 together with Theorem 6.11 in the case of a 2-gerbe $\mathbf{G}$, which is a particular case of a generalized 2-gerbe. □

Recall that 2-dimensional cohomology of $\mathbf{S}$ with coefficients in a 2-gerbe $\mathbf{G}$ is given by the formula

$$\mathrm{H}^2(\mathbf{S};\mathbf{G}) = \mathbf{\Pi_0}(\mathrm{Tors}^2(\mathbf{G}))$$

where $\mathbf{\Pi_0}$ in this case denotes 'equivalence classes.'

Remark 6.13. By analogy with the case of dimension 1, we interpret Theorem 6.11 to say that, for any 2-gerbe $\mathbf{G}$, the 2-topos $\mathrm{stack}^{\mathbf{G}^{\mathrm{op}}}$ *classifies* $\mathbf{G}$-2-torsors in $\mathbf{S}$. This remark can, in principle, be made precise using the notion of a 2-topos (or a bitiopos) from [36]. To this end we need to isolate a notion of morphism of 2-toposes and the analogue of the notion of a point of a topos. We leave this matter open for the time being.

7. The equivalences lifting property

We begin this section by briefly recalling a definition and a result from [24, 25] about the existence of a Quillen model structure on 2-**Cat**(Set). The definition (Definition 7.1) makes sense for any elementary topos $\mathbf{S}$.

Definition 7.1. A 2-functor $F\colon \mathbf{A} \to \mathbf{B}$ in 2-**Cat**($\mathbf{S}$) is said to have the *equivalence lifting property* ((ELP) for short) if

• Given $A \in A_0$, and an *adjoint equivalence* in $\mathbf{B}$, consisting of $b\colon B \to FA$, $b'\colon FA \to B$, $\beta_1\colon bb' \cong 1$, and $\beta_2\colon b'b \cong 1$, there is an adjoint equivalence in $\mathbf{A}$, consisting of $a\colon A' \to A$, $a'\colon A \to A'$, $\alpha_1\colon aa' \cong 1$ and $\alpha_2\colon a'a \cong 1$, with $FA' = B$, $Fa = b$, $Fa' = b'$, $F\alpha_1 = \beta_1$, and $F\alpha_2 = \beta_2$.

• For any 1-cell $a : A \to A'$ in $\mathbf{A}$ and every invertible 2-cell $\beta\colon b \to Fa$ in $\mathbf{B}$, there is a 1-cell $a'\colon A \to A'$ with $F(a') = b$ and an invertible 2-cell $\alpha\colon a' \to a$ with $F(\alpha) = \beta$.

In fact, the result (Theorem 7.2) shown in [24, 25] only for $\mathbf{S}$ = Set, is valid for any Grothendieck topos $\mathbf{S}$, provided the notions involved are interpreted internally.

Theorem 7.2 (Lack [25]). *Let* $\mathbf{S}$ *be a Grothendieck topos. There is a cofibrantly generated Quillen model structure on the category* 2-**Cat**($\mathbf{S}$) *for which the weak equivalences are the weak* 2*-equivalences, and the fibrations are the* 2*-functors with the* (ELP).

We can now use this result in order to exhibit internal 2-stack completions in the topos $\mathbf{S} = \mathrm{Set}^{\mathbf{2}}$.

Proposition 7.3. *A 2-functor* $F\colon \mathbf{A} \to \mathbf{B}$ *in* **Cat** *satisfying the* (ELP) *is a 2-stack regarded as a 2-category object in the topos* $\mathbf{S} = \mathrm{Set}^{\mathbf{2}}$.

PROOF. By Proposition 3.8 we can already deduce, from condition (2) in Definition 7.1, that the Homs are 1-stacks. The proof of the fact that condition (1) in this definition implies the full 2-stack property is entirely analogous to that of the same theorem given in [11]. The basic fact is that item (1) in Definition 4.6 of a weak

2-equivalence is a geometric notion, hence true for any 2-functor $F\colon \mathbf{A} \to \mathbf{B}$ iff it is 'pointwise true,' that is, if for every object c in $\mathbf{C}$, the functor $F(c)\colon \mathbf{A}(c) \to \mathbf{B}(c)$ is a weak equivalence functor in Set, that is, a strong equivalence functor since, as an elementary topos, Set satisfies the Axiom of Choice. □

Corollary 7.4. *Let $F\colon \mathbf{A} \to \mathbf{B}$ be any 2-category object in* $\mathrm{Set}^{\mathbf{2}}$. *Factor the unique morphism $F \to \mathbf{1}$ into*

$$F \to F^* \to \mathbf{1}$$

where $F \to \mathbf{F}^*$ is a trivial cofibration and $F^* \to \mathbf{1}$ is a fibration for the Quillen model structure on 2-$\mathbf{Cat}(\mathbf{S})$, *where* $\mathbf{S} = \mathrm{Set}^{\mathbf{2}}$. *Then,*

$$F \to F^*$$

is the 2-stack completion of F in $\mathbf{Cat}(\mathbf{S})$.

PROOF. The unique $F^* \to \mathbf{1}$ has the (ELP) and so, by Proposition 7.3, F^* is a 2-stack. Since $F \to F^*$ is a trivial cofibration, it is, in particular, a weak 2-equivalence, hence by Corollary 4.10, it is the 2-stack completion of F. □

An explicit description of the stack completion of a 2-functor $F\colon \mathbf{A} \to \mathbf{B}$ is possible in the manner of the analogous 1-dimensional case [11]. We leave this as an interesting exercise.

8. An outline of higher stack completions

In order to deal with higher dimensions we would first need to choose a suitable notion of n-category [28] and, relative to it, a notion of weak n-equivalence n-functor. The latter is done as follows, by induction.

Definition 8.1. Let $\mathbf{S}$ be a topos. We define the notion of a *weak n-equivalence n-functor* in $\mathbf{S}$ by induction on n.

• ($n = 0$) A 0-functor $F\colon A \to B$ in $\mathbf{S}$ is a weak 0-equivalence if it is an isomorphism.

• ($n \geq 1$) An n-functor $F\colon A \to B$ between n-categories in $\mathbf{S}$ is a weak n-equivalence if

(1) $F_0\colon A_0 \to B_0$ is essentially $(n-1)$-surjective, and
(2) For all $x, x' \in A_0$, the $(n-1)$-functor

$$F_{x,x'}\colon \mathrm{Hom}_A(x,x') \to \mathrm{Hom}_B(Fx, Fx')$$

is a weak $(n-1)$-equivalence of $(n-1)$-categories.

We state the following result without proof.

Lemma 8.2. *Let $e\colon J \to I$ be an epimorphism in $\mathbf{S}$. Then the the n-functor*

$$F^{(n)}{}_e\colon J^{(n)}_e \to I^{(n)}_{\mathrm{dis}}$$

from the n-kernel of e to the disccrete n-category on I is a weak n-equivalence.

The enterprise of dealing with higher dimensional stack completions is well beyond our possibilities here. Nevertheless, we wish to indicate an outline of how to proceed once these first steps are done.

Remark 8.3. • The construction of the $(n+1)$-stack completion of a generalized $(n+1)$-gerbe $\mathbf{C}$ as given by the Yoneda embedding

$$\mathbf{C} \xrightarrow{Y} \mathbf{LR}(n\text{-}\mathbf{Stack}^{\mathbf{C}^{\mathrm{op}}})$$

requires proving first that the $(n+1)$-fibration cod: n-**Stack** $\to$ **S** be an $(n+1)$-stack. The complicated coherence conditions involved constitute a technical obstruction to such a program. However, assuming this to be the case, it would follow from it that the above is the $(n+1)$-stack completion of $\mathbf{C}$.

• By contrast, we conjecture that, in higher dimensions, a construction of the n-stack completion of a generalized n-gerbe by means of a Quillen model structure on n-**Gerb** exists, whose weak equivalences are the (special) weak n-equivalence functors, and whose fibrant objects are the 'strong' n-stacks.

• We have seen that, in the cases $n = 1$ and $n = 2$, there are, in principle, two different constructions of the n-stack completion of an n-gerbe. The first, for **S** a Grothendieck topos, is given by a suitable Quillen model structure on n-**Cat**(**S**). This is the one that relates to *homotopy* and gives a precise meaning to the assertion that **S** satisfies an axiom of n-stack completions. The second one carves out of a larger n-stack by cutting down to the locally representables. This is the one that exhibits an 'n-topos' as classifier of the n-torsors, and that relates to *nonabelian cohomology*. A comparison between the two would give rise to an appropriate classification and Morita equivalence theorems.

• Alternative outlines of nonabelian higher cohomology have been given in [13, 14,37], with similar conclusions to those made in Remark 8.3 for higher dimensions.

Acknowledgements. The authors are grateful to the anonymous referee for his useful suggestions on the presentation of the results in dimension 1, which we adopted, and for his appreciation of their work. In addition, they thank George Janelidze and Mike Shulman for valuable remarks leading to some changes in a previous version of this paper. This paper is based on a lecture given by the first-named author at the International Category Theory Meeting 2008 (CT'08), Calais, France, June 2008. Research partially supported from an NSERC individual grant to the first-named author, which included her participation at CT'08 and a visit to McGill University by the second-named author during the month of November 2008. Diagrams typeset with Michael Barr's diagxy package for xy-pic.

References

1. M. Barr, *Exact categories*, Exact Categories and Categories of Sheaves, Lecture Notes in Math., vol. 236, Springer, Berlin, 1971, pp. 1–120.
2. J. Beck, Cornell University, 1967, untitled manuscript.
3. J. Bénabou, *Fibrations petites et localement petites*, C. R. Acad. Sci. Paris Sér. A-B **281** (1975), no. 21, A897–A900.
4. J. Bénabou and J. Roubaud, *Monades et descente*, C. R. Acad. Sci. Paris Sér. A-B **270** (1970), A96–A98.
5. N. Bourbaki, *Éléments de mathématique*: *Algèbre commutative*, Hermann, Paris, 1943.
6. L. Breen, *On the classification of 2-gerbes and 2-stacks*, Astérisque **225** (1994).
7. M. Bunge, *Stack completions and Morita equivalence for categories in a topos*, Cahiers Topologie Géom. Différentielle **20** (1979), no. 4, 401–436.
8. ———, *An application of descent to a classification theorem for toposes*, Math. Proc. Cambridge Philos. Soc. **107** (1990), no. 1, 59–79.
9. ———, *Galois groupoids and covering morphisms in topos theory*, Galois Theory, Hopf Algebras, and Semiabelian Categories (B. Pareigis and W. Tholen, eds.), Fields Inst. Commun., vol. 43, Amer. Math. Soc., Providence, RI, 2004, pp. 131–161.

10. M. Bunge and S. Lack, *van Kampen theorems for toposes*, Adv. Math. **179** (2003), no. 2, 291–317.
11. M. Bunge and R. Paré, *Stacks and equivalence of indexed categories*, Cahiers Topologie Géom. Différentielle **20** (1979), no. 4, 373–399.
12. J. Duskin, *Variations on Beck's tripleability criterion*, Reports of the Midwest Category Seminar, III, Lecture Notes in Math., vol. 106, Springer, Berlin, 1969, pp. 74–129.
13. J. Duskin, *An outline of nonabelian cohomology in a topos*. I: *The theory of bouquets and gerbes*, Cahiers Topologie Géom. Différentielle **23** (1982), no. 2, 165–191.
14. ———, *An outline of a theory of higher-dimensional descent*, 1989, unpublished.
15. J. Giraud, *Cohomologie non abélienne*, Grundlehren Math. Wiss., vol. 179, Springer, Berlin, 1971.
16. A. Grothendieck, *Technique de descente et théorèmes d'existence en géometrie algébrique. I: Généralités. Descente par morphismes fidèlement plats*, Séminaire Bourbaki, Vol. 5, Soc. Math. France, Paris, 1995, Exp. No. 190, 299–327.
17. A. Grothendieck, *La longue marche à travers la théorie de Galois*. Tome 1: *Transcription d'un manuscrit inédit*, with edited and with a foreword by J. Malgoire, Département des Sciences Mathématiques, Université Montpellier II, Montpellier, 1995.
18. C. Hermida, *Some properties of* **Fib** *as a fibred 2-category*, J. Pure Appl. Algebra **134** (1999), no. 1, 83–109.
19. ———, *From coherent structures to universal properties*, J. Pure Appl. Algebra **165** (2001), no. 1, 7–61.
20. ———, *Descent on 2-fibrations and strongly 2-regular 2-categories*, Appl. Categ. Structures **12** (2004), no. 5-6, 427–459.
21. M. Hovey, *Model categories*, Math. Surveys Monogr., vol. 63, Amer. Math. Soc., Providence, RI, 1999.
22. P. T. Johnstone, *Topos theory*, Vol. 10, Academic Press, London, 1977. London Math. Soc. Monogr. Ser.
23. A. Joyal and M. Tierney, *Strong stacks and classifying spaces*, Category Theory (Como, 1990) (A. Carboni, M. C. Pedicchio, and G. Rosolini, eds.), Lecture Notes in Math., vol. 1488, Springer, Berlin, 1991, pp. 213–236.
24. S. Lack, *A Quillen model structure for 2-categories*, K-Theory **26** (2002), no. 2, 171–205.
25. ———, *A Quillen model structure for bicategories*, K-Theory **33** (2004), no. 3, 185–197.
26. W. F. Lawvere, *Teoria delle categorie sopra un topos di base*, Universitá di Perugia, 1973.
27. ———, *Logic of topoi: inside and outside*, Université de Montréal, 1974.
28. T. Leinster, *A survey of definitions of n-category*, Theory Appl. Categ. **10** (2002), 1–70.
29. C. Mulvey, *Intuitionistic algebra and representations of rings*, Mem. Amer. Math. Soc. **148** (1974), 3–57.
30. L. Mauri and M. Tierney, *Two-descent, two-torsors and local equivalence*, J. Pure Appl. Algebra **143** (1999), no. 1-3, 313–327.
31. R. Paré and D. Schumacher, *Abstract families and the adjoint functor theorems*, Indexed Categories and Their Applications, Lecture Notes in Math., vol. 661, Springer, Berlin, 1978, pp. 1–125.
32. D. G. Quillen, *Homotopical algebra*, Lecture Notes in Math., vol. 43, Springer, Berlin, 1967.
33. R. Street, *Fibrations and Yoneda's lemma in a 2-category*, Category Seminar (Sydney, 1972/1973) (G. M. Kelly, ed.), Lecture Notes in Math., vol. 420, Springer, Berlin, 1974, pp. 104–133.
34. ———, *Fibrations in bicategories*, Cahiers Topologie Géom. Différentielle **21** (1980), no. 2, 111–160.
35. ———, *Conspectus of variable categories*, J. Pure Appl. Algebra **21** (1981), no. 3, 307–338.
36. ———, *Characterizations of bicategories of stacks*, Category Theory (Gummersbach, 1981) (K. H. Kamps, D. Pumplün, and W. Tholen, eds.), Lecture Notes in Math., vol. 962, Springer, Berlin, 1982, pp. 282–291.
37. ———, *Descent theory*, Oberwolfach, 1995, unpublished notes.
38. R. Street and D. Verity, *The comprehensive factorization and torsors*, Theory Appl. Categ. **23** (2010), no. 3, 42–75.

Department of Mathematics and Statistics, McGill University, Montréal, QC H3A 2K6, Canada
E-mail address: marta.bunge@mcgill.ca

School of Computing, Queen's University, Kingston, ON K7L 3N6, Canada
E-mail address: chermida@cs.queensu.ca

Centre de Recherches Mathématiques
CRM Proceedings and Lecture Notes
Volume **53**, 2011

Purity and Pure-Injectivity for Topological Modules

Clint Enns and Thomas G. Kucera

Dedicated to Professor Mihály Makkai on the occasion of his seventieth birthday

ABSTRACT. A logic suitable for topological structures was introduced by McKee and Sgro in the 1970s and developed further by Garavaglia, Ziegler and Kucera. This logic satisfies the compactness theorem, a Löwenheim–Skolem theorem, and a Lindström theorem.

The model theory of modules is very rich with strong connections to standard questions of algebra, primarily based on the partial elimination of quantifiers down to positive primitive formulas. Theories of topological modules also have a partial elimination of quantifiers, down to "topological positive primitive formulas," so there is some hope to see if a development of these theories can be made that parallels the development of the model theory of ordinary modules.

In his M.Sc. thesis, the first author investigated whether there are good definitions of "pure embedding" and of "pure-injective module" for topological modules in this context. The answer is "mostly not." We report on these results and related results from the literature here.

Topology is innately a higher order subject: statements about topological properties involve quantifications over individuals, over open sets, and over families of open sets. Nonetheless, in the mid to late 1970s, several groups of people realized independently that many concepts of topological interest could be expressed in a first-order context. The language $\mathcal{L}_t$ was introduced by McKee [21–23] and Sgro [30]; and Garavaglia developed these ideas much further [10]. Meanwhile Martin Ziegler had also been studying similar problems [31], and, with Jörg Flum, published a comprehensive overview of the subject in 1980 [9]. The second author developed some of these ideas further in his Ph.D. thesis [16], in particular aspects of stability theory and applications to theories of topological modules [17]. But in spite of limited attempts in [17] and later efforts by Majewski [20], no really

2010 *Mathematics Subject Classification.* Primary 03C80, 13J99; Secondary 03C85, 03C07, 22A99.

This paper is based in part on work presented by Mr. Enns in his M.Sc. thesis, prepared under the direction of Dr. Kucera.

This paper deals in part with problems left unresolved in the Ph.D. thesis of the second author, prepared under the direction of Dr. Makkai, and submitted to McGill University in 1984.

This is the final form of the paper.

satisfactory formulations of the concepts of "elementary extension" or "saturated model" have been made. In particular, neither author offered any examples.

An extensive list of the simpler topological properties can be axiomatized in the language $\mathcal{L}_t$, but in spite of all this, it is quite limited in its expressive power, and the question arises as to what sorts of "everyday" mathematics can be done in this context. That is the question that concerns us here; in particular, we ask "Is there an interesting and/or useful model theory for topological modules (parallel to the very successful model theory of ordinary modules) that can be developed in $\mathcal{L}_t$?"

The answer is "Almost certainly not."

A question like this can't be answered definitively by some sort of incompleteness theorem, or by specific counterexamples, but only by the cumulative weight of evidence. We show that a certain set of core ideas in the model theory of ordinary modules cannot be expected to generalize to topological modules in any natural way. Our investigation of this question is completely tied to the expressive power of the languages $\mathcal{L}_t$ and $\mathcal{L}_m$; and we do not rule out the possibility that alternative approaches might be more profitable. We make some comments about these approaches at the end.

Part of our motivation is this: it is easier for an embedding of ordinary modules to be pure than it is for it to be elementary; it is easier for a module to be pure-injective than it is for it to be saturated. Thus exploring the nature or feasibility of these concepts for topological modules may shed light on the problems associated with defining the more general concepts for arbitrary theories in $\mathcal{L}_t$, but in a context where we might have more powerful tools at our disposal.

Convention A. Structures for some first-order language are denoted by a script letter $\mathcal{M}, \mathcal{A}, \mathcal{M}_0$, etc; and their underlying sets are consistently denoted by the corresponding uppercase Roman letter M, A, M_0, etc.

Convention B. $\bar{v}, \overline{m}$ ($\overline{m} \in M$), $\overline{U}$, etc., represent finite tuples of variables, individuals (in M), sets, etc. Operations (such as $f(\overline{m}), \overline{U} \cap M$) applied to a tuple are to be interpreted as applying componentwise.

1. Model theory of modules

What are some of the basic ideas that make the model theory of modules so successful? It is these ideas for which we will want to develop parallels in the model theory of topological modules, if possible.

Definition 1.1. Let R be a ring with unit. The *language* ${}_R\mathcal{L}$ *of R-modules* has the symbols $+, -$, and $\mathbf{0}$ of the usual language of abelian groups, as well as for each $r \in R$ a unary operation symbol $\mathbf{r}$, intended to represent scalar multiplication by r.

It is easy to see how to axiomatize "left (or right) module over R" in this language; cf. Eklof and Sabbagh [7]. Building the ring into the language rather than axiomatizing R-modules as two-sorted structures avoids including the already known-to-be-complicated theory of rings into the study of the model theory of modules. Since the ring in question is seldom important for our arguments, we rarely mention it. All results apply equally to left and to right R-modules.

Convention C. All modules are left modules over some fixed ring R, unless otherwise stated. We do not distinguish between abelian groups and (left) $\mathbb{Z}$-modules.

Using this language, a very rich and fruitful model theory of modules has been developed. For all basic definitions and references to the early history of the subject, we refer the reader to Prest's book [28]. A good sense of how the subject area has developed and become entwined with contemporary algebra — and in many ways, has separated from its roots in model theory — over the last 30 years can be obtained from Prest's recent book [29].

Definition 1.2. A *positive primitive formula* (pp formula) is a formula of the form $\exists w_1 \ldots \exists w_n \bigwedge_{j<m} \alpha_j$, where $n \geq 0$, $m \geq 1$, and each α_j is an atomic formula.

Note that up to logical equivalence, the set of positive primitive formulas is closed under existential quantification and conjunctions. For languages and theories of modules, we may assume that each atomic formula is of the form $\mathbf{t} = \mathbf{0}$ for some term $\mathbf{t}$ of ${}_R\mathcal{L}$, specifically of the form $\mathbf{t} = \mathbf{r}_1 u_1 + \cdots + \mathbf{r}_k u_k$ for distinct variables $u_1, \ldots, u_k$.

Here are some of the basic properties of pp formulas in modules; these may all be found in Prest [28, Chapter 2; 2.1, 2.2, 2.7, 2.10].

Proposition 1.3. *Let $\mathcal{M}$ and $\mathcal{N}$ be left R-modules, $f \colon \mathcal{M} \to \mathcal{N}$ an R-module homomorphism, and φ and ψ pp formulas with the same free variables, say $v_0, \ldots, v_{n-1}$.*

(1) *"pp formulas define subgroups": $\varphi[M] = \{\overline{m} \in M^n : \mathcal{M} \models \varphi[\overline{m}]\}$ is a subgroup of $\mathcal{M}^n$.*

(2) *"pp formulas with parameters are empty or define a coset of a definable subgroup": if $\bar{a} = a_j, \ldots, a_{n-1}$ is fixed in M^{n-j}, then $\varphi[M, \bar{a}] = \{\overline{m} \in M^j : \mathcal{M} \models \varphi[\overline{m}, \bar{a}]\}$ is either empty or a coset of the definable subgroup $\varphi[M, \overline{0}]$.*

(3) *"the sum of two pp-definable sets is pp-definable": given φ and ψ, we can find a pp formula χ such that $\chi[M, \bar{a}, \bar{b}] = \varphi[M, \bar{a}] + \psi[M, \bar{b}]$.*

(4) *"pp formulas factor across direct sums": if $\bar{a} \in M \oplus N$, (with a typical element $\langle a_j^0, a_j^1 \rangle$) then $\mathcal{M} \oplus \mathcal{N} \models \varphi[\bar{a}]$ if and only if $\mathcal{M} \models \varphi[\bar{a}^0]$ and $\mathcal{N} \models \varphi[\bar{a}^1]$.*

(5) *"Homomorphisms preserve pp formulas": if $\mathcal{M} \models \varphi[\bar{a}]$ then $\mathcal{N} \models \varphi[f(\bar{a})]$ (equivalently, $f[\varphi[M]] \subseteq \varphi[N]$).*

A homomorphism which reflects pp formulas is called *pure*, in which case it is an embedding. Clearly elementary embeddings are pure; and if $\mathcal{A}$ is a direct summand of $\mathcal{B}$, the natural inclusion of $\mathcal{A}$ in $\mathcal{B}$ is pure.

Theorem 1.4 (Baur [3], Monk [24], "pp elimination of quantifiers"). *Given a complete theory T of R-modules, every formula of ${}_R\mathcal{L}$ is equivalent modulo T to a Boolean combination of pp formulas.*

Thus, in any module $\mathcal{M}$, a definable subset of M^n (definable with parameters) is a Boolean combination of cosets of subgroups of M^n definable without parameters. It is this result that makes the model theory of modules so profitable algebraically: the definable sets are essentially objects of module theory; and it therefore turns out that a very wide range of model theoretic concepts have essentially algebraic formulations, and conversely. In fact considering the definable cosets only (rather than their complements as well) turns out to be sufficient for most model-theoretic purposes. For instance, elementary embeddings are clearly pure; and in many arguments it suffices to use pure embeddings instead of elementary embeddings. Similarly sufficiently saturated models are clearly pure-injective, and in many model-theoretic arguments pure-injective modules suffice.

In fact there is deeper theory underlying all these results. Each positive primitive formula of ${}_R\mathcal{L}$ defines a functor from the category of (left) R-modules to the category of abelian groups. It turns out that these functors are well behaved, and that many concepts of the model theory of modules have their most natural expression in the context of functor categories. In fact, a significant part of the development of the model theory of modules in recent years has gone in this direction and explicit references to model theory can disappear entirely. It would lead us too far astray to attempt to provide any significant details; instead we refer the reader to Prest [29]. Nonetheless, we are left with a strong motivation to understand the category of topological modules, and the functors on it.

There are some basic properties of direct sums and direct products which are important in the model theory of modules:

Theorem 1.5 (see, for instance, Prest [28, Section 2.5]). *Let $\mathcal{M}$ and $\mathcal{N}$ be R-modules, and $\langle \mathcal{M}_i \rangle_{i\in I}$ a family of R-modules.*

(1) *If $\mathcal{M} \prec \mathcal{N}$ then $\mathcal{M}$ is elementarily equivalent to $\mathcal{M} \oplus (\mathcal{N}/\mathcal{M})$.*
(2) $\bigoplus_{i\in I} \mathcal{M}_i \equiv \prod_{i\in I} \mathcal{M}_i$
(3) *For any infinite I, $\mathcal{M}^{(I)} \equiv \mathcal{M}^{(\aleph_0)} \equiv \mathcal{M}^{\aleph_0} \equiv \mathcal{M}^I$*
(4) $\bigoplus_{i\in I} \mathcal{M}_i \preceq \prod_{i\in I} \mathcal{M}_i$

These results, together with the basic results of homological algebra relating the Hom functor with the direct sum and the direct product, begin to explain the power and significance of direct sums and direct products in the model theory of modules.

Definition 1.6. Let $\mathcal{N}$ be an R-module.

(1) $\mathcal{N}$ is *pure-injective* if it is injective over pure embeddings, that is, whenever $e\colon \mathcal{A} \to \mathcal{B}$ is a pure embedding of R-modules and $f\colon \mathcal{A} \to \mathcal{N}$ is a homomorphism, there is a homomorphism $\hat{f}\colon \mathcal{B} \to \mathcal{N}$ lifting f, $f = \hat{f} \circ e$.

(2) $\mathcal{N}$ is *equationally compact* if every system Σ of R-linear equations over N which is finitely solvable in $\mathcal{N}$ is actually solvable in $\mathcal{N}$.

Here Σ may consist of infinitely many equations, may involve infinitely many variables, and the equations may have nonzero constant terms from N. "Finitely solvable" means that every finite subset of Σ has a solution in $\mathcal{N}$.

(3) $\mathcal{N}$ is *pp-compact* if every set Φ of pp formulas over N which is finitely satisfiable in $\mathcal{N}$ is actually satisfiable in $\mathcal{N}$.

Here Φ may consist of infinitely many pp formulas, may involve infinitely many variables, and the formulas may have arbitrary parameters from N. "Finitely satisfiable" means that every finite subset of Φ is realized in $\mathcal{N}$.

Pure injectivity is a natural algebraic concept. Clearly injective modules are pure-injective, so $\mathbb{Q}$ and the Prüfer p-groups $\mathbb{Z}(p^\infty)$ are all pure-injective. Equational compactness and pp-compactness are natural model-theoretic concepts. Clearly $|R|^+$-saturated models are pp-compact; and pp compact structures are equationally compact. If $\mathcal{N}$ carries a compatible compact Hausdorff topology, then the solution sets of linear equations are closed sets, so such a module is clearly equationally compact. So for instance the circle group $\mathbb{R}/\mathbb{Z}$ and the group of p-adic numbers (p prime) are each equationally compact. Clearly any finite module is pp-compact.

Theorem 1.7. *Let $\mathcal{N}$ be an R-module. The following are equivalent:*
(1) *$\mathcal{N}$ is pure-injective;*
(2) *$\mathcal{N}$ is equationally compact;*
(3) *$\mathcal{N}$ is pp-compact.*

The equivalence between pure-injectivity and pp-compactness, coupled with the pp elimination of quantifiers, is what makes pure-injectivity a sufficient replacement for saturation in many arguments. Some of that flavour may be seen in the two arguments that we will outline below.

It seems clear that since the topological language involves references to open sets as well as to individuals that there will be no natural extension of the concept of equational compactness to the context of topological modules. But the other two concepts do hold some promise of generalization, and what we would like to see then is that the equivalence between them should still hold. For ordinary modules, the arguments in both directions are natural, and we provide brief sketches.

PP-COMPACTNESS IMPLIES PURE-INJECTIVITY. Let $e\colon \mathcal{A} \to \mathcal{B}$ be a pure embedding of R-modules and let $f\colon \mathcal{A} \to \mathcal{N}$ be a homomorphism into pp-compact $\mathcal{N}$. Establish a bijection $b \mapsto v_b$ between the elements of B and a set $\mathcal{V}$ of new, distinct variables. Let Φ be the "pp-type" of B over A in $\mathcal{B}$: the set of all pp formulas with parameters from A and free variables from $\mathcal{V}$ satisfied in $\mathcal{B}$ by assigning to each variable its corresponding element of B. Since e is pure, it is easy to see that Φ is finitely satisfiable in $\mathcal{A}$. Now replace each parameter $a \in A$ by $f(a) \in N$ to get a set of pp formulas $f(\Phi)$ with parameters in N. Since f is a homomorphism, it is easy to see that $f(\Phi)$ is finitely satisfiable in $\mathcal{N}$. Since $\mathcal{N}$ is pp-compact, $f(\Phi)$ is satisfied in $\mathcal{N}$, say by $\langle n_b \rangle_{b \in B}$. Define $\hat{f}\colon \mathcal{B} \to \mathcal{N}$ by sending each $b \in B$ to the corresponding element n_b of N. It is easy to check that $\hat{f}$ is a module homomorphism satisfying $f = \hat{f} \circ e$. □

PURE-INJECTIVITY IMPLIES PP-COMPACTNESS. Let Φ be a finitely satisfiable set of pp formulas over the pure-injective module $\mathcal{N}$. Then Φ is certainly consistent, so there is an elementary extension $\mathcal{N}'$ of $\mathcal{N}$ in which Φ has a solution. Elementary extensions are pure. Let $f\colon \mathcal{N} \to \mathcal{N}$ be the identity map. By pure-injectivity, f lifts to $\hat{f}\colon \mathcal{N}' \to \mathcal{N}$. The image of the solution of Φ in $\mathcal{N}'$ under $\hat{f}$ is clearly a solution of Φ in $\mathcal{N}$.

This "short proof" uses much too heavy machinery — full compactness to construct an elementary extension with a desired property. Prest [29, Theorem 4.1.4] takes a somewhat more elementary approach, showing that whenever Φ is a finitely satisfiable set of pp formulas over a module $\mathcal{M}$, the diagonal embedding of $\mathcal{M}$ into a suitably chosen reduced product $\mathcal{M}^I/\mathcal{F}$ is a pure embedding of $\mathcal{M}$ into a module in which Φ has a solution. □

It is these proofs that we would like to see generalize to the context of topological modules in the language $\mathcal{L}_t$. A test of a possible definition of purity for topological modules would be whether it had associated concepts of pure-injectivity and pp-compactness that were related in a useful way.

2. Topological abelian groups and modules

A *topological abelian group* is a structure $\langle \mathcal{A}, \tau \rangle$ where $\mathcal{A} = \langle A, +, -, 0 \rangle$ is an abelian group and τ is a topology under which $+$ and $-$ are continuous. Virtually all of our interest is in Hausdorff topologies: for groups this is equivalent to $\{0\}$ being closed.

Convention D. All topological abelian groups and modules are assumed to be Hausdorff.

This is no great restriction, since "Hausdorff" can be axiomatized by a sentence of the language $\mathcal{L}_t$ to be introduced later.

For each fixed ring R, a *topological R-module* is a structure $\langle \mathcal{M}, \sigma \rangle$ such that $\mathcal{M}$ is an R-module and τ is a topology on the underlying set M of $\mathcal{M}$ under which $+$, $-$ and each map $M \to M : m \mapsto r.m$ are all continuous. This is a restricted case of what is normally considered in the literature: in general, R should be a topological ring, and scalar multiplication should be a continuous map $R \times M \to M$. However the usual language for R-modules incorporates R into the language rather than in the structures themselves, so there is no way to express topological properties of R in the corresponding topological languages. Equivalently, we consider only topological modules over discrete topological rings.

For any two topological structures of the same type, a *topological embedding* is an embedding which is also a homeomorphism onto a subspace of the codomain. That is, the embedding must be continous and open onto its image. The condition "open onto its image" is important in and of itself; a continuous map between two spaces is called *proper* if it is open onto its image. One particular reason for the usefulness of this concept comes from the duality theory of locally compact abelian groups: the Pontryagin dual of an exact sequence of locally compact abelian groups need not be exact; but the dual of a proper exact sequence is proper exact (see, for instance, Armacost [1, [Proposition 9.14]].)

This raises the whole question of what the category **TopAb** of topological abelian groups or the category of topological modules might be, and what its properties are. The maps are just the continuous module homomorphisms. However, the resulting category is not an abelian category; and many associated concepts either fail to exist at all in general in **TopAb**, or are "unnatural" in one way or another. This gives us the first indication that the model theory of topological abelian groups will likely be very different from that of ordinary abelian groups and modules.

(1) Neither monomorphisms nor epimorphisms need be *normal* (that is, the kernel or cokernel of some map, respectively);
(2) Given a monomorphism $f\colon \langle \mathcal{A}, \sigma \rangle \to \langle \mathcal{B}, \tau \rangle$, it may not be possible to extend this to a short exact sequence $0 \to A \xrightarrow{f} B \xrightarrow{g} C \to 0$;
(3) There appears to be no natural definition of tensor product.

For instance, if σ' is a group topology on $\mathcal{A}$ strictly finer than σ, then the identity map id: $\langle \mathcal{A}, \sigma' \rangle \to \langle \mathcal{A}, \sigma \rangle$ is a continuous homomorphism which is both mono and epi, but is neither normal mono nor normal epi; and this monomorphism cannot be extended to a short exact sequence. We might hope to improve the situation by insisting that all the maps be proper; but this is doomed to failure as the composition of proper maps need not be proper.

Nonetheless, there is some natural homological algebra in the categories of topological abelian groups or topological R-modules. Given two topological left R-modules $\langle \mathcal{A}, \sigma \rangle$ and $\langle \mathcal{B}, \tau \rangle$, we let

$$\mathrm{CHom}(\mathcal{A}, \mathcal{B}) = \{f \in \mathrm{Hom}(\mathcal{A}, \mathcal{B}) : f \text{ is continuous}\}.$$

The pointwise sum of two continuous homomorphisms is continuous, so $\mathrm{CHom}(\mathcal{A}, \mathcal{B})$ is a subgroup of $\mathrm{Hom}(\mathcal{A}, \mathcal{B})$. The *compact-open topology* on this subgroup is the topology having as a sub-basis of open sets all sets of the form (K, U) where K is a compact subset of $\mathcal{M}$ and U is open in $\mathcal{N}$. Then $\mathrm{CHom}(\mathcal{A}, \mathcal{B})$ endowed with the compact-open topology is a topological abelian group. There are three standard theorems relating $\mathrm{Hom}(-,-)$ to the direct sum and direct product of modules. Parallel, but not identical, results for $\mathrm{CHom}(-,-)$, the Tychonoff product of topological abelian groups, and the coproduct of topological abelian groups, can be found in exercises 2.10.3 and 2.10.4, in Dikranjan et al. [6, pp. 66 – 69]. These all generalize naturally to topological modules.

3. Weak structures for second-order logic

Henkin [11] introduced a method of approximating higher order logics in first-order logic, now usually called the *method of weak structures.* Associated with any first-order language $\mathcal{L}$ there is a second-order language, consisting of $\mathcal{L}$ augmented by a new, infinite set of variables (called *set variables*), two new quantifiers (universal and existential) that bind set variables, equality $=_2$ between set variables, and a new binary relation symbol $\in$ between individuals and sets. In the second-order logic of $\mathcal{L}$ the set quantifiers are intended to range over all subsets of the domain of an $\mathcal{L}$-structure. Syntactically the same is the two-sorted expansion of $\mathcal{L}$ called $\mathcal{L}_2$. A *weak structure* for the second-order logic of $\mathcal{L}$ is a two-sorted structure where the first sort is just an ordinary $\mathcal{L}$-structure and the second sort is some collection of subsets of the domain of the first sort. Now in two-sorted logic the set quantifiers are interpreted as ranging over the second sort, and $\in$ is interpreted as a relation between individuals of the first sort and individuals of the second sort. The logic of weak structures is just ordinary, two-sorted, first-order, logic.

Consider the sentence

$$\textbf{ext}: \qquad \forall X \forall Y \big(X =_2 Y \leftrightarrow \forall v(v \in X \leftrightarrow v \in Y)\big)$$

If the $\mathcal{L}_2$-structure $\langle \mathcal{M}, A \rangle$ is a model of **ext**, then we can replace each $a \in A$ by

$$\hat{a} = \{m \in M : \langle \mathcal{M}, A \rangle \models m \in A\}$$

to get a weak structure for the second-order logic of $\mathcal{L}$ which is isomorphic as an $\mathcal{L}_2$-structure to $\langle \mathcal{M}, A \rangle$. The study of weak models of **ext** is the desired first-order approximation to the study of the full second-order logic of $\mathcal{L}$.

4. Languages for topological modules

The concept of "topology on a set X" is not even second-order over X ("the union of an arbitrary family of open sets is open"); however, the concept of "basis for a topology on a set X" is axiomatizable in weak structures:

$$\begin{aligned}\textbf{bas}: \quad & \forall v \exists V (v \in V) \\ & \wedge \forall v \forall U \forall V \big[(v \in U \wedge v \in V) \to \exists W (v \in W) \wedge \forall w \big(w \in W \to (w \in U \wedge w \in V)\big)\big]\end{aligned}$$

If $\langle \mathcal{M}, \sigma \rangle$ is a weak structure for $\mathcal{L}$, then it is a model of **bas** iff σ is a basis for a topology on the domain of $\mathcal{M}$. This is what suggested to the various authors mentioned in the introductory paragraphs that one should seek out the formulas φ of $\mathcal{L}_2$ that are *invariant with respect to basis*: formulas such that whenever σ and σ' are two bases for the same topology on the domain of $\mathcal{M}$,

$$\langle \mathcal{M}, \sigma \rangle \models \varphi \iff \langle \mathcal{M}, \sigma' \rangle \models \varphi.$$

These are precisely the formulas of $\mathcal{L}_{\mathrm{t}}$.

Definition 4.1. Let φ be a formula of $\mathcal{L}_2$ written in *negation normal form*, that is, it is written using only the propositional connectives $\neg, \wedge$, and $\vee$, and each occurrence of $\neg$ is attached to an atomic formula. φ is *positive* in U if each occurrence of U is in a negation-free atomic formula $\mathrm{t} \in U$, and φ is *negative* in V if each occurrence of V is in a negated atomic formula $\neg(\mathrm{t} \in V)$.

Definition 4.2. The set of all $\mathcal{L}_{\mathrm{t}}$ formulas is the smallest set $\mathcal{W}$ of $\mathcal{L}_2$ formulas containing the atomic $\mathcal{L}$ formulas and the atomic $\mathcal{L}_2$-formulas of the form $\mathbf{t} \in V$, closed under $\neg, \wedge$, and existential quantifiers of the first sort; such that if φ is in $\mathcal{W}$, $\mathbf{t}$ is an $\mathcal{L}$-term, V is a variable of the second sort, and φ is positive in V, then $\forall V(\mathbf{t} \in V \to \varphi)$ is in $\mathcal{W}$; and such that if φ is in $\mathcal{W}$, $\mathbf{t}$ is an $\mathcal{L}$-term, V is a variable of the second sort, and φ is negative in V, then $\exists V(\mathbf{t} \in V \wedge \varphi)$ is in $\mathcal{W}$.

Note that the atomic $\mathcal{L}_2$-formulas of the form $V_i = V_j$ *are not* formulas of $\mathcal{L}_{\mathrm{t}}$.

We abbreviate $\forall V(\mathbf{t} \in V \to \varphi)$ as $(\forall V)_{\mathbf{t}}\varphi$ and $\exists V(\mathbf{t} \in V \wedge \varphi)$ as $(\exists V)_{\mathbf{t}}\varphi$, intended to be read as "for every open neighborhood V of $\mathbf{t}$, φ holds" and "for some open neighborhood V of $\mathbf{t}$, φ holds" respectively.

Theorem 4.3.
(1) (McKee [23]) *$\mathcal{L}_{\mathrm{t}}$ formulas are invariant with respect to basis;*
(2) (McKee [21,23], Garavaglia [10]) *An $\mathcal{L}_2$ sentence is invariant with respect to basis iff it is $\mathcal{L}_2$-equivalent to an $\mathcal{L}_{\mathrm{t}}$ sentence in topological structures.*

Since $\mathcal{L}_2$ is an ordinary first-order logic, we immediately obtain the compactness theorem and the Löwenheim – Skolem theorem for $\mathcal{L}_{\mathrm{t}}$ ([10,31]); and Ziegler showed that in fact $\mathcal{L}_{\mathrm{t}}$ satisfies a Lindström's theorem for topological structures.

Garavaglia [10] and Flum and Ziegler [9] enumerate many properties than can be expressed in $\mathcal{L}_{\mathrm{t}}$, amongst which: continuity and openness of functions (but not closedness), openness or closedness of definable sets, the closure of a definable set, separation axioms T_0 through $T_{2\frac{1}{2}}$ and regularity, so T_3 (but no higher), and discrete or trivial topologies.

It follows from the Löwenheim – Skolem theorem for $\mathcal{L}_{\mathrm{t}}$ that properties such as "normal," "connected," "compact," or "locally compact" are not expressible in $\mathcal{L}_{\mathrm{t}}$, even by a set of sentences.

The topology on a topological abelian group A or module is special: for each $a \in A$ the translation $x \mapsto a + x$ is a homeomorphism and so the topology is completely determined by the set of (open) neighborhoods of 0: $U \subseteq A$ is open iff for every $a \in U$ there is a neighborhood N of 0 such that $a + N \subseteq U$. Topologies of this sort can be described in terms of monotone systems:

Definition 4.4. A *(k-)monotone system* on a set A is a collection ν of nonempty subsets of A (of A^k) such that $U \in \nu$ and $U \subseteq V \subseteq A$ ($U \subseteq V \subseteq A^k$) implies that $V \in \nu$.

If μ is any collection of nonempty subsets of A then $\hat{\mu}$ denotes the smallest monotone system on A containing μ.

The set of all neighborhoods of 0 in a topological module forms a 1-monotone system.

Flum and Ziegler [9, [pp. 52 et seq.] develop the basic theory of a logic $\mathcal{L}_{\mathrm{m}}$ for monotone systems completely parallel to the development of $\mathcal{L}_{\mathrm{t}}$; in particular $\mathcal{L}_{\mathrm{m}}$ satisfies the compactness theorem and Löwenheim – Skolem theorem, and a Lindström's Theorem.

The language $\mathcal{L}_{\mathrm{m}}$ is defined just as the language $\mathcal{L}_{\mathrm{t}}$, except that the set quantifiers now just appear as $\forall V$ or as $\exists V$, instead of as $(\forall V)_t$ or as $(\exists V)_t$. The appropriate analogue of Theorem 4.3 holds, cf. Flum and Ziegler [9, Part I: 8.5, 8.6, 8.7].

The concept of "topological abelian group" ("topological left R-module") is easy to axiomatize in the languages $\mathcal{L}_{\mathrm{t}}$ and $\mathcal{L}_{\mathrm{m}}$. In addition, for any formula $\varphi(\bar{u}, \overline{V})$ of $\mathcal{L}_{\mathrm{t}}$ one can construct a formula $\varphi'(\bar{u}, \overline{V})$ of $\mathcal{L}_{\mathrm{m}}$ such that for all topological modules $\langle \mathcal{A}, \sigma \rangle$ and corresponding neighborhood basis μ at 0, all $\bar{a}$ in A and all open neighborhoods $\overline{\mathrm{U}}$ of 0 in $\langle \mathcal{A}, \sigma \rangle$,

$$\langle \mathcal{A}, \sigma \rangle \models \varphi[\bar{a}, \overline{\mathrm{U}}] \quad \Leftrightarrow \quad \langle \mathcal{A}, \mu \rangle \models \varphi'[\bar{a}, \overline{\mathrm{U}}],$$

and conversely. As a result, all the basic definitions and theorems for the model theory of topological modules in $\mathcal{L}_{\mathrm{t}}$ carry over, *mutatis mutandis*, to the model theory of topological modules in $\mathcal{L}_{\mathrm{m}}$, and we will not state them explicitly.

We say that two $\mathcal{L}_{\mathrm{t}}$ structures are $\mathcal{L}_{\mathrm{t}}$-elementary equivalent ("$\equiv_{\mathrm{t}}$") if they satisfy exactly the same $\mathcal{L}_{\mathrm{t}}$ sentences; and we say that two $\mathcal{L}_{\mathrm{m}}$ structures are $\mathcal{L}_{\mathrm{m}}$-elementary equivalent ("$\equiv_{\mathrm{m}}$") if they satisfy exactly the same $\mathcal{L}_{\mathrm{m}}$ sentences. Clearly this is a weaker concept than elementary equivalence as two-sorted structures.

Just as for the model theory of ordinary modules, there is a partial elimination of quantifiers for theories of topological modules.

Definition 4.5 (Garavaglia [10]**).** Let $\mathcal{L} = {}_R\mathcal{L}$, the usual language for left R-modules.

A *topological positive primitive formula* (tppf) is a formula of $\mathcal{L}_{\mathrm{t}}$ of the form

$$Q_1 Q_2 \dots Q_n \varphi$$

where φ is a conjunction of atomic $\mathcal{L}_{\mathrm{t}}$ formulas (hence, without loss of generality, of the form $\mathbf{t} = 0$ or of the form $\mathbf{t} \in V$ for some term $\mathbf{t}$ of $\mathcal{L}$), and each Q_i is a quantifier $\exists u$ or a quantifier $(\forall V)_0$ for some variables u, V of the first and second sort respectively.

Clearly the translation between $\mathcal{L}_{\mathrm{t}}$ and $\mathcal{L}_{\mathrm{m}}$ described previously takes topological positive primitive formulas to monotone positive primitive formulas, and conversely.

Theorem 4.6 (Garavaglia [10]**, Kucera** [17, Corollary 3.7]**).** *Let $\langle \mathcal{M}, \sigma \rangle$ be a topological left R-module.*

Let $\varphi(\overline{x})$ be a formula of $\mathcal{L}_{\mathrm{t}}$ with only individual variables free. Then there is $\psi(\overline{x})$ with the same free variables, ψ a Boolean combination of topological positive primitive formulas, such that for all $\overline{m} \in M$, $\langle \mathcal{M}, \sigma \rangle \models \varphi[\overline{m}]$ if and only if $\langle \mathcal{M}, \sigma \rangle \models \psi[\overline{m}]$.

It is this theorem that gives us some hope that there might be a reasonable and useful "model theory of topological modules." We need to wait until we have some more material on saturated modules before we can show (Corollary 5.3) that tppfs define subgroups, as they do in ordinary theories of modules. Although historically it has been the language $\mathcal{L}_t$ which has received the most attention, the language $\mathcal{L}_m$ is often more convenient (as Flum and Ziegler noted); and we will see in the next sections at least one reason for preferring it.

5. Model theory of topological modules in two sorted logic

The general philosophy behind the "weak structures" approach to second-order logic, coupled with the characterization of invariant formulas, suggests that we should ascribe to a topological module $\langle \mathcal{M}, \tau \rangle$ some model theoretic property whenever $\langle \mathcal{M}, \beta \rangle$ has that property in two-sorted first-order logic, for some basis β for τ. Similarly, if ν is the set of all neighborhoods of 0, we ascribe a property to the monotone structure $\langle \mathcal{M}, \nu \rangle$ if $\langle \mathcal{M}, \mu \rangle$ has that property in two-sorted first order logic for some basis μ for ν.

In particular, $\langle \mathcal{M}, \sigma \rangle$ has cardinality κ means that M has cardinality κ and σ has a basis of cardinality κ. We say that $\langle \mathcal{M}, \sigma \rangle$ is (κ-)saturated if and only if $\langle \mathcal{M}, \sigma' \rangle$ is (κ-)saturated as a two-sorted structure for some basis σ' of σ. Many of the key results of [10] and [17] were obtained by using structures that were saturated [or special] in this way. Note that while saturation is a property of the topological structure, it is not a property that is invariant with respect to basis. Clearly if $\langle \mathcal{M}, \sigma \rangle$ is κ-saturated as a two-sorted structure then $\mathcal{M}$ is saturated as an $\mathcal{L}$-structure.

Proposition 5.1 (Flum and Ziegler [9, Part I: Lemma 4.7]**).** *Let $\langle \mathcal{M}, \sigma \rangle \equiv_t \langle \mathcal{N}, \tau \rangle$ be saturated structures of the same cardinality. Then they are topologically isomorphic.*

It is worthwhile to re-iterate exactly what the hypotheses mean: As topological structures, $\langle \mathcal{M}, \sigma \rangle$ and $\langle \mathcal{N}, \tau \rangle$ satisfy the same $\mathcal{L}_t$ sentences. We can find bases σ' and τ' for σ and τ respectively, such that $|M| = |\sigma'| = |N| = |\tau'| = \kappa$ and $\langle \mathcal{M}, \sigma' \rangle$ and $\langle \mathcal{N}, \tau' \rangle$ are κ-saturated as two-sorted structures.

Flum and Ziegler prove this result only for $\kappa = \aleph_0$, but a straightforward modification of their back-and-forth argument proves the same result for saturated structures of any infinite cardinality. In particular, in constructing the back-and-forth system, one must consider partial homeomorphisms of cardinality less than κ, and then use the κ-saturation to show that the resulting system is closed under unions of chains of cardinality less than κ.

Clearly if $\langle \mathcal{M}, \sigma \rangle$ is a topological module, σ is some basis for τ, and $\langle \mathcal{M}, \sigma \rangle \equiv_2 \langle \mathcal{M}', \sigma' \rangle$ or $\langle \mathcal{M}, \sigma \rangle \preceq_2 \langle \mathcal{M}', \sigma' \rangle$ or $\langle \mathcal{M}', \sigma' \rangle \preceq_2 \langle \mathcal{M}, \sigma \rangle$ as two sorted structures, then $\langle \mathcal{M}, \sigma \rangle$ and $\langle \mathcal{M}', \sigma' \rangle$ are $\mathcal{L}_t$-elementarily equivalent. This leads to some interesting constructions; in particular see Example 5.7.

Note that in spite of the general principle enunciated above, it would *not* be useful to try to define "elementary substructure" by saying that "$\langle \mathcal{M}, \sigma \rangle$ is an elementary substructure of $\langle \mathcal{N}, \tau \rangle$ iff for some basis α of σ and some basis β of τ, $\langle \mathcal{M}, \alpha \rangle \preceq_2 \langle \mathcal{N}, \beta \rangle$," for then the resulting relation might not even be transitive.

A standard simplifying assumption of ordinary one-sorted model theory is that an embedding can always be replaced by an inclusion of structures. Of course we can make the same replacement for a two-sorted embedding; but not for embeddings between weak structures. The individuals of the second sort of a weak structure are constrained to be subsets of the first sort. The best that we can do is this: if $f\colon \langle \mathcal{M}, \alpha \rangle \to \langle \mathcal{N}, \beta \rangle$ is an embedding of weak structures, then we may assume without loss of generality that on the first sort, f is an inclusion of structures $\mathcal{M} \subseteq \mathcal{N}$. On the second sort, f is a one-to-one map identifying certain subsets of M (the elements of α) with certain subsets of N (which are elements of β). The fact that f respects atomic formulas implies that for each $A \in \alpha$, $f(A) \cap M = A$. But note that in general, $f(A) \supsetneq f[A] = \{f(m) : m \in A\}$. So in particular if these are topological structures, then α will be contained in the subspace topology induced from β, but this subspace topology might be much finer than α. An important example of this is seen in Corollary 5.4. Further developments of these ideas are seen in the discussion of continuous homomorphisms in Proposition 6.1 and following.

It helps to consider a concrete example, for instance the natural inclusion of the rational group as a subspace of the real group, both with the Euclidean topology. If we wish to consider this embedding as a two-sorted embedding, we have to map open sets of $\mathbb{Q}$ to open sets of $\mathbb{R}$. There is no canonical way of doing this for all open sets of $\mathbb{Q}$. It is however possible to do so for particular choices of basis: we could take instead the basis for the Euclidean topology on $\mathbb{Q}$ consisting of all bounded intervals with rational endpoints, and match each such interval with the corresponding interval in $\mathbb{R}$.

Theorem 5.2. (1) (Flum and Ziegler [9, Part II, Lemma 1.18]) *Let $\langle \mathcal{M}, \sigma \rangle$ be a topological module with a basis μ for the neighborhoods at 0 such that $\langle \mathcal{M}, \mu \rangle$ is $\aleph_1$-saturated. Then τ is closed under countable intersections.*

(2) (Garavaglia [10]) *Let $\langle \mathcal{M}, \sigma \rangle$ be a topological module with a basis μ for the neighborhoods at 0 such that $\langle \mathcal{M}, \mu \rangle$ is $\aleph_1$-saturated Then $\hat{\mu}$ has a basis at 0 consisting of subgroups.*

The following is folklore, but has not been stated before with a clear proof.

Corollary 5.3 (Enns [8, Corollary 4.29]**).** *If $\varphi(\bar{v})$ is a topological positive primitive formula with no set variables free, then φ defines a subgroup of $\mathcal{M}^n$ (for some n) for any topological module $\langle \mathcal{M}, \sigma \rangle$. If $\varphi(\bar{v}, \bar{a})$ is a is a tppf with no set variables free and parameters from M, then it is either unsatisfiable in $\langle \mathcal{M}, \sigma \rangle$, or defines a coset of the subgroup defined by $\varphi(\bar{v}, \bar{0})$.*

PROOF. Let $\langle \mathcal{N}, \tau \rangle$ be an $\aleph_1$-saturated two-sorted elementary extension of $\langle \mathcal{M}, \sigma \rangle$. Then τ has a basis τ_0 at 0 consisting of subgroups of $\mathcal{M}$. Consider the structure $\langle \mathcal{N}, \tau_0 \rangle$. Atomic formulas $\mathbf{t} = 0$ and $\mathbf{t} \in A$ (for any set parameter $A \in \tau_0$) clearly define subgroups since terms are linear. Then the conjunction of atomic formulas (with set parameters) defines an intersection of subgroups; individual existential quantification defines a projection of a subgroup; and universal set quantification defines an intersection of a family of subgroups. Hence any tppf even with set parameters defines a subgroup of $\mathcal{N}^n$ in the structure $\langle \mathcal{N}, \tau_0 \rangle$. In particular, tppfs without set parameters define subgroups. But to say that a definable set is a subgroup is itself an $\mathcal{L}_t$-formula, so by the invariance of $\mathcal{L}_t$ formulas with respect to basis, they also define subgroups in $\langle \mathcal{N}, \tau \rangle$ and hence in $\langle \mathcal{M}, \sigma \rangle$. □

We cannot conclude anything in general about the topological properties of definable subgroups. Noting that open subgroups are closed, we see that although atomic formulas (even with open subgroup parameters) define closed subgroups, and both conjunctions and universal set quantifiers ($(\forall V)_0$) define intersections of closed subgroups, hence closed; it is not true in general that a projection ($(\exists w)$) of a closed subgroup is closed. For instance, consider $\mathbb{Z}$ with the p-adic topology, p a prime. For primes $q \neq p$, $q\mathbb{Z}$ is definable (by $\exists w(qw = v)$), and is not closed; in fact it is a dense subgroup.

Corollary 5.4 (Enns [8, Corollary 2.10]**).** *Let $\langle \mathcal{M}, \sigma\rangle$ be a topological module with a basis μ for the neighborhoods at 0 such that $\langle \mathcal{M}, \mu\rangle$ is $\aleph_1$-saturated. Let $\mathcal{M}_0$ be a countable submodule of $\mathcal{M}$. Then the subspace topology on M_0 is the discrete topology.*

PROOF. For each $0 \neq b \in M_0$, $U_b = M_0 \setminus \{b\}$ is an open neighborhood of 0 (in M_0) since τ is Hausdorff. Hence $\{0\} = \bigcap_{0\neq b \in M_0} U_b$ is open since M_0 is countable. Thus the subspace topology on M_0 is discrete. □

Note that this argument applies to any kind of $\aleph_1$-saturated topological structure, not just to topological modules. So in general we are warned away from using two-sorted embeddings for a study of topological structures, as they will not necessarily be embeddings onto subspaces.

Definition 5.5 (Flum and Ziegler [9, Part II, 2.3]**).** A topological abelian group is *locally pure* if for each $n > 0$ it satisfies the following $\mathcal{L}_{\mathrm{m}}$ sentence:

$$\forall U \exists V \forall u \exists v\, [n \centerdot u \in V \to (v \in U \wedge n \centerdot u = n \centerdot v)]$$

Abelian groups with the discrete or the trivial topology are vacuously locally pure; groups with a topology induced by a linear order are locally pure; and the group of integers with any n-adic topology ($n \geq 2$) is locally pure.

Lemma 5.6 (Flum and Ziegler [9, Part II, 2.4]**).** *An $\aleph_1$-saturated locally pure topological abelian group has a basis at 0 consisting of pure subgroups.*

Example 5.7. Consider the rational group $\mathbb{Q}$ with the usual Euclidean topology $\mathcal{E}$. Flum and Ziegler [9, Part II, 2.9] give a complete axiomatization for the $\mathcal{L}_{\mathrm{m}}$ theory of $\langle \mathbb{Q}, \mathcal{E}_0\rangle$ (where $\mathcal{E}_0$ is the set of all Euclidean neighborhoods of 0 in $\mathbb{Q}$):

"torsion-free, divisible, nonzero, Hausdorff, nondiscrete, and locally pure."

(Note that Flum and Ziegler fail to mention "nondiscrete.")

It follows immediately that the group of the reals with the Euclidean topology is $\mathcal{L}_{\mathrm{m}}$-equivalent to $\langle \mathbb{Q}, \mathcal{E}_0\rangle$; and similarly for any $\mathbb{Q}^n$ with the Euclidean topology.

Now the model theory of the group $\mathbb{Q}$ is very well understood, and quite simple. The groups elementarily equivalent to $\mathbb{Q}$ are precisely the nonzero $\mathbb{Q}$-vector spaces; so the theory is $\aleph_1$-categorical. A model is κ-saturated iff its dimension as a vector space is $\geq \kappa$.

Consider some two-sorted elementary extension $\langle \mathcal{M}, \mu\rangle$ of $\langle \mathbb{Q}, \mathcal{E}_0\rangle$ which is $\aleph_1$-saturated. Then the monotone system generated by μ has a basis μ' of open, pure, subgroups by Lemma 5.6. As well as **bas**, $\langle \mathcal{M}, \mu'\rangle$ satisfies sentences of two-sorted logic asserting that each $U \in \mu$ is an open, pure, subgroup. By the downward Löwenheim – Skolem theorem for two-sorted logic, $\langle \mathcal{M}, \mu'\rangle$ has a countable elementary submodule $\langle \mathcal{M}_0, \mu_0\rangle$, that is, the underlying set of $\mathcal{M}_0$ is countable and μ_0 is countable. This two-sorted structure is isomorphic to a weak structure $\langle \mathcal{M}_0, \mu_0' \rangle$. It follows that $\langle \mathcal{M}_0, \mu_0' \rangle$ is a countable topological abelian group $\mathcal{L}_\mathrm{m}$-equivalent to the rational group, and with a neighborhood basis at 0 consisting of open pure subgroups. Now a pure subgroup of a $\mathbb{Q}$-vector space is just a vector subspace; and a Hausdorff topology on a finite dimensional vector space generated by subspaces will be discrete. So $\mathcal{M}_0 \cong \mathbb{Q}^{(\aleph_0)}$, and μ_0 contains no finite dimensional subspaces. There are many possibilities for such, amongst which a natural example is the set μ_cof of all subspaces of finite codimension. This gives a natural countable model of the $\mathcal{L}_\mathrm{m}$ theory of $\langle \mathbb{Q}, \mathcal{E}_0 \rangle$ which looks quite different from the Euclidean topology on $\mathbb{Q}$ or $\mathbb{R}$, and which does not have $\langle \mathbb{Q}, \mathcal{E}_0 \rangle$ as a subspace. Although we have retained $\mathcal{L}_\mathrm{m}$-elementary equivalence at all stages, none of the two-sorted constructions preserve topologies. In fact, any finite dimensional subspace is discrete. There is a wide variety of similar models. Let $\mathcal{F}$ be any filter on the collection of all vector subspaces of $\mathbb{Q}^{(\omega)}$ containing μ_cof. Then $\mathcal{F}$ generates a group topology on $\mathbb{Q}^{(\omega)}$ that is $\mathcal{L}_\mathrm{m}$ equivalent to $\langle \mathbb{Q}, \mathcal{E}_0 \rangle$.

We make no claim as to what the κ-saturated elementary extensions of $\langle \mathbb{Q}, \mathcal{E}_0 \rangle$ for infinite κ might be; but we do note that $\mathbb{Q}^{(\kappa)}$ with the topology generated by subspaces of codimension less than κ is $\mathcal{L}_\mathrm{m}$-elementarily equivalent to $\langle \mathbb{Q}, \mathcal{E}_0 \rangle$, with a basis at 0 consisting of pure subgroups which is closed under intersections of families of size less than κ.

6. Embeddings and purity relative to $\mathcal{L}_\mathrm{t}$ or $\mathcal{L}_\mathrm{m}$

In the preceding section we saw some use for two-sorted maps between topological structures. This is unnatural from a topological standpoint. Indeed, the continuity of a map between topological structures is not defined in terms of taking open sets forward, but rather in terms of bringing open sets back. It is in this context of continuous homomorphisms that we would hope to find natural definitions of pure embedding.

There is a natural analogue to the homomorphism property for ordinary pp formulas (cf. Prest [28, Lemma 2.7]).

Proposition 6.1 (Enns [8, Lemma 4.24]**).** *Let $\langle \mathcal{M}, \sigma \rangle$ and $\langle \mathcal{N}, \tau \rangle$ be topological modules and let $f \colon M \to N$ be a continuous homomorphism. Then for every topological positive primitive formula $\varphi(\bar{v}, \overline{V})$, every $\overline{m} \in M$, and every $\overline{U} \in \tau$,*

$$\langle \mathcal{M}, \sigma \rangle \models \varphi[\overline{m}, f^{-1}[\overline{U}]] \implies \langle \mathcal{N}, \tau \rangle \models \varphi\left[f(\overline{m}), \overline{U}\right].$$

PROOF. This follows by a simple induction on the length of the quantifier prefix of the topological positive primitive formula φ. □

Note the peculiar role played by the variables of the second sort: they are not "carried upwards," but rather downwards. As a special case, if $\varphi(\bar{v})$ is a tppf with no set variables free and f is a continuous homomorphism, then $\langle \mathcal{M}, \sigma \rangle \models \varphi[\overline{m}] \implies \langle \mathcal{N}, \tau \rangle \models \varphi[f(\overline{m})]$. We do not know how to characterize those R-module homomorphisms which always preserve tpp formulas with no set variables free. For later reference, we will call these *$\mathcal{L}_\mathrm{t}$-homomorphisms.*

Corollary 6.2. *Let $\langle \mathcal{M}, \sigma \rangle$ and $\langle \mathcal{N}, \tau \rangle$ be topological modules with $\langle \mathcal{M}, \sigma \rangle$ a submodule and a subspace of $\langle \mathcal{N}, \tau \rangle$. Then for all $\overline{m} \in M$ and open sets $\overline{U} \in \tau$.*

$$\langle \mathcal{M}, \sigma \rangle \models \varphi[\overline{m}, \overline{U} \cap M] \implies \langle \mathcal{N}, \tau \rangle \models \varphi[f(\overline{m}), \overline{U}].$$

The fact that homomorphisms preserve pp formulas upwards is one motivation for the definition of purity: a pure embedding is one that reflects pp formulas downwards. However there is an immediate problem to making the obvious generalization to topological modules. Consider the case of the very simple topological positive primitive formula $\psi(V) = \exists u\,(u \in V)$ with only the set variable V free.

Proposition 6.3 (Enns [8, Theorem 4.35]). *Let $\langle \mathcal{M}, \sigma\rangle$ and $\langle \mathcal{N}, \tau\rangle$ be topological modules and $f\colon M \to N$ a continuous homomorphism. If for every $U \in \tau$,*

$$\langle \mathcal{N}, \tau\rangle \models (\exists u\,(u \in V))[U] \implies \langle \mathcal{M}, \sigma\rangle \models (\exists u\,(u \in V))\big[f^{-1}[U]\big],$$

then $f[M]$ is dense in N.

PROOF. Clearly $\langle \mathcal{N}, \tau\rangle \models (\exists u\,(u \in V))[U]$ just asserts that U is a nonempty open subset of N; and then the assumption implies that $f^{-1}[U]$ is a nonempty open subset of M. But $f^{-1}[U]$ is nonempty iff $f[M] \cap U$ is nonempty. Hence $f[M]$ is dense in N. □

Clearly if we wish to have a useful model theory of topological modules, this is much too strong a restriction to impose. But there is an easy way out of this particular problem; one that has been alluded to before. If we use the monotone language rather than the topological language, then the above proof disappears, as the formula $\exists u\,(u \in V)$ is trivially true of any neighborhood of 0.

However a slightly more complicated formula points out some serious problems with carrying through set parameters even in $\mathcal{L}_{\mathrm{m}}$-formulas. If X is a set definable in a topological module, say by a formula $\psi(v)$, then the closure of X is also definable, by the $\mathcal{L}_{\mathrm{m}}$-formula $\mathbf{cl}\,\psi(v) = \forall W \exists w\big(w - v \in W \wedge \psi(w)\big)$. In particular, if ψ is a monotone positive primitive formula, so is $\mathbf{cl}\,\psi(v)$. If we allow set parameters we encounter the following problem. Consider the monotone positive primitive formula

$$\psi(v, V) = \forall W \exists w (w \in W \wedge w - v \in V),$$

that is, "$v \in \mathbf{cl}\, V$."

Proposition 6.4. *Let $\langle \mathcal{M}, \mu\rangle$ and $\langle \mathcal{N}, \nu\rangle$ be topological modules (considered as monotone structures) and $f\colon M \to N$ a nonzero, continuous homomorphism. Then f reflects all instances of "$v \in \mathbf{cl}\, V$" if and only if f is an open map.*

PROOF. Suppose that W is a nonempty open set of the topology generated by μ, and that $f[W]$ is *not* open in $\langle \mathcal{N}, \nu\rangle$, that is, that $N \setminus f[W]$ is *not* closed. So we can chose an element m of $\mathbf{cl}_N(N \setminus f[W]) \cap f[W] \neq \varnothing$. Now for any $k \in f[M]$, $n \mapsto k+n$ is a homeomorphism of N onto itself sending $f[M]$ to $f[M]$ and $N \setminus f[M]$ to $N \setminus f[M]$. So we can assume without loss of generality by a suitable translation of W that $f(m) \neq 0$. Since Hausdorff groups are regular, we can choose a *closed* neighborhood $A \in \nu$ of 0 such that $f(m) \notin A$. Set $U = A \cup (N \setminus f[W])$, a neighborhood of 0.

Then trivially $f(m) \in \mathbf{cl}_N(U)$. Suppose that $m \in \mathbf{cl}_M(f^{-1}[U])$. We have

$$\begin{aligned}\mathbf{cl}_M(f^{-1}[U]) &= \mathbf{cl}_M(f^{-1}[A \cup (N \setminus f[W])]) \\ &\subseteq \mathbf{cl}_M(f^{-1}[A]) \cup \mathbf{cl}_M(M \setminus W) = \mathbf{cl}_M(f^{-1}[A]) \cup (M \setminus W)\end{aligned}$$

(the latter since $M \setminus W$ is closed). Thus $m \in \mathbf{cl}_M(f^{-1}[A])$ since $m \in W$ and so $f(m) \in \mathbf{cl}_N(A) = A$ by the continuity of f, contrary to the choice of A.

Conversely suppose that f is an open map. We want to see that for all $U \in \nu$ and all $m \in M$, $f(m) \in \mathbf{cl}_N(U) \Rightarrow m \in \mathbf{cl}_M(f^{-1}[U])$. Since in particular $f[M]$

is an open submodule, it is clopen. Hence for any $A \subseteq N$, $\mathbf{cl}_N(A) \cap f[M] = \mathbf{cl}_N(A \cap f[M])$. So $f(m) \in \mathbf{cl}_N(U)$ implies $f(m) \in \mathbf{cl}_N(U \cap f[M])$. Let W be open in the topology of $\langle \mathcal{M}, \mu \rangle$ with $m \in W$, so that $f[W]$ is open in the topology of $\langle \mathcal{N}, \nu \rangle$. Hence, since $f(m) \in f[W]$, $f[W] \cap (U \cap f[M]) \neq \varnothing$. Thus $W \cap f^{-1}[U] \neq \varnothing$, and so $m \in \mathbf{cl}_M(f^{-1}[U])$. □

So in spite of the promise of Lemma 6.1, we must abandon the idea of considering formulas with free set variables in any definition of topological purity.

Therefore the only decision remaining is what sort of topological constraints to put on the inclusion of $\langle \mathcal{M}, \mu \rangle$ as a submodule of $\langle \mathcal{N}, \nu \rangle$: either there is no relation between μ and ν; or $\mu \supseteq \nu \restriction M$; or $\mu = \nu \restriction M$. Now based on the types of arguments developed by the second author in [17] (e.g., the development of the basics of stability theory) it appears that there might be some sort of interesting "pure" model theory that can be developed under the weakest of these alternatives. In the same paper the second author proposed a definition of $\mathcal{L}_t$-elementary embedding that placed no topological constraints on the maps, and only required that formulas with only individual variables free be preserved and reflected. At about the same time, Majewski [20] put forward a similar definition for elementary embedding, but requiring that the maps in question be topological embeddings. Neither author offered a single example. It is hard to see how such a development would contain much of topological interest. It appears that the most natural definition would be to take the third alternative, namely that $\langle \mathcal{M}, \mu \rangle$ should be a subspace of $\langle \mathcal{N}, \nu \rangle$, but we do explore two possible definitions.

Definition 6.5. Let $\langle \mathcal{M}, \sigma \rangle$ and $\langle \mathcal{N}, \tau \rangle$ be topological modules with $\mathcal{M}$ a submodule of $\mathcal{N}$.

(1) $\langle \mathcal{M}, \sigma \rangle$ is *weakly $\mathcal{L}_t$-pure* in $\langle \mathcal{N}, \tau \rangle$ if for all tppfs $\varphi(\bar{v})$ with no set variables free, and all $\overline{m} \in M$,

$$\langle \mathcal{M}, \sigma \rangle \models \varphi[\overline{m}] \iff \langle \mathcal{N}, \tau \rangle \models \varphi[\overline{m}].$$

(2) $\langle \mathcal{M}, \sigma \rangle$ is *$\mathcal{L}_t$-pure* in $\langle \mathcal{N}, \tau \rangle$ if it is weakly $\mathcal{L}_t$-pure and in addition, σ is the subspace topology induced by τ.

An intermediate case would only require that $\sigma \supseteq \tau \restriction M$.

We choose *not* to call this a "topological(ly) pure submodule." Several people, most notably Peter Loth, have been exploring various concepts of purity in the context of locally compact abelian groups. We will discuss this further in Section 9.

An immediate problem for any of the potential definitions of pure-embedding for topological modules considered here is that of *existence*: does an arbitrary topological module have a proper pure extension? does it have a pure extension satisfying some reasonable conditions? What we have in mind here is the problem of finding some substitute for the use of elementary extensions or the use of reduced products in the proof of "pure-injectivity implies pp-compactness," Theorem 1.7. We have some limited results, the two following and Corollary 7.6.

Proposition 6.6. *Let $\langle \mathcal{M}, \sigma \rangle$ be a topological module and α a basis for σ. Suppose that the weak structure $\langle \mathcal{N}, \beta \rangle$ is a two-sorted elementary extension of $\langle \mathcal{M}, \alpha \rangle$, and let τ be the topology generated by β as a basis. Then $\langle \mathcal{N}, \tau \rangle$ is a weak $\mathcal{L}_t$-pure extension of $\langle \mathcal{M}, \sigma \rangle$.*

PROOF. For $\overline{m} \in M$ and tpp formula φ,

$$\begin{aligned}\langle \mathcal{N}, \tau \rangle \models_t \varphi[\overline{m}] &\Leftrightarrow \langle \mathcal{N}, \beta \rangle \models_2 \varphi[\overline{m}] && \text{(invariance with respect to basis)}\\ &\Leftrightarrow \langle \mathcal{M}, \alpha \rangle \models_2 \varphi[\overline{m}] && \text{(two-sorted elementary substructure)}\\ &\Leftrightarrow \langle \mathcal{M}, \sigma \rangle \models_t \varphi[\overline{m}] && \text{(invariance with respect to basis)}\end{aligned}$$ □

We have one situation in which we can recognize $\mathcal{L}_t$-purity, theories T of topological modules which have complete elimination of set quantifiers from tppfs. "Elimination" here means that for every tppf φ of $\mathcal{L}_t$, there is a ppf $\hat{\varphi}$ of ${}_R\mathcal{L}$ such that $T \models \varphi \leftrightarrow \hat{\varphi}$. Such theories include the theory of $\langle \mathbb{Q}, \mathcal{E} \rangle$ already discussed, and any theory with a compact model, cf. Kucera [17, Theorem 3.13]. For the latter sort of theory, $\hat{\varphi}$ is constructed explicitly from φ, and $\models_t \hat{\varphi} \to \varphi$ holds in all topological modules, so the result is a little sharper.

Theorem 6.7. *Suppose $\langle \mathcal{M}, \sigma \rangle$ is a subspace of $\langle \mathcal{N}, \tau \rangle$ such that $\mathcal{M}$ is a pure submodule of $\mathcal{N}$.*

If $\langle \mathcal{N}, \tau \rangle$ is $\mathcal{L}_t$-equivalent to a compact module, or if $\langle \mathcal{M}, \sigma \rangle$ and $\langle \mathcal{N}, \tau \rangle$ are both models of an $\mathcal{L}_t$-theory such that every tppf with only individual variables free is equivalent, modulo **T**, *to a positive primitive formula, then $\langle \mathcal{M}, \sigma \rangle$ is $\mathcal{L}_t$-pure in $\langle \mathcal{N}, \tau \rangle$.*

PROOF. We only need check that tppfs are reflected downwards. But this is easy: if φ is a topological positive primitive formula with only individual variables free, $\langle \mathcal{N}, \tau \rangle$ satisfies $\forall \overline{v}(\varphi \leftrightarrow \hat{\varphi})$, and in either case $\langle \mathcal{M}, \sigma \rangle$ satisfies $\forall \overline{v}(\hat{\varphi} \to \varphi)$. If $\overline{m} \in M$ and $\langle \mathcal{N}, \tau \rangle \models \varphi[\overline{m}]$, then $\langle \mathcal{N}, \tau \rangle \models \hat{\varphi}[\overline{m}]$, and in particular $\mathcal{N} \models \hat{\varphi}[\overline{m}]$ since $\hat{\varphi}$ is a formula of ${}_R\mathcal{L}$. Thus since $\mathcal{M}$ is pure in $\mathcal{N}$, $\mathcal{M} \models \hat{\varphi}[\overline{m}]$. Thus in either case $\langle \mathcal{M}, \sigma \rangle \models \hat{\varphi}[\overline{m}]$, and so $\langle \mathcal{M}, \sigma \rangle \models \varphi[\overline{m}]$. □

So, for instance, $\langle \mathbb{Q}, \mathcal{E} \rangle$ is $\mathcal{L}_t$-pure in $\langle \mathbb{R}, \mathcal{E} \rangle$. In fact, since $\mathbb{Q}$ is an injective abelian group, any embedding of $\mathbb{Q}$ onto a Euclidean subspace of a model of the $\mathcal{L}_t$-theory of $\langle \mathbb{Q}, \mathcal{E} \rangle$ is $\mathcal{L}_t$-pure in our sense. Since the two structures are in fact $\mathcal{L}_t$-equivalent, it follows by the tpp elimination of quantifiers that this is an elementary embedding in the sense of Majewski [20].

Although this theorem allows us to recognize $\mathcal{L}_t$-pure extensions in certain situations, it does not necessarily help to construct them. We could, for instance, reduce a set of tppfs with no set variables free to a set of equivalent pp formulas, and then use that set to construct a useful extension of some sort by the usual methods of the model theory of modules. However this method would not help to construct an appropriate topology on that extension. Indeed, Corollary 5.4 ensures that sometimes it will not be possible to do so, for a countable nondiscrete module *cannot* be a subspace of an $\aleph_1$-saturated (two-sorted) elementary extension.

7. The role of products and coproducts

Suppose that for each $i \in I$, $\mathcal{X}_i = \langle X_i, \tau_i \rangle$ is a topological space, and let $\mathbb{X} = \prod_{i \in I} X_i$. There are (at least) two topologies of interest on the Cartesian product: the *Tychonoff topology* τ_{prod} which has as a basis all sets of the form $\prod_{i \in I} U_i$, where $U_i \in \tau_i$ for all $i \in I$ and $U_i = X_i$ for all but finitely many $i \in I$; and the *box topology* τ_{box} which has as a basis all sets of the form $\prod_{i \in I} U_i$, where $U_i \in \tau_i$ for all $i \in I$. Clearly for infinite I, the box topology is strictly finer than the Tychonoff topology. It is a fundamental fact that the Tychonoff topology makes the Cartesian product into the direct product in the categorical sense, both in the category of topologies and in the categories of topological abelian groups or topological modules. Thus the projection maps are continuous (and open).

Coproducts in the category of topological spaces are just disjoint unions; but in the category of topological abelian groups the situation is quite different. Let $(\langle \mathcal{M}_i, \tau_i \rangle)_{i \in I}$ be a family of topological abelian groups (modules). The *coproduct topology* τ_{coprod} is defined to be the finest module topology on the direct sum $\bigoplus_{i \in I} \mathcal{M}_i$ which makes all the canonical inclusions $\varepsilon_i \colon \mathcal{M}_i \to \bigoplus_{i \in I} \mathcal{M}_i$ continous; since clearly both the restrictions from the direct product of the Tychonoff topology and of the box topology do so, the coproduct topology is well defined and at least as fine as the restriction of the box topology. (The set of Hausdorff module topologies on a fixed module $\mathcal{M}$ is a complete lattice.) In fact, the coproduct topology coincides with the box topology if I is countable; but it may be be finer if I is uncountable, cf. Nickolas [26]. It is also clear that for finite coproducts, the coproduct topology coincides with the Tychonoff topology. The basic ideas date back to the 1940s, cf. S. Kaplan [14], but the first complete (abstract) description of the coproduct topology was given by P. J. Higgins in 1977 [13], with the first explicit description being given by Chasco and Domínguez [5]; and Enns has a new and perhaps simpler way of deriving this description in his M.Sc. thesis [8].

Theorem 7.1 (Enns [8, Theorem 4.36]**).** *Let $(\langle \mathcal{M}_i, \tau_i \rangle)_{i=1}^n$ be topological modules and let $\varphi(\bar{v}, \overline{V})$ be a topological positive primitive formula. For each $i = 1, \ldots, n$ let $\bar{a}^i$ be a tuple in M_i compatible with $\bar{v}$, and let $\overline{U}^i$ be a tuple in τ_i compatible with $\overline{V}$. Let $\mathbf{a}$ be the corresponding tuple formed pointwise in $\prod_{i=1}^n M_i$ and let $\mathbf{U}$ be the tuple in the product topology with components $(\overline{U}^1)_j \times \cdots \times (\overline{U}^n)_j$. (So each such component is a member of the standard basis for the product topology).*

Then

$$\prod_{i=1}^{n} \langle \mathcal{M}_i, \tau_i \rangle \models \varphi[\mathbf{a}, \mathbf{U}]$$

if and only if

$$\langle \mathcal{M}_i, \tau_i \rangle \models \varphi[\bar{a}^i, \overline{U}^i] \quad \text{for each } i,\ i = 1, \ldots, n.$$

PROOF. The proof is a routine induction on the complexity of topological positive primitive formulas. □

The "correct" generalization of this theorem would be to infinite products; or perhaps to infinite coproducts. But neither of these can work. The specific form of $\mathbf{U}$ is required for the induction to carry through; and in the infinite case the sets of this form are precisely a base for the box topology. So the theorem may fail for infinite products, since in general the sets $\prod_{i \in I} \overline{U}^i{}_j$ are not open in the product topology; and the theorem may fail for infinite coproducts since in general the sets $\prod_{i \in I} \overline{U}^i{}_j$ do not form a base for the coproduct topology. For what it is worth, this theorem does carry through for arbitrary products when endowed with the box product topology; and thus by the result [26] of Nickolas already cited, for countable direct sums as well.

Flum and Ziegler [9] develop a general theory of back-and-forth arguments for $\mathcal{L}_t$; and using these methods in particular prove the following theorem:

Theorem 7.2 ([9, Part I, Theorem 6.1]). *Suppose that $\langle \mathcal{M}_i, \sigma_i \rangle$ and $\langle \mathcal{N}_i, \tau_i \rangle$ are $\mathcal{L}_{\mathrm{t}}$-elementarily equivalent $\aleph_0$-saturated topological R-modules for each $i \in I$. Then*

$$\prod_{i \in I} \langle \mathcal{M}_i, \sigma_i \rangle \equiv_{\mathrm{t}} \prod_{i \in I} \langle \mathcal{N}_i, \tau_i \rangle$$

We may remove the requirement that the modules be $\aleph_0$-saturated if the language is finite, for instance in the case of topological abelian groups.

It is not known if this result holds in complete generality.

By similar methods, Enns proved in his thesis:

Theorem 7.3 ([8, Theorem 4.38]). *Suppose that $\langle \mathcal{M}_i, \sigma_i \rangle$ and $\langle \mathcal{N}_i, \tau_i \rangle$ are $\mathcal{L}_{\mathrm{t}}$-elementarily equivalent $\aleph_0$-saturated topological R-modules for each $i \in I$. Then*

$$\left\langle \bigoplus_{i \in I} \mathcal{M}_i, \sigma_{\mathrm{coprod}} \right\rangle \equiv_{\mathrm{t}} \left\langle \bigoplus_{i \in I} \mathcal{N}_i, \tau_{\mathrm{coprod}} \right\rangle.$$

Once again, we may remove the requirement that the modules be $\aleph_0$-saturated for the case of finite languages, in particular for topological abelian groups; and we do not know if the result holds in complete generality.

Finally we consider the relationship between the direct sum and the direct product. We get a result, but not a direct generalization of the corresponding Theorem 1.5 for ordinary modules. This result depends on an appropriate choice of basis in order to allow the use of set parameters.

The proof consists almost entirely in choosing an efficient notation to describe the various elements and sets involved. Let $(\langle \mathcal{M}_i, \sigma_i \rangle)_{i \in I}$ be topological modules. Let $\mathcal{M}_{\oplus} = \bigoplus_{i \in I} \mathcal{M}_i$; let $\mathcal{M}_{\Pi} = \prod_{i \in I} \mathcal{M}_i$. Let $\bar{a} \in \mathcal{M}_{\oplus}$, $\overline{U} \in \beta$, where β is the standard basis for the Tychonoff product topology. The standard projection and inclusion maps are π_i and ε_i respectively. The *support* of $a \in \mathcal{M}_{\oplus}$ is $\operatorname{supp}(a) = \{i \in I : \pi_i(a) \neq 0\}$; the *support* of $U \in \beta$ is $\operatorname{supp}(U) = \{i \in I : \pi_i[U] \neq M_i\}$; and the *support* of $\bar{a}\overline{U}$, $\operatorname{supp}(\bar{a}\overline{U})$, is the union of the supports of the elements in $\bar{a}$ and the sets in $\overline{U}$. Clearly $\operatorname{supp}(\bar{a}\overline{U})$ is a finite subset of I. Given $I_{\mathrm{C}} \subseteq I$ and $c \in \mathcal{M}_{\Pi}$, let $c \restriction I_0$ be the element d defined by $d_i = c_i$ if $i \in I_0$; otherwise let $d_i = 0$. Clearly if I_0 is finite, then $d \in \mathcal{M}_{\oplus}$. If $\bar{c} = c^0, c^1, \ldots$ is a tuple in $\mathcal{M}_{\Pi}$, let $\bar{c} \restriction I_0 = c^0 \restriction I_0, c^1 \restriction I_0, \ldots$.

Lemma 7.4 (Enns [8, Lemma 4.39]**).** *Let $\psi(\bar{v}, \bar{u}, \overline{V})$ be a topological positive primitive formula, and let all other notation be as in the preceding paragraph. Let $I_0 = \operatorname{supp}(\bar{a}\overline{U})$. Then*

$$\langle \mathcal{M}_{\Pi}, \tau_{\mathrm{prod}} \rangle \models \psi[\bar{a}, \bar{c}, \overline{U}] \implies \langle \mathcal{M}_{\Pi}, \tau_{\mathrm{prod}} \rangle \models \psi[\bar{a}, \bar{c} \restriction I_0, \overline{U}].$$

PROOF. Obvious. Note that in computing the value of a term, which is done component-wise in the product, for i outside the support of $\bar{a}\overline{U}$ the value only depends on $\bar{c}$, not on $\bar{a}$, and membership is trivial because the corresponding open set is all of M_i. Therefore replacing the i-components of the entries in $\bar{c}$ by 0 cannot affect the truth of atomic formulas, and similarly for conjunctions of such. Then the result follows easily by induction on the quantifier complexity of the topological positive primitive formula ψ. The case of universal set quantifiers is immediate since the hypothesis is taken over a basis for the product topology; and the case for existential individual quantifiers follows easily. □

Theorem 7.5 (Enns [8, Theorem 4.40]**).** *Continuing the notation of the lemma and the paragraph preceding it, let $\psi(\bar{v}, \overline{V})$ be a topological positive primitive formula. Then*

$$\langle \mathcal{M}_\oplus, \tau_{\text{prod}} \restriction \mathcal{M}_\oplus \rangle \models \psi[\bar{a}, \overline{U} \cap \mathcal{M}_\oplus] \iff \langle \mathcal{M}_\Pi, \tau_{\text{prod}} \rangle \models \psi[\bar{a}, \overline{U}].$$

PROOF. The forward direction follows by Proposition 6.1 since we have put the subspace topology on $\mathcal{M}_\oplus$. The converse follows immediately by Lemma 7.4. □

Corollary 7.6. *With the same notational conventions, $\langle \mathcal{M}_\oplus, \tau_{\text{prod}} \restriction \mathcal{M}_\oplus \rangle$ is $\mathcal{L}_t$-pure in $\langle \mathcal{M}_\Pi, \tau_{\text{prod}} \rangle$.*

Note that we do not and cannot get a nice relation between the coproduct and the product. For an arbitrary coproduct of discrete modules is discrete, but the corresponding product is not.

8. Pure-injectivity

We seek some suitable definition of pure-injectivity for topological modules in the form of "injective over topologically pure embeddings." We have already discussed a range of candidates for "topological pure embedding," with the two alternatives of Definition 6.5 perhaps showing the most promise.

Immediately we see another obstruction to the use of free set variables in tpp formulas:

Proposition 8.1 (Enns [8, Corollary 4.27]**).** *Let $\varphi(\bar{v}, \overline{V})$ be a topological positive primitive formula. Form $\varphi^\circ(\bar{v})$ from φ by deleting every conjunct of the open matrix of φ of the form "$\mathbf{t} \in V$", V a free set variable of φ. Then φ is satisfiable in $\langle \mathcal{N}, \tau \rangle$ if an only if for some $\bar{a}$ in N, $\langle \mathcal{N}, \tau \rangle \models \varphi[\bar{a}, \overline{M}]$, equivalently $\langle \mathcal{N}, \tau \rangle \models \varphi^\circ[\bar{a}]$.*

PROOF. All occurrences of set variables in a topological positive primitive formula are positive occurrences of some atomic subformula "$\mathbf{t} \in V$." So if such a formula is true at a given assignment of values, it remains true upon replacing the value of V by any larger open set. □

Thus the satisfiability of a set of topological positive primitive formulas does not depend on the free set variables appearing in that set.

In the end, it appears not to matter much what the precise forms of the definitions are: there are strong barriers to proving either direction of the equivalence "topological pure-injective if and only if tpp-compact"; at least by anything approaching the standard methods of the model theory of modules.

8.1. Pure injective implies tpp-compact? First consider the case where $\langle \mathcal{N}, \tau \rangle$ is injective over topologically pure maps in some sense. The "short" proof that pure-injective implies pp-compact for ordinary modules uses quite heavy machinery, namely the existence of sufficiently saturated elementary extensions. Reduced products in two-sorted logic can be used just as they are for ordinary modules, but the embedding constructed will in general be onto a discrete subspace, and therefore not even continuous. So we cannot carry through the proof of Section 1 for $\mathcal{L}_t$ pure embeddings or any similar concept that demands the topologies of the two spaces be related in some way.

For injectivity over weakly $\mathcal{L}_t$-pure embeddings, we can imitate the proof of Section 1 as long as we only demand that the homomorphisms we consider preserve topological positive primitive formulas; there is nothing that we can use to force the lifted homomorphism $\hat{f}$ to be continuous. In particular, although continuity of definable functions within a structure is axiomatizable in $\mathcal{L}_t$, we cannot use formulas or sets of formulas of $\mathcal{L}_t$ to force a homomorphism between structures to be continuous.

8.2. tpp-compact implies pure-injective? Now we consider whether various forms of tpp-compactness might imply some form of injectivity over pure maps for topological modules. So suppose that $\langle \mathcal{N}, \tau \rangle$ is tpp-compact, that $e \colon \langle \mathcal{A}, \alpha \rangle \to \langle \mathcal{B}, \beta \rangle$ is a pure embedding, and that $f \colon \mathcal{A} \to \mathcal{N}$ is a homomorphism. At the very least we assume that e reflects tppfs and that f preserves tppfs (with no set variables free). Then we can imitate the usual proof for ordinary modules: we enumerate B by a set of new, distinct variables $\mathcal{V}$, and let Φ be the set of all tppfs in these variables and with parameters from A satisfied by B. Since e reflects tpp formulas, Φ is finitely satisfiable in $\mathcal{A}$, and since f preserves tpp formulas, the natural image $f[\Phi]$ of Φ in $\mathcal{N}$ is also finitely satisfiable. So if $\langle \mathcal{N}, \tau \rangle$ is compact with respect to sets of tpp formulas with no free variables, $f[\Phi]$ has a solution in $\langle \mathcal{N}, \tau \rangle$ and so there is a map $\hat{f} \colon \langle \mathcal{B}, \beta \rangle \to \langle \mathcal{N}, \tau \rangle$ induced by this solution. Clearly by construction $\hat{f}$ preserves tpp formulas, so in particular $\hat{f}$ is a homomorphism. But that is the only condition that we can impose on $\hat{f}$ by tpp formulas, even if we assume that e is a topological embedding and that f is continuous. There is no sense in which the topology on $\mathcal{B}$ is determined by Φ.

Ignoring for the moment the difficulties already noted in including set parameters anywhere in our definitions, we might ask what could be gained by allowing set parameters. In order to carry set variables through the above argument, it is necessary to assume right from the start that e is a topological embedding and that f is proper. Then $\hat{f}$ would be constructed as a two sorted map $\hat{f}_0 \colon \mathcal{B} \to \mathcal{N}$ and $\hat{f}_1 \colon \beta \to \tau$. Since $\hat{f}$ preserves tpp formulas, in particular "$v \in V$," we obtain immediately that for all nonempty $U \in \beta$, $\hat{f}_0[U] \subseteq \hat{f}_1(U)$, that is, $\hat{f}_0[U] = \hat{f}_1(U) \cap \hat{f}_0[M]$. So $\hat{f}_0$ is open onto its image. But we still have no tools available to conclude that $\hat{f}_0$ is continuous.

9. Alternative approaches

In [2] Paul Bankston introduced a notion of ultraproduct for topological spaces; and has expanded on this idea and used techniques associated with it in subsequent papers over the years. This gives us a sort of model theory of topological spaces without the bother of considering any sort of formal language.

In 1976, C. Ward Henson [12] introduced a logic suitable for the study of Banach spaces. In addition to the language of $\mathbb{Q}$-vector spaces, the appropriate language includes predicates for the closed unit ball and for the elements of norm greater than or equal to one. Notions of "approximate satisfaction" (of positive bounded formulas) and "approximate equivalence" grow naturally out of the metric structure of the Banach spaces. This line of study has proved to be very profitable in the study of Banach spaces, but this sort of logic has little to say about topological modules in general. Closely related to this is the *continuous logic* of Ben Yaacov, Henson, et al., cf. [4] which gives a good model theory for structures endowed with a metric.

Pillay [27] studied topological structures $\langle \mathcal{M}, \sigma \rangle$ where the topology is uniformly definable in some first-order language, that is, there is a formula $\beta(v, \bar{u})$ of the language of $\mathcal{M}$ such that $\{\beta[M, \overline{m}] : \overline{m} \in M\}$ is a basis for σ. Many of the structures we consider here do not fall into this class. Pillay obtains especially nice results for structures where every definable set is a Boolean combination of definable open sets.

There is a rich history of the study of the class of locally compact abelian groups (LCA), motivated by the Pontryagin duality theorm. In 1967 Moskowitz [25] considered to what extent homological algebra can be developed inside LCA. As we have already noted, this category is not abelian, so there are restrictions on what can be accomplished. Moskowitz obtains general results for LCA, but often by restricting his attention to cases where the maps are proper and where one or more of the abelian groups in questions is compactly generated. This opens up the possibility of imitating the development of the model theory of modules in a functorial context, as in Prest [29], with the same sorts of restrictions as considered by Moskowitz. This would *not* need references to the syntax and semantics of any particular language.

We have already made reference to the recent work of Peter Loth and others on definitions of purity and pure-injectivity in LCA. Khan [15] first introduced a concept of purity for topogical abelian groups in 1973; the book of Armacost [1, Chapter 7] of 1981 contains this and much else that is relevant. More recently, Peter Loth has reconsidered the definition of Khan and several of his own; these definitions of purity in LCA are all distinct. See [18, 19] for some of Loth's work most closely connected to the ideas that we explore here, and for further references.

Definition 9.1 (Khan [15]**).** Let $\langle \mathcal{A}, \sigma \rangle$ be a locally compact abelian group. A closed subgroup H of A is *t-pure* in $\langle \mathcal{A}, \sigma \rangle$ if for all closed subgroups $K \supseteq H$ such that K/H is compactly generated, H splits from K.

Theorem 9.2 (Khan [15]**).** *Let $\langle \mathcal{A}, \sigma \rangle$ be a locally compact abelian group and H a closed subgroup of $\langle \mathcal{A}, \sigma \rangle$.*

(1) *If H is a topological direct summand of $\langle \mathcal{A}, \sigma \rangle$ then H is t-pure in $\langle \mathcal{A}, \sigma \rangle$.*
(2) *If H is t-pure in $\langle \mathcal{A}, \sigma \rangle$ then H is pure in $\langle \mathcal{A}, \sigma \rangle$.*

Thus "t-purity" as defined by Khan extends algebraic purity for abelian groups and shares the important property (1) with it. There appears to be no easy correspondence between the notions of purity relative to $\mathcal{L}_t$ that we have introduced, and Khan's notion of t-purity.

Definition 9.3 (Loth [19]**).** A proper short exact sequence

$$0 \to A \xrightarrow{\alpha} B \xrightarrow{\beta} C \to 0$$

of locally compact abelian groups is *t-pure* if $\alpha[A]$ is a *topologically pure subgroup* of B, that is, if

$$\mathbf{cl}_B(n\alpha[A]) = \mathbf{cl}_B(nB) \cap \alpha[A]$$

for every positive integer n.

By taking $n = 1$ it follows that A must be closed in B.

Definition 9.4 (Loth [19]**).** A proper short exact sequence

$$0 \to A \to B \to C \to 0$$

of locally compact abelian groups is *topologically pure* if

$$0 \to \mathbf{cl}_A(nA) \to \mathbf{cl}_B(nB) \to \mathbf{cl}_C(nC) \to 0$$

is proper exact for all positive integers n.

Since the sequences considered are proper, the image of A is locally compact and hence closed in B. Thus a topologically pure exact sequence is t-pure. But Loth [18, Example 3.5] gives an example of a t-pure exact sequence which is not topologically pure. In addition ([18, Example 2.4]) shows that a proper exact sequence that is algebraically pure need not be topologically pure, and that a proper exact sequence which is topologically pure need not be algebraically pure. It follows in particular by Theorem 9.2 that neither of Loth's concepts imply t-purity in the sense of Khan.

In regards to topological purity, the sets nA are pp-definable subgroups; and we have already noted that the closure of a tpp-definable subgroup is again tpp-definable (see the discussion preceding Proposition 6.4). Thus "topological purity" demands only that a restricted class of tpp formulas be reflected.

Loth [19] defines *t-pure-injectivity* in the class of locally compact abelian groups in the usual way from (his version of) t-purity. In this paper, Theorem 2.7 shows that the t-pure-injective locally compact abelian groups have a rather restricted structure: they are a topological direct sum $\mathbb{R}^n \oplus \mathbb{T}^\kappa \oplus G$, where n is a nonnegative integer, κ is a cardinal, and G is a"topological torsion group." We refer the reader to Loth's papers for the details.

10. Summary

We have considered the possibility of developing a useful model theory of topological modules (useful for the purposes of topological algebra) in the languages $\mathcal{L}_\mathrm{t}$ or $\mathcal{L}_\mathrm{m}$.

- Any approach to this problem that involves the use of open-set parameters, or imposes topological constraints on the morphisms between structures, seems unlikely to produce anything useful.
- If we consider the category of topological modules with module homomorphisms that preserve topological positive primitive formulas, we can get some results of a model-theoretic nature, but which appear to have little of topological interest.
- Nothing we have discussed here places any limits on what might be accomplished by other means. In particular, it places no restrictions on what might be accomplished within the (nonaxiomatizable) class of locally compact abelian groups or modules.

References

1. D. L. Armacost, *The structure of locally compact abelian groups*, Monogr. Textbooks Pure Appl. Math., vol. 68, Marcel Dekker Inc., New York, 1981.
2. P. Bankston, *Ultraproducts in topology*, General Topology and Appl. **7** (1977), no. 3, 283–308.
3. W. Baur, *Elimination of quantifiers for modules*, Israel J. Math. **25** (1976), no. 1-2, 64–70.
4. I. Ben Yaacov, A. Berenstein, C. W. Henson, and A. Usvyatsov, *Model theory for metric structures*, Model Theory with Applications to Algebra and Analysis. Vol. 2 (Z. Chatzidakis, D. Macpherson, A. Pillay, and A. Wilkie, eds.), London Math. Soc. Lecture Note Ser., vol. 350, Cambridge Univ. Press, Cambridge, 2008, pp. 315–427.
5. M. J. Chasco and X. Domínguez, *Topologies on the direct sum of topological abelian groups*, Topology Appl. **133** (2003), no. 3, 209–223.

6. D. N. Dikranjan, I. R. Prodanov, and L. N. Stoyanov, *Topological groups: Characters, dualities and minimal group topologies*, Monogr. Textbooks Pure Appl. Math., vol. 130, Marcel Dekker Inc., New York, 1990.
7. P. Eklof and G. Sabbagh, *Model-completions and modules*, Ann. Math. Logic **2** (1970/1971), no. 3, 251–295.
8. C. Enns, *Pure embeddings and pure-injectivity for topological modules*, Master's Thesis, University of Manitoba, 2009.
9. J. Flum and M. Ziegler, *Topological model theory*, Lecture Notes in Math., vol. 769, Springer, Berlin, 1980.
10. S. Garavaglia, *Model theory of topological structures*, Ann. Math. Logic **14** (1978), no. 1, 13–37.
11. L. Henkin, *Completeness in the theory of types*, J. Symbolic Logic **15** (1950), 81–91.
12. C. W. Henson, *Nonstandard hulls of Banach spaces*, Israel J. Math. **25** (1976), no. 1-2, 108–144.
13. P. J. Higgins, *Coproducts of topological abelian groups*, J. Algebra **44** (1977), no. 1, 152–159.
14. S. Kaplan, *Extensions of the Pontrjagin duality.* I: *Infinite products*, Duke Math. J. **15** (1948), 649–658.
15. M. A. Khan, *A notion of purity in locally compact groups*, J. Sci. (Karachi) **2** (1973), no. 1, 128–130.
16. T. G. Kucera, *Applications of stability-theoretical methods to theories of modules*, Ph.D. thesis, McGill University, 1984.
17. ______, *Stability theory for topological logic, with applications to topological modules*, J. Symbolic Logic **51** (1986), no. 3, 755–769.
18. P. Loth, *Topologically pure extensions*, Abelian groups, rings and modules (Perth, 2000) (A. V. Kelarev, R. Göbel, K. M. Rangaswamy, P. Schultz, and C. Vinsonhaler, eds.), Contemp. Math., vol. 273, Amer. Math. Soc., Providence, RI, 2001, pp. 191–201.
19. ______, *On t-pure and almost pure exact sequences of LCA groups*, J. Group Theory **9** (2006), no. 6, 799–808.
20. M. Majewski, *Elementary extensions of topological models in L_t language*, Studia Logica **46** (1987), no. 3, 255–264.
21. T. A. McKee, *Some applications of model theory to topology*, Ph.D. thesis, University of Wisconsin, 1974.
22. ______, *Infinitary logic and topological homeomorphisms*, Z. Math. Logik Grundlagen Math. **21** (1975), no. 5, 405–408.
23. ______, *Sentences preserved between equivalent topological bases*, Z. Math. Logik Grundlagen Math. **22** (1976), no. 1, 79–84.
24. L. G. Monk, *Elementary-recursive decision procedures*, Ph.D. thesis, University of California, Berkeley, 1975.
25. M. Moskowitz, *Homological algebra in locally compact abelian groups*, Trans. Amer. Math. Soc. **127** (1967), 361–404.
26. P. Nickolas, *Coproducts of abelian topological groups*, Topology Appl. **120** (2002), no. 3, 403–426.
27. A. Pillay, *First order topological structures and theories*, J. Symbolic Logic **52** (1987), no. 3, 763–778.
28. M. Prest, *Model theory and modules*, London Math. Soc. Lecture Note Ser., vol. 130, Cambridge Univ. Press, Cambridge, 1988.
29. ______, *Purity, spectra and localisation*, Encyclopedia Math. Appl., vol. 121, Cambridge Univ. Press, Cambridge, 2009.
30. J. Sgro, *Completeness theorems for topological modules*, Ph.D. thesis, University of Wisconsin, 1974.
31. M. Ziegler, *A language for topological structures which satisfies a Lindström-theorem*, Bull. Amer. Math. Soc. **82** (1976), no. 4, 568–570.

Department of Mathematics, University of Manitoba, Winnipeg, MB R3T 2N2, Canada

E-mail address, C. Enns: `clint.enns@gmail.com`

E-mail address, T. G. Kucera: `tkucera@cc.umanitoba.ca`

Centre de Recherches Mathématiques
CRM Proceedings and Lecture Notes
Volume **53**, 2011

Model Theory vs. Categorical Logic: Two Approaches to Pretopos Completion (a.k.a. T^{eq})

Victor Harnik

Dedicated to Michael Makkai on his 70th birthday

Introduction

It was a sunny day in Jerusalem, in the fall of 1980, when Michael Makkai approached me with a question that sounded more like a declaration. "Did you realize that T^{eq} is nothing but the pretopos completion?" Mihaly (pronounced 'Mee-hi') as many of his friends, including myself, call him, spoke with certain excitement, but that was not unusual with him, when the subject was mathematics. This was the first month of a special year on Model Theory organized by Saharon Shelah at the Institute for Advanced Studies of the Hebrew University.

I knew what Shelah's T^{eq} was, but I had no idea about pretoposes. I realized, however, that a remarkable event took place. The very same new mathematical concept came up in two, seemingly totally unrelated, pieces of mathematical research, one of Shelah and the other of Makkai and Reyes. Both works were developed by what we call *theory builders.* In the tradition of good mathematics, both works were seeking to discover *structure.*

Saharon Shelah worked for years to understand the ways in which mathematical models of given theories can be described. His work drastically changed the scene of Model Theory. Especially striking are his results concerning *first order countable complete theories.* He showed that each such theory T belongs to one of two kinds: *either* T has 'many models', meaning that in every uncountable cardinality λ the number $I(\lambda, T)$ of nonisomorphic models of T has the maximal possible value 2^λ, *or,* there is a good *structure theory* for the models of T, meaning that the isomorphism type of each model is determined by a system of *cardinal invariants* (a simplest example being the vector spaces over the field of rationals, which are described by one cardinal, i.e. the dimension). As a corollary, Shelah proved the so called Morley conjecture saying that, for T as above, the function $I(\lambda, T)$ is nondecreasing for uncountable values of λ. Shelah's work goes way beyond the first order countable case and gives much information about the class of models of uncountable first order theories, or of theories in infinitary or higher order logics and, even, more

2010 *Mathematics Subject Classification.* Primary 03G03; Secondary 03C45.
This is the final form of the paper.

general classes of structures called by Shelah '*abstract elementary*.' The first detailed account of his impressive work, was published in [21] and substantial additions, including the proof of Morley's conjecture, appeared in the second edition [23]. For an illuminative, non technical description, see [22]. For the results on abstract elementary classes, see [24] and [25].

Shelah had to overcome a great deal of technical difficulties, using imaginative, ingenious ideas. One of these is connected to the present paper. To every first order theory T, he attached an extension T^{eq}, a theory formulated in a richer language. The two theories are closely connected, but Shelah could use T^{eq} to prove new statements about T itself. It is this construction that appeared, in another disguise, in the work of Makkai and Reyes.

Makkai arrived in Montreal in the fall of 1973 and met there Gonzalo Reyes, his friend from the days of a visit in Berkeley. They both were trained as model theorists, but in Montreal is hard to avoid exposure to Category Theory. At the time of Makkai's arrival, Reyes was already won over to categories, especially due to intensive discussions with André Joyal. Mihaly listened, with great reluctance at first, to Gonzalo's 'preaching' but soon found himself engaged in vivid discussion that led, finally, to a massive work summed up in [15].

The main point of this work is that categories, as underlying structures, are much more prevalent in logic and model theory than previously thought. Not only do the models of a given theory form a category (everyone knew that) but the *first order theories* themselves are, or can be seen as, *categories*. The authors take care to emphasize that this idea has been reached under the influence of the work of Lawvere and discussions with Joyal (as well as a previous work of Joyal and Reyes). However, the extent to which this point of view can be elaborated, is the main novelty of this book. Theories are categories and models become functors into the category of sets. Interpretations of theories into each other become also functors between the corresponding categories and can be seen as *generalized* models. An intimate relation between categorical logic and work of Grothendieck in algebraic geometry, esp. the theory of *topoi*, is established.

Each first order theory T is identified by Makkai and Reyes with a category $\mathbb{T}$ of a kind called *logical* (actually, if we work in the framework of *full classical logic*, as model theorists do, $\mathbb{T}$ is what we call a *Boolean* logical category). Makkai and Reyes notice that the class of categories called *pretoposes* by Grothendieck, is a subclass of the logical categories. The pretoposes enjoy a special property called by them *conceptual completeness*. They also prove that each logical category has a closely related pretopos called its *pretopos completion*.

What Makkai was saying in the sunny fall day of Jerusalem was that **the pretopos completion of the logical category $\mathbb{T}$, identified with T, is nothing but the logical category $\mathbb{T}^{\mathrm{eq}}$ which is identified with the first order theory T^{eq}**.

When I was invited to take part in the *Makkaifest*, I seized the opportunity to clarify to myself and convey to others, the two approaches to the same concept that emerged, more or less at the same time, in two contexts that are so vastly different. This expository paper is based on the talk that I gave at the Makkaifest. I thought that this would be a suitable homage to Mihaly on his 70th birthday. It is, as a matter of fact, an homage to the three men involved, Makkai, Reyes and Shelah and to their important contributions to mathematics.

The paper is addressed to a mixed audience of model and category theorists, hoping to give each group a flavor of the way of thinking of the other.

I assume that the prospective reader has some basic knowledge in both fields and yet, I devote Section 1 to preliminaries in both, category theory and model theory. They are brought mainly to establish nomenclature and notations. A special feature is the description of *multisorted* logic which, although very natural, is seldom used in model theory.

In Section 2, I describe the construction of T^{eq} and give a rough, but hopefully understandable, explanation of its usefulness. In Section 3, I spell out the close connection between the category of models of T and those of T^{eq}.

Actually, the second half of Section 3 as well as Sections 4 and 5, can be seen as a *model-theorist's view of the category-theoretical thinking.* To say it in a nutshell, category theory takes with utmost seriousness the universally accepted mathematical principle that *isomorphic structures are the same.* I hope to be able to soften somewhat the reluctance that most of us, model theorists, feel (precisely as Makkai felt initially) when confronted with the mixture of category theory and logic.

Section 4 presents the 'theories as categories' point of view and Section 5 deals with pretoposes. The final Section 6 contains remarks, a brief description of subsequent developments and a discussion concerning the relevance of conceptual completeness to the problem of giving the widest possible definition to the notion of interpretation of theories.

Acknowledgements. First of all, I want to thank Mihaly for his longtime friendship and endless hours of exciting discussions on mathematics. In particular, I thank him for the many explanations concerning the subject of this paper.

I benefited from useful comments and information provided by Bradd Hart, Wilfrid Hodges, Moshe Kamensky and Anand Pillay.

The few diagrams and pictures were drawn using Michael Barr's diagram package.

1. Preliminaries

In this section we describe our terminology and notations. The presentation will be succinct, as we assume the reader to be familiar with the basic concepts of category theory, on one hand, and (first order) model theory, on the other.

Categories. A category $\mathbb{A}$ has *objects* and *morphisms.* We write $f\colon X \to Y$, or $X \xrightarrow{f} Y$, to indicate that f is an $\mathbb{A}$-morphism with *domain* and *codomain* the $\mathbb{A}$-objects X and Y, respectively. Likewise, $F\colon \mathbb{A} \to \mathbb{B}$ and $\mathbb{A} \xrightarrow{F} \mathbb{B}$ will indicate that F is a *functor* from category $\mathbb{A}$ to category $\mathbb{B}$. The partial operation of *composition* of morphisms (or of functors) will be denoted by $\circ$.

For $\mathbb{A}$-objects X and Y, we will let $\mathbb{A}(X, Y)$ be the set of $\mathbb{A}$-morphisms from X to Y (another customary notation is $\mathrm{Hom}_{\mathbb{A}}(X, Y)$).

Let $\mathbb{S}\mathrm{et}$ be the *category* of sets whose class of objects is the class of all sets and whose morphisms are the functions between sets. Thus, the set of functions between given sets A and B, which is designated by logicians as B^A, is also denoted as $\mathbb{S}\mathrm{et}(A, B)$ by category theorists. $\mathbb{S}\mathrm{et}$ is a *large* category, i.e., one whose collection of objects is a *proper* class (but the collection of morphisms between any two given objects is a *set*).

A morphism $j\colon X \to Y$ in a category $\mathbb{A}$ is called an *isomorphism* iff it has an *inverse* $k\colon Y \to X$, meaning that both $\mathbb{A}$-morphisms $k \circ j\colon X \to X$ and $j \circ k\colon Y \to Y$ are identity morphisms. Notice that in the category $\mathbb{S}$et, a morphism is an isomorphism iff it is a bijection.

Two objects X and Y are called isomorphic *in* $\mathbb{A}$ iff there exists an $\mathbb{A}$-morphism $j\colon X \to Y$ which is an isomorphism. Likewise, two $\mathbb{A}$-morphisms $X \xrightarrow{f} Y$ and $X' \xrightarrow{f'} Y'$ are called isomorphic in $\mathbb{A}$ if we have isomorphisms $X \xrightarrow{j} X'$ and $Y \xrightarrow{k} Y'$ such that the diagram

$$\begin{array}{ccc} X & \xrightarrow{f} & Y \\ {\scriptstyle j}\big\downarrow & & \big\downarrow{\scriptstyle k} \\ X' & \xrightarrow[f']{} & Y' \end{array}$$

commutes (meaning that $k \circ f = f' \circ j$). We define in a similar way isomorphism of arbitrary (finite) *diagrams* in $\mathbb{A}$.

This much for category theory preliminaries, for the moment.

We will use the '$\mapsto$' notation, customary both in logic and category theory, for function description (e.g., '$x \mapsto x^2$' will denote the function $f(x) = x^2$).

Model theory of first order languages. The symbols of a *first order language* are of two kinds: logical and extralogical. The *logical* symbols are common to all first order languages and consist of an infinite sequence of variables as well as of logical connectives (e.g., conjunction or negation), quantifiers, the equality symbol and auxiliary symbols (such as parentheses and commas). The quantifiers and equality symbols operate on variables. What makes the language *first order* is the insistence that variables denote individual elements (rather than more complex objects such as, say, *sets* of elements).

The *extralogical* symbols are peculiar to each particular first order language and consist of sequences of *relation* (or *predicate*), *function* and *individual constant* symbols. Each relation and function symbol has its *arity* (i.e., number of arguments).

A given first order language L allows us to write *formulas* $\varphi(x_1, \ldots, x_n)$ expressing properties of elements (denoted by $x_1, \ldots, x_n$) of any *L-structure.* An L-structure M is specified by a set of elements $d(M)$, the *domain* of M, together with an *interpretation* $\sigma(M)$, *over* $d(M)$, of any extralogical symbol σ of L (e.g., if f is an n-ary function symbol of M, then $f(M)$ is a function $f(M)\colon d(M)^n \to d(M)$).

We will employ the vector notation $\vec{x} = (x_1, \ldots, x_n)$ to denote finite sequences. This will allow abbreviated notations. For example, formulas will appear as $\varphi(\vec{x})$ and $\vec{a} \in d(M)$, will mean that $a_i \in d(M)$, for $1 \leqslant i \leqslant n$, where $\vec{a} = (a_1, \ldots, a_n)$.

We use the notation $M \vDash \varphi[\vec{a}]$ to indicate the fact that a given sequence of elements $\vec{a} \in d(M)$ satisfies, in the structure M, the property expressed by the formula $\varphi(\vec{x})$. This satisfaction relation allows us to define the interpretation of $\varphi(\vec{x})$ in M to be the n-ary relation

$$\varphi(M) = \{\vec{a} \in d(M) : M \vDash \varphi[\vec{a}]\} \subset d(M)^n.$$

Let us mention that what we just described is the most common way of interpreting logical formulas and is called the *classical* interpretation. The first order logic with this interpretation is called *classical first order logic.* There are alternative ways of interpreting logical formulas, the most notable one being the *intuitionistic* interpretation. In this paper, we deal with classical logic only.

A *sentence* φ (i.e., a formula with no free variables) is either *true* or *false* in M, and we denote this as $M \models \varphi$ and $M \not\models \varphi$, respectively. φ is called *valid* if it is true in every L-structure, and we denote this as $\models \varphi$.

A first order *theory* T in the language L is a set of L-sentences. A *model* of T is an L-structure M such that $M \models \varphi$ for all $\varphi \in T$. We use $M \models T$ to denote the fact that M is a model of T.

If T is a first order theory and φ a sentence in its language, we say that T implies φ, and denote this as $T \models \varphi$, iff $M \models \varphi$ for all $M \models T$ (we also say that φ is a T-theorem).

Here is a very simple and illustrative example of a first order theory $\mathcal{E}$ that will accompany our exposition. The language of $\mathcal{E}$ has a binary relation symbol E as its unique extralogical symbol. The axioms of $\mathcal{E}$ will say that: (a) E is an equivalence relation, (b) E has precisely two equivalence classes, and (c) both classes are infinite. Fact (b) is expressed by the first order axiom

$$\exists x_1 \exists x_2 (\neg E(x_1, x_2) \wedge \forall y (E(x_1, y) \vee E(x_2, y))).$$

Fact (c) is expressed by infinitely many first order axioms (for each $n > 0$, one axiom stating that every equivalence class has at least n distinct elements).

A model of $\mathcal{E}$ can be represented schematically as a rectangle split into two classes by a vertical line.

$M :$

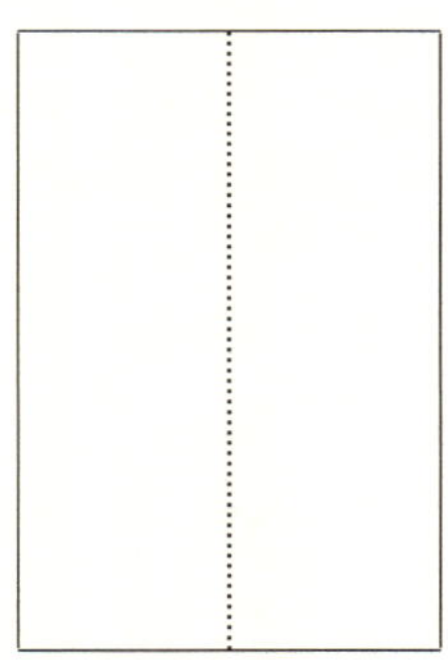

$E(M) \subset d(M) \times d(M)$

The models of a first order theory T form a large category $\operatorname{Mod} T$ whose objects are the models of T and morphisms are the *elementary maps* between models. An elementary map $h\colon M \to N$ is a function $h\colon d(M) \to d(N)$ which preserves satisfaction of formulas, i.e., for any L-formula $\varphi(\vec{x})$ and $\vec{a} \in d(M)$, $M \models \varphi[\vec{a}]$ implies $N \models \varphi[h\vec{a}]$ (where $h\vec{a} = (ha_1, \ldots, ha_n)$, if $\vec{a} = (a_1, \ldots, a_n)$).

Multisorted first order languages. Many of the 'everyday' mathematical structures are more conveniently described as having a domain that is a disjoint union of sets of elements of various *sorts*. For instance, a vector space is naturally construed as a structure with two sorts of elements: scalars and vectors. Also, a category is a structure with two sorts of elements: objects and morphisms. In the model theoretical practice, this situation is dealt with by regarding the various sorts as unary predicates. For our purposes, however, it is more convenient to extend the concepts discussed sofar to *multisorted* languages and structures.

A multisorted language L is based on a set $\mathcal{D} = \mathcal{D}(L)$ of *sort* symbols (the *unisorted* languages that we described above, correspond to the case of a singleton set $\mathcal{D}$). Any L-structure M will have a set $d(M)$ of elements of sort d, for any $d \in \mathcal{D}(L)$. If $\vec{d} = (d_1, \ldots, d_n)$ is a finite sequence of sort symbols, then we denote $\vec{d}(M) = d_1(M) \times \cdots \times d_n(M)$.

The variables of L are sorted. For each $d \in \mathcal{D}$, we have an infinite sequence of variables of sort d. If x is a such, we denote this as $x : d$ and we also write $\vec{x} : \vec{d}$ to mean that $x_i : d_i$ for all $1 \leqslant i \leqslant n$, where $\vec{x} = (x_1, \ldots, x_n)$, $\vec{d} = (d_1, \ldots, d_n)$.

The extralogical symbols are also sorted, and so are their interpretations. Thus, each n-ary function symbol f has arguments of given sorts $\vec{d} = (d_1, \ldots, d_n)$ and values of a sort d_0. We denote this as $f\colon \vec{d} \to d_0$. The interpretation of f in an L-structure M will be a function $f(M)\colon \vec{d}(M) \to d_0(M)$. We let the reader figure out what do we mean by sorting and interpretation of relation and individual constant symbols, and what are the multisorted L-formulas and their interpretations in L-structures. We define in the obvious way the notion of a multisorted first order theory T, its models as well as the concept of elementary map. Thus, we can talk about the category of models $\operatorname{Mod} T$, also in the multisorted context.

2. The theory T^{eq}

Shelah associates with every theory T in a first order language L, a theory T^{eq} in a language L^{eq}. The language L^{eq} is multisorted, even if L was unisorted.

We start the description of L^{eq} with a preliminary definition. An L-formula $\varepsilon(\vec{x}, \vec{y})$ with $\vec{x}, \vec{y} : \vec{d} = (d_1, \ldots, d_n) \in \mathcal{D}^n(L)$, is called an *equivalence formula of* T iff for every $M \vDash T$, the relation $\varepsilon(M) \subset \vec{d}(M) \times \vec{d}(M)$ is an equivalence relation.

Before going any further, let me remark that any theory T has a plethora of equivalence formulae. Indeed, if $\varphi(\vec{x}, \vec{z})$ is any L-formula with $\vec{x} : \vec{d}$, $\vec{z} : \vec{d'}$, then the following is an equivalence formula:

$$\varepsilon(\vec{x}, \vec{y}) := \forall \vec{z}\big(\varphi(\vec{x}, \vec{z}) \leftrightarrow \varphi(\vec{y}, \vec{z})\big)$$

(notice that this is an equivalence formula in a strong sense, since $\varepsilon(M)$ is an equivalence relation for *all* L-structures M, not just those that are models of T). Actually, it is an easy exercise to show that any equivalence formula of T is equivalent, *in* T, to one of this form.

The language L^{eq} will be an extension of L. It will have, in addition to the symbols of L, a new sort symbol d_ε for every equivalence formula $\varepsilon(\vec{x}, \vec{y})$ of T, as well as a function symbol $f_\varepsilon\colon \vec{d} \to d_\varepsilon$ (where we assume that $\vec{x}, \vec{y} : \vec{d}$).

The theory T^{eq} is, in turn, an extension of T. Its axioms are those of T and, with every equivalence formula $\varepsilon(\vec{x}, \vec{y})$, also the following two axioms saying that the elements of d_ε are (in one-to-one correspondence with) the equivalence classes of the relation defined by $\varepsilon(\vec{x}, \vec{y})$ and that f_ε sends each tuple to its equivalence class:

$$\forall z \exists \vec{x} f_\varepsilon(\vec{x}) = z$$
$$\forall \vec{x} \forall \vec{y}\big(\varepsilon(\vec{x}, \vec{y}) \leftrightarrow f_\varepsilon(\vec{x}) = f_\varepsilon(\vec{y})\big)$$

(it is understood that $\vec{x}, \vec{y} : \vec{d}$ and $z : d_\varepsilon$).

The fact that T^{eq} is an extension of T implies that every model $N \vDash T^{\mathrm{eq}}$ has a *restriction* to a model $N_0 = N{\restriction}L \vDash T$; this is the L-structure obtained from N by ignoring the interpretations of those symbols of L^{eq} that are additional to those of

L. It follows that for all L-sentences φ, if $T \models \varphi$ (i.e., φ is true in all models of T) then $T^{\mathrm{eq}} \models \varphi$ as well.

In the case of T^{eq}, we can go also in the converse direction. Every model $M \models T$ has a *canonical* extension to $M^{\mathrm{eq}} \models T^{\mathrm{eq}}$ in which the interpretations of the additional symbols are:

$$d_\varepsilon(M^{\mathrm{eq}}) = \{\vec{a}/\varepsilon(M) : \vec{a} \in \vec{d}(M)\} \quad (= \text{the set of equivalence classes of } \varepsilon(M))$$
$$f_\varepsilon(\vec{a}) = \vec{a}/\varepsilon(M).$$

Recall our illustrative example $\mathcal{E}$, the theory of an equivalence relation E with two equivalence classes, both infinite. $E(x, y)$ is an equivalence formula of $\mathcal{E}$. If $M \models \mathcal{E}$, then the sort $d_E(M^{\mathrm{eq}})$ will have two elements and the function $f_E(M^{\mathrm{eq}})$ (denoted f_E in the figure below) will send the elements of the left class of M to one of these and those of the right class to the other.

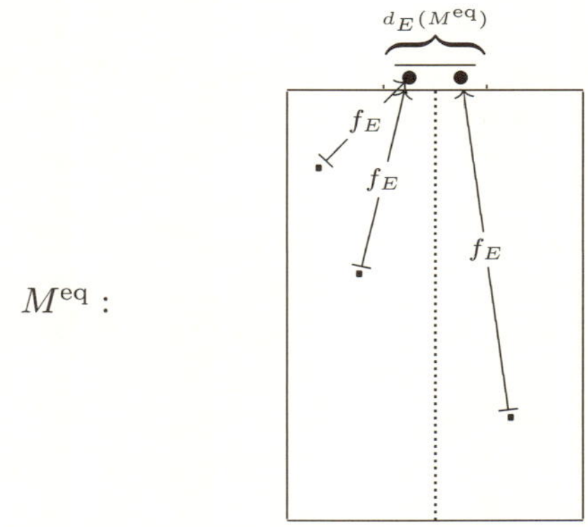

The existence of the canonical extension of models of T implies that T^{eq} is a *conservative* extension of T. This means that T^{eq} has no new theorems expressible in the original language L. In other words, if φ is an L-sentence such that $T^{\mathrm{eq}} \models \varphi$, then $T \models \varphi$ as well (indeed, if $M \models T$ and φ is a T^{eq}-theorem, then $M^{\mathrm{eq}} \models \varphi$ and, as the truth of the L-sentence φ in M^{eq} depends only on the interpretations of the L-symbols, we have also that $M^{\mathrm{eq}}{\restriction}L \models \varphi$; however, $M^{\mathrm{eq}}{\restriction}L = M$).

Remark. Actually, Shelah's construction in [21] was a little different, because he stuck to unisorted languages. Therefore, T^{eq} is conceived by him as a unisorted theory having a unary predicate (rather than a sort symbol) P_ε, for each equivalence formula ε and additional axioms stating, among others, that distinct P_εs denote disjoint sets. To insure that M^{eq} is an extension of M, the domain of M is identified with $P_=(M^{\mathrm{eq}})$, where $P_=$ is the predicate associated with the equivalence formula $x = y$.

For all practical purposes, Shelah's construction is the same with the one that we described above, but one should be aware of the fact that the domain of his M^{eq} is allowed to contain a set of elements that do not belong to any of the sets $P_\varepsilon(M^{\mathrm{eq}})$. This set is totally arbitrary and its elements are never looked at, they are just "sitting idle." As a result, M^{eq} is not uniquely defined from M. This is just an aesthetic flaw, which is avoided when we switch to multisorted languages as suggested by Makkai. This suggestion (prompted by his work with Reyes [15]) was adopted by other authors as well (e.g., [1]).

The usefulness of T^{eq}. Shelah showed that the extension of T and $M \models T$ to T^{eq} and M^{eq} is a very useful device. By skilfully switching between T and T^{eq}, he manages to obtain results pertaining to T, that he could not obtain otherwise. In a joint paper with Leo Harrington ([6]), we offer an explanation for the efficacy of Shelah's construction. The details are quite technical, but I will try to give a rough outline that will, hopefully, convey the flavor.

A central notion of model theory is that of a *type of an element over a set.* If M is an L-structure (where, for simplicity, we assume L to be a unisorted language), $A \subset d(M)$ a subset of its domain and $b \in d(M)$ an element, then the type of b over A is the list of all first order properties of b that are expressible in the language (usually denoted as $L(A)$) obtained from L by adding individual constants that denote the elements of A. More formally, the type of b over A *in* M is defined as

$$\mathrm{tp}^M(b, A) = \{\varphi(x, \vec{a}) : \varphi(x, \vec{y}) \in L, \vec{a} \in A, M \models \varphi[b, \vec{a}]\}$$

(following a common practice of model theorists, we do not distinguish between an element of A and the individual constant of $L(A)$ that denotes it). In the sequel, we will refer to $\mathrm{tp}^M(b, A)$ as the *full* type of a over A (because it incorporates *all* $L(A)$-properties).

Shelah came to the conclusion that the notion of full type in the given theory is too crude. To remedy this, he sometimes "steps out of T" and looks at full types in T^{eq}. As we point out in [6], we could achieve the same results by staying inside T, but speaking also about *partial* types that list only *some* of the first order properties of an element over a given set. Rather than giving any precise definition, we'll examine a simple example based on the theory $\mathcal{E}$ that accompanies our exposition.

Assume that $M \models \mathcal{E}$, $A \subset d(M)$ and $b_1, b_2, b_3 \in d(M)$ are equivalent elements such that $b_1, b_2 \notin A$ but $b_3 \in A$. Then we have $\mathrm{tp}^M(b_1, A) = \mathrm{tp}^M(b_2, A) \neq \mathrm{tp}^M(b_3, A)$, the reason for the inequality being that while b_1, b_2 are distinct from all elements of A, b_3 equals one of them. Yet, these three elements have something in common, as they belong to the same equivalence class. We can express this by saying that the *partial* types $\mathrm{tp}^M_E(b_i, A)$ obtained by restricting ourselves to formulas using only E (and *no* equality symbol $=$) are the same for $i = 1, 2, 3$. Rather than doing that, Shelah "works in T^{eq}" and notices that the *full* types $\mathrm{tp}^{M^{\mathrm{eq}}}(f_E(M^{\mathrm{eq}})(b_i), A)$ are the same (simply, because the elements $f_E(M^{\mathrm{eq}})(b_i)$, for $i = 1, 2, 3$, are identical).

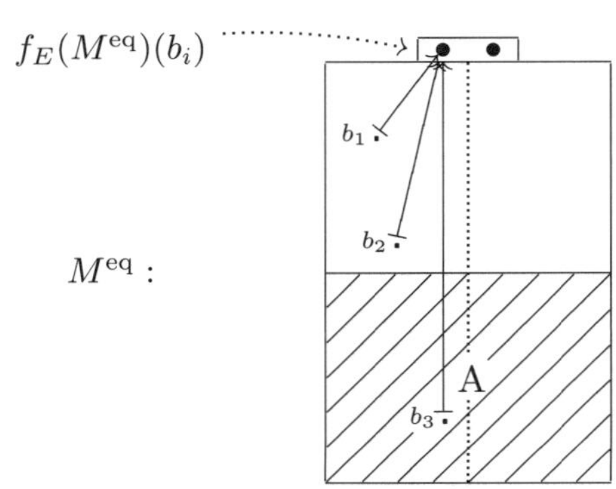

Roughly speaking, the T^{eq}-*full* type $tp^{M^{\mathrm{eq}}}(f_E(M^{\mathrm{eq}})(b_i), A)$ that Shelah prefers to consider, is essentially the same as the *partial* type $tp_E^M(b_i, A)$ that we suggest.

We conclude that, after having elaborated the main properties of the notion of full type in a first order theory T, we can proceed in two ways. *Either* work with full types in the extended theory T^{eq} and then return to the original theory T (this is Shelah's option), *or*, stay within T and use a suitably devised notion of *partial* type which is shown to possess the main properties of full types (this is our alternative path suggested in [6]). Shelah's way is conceptually simpler and is universally accepted. Ours is offered as an explanation of what happens when we work in T^{eq}.

3. The models of T^{eq} are essentially the same with those of T

The theory T^{eq} is, as we saw, a non trivial, significant extension of T. Yet, the two theories are closely related. As we noted, each model of T can be canonically extended to one of T^{eq}, a fact that implies that the extension is conservative. *Moreover*, as it turns out, T^{eq} has no other models besides these canonical extensions. To be more precise, every model of T^{eq} is *isomorphic* to the canonical extension of a model of T. Indeed, if $M \vDash T^{\mathrm{eq}}$ and if $M_0 = M{\restriction}L$ is the restriction of M to a model of T, then a moment of thought will show that M is isomorphic to $M_0^{\mathrm{eq}} = (M{\restriction}L)^{\mathrm{eq}}$. Thus, we may say that T and T^{eq} have *essentially* the same models. This fact can be stated more perspicuously in terms of the categories of models.

The first observation is that the function $M \mapsto M^{\mathrm{eq}}$, that associates with each model of T its canonical extension to a model of T^{eq}, can be extended to a *functor*

$$\operatorname{Mod} T \xrightarrow{(-)^{\mathrm{eq}}} \operatorname{Mod} T^{\mathrm{eq}}$$

by sending any map $h\colon M \to N$ in $\operatorname{Mod} T$ to its extension $h^{\mathrm{eq}}\colon M^{\mathrm{eq}} \to N^{\mathrm{eq}}$ defined by $h^{\mathrm{eq}}(\vec{a}/\varepsilon(M)) = h\vec{a}/\varepsilon(N)$ (of course, we mean here that $h\vec{a} = (ha_1, \dots, ha_k)$, if $\vec{a} = (a_1, \dots, a_k)$).

It is easy to see that, for any two models $M, N \vDash T$, the map $h \mapsto h^{\mathrm{eq}}$ is actually a bijection between the set $\operatorname{Mod} T(M, N)$ of elementary maps from M to N and the set $\operatorname{Mod} T^{\mathrm{eq}}(M^{\mathrm{eq}}, N^{\mathrm{eq}})$ of elementary maps between the corresponding canonical extensions. In the terminology of category theorists, this means that $(-)^{\mathrm{eq}}\colon \operatorname{Mod} T \to \operatorname{Mod} T^{\mathrm{eq}}$ is a *full and faithful* functor. In general, a functor $F\colon \mathbb{C} \to \mathbb{D}$ between categories $\mathbb{C}$ and $\mathbb{D}$ is called *full and faithful* iff for any two objects A, B of $\mathbb{C}$, F is a bijection between the set $\mathbb{C}(A, B)$ of $\mathbb{C}$-morphisms from A to B and the corresponding set $\mathbb{D}(FA, FB)$ of $\mathbb{D}$-morphisms.

Is $(-)^{\mathrm{eq}}$ an isomorphism of categories? In other words, is it also bijective on objects? Not quite. The functor $(-)^{\mathrm{eq}}$ happens to be one-to-one on objects but it is *not* surjective. However, it is not *far* from being surjective, as we remarked already, because every model $M \vDash T^{\mathrm{eq}}$ is *isomorphic* to the image $(M{\restriction}L)^{\mathrm{eq}}$ of its restriction to a model of T. We say that $(-)^{\mathrm{eq}}$ is *essentially surjective on objects.* In general, $F\colon \mathbb{C} \to \mathbb{D}$ is called essentially surjective on objects iff every object B of $\mathbb{D}$ is isomorphic (*in* $\mathbb{D}$) to one of the form FA for some $\mathbb{C}$-object A.

In the terminology of category theorists, a functor, such as $(-)^{\mathrm{eq}}$, which is full and faithful and essentially surjective on objects is called an *equivalence of categories.* This is so, because if we have an equivalence of categories $F\colon \mathbb{C} \to \mathbb{D}$ then the categories $\mathbb{C}$ and $\mathbb{D}$ (are called *equivalent categories* and) are the same from the point of view of category theory. Equivalent categories differ from each other only in the number of isomorphic copies that each object has. A most illuminative example of equivalent categories is the category $\mathbb{FG}\mathrm{rp}$ of finite groups (a large category) on one hand and the category $\mathbb{G}_0$ of finite groups whose domain is a set of natural numbers (a small category!) on the other.

Equivalence of categories is a symmetric relation, so we must have also an equivalence functor from $\mathrm{Mod}\,T^{\mathrm{eq}}$ to $\mathrm{Mod}\,T$. Indeed, the map $M \mapsto M{\restriction}L$ can be extended to a *restriction functor* $(-){\restriction}L\colon \mathrm{Mod}\,T^{\mathrm{eq}} \to \mathrm{Mod}\,T$, which is also an equivalence of categories.

Let us introduce the following terminology. If T' is a first order theory in a language L' that extends the L-theory T, we will say that T' is a *tight* extension of T iff the restriction functor $(-){\restriction}L\colon \mathrm{Mod}\,T' \to \mathrm{Mod}\,T$ is an equivalence of categories.

Thus, T^{eq} is a tight extension of T. It has been devised by Shelah with a definite purpose in mind, and it allowed him to overcome certain difficulties. It is, therefore, a *significant* extension. It seems not unreasonable to expect that in the future, someone will extend T^{eq} further, in a significant but still tight fashion, in order to solve some other problem. Or, maybe, one will find a significant but tight extension T' which is inconsistent with T^{eq}, it will go in "another direction," so to speak. Is this possible at all? Or, maybe, T^{eq} is a *largest* tight extension of T?

To begin with, we must formulate these questions in a precise mathematical form. What is an extension of a theory? A simple minded answer to this question would say that a theory T' in a language L' is an extension of an L-theory T if all symbols of L are also symbols of L' and all axioms of T are axioms of T'. This definition, which is adequate for qualifying T^{eq} to be an extension of T, appears to be too restrictive when compared to our intuition arising from mathematical practice. We would like to say that a theory T' is an extension of T when the language of the latter is not necessarily a sub-language of the former but, rather, it can be *translated* into the language of T', such that the translations of the axioms of T become theorems of T'. Two wellknown examples will clarify what we mean.

The field of complex numbers can be defined from that of reals, when each complex number is identified with an ordered pair of reals. This procedure can be generalized to produce an *algebraically closed field* from any *real closed one* (a real closed field is one that satisfies all the first order properties of the field of real numbers). Thus, we may regard the theory of real closed fields as an extension of the (first order) theory of algebraically closed fields (by the way, this would be an instance of a *non tight* extension). An alternative terminology would be to say that there is a *translation* of the theory of algebraically closed fields into that of real closed ones.

Another important example: the standard construction of a model of plane geometry from the field of reals. Here, every geometrical point is identified with an ordered pair of reals and every line with an ordered triple of reals (the coefficients of a "canonical" equation of the line). This construction is easily generalized to all real closed fields showing that the theory of real closed fields is an extension of elementary geometry (i.e. the theory consisting of all *first order sentences* that are true in the Euclidean plane). This would be an instance of a *tight* extension.

One can make up a more general definition of the notion of extension of a first order theory, in the spirit of these two examples. Such a definition requires the existence of a map $\varphi \mapsto \varphi'$ that translates formulas of T into those of T', such that sentences are mapped to sentences and theorems to theorems (meaning that $T' \models \varphi'$

whenever $T \models \varphi$). Also, this map should allow us to extract a model of T from any one of T'. Definitions of this kind have been produced, the classical reference being [26]. According to the customary terminology, the map $\varphi \mapsto \varphi'$ is called an *interpretation* of T into T'. Hence, authors mostly say that "T is interpretable in T'", rather than "T' is an extension of T." Even [15] uses this language. We will discuss the notion of interpretation of theories and its generalizations in the concluding Section 6.

We can go on and give a general definition of the notion of *tight* extension (using, of course, some category theory). But then, what is a *significant* extension? It seems that model theory, in its traditional form, is not the appropriate context to ask and solve these kind of questions in precise mathematical terms. It is at this point that category theory enters naturally, in a more substantial way.

4. Theories as categories

The work of Makkai and Reyes ([15]) offers a novel, fresh look at the main concepts of model theory. It starts with the innocent looking observation that, given a first order language L, each L-structure M carries with it a function that associates with every L-formula $\varphi(\vec{x})$ its interpretation $\varphi(M)$, which is just an abstract set, i.e. an object of the category $\mathbb{S}\mathrm{et}$ (which we described in Section 1). Moreover, this function contains all the information needed to describe the structure M, because it specifies the interpretations of all sort, relation, function and individual constant symbols (indeed, if d is a sort symbol, then $d(M) = \varphi(M)$, where $\varphi(x)$ is the formula $x = x$, with $x : d$). Thus, we may *identify* M with the function $\varphi(\vec{x}) \mapsto \varphi(M)$. Actually, Makkai and Reyes go further and identify the structures that happen to be models of a first order theory T, with functions that are *functors*. The first task of the program they developed in [15] can be formulated as follows:

To associate with every first order L-theory T, a category $\mathbb{T}$ whose objects are (essentially) *the L-formulas such that, for any model $M \models T$, the function $\varphi(\vec{x}) \mapsto \varphi(M)$ can be extended to a* functor $\mathbb{T} \xrightarrow{M} \mathbb{S}\mathrm{et}$.

The construction of $\mathbb{T}$ is very easy to describe:

The objects of $\mathbb{T}$ will not be formulas but, rather, *equivalence* classes $[\varphi(\vec{x})]_T = \varphi(\vec{x})/\!\sim_T$ of formulas with respect to the equivalence relation defined by

$$\varphi(\vec{x}) \sim_T \varphi'(\vec{x'}) \text{ iff } T \models \forall \vec{x}(\varphi(\vec{x}) \leftrightarrow \varphi'(\vec{x}))$$

where we assume that $\vec{x}, \vec{x'} : \vec{d}$ (notice that in the rightmost formula we changed $\vec{x'}$ to $\vec{x}$, in the argument of φ'). Whenever T is clear from the context, we just write $[\varphi(\vec{x})]_T = [\varphi(\vec{x})]$.

A $\mathbb{T}$-morphism between the objects $[\varphi(\vec{x})]$ and $[\psi(\vec{y})]$ is $[\varphi(\vec{x})] \xrightarrow{[\chi(\vec{x},\vec{y})]} [\psi(\vec{y})]$, where $\chi(\vec{x}, \vec{y})$ is a formula for which

$$T \models \forall \vec{x}, \vec{y}(\chi(\vec{x}, \vec{y}) \to \varphi(\vec{x}) \wedge \psi(\vec{y})) \wedge \forall \vec{x}(\varphi(\vec{x}) \to \exists! \vec{y}(\psi(\vec{y}) \wedge \chi(\vec{x}, \vec{y})))$$

(in other words, in every model $M \models T$, the formula $\chi(\vec{x}, \vec{y})$ defines the graph of a function $\varphi(M) \to \psi(M)$).

As formulas that are equivalent in the theory T have the same interpretation in any of its models, we see that for $M \models T$, we have a well defined function $[\varphi(\vec{x})] \mapsto \varphi(M)$, which extends to a functor $\mathbb{T} \xrightarrow{M} \mathbb{S}\mathrm{et}$, in an obvious way. It is this functor that is identified, by Makkai and Reyes, with the model M itself.

Boolean logical categories. Which categories are $\mathbb{T}$ for some first order theory T? A better formulation of this question, in the spirit of category theory, is: which categories are *equivalent to* $\mathbb{T}$ for some T? Makkai and Reyes set out to discover the properties that characterize these categories. They made the crucial observation that all these properties are shared by the category $\mathbb{S}$et. Taking this as a guiding principle, they defined a notion of *Boolean logical category*, in short *b.l.c.* or *b.l. category*, such that the following conditions are met:

(a) $\mathbb{S}$et *is a Boolean logical category.*
(b) $\mathbb{T}$ *is a b.l.c., for all first order theories* T.
(c) *If* $\mathbb{C}$ *is a* small *b.l.c. then it is equivalent to* $\mathbb{T}$, *for some first order theory* T.

Remark. Statement (c) emphasizes the fact that the category that we associate with a given first order theory T does not have to be determined uniquely, but only *uniquely up to equivalence of categories.* This is in accordance with the categorical point of view and it leaves us with some degree of freedom, when deciding how we are going to define the category $\mathbb{T}$. As a matter of fact, Makkai and Reyes' definition in [15] is different from the one that we gave here. They take the objects of $\mathbb{T}$ to be the L-formulas $\varphi(\vec{x})$ themselves, rather than equivalence classes of such. Their morphisms, however, are equivalence classes $[\chi(\vec{x},\vec{y})]$ of formulas that define functions; this is necessary, in order to insure that the associativity and identity axioms hold for compositions of morphisms.

Rather than going into the technical details of Makkai and Reyes' definition, we will just say that a Boolean logical category is one which is closed under certain *category-theoretical* operations, some of which are also required to have a special property (called "stability under pullbacks"). Each of these category-theoretical operations is a *substitute* for a logical one. A *logical* category is one that is closed only under the category-theoretical operations that substitute the logical operations of *disjunction*, *conjunction* and *existential quantification*, while a *Boolean* logical category is one that is closed under the substitute of *negation* as well.

A digression: category-theoretical operations. Our last remarks may sound somewhat cryptical and they call for some clarifications. What are category-theoretical operations? What does it mean that they substitute certain logical operations?

A word of warning: category-theoretical operations are *not* operations in the customary sense of the word, because they are not uniquely defined but defined *uniquely up to isomorphism.*

This state of affairs is an outcome of the categorical way of thinking, according to which, category-theoretical properties should be *preserved* and *reflected* by equivalence functors. By this we mean that if F is an equivalence functor then a diagram Γ satisfies a given category theoretical property *iff* so does $F\Gamma$ (category theorists say that a property is reflected by a functor F iff whenever $F\Gamma$ has that property, so does Γ). This principle implies that in category theory, *we cannot define objects uniquely.* Indeed, if X and X' are two objects that are isomorphic in a category $\mathbb{A}$, it is an easy exercise to see that there exists an equivalence functor $F\colon \mathbb{A} \to \mathbb{A}$ which is the identity on all $\mathbb{A}$-objects *except* for X', which is mapped to X, i.e., $FX' = X$. Thus, any category-theoretical property enjoyed by X will be shared by X' as well.

Remark. A syntactical characterization of the first order formulae that express category-theoretical properties has been obtained, independently, by Freyd [5] and Blanc [3].

Let us describe a paradigmatic example of a category-theoretical operation, which is also highly relevant to our context. It is probably known to most readers of this paper. The *pullback* of a diagram $X \xrightarrow{f} A \xleftarrow{g} Y$ is a diagram

$$\begin{array}{ccc} P & \xrightarrow{q} & X \\ {\scriptstyle r}\downarrow & & \downarrow{\scriptstyle f} \\ Y & \xrightarrow[g]{} & A \end{array}$$

which commutes and for all $X \xleftarrow{h} Z \xrightarrow{k} Y$ such that $f \circ h = g \circ k$ there is a *unique* $Z \xrightarrow{s} P$ such that $q \circ s = h$, $r \circ s = k$.

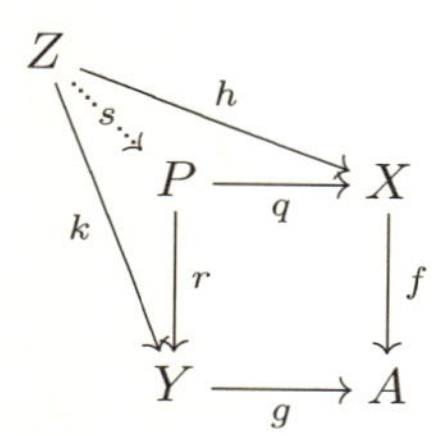

A given diagram may have many pullbacks, however they are all isomorphic to each other (by an isomorphism keeping the original diagram fixed). Yet, it is customary to speak about the *pullback operation* which associates with the given diagram $X \xrightarrow{f} A \xleftarrow{g} Y$ its pullback $X \xleftarrow{q} P \xrightarrow{r} Y$, in spite of the fact that this diagram is not uniquely defined (sometimes, one might even say, carelessly, that the object P alone is *the* pullback of the given diagram!).

A particularly simple example of a pullback occurs in $\mathbb{S}\mathrm{et}$, when $X, Y \subset A$ and f, g are the *inclusion* maps, in which case we may take $P = X \cap Y$ with q, r being inclusion maps as well. Thus, the concept of pullback can be seen, not only as a substitute, but as a *generalization* of the *set-theoretical* operation of intersection of subsets of a given set A. It so happens that the same category-theoretical operation of pullback also generalizes the logical operation of *conjunction.* This fact is very important, as we shall see shortly, and is not hard to verify (hint: given formulas $\varphi(x), \psi(x)$ with $x : d$, the object $[\varphi(x) \wedge \psi(x)]$ of $\mathbb{T}$ is, carelessly speaking, the pullback of the diagram

$$[\varphi(x)] \xrightarrow{[\varphi(x)\wedge x=y]} [y = y] \xleftarrow{[\psi(x)\wedge x=y]} [\psi(x)],$$

where $y : d$; remember that $y = y$ is a formula whose interpretation in any model M is the set $d(M)$ of elements of sort d in M).

We conclude this digression with a remark that may look intriguing to some. The identity function from the set of objects of a category to itself is *not* a category theoretical operation! Question: what would be a category-theoretical generalization of this operation?

Natural transformations. We now return to our main theme. Assume that $M, N \models T$ are models of T. If we now think of M and N as functors from $\mathbb{T}$ to $\mathbb{S}\text{et}$, it is only natural to ask what is the *category theoretical meaning* of an *elementary map* $h\colon M \to N$. It turns out that h is, essentially, what category theorists call a *natural transformation* between the functors M and N.

In general, given categories $\mathbb{C}, \mathbb{D}$ and functors $F, G\colon \mathbb{C} \to \mathbb{D}$, a *natural transformation* η between F and G, also denoted as $\eta\colon F \to G$, is an indexed family of $\mathbb{D}$-morphisms

$$\{FX \xrightarrow{\eta_X} GX : X \text{ an object of } \mathbb{C}\},$$

such that for every $\mathbb{C}$-morphism $X \xrightarrow{f} Y$, the following diagram (in the category $\mathbb{D}$) commutes

$$\begin{array}{ccc} FX & \xrightarrow{\eta_X} & GX \\ {\scriptstyle Ff}\downarrow & & \downarrow{\scriptstyle Gf} \\ FY & \xrightarrow[\eta_Y]{} & GY \end{array}$$

This notion, defined by Eilenberg and Mac Lane in a seminal paper [4], is surprisingly little known among mathematicians from other fields. It is one of the most important concepts of category theory. The authors just mentioned go as far as to say that "category" has been defined in order to define "functor" and "functor" has been defined in order to define "natural transformation" (this observation is quoted in [10, p. 18]).

We are now in a position that allows us to explain why an elementary map $h\colon M \to N$ can be regarded as a natural transformation. As h preserves the interpretations of first order formulas, we see that for any $\varphi(\vec{x})$, the restriction of h to the set $\varphi(M)$ takes its values in $\varphi(N)$. Obviously, this restriction of h depends on $[\varphi(\vec{x})]$, rather than on $\varphi(\vec{x})$ alone; therefore, we may denote it as $h_{[\varphi]}$. Thus, h determines an indexed family $\{\varphi(M) \xrightarrow{h_{[\varphi]}} \varphi(N) : \varphi \text{ an } L\text{-formula}\}$ of functions (i.e., $\mathbb{S}\text{et}$-morphisms), which is easily seen to be a natural transformation between the functors $M, N\colon \mathbb{T} \to \mathbb{S}\text{et}$ (one should keep in mind that the sets $\varphi(M), \varphi(N)$ are the images, under M and N, of the object $[\varphi(\vec{x})]$ of the category $\mathbb{T}$). Conversely, any natural transformation between the *functors* M and N originates in this way from an elementary map between the *models* M and N.

Categories of functors. With our new point of view, the category $\operatorname{Mod} T$ of models of T appears to be one whose objects are functors and morphisms are natural transformations. This is just a particular example of a *category of functors*. It is a remarkable fact that, given any two categories $\mathbb{C}$ and $\mathbb{D}$ we have a category whose objects are the functors from $\mathbb{C}$ to $\mathbb{D}$ and morphisms are the natural transformations between these functors. This category of functors is denoted as $\mathbb{D}^{\mathbb{C}}$.

Having this new concept, we may now say that *the models of the first order theory T are objects of the functor category $\mathbb{S}\text{et}^{\mathbb{T}}$ and the elementary maps between models of T are morphisms of $\mathbb{S}\text{et}^{\mathbb{T}}$*. Actually, for given models $M, N \models T$, the elementary maps from M to N are *precisely* the $\mathbb{S}\text{et}^{\mathbb{T}}$-morphisms between M and N (i.e., the natural transformations between the functors $M, N\colon \mathbb{T} \to \mathbb{S}\text{et}$).

In the terminology of category theory, the situation that we just presented is described by saying that $\operatorname{Mod} T$ is a *full subcategory* of the functor category $\mathbb{S}\text{et}^{\mathbb{T}}$. In general, a category $\mathbb{C}$ is a *subcategory* of $\mathbb{D}$ iff it is a *substructure* of $\mathbb{D}$, in the model theoretic sense; this means that the objects and morphisms of $\mathbb{C}$ are objects and morphisms of $\mathbb{D}$ and, moreover, morphisms of $\mathbb{C}$ have the same domain and codomain and compose in the same way in $\mathbb{D}$ as in $\mathbb{C}$. We say that $\mathbb{C}$ is a *full* subcategory iff in addition, for any two objects X, Y of $\mathbb{C}$, $\mathbb{C}(X, Y) = \mathbb{D}(X, Y)$; this means that all $\mathbb{D}$-morphisms from X to Y are already $\mathbb{C}$-morphisms between the same domain and codomain.

It is quite obvious, as we shall explain in a moment, that $\operatorname{Mod} T$ is a *proper* subcategory of $\mathbb{S}\mathrm{et}^{\mathbb{T}}$. In other words, there are functors from $\mathbb{T}$ to $\mathbb{S}\mathrm{et}$ that are *not* (identifiable with) models of T. In fact, there are many such functors. This leads to a very natural question: which functors from a b.l. category to $\mathbb{S}\mathrm{et}$ can be seen as models?

Logical functors. Any model $M \vDash T$, seen as a functor, should satisfy some obvious conditions. For example, M sends the objects $[\varphi(\vec{x})], [\psi(\vec{x})], [\varphi(\vec{x}) \wedge \psi(\vec{x})]$ of $\mathbb{T}$, to the sets $\varphi(M), \psi(M), \varphi \wedge \psi(M)$, respectively. These sets should satisfy

$$\varphi \wedge \psi(M) = \varphi(M) \cap \psi(M).$$

In other words, the functor M must transfer the *logical* operation $\wedge$ into the *set theoretical* operation $\cap$. As it so happens that both these operations are generalized by the same category-theoretical pullback operation, it is enough to require M to *preserve pullbacks.* This is just one condition that a functor $\mathbb{T} \to \mathbb{S}\mathrm{et}$ must satisfy in order to be a model of T.

Following this line of thought, Makkai and Reyes defined the notion of *logical functor* between logical categories as being one that preserves the category-theoretical operations under which the *logical* categories are closed. One would be tempted to define also a notion of *Boolean* logical functor. This is not necessary, however. As it turns out, a logical functor whose domain and codomain happen to be *Boolean* logical categories, preserves also the additional, negation-related operations that characterize the Boolean logical categories.

Makkai and Reyes were able to show:

If M is a model of T then the functor $M\colon \mathbb{T} \to \mathbb{S}\mathrm{et}$ which is identified with M is a logical functor.

How about the converse of this statement? Let us call, just for a moment, the functors from $\mathbb{T}$ to $\mathbb{S}\mathrm{et}$ that are identified with models of T by the name "*standard* functors." Is any logical functor from $\mathbb{T}$ to $\mathbb{S}\mathrm{et}$ a standard one? *Essentially* it is. *Essentially,* but not precisely! In fact, any logical functor $N\colon \mathbb{T} \to \mathbb{S}\mathrm{et}$ is *isomorphic* to a standard one. This isomorphism takes place in the functor category $\mathbb{S}\mathrm{et}^{\mathbb{T}}$, meaning that there is a standard functor M such that there exist natural transformations $M \xrightarrow{\eta} N$ and $N \xrightarrow{\theta} M$ which are inverses of each other.

The remarkable statement just made can be said more elegantly and usefully as follows. Let $\operatorname{Mod}' T$ be the full subcategory of $\mathbb{S}\mathrm{et}^{\mathbb{T}}$ whose objects are the *logical* functors from $\mathbb{T}$ to $\mathbb{S}\mathrm{et}$. Then $\operatorname{Mod} T$ is a full subcategory of $\operatorname{Mod}' T$. This means that we have an *embedding* functor $\iota\colon \operatorname{Mod} T \to \operatorname{Mod}' T$ (i.e., the functor that maps the objects and morphisms of $\operatorname{Mod} T$ to themselves). This functor is full and faithful, of course, and the statement that we made above means that it is also essentially surjective on objects. Hence:

The embedding functor $\iota\colon \operatorname{Mod} T \to \operatorname{Mod}' T$ is an equivalence of categories.

Thus, the category of models of T is the same, from the category theoretical point of view, as the *category of logical functors* from $\mathbb{T}$ to $\mathbb{S}$et.

From now on we will denote this latter category as $\operatorname{Mod} T$ *(rather than* $\operatorname{Mod}' T$*) and, when saying that* M *is a model of* T*, we will mean that* $M\colon \mathbb{T} \to \mathbb{S}\mathrm{et}$ *is a logical functor (and will denote this as* $M \models T$*).*

Remark. The fact that the category theoretical characterization of the standard functors is only up to isomorphism, should not come as a surprise. In order to be standard, a logical functor M must satisfy certain *set theoretical* properties that cannot be formulated in category theoretical terms. For example, for all $\bar{L}$-formulas $\varphi(x)$ with $x : d$, M should map the object $[\varphi(x)]$ of $\mathbb{T}$ to a *subset* $\varphi(M)$ of $d(M)$ (remember that $d(M)$ is the image under M of the object $[x = x]$). Some of these $\varphi(M)$s should be *proper* subsets of $d(M)$. However, for objects X, Y of $\mathbb{S}$et, there is no way to express, in the language of category theory, the fact that X is a proper subset of Y, simply because this statement is *not* preserved by equivalence functors. The most that we can say, using the concepts of $\mathbb{S}$et-morphism (i.e. function) and morphism composition, is that there exists a *monomorphism* from X into Y (a morphism $f\colon X \to Y$ in a category is called a monomorphism iff $f \circ g = f \circ h$ implies $g = h$, for any two morphisms $f, g\colon Z \to X$; a morphism in $\mathbb{S}$et is a monomorphism iff it is a one-to-one function).

Extensions. We reached a position in which we can produce a precise mathematical definition for the notion of extension of a theory.

Definition 4.1. A theory T_1 is called an *extension* of T, iff there is a logical functor $I\colon \mathbb{T} \to \mathbb{T}_1$. The functor I is called an extension functor or, simply, an extension.

This notion is certainly broad enough to incorporate the one that we outlined roughly at the end of Section 3.

The obvious example, in our context, is that of T^{eq} being an extension of T. The extension functor $I\colon \mathbb{T} \to \mathbb{T}^{\mathrm{eq}}$ is nothing but the *embedding* functor defined, simply, as $I([\varphi(\vec{x})]_T) = [\varphi(\vec{x})]_{T^{\mathrm{eq}}}$ for objects and $I([\chi(\vec{x}, \vec{y})]_T) = [\chi(\vec{x}, \vec{y})]_{T^{\mathrm{eq}}}$ for morphisms.

At this point, we have a very simple and straightforward mathematical definition of *significance* of an extension, a notion that was not clear how to define in the framework of model theory.

Definition 4.2. An extension $I\colon \mathbb{T} \to \mathbb{T}_1$ is called *significant* iff I is *not* an equivalence of categories.

In Section 3 we made the *intuitive* statement that T^{eq} is a significant extension of T. This can be now stated precisely, by saying that the embedding functor $I\colon \mathbb{T} \to \mathbb{T}^{\mathrm{eq}}$ is not an equivalence of categories. Well, this is true for *most* first order theories, but not for *all* of them.

Definition 4.3. We say that a theory T has *elimination of imaginaries* iff the embedding functor $I\colon \mathbb{T} \to \mathbb{T}^{\mathrm{eq}}$ is an equivalence of categories.

This notion has been defined, in model theoretical terms, by Poizat [19] (or [20], in English translation). His definition says, roughly, that each equivalence relation of T has a first order definable 'quotient function' (i.e., a function under which, any two tuples have the same value iff they are equivalent). The explanation of this name rests on the fact that Shelah called the equivalence classes $\vec{a}/\varepsilon(M)$ "imaginary elements of the model M."

Theories with elimination of imaginaries do exist. In fact:

Claim 4.4. *For any first order theory T, the theory T^{eq} has elimination of imaginaries. In other words, the embedding functor $I\colon \mathbb{T}^{\mathrm{eq}} \to (\mathbb{T}^{\mathrm{eq}})^{\mathrm{eq}}$ is an equivalence of categories.*

This easily seen fact was certainly known to Shelah, but it seems that Poizat was the one who first stated it explicitly.

How can we define, in category theoretical terms, the notion of *tight* extension? To do this, we have to discuss first the notion of *restriction* of a model. We start with the observation that each model N of T^{eq} (i.e., logical functor $N\colon \mathbb{T}^{\mathrm{eq}} \to \mathbb{S}\mathrm{et}$) yields one of T by functor composition, namely $N \circ I\colon \mathbb{T} \to \mathbb{S}\mathrm{et}$, which is easily seen to be the logical functor that is identified with the restriction $N{\restriction}L \models T$.

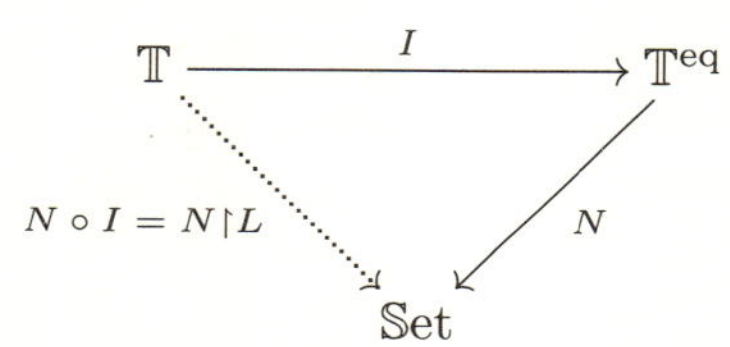

More generally, for *any* extension $I\colon \mathbb{T} \to \mathbb{T}_1$, we have a function $N \mapsto N \circ I$ from the objects of $\operatorname{Mod} T_1$ to those of $\operatorname{Mod} T$. Moreover, this function can be extended to a *functor* $(-) \circ I\colon \operatorname{Mod} T_1 \to \operatorname{Mod} T$, as it follows by basic category theoretical facts (for the uninitiated, it is an easy but rewarding exercise to work out the definition of $\eta \circ I\colon M \circ I \to N \circ I$ for any natural transformation $\eta\colon M \to N$). It is this functor, denoted as $I^* = (-) \circ I\colon \operatorname{Mod} T_1 \to \operatorname{Mod} T$ by Makkai and Reyes, that we see as the *generalized* restriction functor *induced* by I.

Having this concept at hand, the following definition is straightforward.

Definition 4.5. An extension $I\colon \mathbb{T} \to \mathbb{T}_1$ is called *tight* iff the restriction functor $I^*\colon \operatorname{Mod} T_1 \to \operatorname{Mod} T$ is an equivalence of categories.

We can now formulate rigorously the questions raised at the end of Section 3. Does T^{eq} have a tight and significant extension? If not, is T^{eq} a largest tight extension of T? In other words, is T^{eq} an extension of any other tight extension of T? In the next section, we describe the answer given by Makkai and Reyes.

5. The pretopos completion

As we mentioned in the introduction, Makkai and Reyes discovered that the categories called by Grothendieck *pretoposes* form a subclass of the logical categories and enjoy a remarkable property. They also defined a category-theoretical operation that attaches to every logical category a closely related pretopos called its *pretopos completion.* Makkai's remark in the fall of 1980, was that this operation generalizes the construction of T^{eq}. Indeed, if T is a first order complete theory with infinite models (as are the ones considered by Shelah), then the pretopos completion of the category $\mathbb{T}$ turns out to be (equivalent to) the category $\mathbb{T}^{\mathrm{eq}}$. These findings will be presented, in some detail, in this section.

As a first step of our exposition, let us ask when is a Boolean logical category equivalent to one of the form $\mathbb{T}^{\mathrm{eq}}$, for a first order theory T. Of course, Makkai and Reyes did not ask themselves this question because they were not aware of Shelah's construction at the time. Yet, we will follow their guiding principle namely, to seek category-theoretical properties, shared by the category $\mathbb{S}\mathrm{et}$ (which is, itself, a pretopos), that will supply the desired characterization. The direction of inquiry should be clear. The categories that we are looking for, should be closed also under category-theoretical operations that generalize the set theoretical operations of taking quotients of equivalence relations. We are faced again with the same problem: how can we describe complex set theoretical constructions, using only category-theoretical concepts? At first sight, the language of category theory seems more restricted than the one of set theory and, in fact, *it is* more restricted (remember that we cannot express even a simple fact such as an object of $\mathbb{S}\mathrm{et}$ being a subset of another). *One can*, nevertheless, *capture a great deal of the richness of the universe of sets*, using this limited language. How to accomplish this task, is a challenging problem well worth pursuing, as it turns out. If this project is pushed far enough, one reaches the notion of *elementary topos*, the subject of an active and vibrant research field. The category $\mathbb{S}\mathrm{et}$ is a topos, but there are many other interesting instances and it would be a mistake to think of topos theory as generalized set theory only. Actually, the origins of this theory are in the work of Grothendieck (to whom the name '*topos*' is due) in algebraic geometry. To get an idea of the broadness of the subject, one could consult, for example, [9] (which ends with an extensive bibliography; the literature of the field is so vast, that one could not provide a single reference without doing injustice to many others).

The properties needed to define pretoposes fall way short of the axioms for an elementary topos. Even less is required to characterize the categories equivalent to some $\mathbb{T}^{\mathrm{eq}}$. We can define category theoretical notions of *equivalence relation* over an object and *quotient* of an equivalence relation, that generalize the set theoretical concepts with the same name. Once this done, we can state:

Claim 5.1. *A small category $\mathbb{C}$ is equivalent to one of the form $\mathbb{T}^{\mathrm{eq}}$ for some first order theory T iff $\mathbb{C}$ is a Boolean logical category which is closed under the operation of taking quotients of equivalence relations.*

Notice that the "only if" direction namely, the fact that $\mathbb{T}^{\mathrm{eq}}$ is closed under quotients of equivalence relations, is established by Claim 4.4.

In order to define the notion of pretopos, we need two more concepts. The first of these is basic and well known.

An object S of a category $\mathbb{C}$ is called *initial*, if for any object X of $\mathbb{C}$, there is precisely one $\mathbb{C}$-morphism $S \to X$. The empty set $\varnothing$ is the unique initial object of the category $\mathbb{S}\mathrm{et}$. An object $[\varphi(\vec{x})]$ of the logical category $\mathbb{T}$ is initial iff $\varphi(\vec{x})$ is a formula which is a contradiction in the theory T (i.e. $T \vDash \forall\vec{x}\neg\varphi(\vec{x})$).

We shall say that two $\mathbb{C}$-objects X and Y *coexist disjointly in* $\mathbb{C}$ iff there exist in $\mathbb{C}$ a diagram $X \xrightarrow{g} A \xleftarrow{h} Y$ where g, h are monomorphisms, which has a pullback

$$\begin{array}{ccc} S & \xrightarrow{\;q\;} & Y \\ {\scriptstyle r}\big\downarrow & & \big\downarrow{\scriptstyle h} \\ X & \xrightarrow[\;g\;]{} & A \end{array}$$

with S an initial object of $\mathbb{C}$. In the category $\mathbb{S}\mathrm{et}$, this condition means that g and h are bijections whose images $g(X), h(X)$ are *disjoint* subsets of A. It follows easily than any two objects of $\mathbb{S}\mathrm{et}$ coexist disjointly.

Definition 5.2. A *pretopos* is a logical category $\mathbb{C}$ which is closed under the operation of taking quotients of equivalence relations and in which any two objects coexist disjointly. A *Boolean pretopos* is a pretopos which is a Boolean logical category.

Remark. This definition is different from, but equivalent to, Grothendieck's original or the Makkai and Reyes definition.

By Claim 5.1, the categories of the form $\mathbb{T}^{\mathrm{eq}}$ have to satisfy only the first of the two conditions from the definition of a pretopos. Yet, as we already hinted in the first paragraph of this section, in the cases that are of interest for the present paper, the category $\mathbb{T}^{\mathrm{eq}}$ *is* a Boolean pretopos. We are now going to explain this.

If d is a sort symbol of a first order theory T, we say that d *is nonempty* if the set $d(M)$ is nonempty for every model $M \models T$ (in other words, $T \models \exists x(x = x)$, where $x : d$). In unisorted logic, we assume tacitly that the unique sort is nonempty, but in the many sorted context, we are not bound by any nonemptiness assumptions. Let us say that T is a *proper* theory if all its sorts are nonempty and, for some sort symbol d_0, the set $d_0(M)$ has at least two elements for every model $M \models T$ (i.e., $T \models \exists u \exists v(u \neq v)$, where $u, v : d_0$). The theories studied by Shelah are certainly proper, as they are unisorted and have only infinite models. An important simple observation is that *if T is proper, then so is T^{eq}.*

Theorem 5.3. *If T is a proper first order theory, then $\mathbb{T}^{\mathrm{eq}}$ is a Boolean pretopos.*

PROOF (SKETCH). We know that $\mathbb{T}^{\mathrm{eq}}$ is a Boolean logical category closed under quotients of equivalence relations, so, all we have to show is that any two objects coexist disjointly. Let $X = [\varphi(\vec{x})]$, $Y = [\psi(\vec{y})]$ be two objects in $\mathbb{T}^{\mathrm{eq}}$ with $\vec{x} : \vec{d^1}$, $\vec{y} : \vec{d^2}$. As we know, by Claim 4.4, that $\mathbb{T}^{\mathrm{eq}}$ is equivalent to $(\mathbb{T}^{\mathrm{eq}})^{\mathrm{eq}}$, it will be enough to show that X and Y coexist disjointly in this latter category. Let $\vec{d}$ be the sequence $(d_0, d_0, \vec{d^1}, \vec{d^2})$ of sort symbols (where d_0 is the sort that shows that T is proper). With $u, v : d_0$ two distinct variables of the sort d_0, let us denote, for brevity, $\vec{z} = (u, v, \vec{x}, \vec{y}) : \vec{d}$ and take $\vec{z'} = (u', v', \vec{x'}, \vec{y'}) : \vec{d}$ to be a sequence of similar variables *distinct* from their nonprimed counterparts. Let $\varepsilon(\vec{z}, \vec{z'})$ be the formula

$$(u = v \wedge u' = v' \wedge \vec{x} = \vec{x'}) \vee (u \neq v \wedge u' \neq v' \wedge \vec{y} = \vec{y'}).$$

This is an equivalence formula of T^{eq}, hence the theory $(T^{\mathrm{eq}})^{\mathrm{eq}}$ has a new sort symbol d_ε and a function symbol $f_\varepsilon \colon \vec{d} \to d_\varepsilon$ which is the quotient map of ε. Remember that we are seeking an object A and monomorphisms $X \xrightarrow{g} A \xleftarrow{h} Y$, *in* $(T^{\mathrm{eq}})^{\mathrm{eq}}$, showing the disjoint coexistence of $X = [\varphi(\vec{x})]$ and $Y = [\psi(\vec{y})]$. We take $A = [w = w]$, where $w : d_\varepsilon$ (in other words, A is the extension of the sort d_ε), while g, h will be the maps

$$[\varphi(\vec{x})] \xrightarrow{[\gamma]} [w = w] \xleftarrow{[\chi]} [\psi(\vec{y})]$$

where γ and χ are the formulas

$$\varphi(\vec{x}) \wedge \exists u \exists v \exists \vec{y}(u = v \wedge f_\varepsilon(\vec{z}) = w) \text{ and } \psi(\vec{y}) \wedge \exists u \exists v \exists \vec{x}(u \neq v \wedge f_\varepsilon(\vec{z}) = w),$$

respectively.

We let the reader verify that these data show the disjoint coexistence of X and Y. □

Remarks. (1) Notice that the fact that our theories are in the *full* classical logic is crucial here, as the negation and the validity of the classical tautology $u = v \vee u \neq v$ are used in an essential way (by the *full* logic, we mean the one that is allowed to employ all connectives, including negation, and quantifiers, including the universal one; as a result, theories in the full logic are, when regarded as categories, *Boolean* logical categories).

(2) The statement of this theorem is not necessarily true if the theory is not *proper*. Indeed, if T is unisorted and all its models have just one element (i.e. $T \models \exists! x(x = x)$), then a moment of thought will reveal that T is, actually, a theory in *propositional* logic. In this case, $T = T^{\mathrm{eq}}$ and $\mathbb{T}^{\mathrm{eq}}$ is *not* a pretopos.

The following theorem is the independent discovery by Makkai and Reyes of a category theoretical generalization of Shelah's construction of Section 2.

Theorem 5.4 ([15]). *Every logical category* $\mathbb{C}$ *has a tight extension* $I\colon \mathbb{C} \to \mathbb{C}'$ *with the following properties*:

(1) $\mathbb{C}'$ *is a pretopos. If* $\mathbb{C}$ *is Boolean, so is* $\mathbb{C}'$.
(2) *If* $K\colon \mathbb{C} \to \mathbb{P}$ *is a logical functor from* $\mathbb{C}$ *into a pretopos* $\mathbb{P}$, *then there is a logical functor* $J\colon \mathbb{C}' \to \mathbb{P}$ *such that* $K = J \circ I$.

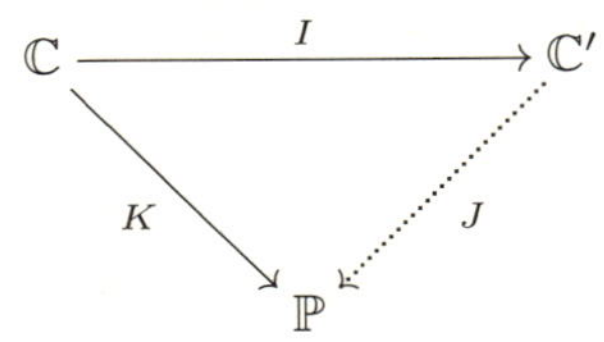

The category $\mathbb{C}'$ or, rather, the extension $I\colon \mathbb{C} \to \mathbb{C}'$ is called, by Makkai and Reyes, *the pretopos completion of* $\mathbb{C}$.

Let us remark that the functor J, of part 2 of Theorem 5.4, is just an object of the functor category $\mathbb{P}^{\mathbb{C}'}$, hence we do not expect it to be *uniquely* defined. Indeed, it is unique *up to an isomorphism* in $\mathbb{P}^{\mathbb{C}'}$.

If $\mathbb{C}$ is a category of the form $\mathbb{T}$ for a *proper* first order theory T, then Shelah's construction of Section 2 furnishes a proof of this theorem. However, Theorem 5.4 is more general, as it covers all logical categories, even not stemming from a proper theory, even not Boolean. Makkai and Reyes' independently found proof is similar to Shelah's construction, the main difference being that it involves an additional step of adding disjoint sums (which is a strong form of disjoint coexistence, used in their definition of pretopos).

Actually, all the results of [15] described in this paper are stated and proven in the general context of *logical* (not necessarily *Boolean*) categories. This is an appropriate place to point out the logical significance of this more general context. The realization that "logical functors are enough", meaning that these functors preserve not only the structure of logical categories, but also the additional structure that occurs in Boolean logical categories and, even, the pretopos structure, focussed the attention of Makkai and Reyes to *coherent logic*, which is the *restricted* fragment of the the *classical* first order logic based only on conjunctions, disjunctions and existential quantification. Coherent logic is carefully described in [15] in an appropriate formal context (known to logicians as *sequent calculus*), and is shown to be related to the concept of logical category in the same way, in which classical first order logic is related to Boolean logical category.

The important result that we present next is the remarkable property of pretoposes that we mentioned in the introduction and in the opening paragraph of the present section. Notice, again, that it refers to arbitrary, not necessarily Boolean, pretoposes.

Theorem 5.5 ([15]). *If $\mathbb{P}$ is a pretopos and $I\colon \mathbb{P} \to \mathbb{P}'$ is a* tight *extension then I is an equivalence of categories.*

When we take a tight extensions of a first order theory, for example the embedding $\mathbb{T} \to \mathbb{T}^{\mathrm{eq}}$, we do not add, in fact, any new models. *However* we do add, *in general*, new *concepts* such as equivalence classes, which now became ordinary *elements* of the models. The theorem that we just presented states that this is not the case if $\mathbb{T}$ was a pretopos, to begin with. Therefore the statement of Theorem 5.5 is called, in [15], **the conceptual completeness of pretoposes**.

The conceptual completeness of *Boolean* pretoposes implies, together with Theorem 5.4, the answer to the question asked at the end of Sections 3 and 4.

Corollary 5.6. *If $\mathbb{T}$ is a Boolean logical category, the embedding $I\colon \mathbb{T} \to \mathbb{T}^{\mathrm{eq}}$ is the* largest *tight extension of $\mathbb{T}$. In other words, for any other tight extension $K\colon \mathbb{T} \to \mathbb{T}_1$ there is a logical functor $\widehat{I}\colon \mathbb{T}_1 \to \mathbb{T}^{\mathrm{eq}}$ such that $\widehat{I} \circ K = I$.*

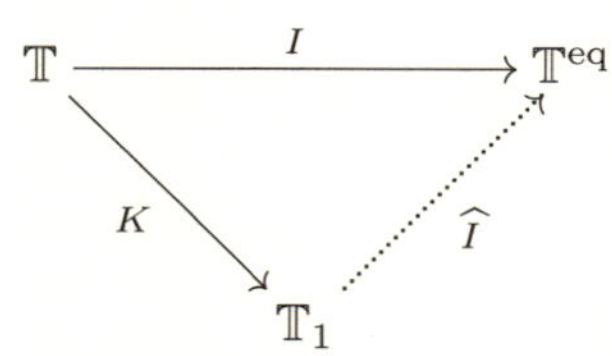

PROOF (SKETCH). Let $I_1\colon \mathbb{T}_1 \to \mathbb{T}_1^{\mathrm{eq}}$ be the embedding of $\mathbb{T}_1$ into its pretopos completion. Then the composition $I_1 \circ K\colon \mathbb{T} \to \mathbb{T}_1$ is a logical functor into a Boolean pretopos, hence, by Theorem 5.4, there exists a logical functor $J\colon \mathbb{T}^{\mathrm{eq}} \to \mathbb{T}_1^{\mathrm{eq}}$ such that $J \circ I = I_1 \circ K$.

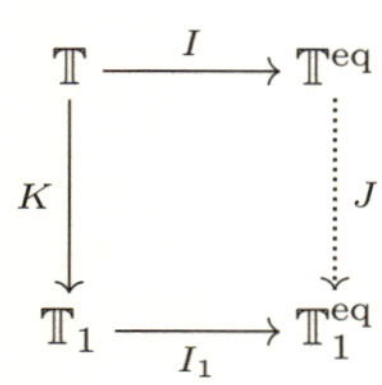

$I_1 \circ K$ is a *tight* extension, as the composition of two such. Therefore, so is $J \circ I$ and as I is tight as well, we conclude that J *is also tight* (the induced restriction functors satisfy $(J \circ I)^* = I^* \circ J^*$ and is an easy exercise to see that if $I^* \circ J^*$ and I^* are equivalence functors, then so is J^*). But then it follows, by Theorem 5.5, that J is an equivalence of categories and so, $\mathbb{T}^{\mathrm{eq}}$ and $\mathbb{T}_1^{\mathrm{eq}}$ *are the same.* Therefore, I_1 is *essentially* a logical functor from $\mathbb{T}_1$ into $\mathbb{T}^{\mathrm{eq}}$! This means that one can modify (using routine techniques) I_1 into an isomorphic logical functor $\widehat{I}\colon \mathbb{T}_1 \to \mathbb{T}^{\mathrm{eq}}$, as desired. □

Remark. Again, one can show that the functor $\widehat{I}$ is unique, up to isomorphism, having the specified property.

6. Concluding remarks

(1) Shelah succeeded to overcome certain difficulties in the development of his classification theory by using an auxiliary construction. He *invented* the extension T^{eq} of a first order theory T.

Makkai and Reyes explored a new framework unveiling the structure behind the categorical approach to logic and model theory. They *discovered* the pretopos completion.

The two contributions, of Shelah on one hand and of Makkai–Reyes on the other, complement each other. The pretopos completion is a concept that *had* to be defined. It lays there as part of the orderly world in which the categories associated with theories and their models live. Once discovered, it is only natural to ask for a context in which this new notion is useful. Shelah's work supplies an instant answer, a gratifying one, to this question.

(2) If $\mathbb{T}$ is a logical category, let $\operatorname{Mod}\mathbb{T}$ be the category of *logical* functors $\mathbb{T} \to \mathbb{S}\mathrm{et}$ (in other words, $\operatorname{Mod}\mathbb{T}$ is the same as the category $\operatorname{Mod}T$ of the models of the coherent theory represented by $\mathbb{T}$). The conceptual completeness Theorem 5.5 says, roughly speaking, that if $\mathbb{T}, \mathbb{T}'$ are pretoposes such that $\operatorname{Mod}\mathbb{T}, \operatorname{Mod}\mathbb{T}'$ "are the same," then so are $\mathbb{T}$ and $\mathbb{T}'$. It seems, therefore, that the category $\operatorname{Mod}\mathbb{T}$ *determines* the pretopos $\mathbb{T}$. Or, using a more "model-theory friendly" way of speech, it seems that for a first order theory T, the category of models $\operatorname{Mod}T$ determines the theory T^{eq}. *Is this really so*?

The answer to this question turns out to be positive, *provided* that we phrase it carefully, as done by Makkai in [13] (for a shorter preliminary version, see [11]). When saying that the pretoposes $\mathbb{T}$ and $\mathbb{T}'$ "are the same," we just mean that they are equivalent categories. However, the meaning of the categories of models "being the same," in the hypothesis of Theorem 5.5, is that they are equivalent by a functor $I^*\colon \operatorname{Mod}\mathbb{T}' \to \operatorname{Mod}\mathbb{T}$ *induced* by a logical functor $I\colon \mathbb{T} \to \mathbb{T}'$. How can we recognize which functors between categories of models are of this, induced, kind?

Makkai points out that the *concrete* category $\operatorname{Mod}\mathbb{T}$ carries some extra structure related to the fact that the class of models of a given first order theory is closed under the ultraproduct operation. He defines an abstract notion of *ultracategory*, which is a category endowed with an extra structure of the kind just mentioned, such that the categories of the form $\operatorname{Mod}T$ are ultracategories (and so is $\mathbb{S}\mathrm{et}$). A functor between ultracategories that preserves this extra structure is called by Makkai an *ultrafunctor*. He shows that the induced functors I^* are ultrafunctors and that they are, up to isomorphism, the only ultrafunctors $\operatorname{Mod}\mathbb{T}' \to \operatorname{Mod}\mathbb{T}$. Moreover, he shows that the pretopos $\mathbb{T}$ is determined by the *ultracategory* $\operatorname{Mod}\mathbb{T}$.

Actually, [13] describes a concrete way of recovering $\mathbb{T}$ from $\operatorname{Mod}\mathbb{T}$. Let Pt be the category whose objects are the pretoposes and morphisms the logical functors between them, and let UC be the category whose objects are the ultracategories and whose morphisms are the ultrafunctors. Then Mod can be construed as a *contravariant* functor from the category Pt of pretoposes to the category UC of ultracategories (a contravariant functor is one that reverses the direction of morphisms, as $I \mapsto I^*$ does). Makkai constructs a *dual* contravariant functor F from UC to Pt such that for any pretopos $\mathbb{T}$, $F\operatorname{Mod}\mathbb{T}$ is a pretopos that is equivalent to $\mathbb{T}$ itself. This result implies Theorem 5.5 and is called by Makkai the *strong conceptual completeness of pretoposes.*

Makkai's work that we just described has had several applications, and I will describe two of them.

The first application is due to Makkai himself and concerns conceptual completeness for *Boolean* pretoposes. For $\mathbb{T}$ a Boolean pretopos, let $\mathrm{Mod}^{\mathrm{iso}}\,\mathbb{T}$ be the category whose objects are the models of $\mathbb{T}$ (i.e., logical functors $M\colon \mathbb{T} \to \mathbb{S}\mathrm{et}$) and morphisms are only the *isomorphisms* between models. In other words, $\mathrm{Mod}^{\mathrm{iso}}\,\mathbb{T}$ is a subcategory of $\mathrm{Mod}\,\mathbb{T}$ with the same objects but considerably fewer morphisms. Using the methods of [13], Makkai proves in [12] that a Boolean pretopos $\mathbb{T}$ can be recovered from the smaller category $\mathrm{Mod}^{\mathrm{iso}}\,\mathbb{T}$ as well. This is a strong conceptual completeness result for Boolean pretoposes which, as Makkai points out, is closely related to an earlier, unpublished, result of Gaifman, a purely model-theoretical definability theorem.

The second application is due to Zawadowski. Given logical categories $\mathbb{T}, \mathbb{T}_1$ and a logical functor $I\colon \mathbb{T} \to \mathbb{T}_1$, let $\mathrm{Mod}_I\,\mathbb{T}$ be the *full* subcategory of $\mathrm{Mod}\,\mathbb{T}$ whose objects are those models $M\colon \mathbb{T} \to \mathbb{S}\mathrm{et}$ of $\mathbb{T}$ that can be *extended* to a model M_1 of $\mathbb{T}_1$ (meaning that $M = M_1 \circ I\colon \mathbb{T} \to \mathbb{S}\mathrm{et}$). In the simplest case, in which the two categories represent theories T, T_1 such that T_1 extends T in the ordinary sense (meaning that the language and the set of theorems of the former includes those of the latter), the model theorist will recognize that the set of objects of $\mathrm{Mod}_I\,\mathbb{T}$ is what is called a PC_Δ-class of structures. In general, the category $\mathrm{Mod}_I\,\mathbb{T}$ has fewer objects than the category $\mathrm{Mod}\,\mathbb{T}$ of *all* the models of $\mathbb{T}$. And yet, Zawadowski uses [13] and proves that in certain cases, $\mathbb{T}$ can be recovered from this proper subcategory of $\mathrm{Mod}\,\mathbb{T}$. Indeed, the main result of [27], implies that *if* $\mathbb{T}, \mathbb{T}_1$ *are* pretoposes *and* $I\colon \mathbb{T} \to \mathbb{T}_1$ *is* conservative, *then* $\mathbb{T}$ *can be recovered from* $\mathrm{Mod}_I\,\mathbb{T}$. This beautiful result, which can be called, aptly, *strong conceptual completeness for* PC_Δ*-classes of pretoposes*, generalizes Makkai's result. When translated into the language of pure category theory, the theorem of Zawadowski is called *the descent theorem for pretoposes* and it affirms a conjecture stated earlier by A. Pitts.

After seing Zawadowski's work, Makkai proved, in [14], a descent theorem for *Boolean* pretoposes as well. For a simpler, heavily model-theoretical proof of both, Zawadowski's and Makkai's, descent theorems, see Ballard and Boshuck's [2].

All the results that we mentioned in this discussion deal with classical logic (either in its full or in its coherent version) and were proven using a mixture of category-theoretical and model-theoretical methods. Andrew Pitts succeeded to prove, in [17], by purely category theoretical methods, a stronger, *constructive* version of conceptual completeness of pretoposes. To be more specific: while the original proof of Makkai – Reyes uses freely the axioms of set theory, as is common in modern mathematics, Pitts' arguments use only the *restricted* set of axioms for an elementary topos with a natural number object (the assumption of having such an object is the topos theoretical version of the axiom of infinity); in particular, this weaker set of axioms does not include an analogue of the axiom of choice, hence the adjective 'constructive' that topos theorists use to design such results.

In a subsequent paper [18], written for a logic inclined audience, Pitts used similar methods to prove a conceptual completeness result for *intuitionistic* first order logic. He states his result in both logical and category-theoretical terms, the latter being what we call conceptual completeness for *Heyting* pretoposes. The validity of a descent theorem for this kind of pretoposes is still an open question.

(3) We mentioned, at the end of Section 3, the notion of *interpretation* of a theory into another, familiar in mathematical logic. As promised there, we are going now to elaborate on this concept, its generalizations and its connection to the notion of extension. This will be our final, and longest, comment and will show that the conceptual completeness of Boolean pretoposes has some relevance to the question whether we have the most general definition.

Given first order languages L, L' and an L'-theory T', an *interpretation of* L *into* T' is a function $\varphi \mapsto \varphi'$ with certain properties that allow us to extract an L-structure M from any model $M' \models T'$.

We are now going to spell out the conditions that the function $\varphi \mapsto \varphi'$ must satisfy in order to be an interpretation of a kind that we call *simple*. This is a non customary name and we use it to distinguish between simple interpretations and the more general kind that we will define later.

We assume first that L, L' are *unisorted* languages. If φ is the L-formula $x = x$ then φ' will be an L'-formula $\delta(\vec{x}')$, with $\vec{x}'$ a k-tuple, that plays a very special role: the elements of the L-structure M extracted from $M' \models T'$ will be the k-tuples of M' that satisfy $\delta(\vec{x}')$. If $\varphi \equiv \varphi(x_1, \ldots, x_n)$ (where we use '$\equiv$', rather than the equality symbol '$=$', to denote identity of formulas) then $\varphi' \equiv \varphi'(\vec{x}'_1, \ldots, \vec{x}'_n)$ with $\vec{x}'_1, \ldots, \vec{x}'_n$ *distinct* k-tuples and we require also that $T' \models \forall \vec{x}'_1 \ldots \forall \vec{x}'_n(\varphi'(\vec{x}'_1, \ldots, \vec{x}'_n) \to \delta(\vec{x}'_i))$ for all $1 \leqslant i \leqslant n$. If $\varphi \equiv R(x_1, \ldots, x_n)$, with R an n-ary relation symbol of L, then the formula $\varphi' \equiv \rho(\vec{x}'_1, \ldots, \vec{x}'_n)$ will define the interpretation $R(M)$ of R in the model M extracted from M' as the set of those n-tuples of k-tuples of elements of M' that satisfy the formula ρ. A very important *simplicity* condition, which makes the interpretation *simple*, requires that if φ is the formula $x = y$ then φ' will be the formula $\vec{x}' = \vec{y}'$ (which is short for $x'_1 = y'_1 \wedge \cdots \wedge x'_k = y'_k$). This condition insures that the equality relation in M will be the *standard* one, i.e. the equality of k-tuples. Next, if $\varphi \equiv f(x_1, \ldots, x_n) = y$, where f is an n-ary function symbol of L, then $\varphi' \equiv \varphi'(\vec{x}'_1, \ldots, \vec{x}'_n, \vec{y}')$ and we require T' to imply that this formula defines $\vec{y}'$ as a function of $(\vec{x}'_1, \ldots, \vec{x}'_n)$. The formula φ' can then be used to define the interpretation $f(M)$ of f in M. The interpretations of individual constants are defined in a similar way. Thus, the values of $\varphi \mapsto \varphi'$ for atomic formulas φ, allow us to extract M from M'. The values of the interpretation for more complex formulas is uniquely determined by additional, very natural, recursive requirements for the function $\varphi \mapsto \varphi'$. It must preserve connectives (e.g., we must have $(\varphi \wedge \psi)' \equiv \varphi' \wedge \psi'$) and translate quantifiers $\forall x, \exists x$ to their versions *relativised* to δ (i.e., $\forall \vec{x}'(\delta(\vec{x})' \to \cdots$ and $\exists \vec{x}'(\delta(\vec{x}') \wedge \cdots)$.

This concludes the definition of a *simple interpretation* of a language L into an L'-theory T', for *unisorted* languages. The extension to the *multisorted* context involves only notational complications, because in this case, we have a formula $x = x$ with $x : d$, for every sort symbol d and this formula is mapped by the simple interpretation to an L'-formula $\delta_d(\vec{x}')$, where the length and the sorting of the sequence of L'-variables $\vec{x}'$ *depend* on d. We skip the obvious details of the extended definition.

If T, T' are theories in the (possibly, multisorted) languages L and L', then a simple interpretation $\varphi \mapsto \varphi'$ of T into T' is a simple interpretation of L into T'

such that for any L-sentence φ, $T \vDash \varphi$ implies that $T' \vDash \varphi'$. It follows, in particular, that for L-formulas φ, ψ, if $\varphi \sim_T \psi$ then $\varphi' \sim_{T'} \psi'$.

Simple interpretations and extensions are essentially the same. Indeed, T has a simple interpretation in T' iff $\mathbb{T}'$ is an extension of $\mathbb{T}$. To be more precise:

Proposition 6.1. *Each simple interpretation $\varphi \mapsto \varphi'$ of T into T' induces an extension functor $I\colon \mathbb{T} \to \mathbb{T}'$ such that I maps any $\mathbb{T}$-morphism $[\varphi]_T \xrightarrow{[\chi]_T} [\psi]_T$ to the $\mathbb{T}'$-morphism $[\varphi']_{T'} \xrightarrow{[\chi']_{T'}} [\psi']_{T'}$. Conversely, every extension functor from $\mathbb{T}$ to $\mathbb{T}'$ is induced by a simple interpretation.*

Remark. Actually, the same extension functor is induced by *several* simple interpretations because, when we construct a simple interpretation inducing a given functor, the value of $\varphi \mapsto \varphi'$ for simple atomic formulas is imposed on us only up to $\sim_{T'}$-equivalence.

Consider a first order theory T in a language L. If T^{eq} is a significant extension of T, as is mostly the case, then $\mathbb{T}$ is *not* an extension of $\mathbb{T}^{\mathrm{eq}}$. This means that T^{eq} has no simple interpretation in T. And yet, every model $M \vDash T$ 'carries' with it its canonical extension M^{eq}. The construction of this canonical extension from M is done, actually, via an *interpretation* in a wider sense in which the elements of the 'extracted' model M^{eq} are interpreted, not as k-tuples in M but, rather, as *equivalence classes* of elements of M. This suggests an obvious generalization of the notion of *simple* interpretation, which we are going to describe.

Again, in order to avoid notational complications, we assume first that the languages are unisorted. An *interpretation* $\varphi \mapsto \varphi'$ of a language L into an L'-theory T' is defined like a simple interpretation with one major modification: the simplicity condition (which said that the interpretation of equality is the standard equality of tuples) is replaced by the following:

If φ is the formula $x = y$ then φ' is a formula $\varepsilon(\vec{x}', \vec{y}')$ such that T' implies that ε defines an equivalence relation on the set of k-tuples satisfying $\delta(\vec{x}')$.

This means that each element of the extracted model M is interpreted as an ε-equivalence class of tuples. As a result, we have to impose additional requirements: T' must imply that ε is a congruence relation with respect to the relation defined by the formula ρ describing the interpretation of a relation symbol R. By this we mean that T' implies that whenever $\varepsilon(\vec{x}'_i, \vec{y}'_i)$ for all $1 \leqslant i \leqslant n$, we also have $\rho(\vec{x}'_1, \ldots, \vec{x}'_n) \leftrightarrow \rho(\vec{y}'_1, \ldots, \vec{y}'_n)$. A similar condition is required for the formula φ describing the interpretation of a function symbol f. For this formula we also require a 'functionality' condition, namely, that T' implies that whenever $\varphi(\vec{x}'_1, \ldots, \vec{x}'_n, \vec{y}')$ and $\varphi(\vec{x}'_1, \ldots, \vec{x}'_n, \vec{z}')$ both hold than we also have $\varepsilon(\vec{y}', \vec{z}')$. With these conditions fulfilled , we obviously can, given any model $M' \vDash T'$, build an L-structure whose elements are ε-equivalence classes of k-tuples of M'. The definition of the notion of interpretation is completed by imposing on complex formulas the same recursive requirements as for simple interpretations. This implies that for any L-formula φ, T' will imply that ε is a congruence relation with respect to the relation defined by φ'.

This definition of interpretation, in the unisorted context, is precisely the one given by Wilfrid Hodges in [7], which contains a thorough treatment of interpretations and their uses in logic, especially in model theory (see also [8], where Hodges graciously traces this definition to Mal′cev, [16, Chapter 15]).

It is a routine exercise to generalize this definition to the multisorted context. Once this done, we can make a precise statement to the effect that T^{eq} *has an interpretation in* T, in fact a very simply described one. Let us call it the *canonical interpretation of* T^{eq} *in* T.

Consider, next, an arbitrary interpretation of an L-theory T in an L'-theory T'. It allows, once given a model $M' \models T'$, to extract a model M of T whose elements are equivalence classes of relations definable in M'. But these equivalence classes are just *elements* of the canonical extension M'^{eq}! This means that the given interpretation of T in T' can be represented as a *simple* interpretation of T in T'^{eq} followed by the *canonical interpretation* of T'^{eq} in T'. Thus, interpretations in T' are essentially the same as *simple* interpretations of T in T'^{eq}. This gives a compact definition of the generalized version of interpretability.

Definition 6.2. A theory T is interpretable in another theory T' iff there exists a logical functor $F\colon \mathbb{T} \to \mathbb{T}'^{\text{eq}}$. Such a functor F will be called an interpretation of T in T'.

Remarks. (1) Of course, every theory T is interpretable in itself via the obvious embedding $I\colon \mathbb{T} \to \mathbb{T}^{\text{eq}}$. We will refer to I as the *canonical* interpretation of T into itself.

(2) If we have interpretations $F\colon \mathbb{T} \to \mathbb{T}'^{\text{eq}}$ of T in T' and $G\colon \mathbb{T}' \to \mathbb{T}''^{\text{eq}}$ of T' in T'', then we would like to define their composition $G * F$ as an interpretation of T in T''. This cannot be done as direct functor composition, because the codomain of F does not equal the domain of G. However, as the canonical interpretation $I'\colon \mathbb{T}' \to \mathbb{T}'^{\text{eq}}$ is the pretopos completion of $\mathbb{T}'$, the functor G can be "lifted," by Theorem 5.4, to a functor $K\colon \mathbb{T}'^{\text{eq}} \to \mathbb{T}''^{\text{eq}}$ (which is unique up to isomorphism) such that $G = K \circ I'$. We then define $G * F = K \circ F$.

We now come to the main point of this comment. Had we reached the *most general* definition of the interpretation concept? Or, maybe, there are some more imaginative ways in which L-structures can be extracted from model of a given theory T'? The conceptual completeness of Boolean pretoposes seems to indicate that whatever means of interpretation one may come up with, they are already present in the pretopos completion of T', namely in T'^{eq}. We may hasten to think that *yes*, Definition 6.2 is most general reasonable one. *However*, here is an example that may cast some doubt on this belief. Let T^{EQ} be the *unisorted* version of T^{eq}, as envisaged by Shelah. Remember that we no longer have sorts d_ε but unary predicates P_ε and that we might have, in a model of T^{EQ} a set of elements which are outside the interpretations of all of these new unary predicates. Well, we would expect T^{EQ} to be interpretable in T and yet, there is no logical functor $F\colon \mathbb{T}^{\text{EQ}} \to \mathbb{T}^{\text{eq}}$. The reason for this is that $\mathbb{T}^{\text{EQ}}$ has an 'all inclusive' object $[x = x]_{T^{\text{EQ}}}$ into which any other object of $\mathbb{T}^{\text{EQ}}$ can be embedded (by a monomorphism), while there is no such in the category $\mathbb{T}^{\text{eq}}$.

We may correct this situation by a slight modification of Definition 6.2. For the purpose of the next definition, let me stress that a *finite* language is one whose set of extralogical symbols is finite.

Definition 6.3. An L-theory T is *locally interpretable* in another theory T' iff for every *finite* sublanguage $L_0 \subset L$, the restricted theory $T{\restriction}L_0$ is interpretable in T'.

It is not hard to see that T^{EQ} is locally interpretable in T. Is local interpretability a reasonable notion? One might argue that the possibility of extracting chunks of models of T from any given model of T', but not the model as a whole, is not really an interpretation. If we adopt this point of view, then we should probably have second thoughts and, contrary to our initial intuition, conclude that T^{EQ} is *not* interpretable in T, in any reasonable sense of the word.

On the other hand, the original use of the concept of interpretation in [26] was for the purpose of proving certain theories to be undecidable. The basic fact is the following:

If T is an undecidable theory which has a recursive *interpretation in T', then T' is undecidable as well.*

This statement remains true if we replace 'interpretation' by 'local interpretation.' From this point of view, local interpretability *is* a reasonable notion. However, this is not very significant, because in most applications of the basic fact, the theory T has a finite language anyhow.

Based on the conceptual completeness of pretoposes, I am ready to venture and propose the following (nonmathematical) conjecture:

For T a first order theory formulated in a finite *language, Definition* 6.2 *provides the most general reasonable notion of interpretability.*

Remark. The referee kindly suggested that I discuss, in this context, the intuitive concept of *bi-interpretability* of theories, as well. We usually say that two theories T and T' are *bi-interpretable* if each of them has an interpretation into the other, such that these two interpretations are, roughly speaking, inverses of each other. Bi-interpretable theories are considered to have the same logical strength. A non trivial example, which we mentioned already, is that of real closed fields vs. elementary geometry.

If we speak about finite-language theories and adopt the conjecture that I just proposed, it seems reasonable to reformulate the rough description above as the following precise mathematical definition (where we use the concepts and notations introduced in the remarks following Definition 6.2): T and T' are bi-interpretable iff we have interpretations $F\colon \mathbb{T} \to \mathbb{T}'^{\mathrm{eq}}$ and $G\colon \mathbb{T}' \to \mathbb{T}^{\mathrm{eq}}$ such that $G * F \simeq I$ (meaning that the two functors $G * F, I\colon \mathbb{T} \to \mathbb{T}^{\mathrm{eq}}$ are isomorphic via a natural transformation) and $F * G \simeq I'$, where I and I' are the canonical interpretations of T and T' into themselves. Not surprisingly, it turns out (as one can see) that *T and T' are bi-interpretable iff the pretoposes $\mathbb{T}^{\mathrm{eq}}$ and $\mathbb{T}'^{\mathrm{eq}}$ are equivalent categories.*

References

1. J. T. Baldwin, *Fundamentals of stability theory*, Perspectives in Mathematical Logic, Springer-Verlag, Berlin, 1988.
2. D. Ballard and W. Boshuck, *Definability and descent*, J. Symbolic Logic **63** (1998), no. 2, 372–378.
3. G. Blanc, *Équivalence naturelle et formules logiques en théorie des catégories*, Arch. Math. Logik Grundlag. **19** (1978/79), no. 3-4, 131–137.
4. S. Eilenberg and S. Mac Lane, *General theory of natural equivalences*, Trans. Amer. Math. Soc. **58** (1945), 231–294.
5. P. Freyd, *Properties invariant within equivalence types of categories*, Algebra, Topology, and Category Theory: A Collection of Papers in Honor of Samuel Eilenberg (A. Heller and M. Tierney, eds.), Academic Press, New York, 1976, pp. 55–61.

6. V. Harnik and L. Harrington, *Fundamentals of forking*, Ann. Pure Appl. Logic **26** (1984), no. 3, 245–286.
7. W. Hodges, *Model theory*, Encyclopedia Math. Appl., vol. 42, Cambridge Univ. Press, Cambridge, 1993.
8. ______, *A shorter model theory*, Cambridge Univ. Press, Cambridge, 1997.
9. F. W. Lawvere, *Comments on the development of topos theory*, Development of Mathematics 1950–2000 (J.-P. Pier, ed.), Birkhäuser, Basel, 2000, pp. 715–734.
10. S. Mac Lane, *Categories for the working mathematician*, 2nd ed., Grad. Texts in Math., vol. 5, Springer, New York, 1998.
11. M. Makkai, *Stone duality for first order logic*, Proceedings of the Herbrand Symposium (Marseille, 1981) (J. Stern, ed.), Stud. Logic Found. Math., vol. 107, North-Holland, Amsterdam, 1982, pp. 217–232.
12. ______, *Ultraproducts and categorical logic*, Methods in Mathematical Logic (Caracas, 1983) (C. A. Di Prisco, ed.), Lecture Notes in Math., vol. 1130, Springer, Berlin, 1985, pp. 222–309.
13. ______, *Stone duality for first order logic*, Adv. in Math. **65** (1987), no. 2, 97–170.
14. ______, *Duality and definability in first order logic*, Mem. Amer. Math. Soc. **105** (1993), no. 503.
15. M. Makkai and G. E. Reyes, *First order categorical logic: Model-theoretical methods in the theory of topoi and related categories*, Lecture Notes in Math., vol. 611, Springer, Berlin, 1977.
16. A. I. Mal′cev, *The metamathematics of algebraic systems. Collected papers: 1936–1967*, Stud. Logic Found. Math., vol. 66, North-Holland, Amsterdam, 1971. Translated, edited, and provided with supplementary notes by B. F. Wells III.
17. A. M. Pitts, *Interpolation and conceptual completeness for pretoposes via category theory*, Mathematical logic and theoretical computer science (College Park, MD, 1984–1985) (D. W. Kueker, E. G. K. López-Escobar, and C. H. Smith, eds.), Lecture Notes in Pure and Appl. Math., vol. 106, Dekker, New York, 1987, pp. 301–327.
18. ______, *Conceptual completeness for first-order intuitionistic logic: an application of categorical logic*, Ann. Pure Appl. Logic **41** (1989), no. 1, 33–81.
19. B. Poizat, *Cours de théorie des modèles: Une introduction à la logique mathématique contemporaine*, Bruno Poizat, Lyon, 1985.
20. ______, *A course in model theory: An introduction to contemporary mathematical logic*, Universitext, Springer, New York, 2000. Translated by M. Klein and revised by the author.
21. S. Shelah, *Classification theory and the number of nonisomorphic models*, Stud. Logic Found. Math., vol. 92, North-Holland, Amsterdam, 1978.
22. ______, *Classification of first order theories which have a structure theorem*, Bull. Amer. Math. Soc. (N.S.) **12** (1985), no. 2, 227–232.
23. ______, *Classification theory and the number of nonisomorphic models*, 2nd ed., Stud. Logic Found. Math., vol. 92, North-Holland, Amsterdam, 1990.
24. ______, *Classification theory for elementary abstract classes*, Stud. Log. (Lond.), vol. 18, Mathematical Logic and Foundations, College Publications, London, 2009.
25. ______, *Classification theory for abstract elementary classes*. Vol. 2, Stud. Log. (Lond.), vol. 20, Mathematical Logic and Foundations, College Publications, London, 2009.
26. A. Tarski, *Undecidable theories*, Stud. Logic Found. Math., North-Holland, Amsterdam, 1953. In collaboration with A. Mostowski and R. M. Robinson.
27. M. W. Zawadowski, *Descent and duality*, Ann. Pure Appl. Logic **71** (1995), no. 2, 131–188.

DEPARTMENT OF MATHEMATICS, UNIVERSITY OF HAIFA, HAIFA 31905, ISRAEL
E-mail address: `harnik@math.haifa.ac.il`

Centre de Recherches Mathématiques
CRM Proceedings and Lecture Notes
Volume **53**, 2011

The Dialectica Monad and Its Cousins

Pieter J. W. Hofstra

ABSTRACT. I give an expositional account of the dialectica construction and some related constructions from the point of view of quantification in fibrations. This allows for concise conceptual formulations of these constructions and explains their universal properties. There are two main points I wish to convey; the first is that the categorical dialectica construction decomposes into two steps, following the quantifier pattern of the original translation; the second is that the categorical embodiment of Skolemization takes the form of a pseudodistributive law between the pseudomonads which freely add universal and existential quantification.

1. Introduction

1.1. Background. The original purpose of Gödel's famous dialectica interpretation [11] was to reduce the problem of proving consistency of first-order arithmetic to the consistency of a simply typed system of computable functionals (called *System T*). The key feature of the translation is that it turns formulae of arbitrary quantifier complexity into formulae of the form $\exists\vec{f}\forall\vec{x}.\alpha(\vec{f},\vec{x})$, where α is quantifier-free, by using the principle of *Skolemization*:

$$\frac{\forall u\exists x.\alpha(u,x)}{\exists f\forall u.\alpha(u,fu)}\ \mathrm{Sk}$$

This principle is a form of choice: it allows us to replace a quantifier combination of the form $\forall\exists$ by one of the form $\exists\forall$ by introducing a choice function f. Because this choice function is a higher-type object, the resulting "Skolemized" formula is no longer first order.

Over the years, several authors have tried to capture the dialectica interpretation in categorical terms. The development which is the main focus of this paper is abstract in nature and aims at defining and analysing "dialectica-like" structures without direct reference to arithmetic. The starting point of this line of research is the PhD work of Valeria de Paiva [7], in which the so-called *dialectica categories* were introduced. Let me briefly sketch the construction. One starts with a category C with finite limits and builds a new category $\mathfrak{Dial}(\mathsf{C})$, the objects of which have

2010 *Mathematics Subject Classification.* 03F10, 03F50, 18C20, 18C50, 18D30.
Partially supported by NSERC discovery grant.
This is the final form of the paper.

the form (X, U, α) where α is a subobject of $X \times U$ in C; such an object is thought of as the formula $\exists x \forall u.\alpha(x, u)$. A morphism from (X, U, α) to (Y, V, β) consists of a pair (f, f_0), where $f\colon X \to Y$ and $f_0\colon X \times V \to U$, subject to the condition[1] that $\alpha(x, f_0(x, v)) \vdash \beta(fx, v)$. This definition of morphism, as explained in loc. cit., is motivated by the way the dialectica interpretation acts on implicational formulae.

The main focus in the original work on the dialectica categories was on the categorical structure of $\mathfrak{Dial}(\mathsf{C})$: one can show that these always have a monoidal structure, and that under additional assumptions on C they form a model of intuitionistic linear logic. De Paiva [8] also considers a variant called the *Girard construction*, and describes how the dialectica category $\mathfrak{Dial}(\mathsf{C})$ can be presented as the result of first applying the Girard construction to C, and then forming the co-Kleisli category for a comonad on this category.

The original construction of the dialectica categories was formulated in terms of the subobject logic of C. However, this can be generalized quite a bit: the case of fibred preorders was explored in detail by Martin Hyland [13], and Bodil Biering's PhD thesis [5] investigates dialectica categories of the form $\mathfrak{Dial}(\mathsf{p})$, where p is an arbitrary cloven fibration.

These constructions raise various questions. What is the precise categorical nature of the dialectica categories? Can they be described in more conceptual terms, for example in terms of universal properties? And in what way does the construction of these categories capture the essential ingredient of Gödel's original translation, namely the principle of Skolemization?

1.2. Goals of the paper. The main aim of this paper is to give an exposition and interpretation of the dialectica construction from a modern categorical perspective, which gives clear and precise answers to the above questions. The dialectica interpretation has somewhat of a reputation of being complicated; however, by giving centre stage to the well-known concepts of monads, simple products and coproducts I hope that what emerges is a conceptually pleasing and accessible picture.

The second goal of the paper is to provide explicit statements and reasonably detailed proofs of facts which are perhaps folklore among some experts in the field but which have, to the best of my knowledge, never been stated or worked out in the literature. The first statement of this kind is that the dialectica construction is really a two-step construction: $\mathfrak{Dial}(\mathsf{p})$ (where p is a fibration) is obtained by first applying the monad which freely adds simple universal quantification and then applying the monad which freely adds simple existential quantification. The second fact is that the dialectica construction (when performed on the level of general fibrations and under the usual assumption that the base category is Cartesian closed) is itself monadic. The main observation in this general categorical setting is that the principle of Skolemization takes the form of a (pseudo-)distributive law between the abovementioned quantification monads.

The paper has been written with a readership familiar with basic category theory in mind. In particular, I assume that the reader is comfortable with adjunctions, monads, categories of algebras for a monad and distributive laws. A standard reference to these topics is the textbook [3]. Since fibrations are omnipresent in the

[1] I use logical notation here; this will be made precise in Section 2.3.

paper I have included a background section on the basic theory of fibred categories, including all the results we need for the rest of the work.

At various places we shall be forced to deal with "up-to-isomorphism" versions of monads and related concepts, i.e., with pseudomonads, pseudo-algebras and pseudodistributive laws. Because this is not the right place to learn about this material (and because those with an interest in the matter will likely already know the basic theory) I have not included any review material of these concepts; a concise treatment containing all definitions which are used in this paper can be found in [6] (see also the references therein); we also mention [10], which is a survey of the theory of pseudodistributive laws.

I should stress however that intimate knowledge of these higher-categorical notions is not necessary in order to follow the paper. First of all, I have made an attempt to organize the material in such a way that it becomes clear why and where certain coherence issues arise and why some of the structures under consideration are not strict. The reader who is interested in the general picture but not in the 2-categorical subtleties can simply pretend that the structure is strict and still follow the main developments. Moreover, for most results I have tried to describe the key ideas for the proof without going into too much distracting detail; and I have delegated more complete proofs involving coherence issues and such to separate sections which can be skipped without losing track of the main story.

I should also point out some things which I have not aimed for in this paper. First of all, the main focus in several papers on the dialectica construction and its variants was on the categorical structure of the dialectica categories and the way in which they model linear logic. While this is clearly important and interesting, it is not within the scope of the current exposition, which focuses solely on the construction itself. Similarly, I will not address variants such as the Diller-Nahm interpretation.

Second, I have not attempted to say much on the history of the subject, nor to be complete in the references. Because of the expositional nature of this paper there appear various results which are either standard, a variation on a standard result, or an explicitation of a result which is implicit in the literature. Whilst trying to avoid cluttering the exposition with too many literature references, I have tried to give credit where that is due, and I hope that I will be forgiven for possible omissions.

1.3. Overview. The paper is structured in the following manner. We begin in Section 2 with a quick review of the theory of fibrations. Readers who are already familiar with fibrations can safely skip this section.

Section 3 deals with notions of quantification in fibrations. We first explain in some detail how quantification is handled on the fibrational level and introduce the main kind of quantification we shall study in the rest of the paper, namely simple (co)products. This is all standard material. Next, in Section 3.2, we consider constructions which freely add quantification to a fibration, starting with the easiest such, namely the families construction. This allows us to reformulate these matters in monadic terms and sets us up for a better understanding of subsequent matters. After that (Section 3.3) we investigate one of the main constructions of interest, namely adding simple existential quantification. Because this is one of the ingredients for the dialectica construction we spell out the details of this construction, and show that it naturally comes equipped with all the structure needed to make it a pseudomonad; a formal proof is given in the Appendix. The dual construction for adding simple universal quantification is briefly described in Section 3.4; this is the second main ingredient for the dialectica construction. Finally, Section 3.5 describes a variation giving rise to a decomposition of the constructions described above into a linear part followed by a co-Kleisli construction.

Section 4 is the heart of the matter: it first introduces the dialectica categories, and describes them in terms of the monads obtained in the previous section. Then we introduce a (pseudo-)distributive law which explains why the dialectica construction can be equipped with the structure of a pseudomonad (Section 4.2), and characterize the algebras for the dialectica monad. The proof of this central fact is given in Section 4.3.

We end (Section 5) with a brief discussion of several aspects which could not be included in this paper and with some suggestions for further exploration.

1.4. Acknowledgments. As often happens, the present paper is really the consequence of the author trying to understand someone else's work. In this case, it was Bodil Biering's PhD thesis [5] which led me to formulate matters in the way presented here, and to explore various sidelines and related issues. I thank Phil Scott and Michael Warren for their stimulating comments, remarks and questions; I've benefited greatly from talking to various people, including Bodil Biering, Lars Birkedal, Richard Garner, Claudio Hermida, Martin Hyland, Pino Rosolini and Alex Simpson. To Valeria de Paiva I'm grateful for urging me to put my thoughts on paper. Finally, I am greatly indebted to the anonymous referees whose thoughtful and detailed reports have been very helpful in improving the focus and presentation of the material.

2. Background

In this section I guide the reader through the basic theory of fibred categories. I should stress that the aim here is not to provide a comprehensive account of the theory, but to introduce just enough to keep this paper self-contained and to give the reader sufficient insight into the material to be able to follow the rest of the paper. A more thorough introductory text to the subject is [24]; for a lively programmatic discussion, see [4]; finally, there is the textbook [16].

2.1. Basic theory and examples. Fibred category theory is a convenient setting for modeling mathematical situations in which one collection of objects depends on, or is *indexed*, or *fibred* over another. We begin by giving a concrete and elementary definition, and then turn to some instructive examples. Along the way we introduce notation and terminology.

Definition 2.1. Let $\mathsf{p}\colon \mathsf{E} \to \mathsf{B}$ be a functor.

(i) A morphism v in E is *vertical* if $\mathsf{p}(v) = 1$

(ii) For an object I in B, the *fibre of* p *over* I is the subcategory of E consisting of those objects X for which $\mathsf{p}X = I$ and with vertical morphisms between them. We denote this category by $\mathsf{p}^{-1}(I)$, or by E_I when p is understood.

If $\mathsf{p}X = I$ we say that X is an object over I. Similarly, if $\mathsf{p}v = f$, we call v a morphism over f.

Definition 2.2 (Cartesian morphism; fibration). (i) Let $m\colon X \to Y$ be a morphism over $\mathsf{p}m = f$. We say that m is *Cartesian* (over f) if for every other map $n\colon Z \to Y$ and every factorization $\mathsf{p}(n) = fg$ there exists a unique morphism v over g for which $mv = n$. Diagrammatically:

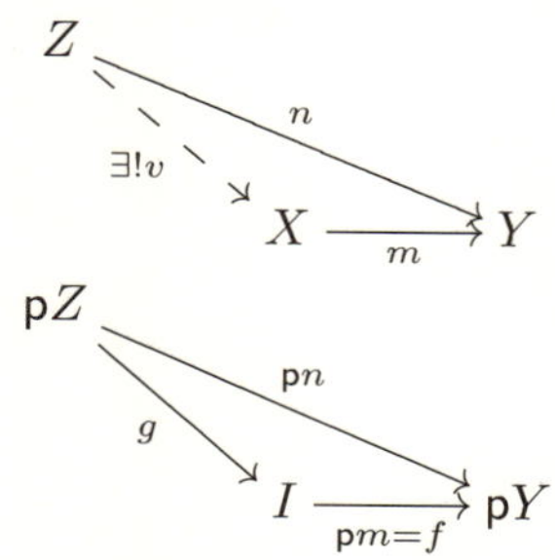

In this case, we call m a *Cartesian lifting* of f

(ii) The functor $\mathsf{p}\colon \mathsf{E} \to \mathsf{B}$ is a *fibration* if every map $f\colon I \to \mathsf{p}Y$ has a Cartesian lifting with codomain Y. We will usually refer to B as the *base category* and to E as the *total category* of p.

A more abstract way of saying this is: $\mathsf{p}\colon \mathsf{E} \to \mathsf{B}$ is a fibration precisely when each induced functor $\mathsf{p}/Y\colon \mathsf{E}/Y \to \mathsf{B}/\mathsf{p}Y$ has a right adjoint right inverse.

The following properties are readily verified from the definition and will frequently be used:

Lemma 2.3. *Let* $\mathsf{p}\colon \mathsf{E} \to \mathsf{B}$ *be a fibration.*

(i) *The composite of two Cartesian maps is again Cartesian.*
(ii) *If two maps m, m' are both Cartesian over f, then there is a unique vertical isomorphism v for which $mv = m'$.*
(iii) *Every morphism in E factors as a vertical map followed by a Cartesian map and this factorization is unique up to unique vertical isomorphism.*
(iv) *When two vertical maps v, v' in E satisfy $mv = mv'$ for some Cartesian map m, then $v = v'$.*
(v) *Any map in E which is both vertical and Cartesian is an isomorphism.*

We now present a couple of standard instructive examples; examples pertaining specifically to logic will be discussed below in Section 2.3.

Examples 2.4. (1) For any pair B, C of categories consider the projection $\mathsf{B} \times \mathsf{C} \to \mathsf{B}$. This is a fibration; maps in $\mathsf{B} \times \mathsf{C}$ of the form $(1, g)$ are vertical, while those of the form $(f, 1)$ are Cartesian. Fibrations of this form are sometimes called *constant*, because all fibres are isomorphic (to C).

(2) A fibration p is called *discrete* when the fibre categories are discrete categories, i.e., sets. For such p, the assignment $I \mapsto \mathsf{p}^{-1}(I)$ is the object part of a presheaf on B. The Grothendieck construction (see Section 2.2 below) provides an inverse to this construction, so that discrete fibrations are essentially the same thing as presheaves.

(3) For any category B, let $\mathsf{B}^{\to}$ be the arrow category whose objects are maps $f\colon X \to Y$ and whose morphisms are commutative squares

$$\begin{array}{ccc} X & \xrightarrow{f} & Y \\ {\scriptstyle m}\downarrow & & \downarrow{\scriptstyle n} \\ X' & \xrightarrow[f']{} & Y' \end{array}$$

The codomain functor cod: $\mathsf{B}^{\to} \to \mathsf{B}$ is a fibration (called the *codomain fibration* on B) precisely when B has pullbacks; the Cartesian maps are the pullback squares. Later we shall consider subfibrations of the codomain fibration.

The following elementary facts are easily verified.

Lemma 2.5. (i) *Fibrations are closed under composition, i.e., when* p *and* q *are fibrations, then so is* pq.
(ii) *Fibrations are preserved by pullback, i.e., when*

$$\begin{array}{ccc} \mathsf{E}' & \xrightarrow{\overline{F}} & \mathsf{E} \\ {\scriptstyle \mathsf{p}'}\downarrow & & \downarrow{\scriptstyle \mathsf{p}} \\ \mathsf{B}' & \xrightarrow[F]{} & \mathsf{B} \end{array}$$

is a pullback square in $\mathfrak{Cat}$ *and* p *is a fibration, then so is* p'.

We end this section with a brief mention of *opfibrations*. A functor $\mathsf{p}\colon \mathsf{E} \to \mathsf{B}$ is an opfibration when $\mathsf{p}^{\mathrm{op}}\colon \mathsf{E}^{\mathrm{op}} \to \mathsf{B}^{\mathrm{op}}$ is a fibration. Explicitly, this means that every map $f\colon \mathsf{p}X \to J$ in B has an *op-Cartesian lifting* $\tilde{f}\colon X \to f_!X$, which is initial in the sense that for any other map $g\colon X \to Y$ and any factorization $\mathsf{p}(g) = mf$ there is a unique map $f_!X \to Y$ over m making

$$\begin{array}{ccc} & & Y \\ & {\scriptstyle g}\nearrow & \uparrow{\scriptstyle \exists!} \\ X & \xrightarrow[\tilde{f}]{} & f_!X \end{array}$$

commute.

2.2. Cloven fibrations. Being a fibration is a property of a functor, as opposed to additional structure: Cartesian liftings are required to exist, but a fibration does not come equipped with a specific choice of Cartesian liftings. The notion of an indexed category, defined below, is structural, and may be regarded as giving a *presentation* of a fibration by explicitly specifying Cartesian liftings.

Definition 2.6. A B-*indexed category* is a pseudofunctor $P\colon \mathsf{B}^{\mathrm{op}} \to \mathfrak{Cat}$. Explicitly, this gives for each object I of B a category $P(I)$, for each $f\colon I \to J$ a functor $P(f)\colon P(J) \to P(I)$ (called a *reindexing functor*), together with coherence isomorphisms $P(f)P(g) \cong P(gf)$ and $P(1) \cong 1$, subject to coherence axioms.

Moreover, P is called *strict* if these coherence isomorphisms are identities (in which case P can be regarded simply as a functor $P : \mathsf{B}^{\mathrm{op}} \to \mathfrak{Cat}$).

Any B-indexed category P gives rise to a fibration over B, as follows.

Construction 2.7 (Grothendieck construction). Let $\int_{\mathsf{B}} P$ be the category with objects (I, x) where $x \in \mathrm{Ob}(P(I))$. A morphism $(I, x) \to (J, j)$ is a pair (f, m) where $f\colon I \to J$ and $m\colon x \to P(f)(y)$. It is easily seen that maps of the form $(f, 1)$ are Cartesian, and that morphisms of the form $(1, m)$ are vertical.

The projection $\pi_P\colon \int_{\mathsf{B}} P \to \mathsf{B}$ is a fibration, and by construction it is endowed with a natural choice of Cartesian liftings: the lift of $f\colon I \to J = \pi_P(J, x)$ may be taken to be $(f, 1)\colon (I, P(f)(x)) \to (J, x)$. Such fibrations are called *cloven* (and a choice of Cartesian liftings is called a *cleavage*). If P is a strict functor, then the cleavage is functorial, meaning that the chosen Cartesian liftings compose and that the chosen lifting of identities are again identities. Fibrations with a functorial cleavage are called *split*.

In case a fibration is equipped with a cleavage, we will denote the chosen Cartesian lifting of $f\colon I \to \mathsf{p}Y$ by $\bar{f}\colon f^*Y \to Y$, and refer to f^*Y as the *reindexing of Y along f*. Similarly, for a vertical morphism $p\colon Y \to Z$, we denote by f^*p the unique mediating map in

$$\begin{array}{ccc} f^*Y & \xrightarrow{\bar{f}} & Y \\ {\scriptstyle f^*p}\downarrow & & \downarrow{\scriptstyle p} \\ f^*Z & \xrightarrow[\bar{f}]{} & Z. \end{array}$$

We shall be sloppy about one aspect concerning cleavages here: even though a cleavage is additional structure on a fibration, we omit it from the notation, since we shall not be dealing with fibrations equipped with different cleavages at the same time.

For a detailed treatment of how the above construction can be extended to a 2-equivalence between indexed categories and cloven fibrations, see [16, 24].

We mention one construction which is easily understood in terms of indexed categories and which we will use later in the paper, namely that of the *opposite* of a fibration; it should not be confused with the opposite of a functor.

Definition 2.8 (Opposite of a fibration). For a given B-indexed category $P\colon \mathsf{B}^{\mathrm{op}} \to \mathfrak{Cat}$ corresponding to a cloven fibration $\int_{\mathsf{B}} P = \mathsf{p}$, composition with the involution $(-)^{\mathrm{op}} : \mathfrak{Cat} \to \mathfrak{Cat}$, gives a new indexed category $(-)^{\mathrm{op}} \circ P : \mathsf{B}^{\mathrm{op}} \to \mathfrak{Cat}$. Now define the fibration $\mathsf{p}^{\mathrm{op}} \stackrel{\mathrm{def}}{=} \int_{\mathsf{B}} (-)^{\mathrm{op}} P$.

The fibre over I of p^{op} is $\mathsf{p}^{-1}(I)^{\mathrm{op}}$, while its Cartesian maps are the same as those of p.

2.3. Logical aspects. We now turn to the logical perspective on fibred categories. We first discuss an important example.

Example 2.9 (Subobject fibration). For a set J, we may consider $\mathcal{P}(J)$, the powerset of J; since $\mathcal{P}(J)$ is partially ordered by inclusion, we may regard it as a posetal category. When $f\colon I \to J$ is a function, we get an induced order-preserving function $f^*\colon \mathcal{P}(J) \to \mathcal{P}(I)$, sending $U \subseteq J$ to $f^*U = \{i \in I | f(i) \in J\}$. This defines a functor

$$\mathcal{P}\colon \mathsf{Set}^{\mathrm{op}} \to \mathfrak{Cat}$$

and hence a fibration denoted $\mathsf{Sub}_{\mathsf{Set}} \to \mathsf{Set}$ called the *subobject fibration*.

Remark 2.10. The above example can be generalized to any category B in which pullbacks of monomorphisms exist: for any object J define $\mathsf{Sub}_{\mathsf{B}}(J)$ to be the partial order of subobjects (equivalence classes of monics) of J, and define reindexing to be given by pulling back subobjects. Then we obtain a subobject fibration $\mathsf{Sub}_{\mathsf{B}} \to \mathsf{B}$. (In the case $\mathsf{B} = \mathsf{Set}$ we used the fact that $\mathcal{P}(I) \cong \mathsf{Sub}(I)$ to pick canonical representatives of subobjects.)

Let us think in logical terms about the subobject fibration $\mathsf{Sub}_{\mathsf{Set}} \to \mathsf{Set}$. An object A in the fibre over I is a subset of I, and hence can be regarded as a *predicate* on I. Similarly, the inclusion relation on subsets may be thought of as *entailments* between such predicates. Finally, for a function $f\colon J \to I$ and a predicate A on I, we may regard the predicate $f^*A = \{j \in J : f(j) \in A\}$ as the result of making the *substitution* $[f(j)/i]$ in $A(i)$.

We generalize this as follows: given a fibration $\mathsf{p}\colon \mathsf{E} \to \mathsf{B}$, we think of the base category B as a category of *types*, or *contexts*. When I is such a type, then the fibre category E_I is thought of as a category of *predicates* with free variables of type I. To stress this viewpoint, we sometimes write $\alpha(i)$ for such an object.

Furthermore, morphisms $\alpha(i) \to \beta(i)$ in the fibre E_I are thought of as *proofs* that α implies β. We denote such a proof by $\phi(i) : \alpha(i) \to \beta(i)$. Because we may have more than one proof of such an implication, the fibre category is generally not a pre-order.

What about Cartesian maps? Well, morphisms in B are thought of as *terms*, or as *context morphisms*; more specifically, a morphism $f\colon I \to J$ may be regarded as a term of type J with free variables from I (in context I). When $\alpha(j)$ is a predicate over J, then the reindexing of $\alpha(j)$ along f is the predicate $\alpha(fi)$ over I obtained by substituting the term f for j in $\alpha(j)$; the Cartesian lifting $\bar{f}$ of f then relates $\alpha(j)$ to its substitution instance $\alpha(fi)$.

$$\begin{array}{ccc} \alpha(fi) & \xrightarrow{\bar{f}} & \alpha(j) \\ & & \\ I & \xrightarrow[f]{} & J \end{array} \qquad\qquad \begin{array}{c} \mathsf{E} \\ \downarrow{\scriptstyle \mathsf{p}} \\ \mathsf{B} \end{array}$$

It is important to keep in mind that this logical notation for fibrations suppresses coherence data. For example, given a predicate $\alpha(j, x)$ in two variables and maps $f\colon I \to J$ and $g\colon Y \to X$, we could first form $\alpha(fi, x)$ and then $\alpha(fi, gy)$, or the other way around. The notation suggests that the two resulting predicates are identical, whereas in general we only have a vertical isomorphism between the two.

Another class of examples which is relevant to the theme of this paper arises by considering fibrations built from the syntax of a theory $\mathbb{T}$ (at least regular). Good concrete examples to keep in mind are Heyting Arithmetic or Gödel's system T.

From such a theory we first build a base category whose objects are the variable contexts of the theory; the morphisms in the base are the (equivalence classes of) provably functional relations of the theory.

Next, we build a fibration over this base by taking the objects in the fibre over a type T to be the formulae of $\mathbb{T}$ with free variable from T. Reindexing is given by substitution. Such a fibration is usually referred to as the *syntactic fibration* associated to the theory $\mathbb{T}$; see [16] for more information.

Remark 2.11. This paper is mainly about quantification in fibrations; however, it seems strange to discuss the logical perspective and not say a word about the interpretation of the propositional connectives. In a nutshell: in order for a fibration to interpret a certain propositional connective, the fibres need to have appropriate categorical structure and this structure needs to be preserved by reindexing. For example, to interpret conjunction the fibres should have Cartesian products, and reindexing should preserve these products. Note that in a posetal fibration this just means that the fibres have binary meets preserved by reindexing. Similarly, disjunction is interpreted using fibrewise coproducts, implication using fibred exponentials, and so on.

2.4. 2-categorical aspects. We now describe how fibrations, as well as fibrations over a fixed base B naturally form a 2-category. We will not worry about size issues, although typically we will have in mind that the base is small. After that, we expore some properties of this 2-category which will be used in the rest of the paper, in particular in connection to coherence matters.

We first define the 2-category $\mathfrak{CFib}$ of cloven fibrations over arbitrary base as follows. The objects of $\mathfrak{CFib}$ are cloven fibrations $\mathsf{p}\colon \mathsf{E} \to \mathsf{B}$. For objects $\mathsf{p}\colon \mathsf{E} \to \mathsf{B}$ and $\mathsf{q}\colon \mathsf{D} \to \mathsf{B}'$, a 1-cell is a commutative square

$$
\begin{array}{ccc}
\mathsf{D} & \xrightarrow{F_0} & \mathsf{E} \\
{\scriptstyle \mathsf{q}}\downarrow & & \downarrow{\scriptstyle \mathsf{p}} \\
\mathsf{B}' & \xrightarrow[F]{} & \mathsf{B}
\end{array}
\tag{2.1}
$$

in which F_0 is a *fibred functor*, in the sense that it sends q-Cartesian maps to p-Cartesian maps. It is *not* required that F_0 preserves the cleavage on the nose.

Finally, given two such 1-cells (F, F_0) and (G, G_0) from q to p, a 2-cell from (F, F_0) to (G, G_0) is a pair of natural transformations $\phi\colon F \to G$, $\phi_0\colon F_0 \to G_0$, where ϕ_0 lies over ϕ, in the sense that for each X in D, the component $(\phi_0)_X$ is sent by p to $\phi_{\mathsf{q}X}$.

There is a forgetful 2-functor $\mathfrak{CFib} \to \mathfrak{Cat}$ which sends $\mathsf{p}\colon \mathsf{E} \to \mathsf{B}$ to the base B. When we take the fibre of this forgetful functor at a base category B, we obtain the 2-category $\mathfrak{CFib}(\mathsf{B})$ of cloven fibrations over B, fibred functors and vertical natural transformations.

Definition 2.12 (Fibred functor, fibred natural transformation). Let $\mathsf{p}\colon \mathsf{E} \to \mathsf{B}$ and $\mathsf{q}\colon \mathsf{D} \to \mathsf{B}$ be functors.

(i) A functor $F\colon \mathsf{E} \to \mathsf{D}$ over B is called a *fibred functor*[2] if it preserves Cartesian maps.

(ii) A natural transformation $\phi : F \Rightarrow F'$ between two functors over B is called *vertical* when all of its components are vertical maps, i.e., when $\mathsf{q}(\phi_X) = 1$ for each object X of E.

To summarize: $\mathfrak{CFib}(\mathsf{B})$ is the 2-category with

0-Cells: cloven fibrations $\mathsf{p}\colon \mathsf{E} \to \mathsf{B}$
1-Cells: fibred functors
2-Cells: vertical natural transformations.

Remark 2.13. The forgetful 2-functor $\mathfrak{CFib}(\mathsf{B}) \to \mathfrak{Cat}/\mathsf{B}$ is in fact 2-monadic: the free (split) fibration on a functor $\mathsf{p}\colon \mathsf{E} \to \mathsf{B}$ is the projection $\mathsf{B}/\mathsf{p} \to \mathsf{B}$; moreover, "fibration structure" (i.e., a cleavage) on a functor $\mathsf{p}\colon \mathsf{E} \to \mathsf{B}$ is precisely a pseudo-algebra structure for this 2-monad (strict algebras correspond to splittings). See [23] for a detailed treatment of this approach to fibrations.

The following elementary fact will often be used:

---

[2]Some authors call this a *Cartesian* functor, but we will not follow this.

Lemma 2.14. *Let F be a functor over B from $\mathsf{p}\colon \mathsf{E} \to \mathsf{B}$ to $\mathsf{q}\colon \mathsf{D} \to \mathsf{B}$, where p, q are cloven. Then F is a fibred functor if and only if each of the unique mediating maps*

$$\begin{array}{ccc} F(f^*\alpha) & & \\ \downarrow & \searrow & \\ f^*(F\alpha) & \longrightarrow & F\alpha \end{array}$$

is an isomorphism.

This encourages us to extend the logical notation for fibrations to include functors. For example, we write $F\alpha(i,x)$ for the image under the fibred functor F of the predicate $\alpha(i,x)$; then the expression $F\alpha(fj,x)$ is ambiguous, but by the above lemma the two possible interpretations are coherently isomorphic.

Similarly, for a fibred transformation $\tau\colon F \Rightarrow G$, we note that the square

$$\begin{array}{ccc} F(f^*\alpha) & \xrightarrow{\tau_{f^*\alpha}} & G(f^*\alpha) \\ {\scriptstyle\cong}\downarrow & & \downarrow{\scriptstyle\cong} \\ f^*(F\alpha) & \xrightarrow[f^*\tau_\alpha]{} & f^*(G\alpha) \end{array}$$

commutes.

3. Quantification in Fibrations

In this section we first explain various notions of quantification in more detail; then we turn to the process of freely adding existential quantification to a given fibration, showing that this process is monadic in a suitable 2-categorical sense. We do the same for universal quantification, which is then a relatively easy dualization. Finally, we discuss the linear variants of these free constructions.

From now on all fibrations are equipped with a cleavage.

3.1. Sums and products. The starting point for the interpretation of the quantifiers in the fibrational setting is the observation that the quantifiers $\forall, \exists$ may be characterized in terms of adjointness relations. Let me explain this: consider a predicate $\alpha(x,y)$ over $X \times Y$. Existentially quantifying over the variable y results in the predicate $\exists y.\alpha(x,y)$, which now only has free variable x and hence must live in the fibre over X. Thus, this form of quantification should be an operation $\exists_Y\colon \mathsf{p}^{-1}(X \times Y) \to \mathsf{p}^{-1}(X)$. The main observation, due to Lawvere, is now that this operation is *left adjoint* to the reindexing functor $\mathsf{p}^{-1}(X) \to \mathsf{p}^{-1}(X \times Y)$ (this functor is also referred to as a *weakening* functor; it introduces a dummy variable). Similarly, universal quantification can be expressed in terms of right adjoints to reindexing functors.

Suppose then that B is a category with (chosen) pullbacks. For an arbitrary map $f\colon I \to J$ in B, we may ask whether the functor $f^*\colon \mathsf{p}^{-1}(J) \to \mathsf{p}^{-1}(I)$ has a left adjoint. If it does, we think of this adjoint as giving *existential quantification along f* and denote it by $\exists_f\colon \mathsf{p}^{-1}(I) \to \mathsf{p}^{-1}(J)$.

However, merely asking that p has left adjoints to reindexing functors is not sufficient: we have to make sure that these left adjoints behave well with respect to substitution. To this end, consider a pullback square in B:

$$(3.1)\qquad \begin{array}{ccc} I & \xrightarrow{f} & J \\ {\scriptstyle h}\downarrow & & \downarrow{\scriptstyle g} \\ V & \xrightarrow[k]{} & W \end{array}$$

Given a predicate on J, we may now either first make the substitution $[f(i)/j]$ and then quantify along h, or we first quantify along g and then substitute $[k(v)/w]$. We want those two to agree. Formally, there is a canonical comparison map

$$\kappa\colon \exists_h f^* \to k^* \exists_g$$

obtained as follows: first take the unit $1 \to g^*\exists_g$, and apply to it the functor f^* to get a map $f^* \to f^*g^*\exists_g$. Then use the canonical isomorphism $h^*k^* \cong f^*g^*$ to get a map $f^* \to h^*k^*\exists_g$. Finally, transpose this along the adjunction $\exists_h \dashv h^*$ to get the desired $\exists_h f^* \to k^*\exists_g$. We say that p satisfies the *Beck – Chevalley Condition* (for the given pullback square) when this mediating map is an isomorphism.

Definition 3.1. We say that a fibration p over a category B with pullbacks has *existential quantification* when all reindexing functors have left adjoints satisfying the Beck – Chevalley Condition (BCC). Dually, p has *universal quantification*, when reindexing functors have right adjoints satisfying the BCC.

For fibrations with existential quantification, observe that for $f\colon I \to J$ in B and $\alpha \in \mathsf{p}^{-1}(I)$, there is a canonical map

$$\tilde{f}\colon \alpha \to \exists_f(\alpha)$$

obtained by precomposing the unit $\alpha \to f^*\exists_f(\alpha)$ with the Cartesian map $\bar{f}\colon f^*\exists_f(\alpha) \to \exists_f(\alpha)$. This map is op-Cartesian over f, a fact which we will exploit later.

We have already seen examples of fibrations with existential quantification: the archetypical such is the codomain fibration $\mathsf{B}^{\to} \to \mathsf{B}$. Given a map $f\colon I \to J$ in B, the left adjoint to the pullback functor $f^*\colon \mathsf{B}/J \to \mathsf{B}/I$ is usually denoted by Σ_f, and acts by composition with f. (Moreover, the codomain fibration has universal quantification whenever the base category is locally Cartesian closed, see, e.g, [16]).

For a different example, consider a complete Heyting algebra (frame, locale) Ω, and consider the assignment $I \mapsto \Omega^I = \mathsf{Set}[I, \Omega]$. Reindexing is given by precomposition. The reindexing functors $f^*\colon \Omega^J \to \Omega^I$ have both adjoints, given by (for $\alpha\colon I \to \Omega$)

$$\exists_f(\alpha)(j) = \bigvee_{f(i)=j} \alpha(i) \qquad \forall_f(\alpha)(j) = \bigwedge_{f(i)=j} \alpha(j)$$

Now that we understand this general form of quantification, it is easy to consider more restricted versions by only asking for adjoints to certain reindexing functors. We will be mostly concerned with reindexing along projections. This is usually referred to as *simple quantification*. Explicitly:

Definition 3.2. A fibration p over a category B with finite products has *simple existential quantification*, or *simple coproducts*, when all weakening functors have left adjoints satisfying the Beck – Chevalley Condition (BCC) for pullback squares of the form

$$\begin{array}{ccc} I\times X & \xrightarrow{\pi_I} & I \\ {\scriptstyle f\times X}\Big\downarrow & & \Big\downarrow{\scriptstyle f} \\ J\times X & \xrightarrow[\pi_J]{} & J. \end{array}$$

Dually, p has *simple universal quantification*, or *simple products*, when weakening functors have right adjoints satisfying the BCC.

Remark 3.3. The idea of selecting a particular class of maps to quantify along (i.e., by specifying a subcategory of $\mathsf{B}^{\to}$ which is pullback-stable) allows us to define quite general notions of quantification; this has been made more precise via notions such as CT-structure and comprehension categories, see [16].

What does it mean for a fibred functor to preserve coproducts? Consider two fibrations with coproducts and a fibred functor F between them. Given a morphism $f: I \to J$ in the base and a predicate α over I, we have a unique vertical comparison map induced by op-Cartesianness of $\tilde{f}$.

$$\begin{array}{ccc} & & F(\exists_f\alpha) \\ & {\scriptstyle F\tilde{f}}\nearrow & \Big\uparrow{\scriptstyle \psi_{f,\alpha}} \\ F\alpha & \xrightarrow[\tilde{f}]{} & \exists_f(F\alpha) \end{array}$$

Definition 3.4. A fibred functor F *preserves coproducts* when each of the comparison morphisms $\psi_{f,\alpha}$ is an isomorphism.

Of course, we have a similar definition for preservation of products, and it is now also clear what is meant by a fibred functor preserving simple (co)products. We introduce the following 2-categories: $\mathfrak{CFib}_{\forall}(\mathsf{B})$ is the 2-category with

0-Cells: Cloven fibrations over B with (chosen) simple products
1-Cells: Simple product-preserving fibred functors
2-Cells: Fibred natural transformations

and similarly $\mathfrak{CFib}_{\exists}(\mathsf{B})$ has

0-Cells: Cloven fibrations over B with (chosen) simple coproducts
1-Cells: Simple coproduct-preserving fibred functors
2-Cells: Fibred natural transformations

We will extend the logical notation for fibrations to include the quantifiers, writing $\exists x.\alpha(i,x)$ for the result of applying $\exists_X$ to $\alpha(i,x)$. Then an expression such as $\exists x.\alpha(fj,x)$ is again ambiguous, but the BCC guarantees that both possible interpretations are coherently isomorphic.

3.2. The families construction. We now turn to the question of how to freely add existential quantification to a given fibration. The answer is easiest for adding all coproducts; we first briefly review this, as it makes the subsequent constructions clear and gives us the opportunity to point out some structural features.

First recall that cod: $\mathsf{B}^{\to} \to \mathsf{B}$ (for B with pullbacks) is a fibration with existential quantification along all maps. In fact, it is the free such on the terminal fibration $1: \mathsf{B} \to \mathsf{B}$ on B: given any fibration with coproducts $\mathsf{E} \xrightarrow{\mathsf{p}} \mathsf{B}$ and any fibred functor F from the terminal to p, there is an induced fibred functor from the codomain fibration to p which sends an arrow $f\colon I \to J$ to $\exists_f(FI)$.

We now exploit this by defining, for general p, a new fibration $\mathfrak{Fam}(\mathsf{p})$ as follows:

$$\begin{array}{ccc} \mathfrak{Fam}(\mathsf{E}) & \longrightarrow & \mathsf{E} \\ \downarrow & & \downarrow{\scriptstyle \mathsf{p}} \\ \mathsf{B}^{\to} & \xrightarrow{\text{dom}} & \mathsf{B} \\ \downarrow{\scriptstyle \text{cod}} & & \\ \mathsf{B} & & \end{array} \qquad \mathfrak{Fam}(\mathsf{p})\colon \mathfrak{Fam}(\mathsf{E}) \to \mathsf{B}$$

Here, the square is a pullback. Thus, we label the objects and arrows of the free structure on the terminal by objects of E; explicitly $\mathfrak{Fam}(\mathsf{E})$ is the category with

Objects: pairs $(I \xrightarrow{f} J, \alpha)$ where $\alpha \in \mathsf{p}^{-1}(I)$

Arrows: an arrow from $(I \xrightarrow{f} J, \alpha)$ to $(H \xrightarrow{g} K, \beta)$ consists of a triple (p, q, ϕ), where $gp = qf$ is a commutative square in B and $\phi\colon \alpha \to \beta$ is a map in E over p.

One may think of an object $(f\colon I \to J, \alpha)$ as the predicate $\exists_f \alpha$; alternatively, one regards it as an I-indexed family of objects of E, whence the name of the construction.

Using the universal property of the pullback, this construction is easily seen to be 2-functorial. Moreover, one can construct a unit and a multiplication map. Explicitly, the unit is given by

$$\eta_{\mathsf{p}}\colon \mathsf{E} \to \mathfrak{Fam}(\mathsf{E}); \qquad \alpha \in \mathsf{p}^{-1}(I) \mapsto (I \xrightarrow{1} I, \alpha)$$

and the multiplication is

$$\mu_{\mathsf{p}}\colon \mathfrak{Fam}^2(\mathsf{E}) \to \mathfrak{Fam}(\mathsf{E}); \qquad (I \xrightarrow{f} J, (J \xrightarrow{g} K, \alpha)) \mapsto (I \xrightarrow{gf} K, \alpha).$$

The monad laws are just the composition laws in B.

Theorem 3.5. *The construction* $\mathsf{p} \mapsto \mathfrak{Fam}(\mathsf{p})$ *extends to a strict* 2*-monad on* $\mathfrak{CFib}(\mathsf{B})$, *which has the KZ property. The* 2*-category of pseudo-algebras is* 2*-equivalent to the* 2*-category of cloven fibrations with chosen coproducts and coproduct-preserving fibred functors.*

Remark 3.6. Let me first comment on the KZ-property: a pseudomonad is said to be *Kock – Zöberlein* (see [19–21]) when pseudo-algebra structures on p are left adjoint to the unit at p. (To show that a monad has this property, it suffices to test it for free algebras). This has useful consequences: first, when p admits a pseudo-algebra structure it is essentially unique. Second, any morphism between algebras is automatically a lax algebra morphism in a unique way, and is a pseudo algebra map whenever the unique mediating 2-cell is an isomorphism. And third, any 2-cell between lax algebra morphisms is automatically an algebra 2-cell. The dual notion is called a co-KZ monad, and is characterized by algebra structure maps being right adjoint to the unit.

Now let me sketch why a pseudo-algebra structure on p amounts to having chosen coproducts in p. A pseudo-algebra structure on p is, by the KZ property, a (fibred!) left adjoint to the unit $\eta_{\mathsf{p}}\colon \mathsf{p} \to \mathfrak{Fam}(\mathsf{p})$. Given such left adjoint K we thus have adjunctions for each fibre

$$\mathsf{E}_I \underset{\eta_I}{\overset{K_I}{\underset{\longrightarrow}{\overset{\longleftarrow}{\bot}}}} \mathfrak{Fam}(\mathsf{E})_I$$

where the counit is an isomorphism, and hence η_I is fully faithful. It is straightforward to verify that the fact that K is a fibred functor translates into the BCC for the local adjunctions $K_I \dashv \eta_I$, in the sense that given a map $f\colon I \to J$ the canonical natural transformation $K_I f^* \to f^* K_J$ is an isomorphism (this map depends on the adjunction morphisms). We now define existential quantification $\exists_f\colon \mathsf{E}_I \to \mathsf{E}_J$ in p along f to be the composite

$$\mathsf{E}_I \xrightarrow{\eta_I} \mathfrak{Fam}(\mathsf{E})_I \xrightarrow{\exists_f} \mathfrak{Fam}(\mathsf{E})_J \xrightarrow{K_J} \mathsf{E}_J; \qquad \alpha \mapsto K(I \xrightarrow{f} J, \alpha).$$

The adjointness $\exists_f \dashv f^*$ for p is now immediate from the following sequence of bijections:

$$\begin{array}{cl} K_J\exists_f\eta_I\alpha \to \beta & \\ \hline \exists_f\eta_I\alpha \to \eta_J\beta & \text{by } K_I \dashv \eta_I \\ \hline \eta_I\alpha \to f^*\eta_J\beta & \text{by } \exists_f \dashv f^* \text{ for } \mathfrak{Fam}(\mathsf{p}) \\ \hline \eta_I\alpha \to \eta_I f^*\beta & \eta \text{ is fibred} \\ \hline \alpha \to f^*\beta & \eta_I \text{ is fully faithful} \end{array}$$

The BCC for the K_I easily implies that for the $\exists_f$. Note that the adjunction morphisms for $K \dashv \eta$ determine the adjunction morphisms of $\exists_f \dashv f^*$.

Conversely, when we have left adjoints to reindexing functors then we can use the same definition

$$K_I\colon \mathfrak{Fam}(\mathsf{E})_I \to \mathsf{E}_I; \qquad (J \xrightarrow{f} I, \alpha) \mapsto \exists_f(\alpha)$$

to locally construct a left adjoint to the unit. The BCC now guarantees that these fit together to form a global adjoint $K \dashv \eta$. Details are straightforward.

3.3. The simple fibration, or adding simple coproducts. Next, we investigate adding simple coproducts to a fibration. We follow the pattern from the previous section. First of all, the codomain fibration gets replaced by a smaller fibration called the *simple fibration* over B. This is defined as follows:

Construction 3.7. The category $\mathfrak{Sum}(\mathsf{B})$ has:

Objects: pairs (I, X), where I and X are objects of B.

Morphisms: a map from (I, X) to (J, Y) is a pair $(f, f_0,)$, where
- $f\colon I \to J$ is a morphism in B
- $f_0\colon I \times X \to Y$ is a morphism in B.

Identities: the identity on (I, X) is the map $(1_I, \pi_X)$.

Composition: given $(f, f_0)\colon (I, X) \to (J, Y)$ and $(g, g_0)\colon (J, Y) \to (K, Z)$, the composite of these maps is defined to be the pair (h, h_0) where

$$h = gf; \qquad h_0 = g_0\langle f\pi_I, f_0\rangle; \qquad h_0(i, x) = g_0(fi, f_0(i, x)).$$

This category is fibred over B via the first projection. The fibre over I is also referred to as the *simple slice* over I. There will be no notational distinction made between $\mathfrak{Sum}(\mathsf{B})$ regarded as a fibration and as category.

We now extend this construction in the canonical way to general fibrations: given $\mathsf{E} \xrightarrow{\mathsf{p}} \mathsf{B}$, consider

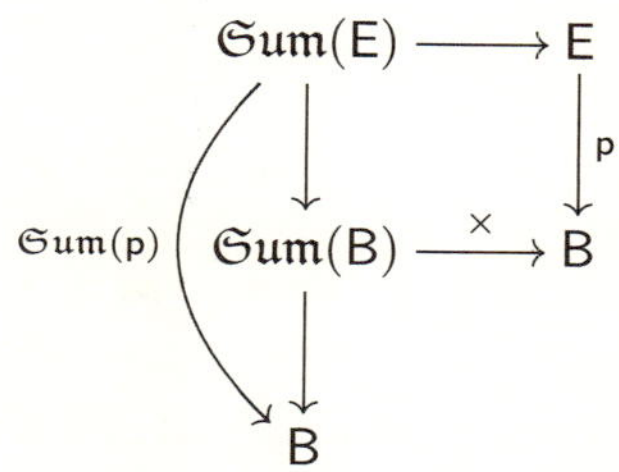

where the square is a pullback. The total category $\mathfrak{Sum}(\mathsf{E})$ may be described explicitly as follows:

Construction 3.8. Let $\mathsf{p}\colon \mathsf{E} \to \mathsf{B}$ be a fibration. The category $\mathfrak{Sum}(\mathsf{E})$ has:

Objects: triples (I, X, α), where I and X are objects of the base B, and where $\alpha \in \mathsf{p}^{-1}(I \times X)$ is an object in the fibre over $I \times X$.

Morphisms: a map from (I, X, α) to (J, Y, β) is a triple (f, F, ϕ), where
- $f\colon I \to J$ is a morphism in B
- $f_0\colon I \times X \to Y$ is a morphism in B
- $\phi = \phi(i, y)\colon \alpha(i, x) \to \beta(f(i), f_0(i, x))$ is a morphism in the fibre over $I \times X$.

The rest of the structure is straightforward, and it is readily verified that $\mathsf{p} \mapsto \mathfrak{Sum}(\mathsf{p})$ is 2-functorial.

Theorem 3.9. *The 2-functor $\mathsf{p} \mapsto \mathfrak{Sum}(\mathsf{p})$ can be endowed in a canonical way with the structure of a pseudomonad on $\mathfrak{CFib}(\mathsf{B})$, which is a strict 2-monad when the product structure on B is strict. This pseudomonad is KZ, and its 2-category of pseudo-algebras is 2-equivalent to the 2-category $\mathfrak{CFib}_{\exists}$.*

Note first that in case products are strict in B, it is easy to see that we get a 2-monad: for then the collection of first projections is closed under composition, so all of the 2-monad structure is obtained by simply restricting that of the families monad.

Formally, to specify a pseudomonad structure on $\mathfrak{Sum}$ means to give the following data:

- a pseudonatural transformation $(\eta, \overline{\eta})\colon 1 \to \mathfrak{Sum}$
- a pseudonatural transformation $(\mu, \overline{\mu})\colon \mathfrak{Sum}^2 \to \mathfrak{Sum}$
- invertible modifications $\lambda\colon \mu \circ \mathfrak{Sum}\,\eta \to 1, \rho\colon \mu \circ \eta\,\mathfrak{Sum} \to 1$ and $\theta\colon \mu \circ \mu\,\mathfrak{Sum} \to \mu \circ \mathfrak{Sum}\,\mu$.

Moreover, this data is supposed to satisfy certain coherence conditions. Here I will simply give a direct description of this structure; in the Appendix the coherence will be discussed in some detail.

First, the unit for $\mathfrak{Sum}$. At a fibration p, this is the map which acts as follows on an object α and a morphism $\phi : \alpha \to \beta$ in E_I:

$$\eta_{\mathsf{p}}\colon \mathsf{E} \to \mathfrak{Sum}(\mathsf{E}); \qquad \alpha \mapsto (I, 1, \alpha'); \qquad \phi \mapsto (1_I, !, \phi')$$

where α' is the result of reindexing α along the unit isomorphism $I \times 1 \to I$, and ϕ' is the unique vertical map making

$$\begin{array}{ccc} \alpha' & \xrightarrow{\overline{\pi_I}} & \alpha \\ {\scriptstyle \phi'}\downarrow & & \downarrow{\scriptstyle \phi} \\ \beta' & \xrightarrow[\overline{\pi_I}]{} & \beta \end{array}$$

commute.

We see right away why we get a complication which was not present in the case of $\mathfrak{Fam}$ (and which also doesn't show up when products are strict): in order to get a well-defined object of $\mathfrak{Sum}(\mathsf{E})$, we need to move to an isomorphic fibre.

As a consequence, the unit is no longer strictly natural, but only pseudonatural: given $F \colon \mathsf{p} \to \mathsf{q}$, consider the square

$$\begin{array}{ccc} \mathsf{E} & \xrightarrow{\eta_{\mathsf{p}}} & \mathfrak{Sum}(\mathsf{E}) \\ {\scriptstyle F}\downarrow & \Downarrow \overline{\eta}_F & \downarrow{\scriptstyle \mathfrak{Sum}(F)} \\ \mathsf{D} & \xrightarrow[\eta_{\mathsf{q}}]{} & \mathfrak{Sum}(\mathsf{D}) \end{array} \tag{3.2}$$

The mediating vertical natural transformation $\overline{\eta}_F$ arises from the fact that F preserves reindexing up to unique vertical isomorphism: thus the component of $\overline{\eta}_F$ at α over I is the unique vertical isomorphism making

$$\begin{array}{ccc} F(\alpha') & & \\ \downarrow & \searrow^{F\overline{\pi_I}} & \\ (F\alpha)' & \xrightarrow[\overline{\pi_I}]{} & F\alpha \end{array}$$

commute (where $\pi_I \colon I \times 1 \to I$).

The $\overline{\eta}_F$ are the components of a natural transformation

$$\begin{array}{ccc} \mathfrak{CFib}(\mathsf{p},\mathsf{q}) & \xrightarrow{\mathfrak{Sum}} & \mathfrak{CFib}(\mathfrak{Sum}(\mathsf{p}), \mathfrak{Sum}(\mathsf{q})) \\ \Vert & \Downarrow \overline{\eta} & \downarrow \\ \mathfrak{CFib}(\mathsf{p},\mathsf{q}) & \longrightarrow & \mathfrak{CFib}(\mathsf{p}, \mathfrak{Sum}(\mathsf{q})) \end{array}$$

making $(\eta, \overline{\eta})$ a well-defined pseudonatural transformation $1 \to \mathfrak{Sum}$.

A similar story can be told for the multiplication. A typical object of $\mathfrak{Sum}^2(\mathsf{E})$ has the form $(I, X, (I \times X, U, \alpha))$, with $\alpha \in \mathsf{E}_{(I \times X) \times U}$. We may write (I, X, U, α) for such an object. The multiplication map μ_{p} acts as follows:

$$\mu(I, X, U, \alpha) = (I, X \times U, \alpha'),$$

where α' is now the reindexing of α along the associativity isomorphism. As for the unit, we can construct a family of natural transformations $\overline{\mu}$ making $(\mu, \overline{\mu})$ a pseudonatural transformation $\mathfrak{Sum}^2 \to \mathfrak{Sum}$.

The modifications λ, ρ, θ up to which the monad laws hold, are now easily seen to be the canonical isomorphisms

$$\lambda \colon (I, 1 \times X, \alpha') \to (I, X, \alpha)$$

$$\rho\colon (I, X \times 1, \alpha') \to (I, X, \alpha)$$
$$\theta\colon (I, X \times (U \times V), \alpha') \to (I, (X \times U) \times V, \alpha).$$

(In each of these, α' refers to the reindexing of α along the appropriate coherence isomorphism.)

3.4. Adding products. We now dualize the above results. First we give an explicit description:

Construction 3.10. Let $\mathsf{p}\colon \mathsf{E} \to \mathsf{B}$ be a cloven fibration. Construct a category $\mathfrak{Prod}(\mathsf{E})$ as follows:

Objects: triples (I, X, α), where I and X are objects of the base B, and where $\alpha \in \mathsf{p}^{-1}(I \times X)$ is an object in the fibre over $I \times X$.

Morphisms: a map from (I, X, α) to (J, Y, β) is a triple (f, f_0, ϕ), where
- $f\colon I \to J$ is a morphism in B
- $f_0\colon I \times Y \to X$ is a morphism in B
- $\phi = \phi(i, y)\colon \alpha(i, f_0(i, y)) \to \beta(f(i), y)$ is a morphism in the fibre over $I \times Y$.

The identity map on (I, X, α) is $(1_I, \pi_X, 1_\alpha)$. The composition of (f, f_0, ϕ): $(I, X, \alpha) \to (J, Y, \beta)$ and $(g, g_0, \psi)\colon (J, Y, \beta) \to (K, Z, \gamma)$ is given by (h, h_0, χ) where

$$h = gf\colon I \to K, \qquad h_0(i, z) = f_0(i, g_0(fi, z))\colon I \times Z \to X$$

and where $\chi(i, z)$ is the composite

$$\alpha(i, h_0(i, z)) = \alpha(i, f_0(i, g_0(fi, z))) \xrightarrow{\phi(i, g_0(fi, z))} \beta(fi, g_0(fi, z)) \xrightarrow{\psi(fi, z)} \gamma(gfi, z).$$

Proposition 3.11. *There is an isomorphism of fibrations*

$$\mathfrak{Prod}(\mathsf{p}) \cong \mathfrak{Sum}(\mathsf{p}^{\mathrm{op}})^{\mathrm{op}}.$$

This isomorphism is natural in p.

Here one has to recall that p^{op} stands for the *fibrewise opposite* of p as described in Definition 2.8.

Proof. The first statement of the theorem is easily checked by hand: both categories have the same objects. In the fibre over I, a map $(I, X, \alpha) \xrightarrow{(1, f_0, \phi)} (I, Y, \beta)$ of $\mathfrak{Sum}(\mathsf{p}^{\mathrm{op}})$ has $f_0\colon I \times X \to Y$ and ϕ: $\beta(i, f_0(i, x)) \to \alpha(i, x)$. But this is precisely a map $(I, Y, \beta) \to (I, X, \alpha)$ in $\mathfrak{Sum}(\mathsf{p})$.

The second statement is left as a straightforward exercise. □

As a consequence, $\mathfrak{Prod}$ is also a pseudomonad on $\mathfrak{CFib}(\mathsf{B})$. (This may of course also be verified directly along the same lines as for $\mathfrak{Sum}$.) We state:

Theorem 3.12. *The assignment* $\mathsf{p} \mapsto \mathfrak{Prod}(\mathsf{p})$ *can be endowed in a canonical way with the structure of a pseudomonad on* $\mathfrak{CFib}(\mathsf{B})$, *which is a strict 2-monad when the product structure on* B *is strict. This pseudomonad is co-KZ, and its 2-category of pseudo-algebras is 2-equivalent to the 2-category* $\mathfrak{CFib}_\forall$.

Because of this result, we are justified to think of an object (I, X, α) of $\mathfrak{Prod}(\mathsf{p})$ as the predicate $\forall x.\alpha(i, x)$.

3.5. Variations. In this section we make a brief excursion to the linear approximations of the monads described above.

Let us first look at the linear variant of the monad $\mathfrak{Sum}$. Explicitly, it is constructed as follows:

Construction 3.13. Let $\mathsf{p}\colon \mathsf{E} \to \mathsf{B}$ be a cloven fibration. Construct a category $\mathfrak{Sum}_L(\mathsf{p})$ as follows:

Objects: triples (I, X, α), where I and X are objects of the base B, and where $\alpha \in \mathsf{p}^{-1}(I \times X)$ is an object in the fibre over $I \times X$.

Morphisms: a map from (I, X, α) to (J, Y, β) is a triple (f, f_0, ϕ), where

- $f\colon I \to J$ is a morphism in B
- $f_0\colon X \to Y$ is a morphism in B
- $\phi = \phi(i, y) : \alpha(u, x) \to \beta(f(i), f_0(x))$ is a morphism in the fibre over $I \times Y$.

It is easy to see that this construction is the result of restricting the simple slice over B to the category with the same objects, but only those maps $(I, X) \to (J, Y)$ for which the second factor $f_0\colon I \times X \to Y$ is *independent* of X, i.e. factors as $f_0'\pi_X$ for some $f_0'\colon X \to Y$.

We have the expected result:

Theorem 3.14. *The assignment* $\mathsf{p} \mapsto \mathfrak{Sum}_L(\mathsf{p})$ *carries the structure of a pseudomonad on* $\mathfrak{CFib}(\mathsf{B})$. *When products in* B *are strict, then* $\mathfrak{Sum}_L$ *is a strict 2-monad.*

All structure is inherited from $\mathfrak{Sum}$, and we have an inclusion of pseudomonads $\mathfrak{Sum}_L \to \mathfrak{Sum}$. However, it should be noted that $\mathfrak{Sum}_L$ does not have the KZ-property.

For the following result which further describes how $\mathfrak{Sum}_L$ relates to $\mathfrak{Sum}$, recall that a fibration p has *equality* when it has left adjoints to contraction functors $\delta^*\colon \mathsf{p}^{-1}(U \times U) \to \mathsf{p}^{-1}(U)$ subject to the Beck–Chevalley condition.

Theorem 3.15. *Assume that* p *has equality. Then there is a comonad* $!_{\mathsf{p}}$ *on* $\mathfrak{Sum}_L(\mathsf{p})$ *such that the co-Kleisli object for this comonad is* $\mathfrak{Sum}(\mathsf{p})$.

PROOF. First note that if p has equality, then so does $\mathfrak{Sum}_L(\mathsf{p})$: this is given by $(I \times I, X, \alpha) \mapsto (I, X, \exists_{\Delta \times X}(\alpha))$.

For the comonad on $\mathfrak{Sum}_L(\mathsf{p})$, we define $!_{\mathsf{p}}(I, X, \alpha) = (I, I \times X, \hat{\alpha})$, where $\hat{\alpha}(i, i', x) = \alpha(i, x) \wedge (i = i')$. Equivalently, $\hat{\alpha} = \exists_{\Delta \times X}(\alpha)$. It is now easily seen that a map $!_{\mathsf{p}}(I, X, \alpha) \to (J, Y, \beta)$ in $\mathfrak{Sum}_L(\mathsf{p})$ is precisely a map $(I, X, \alpha) \to (J, Y, \beta)$ in $\mathfrak{Sum}(\mathsf{p})$. □

What are the (pseudo-)algebras for $\mathfrak{Sum}_L$? These are the fibrations equipped with a notion of quantification (again denoted $\exists$) satisfying the following inference rule:

$$\frac{\alpha(i, x) \to \beta(fi, f_0x)}{\exists x.\alpha(i, x) \to \exists y.\beta(fi, y)}$$

There is a sense in which this structure may be thought of as linear quantification; see the comments in the last section.

We now turn to the dual results:

Construction 3.16. The category $\mathfrak{Prod}_L(\mathsf{p})$ has

Objects: triples (I, X, α), where I and X are objects of the base B, and where $\alpha \in \mathsf{p}^{-1}(I \times X)$ is an object in the fibre over $I \times X$.

Morphisms: a map from (I, X, α) to (J, Y, β) is a triple (f, F, ϕ), where
- $f \colon I \to J$ is a morphism in B
- $f_0 \colon Y \to X$ is a morphism in B
- $\phi = \phi(i, y) : \alpha(i, f_0(y)) \to \beta(f(i), y)$ is a morphism in the fibre over $I \times Y$.

Again, this category is fibred over the base B via the first projection. Because all the structure on $\mathfrak{Prod}$ restricts to $\mathfrak{Prod}_L$ (or using the isomorphism $\mathfrak{Prod}_L(\mathsf{p}) \cong (\mathfrak{Sum}_L(\mathsf{p}^{\mathrm{op}}))^{\mathrm{op}}$), the following is immediate:

Theorem 3.17. *The assignment* $\mathsf{p} \mapsto \mathfrak{Prod}_L(\mathsf{p})$ *carries the structure of a pseudomonad on the category* $\mathfrak{CFib}(\mathsf{B})$ *of fibrations over B.*

Under the assumption that the base category B is Cartesian closed, the inclusion $\mathfrak{Prod}_L(\mathsf{p}) \to \mathfrak{Prod}(\mathsf{p})$ is part of a co-Kleisli adjunction for a comonad ! on $\mathfrak{Prod}(\mathsf{p})$. This comonad is defined by

$$!_{\mathsf{p}}(I, X, \alpha) = (I, X^I, \tilde{\alpha}); \qquad \tilde{\alpha}(i, F) = \alpha(i, Fi),$$

where $\tilde{\alpha}$ is the obvious reindexing of α.

This indeed makes $!_{\mathsf{p}}$ into a fibred functor; moreover, it is indeed a comonad: the counit

$$!_{\mathsf{p}}(I, X, \alpha) = (I, X^I, \tilde{\alpha}) \to (I, X, \alpha)$$

is induced by $X \to X^I$, the transpose of the projection. The comultiplication

$$!_{\mathsf{p}}(I, X, \alpha) = (I, X^I, \tilde{\alpha}) \to (I, (X^I)^I, \tilde{\tilde{\alpha}}) = !_{\mathsf{p}}!_{\mathsf{p}}(I, X, \alpha)$$

is induced by $X^{\pi_1} \colon X^I \to X^{I \times I} \cong (X^I)^I$.

Finally, a map

$$!_{\mathsf{p}}(I, X, \alpha) = (I, X^I, \tilde{\alpha}) \xrightarrow{(f, f_0, \phi)} (J, Y, \beta)$$

in $\mathfrak{Prod}_L(\mathsf{p})$ amounts to $f \colon I \to J, f_0 \colon Y \to X^I, \phi : \tilde{\alpha}(i, f_0 y) \to \beta(fi, y)$; by taking the transpose of F this corresponds to a morphism

$$(I, X, \alpha) \xrightarrow{(f, \hat{f}_0, \phi)} (J, Y, \beta)$$

in $\mathfrak{Prod}(\mathsf{p})$, showing that the latter is indeed the co-Kleisli category for $!_{\mathsf{p}}$.

This gives:

Theorem 3.18. *For each cloven fibration* p *over a Cartesian closed base category, there is a co-Kleisli adjunction*

$$\mathfrak{Prod}(\mathsf{p}) \underset{\longleftarrow}{\overset{!_{\mathsf{p}}}{\underset{\perp}{\longrightarrow}}} \mathfrak{Prod}_L(\mathsf{p})$$

Again, the algebras for the monad $\mathfrak{Prod}_L$ may be roughly described as cloven fibrations equipped with a form of linear simple products; in logical terms, these obey the inference rule

$$\frac{\alpha(i, f_0 x) \to \beta(fi, x)}{\forall x. \alpha(i, x) \to \forall y. \beta(fi, y)}$$

The reason for calling this a linear quantifier is that while the ordinary universal quantifier behaves as a generalized conjunction, this linear quantifier behaves more

TABLE 1. Quantification monads

Monad	Pattern	Inference Rule	Structure
$\mathfrak{Sum}_L$	$I \to J$, $X \to Y$	$\dfrac{\alpha(i,x) \to \beta(fi, f_0x)}{\exists x\alpha(i,x) \to \exists y\beta(fi,y)}$	Linear simple coproducts
$\mathfrak{Sum}$	$I \to J$, $X \to Y$, $I \to Y$	$\dfrac{\alpha(i,x) \to \beta(fi, f_0ix)}{\exists x\alpha(i,x) \to \exists y\beta(fi,y)}$	Simple coproducts
$\mathfrak{Prod}_L$	$I \to J$, $Y \to X$	$\dfrac{\alpha(i, f_0y) \to \beta(fi,y)}{\forall x\alpha(i,x) \to \forall y\beta(fi,y)}$	Linear simple products
$\mathfrak{Prod}$	$I \to J$, $Y \to X$, $J \to X$	$\dfrac{\alpha(i, f_0iy) \to \beta(fi,y)}{\forall x\alpha(i,x) \to \forall y\beta(fi,y)}$	Simple products

like a tensor product, since one does not necessarily have projections (i.e., the instantiation rule $\forall x.\alpha(x) \vdash \alpha(a)$ is generally not valid).

We summarize the four constructions in Table 1.

4. The Dialectica Monad

In this section we consider the dialectica construction and analyze it in terms of the monads considered earlier. We begin by showing directly that the dialectica construction may be decomposed into two steps, following the $\exists\forall$ quantifier pattern suggested by the original interpretation. We then indicate another decomposition via a linear approximation followed by an exponential comonad. After that we come to the main issue, namely the observation that the dialectica construction is itself a (pseudo-)monad, provided the base category is Cartesian closed; to this end, we show that the type-theoretic axiom of choice (i.e., Skolemization) in fact takes the form of a pseudodistributive law of $\mathfrak{Prod}$ over $\mathfrak{Sum}$, thus making the composite into pseudomonad. This explains the categorical content of Skolemization and the universal property of the dialectica construction.

4.1. The dialectica construction. We now introduce the main construction of interest.

Construction 4.1 (Dialectica construction). Let $\mathsf{p}\colon \mathsf{E} \to \mathsf{B}$ be a fibration. Define a category $\mathfrak{Dial}(\mathsf{p})$ as follows:

Objects: quadruples (I, X, U, α), where I, X and U are objects of the base B, and where $\alpha \in \mathsf{p}^{-1}(I \times X \times U)$ is an object in the fibre over $I \times X \times U$.

Morphisms: a map from (I, X, U, α) to (J, Y, V, β) is a quadruple (f, f_0, f_1, ϕ), where

- $f\colon I \to J$ is a morphism in B
- $f_0\colon I \times X \to Y$ is a morphism in B
- $f_1\colon I \times X \times V \to U$ is a morphism in B

- $\phi = \phi(i,x,v)\colon \alpha(i,x,f_1(i,x,v)) \to \beta(f(i), f_0(i,x), v)$ is a morphism in the fibre over $I \times X \times V$.

This is again fibred over B via the first projection. Note that the fibre over 1 is, up to equivalence, just the the total category of $\mathfrak{Prod}(\mathsf{p})$; in the original presentation [7] it was the latter category (for the special case where p was the subobject fibration) which was studied. This category can thus be viewed as a category of *dialectica propositions.*

As a first step towards understanding the structure of the fibration $\mathfrak{Dial}(\mathsf{p})$, we observe that the construction can be decomposed into two steps:

Lemma 4.2. *There is an isomorphism of fibrations, natural in* p:

$$\mathfrak{Dial}(\mathsf{p}) \cong \mathfrak{Sum}\,\mathfrak{Prod}(\mathsf{p}).$$

PROOF. This is an easy direct verification; an object of $\mathfrak{Sum}\,\mathfrak{Prod}(\mathsf{p})$ has the form $(I, X, (I \times X, U, \alpha))$, and we may identify this with (I, X, U, α) as usual. A morphism $(I, X, U, \alpha) \to (J, Y, V, \beta)$ has the form (f, f_0, F), where $f\colon I \to J$, $f_0\colon I \times X \to Y$ and

$$F\colon (I \times X, U, \alpha) \to \langle f\pi_I, f_0\rangle^*(J \times Y, V, \beta)$$

in $\mathfrak{Prod}(\mathsf{p})$. Thus, F consists of maps

$$f_1\colon I \times X \times V \to U; \quad \phi(i,x,v)\colon \alpha(i,x,f_1(i,x,v)) \to \beta(fi, f_0(i,x), v),$$

and hence the tuple (f, f_0, f_1, ϕ) constitutes a map in $\mathfrak{Dial}(\mathsf{p})$. Further details are straightforward and left to the reader. □

Concretely this says that $\mathfrak{Dial}(\mathsf{p})$ arises from p by first adding universal quantification, and then existential quantification. Thus we are justified in thinking of an object (I, X, U, α) as $\exists x \forall u.\alpha(i,x,u)$, and this makes clear how this abstract categorical construction is related to Gödel's original interpretation.

4.2. Skolemization as a distributive law. As already mentioned in the introduction, Skolemization is a syntactic translation which is based on the idea of replacing a quantifier combination of the form $\forall x : X.\exists u : U.\alpha(x,u)$ by one of the form $\exists f : U^X.\forall x : X.\alpha(x, fx)$. Since the existential quantifier ranges over a function type, this construction is only available in the presence of function types. Skolemization is the key ingredient in Gödel's dialectica interpretation, because it tells us how to replace a sequence of alternating quantifiers by a sequence of the form $\exists\vec{f}\forall\vec{x}$.

First, recall from the previous section that the monad $\mathfrak{Prod}$ admits a decomposition into a linear component followed by a co-Kleisli construction. The same idea works for $\mathfrak{Dial}$: we consider an auxiliary fibration $\mathfrak{Dial}_l(\mathsf{p})$ by taking the same objects as $\mathfrak{Dial}(\mathsf{p})$, but only those morphisms for which the third coordinate does not depend on the second (but it may depend on the first). Thus schematically the morphisms have the form

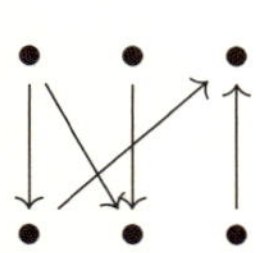

and we may think of objects (I, X, U, α) of $\mathfrak{Dial}_l(\mathsf{p})$ as predicates

$$\begin{bmatrix}\exists x\\ \forall u\end{bmatrix} \alpha(i,x,u),$$

where the branching quantifier indicates that u and x are independent.

How does this help? Well, if we have an expression with two such quantifiers, the quantifier independence allows us to contract it as follows:

$$\begin{bmatrix}\exists x\\ \forall u\end{bmatrix}\begin{bmatrix}\exists y\\ \forall v\end{bmatrix}\alpha(i,x,u,y,v)\mapsto\begin{bmatrix}\exists xy\\ \forall uv\end{bmatrix}\alpha(i,x,u,y,v).$$

This operation may be recast in terms of fibrations as a fibred functor

$$\mathfrak{Dial}_i^2(p)\to\mathfrak{Dial}_i(\mathsf{p});\qquad (I,X,U,Y,V,\alpha)\mapsto(I,X\times Y,U\times V,\alpha)$$

(the reader can easily verify that this assignment works on the level of morphisms precisely because of the independence conditions) which is the component at p of a multiplication on $\mathfrak{Dial}_i$. One can now work out that $\mathfrak{Dial}_i$ can be equipped with the structure of a pseudomonad.

Now back to the general case of $\mathfrak{Dial}(\mathsf{p})$. Here, we cannot use this formula for the multiplication directly. Instead, we need to first break the quantifier dependence by using the axiom of choice.

Lemma 4.3. *Let* B *be a Cartesian closed category. Then the inclusion of fibrations* $\mathcal{I}_{\mathsf{p}}\colon \mathfrak{Dial}_i(\mathsf{p})\to\mathfrak{Dial}(\mathsf{p})$ *has a fibred right adjoint.*

PROOF. This is a straightforward calculation along the same lines as for $\mathfrak{Prod}_L$. Explicitly, the right adjoint $\mathcal{R}_{\mathsf{p}}$ sends an object (I,X,U,α) to $(I,X,U^X,\check{\alpha})$, where $\check{\alpha}(i,x,f)=\alpha(i,x,fx)$. Then we have a bijective correspondence of maps in $\mathfrak{Dial}_i(\mathsf{p})$ from (J,Y,V,β) to $(I,X,U^X,\check{\alpha})$ and maps in $\mathfrak{Dial}(\mathsf{p})$ from (J,Y,V,β) to (I,X,U,α). Details are left to the reader. □

We may also prove (although we shall not need this) that $\mathfrak{Dial}(\mathsf{p})$ is in fact the Kleisli category for the induced monad $!_{\mathsf{p}}$ on $\mathfrak{Dial}_i(\mathsf{p})$. Finally, we note that the condition that B is Cartesian closed is not only sufficient but is also necessary for this result. (Just take p to be the identity fibration on B.)

Consider now the following map $\lambda_{\mathsf{p}}:\mathfrak{Prod}\,\mathfrak{Sum}(\mathsf{p})\to\mathfrak{Sum}\,\mathfrak{Prod}(\mathsf{p})$:

$$(I,X,U,\alpha)\mapsto(I,U^X,X,\hat{\alpha}),\qquad \hat{\alpha}(i,f,x)=\alpha(i,x,fx).$$

Thus λ_{p} is the composite of the right adjoint $\mathcal{R}_{\mathsf{p}}$ and the twist map.

Theorem 4.4. *When* B *is Cartesian closed, the maps* λ_{p} *are the components of a pseudonatural transformation*

$$\lambda\colon\ \mathfrak{Prod}\,\mathfrak{Sum}\to\mathfrak{Sum}\,\mathfrak{Prod}$$

which underlies a pseudodistributive law of $\mathfrak{Prod}$ *over* $\mathfrak{Sum}$.

PROOF. A direct verification is long and tedious, but straightforward (the many coherence conditions pose no real difficulty, as the coherence morphisms are all defined using uniqueness properties). In Section 4.3 we give a more conceptual proof using the fact that pseudodistributive laws of $\mathfrak{Prod}$ over $\mathfrak{Sum}$ correspond to liftings of $\mathfrak{Sum}$ to the category of algebras for $\mathfrak{Prod}$. □

Remark 4.5. Even when $\mathfrak{Sum}$ and $\mathfrak{Prod}$ happen to be strict, this pseudodistributive law will not be strict, unless (i) the Cartesian closed structure on B is strict and (ii) the fibrations involved are split. The reason is that the coherence isomorphisms involved in the distributive law arise as comparison morphisms between various ways of reindexing along structure maps for the Cartesian closed structure on the base.

The immediate consequence of Theorem 4.4 is of course that the composite $\mathfrak{Dial}$ can also be equipped with the structure of a pseudomonad, and that the multiplication is given by

$$\mathfrak{Dial}^2(p) \to \mathfrak{Dial}(\mathsf{p}); \qquad (I, X, U, Y, V, \alpha) \mapsto (I, X \times Y^U, U \times V, \hat{\alpha})$$

where of course $\hat{\alpha}(i, x, f, u, v) = \alpha(i, x, u, fx, v)$.

Theorem 4.6. *When* B *is Cartesian closed, the dialectica construction is (the underlying 2-functor of) a pseudomonad on* $\mathfrak{CFib}$. *The pseudo-algebras for this monad are the cloven fibrations which have (chosen) simple products and simple coproducts satisfying the distributivity condition*

$$\forall u \exists x. \alpha(i, u, x) \cong \exists f \forall u. \alpha(i, fu, u). \tag{4.1}$$

Indeed, when $\mathsf{p}\colon \mathsf{E} \to \mathsf{B}$ is a (pseudo-)algebra for $\mathfrak{Dial}$, then the following diagram must commute up to natural isomorphism:

$$\begin{array}{ccccc}
\mathfrak{Prod}\,\mathfrak{Sum}(\mathsf{E}) & & \xrightarrow{\quad\lambda\quad} & & \mathfrak{Sum}\,\mathfrak{Prod}(\mathsf{E}) \\
{\scriptstyle \mathfrak{Prod}(\exists)}\big\downarrow & & & & \big\downarrow{\scriptstyle \mathfrak{Sum}(\forall)} \\
\mathfrak{Prod}(\mathsf{E}) & \xrightarrow[\forall]{} & \mathsf{E} & \xleftarrow[\exists]{} & \mathfrak{Sum}(\mathsf{E})
\end{array}$$

Chasing this diagram both ways around for an object (I, X, U, α) results in a predicate $\forall u \exists x. \alpha(i, u, x)$ on the one hand, and $\exists f \forall x. \alpha(i, fx, x)$ on the other.

We remark in passing that there is a dual result, which comes for free now: there is also a pseudodistributive law

$$\kappa\colon \mathfrak{Sum}\,\mathfrak{Prod} \to \mathfrak{Prod}\,\mathfrak{Sum}$$

making the composite $\mathfrak{Prod}\,\mathfrak{Sum}$ into a pseudomonad whose algebras are fibrations with (chosen) simple products and coproducts satisfying the distributivity condition

$$\exists x \forall u. \alpha(i, x, u) \cong \forall f \exists x. \alpha(i, fx, x). \tag{4.2}$$

4.3. Proof of main result. We now give a proof that $\mathfrak{Dial}$ carries a pseudomonad structure by verifying that $\mathfrak{Prod}$ distributes over $\mathfrak{Sum}$ (throughout, we of course assume that B is Cartesian closed). As is well-known (see, e.g., [6]), to give a distributive law of $\mathfrak{Prod}$ over $\mathfrak{Sum}$ is equivalent to giving a lifting of $\mathfrak{Sum}$ to the 2-category of pseudo-algebras for $\mathfrak{Prod}$. In our case, the fact that the monad $\mathfrak{Sum}$ has the KZ property implies that such a lifting is essentially unique. It has been shown[3] that to give this lifting it suffices to prove the following three lemmata:

Lemma 4.7. *When a fibration* $\mathsf{p}\colon \mathsf{E} \to \mathsf{B}$ *has (chosen) simple products, then so does* $\mathfrak{Sum}(\mathsf{p})$.

Lemma 4.8. *When* p *has (chosen) simple products, then the unit* $\eta\colon \mathsf{p} \to \mathfrak{Sum}(\mathsf{p})$ *preserves them.*

Lemma 4.9. *When* p *has (chosen) simple products and* $H\colon \mathsf{p} \to \mathfrak{Sum}(\mathsf{q})$ *preserves them, then so does the extension* $H^+\colon \mathfrak{Sum}(\mathsf{p}) \to \mathfrak{Sum}(\mathsf{q})$.

[3]I am grateful to the referee for pointing me to this result, of which the preorder enriched version can be found in [15].

In the proofs of these lemmas, we shall suppress some coherence data arising from the reindexing of the fibrations involved and the fibred functors between them in order to keep things readable.

PROOF OF LEMMA 4.7. Assume p has chosen simple products, i.e., right adjoints $\forall_{I,X}$ to weakening functors $\pi_I^*\colon \mathsf{E}_I \to \mathsf{E}_{I\times X}$ subject to the BCC. We need to show that $\mathfrak{Sum}(\mathsf{p})$ has the same structure. Let

$$\forall_{I,X}\colon\ \mathfrak{Sum}(\mathsf{E})_{I\times X} \to \mathfrak{Sum}(\mathsf{E})_I$$

be defined by

$$(I\times X, U, \alpha(i,x,u)) \mapsto (I, U^X, \forall x.\alpha(i,x,fx))$$

where f is of type U^X and where we write $\forall x$ instead of the more cumbersome $\forall_{I,X}$.

This mapping is right adjoint to weakening: a morphism

$$(I, V, \beta(i,v)) \to (I, U^X, \forall x.\alpha(i,x,fx))$$

consists of

$$f\colon I\times V \to U^X \qquad\text{and}\qquad \phi(i,v)\colon \beta(i,v) \to \forall x.\alpha(i,x,f(i,v)(x)).$$

Such data is in 1-1 correspondence with maps

$$\hat{f}\colon (I\times X)\times V \to U \qquad\text{and}\qquad \hat{\phi}(i,x,v)\colon \beta(i,v) \to \alpha(i,x,\hat{f}(i,x,v)),$$

i.e., with morphisms $\pi^*(I,X,\beta(i,v)) = (I\times X, V, \beta(i,v)) \to (I\times X, U, \alpha(i,x,u))$.

For the BCC, consider a morphism $h\colon I\to J$, and an object $(I\times X, U, \alpha(i,x,u))$ over I, and consider the diagram

$$\begin{array}{ccccc}
J\times U^X & \xleftarrow{\ \pi\ } & J\times U^X\times X & \xrightarrow{\ v\ } & J\times X\times U\\
\Big\downarrow{\scriptstyle h\times U^X} & & \Big\downarrow{\scriptstyle h\times U^X\times X} & & \Big\downarrow{\scriptstyle (h\times X)\times U}\\
I\times U^X & \xleftarrow[\ \pi\ ]{} & I\times U^X\times X & \xrightarrow[\ w\ ]{} & I\times X\times U.
\end{array}$$

Here, the left-hand square is a pullback, and v, w are the obvious maps built from the evaluation and pairing, so the right-hand square commutes.

Using the coherence for reindexing we get a canonical isomorphism

$$(h\times U^X\times X)^* v^*\alpha \to w^*(h\times X\times U)^*\alpha$$

Applying $\forall_X$ to this isomorphism and precomposing with the canonical isomorphism from the BCC for p gives

$$(h\times U^X)^*\forall_X v^*\alpha \to \forall_X w^*(h\times X\times U)^*\alpha.$$

It is now readily verified that this map is the third component of the canonical map

$$h^*\forall_X(I\times X, U,\alpha) \to \forall_X(h\times X)^*(I\times X, U, \alpha)$$

so that the BCC holds for the product in $\mathfrak{Sum}(\mathsf{p})$ as we have defined it. □

PROOF OF LEMMA 4.8. Consider $\alpha\in \mathsf{E}_{I\times X}$. One the one hand we have

$$\eta\forall_X\alpha = (I, 1, \pi_I^*\forall_X\alpha)$$

and on the other hand we have

$$\forall_X\eta(\alpha) = \forall_X(I\times X, 1, \pi_{I\times X}^*\alpha) = (I, 1^X, \forall_X v^*\pi_{I\times X}^*\alpha)$$

where $v\colon I \times 1^X \times X \to I \times X \times 1$ is again the canonical map. These two objects are canonically isomorphic, since we may use the BCC for p for the pullback in the diagram below.

$$\begin{array}{ccccc} I \times 1^X & \xleftarrow{\pi} & I \times 1^X \times X & \xrightarrow{v} & I \times X \\ \cong\downarrow & & \downarrow\cong & \nearrow_{=} & \\ I & \xleftarrow[\pi]{} & I \times X & & \end{array}$$

□

Proof of Lemma 4.9. Given a product-preserving map $H\colon \mathsf{p} \to \mathfrak{Sum}(\mathsf{q})$; we will write, for $\alpha \in \mathsf{E}_I$,

$$H(\alpha) = (I, H_0[\alpha], H_1[\alpha])$$

First, to say that H is a fibred functor means that for a map $f\colon J \to I$ there are canonical isomorphisms

$$\begin{aligned} I \times H_0[f^*\alpha] &\cong I \times H_0[\alpha] \\ H_1[f^*\alpha] &\cong (f \times 1)^* H_1[\alpha] \end{aligned}$$

To say that H preserves simple products means that there are canonical isomorphisms

$$\begin{aligned} I \times H_0[\forall x.\alpha] &\cong I \times H_0[\alpha]^X \\ H_1[\forall x.\alpha] &\cong \forall x.\widehat{H_1[\alpha]} \end{aligned}$$

where $\widehat{H_1[\alpha]}$ is the result of reindexing $H_1[\alpha]$ along $I \times H_0[\alpha]^X \times X \to I \times X \times H_0[\alpha]$.

Next, the extension $H^\dagger$ is defined as $\mu_{\mathsf{q}}\,\mathfrak{Sum}(H)$. Explicitly,

$$H^\dagger(I, X, \alpha) = (I, X \times H_0[\alpha], H_1[\alpha])$$

where we suppress the reindexing along the associativity isomorphism. To show that $H^\dagger$ preserves products, we consider an object $(I \times X, U, \alpha)$, and compute both $\forall_X H^\dagger(I \times X, U, \alpha)$ and $H^\dagger \forall_X (I \times X, U, \alpha)$. We have

$$\begin{aligned} H^\dagger \forall_X (I \times X, U, \alpha) &= H^\dagger(I, U^X, \forall x.\hat{\alpha}) && \text{by def. of } \forall \text{ in } \mathfrak{Sum} \\ &= (I, U^X \times H_0[\forall x.\hat{\alpha}], H_1[\forall x.\hat{\alpha}]) && \text{by def. of } H^\dagger \\ &\cong (I, U^X \times H_0[\hat{\alpha}]^X, \forall x.\widehat{H_1[\hat{\alpha}]}) && H \text{ preserves } \forall \\ &\cong (I, (U \times H_0[\alpha])^X, \forall x.\widehat{H_1[\alpha]}) && \\ &= \forall_X (I \times X, U \times H_0[\alpha], H_1[\alpha]) && \text{by def. of } \forall \text{ in } \mathfrak{Sum} \\ &= \forall_X H^\dagger(I \times X, U, \alpha) && \text{by def. of } H^\dagger. \end{aligned}$$

The isomorphisms in the middle arise as follows: for the first one, we use the fact that H preserves reindexing and simple products. For the second, it seems as if we've lost a hat in the process, but consider the diagram

$$\begin{array}{ccccc} I \times (U \times H_0[\alpha])^X & \longleftarrow & I \times (U \times H_0[\alpha])^X \times X & \xrightarrow{v} & I \times X \times U \times H_0[\alpha] \\ \cong\downarrow & & \cong\downarrow & & \uparrow w \\ I \times U^X \times H_0[\hat{\alpha}]^X & \longleftarrow & I \times U^X \times H_0[\hat{\alpha}]^X \times X & \xrightarrow[v']{} & I \times U^X \times X \times H_0[\hat{\alpha}] \end{array}$$

The object $H_1[\alpha]$ lives over the top right corner; reindexing along w and then along v' gives $H_1[\hat{\alpha}]$ (because H commutes with reindexing) and then $\widehat{H_1[\hat{\alpha}]}$. When we reindex the latter along the canonical isomorphism in the middle, we get something isomorphic to $\widehat{H_1[\alpha]}$. Finally, an application of the BCC for the left-hand square gives the desired isomorphism. □

5. Concluding thoughts

We end with a few loose ends, and some possible avenues for further research.

The Chu construction. There is a cousin of the dialectica construction which we haven't mentioned yet, namely the Chu construction (see, e.g., [2]); this construction has been compared to the dialectica interpretation in [9] because of certain formal similarities. One can define, for an arbitrary fibration p, a category $\mathsf{Chu}(\mathsf{p})$, by taking the subcategory of $\mathfrak{Sum}_L(\mathsf{p})$ on the same objects, but only on those morphisms (f, f_0, ϕ) for which the third component is an isomorphism. At present, I'm not sure whether this is a fruitful course of action: for one thing, it is not clear whether it gives a nice universal property of the Chu construction along the same lines as for the other constructions. I should point out that a completely different approach to finding a universal property of the Chu construction is given in a paper by Dusko Pavlovic [22].

The case of locally Cartesian closed categories. In this paper we have considered the simple (in the sense of simply typed) version of the dialectica construction, mainly because of the fact that this is how the construction appears in the literature. However, in [14] a dependently typed version is suggested. In this case, one works over a locally Cartesian closed base category, in which case one can use the families monad and its dual.

Linear quantifiers. We have been brief about the monads for "linear" quantification and in particular about their algebras. This is partly because it requires a bit more work to describe these structures in detail, and also because at this point it is not fully clear what role they should play. One thing which is worth mentioning is that there is a connection with independence-friendly and branching quantifiers; the linear quantifiers introduced here may be viewed as an instance of these. An example of a fibration which has this structure arises from work by Abramsky and Väänänen on the Hodges semantics for independence-friendly logic [1]. However, at present there is no general categorical treatment of branching quantifiers available, and this will be left for future work.

Bicompletion of a fibration. The dialectica construction combines the monads for simple products and coproducts by means of a distributive law between them. However, there are other ways of combining monads than by using distributive laws. Here, we would like to mention that there is another pseudomonad on $\mathfrak{CFib}(\mathsf{B})$, which takes a fibration p and produces a fibration $\Lambda(\mathsf{p})$, the free bicompletion of p. For this construction, Cartesian closure of the base category is not needed. The pseudo-algebras for Λ are the fibrations which have (chosen) simple products and coproducts. Roughly, $\Lambda(\mathsf{p})$ is constructed by iterating the monads $\mathfrak{Sum}$ and $\mathfrak{Prod}$ and by taking a suitable colimit. Explicitly, an object of $\Lambda(\mathsf{p})$ is a tuple $(I, X_1, X_2, \dots, X_n, \alpha)$, where α is an object of $\mathsf{p}^{-1}(I \times X_1 \times \cdots \times X_n)$. We think of such an object as the predicate $\exists x_1 \forall x_2 \dots \forall x_n . \alpha(i, x_1, \dots, x_n)$. The morphisms in $\Lambda(\mathsf{p})$ are generated by those of the iterates $(\mathfrak{Sum}\,\mathfrak{Prod})^n(\mathsf{p})$, together with newly added isomorphisms forcing the resulting category to have the desired structure, e.g., $(I, X_1, 1, 1, X_2, \alpha) \cong (I, X_1, X_2, \alpha)$, etc. A more detailed description and analysis of this construction is in preparation.

The free bicompletion of an ordinary category was first studied in [17], where it was characterized in terms of its so-called *softness* property.

Appendix A. Coherence for $\mathfrak{Sum}$

This appendix establishes the coherence for the pseudomonad structure on $\mathfrak{Sum}$. Let me say first that a direct verification is not difficult; the work is greatly reduced by making use of that the multiplication is left adjoint to the unit. Another approach would be to replace $\mathfrak{Sum}(\mathsf{p})$ by its closure in $\mathfrak{Fam}(\mathsf{p})$; the coherence for the resulting pseudomonad would then be automatic, and one would only need to show that it is equivalent to $\mathfrak{Sum}$.

However, one may still ask for a more conceptual explanation; part of the motivation here is that there is a sense in which $\mathfrak{Sum}$ is a club[4] (see [18]), i.e., determined by its action on the terminal fibration; moreover, one would hope that coherence for $\mathfrak{Sum}$ is a formal consequence of the coherence at the terminal object. This should also help in dealing with pseudomonads on $\mathfrak{CFib}(\mathsf{B})$ similar to $\mathfrak{Sum}$ (but perhaps not satisfying the KZ property).

In what follows I heavily use the fact that the forgetful 2-functor $\mathfrak{CFib} \to \mathfrak{Cat}$ is a 2-fibration (see [12] for detailed discussion). In particular, we can lift natural transformations from $\mathfrak{Cat}$ to $\mathfrak{CFib}$ as follows: consider a fibration $\mathsf{E} \xrightarrow{p} \mathsf{B}$, functors $L, R \colon \mathsf{B}' \to \mathsf{B}$ in the base, and a natural transformation $\phi \colon L \Rightarrow R$. Now suppose we have a lifting $\overline{R}$ of R, as in

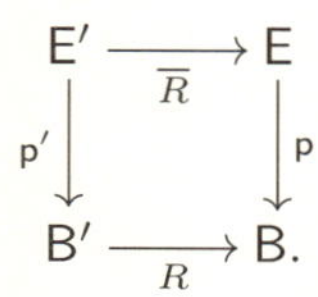

Then we may define a lift of L by letting $\tilde{L}(\alpha) = \phi_I^*(\alpha)$, where $\mathsf{p}'(\alpha) = I$. Moreover, for each object α over I we get a Cartesian morphism $\bar{\phi}_{(I,\alpha)} \colon \phi_I^*(\alpha) \to \alpha$ over ϕ_I. These assemble to give a natural transformation $\bar{\phi} \colon \tilde{L} \Rightarrow \overline{R}$ over ϕ, as in

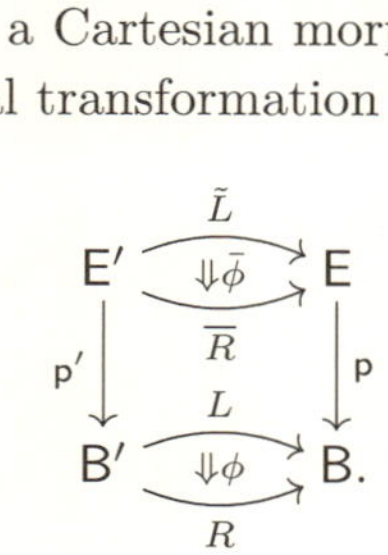

We call $\bar{\phi}$ the *Cartesian lifting* of ϕ. Note that by construction all components of $\bar{\phi}$ are Cartesian. The assignment $\phi \mapsto \bar{\phi}$ has various other good naturality properties, which we won't spell out.

---

[4]As far as I'm aware, no precise definition of pseudoclub exists.

Instead of just lifting a single natural transformation we can also lift entire pasting diagrams; for example, a diagram

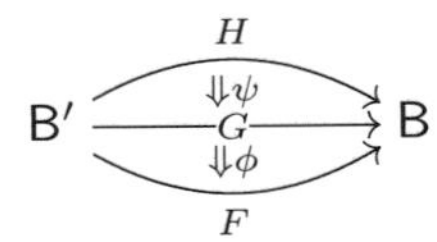

can be lifted by first lifting ϕ, and then lifting ψ. One can also lift the composite $\psi\phi$ in one single step, and the two outcomes differ by a unique vertical transformation, whose component at an object α over I is simply the unique vertical map mediating between $\psi_I^*\phi_I^*(\overline{F}\alpha)$ and $(\phi\psi)_I^*(\overline{F}(\alpha))$. A similar observation can be made for whiskering ϕ on either side. The result of this is, that, up to unique vertical natural isomorphism, the lifting of a pasting diagram is well-defined.

We are now in a position to explain in which sense the coherence for $\mathfrak{Sum}$ comes for free. To this end, I will describe data on the terminal object from which one can construct a pseudomonad on $\mathfrak{CFib}(\mathsf{B})$.

Datum 1. A (split) fibration $\mathsf{t}\colon \mathfrak{T}(\mathsf{B}) \to \mathsf{B}$.

Datum 2. A functor $l\colon \mathfrak{T}(\mathsf{B}) \to \mathsf{B}$ (which is not supposed to be a fibred functor).

Given these, we may to extend $\mathfrak{T}$ to a 2-functor on $\mathfrak{CFib}$ via pullback:

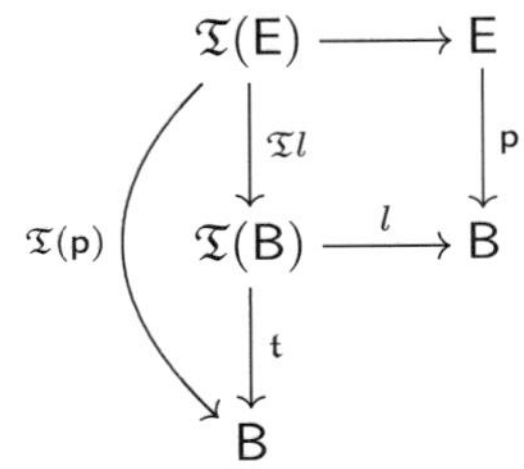

just as we did for $\mathfrak{Sum}$ and $\mathfrak{Fam}$.

Datum 3. Fibred functors $\eta_\mathsf{B}\colon \mathsf{B} \to \mathfrak{T}(\mathsf{B})$ and $\mu_\mathsf{B}\colon \mathfrak{T}^2(\mathsf{B}) \to \mathfrak{T}(\mathsf{B})$.

Datum 4. Natural isomorphisms (not fibred)

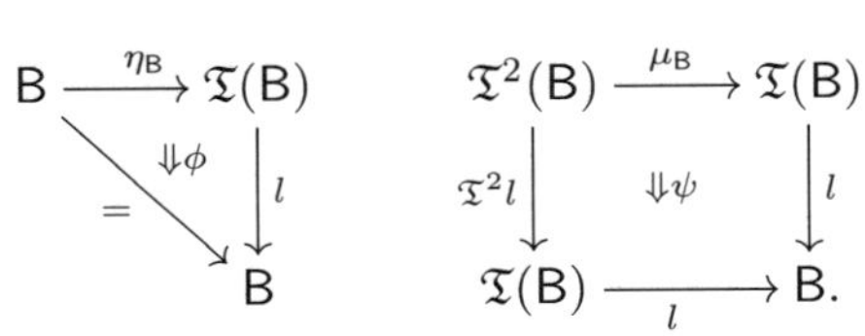

These allow us, using the lifting properties of natural transformations, to extend η_B and μ_B to pseudonatural transformations $(\eta, \overline{\eta})\colon 1 \to \mathfrak{T}, (\mu, \overline{\mu}) : \mathfrak{T}^2 \to \mathfrak{T}$. More explicitly, here is how we get the unit at $\mathsf{p} : \mathsf{E} \to \mathsf{B}$. We lift the natural transformation $\phi\colon l\eta_\mathsf{B} \to 1_\mathsf{B}$ to a natural transformation $\overline{\phi}_\mathsf{E}\colon \tilde{L} \to 1_\mathsf{E}$. Now factorize this using the fact that the right-hand square in the diagram below is a pullback:

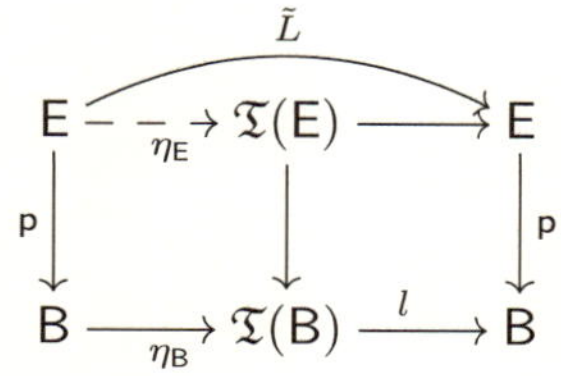

The left-hand square is in general not a pullback. We now obtain (using the lifting properties of natural transformations) a natural transformation $\overline{\eta}_F$ as in

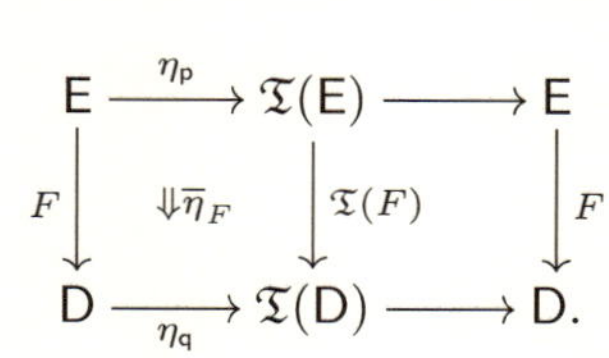

making $(\eta, \overline{\eta})$ a pseudonatural transformation. Completely analoguous is the construction of μ and $\overline{\mu}$ from μ_{B} giving $(\mu, \overline{\mu})\colon \mathfrak{T}^2 \to \mathfrak{T}$.

Finally, we specify the data needed to ensure that we can endow $\mathfrak{T}$ with the structure of a pseudomonad.

Datum 5. Fibred natural isomorphisms

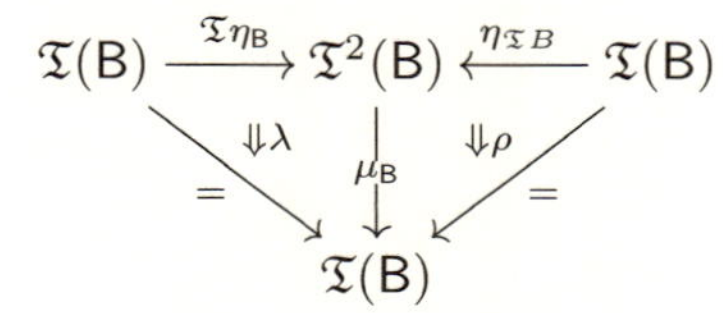

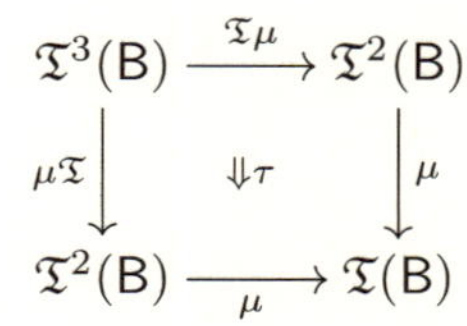

subject to the usual coherence conditions for the modifications up to which the monad laws hold, and in addition three conditions relating λ, ρ and τ to ϕ and ψ:

(a)

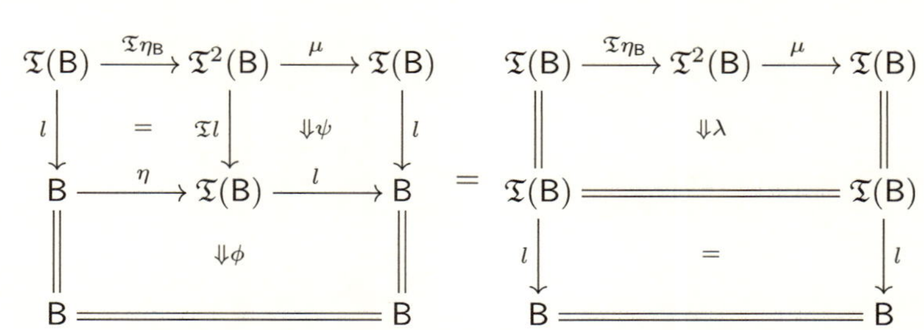

(b)

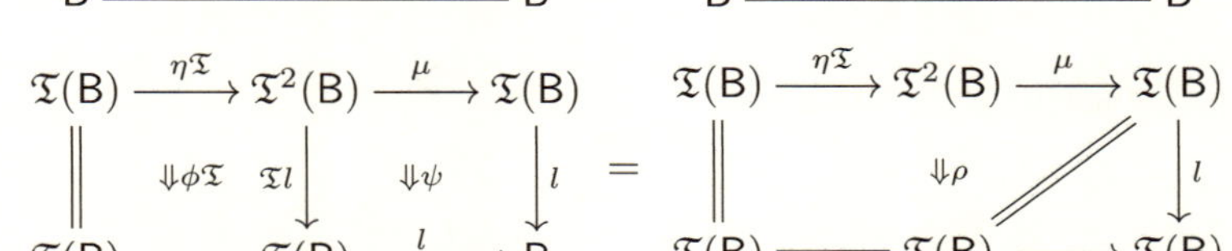

(c)

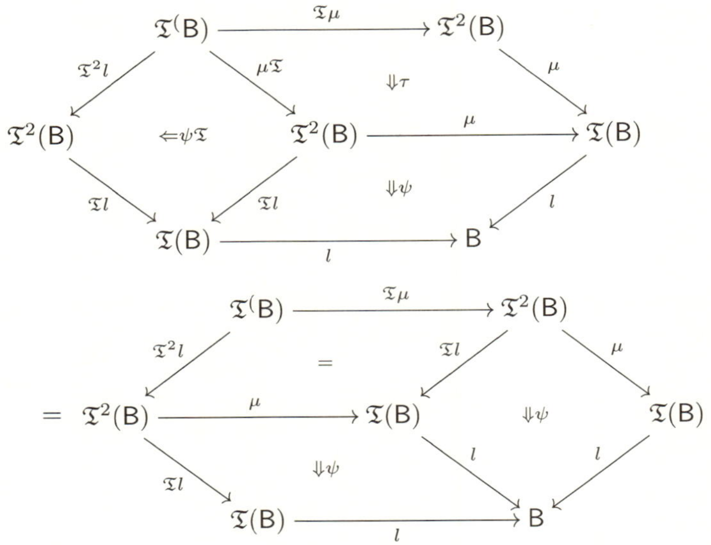

In the case of $\mathfrak{Sum}$, all of the isomorphisms $\lambda, \rho, \tau, \phi, \psi$ are simply coherence isomorphisms for the product structure on B, and the coherence conditions are immediate.

We can now lift λ, ρ and τ. The point is here that we know that λ, ρ and τ fit into pasting diagrams whose composite is equal to a pasting diagram involving ψ, ϕ, and thus that we can find liftings which are of the right type (i.e. have the correct domain and codomain). It now also follows immediately by uniqueness of such liftings that the coherence conditions are inherited from those at B.

This gives the main observation of this section:

Theorem A.1. *The data* $1-5$ *described above, induces a pseudomonad structure on* $\mathfrak{T}$. *The underlying pseudofunctor is strict, and when each of* $\phi, \psi, \lambda, \rho$ *and* τ *are strict then* $(\mathfrak{T}, \eta, \mu)$ *is a strict* 2*-monad.*

References

1. S. Abramsky and J. Väänänen, *From IF to BI: A tale of dependence and separation*, Synthese **167** (2009), no. 2, 207–230.
2. M. Barr, *The Chu construction*, Theory Appl. Categ. **2** (1996), no. 2, 17–35.
3. M. Barr and C. Wells, *Toposes, triples and theories*, Grundlehren Math. Wiss., vol. 278, Springer, New York, 1985.
4. J. Bénabou, *Fibered categories and the foundations of naive category theory*, J. Symbolic Logic **50** (1985), no. 1, 10–37.
5. B. Biering, *Dialectica interpretations: A categorical analysis*, Ph.D. Dissertation, IT University of Copenhagen, 2008.
6. E. Cheng, J. M. E. Hyland, and J. Power, *Pseudo-distributive laws*, Electronic Notes in Theoretical Computer Science **83** (2004).
7. V. C. V. de Paiva, *The Dialectica categories*, Categories in Computer Science and Logic (Boulder, CO, 1987) (J. W. Gray and A. Scedrov, eds.), Contemp. Math., vol. 92, Amer. Math. Soc., Providence, RI, 1989, pp. 47–62.
8. ———, *A Dialectica-like model of linear logic*, Category Theory and Computer Science (Manchester, 1989) (D. H. Pitt, D. E. Rydeheard, P. Dybjer, A. M. Pitts, and A. Poigné, eds.), Lecture Notes in Comput. Sci., vol. 389, Springer, Berlin, 1989, pp. 341–356.
9. ———, *Dialectica and Chu constructions: cousins?*, Theory Appl. Categ. **17** (2007), 127–152.

10. N. Gambino, *On the coherence conditions for pseudo-distributive laws* (2009), available at arXiv:0907.1359.
11. K. Gödel, *Über eine bisher noch nicht benützte Erweiterung des finiten Standpunktes*, Dialectica **12** (1958), 280–287.
12. C. Hermida, *Some properties of* **Fib** *as a fibred 2-category*, J. Pure Appl. Algebra **134** (1999), no. 1, 83–109.
13. J. M. E. Hyland, *Proof theory in the abstract*, Ann. Pure Appl. Logic **114** (2002), no. 1-3, 43–78.
14. ———, *Fibrations in logic*, Category Theory 2007 (Coimbra, 2007), invited lecture.
15. ———, *Some reasons for generalizing domain theory*, Math. Structures Comput. Sci. **20** (2010), no. 2, 239–265.
16. B. Jacobs, *Categorical logic and type theory*, Stud. Logic Found. Math., vol. 141, North-Holland, Amsterdam, 1999.
17. A. Joyal, *Free bicomplete categories*, C. R. Math. Rep. Acad. Sci. Canada **17** (1995), no. 5, 219–224.
18. G. M. Kelly, *Doctrinal adjunction*, Category Seminar (Sydney, 1972/1973) (G. M. Kelly, ed.), Lecture Notes in Math., vol. 420, Springer, Berlin, 1974, pp. 257–280.
19. G. M. Kelly and S. Lack, *On property-like structures*, Theory Appl. Categ. **3** (1997), no. 9, 213–250.
20. A. Kock, *Monads for which structures are adjoint to units*, J. Pure Appl. Algebra **104** (1995), no. 1, 41–59.
21. F. Marmolejo, *Doctrines whose structure forms a fully faithful adjoint string*, Theory Appl. Categ. **3** (1997), no. 2, 24–44.
22. D. Pavlovic, *Chu* I: *cofree equivalences, dualities and *-autonomous categories*, Math. Structures Comput. Sci. **7** (1997), no. 1, 49—73.
23. R. Street, *Fibrations in bicategories*, Cahiers Topologie Géom. Différentielle **21** (1980), no. 2, 111–160.
24. T. Streicher, *Fibred categories à la Bénabou* (1999), lecture notes, available at http://www.mathematik.tu-darmstadt.de/~streicher/.

Department of Mathematics and Statistics, University of Ottawa, 585 King Edward Avenue, Ottawa, ON K1N 6N5, Canada
E-mail address: phofstra@uottawa.ca

Centre de Recherches Mathématiques
CRM Proceedings and Lecture Notes
Volume **53**, 2011

A Categorical Approach to Internality

Moshe Kamensky

ABSTRACT. Model theoretic internality provides conditions under which the group of automorphisms of a model over a reduct is itself a definable group. In this paper we formulate a categorical analogue of the condition of internality, and prove an analogous result on the categorical level. The model theoretic statement is recovered by considering the category of definable sets.

1. Introduction

In model theory, one works with an abstract notion of a structure, or a model. Such a structure can be a set $\mathcal{M}$ (or a collection of sets), together with a distinguished collection of subsets of $\mathcal{M}$ and its powers, the definable subsets. An example could be an algebraically closed field, with the definable sets being the field operations, solutions to equations derived from them (polynomial equations).

There is a natural notion of maps of such structures, and so one may form the group $\mathrm{Aut}(\mathcal{M})$ of automorphisms (invertible such self-maps) of $\mathcal{M}$. This group and certain subgroups of it are an important tool in the study of definable sets and other model-theoretic objects. In general, the elements of this group have nothing to do with the elements of the structure $\mathcal{M}$ itself. However, under special circumstances, certain such automorphism groups are *themselves definable*: there is a canonical correspondence between the elements of the group and a definable subset in $\mathcal{M}$, and the (graph of the) group operation is also definable. This condition is known as internality. Internality also provides (under suitable assumptions) a definable Galois theory, that provides a correspondence between certain intermediate structures and definable subgroups.

Internality was first discovered in Zil′ber [15], where it was used to study strongly minimal sets. It can be used in an abstract setting to study the relation between definable sets that are "close" to each other. However, it was also discovered in Poizat [13] that, when applied to a particular theory (differentially closed fields), the definable Galois group and the associated Galois theory specialize to a classical construction, the Galois group (and Galois theory) of linear differential

2010 *Mathematics Subject Classification.* 03C40, 03G30, 18C10.

Key words and phrases. model theoretic internality, elimination of imaginaries, closed categories.

Supported by NSF grant 1001741.

This is the final form of the paper.

equations. It has since been applied to provide and study Galois theory in various situations (e.g., Pillay [12]). A more detailed review of internality, along with some references appears in Section 3.

Whereas the original definition of internality appeared under rather specific model theoretic assumptions (strongly minimal structures), it was subsequently generalized and simplified. In Hrushovski [3] the construction appeared without any restrictions on the theories in question.

It is shown in Kamensky [7], that a certain categorical construction, the algebraic group associated with a Tannakian category, can also be viewed as an instance of this formalism. Since the theory of internality no longer depends on "stability theoretic" assumption, it was natural to ask whether it is possible to go in the other direction, and recover the theory of internality (or some variant) from a categorical statement.

The present paper is an attempt to answer this question. In the spirit of Makkai and Reyes [10], we replace a theory $\mathcal{T}$ by its category of definable sets, and interpretations and models by suitable functors. We do not use the precise characterization provided by Makkai and Reyes [10], since our result holds in greater generality, but the motivation comes from there. On the other hand, though the category of definable sets is not a topos, when the theory eliminates imaginaries it is rather close being that, and we use some ideas from topos theory, and more generally from category theory, in the style of Mac Lane and Moerdijk [9], that remain relevant in this case.

The main results of this paper appear in Sections 4 and 5. In Section 4, we make some observations on the category of definable sets of a theory (some of these appear in Makkai and Reyes [10]), and formulate categorical versions of stably embedded definable sets, and internality. We then prove a weak version of the existence of definable Galois groups in the categorical context (Corollary 4.13). In Section 5 we introduce additional assumptions on the data, that allow us to prove the full analogy to the model theoretic case. In Section 6 we provide a sketch of two possible applications of these theorems.

The main point of this paper is to explicate some analogies between model theoretic notions and constructions with categorical counterparts. It is therefore my hope that this paper is readable by anyone with minimal knowledge in these areas. To this end, the paper includes two sections which are almost completely expository. Section 2 reviews some basic notions from category theory, and discusses the categorical assumptions we would like to make. The only new result there is Proposition 2.7 (which is a direct application of ends, and is possibly known) and its corollary. Section 3 contains a brief, and rather informal review of basic model theory, as well as internality and the associated Galois group. Hopefully, there are no new results there.

2. Categorical background

We review some notions from category theory that will be used in the description of internality. The classical reference for these is Mac Lane [8], but Wikipedia [14] is mostly sufficient. All categories in this paper are (essentially) small.

A contravariant functor from a category $\mathcal{C}$ to the category of sets will be called a *presheaf* (on $\mathcal{C}$). The functor that assigns to an object $\mathbf{X}$ the presheaf

$\mathbf{Y} \mapsto \mathrm{Hom}(\mathbf{Y}, \mathbf{X})$ it represents is called the *Yoneda embedding*, and the Yoneda lemma states that this embedding is fully faithful (induces a bijection on sets of morphisms). Hence we also denote by $\mathbf{X}$ the presheaf it represents, and view $\mathcal{C}$ as a subcategory of its presheaf category whenever this is convenient. More generally, the Yoneda lemma states that $\mathcal{F}(\mathbf{X}) = \mathrm{Hom}(\mathbf{X}, \mathcal{F})$ for any presheaf $\mathcal{F}$.

All categories we consider, with the exception of index categories, will have all finite (inverse) limits. If $\mathcal{C}$ is such a category, we denote by $\mathbf{1}$ the terminal object (empty product) of $\mathcal{C}$, and by Γ (global sections) the functor $\mathbf{X} \mapsto \mathrm{Hom}(\mathbf{1}, \mathbf{X})$.

2.1. Pro- and Ind-objects. Our categories will not, usually, be closed under infinite limits or colimits. However, certain such limits appear naturally in what follows.

We recall that a *filtering category* is a category $\mathcal{I}$ in which any two objects map into a common object, and for any two maps $f, g\colon \mathbf{A} \to \mathbf{B}$ there is some map $h\colon \mathbf{B} \to \mathbf{C}$ with $h \circ f = h \circ g$. A *filtering system* in a category $\mathcal{C}$ is a functor from a filtering category to $\mathcal{C}$. The category $\mathrm{Ind}(\mathcal{C})$ of *ind-objects* of $\mathcal{C}$ is a category that contains $\mathcal{C}$ as a full subcategory, admits the colimits of all filtering systems, and satisfies

$$\mathrm{Hom}\big(\mathbf{X}, \mathrm{Ind}_{\mathcal{I}}(\mathbf{Y}_i)\big) = \varprojlim_{\mathcal{I}}\big(\mathrm{Hom}(\mathbf{X}, \mathbf{Y}_i)\big) \tag{1}$$

where $\mathbf{X}$ and $\mathbf{Y}_i$ are objects of $\mathcal{C}$, and $\mathrm{Ind}_{\mathcal{I}}(\mathbf{Y}_i)$ is the colimit of the system in $\mathrm{Ind}(\mathcal{C})$ (note that if $\mathcal{C}$ itself happens to admit the colimit $\varinjlim_{\mathcal{I}} \mathbf{Y}_i$ of this system, this object will not be isomorphic to $\mathrm{Ind}_{\mathcal{I}}(\mathbf{Y}_i)$ in $\mathrm{Ind}(\mathcal{C})$). A construction of this category appears in Grothendieck et al. [2].

The category $\mathrm{Pro}(\mathcal{C})$ of pro-objects is the dual notion: it is the opposite of the category of ind-objects of the opposite of $\mathcal{C}$. Thus it consists of inverse limits $\mathrm{Pro}(\mathbf{X}_i)$ of cofiltering systems in $\mathcal{C}$, with $\mathcal{C}$ embedded as a full subcategory, and satisfying

$$\mathrm{Hom}(\mathrm{Pro}_{\mathcal{I}}(\mathbf{X}_i), \mathbf{Y}) = \varprojlim_{\mathcal{I}}\big(\mathrm{Hom}(\mathbf{X}_i, \mathbf{Y})\big) \tag{2}$$

with similar conventions to the above.

Because of the above properties, the objects of $\mathcal{C}$ will be called *compact* when considered as objects of $\mathrm{Ind}(\mathcal{C})$ or $\mathrm{Pro}(\mathcal{C})$. A functor (or a presheaf) will be called pro-representable (ind-representable) if it is represented by a pro- (ind-)object. In other words, it is a filtered colimit of representable functors (presheaves).

By a *pro-group* (resp. a pro-monoid, etc.) in $\mathcal{C}$ we mean a pro-object of $\mathcal{C}$ that can be represented by an inverse filtering system of group (resp. monoid) objects of $\mathcal{C}$. Such a pro-group is a group object in $\mathrm{Pro}(\mathcal{C})$, but the converse is not true, in general: there may be group objects in $\mathrm{Pro}(\mathcal{C})$ which are not of this form.

2.2. Closed and Ind-closed categories. Let $\mathcal{C}$ be a category with finite inverse limits. Given two objects $\mathbf{X}$ and $\mathbf{Y}$ of $\mathcal{C}$, we denote by $\underline{\mathrm{Hom}}(\mathbf{X}, \mathbf{Y})$ the presheaf $\mathbf{Z} \mapsto \mathrm{Hom}(\mathbf{Z} \times \mathbf{X}, \mathbf{Y})$. There is, therefore, a natural evaluation map $\underline{\mathrm{ev}}\colon \underline{\mathrm{Hom}}(\mathbf{X}, \mathbf{Y}) \times \mathbf{X} \to \mathbf{Y}$, where $\mathbf{X}$ and $\mathbf{Y}$ are identified with the presheaf they represent. Applying the definition with $\mathbf{Z} = \mathbf{1}$ (and using Yoneda's lemma), we see that $\Gamma\big(\underline{\mathrm{Hom}}(\mathbf{X}, \mathbf{Y})\big) = \mathrm{Hom}(\mathbf{X}, \mathbf{Y})$.

The category $\mathcal{C}$ is called (Cartesian) *closed* (Mac Lane [8, Section IV.6]) if all such $\underline{\mathrm{Hom}}$ presheaves are representable (in which case the representing object is denoted by the same symbol, and is called the *internal* Hom). In other words, the functor $\mathbf{Z} \mapsto \mathbf{Z} \times \mathbf{X}$ has a right adjoint $\mathbf{Y} \mapsto \underline{\mathrm{Hom}}(\mathbf{X}, \mathbf{Y})$. By the Yoneda lemma, the evaluation map is also in $\mathcal{C}$ in this case.

2.3. Definition. We will say that a category $\mathcal{C}$ with finite inverse limits is *ind-closed* if $\underline{\mathrm{Hom}}(\mathbf{X}, \mathbf{Y})$ is ind-representable for all objects $\mathbf{X}$ and $\mathbf{Y}$.

Note that $\mathrm{Ind}(\mathcal{C})$ again has all finite inverse limits, but the condition does not imply that $\mathrm{Ind}(\mathcal{C})$ is closed. Instead, we know that $\underline{\mathrm{Hom}}(\mathbf{X}, \mathbf{Y})$ exists (and is constructed in the obvious way) whenever $\mathbf{X}$ is compact. Again, the evaluation map is in $\mathrm{Ind}(\mathcal{C})$.

It may sometimes be difficult to determine whether a category is closed. However, for ind-closed, we have the following formal criterion.

2.4. Proposition. *Assume that $\mathcal{C}$ has all finite limits and colimits. Then it is ind-closed if and only if, for any object $\mathbf{X}$ of $\mathcal{C}$, the functor $\mathbf{Z} \mapsto \mathbf{Z} \times \mathbf{X}$ commutes with finite colimits.*

PROOF. The "Grothendieck construction" presents every presheaf as a colimit of representable presheaves (Mac Lane and Moerdijk [9, Section VII]). The system produced by the construction is filtering if the presheaf is left-exact (takes finite colimits to the corresponding inverse limits). Hence in a category that has finite colimits, the ind-representable presheaves are precisely the left-exact ones (it is clear that any ind-representable presheaf is left-exact).

Since any representable presheaf is left-exact, $\underline{\mathrm{Hom}}(\mathbf{X}, \mathbf{Y})$ is left-exact if $- \times \mathbf{X}$ preserves finite colimits.

Conversely, if $\mathcal{C}$ is ind-closed, then $- \times \mathbf{X}$ commutes with colimits in $\mathrm{Ind}(\mathcal{C})$, since it has a right adjoint there. The inclusion of $\mathcal{C}$ in $\mathrm{Ind}(\mathcal{C})$ preserves finite colimits, so the result follows. □

In our case, the requirement that finite colimits exist is too strong. It will be replaced by a more direct condition, elimination of imaginaries, ensuring that the system produced by the Grothendieck construction is filtering (see Proposition 4.2).

2.5. Closed functors. If $\mathcal{F}\colon \mathcal{C} \to \mathcal{D}$ is any functor, it extends by composition to a functor $\mathcal{F}\colon \mathrm{Ind}(\mathcal{C}) \to \mathrm{Ind}(\mathcal{D})$. If $\mathcal{C}$ is ind-closed, and $\mathcal{F}$ preserves finite inverse limits, we may apply $\mathcal{F}$ to the evaluation map, to obtain a map $\mathcal{F}(\underline{\mathrm{Hom}}(\mathbf{X}, \mathbf{Y})) \times \mathcal{F}(\mathbf{X}) \to \mathcal{F}(\mathbf{Y})$. This induces a map $\mathcal{F}(\underline{\mathrm{Hom}}(\mathbf{X}, \mathbf{Y})) \to \underline{\mathrm{Hom}}(\mathcal{F}(\mathbf{X}), \mathcal{F}(\mathbf{Y}))$, and we say that such a functor $\mathcal{F}$ is *closed* if this map is an isomorphism.

2.6. Endomorphisms of left exact functors. We recall that a map (natural transformation) θ from a functor $\mathcal{F}\colon \mathcal{C} \to \mathcal{D}$ to another functor $\mathcal{G}\colon \mathcal{C} \to \mathcal{D}$ consists of a collection morphisms $\theta_{\mathbf{X}}\colon \mathcal{F}(\mathbf{X}) \to \mathcal{G}(\mathbf{X})$, one for each object $\mathbf{X}$ of $\mathcal{C}$, commuting with all maps that come from $\mathcal{C}$.

Let $\mathcal{C}$ and $\mathcal{D}$ be two categories with finite inverse limits, and let $\mathcal{F}\colon \mathcal{C} \to \mathcal{D}$ be a left exact functor. For an object $\mathbf{T}$ of $\mathcal{D}$, we denote by $\mathrm{End}_{\mathbf{T}}(\mathcal{F})$ the set of natural transformations from $\mathbf{T} \times \mathcal{F}$ to $\mathcal{F}$ (where $(\mathbf{T} \times \mathcal{F})(\mathbf{X}) = \mathbf{T} \times \mathcal{F}(\mathbf{X})$). Each such natural transformation induces a transformation of $\mathbf{T} \times \mathcal{F}$ to itself, and therefore $\mathbf{T} \mapsto \mathrm{End}_{\mathbf{T}}(\mathcal{F})$ is a presheaf of monoids on $\mathcal{D}$. We denote by $\mathrm{Aut}_{\mathbf{T}}(\mathcal{F})$ the subset of $\mathrm{End}_{\mathbf{T}}(\mathcal{F})$ consisting of transformations where the induced endomorphism above is invertible. Thus it is a presheaf of groups on $\mathcal{D}$.

More generally, if $\mathcal{I}\colon \mathcal{C}_0 \to \mathcal{C}$ is another functor, we denote by $\mathrm{End}_{\mathbf{T}}(\mathcal{F}/\mathcal{I})$ and $\mathrm{Aut}_{\mathbf{T}}(\mathcal{F}/\mathcal{I})$ the subsets of $\mathrm{End}_{\mathbf{T}}(\mathcal{F})$ and $\mathrm{Aut}_{\mathbf{T}}(\mathcal{F}/\mathcal{I})$ consisting of those endomorphisms whose "restriction" to the composition $\mathcal{F} \circ \mathcal{I}$ is the identity. These are again presheaves on $\mathcal{D}$. The main observation in this section is that if $\mathcal{D}$ is closed, these presheaves are represented by a pro-monoid and a pro-group, respectively, of $\mathcal{D}$.

2.7. Proposition. *Let $\mathcal{I}\colon \mathcal{C}_0 \to \mathcal{C}$ and $\mathcal{F}\colon \mathcal{C} \to \mathcal{D}$ be functors. Assume that $\mathcal{D}$ admits finite inverse limits, and that $\underline{\mathrm{Hom}}(\mathcal{F}(\mathbf{X}), \mathcal{F}(\mathbf{Y}))$ exists for all objects $\mathbf{X}$ and $\mathbf{Y}$ of $\mathcal{C}$. Then the functors $\mathbf{T} \mapsto \mathrm{End}_{\mathbf{T}}(\mathcal{F}/\mathcal{I})$ and $\mathbf{T} \mapsto \mathrm{Aut}_{\mathbf{T}}(\mathcal{F}/\mathcal{I})$ are pro-representable, by a pro-monoid $\mathbf{M}$ and pro-group $\mathbf{G}$, respectively.*

If $\mathcal{H}\colon \mathcal{D} \to \mathcal{E}$ is any closed *functor (at least when restricted to the essential image of $\mathcal{F}$), then $\mathcal{H}(\mathbf{M})$ and $\mathcal{H}(\mathbf{G})$ represent $\mathbf{T} \mapsto \mathrm{End}_{\mathbf{T}}(\mathcal{H} \circ \mathcal{F}/\mathcal{I})$ and $\mathbf{T} \mapsto \mathrm{Aut}_{\mathbf{T}}(\mathcal{H} \circ \mathcal{F}/\mathcal{I})$, respectively.*

The proof is a direct consequence of the end construction, as explained below.

2.8. Corollary. *Let $\mathcal{I}\colon \mathcal{C}_0 \to \mathcal{C}$ be a functor into an ind-closed category $\mathcal{C}$. Then there are pro-monoid $\mathbf{M}$ and pro-group $\mathbf{G}$ in* $\mathrm{Ind}(\mathcal{C})$*, such that for any closed functor $\mathcal{H}\colon \mathcal{C} \to \mathcal{E}$,* $\mathrm{End}(\mathcal{H}/\mathcal{I}) = \Gamma(\mathcal{H}(\mathbf{M}))$ *and* $\mathrm{Aut}(\mathcal{H}/\mathcal{I}) = \Gamma(\mathcal{H}(\mathbf{G}))$.

PROOF. The embedding $\mathcal{F}$ of $\mathcal{C}$ into $\mathcal{D} = \mathrm{Ind}(\mathcal{C})$ satisfies the condition of the proposition. □

In the model theoretic application, $\mathcal{H}$ will be extension of constants (base-changes), so $\Gamma \circ \mathcal{H}$ is "taking $\mathcal{N}$-points" for some model $\mathcal{N}$ (or simply a model, in the language of Makkai and Reyes [10]; see Section 4.6).

2.9. Ends. The following construction is presented in Mac Lane [8, Section IX.5]. Let $\mathcal{S}\colon \mathcal{C}^{\mathrm{op}} \times \mathcal{C} \to \mathcal{D}$ be a bifunctor (so for any object $\mathbf{X}$ of $\mathcal{C}$, $\mathcal{S}(\mathbf{X}, -)$ and $\mathcal{S}(-, \mathbf{X})$ are a functor and a presheaf, resp., with values in $\mathcal{D}$). The *end* of $\mathcal{S}$ is an object e of $\mathcal{D}$, together with a morphism $p_{\mathbf{X}}\colon e \to \mathcal{S}(\mathbf{X}, \mathbf{X})$ for each object $\mathbf{X}$ of $\mathcal{C}$, such that for any morphism $f\colon \mathbf{X} \to \mathbf{Y}$ of $\mathcal{C}$, the diagram

$$\begin{array}{ccc} e & \xrightarrow{p_{\mathbf{X}}} & \mathcal{S}(\mathbf{X}, \mathbf{X}) \\ {\scriptstyle p_{\mathbf{Y}}}\downarrow & & \downarrow{\scriptstyle \mathcal{S}(\mathbf{X}, f)} \\ \mathcal{S}(\mathbf{Y}, \mathbf{Y}) & \xrightarrow[\mathcal{S}(f, \mathbf{Y})]{} & \mathcal{S}(\mathbf{X}, \mathbf{Y}) \end{array} \tag{3}$$

commutes, and which is universal with these properties.

2.10. Example. Let $\mathcal{F}, \mathcal{G}\colon \mathcal{C} \to \mathcal{D}$ be two functors, and let $\mathcal{S}\colon \mathcal{C}^{\mathrm{op}} \times \mathcal{C} \to \mathcal{S}et$ be the functor $(\mathbf{X}, \mathbf{Y}) \mapsto \mathrm{Hom}(\mathcal{F}(\mathbf{X}), \mathcal{G}(\mathbf{Y}))$. The set $\mathrm{Hom}(\mathcal{F}, \mathcal{G})$ of natural transformations between $\mathcal{F}$ and $\mathcal{G}$ maps to each $\mathcal{S}(\mathbf{X}, \mathbf{X})$ by restriction, and the commuting diagrams as above amount to the statement that a collection of such restrictions amalgamates to a natural transformation. Hence $\mathrm{Hom}(\mathcal{F}, \mathcal{G})$ is the end of $\mathcal{S}$.

The end of a functor is in a natural way an inverse limit of diagrams as above. Hence it exists if the category $\mathcal{D}$ admits all inverse limits. The proof of Proposition 2.7 is a repetition of the above example internally:

PROOF OF PROPOSITION 2.7. Assume first that $\mathcal{C}_0$ is the empty category. By assumption, we have a functor $\mathcal{S}\colon \mathcal{C}^{\mathrm{op}} \times \mathcal{C} \to \mathcal{D}$, $\mathcal{S}(\mathbf{X}, \mathbf{Y}) = \underline{\mathrm{Hom}}(\mathcal{F}(\mathbf{X}), \mathcal{F}(\mathbf{Y}))$.

Since the category $\mathrm{Pro}(\mathcal{D})$ admits all filtered inverse limits, and $\mathcal{D}$ (and therefore $\mathrm{Pro}(\mathcal{D})$) admits all finite inverse limits, $\mathrm{Pro}(\mathcal{D})$ admits all inverse limits (Mac Lane [8, Section IX.1]). Therefore, the end $\mathbf{M}$ of $\mathcal{S}$ exists. The fact that $\mathbf{M}$ represents $\mathbf{T} \mapsto \mathrm{End}_{\mathbf{T}}(\mathcal{F})$ follows directly from the definitions.

The fact that $\mathbf{M}$ can be given a system of monoid objects follows from the fact that $\mathrm{End}_{\mathbf{T}}(\mathcal{F})$ is the filtered inverse limit of $\mathrm{End}_{\mathbf{T}}(\mathcal{F}_\alpha)$, where $\mathcal{F}_\alpha$ are restrictions of $\mathcal{F}$ to finite subcategories $\mathcal{C}_\alpha$ of $\mathcal{C}$.

When $\mathcal{C}_0$ is not empty, the pro-monoid in the theorem is the pullback of $\mathbf{M}$ constructed above along the product of the maps $\mathbf{1} \to \underline{\mathrm{Hom}}\left(\mathcal{F}(\mathcal{I}(\mathbf{X})), \mathcal{F}(\mathcal{I}(\mathbf{X}))\right)$ corresponding to the identity $\mathcal{F}(\mathcal{I}(\mathbf{X})) \to \mathcal{F}(\mathcal{I}(\mathbf{X}))$. Hence it is again of the same type. Likewise, the object $\mathbf{G}$ is a suitable pullback.

The last statement follows since in this case, by assumption, the functors $\mathcal{H} \circ \mathcal{S}$ and $(\mathbf{X}, \mathbf{Y}) \mapsto \underline{\mathrm{Hom}}\left(\mathcal{H}(\mathcal{F}(\mathbf{X})), \mathcal{H}(\mathcal{F}(\mathbf{Y}))\right)$ are isomorphic, hence so are their ends. $\square$

3. Model theory and internality

In this section we summarize some basic notions from model theory. Most of them are classical, and can be found, for example, in Marker [11]. The exception is the notion of internality, which is the central notion for us, and which is recalled here following Hrushovski [4]. The first three subsections are illustrated in Example 3.4.

3.1. Theories, models and definable sets. Logic is concerned with studying mathematical theories as mathematical objects. Thus, the data of a *theory* consists of a formal language, together with a collection of statements (axioms) written in this language. A *model* of the theory $\mathcal{T}$ is a structure consisting of an interpretation of the symbols in the language of $\mathcal{T}$, in which all the axioms of $\mathcal{T}$ hold. For example, the theory of fields, and the theory ACF of algebraically closed fields can both be written in a language constructed of the symbols $(0, 1, +, -, \cdot)$, and a model of ACF is a particular algebraically closed field (see Example 3.4).

A *formula* is written in the same formal language, but has *free variables*, into which elements of the model can be plugged. Any such formula $\phi(x_1, \dots, x_n)$ (where $x_1, \dots, x_n$ contain all free variables of ϕ) thus determines a subset $\phi(\mathcal{M})$ of $\mathcal{M}^n$, for any model $\mathcal{M}$, namely, the set of all tuples $\bar{a}$ for which $\phi(\bar{a})$ holds. Two formulas ϕ and ψ are *equivalent* if $\phi(\mathcal{M}) = \psi(\mathcal{M})$ for all models $\mathcal{M}$. An equivalence class under this relation is called a *definable set*. Thus, if $\mathbf{X}$ is a definable set, there is a well defined set $\mathbf{X}(\mathcal{M})$ for any model $\mathcal{M}$ (but the syntactic information is lost). If we have a fixed model $\mathcal{M}$ in mind, the set $\mathbf{X}(\mathcal{M})$ is also called definable.

If $\mathcal{M}$ is the model of some theory, the set of all statements (in the underlying language) that are true in $\mathcal{M}$ is a theory $\mathcal{T}(\mathcal{M})$, and $\mathcal{M}$ is a model of $\mathcal{T}(\mathcal{M})$.

We will assume all our theories to be multisorted, i.e., the variables of a formula can take values in any number of disjoint sets. In fact, if $\mathbf{X}$ and $\mathbf{Y}$ are sorts, we view $\mathbf{X} \times \mathbf{Y}$ as a new sort. In particular, any definable set is a subset of some sort (for example, in ACF the sorts correspond to the affine spaces $\mathbb{A}^n$). By a statement such as $a \in \mathcal{M}$ we will mean that a is an element of one of the sorts. Likewise, a subset of $\mathcal{M}$ is collection of subsets of $\mathbf{X}(\mathcal{M})$, one for each sort $\mathbf{X}$.

3.2. Parameters and definable closure. Let $\mathcal{M}$ be a model of a theory $\mathcal{T}$, $\mathbf{X}$ and $\mathbf{Y}$ definable sets. A function $f\colon \mathbf{X}(\mathcal{M}) \to \mathbf{Y}(\mathcal{M})$ is *definable* if its graph is definable. An element $b \in \mathcal{M}$ is definable over another element a if $f(a) = b$ for some definable function f. The *definable closure* $\mathrm{dcl}(\mathcal{A})$ of a subset $\mathcal{A}$ of $\mathcal{M}$ is the set of elements definable over some tuple a of elements of $\mathcal{A}$. The set $\mathcal{A}$ is *definably closed* if $\mathcal{A} = \mathrm{dcl}(\mathcal{A})$. We denote by $\mathbf{X}(\mathcal{A})$ the set $\mathbf{X}(\mathcal{M}) \cap \mathrm{dcl}(\mathcal{A})$.

More generally, we may consider formulas $\phi(x, a)$ with parameters in $\mathcal{A}$, obtained by plugging a tuple a from $\mathcal{A}$ in a regular formula $\phi(x, y)$. These determine subsets in models containing $\mathcal{A}$, give rise to definable sets over $\mathcal{A}$. If $\mathcal{T} = \mathcal{T}(\mathcal{M})$ is the theory of $\mathcal{M}$, it is possible to expand $\mathcal{T}$ by constant symbols and suitable axioms to obtain a theory $\mathcal{T}_{\mathcal{A}}$, whose models are the models of $\mathcal{T}$ containing $\mathcal{A}$, and whose definable sets the $\mathcal{A}$-definable sets of $\mathcal{T}$.

An *elementary map* from a model $\mathcal{M}$ to another model $\mathcal{N}$ is a map α from $\mathcal{M}$ to $\mathcal{N}$ such that for each definable set $\mathbf{X}$, $\mathbf{X}(\mathcal{M}) = \alpha^{-1}(\mathbf{X}(\mathcal{N}))$. Such a map is necessarily an embedding, and when it is fixed, we identify elements of $\mathcal{M}$ with their image in $\mathcal{N}$. An *automorphism* of $\mathcal{M}$ is an invertible elementary map. The automorphism α of $\mathcal{M}$ is over $\mathcal{A} \subseteq \mathcal{M}$ if α fixes each element of $\mathcal{A}$.

An automorphism over $\mathcal{A}$ will also fix pointwise any element of $\mathrm{dcl}(\mathcal{A})$. Hence $\mathrm{dcl}(\mathcal{A})$ is contained in the set $\mathcal{M}^G$ of elements of $\mathcal{M}$ fixed pointwise by all elements of the group $G = \mathrm{Aut}(\mathcal{M}/\mathcal{A})$ of automorphisms of $\mathcal{M}$ fixing $\mathcal{A}$ pointwise. This inclusion might be strict, in general, but any model admits an elementary embedding in a *homogeneous* model, where this inclusion becomes an equality whenever the cardinality of $\mathcal{A}$ is smaller than that of the model.

3.3. Imaginaries and interpretations. A *definable equivalence relation* on a definable set $\mathbf{X}$ is a definable subset of $\mathbf{X} \times \mathbf{X}$ that determines an equivalence relation in any model. The theory *eliminates imaginaries* if any equivalence relation has a quotient, i.e., any equivalence relation $\mathbf{E}(x, y)$ can be represented as $f(x) = f(y)$ for some definable function f on $\mathbf{X}$.

A *definition* of a theory $\mathcal{T}_1$ in another theory $\mathcal{T}_2$ is specified by assigning, to each definable set $\mathbf{X}$ of $\mathcal{T}_1$ a definable set $\mathbf{X}_2$ of $\mathcal{T}_2$, such that for any model $\mathcal{M}_2$ of $\mathcal{T}_2$, the sets $\mathbf{X}_2(\mathcal{M}_2)$ comprise a model $\mathcal{M}_1$ of $\mathcal{T}_1$ (when interpreted as a $\mathcal{T}_1$ structure in the obvious way). The expansion $\mathcal{T}_{\mathcal{A}}$ of $\mathcal{T}$ mentioned above is an example of a definition of $\mathcal{T}$ in $\mathcal{T}_{\mathcal{A}}$.

For any theory $\mathcal{T}$ there is a theory $\mathcal{T}^{\mathrm{eq}}$ and a definition of $\mathcal{T}$ in $\mathcal{T}^{\mathrm{eq}}$, where $\mathcal{T}^{\mathrm{eq}}$ eliminates imaginaries, and the definition is universal with this property. Any model of $\mathcal{T}$ can be expanded uniquely to a model of $\mathcal{T}^{\mathrm{eq}}$. One can often assume that that a theory admits elimination of imaginaries via this process. Through the rest of this section, we assume that our theories eliminate imaginaries.

3.4. Example. We illustrate some of the notions discussed above for the theory ACF of algebraically closed fields. All the fact in this example appear in Marker [11].

As mentioned above, this theory is given in the language determined by functions symbols $+, -, \cdot$, and constant symbols 0 and 1. The properties of an algebraically closed field can be expressed by (infinitely many) statements in this language. For instance, commutativity of addition is written as $\forall x \forall y (x + y = y + x)$.

An example of a model is the field $\mathbb{C}$ of complex numbers. The full theory $\mathcal{T}(\mathbb{C})$ of the complex numbers as a field turns out to be ACF together with the (infinitely many) axioms stating that the characteristic is 0.

The sorts in this theory are all Cartesian powers of the field sort, hence can be identified with $\mathbb{A}^n$, the affine n-space. A formula consists of quantifiers ($\forall$ and $\exists$) applied to Boolean combinations of polynomial equations. For example $\exists y(y^2 = x)$. Such a formula determines a subset of some $\mathbb{A}^n$ (where n is the number of free variables). Whereas the formulas may have an arbitrary number of quantifiers, the subset it defines can also be defined by a formula without quantifiers. In other words, the definable sets can all be represented by quantifier-free formulas, i.e., they are the constructible subsets.

An automorphism is a bijection of the field with itself that preserves the operations, 0 and 1. Hence it is the same as a field automorphism. A set of parameters is a subset of some algebraically closed field. Since the field operations are definable, a definably closed set is a field, which is also perfect, since extracting pth roots (when $p > 0$ is the characteristic) is again a definable function. By considering the automorphism group of the model, we find that perfect fields are precisely the definably closed sets: The definable closure of a set $\mathcal{A}$ is contained in the fixed field of the group of automorphisms of the algebraically closed field fixing $\mathcal{A}$. This fixed field is, by field theory, the perfect field generated by $\mathcal{A}$.

Let $\mathbf{X}$ be the definable set of pairs (x, y). The definable subset $\mathbf{E}\big((x, y), (z, w)\big)$ of $\mathbf{X}\times\mathbf{X}$ given by $(x = z \wedge y = w) \vee (x = w \wedge y = z)$ is an equivalence relation, which is definable not only in ACF, but in any theory (since the formula mentions only equality). In the theory of equality (with no other symbols) this relation does not have a quotient. However, if $\mathbf{X}$ is a field, the definable function $f(x, y) = (x+y, xy)$ is a quotient. In fact, ACF eliminates imaginaries.

3.5. Internality. Let $\mathcal{T}$ be an expansion of $\mathcal{T}_0$ (so we have an interpretation of $\mathcal{T}_0$ in $\mathcal{T}$), and let $\mathcal{M}$ be a model of $\mathcal{T}$, with restriction $\mathcal{M}_0$ to $\mathcal{T}_0$. An automorphism of $\mathcal{M}$ over $\mathcal{M}_0$ consists certain bijections from $\mathbf{X}(\mathcal{M})$ to itself, for each definable set $\mathbf{X}$ of $\mathcal{T}$. These bijections will rarely come from *definable* bijections in $\mathcal{T}$ (even with parameters). Internality, which is the main notion of interest for us, provides a situation where all such automorphisms are, in fact, definable (with parameters in $\mathcal{M}$).

A definable set $\mathbf{X}$ in a theory $\mathcal{T}$ is *stably embedded* if for any model $\mathcal{M}$ of $\mathcal{T}$, any $\mathcal{M}$-definable subset of $\mathbf{X}^n$ (for any n) is definable over $\mathbf{X}(\mathcal{M})$. An interpretation of a theory $\mathcal{T}_0$ in $\mathcal{T}$ is a *stable embedding* if any definable set of $\mathcal{T}_0$ is stably embedded when viewed as a definable set in $\mathcal{T}$.

Given a stable embedding of $\mathcal{T}_0$ in $\mathcal{T}$, let $\mathcal{M}$ be a model of $\mathcal{T}$ of cardinality bigger than the language, and let $\mathcal{M}_0$ be the induced model of $\mathcal{T}_0$. Then $\mathcal{T}$ is an *internal cover* of $\mathcal{T}_0$ if for any such model there is a small subset $\mathcal{A}$ of $\mathcal{M}$ (smaller than the cardinality of $\mathcal{M}$), such that $\mathrm{dcl}(\mathcal{M}_0 \cup \mathcal{A}) = \mathcal{M}$. A set $\mathcal{A}$ of this form can (and will) always be taken to be definably closed, and will be called a set of *internality parameters*. We will assume for simplicity that such an $\mathcal{A}$ can be found with $\mathcal{A}_0 = \mathcal{M}_0 \cap \mathcal{A}$ contained in the definable closure of 0 in $\mathcal{T}_0$ (the theory works also without this assumption, but then one has to deal with definable groupoids, rather than groups; cf. Hrushovski [4]).

3.6. Example. Let $\mathcal{T}$ be the theory generated by two sorts, $\mathbf{L}$ and $\mathbf{X}$, and stating that $\mathbf{L}$ is an algebraically closed field, and $\mathbf{X}$ is a vector space over $\mathbf{L}$, of a fixed dimension n (in the natural language for such a structure). Then the theory $\mathcal{T}_0$ of algebraically closed fields is stably embedded in $\mathcal{T}$ via $\mathbf{L}$, and $\mathcal{T}$ is an internal cover of $\mathcal{T}_0$, since in any model $\mathcal{M}$ of $\mathcal{T}$, the definable closure of any basis of $\mathbf{X}(\mathcal{M})$ over $\mathbf{L}(\mathcal{M})$ is a set of internality parameters.

The internal covers are interesting because of the following theorem, which was first proved by Zil′ber ([15]) in the context of strongly minimal structures. The version here was proved in Hrushovski [3, Appendix B], but was reformulated in (approximately) our language in Hrushovski [4].

3.7. Theorem. *Let $\mathcal{T}$ be an internal cover of $\mathcal{T}_0$. There is a pro-group* $\mathbf{G}$ *in* $\mathcal{T}$, *together with a definable action* $m_Q\colon \mathbf{G} \times \mathbf{Q} \to \mathbf{Q}$ *of* $\mathbf{G}$ *on every definable set* $\mathbf{Q}$ *of* $\mathcal{T}$, *such that for any model* $\mathcal{M}$ *of* $\mathcal{T}$, $\mathbf{G}(\mathcal{M})$ *is identified with* $\mathrm{Aut}(\mathcal{M}/\mathcal{M}_0)$ *through this action (with* $\mathcal{M}_0$ *the restriction of* $\mathcal{M}$ *to* $\mathcal{T}_0$*).*

Another part of this theorem, which will not be discussed in this paper, provides a Galois correspondence. As mentioned in the introduction, Poizat [13] developed this Galois correspondence to recover the Galois theory of a linear differential equation. Other applications appear in Pillay [12], Hrushovski [3] and Kamensky [5]. In Kamensky [7], this theorem is the main tool in a model theoretic proof of the Tannakian formalism.

4. Categorical internality

Let $\mathcal{T}$ be a first order theory. The collection of definable sets, with definable maps between them, forms a category, $\mathcal{D}_{\mathcal{T}}$. Our aim in this section is to reformulate some of the notions and results of Section 3, and especially the notion of internality, in terms of the category $\mathcal{D}_{\mathcal{T}}$. We in fact work with more general categories, and obtain a weaker result than Theorem 3.7. In Section 5, we introduce additional assumptions on the category, and deduce the full strength of the theorem.

4.1. The category of definable sets. The structure of the category $\mathcal{D}_{\mathcal{T}}$ was described in Makkai and Reyes [10]. We will not need the full characterization, since our result holds under more general and simpler assumptions. We mention, however, that a category of the form $\mathcal{D}_{\mathcal{T}}$ admits all finite inverse limits, and finite coproducts.[1] A *logical functor* between such categories preserves these limits (and has additional properties). When $\mathcal{T}$ eliminates imaginaries, $\mathcal{D}_{\mathcal{T}}$ is a special kind of a *pre-topos*, in the sense of Grothendieck et al. [2]. In fact, the construction of $\mathcal{T}^{\mathrm{eq}}$ appeared first in Makkai and Reyes [10], as a universal embedding of a logical category in a pretopos.

The categories $\mathrm{Pro}(\mathcal{D}_{\mathcal{T}})$ and $\mathrm{Ind}(\mathcal{D}_{\mathcal{T}})$ were discussed in Kamensky [6]. Essentially the results are that one may evaluate $\mathcal{M}$ points of a pro- or ind- definable set by evaluating them for each member in a system, and taking the corresponding limit. Definable maps between such sets are then precisely those that come from maps in the corresponding category, and if the model is sufficiently saturated, a definable map is an isomorphism if it is bijective on $\mathcal{M}$-points.

We denote by $\mathbf{1}$ and $\mathbf{2}$ the definable sets with 1 and 2 (named) elements. These are the final object in $\mathcal{D}_{\mathcal{T}}$ and its coproduct with itself, respectively.

4.2. Proposition. *The theory $\mathcal{T}$ eliminates imaginaries if and only if $\mathcal{D}_{\mathcal{T}}$ is ind-closed* (*in the sense of Definition* 2.3).

Proof. Assume $\mathcal{D}_{\mathcal{T}}$ is ind-closed, and let $\mathbf{E} \subseteq \mathbf{X} \times \mathbf{X}$ be an equivalence relation. The corresponding characteristic function $\chi_{\mathbf{E}}\colon \mathbf{X} \times \mathbf{X} \to \mathbf{2}$ corresponds, by assumption, to some $t\colon \mathbf{X} \to \mathcal{P}(\mathbf{X}) := \underline{\mathrm{Hom}}(\mathbf{X}, \mathbf{2})$. If $\mathcal{P}(\mathbf{X})$ is represented by some system $(\mathbf{Y}_a, f_{a,b})$, the map t is represented by some $t_a\colon \mathbf{X} \to \mathbf{Y}_a$, and so

[1] We allow disjoint unions of definable sets as a basic logical operation. This is automatic if there are at least two constant symbols, or if $\mathcal{T}$ eliminates imaginaries, and is convenient otherwise. It implies that $\mathcal{D}_{\mathcal{T}}$ has finite coproducts.

the relation $\mathbf{E}(x, y)$ is equivalent to the intersection over all b of the conditions $f_{a,b}(t_a(x)) = f_{a,b}(t_a(y))$, hence, by compactness, by one such condition. The map $f_{a,b} \circ t_a$ onto its image is then the quotient.

Conversely, assume that $\mathcal{T}$ eliminates imaginaries, and let $\mathbf{X}$ and $\mathbf{Y}$ be two definable sets. Let $\mathcal{I}$ be the category whose objects are definable maps $\mathbf{Z} \times \mathbf{X} \to \mathbf{Y}$, and whose morphisms are definable maps of the $\mathbf{Z}$ coordinate that make the obvious diagram commute. For each such object, we view the elements of $\mathbf{Z}$ as maps from $\mathbf{X}$ to $\mathbf{Y}$. Let $\bar{\mathbf{Z}}$ be the quotient of $\mathbf{Z}$ by the definable equivalence relation saying that $z_1, z_2 \in \mathbf{Z}$ define the same map. The quotient map determines a map in $\mathcal{I}$ from $\mathbf{Z} \times \mathbf{X} \to \mathbf{Y}$ to $\bar{\mathbf{Z}} \times \mathbf{X} \to \mathbf{Y}$, which coequalizes any two maps of $\mathcal{I}$ into its domain. Since $\mathcal{I}$ also inherits the coproducts from $\mathcal{D}_\mathcal{T}$, it has all finite colimits, and is therefore filtering category. The "forgetful" functor that assigns to an object $\mathbf{Z} \times \mathbf{X} \to \mathbf{Y}$ of $\mathcal{I}$ the definable set $\mathbf{Z}$ in $\mathcal{T}$ is therefore a filtering system in $\mathcal{D}_\mathcal{T}$. It is clear that this system represents $\underline{\mathrm{Hom}}(\mathbf{X}, \mathbf{Y})$. □

4.3. Remark. The system used in the second part of the above proof is the one used in the Grothendieck construction, to show that any presheaf is a colimit of sheaves. The existence of quotients produces the coequalizers in this system, despite the fact that the original category $\mathcal{D}_\mathcal{T}$ might not have admitted them.

A model $\mathcal{M}$ of $\mathcal{T}$ determines a functor $\mathbf{X} \mapsto \mathbf{X}(\mathcal{M})$ from $\mathcal{D}_\mathcal{T}$ to the category of sets, which preserves all inverse limits, coproducts, and images. Conversely, any such functor is (isomorphic to) a model. An elementary map from $\mathcal{M}$ to $\mathcal{N}$ as models is the same as a map of the corresponding functors. A subset $\mathcal{A}$ of $\mathcal{M}$ is definably closed if and only if it defines a left-exact subfunctor of $\mathcal{M}$ that preserves coproducts.

Similarly, a definition of the theory $\mathcal{T}_1$ in another theory $\mathcal{T}_2$ is the same as a logical functor $\mathcal{I}$ (in the same sense as above) from $\mathcal{D}_{\mathcal{T}_1}$ to $\mathcal{D}_{\mathcal{T}_2}$. Given a model $\mathcal{M}$ of $\mathcal{T}_2$ (viewed as a functor), the corresponding model of $\mathcal{T}_1$ is $\mathcal{M} \circ \mathcal{I}$ (See also Makkai and Reyes [10, Section 7]). We will from now on identify models with the corresponding functors (so that $\mathbf{X}(\mathcal{M}) = \mathcal{M}(\mathbf{X})$ for a definable set $\mathbf{X}$ and a model $\mathcal{M}$), and likewise for definably closed sets and definitions.

The following result is essentially the same as Makkai and Reyes [10, Theorem 7.1.8].

4.4. Proposition. *Let $\mathcal{I}: \mathcal{D}_{\mathcal{T}_1} \to \mathcal{D}_{\mathcal{T}_2}$ be a definition, where $\mathcal{T}_1$ eliminates imaginaries, and let $\mathcal{I}^*$ be the induced functor from models of $\mathcal{T}_2$ to models of $\mathcal{T}_1$ (obtained by composition). Then $\mathcal{I}$ is an equivalence of categories if and only if $\mathcal{I}^*$ is.*

PROOF. Assume that $\mathcal{I}^*$ is an equivalence. We will show that $\mathcal{H}$ is essentially surjective. Let $\mathbf{X}$ be a definable set of $\mathcal{T}_2$, and assume first that $\mathbf{X} \subseteq \mathcal{I}(\mathbf{Y})$ for some definable set $\mathbf{Y}$ of $\mathcal{T}_1$. Then for any two models $\mathcal{M}$ and $\mathcal{N}$ of $\mathcal{T}_2$ such that $\mathcal{M} \circ \mathcal{I} = \mathcal{N} \circ \mathcal{I}$, we have $\mathcal{M}(\mathbf{X}) = \mathcal{N}(\mathbf{X})$ (by equivalence). Hence by Beth's definability theorem, $\mathbf{X} = \mathcal{I}(\mathbf{X}_0)$ for some $\mathbf{X}_0$.

We next claim that for any model $\mathcal{M}_2$ of $\mathcal{T}_2$, any element of $\mathcal{M}_2$ is definable over the restriction $\mathcal{M}_1 = \mathcal{M}_2 \circ \mathcal{I}$. Indeed, $\mathcal{M}_2$ embeds into a sufficiently saturated model $\mathcal{N}_2$, and if there is an automorphism of $\mathcal{N}_2$ that fixes pointwise $\mathcal{M}_1$ but not $\mathcal{M}_2$, we get two different embeddings of $\mathcal{M}_2$ in $\mathcal{N}_2$, which agree on $\mathcal{M}_1$, contradicting the equivalence.

It follows, by compactness, that any definable set $\mathbf{Y}$ of $\mathcal{T}_2$ is the image of a definable map from a definable set $\mathcal{I}(\mathbf{X})$ that comes from $\mathcal{T}_1$. By the previous point, the kernel of this map also comes from $\mathcal{T}_1$, hence, since $\mathcal{T}_1$ eliminates imaginaries, $\mathbf{Y}$ itself comes from $\mathcal{T}_1$ as well.

This shows that $\mathcal{I}$ is essentially surjective. It is full since any map is a definable set, and it is faithful since for equality of definable sets is first order. Consequently, $\mathcal{I}$ is an equivalence. The other direction is trivial. □

As indicated in Section 3.2, one often considers sets definable with parameters in a set $\mathcal{A}$. These can be viewed as definable sets in an expansion $\mathcal{T}_{\mathcal{A}}$ of $\mathcal{T}$. The expansion corresponds to a functor $\mathcal{I}_{\mathcal{A}}\colon \mathcal{D}_{\mathcal{T}} \to \mathcal{D}_{\mathcal{T}_{\mathcal{A}}}$.

4.5. Proposition. *Assume $\mathcal{T}$ eliminates imaginaries, and let $\mathcal{M}$ be a model of $\mathcal{T}$. Then for any definably closed $\mathcal{A} \subseteq \mathcal{M}$, the functor $\mathcal{I}_{\mathcal{A}}$ is closed (in the sense of Section* 2.5).

PROOF. A definable map f in $\mathcal{T}_{\mathcal{A}}$ from $\mathbf{Z}$ to $\mathcal{I}_{\mathcal{A}}(\underline{\mathrm{Hom}}(\mathbf{X},\mathbf{Y}))$ is given by formulas $\phi(w,z) \subseteq \mathbf{W}\times\mathbf{Z}$ and $\mathbf{F}\subseteq \mathbf{W}\times\mathbf{Z}\times\mathcal{I}_{\mathcal{A}}(\underline{\mathrm{Hom}}(\mathbf{X},\mathbf{Y}))$, such that $\mathbf{Z}=\phi(a,-)$, $f=\mathbf{F}(a,-,-)$, $a\in\mathbf{W}(\mathcal{A})$ and $\mathbf{F},\mathbf{W},\phi$ are in $\mathcal{T}$. There is a $\mathcal{T}$ definable subset $\mathbf{W}_0\subseteq\mathbf{W}$ such that $\mathbf{F}(w,-,-)$ is a function from $\phi(w,-)$ to $\underline{\mathrm{Hom}}(\mathbf{X},\mathbf{Y})$ for all $w\in\mathbf{W}_0$. We may assume that $\mathbf{W}=\mathbf{W}_0$, so that $\mathbf{F}$ gives a $\mathcal{T}$ definable function from ϕ to $\underline{\mathrm{Hom}}(\mathbf{X},\mathbf{Y})$. Thus it corresponds to a map $\tilde{F}\colon \phi\times\mathbf{X}\to\mathbf{Y}$. The restriction $\tilde{f}\colon \mathbf{Z}\times\mathbf{X}\to\mathbf{Y}$ of this map to $\phi(a,-)\times\mathbf{X}$ corresponds to the original map f.

Reversing this argument gives, for each map $\tilde{f}\colon \mathbf{Z}\times\mathbf{X}\to\mathbf{Y}$ in $\mathcal{T}_{\mathcal{A}}$ a corresponding map $f\colon \mathbf{Z}\to\mathcal{I}_{\mathcal{A}}(\underline{\mathrm{Hom}}(\mathbf{X},\mathbf{Y}))$. This shows that $\mathcal{I}_{\mathcal{A}}(\underline{\mathrm{Hom}}(\mathbf{X},\mathbf{Y}))$ represents the correct functor in $\mathcal{T}_{\mathcal{A}}$, and therefore coincides with $\underline{\mathrm{Hom}}(\mathcal{I}_{\mathcal{A}}(\mathbf{X}),\mathcal{I}_{\mathcal{A}}(\mathbf{Y}))$.

(More abstractly, $\mathcal{T}_{\mathcal{A}}$ can be viewed as a full subcategory of $\mathrm{Pro}(\mathcal{T})$, and $\mathcal{I}_{\mathcal{A}}$ as the inclusion of $\mathcal{T}$ in $\mathrm{Pro}(\mathcal{T})$. This inclusion is always closed.) □

4.6. Since the definable closure of the empty set can be identified with the set of one element definable subsets, we get that for each definable set, $\mathbf{X}(0)=\Gamma(\mathbf{X})=\mathrm{Hom}(\mathbf{1},\mathbf{X})$. Likewise, for any subset $\mathcal{A}$ of a model $\mathcal{M}$ we have

$$\mathcal{A}(\mathbf{X})=\mathbf{X}(\mathcal{A})=\Gamma_{\mathcal{A}}(\mathcal{I}_{\mathcal{A}}(\mathbf{X})) \tag{4}$$

(where $\Gamma_{\mathcal{A}}$ is global sections in $\mathcal{D}_{\mathcal{T}_{\mathcal{A}}}$). In other words, $\mathcal{A}=\Gamma_{\mathcal{A}}\circ\mathcal{I}_{\mathcal{A}}$. We deduce that $\underline{\mathrm{Hom}}(\mathbf{X},\mathbf{Y})$ gives the expected result when evaluated on parameters.

4.7. Corollary. *Assume $\mathcal{T}$ eliminates imaginaries. For any definably closed set $\mathcal{A}$ and any definable sets $\mathbf{X}$ and $\mathbf{Y}$, the evaluation map identifies $\underline{\mathrm{Hom}}(\mathbf{X},\mathbf{Y})(\mathcal{A})$ with the set of $\mathcal{A}$-definable maps from $\mathbf{X}$ to $\mathbf{Y}$. In particular, $\mathcal{P}(\mathbf{X})(\mathcal{A})$ is the set of $\mathcal{A}$-definable subsets of $\mathbf{X}$.*

PROOF. According to Proposition 4.5 and the above remark,

$$\begin{aligned}\underline{\mathrm{Hom}}(\mathbf{X},\mathbf{Y})(\mathcal{A}) &= \Gamma_{\mathcal{A}}\left(\mathcal{I}_{\mathcal{A}}(\underline{\mathrm{Hom}}(\mathbf{X},\mathbf{Y}))\right)\\ &= \mathrm{Hom}_{\mathcal{A}}\left(1,\underline{\mathrm{Hom}}(\mathbf{X},\mathbf{Y})\right)=\mathrm{Hom}_{\mathcal{A}}(\mathbf{X},\mathbf{Y}).\end{aligned}$$ □

Finally, we can characterize the stable embeddings.

4.8. Proposition. *Let $\mathcal{I}\colon \mathcal{D}_{\mathcal{T}_1}\to\mathcal{D}_{\mathcal{T}_2}$ be an interpretation, where both theories eliminate imaginaries. Then the interpretation is stably embedded if and only if $\mathcal{I}$ is closed.*

PROOF. Assume that the interpretation is stably embedded, let $\mathbf{X}, \mathbf{Y}$ be definable sets in $\mathcal{T}_1$, and let $\mathcal{M}$ be a model of $\mathcal{T}_2$. The map $t\colon \mathcal{I}(\underline{\mathrm{Hom}}(\mathbf{X},\mathbf{Y})) \to \underline{\mathrm{Hom}}(\mathcal{I}(\mathbf{X}),\mathcal{I}(\mathbf{Y}))$ determines a map

$$t\colon \mathcal{I}(\underline{\mathrm{Hom}}(\mathbf{X},\mathbf{Y}))(\mathcal{M}) \to \underline{\mathrm{Hom}}(\mathcal{I}(\mathbf{X}),\mathcal{I}(\mathbf{Y}))(\mathcal{M})$$

By Corollary 4.7, the codomain is the set of $\mathcal{M}$-definable maps from $\mathcal{I}(\mathbf{X})$ to $\mathcal{I}(\mathbf{Y})$ in $\mathcal{T}_2$, while the domain is the set of definable maps from $\mathbf{X}$ to $\mathbf{Y}$ in $\mathcal{T}_1$ with parameters from the model $\mathcal{M} \circ \mathcal{I}$ (and t assigns to each map the same map viewed as a map between $\mathcal{I}(\mathbf{X})$ and $\mathcal{I}(\mathbf{Y})$). Since $\mathcal{T}_1$ asserts that all maps in $\underline{\mathrm{Hom}}(\mathbf{X},\mathbf{Y})$ are distinct, the same has to hold in $\mathcal{T}_2$. Therefore, t is injective. Since $\mathcal{T}_1$ is stably embedded, any $\mathcal{M}$-definable map between $\mathcal{I}(\mathbf{X})$ and $\mathcal{I}(\mathbf{Y})$ is definable with parameters in $\mathcal{M} \circ \mathcal{I}$, hence t is surjective. Since t is a bijection in every model, t is an isomorphism.

Assume conversely that $\mathcal{I}$ is closed, let $\mathbf{X}$ be a definable set in $\mathcal{T}_1$, and let $\mathcal{M}$ be a model of $\mathcal{T}_2$. Any $\mathcal{M}$-definable subset $\mathbf{Z} \subseteq \mathcal{I}(\mathbf{X})$ (in $\mathcal{T}_2$), corresponds to an $\mathcal{M}$-point of $\mathcal{P}(\mathcal{I}(\mathbf{X}))$ (by Corollary 4.7), hence to an $\mathcal{M} \circ \mathcal{I}$ point of $\mathcal{P}(\mathbf{X})$, which can be used to define $\mathbf{Z}$. □

4.9. Internality. We now return to discuss categories $\mathcal{C}$ which are ind-closed and admit finite coproducts. Thus, $\mathcal{C} = \mathcal{D}_{\mathcal{T}}$ for a theory $\mathcal{T}$ with elimination of imaginaries is an example. Our aim is to formulate the notion of internality in Section 3.5 in this setting. From Proposition 4.8, we already have the notion of a stably embedded interpretation: This is simply a closed functor between two such ind-closed categories.

Let $\mathcal{T}$ be an internal cover of $\mathcal{T}_0$ (both eliminating imaginaries), and let $\mathbf{X}$ be a definable set in $\mathcal{T}$. By assumption (and compactness), there is a parameter a, and an a-definable surjective map f_a from some sort in $\mathcal{T}_0$ onto $\mathbf{X}$. Since $\mathcal{T}_0$ eliminates imaginaries, we may assume f_a to be bijective. The domain $\mathbf{X}_0$ of f_a is a definable set in $\mathcal{T}_0$, hence (since $\mathcal{T}_0$ is stably embedded) it can be defined with a (canonical) parameter from a model of $\mathcal{T}_0$, and therefore (because of the assumption in the end of Section 3.5) with no parameters at all.

Thus, after fixing suitable parameters of internality, we may associate to each definable set $\mathbf{X}$ in $\mathcal{T}$, a definable set $\mathbf{X}_0$ in $\mathcal{T}_0$. Since this process commutes with products and inclusions and complements, this is an interpretation of $\mathcal{T}$ in $\mathcal{T}_0$. Furthermore, this is a stably embedded interpretation: any subset of $\mathbf{X}_0$ definable with parameters can be translated via the same function f_a to a subset definable with parameters of $\mathbf{X}$. This motivates the following definition.

4.10. Definition. Let $\mathcal{C}_0$ be an ind-closed category. An *internal cover* of $\mathcal{C}_0$ consists of an ind-closed category $\mathcal{C}$, together with closed functors $\mathcal{I}\colon \mathcal{C}_0 \to \mathcal{C}$ and $\mathcal{F}\colon \mathcal{C} \to \mathcal{C}_0$, and an isomorphism of $\mathcal{F} \circ \mathcal{I}$ with the identity functor (we will ignore this isomorphism in the notation and assume that $\mathcal{F} \circ \mathcal{I}$ is the identity functor).

4.11. Remark. The assumption that $\mathcal{F} \circ \mathcal{I}$ is the identity is equivalent (under the other conditions) to $\mathcal{F}$ being "internally left adjoint" to $\mathcal{I}$, in the sense that the canonical map $\underline{\mathrm{Hom}}(\mathcal{F}(\mathbf{X}),\mathbf{Y}_0) \to \mathcal{F}(\underline{\mathrm{Hom}}(\mathbf{X},\mathcal{I}(\mathbf{Y}_0)))$ is an isomorphism (this can be seen by applying the isomorphism with $\mathbf{X} = \mathbf{1}$). The "unit of adjunction" obtained by setting $\mathbf{Y}_0 = \mathcal{F}(\mathbf{X})$ (which does not formally exist, in general) is, in the model theoretic setting, the collection of maps f_a described above (which "do not exist" because they require parameters).

We now wish to prove an analogue of Theorem 3.7. According to Section 4.6, a "model" of $\mathcal{C}_0$ is given by a composition of functors $\Gamma_{\mathcal{D}} \circ \mathcal{H}$, where $\mathcal{H}$ is a closed functor into an ind-closed category $\mathcal{D}$, and $\Gamma_{\mathcal{D}}$ is the functor $\mathrm{Hom}(\mathbf{1}, -)$ in $\mathcal{D}$. A model of $\mathcal{C}$ is thus given by $\Gamma_{\mathcal{D}} \circ \mathcal{H} \circ \mathcal{F}$. We are interested in automorphisms of this functor that fix $\mathcal{I}$. We also know from Corollary 2.8 that the group $\mathrm{Aut}(\mathcal{H} \circ \mathcal{F}/\mathcal{I})$ of "definable" automorphisms comes from a pro-group, but $\Gamma_{\mathcal{D}}$ is far from being closed. Regardless, we have the following statement.

4.12. Theorem. *Let $\mathcal{I}\colon \mathcal{C}_0 \to \mathcal{C}, \mathcal{F}\colon \mathcal{C} \to \mathcal{C}_0$ be an internal cover (so $\mathcal{C}_0$ and $\mathcal{C}$ are ind-closed). Then for any left-exact $\mathcal{H}\colon \mathcal{C}_0 \to \mathcal{D}$, the map of groups*

$$\mathrm{Aut}(\mathcal{H} \circ \mathcal{F}/\mathcal{I}) \to \mathrm{Aut}(\Gamma_{\mathcal{D}} \circ \mathcal{H} \circ \mathcal{F}/\mathcal{I})$$

obtained by applying $\Gamma_{\mathcal{D}}$ is an isomorphism.

We note that $\mathcal{H}$ need not be closed in this theorem.

PROOF. Let θ be an automorphism of $\mathcal{R} = \Gamma_{\mathcal{D}} \circ \mathcal{H} \circ \mathcal{F}$ fixing $\mathcal{R} \circ \mathcal{I}$, let $\mathbf{Q}$ be an object of $\mathcal{C}$, and set $\mathbf{X} = \underline{\mathrm{Hom}}\big(\mathbf{Q}, \mathcal{I}(\mathcal{F}(\mathbf{Q}))\big)$. Thus there is an evaluation map $\underline{\mathrm{ev}}\colon \mathbf{X} \times \mathbf{Q} \to \mathcal{I}\big(\mathcal{F}(\mathbf{Q})\big)$. By the remark above, we have $\mathcal{F}(\mathbf{X}) = \underline{\mathrm{Hom}}\big(\mathcal{F}(\mathbf{Q}), \mathcal{F}(\mathbf{Q})\big)$, and hence an evaluation map $\mathcal{F}(\mathbf{X}) \times \mathcal{F}(\mathbf{Q}) \to \mathcal{F}(\mathbf{Q})$. Since $\mathcal{R}$ is left exact, we thus get a map

$$\mathcal{R}(\underline{\mathrm{ev}})\colon \mathcal{R}(\mathbf{X}) \times \mathcal{R}(\mathbf{Q}) \to \mathcal{R}(\mathbf{Q})$$

In particular, each element of $\mathcal{R}(\mathbf{X})$ determines a map of $\mathcal{R}(\mathbf{Q})$ to itself. We claim that $\theta_{\mathbf{Q}}$ is represented by an element of $\mathcal{R}(\mathbf{X})$, $g\colon \mathbf{1} \to \mathcal{H}\big(\mathcal{F}(\mathbf{X})\big)$. This will show that θ comes from an automorphism of $\mathcal{H} \circ \mathcal{F}$, namely the composition

$$\mathcal{H}\big(\mathcal{F}(\mathbf{Q})\big) = \mathbf{1} \times \mathcal{H}\big(\mathcal{F}(\mathbf{Q})\big) \xrightarrow{g \times \mathrm{Id}} \mathcal{H}\big(\mathcal{F}(\mathbf{X})\big) \times \mathcal{H}\big(\mathcal{F}(\mathbf{Q})\big) \to \mathcal{H}\big(\mathcal{F}(\mathbf{Q})\big)$$

We first note that there is a map $\mathbf{1} \to \mathcal{F}(\mathbf{X})$ in $\mathcal{C}_0$ corresponding to the identity on $\mathcal{F}(\mathbf{Q})$. Applying $\Gamma \circ \mathcal{H}$, we deduce that the identity map is represented by an element 1 in $\mathcal{R}(\mathbf{X})$. Since $\mathbf{X}, \mathbf{Q}, \mathcal{I}\big(\mathcal{F}(\mathbf{Q})\big)$ and the evaluation map are from $\mathcal{C}$, the evaluation map $\mathcal{R}(\underline{\mathrm{ev}})$ is preserved by θ. Also, $\theta_{\mathcal{I}(\mathcal{F}(\mathbf{Q}))}$ is the identity. Hence, for all $q \in \mathcal{R}(\mathbf{Q})$, we get $\mathcal{R}(\underline{\mathrm{ev}})\big(\theta_{\mathbf{X}}(1), \theta_{\mathbf{Q}}(q)\big) = \mathcal{R}(\underline{\mathrm{ev}})(1, q) = q$. It follows that $\theta_{\mathbf{X}}(1)$, as an endomorphism of $\mathcal{R}(\mathbf{Q})$ is invertible, with inverse $\theta_{\mathbf{Q}}$. This complete the proof of the surjectivity. The injectivity is similar, using the composition map instead of the evaluation. □

4.13. Corollary. *Let $\mathcal{I}\colon \mathcal{C}_0 \to \mathcal{C}, \mathcal{F}\colon \mathcal{C} \to \mathcal{C}_0$ be an internal cover. There is a pro-group $\mathbf{G}$ in $\mathrm{Ind}(\mathcal{C})$, and an action of $\mathbf{G}$ on each object of $\mathcal{C}$, such that for any closed functor $\mathcal{H}\colon \mathcal{C}_0 \to \mathcal{D}$ into an ind-closed category, $\Gamma_{\mathcal{D}}\big(\mathcal{H}(\mathcal{F}(\mathbf{G}))\big) = \mathrm{Aut}(\Gamma_{\mathcal{D}} \circ \mathcal{H} \circ \mathcal{F}/\mathcal{I})$.*

PROOF. By Corollary 2.8 we have such a $\mathbf{G}$ with the property that

$$\Gamma_{\mathcal{D}}\big(\mathcal{H}(\mathcal{F}(\mathbf{G}))\big) = \mathrm{Aut}(\mathcal{H} \circ \mathcal{F}/\mathcal{I})$$

By Theorem 4.12, the last group is equal to $\mathrm{Aut}(\Gamma_{\mathcal{D}} \circ \mathcal{H} \circ \mathcal{F}/\mathcal{I})$. □

As explained above, an internal cover of theories gives rise to an internal cover in our sense. Applying the corollary with $\mathcal{H}$ an expansion by constants in a model recovers Theorem 3.7, but with a pro-group in ind-definable, rather than definable sets.

5. Recovering a pro-group

The model theoretic statement in Theorem 3.7 produces a pro-definable group, i.e., a pro-group in the category $\mathcal{C}$ of definable sets, whereas the more general statement of Corollary 4.13 provides only a pro-group in $\mathrm{Ind}(\mathcal{C})$. This difference seems unavoidable in general, but in this section we would like to formulate additional conditions that ensure the stronger statement (these conditions are easily seen to hold in the first order setting).

The idea, which appears in the model theoretic proof, is that the objects $\mathbf{X} = \underline{\mathrm{Hom}}(\mathbf{Q}, \mathbf{Q})$ which participate in the construction of $\mathbf{G}$ are unnecessarily big. If $\mathbf{Y}$ is an object in a system that represents $\mathbf{X}$, it will not, in general, be closed under composition. However, if $f_{\mathbf{Q}}$ is an "element" of $\mathbf{Y}$ that is part of a natural automorphism, then composition with it remains within $\mathbf{Y}$, since the map from $\mathbf{Y}$ to $\mathbf{X}$ is preserved by the natural automorphism. Hence, to construct the natural automorphisms, it is possible to replace $\mathbf{X}$ by a compact subobject.

For this to work, we need to know that the internal Homs can be constructed from compact subobjects. This is captured by the following definitions.

5.1. Definition. A map $f\colon \mathbf{X} \to \mathbf{Y}$ in $\mathrm{Ind}(\mathcal{C})$ is called *proper* if for any $g\colon \mathbf{Z} \to \mathbf{Y}$ with $\mathbf{Z}$ compact, the pullback $\mathbf{X} \times_{\mathbf{Y}} \mathbf{Z}$ is also compact.

An ind-object $\mathbf{Y}$ is called *semi-strict* if any map from a compact object to $\mathbf{Y}$ is proper.

For example, if $\mathbf{Y}$ is compact, then $f\colon \mathbf{X} \to \mathbf{Y}$ is proper if and only if it is in $\mathcal{C}$.

Recall that an ind-object $\mathbf{Y}$ is *strict* if any map $\mathbf{X} \to \mathbf{Y}$, with $\mathbf{X}$ compact, factors as $\mathbf{X} \to \mathbf{Z} \xrightarrow[i]{\mathbf{Y}}$, with $\mathbf{Z}$ compact and i a monomorphism (equivalently, $\mathbf{Y}$ can be represented by a system of monomorphisms).

5.2. Lemma. *Let* $\mathbf{Y}$ *be an ind-object.*

(1) *If* $\mathbf{Y}$ *is strict, then it is semi-strict*
(2) $\mathbf{Y}$ *is semi-strict if and only if for any two ind-objects* $\mathbf{X}$ *and* $\mathbf{Z}$ *over* $\mathbf{Y}$, *the map* $\mathbf{X} \times_{\mathbf{Y}} \mathbf{Z} \to \mathbf{X} \times \mathbf{Z}$ *is proper.*

The converse of (1) seems to require additional assumptions (for example, quotients by equivalence relations).

PROOF. (1) Since $\mathbf{Y}$ is strict, we may assume that both arrows from compact objects $\mathbf{X}$ and $\mathbf{Z}$ factor through the same monomorphism $\mathbf{W} \to \mathbf{Y}$, with $\mathbf{W}$ compact. Hence $\mathbf{X} \times_{\mathbf{Y}} \mathbf{Z} = \mathbf{X} \times_{\mathbf{W}} \mathbf{Z}$ is compact as well.

(2) This immediately reduces to the case when $\mathbf{X}$ and $\mathbf{Z}$ are compact, in which case we need to show that $\mathbf{X} \times_{\mathbf{Y}} \mathbf{Z}$ is compact, which is precisely the definition of $\mathbf{Y}$ being semi-strict. □

For two objects $\mathbf{X}$ and $\mathbf{Y}$, let $\underline{\mathrm{Bi}}(\mathbf{X}, \mathbf{Y})$ be the ind-object consisting of invertible maps from $\mathbf{X}$ to $\mathbf{Y}$, and let $\underline{\mathrm{Bi}}(\mathbf{X}) = \underline{\mathrm{Bi}}(\mathbf{X}, \mathbf{X})$ (this is a group ind-object).

5.3. Theorem. *In the situation of Corollary* 4.13, *assume in addition*:

(1) *For any two objects* $\mathbf{X}$ *and* $\mathbf{Y}$ *of* $\mathcal{C}$, *the ind-object* $\underline{\mathrm{Hom}}(\mathbf{X}, \mathbf{Y})$ *is semi-strict.*
(2) *In* $\mathcal{C}_0$, *each* $\underline{\mathrm{Bi}}(\mathbf{X}, \mathbf{Y})$ *is strict.*
(3) *If* $\mathbf{X}$ *is an object of* $\mathcal{C}$ *such that* $\mathcal{F}(\mathbf{X})$ *is compact, then* $\mathbf{X}$ *is compact.*

(4) *For any monomorphism* $k\colon \mathbf{X} \to \mathbf{Y}$ *and object* $\mathbf{Z}$, *all in* $\mathcal{C}_0$, *the induced map* $\underline{\circ\, k}\colon \underline{\mathrm{Hom}}(\mathbf{Z},\mathbf{X}) \to \underline{\mathrm{Hom}}(\mathbf{Z},\mathbf{Y})$ *is proper.*

Then the group $\mathbf{G}$ *of the conclusion of Corollary* 4.13 *is a pro-group in* $\mathcal{C}$.

Note that assumption (3) holds if $\mathcal{F}$ reflects isomorphisms, in the sense that i is an isomorphism if $\mathcal{F}(i)$ is.

PROOF. The proof is analogous to the model theoretic proof. The idea is that by internality, for any object $\mathbf{Q}$ of $\mathcal{C}$ there is an object $\mathbf{C}$ of $\mathcal{C}_0$ with $\underline{\mathrm{Bi}}(\mathbf{Q},\mathbf{C})$ "nonempty." The group $\mathbf{G}$ acts freely on this ind-object, and therefore, by assumption, on a compact subobject $\mathbf{X}$. $\mathbf{G}$ is then obtained by a sequence of pullbacks as in the proof of Proposition 2.7, except that now the functor whose end we are taking has values in $\mathcal{C}$ itself for objects on the diagonal. The pullbacks are again compact by the properness assumption.

More precisely, in the construction of the end, the pro-group $\mathbf{G}$ is constructed as an inverse system of pullbacks of the form $\mathbf{G}_3 = \mathbf{G}_1 \times_{\underline{\mathrm{Hom}}(\mathbf{X}_1,\mathbf{X}_2)} \mathbf{G}_2$, where $\mathbf{G}_i$ is a subgroup of $\underline{\mathrm{Bi}}(\mathbf{X}_i)$, and the product is with respect composition maps with a given morphism $f\colon \mathbf{X}_1 \to \mathbf{X}_2$ in $\mathcal{C}$ ($\mathbf{G}_3$ is then viewed as a subgroup of $\underline{\mathrm{Bi}}(\mathbf{X}_1 \times \mathbf{X}_2)$). Since we are assuming that $\underline{\mathrm{Hom}}(\mathbf{X}_1,\mathbf{X}_2)$ is semi-strict, $\mathbf{G}_3$ will be compact if $\mathbf{G}_1$ and $\mathbf{G}_2$ are. Hence, it is enough to prove that each $\underline{\mathrm{Bi}}(\mathbf{Q})$ contains a compact subgroup $\mathbf{G}_{\mathbf{Q}}$, which is in the system.

Let $\mathbf{C} = \mathcal{I}(\mathcal{F}(\mathbf{Q}))$ (so that $\mathcal{F}(\mathbf{C}) = \mathcal{F}(\mathbf{Q})$), and let $\tilde{t}\colon \mathbf{X} \to \underline{\mathrm{Bi}}(\mathbf{Q},\mathbf{C})$ be a map from a compact object in $\mathcal{C}$, such that the map $\mathbf{1} \to \underline{\mathrm{Hom}}(\mathcal{F}(\mathbf{C}),\mathcal{F}(\mathbf{C}))$ corresponding to the identity map on $\mathcal{F}(\mathbf{C})$ factors through $\mathcal{F}(\tilde{t})$. Thus, we have a corresponding evaluation map $t\colon \mathbf{X} \times \mathbf{Q} \to \mathbf{C}$, and a map $i\colon \mathbf{1} \to \mathcal{F}(\mathbf{X})$, such that $\mathcal{F}(t) \circ (i \times \mathrm{id}_{\mathcal{F}(\mathbf{Q})})$ is the identity. We also have a "composition with the inverse" map $m\colon \mathbf{X} \times \mathbf{X} \to \underline{\mathrm{Bi}}(\mathbf{C})$, given on points $(x,y) \in \mathbf{X} \times \mathbf{X}$ by $m(x,y) = x \circ y^{-1}$. We let $\mathbf{G}_{\mathbf{Q}}$ be the "subgroup" of $\underline{\mathrm{Bi}}(\mathbf{X})$ preserving m; More precisely, $\underline{\mathrm{Bi}}(\mathbf{X})$ embeds in $\underline{\mathrm{Bi}}(\mathbf{X} \times \mathbf{X})$ by acting diagonally, and composition with m gives in total a map $\underline{\mathrm{Bi}}(\mathbf{X}) \to \underline{\mathrm{Hom}}(\mathbf{X} \times \mathbf{X},\mathbf{C})$. $\mathbf{G}_{\mathbf{Q}}$ is the pullback of this map along the map $\breve{m}\colon \mathbf{1} \to \underline{\mathrm{Hom}}(\mathbf{X} \times \mathbf{X},\mathbf{C})$ corresponding to m.

$$
\begin{array}{ccc}
\mathbf{G}_{\mathbf{Q}} & \longrightarrow & \underline{\mathrm{Bi}}(\mathbf{X} \times \mathbf{X}) \\
\big\downarrow & & \big\downarrow{\scriptstyle \underline{\circ m}} \\
\mathbf{1} & \xrightarrow[\breve{m}]{} & \underline{\mathrm{Hom}}(\mathbf{X} \times \mathbf{X},\mathbf{C}).
\end{array}
\tag{5}
$$

The ind-group $\mathbf{G}_{\mathbf{Q}}$ is one of the groups that appear in the system defining the total group of automorphisms. Thus, it is enough to prove that it is compact. For that, it is sufficient, by the assumption, to prove that $\mathcal{F}(\mathbf{G}_{\mathbf{Q}})$ is compact. Hence we may assume that $\mathbf{X}$ is a subobject $k\colon \mathbf{X} \to \underline{\mathrm{Bi}}(\mathcal{F}(\mathbf{C}))$ containing the identity $i\colon \mathbf{1} \to \mathbf{X}$, and $\mathbf{G}_{\mathbf{Q}}$ is a group acting on $\mathbf{X}$. The identity map i and the action μ of $\mathbf{G}_{\mathbf{Q}}$ on $\mathbf{X}$ induce a map $j = \mu \circ (\mathrm{id} \times i)\colon \mathbf{G}_{\mathbf{Q}} \to \mathbf{X}$. The pullback diagram defining $\mathbf{G}_{\mathbf{Q}}$ implies that we have a pullback diagram

$$
\begin{array}{ccc}
\mathbf{G}_{\mathbf{Q}} & \xrightarrow{\breve{\mu}} & \underline{\mathrm{Hom}}(\mathbf{X},\mathbf{X}) \\
{\scriptstyle j}\big\downarrow & & \big\downarrow{\scriptstyle \underline{\circ k}} \\
\mathbf{X} & \xrightarrow[\breve{\circ}]{} & \underline{\mathrm{Hom}}\left(\mathbf{X},\underline{\mathrm{Bi}}(\mathcal{F}(\mathbf{C}))\right).
\end{array}
\tag{6}
$$

Since k is a monomorphism, the vertical map $\underline{\ \circ k}$, induced by composition with k, is by assumption proper. Therefore $\mathbf{G_Q}$, being the pullback over a compact object of a proper map, is itself compact. □

6. Examples

As explained above, Corollary 4.13 and Theorem 5.3 are categorical analogues of the classical model theoretic Theorem 3.7. I now briefly mention two additional examples where the more general results may be applied.

6.1. Partial automorphisms. Instead of considering automorphisms of the full model $\mathcal{M}$, as in Theorem 3.7, one may be interested in automorphisms of a definable set $\mathbf{Q}$ in $\mathcal{T}$ that preserve only part of the structure on $\mathbf{Q}$. This question was studied in Kamensky [5], where one is ultimately interested in automorphisms of a linear difference equation that preserve only the quantifier free formulas.

The categorical version may be applied in this case by taking $\mathcal{C}$ to be the subcategory consisting of the definable sets one wishes to preserve. The condition that this category eliminates imaginaries corresponds to the requirement that the internality datum that witnesses the fact that $\mathrm{dcl}(A \cup \mathcal{M}_0) = \mathcal{M}$, should be preserved.

6.2. Tannakian categories. As mentioned above, the model theoretic theorem can be applied to prove a version of the Tannakian reconstruction theorem. Below I outline the "composition" of the model theoretic argument with the categorical formulation of internality described above. This is not a complete example, but a "work plan," since some of the assumptions may be difficult to verify. The idea is only to show how the language used here is translated in this case.

Let $\Bbbk$ be a field of characteristic 0. Recall (Deligne [1, Section 7]) that if $\widetilde{\mathcal{C}}$ is a (symmetric, rigid, $\Bbbk$-linear) tensor category, one may construct the category $\mathcal{S}ch_{\widetilde{\mathcal{C}}}$ of $\widetilde{\mathcal{C}}$-schemes as the opposite category of the category of (commutative, unital) algebras in $\mathrm{Ind}(\widetilde{\mathcal{C}})$ (such an algebra consists of an ind-object A of $\widetilde{\mathcal{C}}$, together with morphisms $m\colon A\otimes A \to A$ and $u\colon \mathbf{1} \to A$, satisfying obvious axioms). This category admits finite inverse limits, since a pullback corresponds to a tensor product of algebras (over a base), as well as finite coproducts.

Furthermore, there is a faithful functor $\mathbf{X} \mapsto \widehat{\mathbf{X}}$ from $\widetilde{\mathcal{C}}$ to $\mathcal{S}ch_{\widetilde{\mathcal{C}}}$, sending $\mathbf{X}$ to the symmetric algebra on the dual $\check{\mathbf{X}}$. Let $\mathcal{C}$ be the full subcategory of $\mathcal{S}ch_{\widetilde{\mathcal{C}}}$ containing all object of the form $\widehat{\mathbf{X}} \times_{\widehat{\mathbf{Y}}} \widehat{\mathbf{Z}}$, with $\mathbf{X}, \mathbf{Y}$ and $\mathbf{Z}$ objects of $\widetilde{\mathcal{C}}$, as well as their finite coproducts (these are the "finitely presented" $\widetilde{\mathcal{C}}$-schemes). It again admits all finite inverse limits and coproducts. It is possible to show, like in usual algebraic geometry, that $\underline{\mathrm{Hom}}(\mathbf{X}, \mathbf{Y})$ is a strict ind-object for all objects $\mathbf{X}$ and $\mathbf{Y}$ of $\mathcal{C}$.

In particular, we take $\widetilde{\mathcal{C}}_0$ to be the category of finite-dimensional vector spaces over $\Bbbk$, so that $\mathcal{C}_0$ is the category of finitely presented affine schemes over $\Bbbk$. An exact tensor functor between tensor categories induces a left-exact functor between the corresponding categories of schemes, which restricts to a functor between the finitely presented subcategories. Furthermore, the construction of the $\underline{\mathrm{Hom}}$ ind-schemes shows that the resulting functor is a stable embedding. In particular, the (essentially unique) tensor functor from $\widetilde{\mathcal{C}}_0$ to $\widetilde{\mathcal{C}}$ induces a functor $\mathcal{I}\colon \mathcal{C}_0 \to \mathcal{C}$.

If $\widetilde{\mathcal{C}}$ is neutral Tannakian, i.e., we have an exact tensor functor ω to $\widetilde{\mathcal{C}}_0$, this functor again induces a stable embedding $\mathcal{F}\colon \mathcal{C} \to \mathcal{C}_0$, with $\mathcal{F} \circ \mathcal{I}$ isomorphic to the identity. We thus get an internal cover, in the sense of Definition 4.10.

If $\mathbf{T}$ is any (affine) scheme over $\Bbbk$, setting $\mathcal{H}_{\mathbf{T}}$ to be base change by $\mathbf{T}$ (i.e., $\mathcal{H}_{\mathbf{T}}(\mathbf{U}) = \mathbf{T} \times \mathbf{U}$ as a functor from $\Bbbk$-schemes to $\mathbf{T}$-schemes), one gets from Corollary 4.13 that any automorphism of the functor $\mathbf{X} \mapsto \mathcal{F}(\mathbf{X})(\mathbf{T})$ (as a functor from $\mathcal{C}$ to sets) comes from a unique compatible collection of automorphisms of the $\mathbf{T}$-schemes $\mathbf{T} \times \mathcal{F}(\mathbf{X})$; and furthermore, all such automorphisms correspond to the $\mathbf{T}$-points of (with the help of Theorem 5.3, whose assumptions should be verified) a pro-group-scheme over $\Bbbk$, acting on all $\mathcal{F}(\mathbf{X})$. In fact, this group comes from a pro-group in $\mathcal{C}$ (which is called the fundamental group of $\mathcal{C}$ in Deligne [1]).

Note that if $\mathbf{X}$ is the $\widetilde{\mathcal{C}}$-scheme associated to an object $\widetilde{\mathbf{X}}$ of $\widetilde{\mathcal{C}}$, then $\mathcal{F}(\mathbf{X})$ is $\omega(\widetilde{\mathbf{X}})$ with its $\Bbbk$-scheme structure. Hence, if $\mathbf{T} = \operatorname{spec}(A)$ is an affine scheme over $\Bbbk$, then $\mathcal{F}(\mathbf{X})(\mathbf{T}) = \mathcal{F}(\mathbf{X})(A)$ is identified (at least as a set) with $A \otimes_{\Bbbk} \omega(\widetilde{\mathbf{X}})$. Furthermore, the linear space and tensor structures are described by suitable maps between schemes, all of which come from $\mathcal{C}$. Therefore, an automorphism of the functor $\mathbf{X} \mapsto \mathcal{F}(\mathbf{X})(\mathbf{T})$ as a functor to sets is the same as a tensor automorphism of the functor $\mathbf{X} \mapsto A \otimes_{\Bbbk} \omega(\mathbf{X})$.

References

1. P. Deligne, *Catégories tannakiennes*, The Grothendieck Festschrift, Vol. II (P. Cartier, L. Illusie, N. M. Katz, G. Laumon, and K. A. Ribet, eds.), Progr. Math., vol. 87, Birkhäuser, Boston, MA, 1990, pp. 111–195.
2. M. Artin, A. Grothendieck, and J.-L. Verdier (eds.), *Théorie des topos et cohomologie étale des schémas. Tome 1: Théorie des topos*, Lecture Notes in Math., vol. 269, Springer, Berlin, 1972.
3. E. Hrushovski, *Computing the Galois group of a linear differential equation*, Differential Galois Theory (Będlewo, 2001) (T. Crespo and Z. Hajto, eds.), Banach Center Publ., vol. 58, Polish Acad. Sci., Warsaw, 2002, pp. 97–138.
4. ———, *Groupoids, imaginaries and internal covers*, available at `arXiv:math/0603413`.
5. M. Kamensky, *Definable groups of partial automorphisms*, Selecta Math. (N.S.) **15** (2009), no. 2, 295–341.
6. ———, *Ind- and pro- definable sets*, Ann. Pure Appl. Logic **147** (2007), no. 3, 180–186.
7. ———, *Model theory and the Tannakian formalism*, available at `arXiv:0908.0604`.
8. S. Mac Lane, *Categories for the working mathematician*, 2nd ed., Grad. Texts in Math., vol. 5, Springer, New York, 1998.
9. S. Mac Lane and I. Moerdijk, *Sheaves in geometry and logic: A first introduction to topos theory*, Universitext, Springer, New York, 1994.
10. M. Makkai and G. E. Reyes, *First order categorical logic: Model-theoretical methods in the theory of topoi and related categories*, Lecture Notes in Math., vol. 611, Springer, Berlin, 1977.
11. D. Marker, *Model theory: An introduction*, Grad. Texts in Math., vol. 217, Springer, New York, 2002.
12. A. Pillay, *Algebraic D-groups and differential Galois theory*, Pacific J. Math. **216** (2004), no. 2, 343–360.
13. B. Poizat, *Une théorie de Galois imaginaire*, J. Symbolic Logic **48** (1983), no. 4, 1151–1170.
14. Wikipedia, *Outline of category theory*, available at `en.wikipedia.org/wiki/Outline_of_category_theory`.
15. B. I. Zil′ber, *Totally categorical theories: structural properties and the nonfinite axiomatizability*, Model Theory of Algebra and Arithmetic (Karpacz, 1979) (L. Pacholski, J. Wierzejewski, and A. J. Wilkie, eds.), Lecture Notes in Math., vol. 834, Springer, Berlin, 1980, pp. 381–410.

Department of Mathematics, University of Notre Dame, Notre Dame, IN 46556, USA

E-mail address: kamensky.1@nd.edu

URL: mkamensky.notlong.com

Centre de Recherches Mathématiques
CRM Proceedings and Lecture Notes
Volume **53**, 2011

Computable Structures of Scott Rank ω_1^{CK}

Julia F. Knight

This paper is dedicated to Mihaly Makkai

ABSTRACT. Mihaly Makkai made fundamental contributions to infinitary logic. This paper describes work growing out of his 1981 paper entitled "An example concerning Scott heights" [13], in which he constructed an example of an arithmetical structure of Scott rank ω_1^{CK}. Makkai's original example was a structure of an unfamiliar kind — a "group-tree." By reworking the construction slightly, we can make Makkai's example computable [10]. In [3, 5, 6, 18], there are further examples of computable structures of Scott rank ω_1^{CK} of more familiar kinds: trees, undirected graphs, fields (of any desired characteristic), linear orderings, 2-step nilpotent groups, and Boolean algebras.

Makkai's original example had the feature that its computable infinitary theory is $\aleph_0$-categorical. The same is true for the further examples in [3, 5, 18]. J. Millar and Sacks [16] gave an example of a structure $\mathcal{A}$ of Scott rank ω_1^{CK} such that the computable infinitary theory of $\mathcal{A}$ is not $\aleph_0$-categorical, and $\omega_1^{\mathcal{A}} = \omega_1^{\mathrm{CK}}$ — this means that $\mathcal{A}$ lives in a fattening of the least admissible set. It is not known whether there is a computable (or hyperarithmetical) structure of Scott rank ω_1^{CK}, such that the computable infinitary theory is not $\aleph_0$-categorical.

1. Infinitary logic

In this section, we give some background on infinitary logic. We recall the Scott Isomorphism Theorem, and we indicate how Scott's proof gave rise to the notion of *Scott rank.*

1.1. Formulas of $L_{\omega_1\omega}$. Let L be a countable language. In $L_{\omega_1\omega}$, we allow countable disjunctions and conjunctions, but only finite strings of quantifiers. For formulas of $L_{\omega_1\omega}$, we do not have *prenex* normal form — in general, we cannot bring the quantifiers to the front of a formula. However, we can bring the negations inside, giving another normal form. For formulas in this normal form, we have the following useful classification.

2010 *Mathematics Subject Classification.* Primary 54C40, 14E20; Secondary 46E25, 20C20.
Key words and phrases. Infinitary logic, computability, Scott rank.
This is the final form of the paper.

Definition 1.1 (Classification of $L_{\omega_1\omega}$ formulas).

- $\varphi(\overline{x})$ is Σ_0 and Π_0 if it is finitary quantifier-free.
- For $0 < \alpha < \omega_1$,
  - $\varphi(\overline{x})$ is Σ_α if it is a countable disjunction of formulas of the form $(\exists\overline{u})\,\psi(\overline{x},\overline{u})$, where ψ is Π_β for some $\beta < \alpha$,
  - $\varphi(\overline{x})$ is Π_α if it is a countable conjunction of formulas of the form $(\forall\overline{u})\,\psi(\overline{x},\overline{u})$, where ψ is Σ_β for some $\beta < \alpha$.

1.2. Expressive power of $L_{\omega_1\omega}$. We illustrate the expressive power of $L_{\omega_1\omega}$ by some results on axiomatizing "Borel" classes of structures. Let L be a countable language, and let $\mathrm{Mod}(L)$ be the class of L-structures with universe ω. There is a natural topology $\mathrm{Mod}(L)$, generated by basic clopen neighborhoods of the form $\mathrm{Mod}(\sigma)$, where σ is a finitary quantifier-free sentence in the language $L \cup \omega$. The *Borel* subsets of $\mathrm{Mod}(L)$ are obtained from these basic neighborhoods by closing under countable unions and intersections (we get closure under complements for free). The following result is from [12] (see also [9]).

Theorem 1.2 (Lopez-Escobar). *Let $K \subseteq \mathrm{Mod}(L)$, where K is closed under isomorphism. Then K is Borel iff it is axiomatized by a sentence of $L_{\omega_1\omega}$.*

We may classify the Borel sets as follows.

Definition 1.3 (Classification of Borel sets).

- The $\boldsymbol{\Pi}_0$ and $\boldsymbol{\Sigma}_0$ sets are the basic neighborhoods.
- For a countable ordinal $\alpha > 0$,
  - a set is $\boldsymbol{\Sigma}_\alpha$ if it is a countable union of sets each of which is $\boldsymbol{\Pi}_\beta$ for some $\beta < \alpha$.
  - a set is $\boldsymbol{\Pi}_\alpha$ if the complement is $\boldsymbol{\Sigma}_\alpha$.

Vaught [23] proved the following refinement of Lopez-Escobar's theorem.

Theorem 1.4 (Vaught). *Let $K \subseteq \mathrm{Mod}(L)$, where K is closed under isomorphism. Then K is $\boldsymbol{\Sigma}_\alpha$ iff it is axiomatized by a Σ_α sentence.*

A second illustration of the expressive power of $L_{\omega_1\omega}$ is the *Scott Isomorphism Theorem* [21], stated below.

Theorem 1.5 (Scott). *Let L be a countable language. For each countable L-structure $\mathcal{A}$, there is an $L_{\omega_1\omega}$ sentence — a* Scott sentence *— whose countable models are just the isomorphic copies of $\mathcal{A}$.*

In proving Theorem 1.5, Scott defined a family of equivalence relations on tuples in $\mathcal{A}$. He then assigned countable ordinals to the tuples, and to $\mathcal{A}$ itself. This is the origin of the notion of "Scott rank." There are several definitions of Scott rank in use. The one given below is similar to Scott's definition, but not an exact match.

Definition 1.6 (Equivalence relations on tuples). Let $\bar{a}, \bar{b}$ be tuples in $\mathcal{A}$.

(1) $\bar{a} \sim^0 \bar{b}$ if $\bar{a}$ and $\bar{b}$ satisfy the same quantifier-free formulas,
(2) for $\alpha > 0$, $\bar{a} \sim^\alpha \bar{b}$ if for each $\beta < \alpha$, for each $\bar{c}$, there exists $\bar{d}$ such that $\bar{a}, \bar{c} \sim^\beta \bar{b}, \bar{d}$, and for each $\bar{c}$, there exists $\bar{d}$ such that $\bar{a}, \bar{c} \sim^\beta \bar{b}, \bar{d}$.

In Scott's definition, in clause 2, $\bar{c}$ and $\bar{d}$ are single elements instead of tuples.

Definition 1.7 (Scott rank).

- The *Scott rank of a tuple* $\bar{a}$ in $\mathcal{A}$ is the least β such that for all $\bar{b}$, $\bar{a} \sim^\beta \bar{b} \Rightarrow (\mathcal{A}, \bar{a}) \cong (\mathcal{A}, \bar{b})$.

- The *Scott rank of* $\mathcal{A}$, denoted by $SR(\mathcal{A})$, is the least ordinal α greater than the ranks of all tuples in $\mathcal{A}$.

Example 1.8. If $\mathcal{A}$ is an ordering of type ω, then $\mathrm{SR}(\mathcal{A}) = 2$.

PROOF. Let $\bar{a}$ and $\bar{b}$ be tuples in $\mathcal{A}$. Then $\bar{a} \sim^0 \bar{b}$ if $\bar{a}$ and $\bar{b}$ are ordered in the same way, and $\bar{a} \sim^1 \bar{b}$ if they partition the ordering into intervals of the same sizes. This is enough to guarantee that there is an automorphism of $\mathcal{A}$ taking $\bar{a}$ to $\bar{b}$. □

Sacks and J. Millar use a different definition of Scott rank. Let L be a countable language, and let $\mathcal{A}$ be a countable L-structure. Let $(\mathcal{A}_\alpha)_{\alpha<\omega_1}$ be a sequence of expansions of $\mathcal{A}$, where $\mathcal{A}_0 = \mathcal{A}$, $\mathcal{A}_{\alpha+1}$ is the expansion of $\mathcal{A}_\alpha$ with predicates for all of the complete types realized in $\mathcal{A}_\alpha$, and for limit α, $\mathcal{A}_\alpha$ is the limit of the expansions $\mathcal{A}_\beta$, for $\beta < \alpha$, including all of the predicates. The *Scott rank* of $\mathcal{A}$ is the least α such that the structure $\mathcal{A}_\alpha$ is atomic.

While we assigned the linear ordering Scott rank 2, the definition of Sacks and Millar would assign it Scott rank 0. However, for the ranks that we care about, the different definitions agree. We make this statement precise in a comment after Theorem 2.8.

2. Computable structure theory

In this section we say what it means for a structure to be computable (or arithmetical, or hyperarithmetical). Next, we describe the *computable infinitary formulas*, and we give a version of compactness for these formulas. Using this, we set an upper bound on the Scott ranks of computable, or hyperarithmetical, structures.

We adopt the following conventions.

(1) Our structures are countable, with universe a subset of ω.
(2) Our languages are computable; i.e., the set of symbols is computable, and we have an effective procedure that we can apply to any symbol to determine the type (relation, operation) and arity.
(3) We identify a sentence with its Gödel number.
(4) We identify a structure $\mathcal{A}$ with its atomic diagram $D(\mathcal{A})$.

Under these conventions, a structure $\mathcal{A}$ is identified with a subset of ω. We say that $\mathcal{A}$ is *computable* (or *arithmetical*, or *hyperarithmetical*) if $D(\mathcal{A})$ is computable (or arithmetical, or hyperarithmetical).

2.1. Computable infinitary formulas.

Computable infinitary formulas are infinitary formulas in which the infinite disjunctions and conjunctions are over computably enumerable sets. Such formulas are comprehensible even though they are infinitely long.

Example 2.1. There is a natural computable infinitary sentence characterizing the class of Abelian p-groups. We take the conjunction of the finitary axioms for Abelian groups and the following:

$$(\forall x) \bigvee_n p^n x = 0$$

We classify computable infinitary formulas as follows.

- $\varphi(\overline{x})$ is computable Σ_0 and Π_0 if it is finitary quantifier free.
- For a computable ordinal $\alpha > 0$,

- $\varphi(\overline{x})$ is computable Σ_α if it is a c.e. disjunction of formulas of the form $(\exists\overline{u})\,\psi(\overline{x},\overline{u})$, where ψ is computable Π_β for some $\beta < \alpha$,
- $\varphi(\overline{x})$ is computable Π_α if it is a c.e. conjunction of formulas of the form $(\forall\overline{u})\,\psi(\overline{x},\overline{u})$, where ψ is computable Σ_β for some $\beta < \alpha$.

We illustrate the expressive power of computable infinitary formulas by an effective version of Theorem 1.4. Let L be a computable language. The *effective Borel* subsets of $\mathrm{Mod}(L)$ are obtained from the basic clopen neighborhoods by closing under computably enumerable unions and intersections. Again, we get closure under complements for free. The result below is the analogue of Theorem 1.4. The result is given in [22].

Theorem 2.2 (Vanden Boom). *Let $K \subseteq \mathrm{Mod}(L)$, closed under isomorphism. If K is Σ_α in the effective Borel hierarchy, then it is axiomatized by a computable Σ_α sentence.*

In proving Theorem 1.4, Vaught used "Vaught transforms" — quantifiers with the meanings "there exist nonmeagerly many" and "there exist comeagerly many." Vanden Boom's proof is stated directly in terms of forcing.

2.2. Barwise – Kreisel compactness. The usual Compactness Theorem fails for infinitary logic, but there is a version of compactness, due to Barwise, for "admissible" fragments. Here is a special case for computable infinitary formulas. Before Barwise, and unknown to Barwise, Kreisel had given, in an obscure footnote to [11], a version of compactness for ω-logic. The statement below (taken from [1]) is closer to that of Kreisel.

Theorem 2.3 (Barwise – Kreisel compactness). *Let Γ be a Π^1_1 set of computable infinitary sentences. If every Δ^1_1 set $\Gamma' \subseteq \Gamma$ has a model, then Γ has a model.*

Unlike ordinary compactness, Barwise – Kreisel compactness can be used to produce computable structures.

Corollary 2.4 (Existence of computable models). *Let Γ be a Π^1_1 set of computable infinitary sentences. If every Δ^1_1 subset of Γ has a computable model, then Γ has a computable model.*

The next two corollaries say that for computable (or hyperarithmetical) structures, computable infinitary sentences have all of the expressive power that we could wish for.

Corollary 2.5 (Isomorphism). *Suppose $\mathcal{A}$ and $\mathcal{B}$ are computable (or hyperarthmetical) structures satisfying the same computable infinitary sentences. Then $\mathcal{A} \cong \mathcal{B}$.*

Corollary 2.6 (Orbits). *Suppose $\bar{a}$ and $\bar{b}$ are tuples satisfying the same computable infinitary formulas in a computable structure $\mathcal{A}$. Then there is an automorphism of $\mathcal{A}$ taking $\bar{a}$ to $\bar{b}$.*

2.3. Possible Scott ranks for computable structures. Using Corollary 2.6, Nadel [19] proved the following.

Theorem 2.7 (Nadel). *If $\mathcal{A}$ is a computable (or hyperarithmetical) structure, then $\mathrm{SR}(\mathcal{A}) \leq \omega_1^{\mathrm{CK}} + 1$.*

We can say more about the possible Scott ranks for computable (or hyperarithmetical) structures. The result below is well-known.

Theorem 2.8. *Suppose $\mathcal{A}$ is computable (or hyperarithmetical).*

(1) $\mathrm{SR}(\mathcal{A})$ *is computable if there is a computable ordinal α such that the orbits of all tuples are defined by computable Π_α formulas,*
(2) $\mathrm{SR}(\mathcal{A}) = \omega_1^{\mathrm{CK}}$ *if the orbits of all tuples are defined by computable infinitary formulas, but there is no computable ordinal α, as in* (1),
(3) $\mathrm{SR}(\mathcal{A}) = \omega_1^{\mathrm{CK}} + 1$ *if there is some tuple whose orbit is not defined by any computable infinitary formula.*

We saw that different definitions of Scott rank assign different ranks to the linear ordering of type ω. However, the result above is true not just for the definition we chose, but also for the one used by Sacks and J. Millar.

2.4. Computable infinitary theories.

Definition 2.9. Let $\mathcal{A}$ be a structure, and let $\bar{a}$ be a finite tuple in $\mathcal{A}$.

- The *computable infinitary theory* $\mathrm{Th}_\infty(\mathcal{A})$ is the set of computable infinitary sentences true in $\mathcal{A}$.
- The *computable infinitary type* of $\bar{a}$ is the set of computable infinitary formulas satisfied by $\bar{a}$.

Remark. Let $\mathcal{A}$ be a computable (or hyperarithmetical) structure. Suppose $\mathrm{SR}(\mathcal{A}) = \omega_1^{\mathrm{CK}} + 1$, witnessed by the tuple $\bar{a}$. Then the computable infinitary type of $\bar{a}$ is nonprincipal. If $\mathrm{SR}(\mathcal{A}) = \omega_1^{\mathrm{CK}}$, then $\mathcal{A}$ is an atomic model of its computable infinitary theory — the types of all finite tuples are principal.

3. Examples of different Scott ranks

In this section we give examples of computable structures of computable Scott rank, and of Scott rank $\omega_1^{\mathrm{CK}} + 1$. We then describe Makkai's example of Scott rank ω_1^{CK}.

3.1. Computable Scott rank.

Proposition 3.1. *For all computable (or hyperarithmetical) structures $\mathcal{A}$ in the following classes,* $\mathrm{SR}(\mathcal{A})$ *is a computable ordinal.*

- *well orderings*
- *superatomic Boolean algebras*
- *reduced Abelian p-groups.*

3.2. Scott rank $\omega_1^{\mathrm{CK}}+1$. The ordinal ω_1^{CK} has no copy that is computable, or hyperarithmetical. In fact, there is no copy in any fattening of the least admissible set. In his 1966 thesis, Harrison showed the following (see the paper [8]).

Theorem 3.2 (Harrison). *There is a computable ordering of type $\omega_1^{\mathrm{CK}}(1+\eta)$.*

Fact. The Harrison ordering has Scott rank $\omega_1^{\mathrm{CK}} + 1$.

Proof. Any element outside the initial copy of ω_1^{CK} has Scott rank ω_1^{CK}. □

The computable infinitary theory of the Harrison ordering has an atomic model; namely, ω_1^{CK}. This model has Scott rank ω_1^{CK}. For some time after 1966, there were no known computable, or hyperarithmetical, structures of Scott rank ω_1^{CK}. Makkai's 1981 paper [13] filled the gap.

3.3. Makkai's example. Makkai [13] proved the following (see also [17]).

Theorem 3.3 (Makkai). *There is an arithmetical structure of Scott rank ω_1^{CK}.*

Here is the outline of the proof.

- Define a computable operator that takes each tree T isomorphic to a subtree of $\omega^{<\omega}$ to a structure $\mathcal{A}(T)$ such that if the input tree T is "special," then $\mathcal{A}(T)$ will have Scott rank ω_1^{CK}.
- Show that there is an arithmetical special tree.

Below, we describe Makkai's computable operator. Let T be a tree, and let T_n be the set of elements at level n in T.

(1) Let G_n be the Abelian group consisting of the finite subsets of T_n under symmetric difference, $\triangle$. The identity is $\varnothing$. For distinct $n > 0$, we replace $\varnothing$ by distinct elements id_n.
(2) Put a tree structure on $G = \bigcup_n G_n$, with predecessors defined as follows. Suppose $a \in G_{n+1}$, say $a = \{t_1, \ldots, t_k\}$, where t_i has predecessor t_i'. Then $p(a)$ is the sum of the $\{t_i'\}$. Let $p(\mathrm{id}_{n+1}) = \mathrm{id}_n$.
(3) Define $f_a \colon G \to G$, for $a \in G$ as follows. If $a \in G_n$, then for $b \in G_m$, let $k = \min\{m, n\}$, and let a^* and b^* be the result of iterating $p(x)$ until the value is in G_k. Then the value of $f_a(b)$ is $a^* \triangle b^*$ — this is also the value of $f_b(a)$.
(4) Finally, let $\mathcal{A}(T)$ be $(G, (f_a)_{a \in G})$.

The whole process, passing from T to $\mathcal{A}(T)$, is effective. That is, there is a computable operator taking T to (a copy of) $\mathcal{A}(T)$. We must describe the "special" trees. First, for an arbitrary tree $T \subseteq \omega^{<\omega}$, we define *tree rank.*

Definition 3.4 (Ranks of elements of T).

(1) $\mathrm{rk}(x) = 0$ if x is terminal,
(2) for $\alpha > 0$, $\mathrm{rk}(x) = \alpha$ if all successors of x have ordinal rank, and α is the first ordinal greater than all of these ordinals,
(3) $\mathrm{rk}(x) = \infty$ if x does not have ordinal rank.

Definition 3.5 (Rank of T). Let Λ be the top node of T (corresponding to the empty sequence). Then $\mathrm{rk}(T) = \mathrm{rk}(\Lambda)$.

Definition 3.6 (path). For a tree $T \subseteq \omega^{<\omega}$, a *path* is a function $f \in \omega^\omega$ such that for all n, $f|n \in T$.

Fact. $\mathrm{rk}(x) = \infty$ iff there is a path through x.

3.4. Conditions guaranteeing that $\mathrm{SR}(\mathcal{A}(T)) = \omega_1^{CK}$.

Definition 3.7 (Special tree). Let T be a hyperarithmetical tree, and let T_n be the set of elements at level n. We say that T is *special* if for each n, there is a computable bound on the ordinal ranks of elements of T_n but there is no computable bound on the ordinal ranks of elements of T.

The next result says that the special trees are the ones we want.

Proposition 3.8 (Makkai). *If T is special, then $\mathcal{A}(T)$ has Scott rank ω_1^{CK}.*

3.5. Existence of special trees. To complete the proof of Theorem 3.3, we need the following.

Proposition 3.9 (Makkai). *There is an arithmetical special tree T.*

Applying the computable operator to the Makkai's special tree T, we obtain an arithmetical structure of Scott rank ω_1^{CK}. This completes the proof of Theorem 3.3. We will say more about special trees in the next section.

4. Modifying Makkai's example

In this section, we first indicate how to re-work Makkai's example to make it computable. Next, we give a computable tree of Scott rank ω_1^{CK}. We indicate how the tree gives rise to further computable structures of Scott rank ω_1^{CK} in familiar classes.

4.1. Making Makkai's example computable. The following result is from [10].

Proposition 4.1 (K – J. Millar). *There is a computable special tree.*

PROOF SKETCH. Recall that for a tree T, T_n denotes the set of elements at level n. Using an idea of Sacks, we can produce the desired tree as follows.

(1) First, for each computable ordinal α, we unwind a notation for α, slowly, to obtain a computable tree T of computable rank $\geq \alpha$, such that the set of ordinal ranks of elements of T_n has order type at most $\omega \cdot n$.

(2) Then we apply Barwise – Kreisel compactness to get a computable tree T such that $\mathrm{rk}(T) = \infty$, T has no hyperarithmetical path, and the set of *ordinal* ranks of elements of T_n has order type at most $\omega \cdot n$. We can show that this tree is special. □

Corollary 4.2. *There is a computable structure of Scott rank ω_1^{CK}.*

PROOF. Applying Makkai's computable operator to the computable special tree T from Proposition 4.1, we get the desired structure $\mathcal{A}(T)$. □

4.2. A computable tree of Scott rank ω_1^{CK}. In Makkai's construction, the structure $A(T)$, incorporating some group structure, is, in general, more homogeneous than T. In [5], we focus directly on the homogeneity, and we obtain computable structure of Scott rank ω_1^{CK} that is just a tree — there is no group structure.

Definition 4.3 (Rank-homogeneous tree). A tree T is *rank-homogeneous* if

- for all $x \in T_n$ and all computable α, if there exists $y \in T_{n+1}$ with $\mathrm{rk}(y) = \alpha < \mathrm{rk}(x)$, then x has infinitely many successors y' with $rk(y') = \alpha$.
- for all $x \in T_n$ with $\mathrm{rk}(x) = \infty$, x has infinitely many successors y with $\mathrm{rk}(y) = \infty$.

For rank-homogeneous trees, we have isomorphism invariants similar to those for Abelian p-groups. To describe a countable Abelian p-group, up to isomorphism, it is enough to give the Ulm sequence and the dimension of the integer part. For a rank-homogeneous tree, it is enough to give the tree ranks at each level.

Proposition 4.4. *If T and T' are rank-homogeneous trees, and for all n, T_n and T'_n represent the same ordinal ranks, then $T \cong T'$.*

The result below gives conditions under which a computable rank-homogeneous tree has computable Scott rank, or Scott rank ω_1^{CK}, or Scott rank $\omega_1^{CK} + 1$.

Proposition 4.5 (Calvert – K – J. Millar). *Suppose T is a computable (or hyperarithmetical) rank-homogeneous tree.*

- $SR(T) < \omega_1^{CK}$ *if there is a computable bound on the ordinal ranks of tree elements.*
- $SR(T) = \omega_1^{CK}$ *if T is special,*
- $SR(T) = \omega_1^{CK} + 1$ *if for some n, there is no computable bound on the ordinal ranks of elements of T_n.*

PROOF SKETCH. We use the following observations.

• The orbit of a tuple $\overline{x}$ is determined by the isomorphism type of the finite subtree generated by $\overline{x}$, and the tree ranks of the nodes. In particular, the orbit of a single element x is determined by the tree ranks of x and the finite sequence of predecessors.

• If α_n is a bound on the ordinal tree ranks that occur at level n, then for x at level n, we can say that $\text{tr}(x) = \infty$ by saying that $\text{tr}(x) > \alpha_n$. □

Theorem 4.6 (Calvert – K – J. Millar). *There is a computable rank-homogeneous tree T that is special.*

PROOF SKETCH. We produce a computable special tree T', labeled by elements from a Harrison ordering H, so that if $x \in T'$ has label h, then $tr(x) = \alpha$ iff $pred(h)$ has order type α in H, and $\text{tr}(x) = \infty$ iff h comes from the non-well-founded part of H. Using the labels, we can replace T' by a rank-homogeneous tree T, preserving the ordinal ranks that appear at each level. □

Our tree shares with the Harrison ordering a certain "approximability" feature.

Definition 4.7. Suppose $\mathcal{A}$ is a computable structure of noncomputable Scott rank. We say that $\mathcal{A}$ is *strongly computably approximable* if for any Σ_1^1 set S, there is a uniformly computable sequence $(\mathcal{C}_n)_{n\in\omega}$ such that if $n \in S$, then $\mathcal{C}_n \cong \mathcal{A}$, and if $n \notin S$, then $SR(\mathcal{C}_n)$ is computable.

The Harrison ordering is strongly computably approximable — the approximations $\mathcal{C}_n$ are well-ordered. We may take our computable rank-homogeneous special tree to be strongly computably approximable — the approximations have no path.

Corollary 4.8. *There is a computable tree T of Scott rank ω_1^{CK}. Moreover, we may take T to be strongly computably approximable.*

4.3. Further computable structures of Scott rank ω_1^{CK}. In [3], we use the computable tree of Scott rank ω_1^{CK} to produce further structures of Scott rank ω_1^{CK}.

Theorem 4.9 (Calvert – Goncharov – K). *Each of the following classes contains a computable structure of Scott rank ω_1^{CK} (as well as one of Scott rank $\omega_1^{CK} + 1$). Moreover, we may take all of the structures to be strongly computably approximable.*

(1) *undirected graphs,*
(2) *linear orderings,*
(3) *fields of any fixed characteristic,*

Morozov [18] added the class of 2-step nilpotent groups to the list in Theorem 4.9.

Theorem 4.10 (Morozov). *There is a computable 2-step nilpotent group of Scott rank ω_1^{CK} (also computably approximable).*

In [6], we added the class of Boolean algebras.

Theorem 4.11 (Fokina – K – Maher – Melnikov – Quinn). *There is a computable Boolean algebra of Scott rank ω_1^{CK} (also computably approximable).*

The idea behind Theorems 4.9, 4.10, and 4.11 is to show that certain computable transformations between classes of structures preserve Scott rank.

Definition 4.12 (Computable transformation). Let K and K' be classes of structures, all with universe a subset of ω. A *computable transformation* from K to K' is a Turing operator $\Phi = \varphi_e$ such that for each $\mathcal{A} \in K$, there is some $\mathcal{B} \in K'$ such that $\varphi_e^{D(\mathcal{A})} = \chi_{D(\mathcal{B})}$.

If Φ is as in the definition above, we write $\Phi(\mathcal{A})$ for $\mathcal{B}$.

Definition 4.13 (Preservation of Scott rank). Let K and K' be classes of structures (with universe a subset of ω). Let Φ be a computable transformation from K to K'. We say that Φ *preserves Scott rank* provided that for all $\mathcal{A} \in K$, either $\mathcal{A}$ and $\Phi(\mathcal{A})$ have the same Scott rank, or else both Scott ranks are ordinals computable in $\mathcal{A}$.

Remark. A rank-preserving computable operator Φ takes computable members of K to computable members of K'. Supposing that $\mathcal{A}$ is a computable member of K, if $\mathrm{SR}(\mathcal{A})$ is computable, then $\mathrm{SR}(\Phi(\mathcal{A}))$ is also computable, but the Scott ranks need not exactly match.

The following theorem is a combination of several results.

Theorem 4.14. *There are computable transformations that preserve Scott rank between the following pairs of classes*:

(1) *from directed graphs to undirected graphs*,
(2) *from undirected graphs to fields of any desired characteristic*,
(3) *from undirected graphs to linear orderings*,
(4) *from fields of any desired characteristic to 2-step nilpotent groups*,
(5) *from rank-homogeneous trees to Boolean algebras.*

For (1), Marker [15], Nies [20], and others, independently, described the transformation. For (2) and (3), Friedman and Stanley [7] described the transformations. For (1), (2), and (3), the proof that the operators preserve Scott rank is in [3]. For (4), Mal′cev [14] described the transformation. Morozov [18] showed that it preserves Scott rank. For (5), the definition of the computable transformation and the proof that it preserves Scott rank are both in [6]. Below we describe three of the computable transformations.

Definition 4.15. A computable transformation Φ from a class $K \subseteq \mathrm{Mod}(L)$ to a class $K' \subseteq \mathrm{Mod}(L')$ is a *Turing computable embedding* if for all $\mathcal{A}, \mathcal{A}' \in K$, $\mathcal{A} \cong \mathcal{A}'$ iff $\Phi(\mathcal{A}) \cong \Phi(\mathcal{A}')$.

In [7], Friedman and Stanley give a *Borel* embedding of undirected graphs into fields, but in fact, their embedding is computable.

Theorem 4.16 (Friedman – Stanley). *There is a Turing computable embedding of undirected graphs into fields of any desired characteristic.*

Description of embedding. Let $\mathcal{F}$ be an algebraically closed field of the desired characteristic, with algebraically independent elements b_n for $n \in \omega$. For a graph $\mathcal{G}$, let $\Phi(\mathcal{G})$ be the subfield of $\mathcal{F}$ generated by elements of the algebraic closure of b_n, for $n \in \omega$, and $\sqrt{d+d'}$, where for some i and j joined by an edge in $\mathcal{G}$, the elements d, d' are inter-algebraic with b_i, b_j, respectively. (For characteristic 2, we use $(d+d')^{1/3}$ instead of $\sqrt{d+d'}$.) The transformation is 1-1 on isomorphism types since there is a copy of the input graph defined (using computable infinitary formulas) in the output field. □

The following result is from [14] (see also [18]).

Theorem 4.17 (Mal′cev). *There is a Turing computable embedding of fields (or rings) into 2-step nilpotent groups.*

Description of embedding. For a field $\mathcal{F}$, let $\Phi(\mathcal{F})$ be the *Heisenberg group*, consisting of all matrices of the form

$$\begin{bmatrix} 1 & a & b \\ 0 & 1 & c \\ 0 & 0 & 1 \end{bmatrix}$$

with entries $a, b, c \in \mathcal{F}$, under the operation of matrix multiplication.

To show that the embedding is 1-1 on isomorphism types, we note that the field is defined, in a uniform way, in its Heisenberg group. Let $\mathcal{F}$ be a field, and let $\mathcal{H} = (H, *, 1)$ be the Heisenberg group. For a noncommuting pair $u, v \in H$, we get an isomorphic copy of $\mathcal{F}$, with universe equal to the center $Z(H)$, and operations defined as follows:

Addition: $x + y = z$ iff $x * y = z$

Multiplication: $x \times_{u,v} y = z$ iff there exist x', y' such that $[x', u] = 1$, $[y', v] = 1$, $[u, y'] = y$, $[x', v] = x$, and $[x', y'] = z$. □

Theorem 4.18 (Friedman – Stanley). *There is a Turing computable embedding of undirected graphs into linear orderings.*

Description of embedding. Consider $\mathbb{Q}^{<\omega}$ with the lexicographic ordering. Let $(\mathbb{Q}_k)_{k\in\omega}$ be a partition of $\mathbb{Q}$ into dense subsets — $\mathbb{Q}_0$ and $\mathbb{Q}_1$ play special roles. Let $(t_n)_{n\in\omega}$ be a list of the quantifier-free types for tuples in undirected graphs. For a graph $\mathcal{G}$, let $\Phi(\mathcal{G})$ be the restriction of the ordering on $\mathbb{Q}^{<\omega}$ to elements of the form $q_0 r_0 \ldots q_n r_n k$, where for some finite sequence $a_0, \ldots, a_n$ in $\mathcal{G}$, satisfying the type t_m, we have $q_n \in Q_{a_n}$, $r_0, \ldots, r_{n-1} \in Q_0$, $r_n \in Q_1$, and $k < m$. □

Remark. For this embedding, the graph $\mathcal{G}$ is not definable or interpretable in $\Phi(\mathcal{G})$. Still, there is a tight connection between orbits of tuples in $\mathcal{G}$ and in $\Phi(\mathcal{G})$.

5. Categoricity

In this section, we observe that in the various known examples of computable structures of Scott rank ω_1^{CK}, the computable infinitary theory is $\aleph_0$-categorical. We end with an interesting result of J. Millar and Sacks which hints that this might not always be the case.

Let $\mathcal{A}$ be computable (or hyperarithmetical). If $SR(\mathcal{A}) < \omega_1^{CK}$, then $Th_\infty(\mathcal{A})$ is $\aleph_0$-categorical. In fact, $\mathcal{A}$ has a computable infinitary Scott sentence. If $SR(\mathcal{A}) = \omega_1^{CK}+1$, then $Th_\infty(\mathcal{A})$ is not $\aleph_0$-categorical — there is a nonprincipal type witnessing the rank. The interesting case is where $SR(\mathcal{A}) = \omega_1^{CK}$.

5.1. Categoricity for known examples.

Proposition 5.1 (Makkai). *Let $\mathcal{A}$ be Makkai's original example of an arithmetical group tree of Scott rank ω_1^{CK}. Then $/Th_\infty(\mathcal{A})$ is $\aleph_0$-categorical.*

The following result is in [4].

Theorem 5.2 (Calvert – Goncharov – K – J. Millar). (1) *If $\mathcal{T}$ is a computable special rank-homogeneous tree, then* $Th_\infty(\mathcal{T})$ *is $\aleph_0$-categorical.*
(2) *If $\mathcal{A}$ is one of the structures, undirected graph, field, or linear ordering, obtained by applying the rank-preserving computable transformations from* [3] *to a tree $\mathcal{T}$ such that* $Th_\infty(\mathcal{T})$ *is $\aleph_0$-categorical, then* $Th_\infty(\mathcal{A})$ *is $\aleph_0$-categorical.*

Morozov showed the following.

Theorem 5.3 (Morozov). *If $\mathcal{A}$ is a 2-step nilpotent group obtained by applying the Mal′cev transformation to a field $\mathcal{F}$ such that* $Th_\infty(\mathcal{F})$ *is $\aleph_0$-categorical, then* $Th(\mathcal{A})$ *is $\aleph_0$-categorical.*

The following result is from [16].

Theorem 5.4 (J. Millar – Sacks). *There is a structure $\mathcal{A}$ such that* $SR(\mathcal{A}) = \omega_1^{CK}$, $\omega_1^{\mathcal{A}} = \omega_1^{CK}$, *and the theory of $\mathcal{A}$ in the corresponding admissible fragment is not $\aleph_0$-categorical.*

The construction of J. Millar and Sacks involves meta-recursion.

Problem 1. Is there a computable structure $\mathcal{A}$ such that $SR(\mathcal{A}) = \omega_1^{CK}$ and $Th_\infty(\mathcal{A})$ is not $\aleph_0$-categorical?

References

1. C. J. Ash and J. Knight, *Computable structures and the hyperarithmetical hierarchy*, Stud. Logic Found. Math., vol. 144, North-Holland, Amsterdam, 2000.
2. W. Calvert, E. Fokina, S. S. Goncharov, J. F. Knight, O. Kudinov, A. S. Morozov, and V. Puzarenko, *Index sets for classes of high rank structures*, J. Symbolic Logic **72** (2007), no. 4, 1418–1432.
3. W. Calvert, S. S. Goncharov, and J. F. Knight, *Computable structures of Scott rank ω_1^{CK} in familiar classes*, Advances in Logic (Denton, TX, 2004) (S. Gao, S. Jackson, and Y. Zhang, eds.), Contemp. Math., vol. 425, Amer. Math. Soc., Providence, RI, 2007, pp. 49–66.
4. W. Calvert, S. S. Goncharov, J. F. Knight, and J. Millar, *Categoricity of computable infinitary theories*, Arch. Math. Logic **48** (2009), no. 1, 25–38.
5. W. Calvert, J. F. Knight, and J. Millar, *Computable trees of Scott rank ω_1^{CK}, and computable approximation*, J. Symbolic Logic **71** (2006), no. 1, 283–298.
6. E. Fokina, J. F. Knight, C. Maher, A. Melnikov, and S. M. Quinn, *Classes of Ulm type, and relations between the class of rank-homogeneous trees and other classes*, J. Symbolic Logic, to appear.
7. H. Friedman and L. Stanley, *A Borel reducibility theory for classes of countable structures*, J. Symbolic Logic **54** (1989), no. 3, 894–914.
8. J. Harrison, *Recursive pseudo-well-orderings*, Trans. Amer. Math. Soc. **131** (1968), 526–543.
9. H. J. Keisler, *Model theory for infinitary logic. Logic with countable conjunctions and finite quantifiers*, Stud. Logic Found. Math., vol. 62, North-Holland, Amsterdam, 1971.

10. J. F. Knight and J. Millar, *Computable structures of rank ω_1^{CK}*, J. Math. Log., to appear.
11. G. Kreisel, *Set theoretic problems suggested by the notion of potential totality*, Infinitistic Methods (Warsaw, 1959), Pergamon, Oxford, 1961, pp. 103–140.
12. E. G. K. Lopez-Escobar, *An interpolation theorem for denumerably long formulas*, Fund. Math. **57** (1965), 253–272.
13. M. Makkai, *An example concerning Scott heights*, J. Symbolic Logic **46** (1981), no. 2, 301–318.
14. A. I. Mal′cev, *On a correspondence between rings and groups*, Fifteen Papers on Algebra, Amer. Math. Soc. Transl. Ser. 2, vol. 45, 1965, pp. 221–232.
15. D. Marker, *Model theory: An introduction*, Grad. Texts in Math., vol. 217, Springer, New York, 2002.
16. J. Millar and G. E. Sacks, *Atomic models higher up*, Ann. Pure Appl. Logic **155** (2008), no. 3, 225–241.
17. A. S. Morozov, *Groups of computable automorphisms*, Handbook of Recursive Mathematics, Vol. 1 (Yu. L. Ershov, S. S. Goncharov, A. Nerode, J. B. Remmel, and V. W. Marek, eds.), Stud. Logic Found. Math., vol. 138, North-Holland, Amsterdam, 1998, pp. 311–345.
18. ______, notes on the Mal′cev transformation from rings to Heisenberg groups.
19. M. Nadel, *Scott sentences and admissible sets*, Ann. Math. Logic **7** (1974), 267–294.
20. A. Nies, *Undecidable fragments of elementary theories*, Algebra Universalis **35** (1996), no. 1, 8–33.
21. D. Scott, *Logic with denumerably long formulas and finite strings of quantifiers*, Theory of Models (Berkeley, CA, 1963) (J. W. Addison, L. Henkin, and A. Tarski, eds.), North-Holland, Amsterdam, 1965, pp. 329–341.
22. M. Vanden Boom, *The effective Borel hierarchy*, Fund. Math. **195** (2007), no. 3, 269–289.
23. R. Vaught, *Invariant sets in topology and logic*, Fund. Math. **82** (1974/75), 269–294. Collection of articles dedicated to Andrzej Mostowski on his sixtieth birthday, VII.

Department of Mathematics, University of Notre Dame, 255 Hurley Hall, Notre Dame, IN 46556, USA

E-mail address: knight.1@nd.edu

Centre de Recherches Mathématiques
CRM Proceedings and Lecture Notes
Volume **53**, 2011

The Lorentz Category in Special Relativity

Joachim Lambek

ABSTRACT. Physical entities may be represented by quaternions with complex components. Conceiving these as arrows in an additive category with two objects, one may encode the information how they transform under a Lorentz transformation, and handle Maxwell's equations and the related Lorentz force. To cope with the Dirac equation we introduce regular matrix representations of the quaternion units and a third object into the category.

1. Introduction

Special relativity saw the light of day in 1905. Only six years later, two authors independently realized that it could be presented elegantly in terms of *biquaternions*, that is quaternions with complex components. (See Conway [1, 2] and Silberstein [10].)

The idea was to represent physical quantities and operators as biquaternions, together with instructions on how to transform them when the coordinate system was changed by a *Lorentz transformation*, that is a linear transformation which leaves the norm of the biquaternion

$$x = x_0 + \mathrm{i}(\mathrm{i}_1 x_1 + \mathrm{i}_2 x_2 + \mathrm{i}_3 x_3)$$

invariant.

A Lorentz transformation takes the form

$$x \mapsto q x q^{*\dagger}$$

where q is a biquaternion of norm $qq^\dagger = q^\dagger q = 1$, $q^\dagger$ being the quaternion conjugate of q. But it turns out that any biquaternion a of physical significance may be transformed as

$$a \mapsto q^X a q^{Y\dagger},$$

where X and Y are superscripts 1 or $*$, $q^1 = q$ and q^* is the complex conjugate of q. (Later we shall also admit the superscript 0 such that $q^0 = 1$, to deal with Dirac spinors.)

Today we may think of the superscripts as objects of an additive category, called a ring with two objects by Barry Mitchell, and known as a *Morita context* in

2010 *Mathematics Subject Classification.* 18C99, 83A05.
This is the final form of the paper.

ring theory, and the biquaternions as morphisms. Thus $a\colon X \to Y$ is to mean that $a \mapsto q^X a q^{Y\dagger}$ when the coordinate system is changed. In particular

$$X \xrightarrow{a} Y \xrightarrow{b} Z$$

means that

$$ab \mapsto q^X a \underbrace{q^{Y\dagger} q^Y} b q^{Z\dagger} = q^X ab q^{Z\dagger}.$$

Note that $a\colon X \to Y$ implies $a^*\colon *X \to *Y$, since

$$(q^X a q^{Y\dagger})^* = q^{*X} a^* q^{*Y\dagger},$$

but $a^\dagger\colon Y \to X$, since

$$(q^X a q^{Y\dagger})^\dagger = q^Y a^\dagger q^{X\dagger}.$$

(Later we will introduce a third object 0 to help representing spinors.)

2. Some Hermitian biquaternions in physics

Of special interest are *Hermitian* biquaternions a for which $a^* = a^\dagger$, i.e., $a^{*\dagger} = a$, in other words

$$a = a_0 + \mathrm{i}(\mathrm{i}_1 a_1 + \mathrm{i}_2 a_2 + \mathrm{i}_3 a_3),$$

when a_0, a_1, a_2 and a_3 are real and $\mathrm{i}_1, \mathrm{i}_2$ and i_3 such that

$$\mathrm{i}_1^2 = \mathrm{i}_2^2 = \mathrm{i}_3^2 = \mathrm{i}_1\mathrm{i}_2\mathrm{i}_3 = -1$$

are the basic quaternion units.

In particular, the following Hermitian biquaternions played a rôle in the early development:

space-time:

$$x = x_0 + \mathrm{i}(\mathrm{i}_1 x_1 + \mathrm{i}_2 x_2 + \mathrm{i}_3 x_3) = t + \mathrm{i}\mathbf{x},$$

where $x_0 = t$ represents time and $\mathbf{x}$ position in 3-space;

energy-momentum:

$$p = p_0 + \mathrm{i}(\mathrm{i}_1 p_1 + \mathrm{i}_2 p_2 + \mathrm{i}_3 p_3) = \varepsilon + \mathrm{i}\mathbf{p},$$

where $p_0 = \varepsilon$ represents energy and $\mathbf{p}$ momentum;

four-potential:

$$A = A(x) = \varphi + \mathrm{i}(\mathrm{i}_1 A_1 + \mathrm{i}_2 A_2 + \mathrm{i}_3 A_3) = \varphi + \mathrm{i}\mathbf{A},$$

where $e\varphi$ represents the potential energy and $e\mathbf{A}$ the potential momentum, e being the electric charge of a moving particle;

partial differentiation:

$$D = \frac{\partial}{\partial x_0} - \mathrm{i}\left(\mathrm{i}_1 \frac{\partial}{\partial x_1} + \mathrm{i}_2 \frac{\partial}{\partial x_2} + \mathrm{i}_3 \frac{\partial}{\partial x_3}\right) = \frac{\partial}{\partial t} - \mathrm{i}\nabla.$$

All of these are to be thought of as arrows $1 \to *$, thus transforming like $x \mapsto q x q^{*\dagger}$, hence possessing an invariant norm. For example, the Hermitian differential $\mathrm{d}x$ has norm

$$\mathrm{d}s^2 = \mathrm{d}x\, \mathrm{d}x^\dagger = \mathrm{d}x^\dagger\, \mathrm{d}x,$$

where s is called the *interval*, and the energy-momentum p has norm

$$m^2 = pp^\dagger = p^\dagger p,$$

where m is known as the *rest-mass*.

When $m \neq 0$ one requires that also $\mathrm{d}s \neq 0$ and writes

$$p = m\frac{\mathrm{d}x}{\mathrm{d}s} = m\frac{\mathrm{d}t}{\mathrm{d}s}\left(1 + \mathrm{i}\frac{\mathrm{d}\mathbf{x}}{\mathrm{d}t}\right) = m(1 - v^2)^{-1/2}(1 + \mathrm{i}\mathbf{v}),$$

where

$$\mathbf{v} = \mathrm{i}_1\frac{\mathrm{d}x_1}{\mathrm{d}t} + \mathrm{i}_2\frac{\mathrm{d}x_2}{\mathrm{d}t} + \mathrm{i}_3\frac{\mathrm{d}x_3}{\mathrm{d}t}$$

is the classical velocity vector of norm v^2. (We have chosen units so that the velocity of light and the dielectric constant are 1 and Planck's constant is 2π.)

In the absence of an external force, one postulates *conservation* of energy-momentum $\mathrm{d}p/\mathrm{d}s = 0$. In the presence of an electro-magnetic force, the so-called *Lorentz force*, one would expect

$$\frac{\mathrm{d}}{\mathrm{d}s}(p + eA) = 0,$$

A being the electro-magnetic four-potential, eA the potential energy-momentum.

It used to be assumed that there is no way of measuring $\mathbf{A}$, but that $A = \varphi + \mathrm{i}\mathbf{A}$ is only known by the vector part of D^*A:

$$V(D^*A) = -(\mathbf{B} + \mathrm{i}\mathbf{E}),$$

where $\mathbf{B}$ and $\mathbf{E}$ are the magnetic and electric fields respectively. But this is the same as the vector part of $D^*(A - D\theta)$, where $\theta = \theta(x)$ is a scalar, since $D^*D\theta$ is a scalar. Thus, the measurable effect of A appeared to be invariant under the *gauge transformation*

$$A \mapsto A_\theta = A - D\theta.$$

This was traditionally exploited by assuming that the scalar part $D \odot A = 0$. We will refrain from making this assumption here. The gauge transformation also does not affect Maxwell's equation, which asserts that

$$D(\mathbf{B} + \mathrm{i}\mathbf{E}) = DD^*A = -J,$$

where J denotes the *charge-current density.*

3. Working with Hermitian biquaternions

Recalling that $a\colon X \to Y$ implies $a^*\colon X* \to Y*$ and $a^\dagger\colon Y \to X$, we infer that it implies $a^{*\dagger}\colon Y* \to X*$. In particular, if $a\colon 1 \to *$, then

$$a^{*\dagger}\colon 1 = ** \to *,$$

hence also

$$H(a) = \frac{1}{2}(a + a^{*\dagger}),\ a - H(a) = \frac{1}{2}(a - a^{*\dagger})\colon 1 \to *.$$

We call $H(a)$ the *Hermitian part* of a and $a - H(a) = -\mathrm{i}H(\mathrm{i}a)$ the *skew-Hermitian part.*

On the other hand, if $a\colon 1 \to 1$ then $a^\dagger\colon 1 \to 1$, hence also

$$S(a) = \frac{1}{2}(a + a^\dagger),\ V(a) = \frac{1}{2}(a - a^\dagger) : 1 \to 1.$$

We call these the *scalar* and *vector* parts of a respectively.

From the Hermitian biquaternions $a, b\colon 1 \to *$ we can construct the composite

$$1 \xrightarrow{a} * \xrightarrow{b^\dagger} 1,$$

which we will write as

$$ab^{\dagger}: 1 \to 1,$$

the order of multiplication being opposite from the usual categorical convention.

If $a = a_0 + \mathrm{i}\mathbf{a}$ and $b = b_0 + \mathrm{i}\mathbf{b}$, we have

$$ab^{\dagger} = a_0 b_0 - \mathbf{a} \circ \mathbf{b} + \mathbf{a} \times \mathbf{b} + \mathrm{i}(\mathbf{a} b_0 - a_0 \mathbf{b}),$$

where the scalar part

$$a \odot b = S(ab^{\dagger}) = a_0 b_0 - \mathbf{a} \circ \mathbf{b}$$

and the vector part

$$ab - a \odot b = V(ab^{\dagger}) = \mathbf{a} \times \mathbf{b} + \mathrm{i}(\mathbf{a} b_0 - a_0 \mathbf{b})$$

both represent arrows $1 \to 1$. Note that $a \odot b = b \odot a$, but $V(ab^{\dagger}) = -V(ba^{\dagger})$.

If a, b, c are three Hermitian biquaternions $* \to 1$, we have

$$* \xrightarrow{a} 1 \xrightarrow{b^*} * \xrightarrow{c} 1,$$

so that

$$ab^* c \colon * \to 1.$$

A routine calculation shows that its Hermitian part

$$H(ab^* c) = a(b \odot c) - b(c \odot a) + c(a \odot b). \tag{1}$$

This identity first appears in Gürsey [5]. We will apply it in case $a = e(\mathrm{d}x/\,\mathrm{d}s)$, $b = D$ and $c = A$.

4. Conservation of energy-momentum

From (1) we obtain

$$-H\Big(e\frac{\mathrm{d}x}{\mathrm{d}s}D^* A\Big) + e\frac{\mathrm{d}x}{\mathrm{d}s}(D \odot A) - D\Big(e\frac{\mathrm{d}x}{\mathrm{d}s} \odot A\Big) + e\Big(\frac{\mathrm{d}x}{\mathrm{d}s} \odot D\Big)A = 0. \tag{2}$$

Here the first two terms combine to

$$\begin{aligned} -H\Big(e\frac{\mathrm{d}x}{\mathrm{d}s}(D^* A - D \odot A)\Big) &= H\Big(e\frac{\mathrm{d}x}{\mathrm{d}s}(\mathbf{B} + \mathrm{i}\mathbf{E})\Big) \\ &= e\frac{\mathrm{d}\mathbf{x}}{\mathrm{d}s} \circ \mathbf{E} + \mathrm{i}e\Big(\frac{\mathrm{d}\mathbf{x}}{\mathrm{d}s} \times \mathbf{B} + \frac{\mathrm{d}t}{\mathrm{d}s}\mathbf{E}\Big) \\ &= e\frac{\mathrm{d}t}{\mathrm{d}s}\big(\mathbf{v} \circ \mathbf{E} + \mathrm{i}(\mathbf{v} \times \mathbf{B} + \mathbf{E})\big). \end{aligned}$$

We recognize $e(\mathbf{v} \times \mathbf{B} + \mathbf{E})$ as the *classical Lorentz force* $\mathrm{d}\mathbf{p}/\,\mathrm{d}t$, so we must interpret $e(\mathbf{v} \circ \mathbf{E})$ as the *classical Lorentz power* $\mathrm{d}\varepsilon/\,\mathrm{d}t$, and the whole expression as a relativistic version of the Lorentz force $\mathrm{d}p/\,\mathrm{d}s$.

Thus (2) asserts that

$$\frac{\mathrm{d}}{\mathrm{d}s}(p + eA) = \frac{\mathrm{d}p}{\mathrm{d}s} + \frac{\mathrm{d}eA}{\mathrm{d}s} = D\Big(A \odot e\frac{\mathrm{d}x}{\mathrm{d}s}\Big),$$

where $p+eA$ is the kinetic plus potential energy-momentum. At first sight, the right hand side appears puzzling, but it seems to be related to the quantum-mechanical *Bohm – Aharonov* effect. (See Ryder [9]). Anyway, if we write

$$A \odot \frac{\mathrm{d}x}{\mathrm{d}s} = \frac{\mathrm{d}\theta}{\mathrm{d}s}, \quad A_\theta = A - D\theta,$$

we have

$$\frac{\mathrm{d}}{\mathrm{d}s}(p + eA_\theta) = 0$$

and we may as well redefine the 4-potential as A_θ, thus establishing that the total energy momentum is conserved.

There is no point in repeating this process, since

$$\begin{aligned}\frac{\mathrm{d}x}{\mathrm{d}s} \odot A_\theta &= \frac{\mathrm{d}x}{\mathrm{d}s} \odot A - \frac{\mathrm{d}x}{\mathrm{d}s} \odot D\theta \\ &= \frac{\mathrm{d}x}{\mathrm{d}s} \odot A - \frac{\mathrm{d}\theta}{\mathrm{d}s} = 0.\end{aligned}$$

Hence we may as well assume that $A = A_\theta$ in the first place, i.e., $\dot{x} \odot A = 0$, i.e., the potential is "orthogonal" to the current in Minkowski space.

In summary of the above argument, Maxwell's equations show how the electromagnetic field depends on the charge-current density and may be derived from a four-potential, which is determined only up to a gauge transformation. The path of a charged particle is influenced by the Lorentz force due to this field. Aided by a suitable gauge transformation, we may assume that the underlying four-potential is orthogonal to the path of the given particle in Minkowski space. When multiplied by the charge, the four-potential then gives rise to the potential energy-momentum of the charged particle, and the relativistic Lorentz force may be accounted for by the conservation of its total (kinetic plus potential) energy-momentum. Physicists must be familiar with this observation, though I have not seen it mentioned in any of the texts I have consulted.

5. Dirac spinors

In vacuum, the photon, or any particle of mass zero, satisfies the equation $DD^*A = 0$. This has been generalized to

$$DD^*A + m^2A = 0$$

for particles of mass $m \neq 0$, the so-called *Klein – Gordon* equation. This second order differential equation may be replaced by two first order equations, provided $m \neq 0$:

$$D^*A = mA', \quad DA' = -mA. \tag{3}$$

For bosons of spin 1,

$$A\colon 1 \to *, \quad A'\colon * \to *$$

but for fermions of spin $\frac{1}{2}$ we postulate

$$A\colon 1 \to 0, \quad A'\colon * \to 0$$

and then A and A' are called *spinors*. We have now introduced a third object 0 into our category.

The two equations of (3) may be combined into one:

$$D^*\Psi = -m\mathrm{j}\Psi,$$

the so-called Dirac equation, where

$$\Psi = A + \mathrm{j}A'\colon 1 \to 0$$

provided

$$\mathrm{j}^2 = -1, \quad \mathrm{ji} = -\mathrm{ij}.$$

This looks odd when i is the usual complex number, but becomes realistic if left multiplication by i and j is replaced by right multiplication with the basic quaternions i_1 and i_2 respectively.

(Note that $q^0 = 1$, hence the set $\{0, *, 1\}$ behaves like $\mathbb{Z} \bmod 3$ under multiplication. The arrow $\Psi\colon 1 \to 0$ represents the fact that $\Psi \mapsto q\Psi$ under a Lorentz transformation.)

More precisely, we identify the quaternion units i_α ($\alpha = 0$ to 3) with their left regular matrix representation and write j_α for the corresponding right representation.

The Dirac equation then takes the form

$$(D^* + m\mathrm{j}_2)\Psi = 0,$$

where $D^* = (\partial/\partial t) + \mathrm{j}_1\nabla$. (There is no need for introducing the ad hoc Pauli matrices.)

Since the Klein – Gordon equation has the explicit solution

$$A = \cos(x \odot p)\Psi_0,$$

Ψ_0 being a constant matrix, our procedure yields a solution to the Dirac equation:

$$\Psi = \big(\cos(x \odot p) + \eta \sin(x \odot p)\big)\Psi_0,$$

where

$$-\eta = \mathrm{j}_2 \frac{\mathrm{d}x^\dagger}{\mathrm{d}s} = (\mathrm{d}x/\,\mathrm{d}s)j_2.$$

Taking advantage of the fact that $\eta^2 = -1$, we may write this more elegantly thus:

$$\Psi = \exp(\eta(x \odot p))\Psi_0.$$

In the Dirac equation, Ψ may be multiplied by any constant matrix on the right, in particular by a column vector $[c]$, where c is a quaternion with real components c_α, and $[c]$ is the transposed of the row vector (c_0, c_1, c_2, c_3), rendering $\Psi[c]$ into a column vector itself, say of the form $[\psi]$, ψ being a real quaternion. Then

$$\mathrm{i}_\alpha[\psi] = [\mathrm{i}_\alpha \psi], \; j_\alpha[\psi] = [\psi \mathrm{i}_\alpha],$$

hence the Dirac equation with $\Psi = [\psi]$ takes the quaternionic form

$$\frac{\partial \psi}{\mathrm{d}t} + \nabla \psi \mathrm{i}_1 + m\psi \mathrm{i}_2 = 0.$$

We have taken j_α to be the right regular representation of i_α, but we might as well have taken it to be that of $r\mathrm{i}_\alpha r^\dagger$, where r is any quaternion of norm 1. Then the above equation takes the form

$$\frac{\partial \psi}{\partial t} r + \nabla \psi r \mathrm{i}_1 + m\psi r \mathrm{i}_2 = 0,$$

which amounts to replacing ψ by ψr.

References

1. A. W. Conway, *On the applications of quaternions to some recent development of electrical theory*, Proc. Roy. Irish Acad. Sect. A **29** (1911), 1–9.
2. ———, *The quaternion form of relativity*, Philos. Mag. (6) **24** (1912), 208.
3. ———, *Quaternions and quantum mechanics*, Pont. Acad. Sci. Acta **12** (1948), 259–277.
4. R. P. Feynman, R. B. Leighton, and M. Sands, *The Feynman lectures on physics*, Addison-Wesley, Reading, MA, 1963.

5. F. Gürsey, *Contribution to the quaternion formalism in special relativity*, Rev. Fac. Sci. Univ. Istanbul. Sér. A. **20** (1955), 149–171.
6. J. Lambek, *If Hamilton had prevailed: quaternions in physics*, Math. Intelligencer **17** (1995), no. 4, 7–15.
7. ———, *Quaternions and the particles of nature*, Technical Report 96-03, Department of Mathematics, McGill University, 1996.
8. R. Penrose, *The road to reality*, Alfred A. Knopf Inc., New York, 2005. A complete guide to the laws of the universe.
9. L. H. Ryder, *Quantum field theory*, 2nd ed., Cambridge Univ. Press, Cambridge, 1996.
10. L. Silberstein, *Quaternionic form of relativity*, Philos. Mag. (6) **23** (1912), 790–809.
11. ———, *Theory of relativity*, MacMillan, London, 1924.
12. A. Sudbery, *Quantum mechanics and the particles of nature: An outline for mathematicians*, Cambridge Univ. Press, Cambridge, 1986.

Department of Mathematics and Statistics, McGill University, Montréal, QC H3A 2K6, Canada

E-mail address: `lambek@math.mcgill.ca`

Centre de Recherches Mathématiques
CRM Proceedings and Lecture Notes
Volume **53**, 2011

A Model-Theoretic Counterpart to Moishezon Morphisms

Rahim Moosa

ABSTRACT. The notion of being *Moishezon* to a set of types, a natural strengthening of internality motivated by complex geometry, is introduced. Under the hypothesis of Pillay's [6] canonical base property, and using results of Chatzidakis [2], it is shown that if a stationary type of finite U-rank at least two is almost internal to a nonmodular minimal type and *admits a diagonal section*, then it is Moishezon to the set of nonmodular minimal types. This result is inspired by Campana's [1] "first algebraicity criterion" in complex geometry. Other related abstractions from complex geometry, including *coreductions* and *generating fibrations*, are also discussed.

1. Introduction

Compact complex manifolds can be viewed as first order structures in the language where there are predicates for all complex-analytic relations. The corresponding theory admits quantifier elimination and is of finite Morley rank. Many of the techniques and approaches used by complex geometers in studying the bimeromorphic structure of compact complex manifolds can be viewed as specialisations of the methods of finite rank geometric stability theory. But this paper has to do with the other direction; abstracting ideas *from* bimeromorphic geometry *to* stability theory. The goal is to develop new model-theoretic tools that may be useful in the study of other contexts, such as differential-algebraic geometry.

The general model-theoretic theme that we are interested in is the following: Given a finite rank stationary type $p(x)$ in a stable theory T, how does p relate to the nonmodular minimal types of T? If the Zilber trichotomy holds in T (as it does in the theory of compact complex manifolds) then a nonmodular minimal type corresponds to algebraic geometry over an algebraically closed field, and so one is trying to quantify how much of an expansion of algebraic geometry the type $p(x)$ involves. In the different theories the expansion from algebraic geometry is in different directions: the theory of compact complex manifolds adds meromorphic structure while the theory of differentially closed fields in characteristic zero, or separably closed fields in positive characteristic, adds (Hasse – Schmidt) differential-algebraic structure.

2010 *Mathematics Subject Classification.* Primary 03C45, 03C98; Secondary 32J27.
Rahim Moosa was supported by an NSERC Discovery Grant.
This is the final form of the paper.

The notions appearing in this paper were originally developed to prove, under suitable hypotheses on the theory, that every finite rank type has a minimum extension that is internal to the set of nonmodular minimal types. The difference in rank then, between the type and this minimum extension, would be a new way to quantify how far the type is from the "algebraic" part of that theory. In complex geometry such minimum extensions, called *algebraic coreductions*, have been known to exist since the early nineteen eighties (due to Campana [1]). In its first form this paper was devoted to developing the model-theoretic machinery that would allow the complex-geometric constructions to go through in a wide class of stable theories. As it turned out however, around the same time, Chatzidakis [2] found a much more direct proof of this theorem using milder (but similar) assumptions on the theory. Nevertheless, we feel that some of the ideas and arguments that were extracted and abstracted from complex geometry may be of independent interest and may have other applications. We therefore present them here.

The paper is laid out as follows: In Section 2 we introduce a strengthening of internality modelled after the complex-geometric notion of a Moishezon morphism. In Section 3, under the assumption that the theory admits the canonical base property (a stability-theoretic condition that is true of compact complex manifolds [6] and differentially closed fields [7]), we give a criterion for when a finite rank type is Moishezon to the set of nonmodular minimals. We rely heavily here on results of Chatzidakis [2], without which we would have had to assume an apparently stronger (but *a posteriori* equivalent) property introduced by the author and Pillay in [5]. Much of the work in this section involves pushing the techniques of [5] as far as they will go. In any case, our criterion can be viewed as a model-theoretic counterpart of a very special case of Campana's "first algebraicity criterion" from [1]. Finally, in section 4, we discuss the connection to the existence of coreductions, a problem which we also reformulate in terms of the *generating families* that appeared in [5].

We work throughout in $\overline{M}^{\mathrm{eq}}$ where $\overline{M}$ is a sufficiently saturated model of a complete stable theory T. All parameter sets are assumed to be of cardinality less than the cardinality of $\overline{M}$, and all tuples $a, b, \ldots$ will be assumed to be possibly infinite tuples of length strictly less than the cardinality of $\overline{M}$. We will only be concerned with precision up to interalgebraicity. In particular, following [5], by the *canonical base* of a stationary type $p(x) \in S(A)$, which we will denote by $\overline{\mathrm{Cb}}(p)$, we will mean (an enumeration of) the *algebraic* closure of A_0, where A_0 is the smallest definably closed subset of $\mathrm{dcl}(A)$ over which p does not fork and the restriction of p to which is stationary.

2. Moishezon: between internal and algebraic

Suppose $\mathbf{P}$ is a set of partial types and $p(x) \in S(B)$ is a stationary type. Here, and throughout this paper, for any set $C \supseteq B$, $p(x)|C$ denotes the nonforking extension of p to C and $\mathbf{P}_C$ denotes the set of partial types in $\mathbf{P}$ whose domains are contained in C. We also use $\mathbf{P}_C^{\overline{M}}$ to denote the set of solutions $\bigcup_{q\in\mathbf{P}_C} q^{\overline{M}}$.

We will say that $p(x) \in S(B)$ is $\mathbf{P}$-*algebraic* if $p^{\overline{M}} \subseteq \mathrm{acl}(B\mathbf{P}_B^{\overline{M}})$. Recall that a stationary type $p(x) \in S(B)$ is *almost* $\mathbf{P}$-*internal* if there is some $C \supseteq B$ such that $p(x)|C$ is $\mathbf{P}$-algebraic. We wish to introduce a condition that lies strictly in between $\mathbf{P}$-algebraicity and almost $\mathbf{P}$-internality. To motivate our definition we turn to the model theory of compact complex manifolds.

Let $\mathcal{A}$ be the many-sorted first order structure where there is a sort for every reduced and irreducible complex-analytic space and where the basic predicates are the complex-analytic subsets of cartesian powers of sorts. The theory of this structure, denoted by CCM, admits quantifier elimimination and is, sort by sort, of finite Morely rank. One distinguished sort in $\mathcal{A}$ is the projective line $\mathbb{P}$. The complex field is definable on this sort, and in fact the induced structure on $\mathbb{P}$ is bi-interpretable with the pure field structure on the complex numbers.

Working in a saturated elementary extension of $\mathcal{A}$, let us consider what it means for a stationary type $p(x) = \operatorname{stp}(a/b)$ to be $\mathbb{P}$-algebraic or almost $\mathbb{P}$-internal. (Here a and b are assumed to be finite tuples.) If we let $X = \operatorname{loc}(a, b)$ and $Y = \operatorname{loc}(b)$ — where here loc denotes the *locus*, that is, the smallest complex-analytic set containing the tuple in question — then, as is explained in Section 2 of [3], p is the generic type of a generic fibre of the fibration $X \to Y$ induced by the projection map. It follows from results in that paper (see also the discussion in Section 3 of [5]) that

- p is $\mathbb{P}$-algebraic if and only if X meromorphically embeds into $Y \times \mathbb{P}_n$ over Y, for some $n \geq 0$, and
- p is almost $\mathbb{P}$-internal if and only if p is $\mathbb{P}$-internal if and only if there is a complex-analytic space over Y, $Y' \to Y$, such that the fibred product $X \times_Y Y'$ meromorphically embeds into $Y' \times \mathbb{P}_n$ over Y', for some $n \geq 0$.

But in complex geometry there is an intermediate notion that plays a much more important role than either of the above two conditions: a surjective morphism $X \to Y$ is *Moishezon* if X meromorphically embeds into $\mathbb{P}(\mathcal{F})$ over Y, where $\mathcal{F}$ is a coherent analytic sheaf on Y and $\mathbb{P}(\mathcal{F})$ denotes the projective linear space associated to $\mathcal{F}$. If p is $\mathbb{P}$-algebraic then $X \to Y$ is seen to be Moishezon by taking $\mathcal{F}$ to be the free sheaf $\mathcal{O}_Y^n$. The converse fails because not every projective linear space splits. On the other hand, as every projective linear space does split after base change, Moishezon-ness of $X \to Y$ implies that p is almost $\mathbb{P}$-internal. The converse of this implication also fails: the results of [3] show that if $\dim(a/b) = 1$ then p is $\mathbb{P}$-internal, but there exist non-Moishezon fibrations of dimension one, for example of Hopf surfaces. Hence Moishezon-ness of $X \to Y$ lies strictly in between $\mathbb{P}$-algebraicity and almost $\mathbb{P}$-internality of p.

Since we do not as yet see how to formulate the abstract analogue of a projective linear space, we focus instead on the following properties of Moishezon fibrations:

(1) The restriction of a Moishezon morphism to a complex-analytic subspace is again Moishezon.
(2) The composition of Moishezon morphisms is Moishezon.

The following definition in the general stable setting extracts, almost literally, the model-theoretic content of these properties.

Definition 2.1. Suppose $p(x) \in S(Ab)$ is a stationary type and $\mathbf{P}$ is an $\operatorname{Aut}_A(\overline{M})$-invariant set of partial types. We will say that $p(x)$ is $\mathbf{P}$-*Moishezon over* A if whenever $a \models p(x)$ and c is such that $\operatorname{stp}(b/Ac)$ is almost $\mathbf{P}$-internal, then $\operatorname{stp}(a/Ac)$ is almost $\mathbf{P}$-internal.

Note that to verify $\mathbf{P}$-Moishezon-ness it suffices to prove that for *some* $a \models p(x)$, whenever c is such that $\operatorname{stp}(b/Ac)$ is almost $\mathbf{P}$-internal, then $\operatorname{stp}(a/Ac)$ is also.

Remark 2.2. This is a somewhat tentative definition. For one thing, it is a variant of a condition that was implicitly considered by Pillay and the author in [5]:

($*$) Whenever $a \models p(x)$, c is such that $\operatorname{stp}(b/Ac)$ is almost $\mathbf{P}$-internal, *and* $a \perp_{Ab} c$, then $\operatorname{stp}(ab/Ac)$ is almost $\mathbf{P}$-internal.

Note that since internality is preserved by concatenation, $\mathbf{P}$-Moishezon-ness is a strengthening of ($*$). Even so, it is not strong enough to precisely capture the complex-geometric notion. While the generic type of a Moishezon morphism in CCM is $\mathbb{P}$-Moishezon (this is Example 2.3 below), not all $\mathbb{P}$-Moishezon types arise in that way: Suppose Y is a strongly minimal compact complex manifold $f\colon X \to Y$ is a fibration with $\dim(X) = \dim(Y) + 1$. Then f need not be Moishezon, but the generic type of f will be $\mathbb{P}$-Moishezon. Indeed, if a is generic in X then $\operatorname{tp}\big(a/f(a)\big)$ is $\mathbb{P}$-internal (by one-dimensionality, see [3]), and as the only $\mathbb{P}$-internal extensions of $\operatorname{tp}\big(f(a)\big)$ are algebraic extensions, $\operatorname{tp}\big(a/f(a)\big)$ is in fact $\mathbb{P}$-Moishezon.

Example 2.3. *Working in* CCM, *suppose X and Y are irreducible compact complex spaces, and $f\colon X \to Y$ is a Moishezon morphism. If $a \in X$ is generic, then* $\operatorname{stp}\big(a/f(a)\big)$ *is $\mathbb{P}$-Moishezon over $\varnothing$.*

Proof. Let $b := f(a)$ and suppose $\operatorname{stp}(b/c)$ is $\mathbb{P}$-internal. (Recall that in CCM, almost internality and internality coincide.) Let $U := \operatorname{loc}(a,b,c)$, $V := \operatorname{loc}(b,c)$, and $W := \operatorname{loc}(c)$. Then, as $\operatorname{stp}(bc/c)$ is $\mathbb{P}$-internal, there exists $W' \to W$ such that $V \times_W W' \to W'$ meromorphically embeds into $W' \times \mathbb{P}_n$ over W'. In particular, $V \times_W W' \to W'$ is Moishezon.

On the other hand, since f is Moishezon, so is $\Gamma(f) \to Y$, where $\Gamma(f) = \operatorname{loc}(a,b)$ is the graph of f. Moishezon morphisms are preserved by base change. Hence $\Gamma(f) \times_Y V \to V$ is Moishezon. Note that $Y = \operatorname{loc}(b)$. As U is contained in $\Gamma(f) \times_Y V$ over V, $U \to V$ is also Moishezon. Hence $U \times_V (V \times_W W') = U \times_W W' \to V \times_W W'$ is Moishezon.

The composition of Moishezon morphisms is Moishezon. So $U \times_W W' \to W'$ is Moishezon. Hence there exists $W'' \to W'$ such that $U \times_W W'' \to W''$ meromorphically embeds in $W'' \times \mathbb{P}_m$ over W'' for some m. It follows that the generic type of $U \to W$, which is $\operatorname{stp}(ab/c)$, is $\mathbb{P}$-internal. Hence, $\operatorname{stp}(a/c)$ is $\mathbb{P}$-internal, as desired. $\square$

The following properties of $\mathbf{P}$-Moishezon-ness are immediate consequences of the of the definitions.

Proposition 2.4. *Suppose $p(x) \in S(Ab)$ is stationary and $\mathbf{P}$ is an $\operatorname{Aut}_A(\overline{M})$-invariant set of partial types.*

(a) *If p is $\mathbf{P}_A$-algebraic then it is $\mathbf{P}$-Moishezon over A. In particular, if p is almost $\mathbf{P}_A$-internal then some nonforking extension of p is $\mathbf{P}$-Moishezon over A.*[1]

(b) *If p is $\mathbf{P}$-Moishezon over A then it is almost $\mathbf{P}$-internal.*

(c) *If $p \restriction A$ is stationary and almost $\mathbf{P}$-internal then p is $\mathbf{P}$-Moishezon over A.*

(d) *If p is $\mathbf{P}$-Moishezon over A then every extension of p is also.*

[1] Note that, unlike almost $\mathbf{P}$-internality, having a $\mathbf{P}$-Moishezon nonforking extension does not imply being $\mathbf{P}$-Moishezon.

(e) *If p is* $\mathbf{P}$*-Moishezon over A and $a' \in \operatorname{acl}(Aba)$ for some $a \models p(x)$, then* $\operatorname{stp}(a'/Ab)$ *is* $\mathbf{P}$*-Moishezon over A.*

(f) *If* $\operatorname{stp}(a/Ab)$ *and* $\operatorname{stp}(a'/Ab)$ *are* $\mathbf{P}$*-Moishezon over A, then* $\operatorname{stp}(aa'/Ab)$ *is also* $\mathbf{P}$*-Moishezon over A.*

(g) *If* $\operatorname{stp}(a/Ab)$ *and* $\operatorname{stp}(b/Ac)$ *are* $\mathbf{P}$*-Moishezon over A, then so is* $\operatorname{stp}(a/Ac)$.

(h) *If* $\operatorname{stp}(a/Abc)$ *is* $\mathbf{P}$*-Moishezon over A then it is* $\mathbf{P}$*-Moishezon over Ab.*

PROOF. For part (a) suppose $\operatorname{stp}(b/Ac)$ is almost $\mathbf{P}$-internal. So there exists $C \supseteq Ac$ such that $b \mathop{\smile\!\!\!\!\!\mid}_{Ac} C$ and $b \in \operatorname{acl}(C\mathbf{P}_C^{\overline{M}})$. Let $a \models p(x)|Cb$. Then $a \mathop{\smile\!\!\!\!\!\mid}_{Ac} C$ and, since $p^{\overline{M}} \subseteq \operatorname{acl}(Ab\mathbf{P}_A^{\overline{M}})$, we have $a \in \operatorname{acl}(C\mathbf{P}_C^{\overline{M}})$. So $\operatorname{stp}(a/Ac)$ is almost $\mathbf{P}$-internal. Hence $p(x) = \operatorname{tp}(a/Ab)$ is $\mathbf{P}$-Moishezon over A.

In particular, if p is almost $\mathbf{P}_A$-internal then from some $b' \supseteq b$, the nonforking extension $q(x) := p(x)|Ab'$ is $\mathbf{P}_A$-algebraic, and hence $\mathbf{P}$-Moishezon over A.

For part (b) just take $c = b$ in the definition of $\mathbf{P}$-Moishezon.

Part (c). Suppose $p \restriction A$ is stationary and almost $\mathbf{P}$-internal. Then every extension of $p \restriction A$ is almost $\mathbf{P}$-internal. In particular $\operatorname{stp}(a/Ac)$ is almost $\mathbf{P}$-internal for any $a \models p(x)$ and any c such that $\operatorname{stp}(b/Ac)$ is almost $\mathbf{P}$-internal. That is, $p(x) = \operatorname{tp}(a/Ab)$ is $\mathbf{P}$-Moishezon over A.

Part (d). Suppose $q \in S(Ab')$ is an extension of p, where $b' \supseteq b$. Suppose $\operatorname{stp}(b'/Ac)$ is $\mathbf{P}$-internal and let $a \models q(x)$. Then $a \models p(x)$ and $\operatorname{stp}(b/Ac)$ is $\mathbf{P}$-internal, so that $\operatorname{stp}(a/Ac)$ is almost $\mathbf{P}$-internal, as desired.

Part (e). If $\operatorname{stp}(b/Ac)$ is almost $\mathbf{P}$-internal then by the $\mathbf{P}$-Moishezon-ness of $\operatorname{stp}(a/Ab)$ over A, $\operatorname{stp}(a/Ac)$ is also almost $\mathbf{P}$-internal. So $\operatorname{stp}(ab/Ac)$ is almost $\mathbf{P}$-internal. Since $a' \in \operatorname{acl}(Aba)$, $\operatorname{stp}(a'/Ac)$ is almost $\mathbf{P}$-internal, as desired.

Part (f). If $\operatorname{stp}(b/Ac)$ is almost $\mathbf{P}$-internal then by $\mathbf{P}$-Moishezon-ness both $\operatorname{stp}(a/Ac)$ and $\operatorname{stp}(a'/Ac)$ are almost $\mathbf{P}$-internal. Hence, $\operatorname{stp}(aa'/Ac)$ is almost $\mathbf{P}$-internal, as desired.

Part (g). If $\operatorname{stp}(c/Ad)$ is almost $\mathbf{P}$-internal then by the $\mathbf{P}$-Moishezon-ness of $\operatorname{stp}(b/Ac)$ over A, $\operatorname{stp}(b/Ad)$ is also almost $\mathbf{P}$-internal. But then, by the $\mathbf{P}$-Moishezon-ness of $\operatorname{stp}(a/Ab)$ over A, $\operatorname{stp}(a/Ad)$ is almost $\mathbf{P}$-internal, as desired.

Finally, to prove part (h), suppose $\operatorname{stp}(c/Abd)$ is almost $\mathbf{P}$-internal. Then so is $\operatorname{stp}(bc/Abd)$, and so, by the $\mathbf{P}$-Moishezon-ness of $\operatorname{stp}(a/Abc)$, $\operatorname{stp}(a/Abd)$ is almost $\mathbf{P}$-internal, as desired. □

3. A criterion for being Moishezon to the nonmodular minimal types

In this section we take $\mathbf{P}$ to be the set of all nonmodular minimal types. An essential part of understanding the structure of finite-rank definable sets in a stable theory is the task of determining whether a given type bears any relation to $\mathbf{P}$. For example, is it nonorthogonal to $\mathbf{P}$? If so, is it almost internal to $\mathbf{P}$? In this section we are interested in the question: When is an almost $\mathbf{P}$-internal type actually $\mathbf{P}$-Moishezon? The following algebraicity criterion in complex geometry will serve as a template for us:

Theorem 3.1 (Campana [1, Théorème 2]**).** *Suppose X is a compact complex space of Kähler-type, and $f\colon X \to Y$ is a surjective morphism whose fibres are Moishezon.*[2] *If there exists an analytic subspace $A \subseteq X$ such that $f|_A\colon A \to Y$ is*

[2]Recall that a compact complex space A is *Moishezon* if the morphism $f\colon A \to \{\text{pt.}\}$ is Moishezon. Equivalently, A is bimeromorphic to a projective algebraic variety.

surjective and Moishezon, then f is Moishezon. In particular, if f has a meromorphic section then it is Moishezon.

Actually, we have stated the theorem here in a more general way than it appears in [1] where Campana assumes in addition that Y is Moishezon and then concludes that X is Moishezon. But the proof he gives there goes through word for word in the relative setting. The model-theoretic significance of the Kähler assumption has to do with saturation in an appropriate language, for details about which we refer the reader to [4]. Let us only mention that if X is Kähler then generic type of f is $\mathbb{P}$-internal if and only if the general fibres of f are Moishezon. So the above theorem is about passing from $\mathbb{P}$-internality to Moishezon-ness. To find a model-theoretic counterpart to this theorem (or more accurately, to the "in particular" clause) we begin by approximating the notion of a section as follows.

Definition 3.2. Let us say that a complete stationary type p *admits a diagonal section* if there exists $a \in \operatorname{acl}(\operatorname{dom}(p))$ such that $a \models p \restriction_{\overline{\operatorname{Cb}}(p)}$.

Remark 3.3. Every stationary type has a nonforking extension that admits a diagonal section: if $a \models p(x) \in S(B)$ then the nonforking extension $p|Ba$ admits a diagonal section, namely a itself. This is analogous to the fact that for every morphism $X \to Y$, the diagonal map is a section to $X \times_Y X \to X$.

What one might hope for, at least in the finite U-rank setting, is that if p is $\mathbf{P}$-internal and admits a diagonal section then it is $\mathbf{P}$-Moishezon (over the empty set). We will in fact prove something like this in Theorem 3.7 below, but only under the additional hypothesis of the "canonical base property," which we now discuss in the following digression.

3.1. The canonical base property. Motivated by complex-geometric results due to Campana and Fujiki from the nineteen eighties, Pillay [6] introduced the following condition: T has the *canonical base property* (CBP) if whenever $U(a/A) < \omega$ and $b = \overline{\operatorname{Cb}}(a/Ab)$ then $\operatorname{stp}(b/Aa)$ is almost $\mathbf{P}$-internal.[3] The complex-geometric results alluded to above more or less directly imply that CCM has the CBP. In [7] Pillay and Ziegler show that differentially closed fields have the CBP. Indeed, no stable theories have been shown to fail the CBP. Again motivated by the complex-geometric situation, Pillay and the author introduced an apparent strengthening of this condition in [5], which we called the *uniform canonical base property* (UCBP), and which essentially says that whenever $U(a/A) < \omega$ and $b = \overline{\operatorname{Cb}}(a/Ab)$ then $\operatorname{stp}(b/Aa)$ is $\mathbf{P}$-Moishezon. "Essentially" because we were working with a slightly weaker variant of Moishezon-ness, namely condition $(*)$ of Remark 2.2. In any case, we pointed out that CCM has the UCBP, and we asked whether or not the CBP always implies the UCBP. Recently, Chatzidakis has shown that this is indeed the case. In terms of Moishezon-ness, her theorem can be stated as follows:

Theorem 3.4 (Chatzidakis [2, Theorem 2.9]**).** *Suppose T has the CBP. If $U(a/A) < \omega$ and $b = \overline{\operatorname{Cb}}(a/Ab)$, then* $\operatorname{stp}(b/Aa)$ *is* $\mathbf{P}$*-Moishezon over A.*

One of the ingredients that goes into proving Theorem 3.4 is the following consequence of the CBP, also due to Chatzidakis [2]: if $U(a/A) < \omega$ and $b = \overline{\operatorname{Cb}}(a/Ab)$, then $\operatorname{tp}(b/\operatorname{acl}(Aa) \cap \operatorname{acl}(Ab))$ is almost $\mathbf{P}$-internal. In [5] the author and

[3]Here $\mathbf{P}$ is still the set of all nonmodular minimal types, but in fact, by the results of [2], it changes nothing if we replace $\mathbf{P}$ by the set of *all* minimal types (see Remark 1.1(b) of [5]).

Pillay gave a more geometrically motivated proof of this fact (using the UCBP). Given Theorem 3.4, one is tempted to ask whether or not $\operatorname{tp}(b/\operatorname{acl}(Aa)\cap\operatorname{acl}(Ab))$ is in fact **P**-Moishezon over A. While this is too much to ask, our arguments in [5] can be stretched to give:

Lemma 3.5. *Suppose the CBP holds for T, $U(a/A)<\omega$, $b=\overline{\operatorname{Cb}}(a/Ab)$ and $a=\overline{\operatorname{Cb}}(b/Aa)$. Then $\operatorname{tp}(b/\operatorname{acl}(Aa)\cap\operatorname{acl}(Ab))|Aa$ is **P**-Moishezon over A.*

PROOF. While this is stronger than the statement of Proposition 4.4 in [5] (which only concludes that $\operatorname{tp}(b/\operatorname{acl}(Aa)\cap\operatorname{acl}(Ab))$ is almost **P**-internal), it actually follows from the proof given there. Let d be such that $Ad=\operatorname{acl}(Aa)\cap\operatorname{acl}(Ab)$.

By Theorem 3.4, both $\operatorname{stp}(a/Ab)$ and $\operatorname{stp}(b/Aa)$ are **P**-Moishezon over A. Now, recursively define $a=a_0,a_1,\dots$ and $b=b_0,b_1,\dots$ such that

(i) $a_{i+1}\models\operatorname{stp}(a_i/Ab_i)|Ab_ia$
(ii) $b_{i+1}\models\operatorname{stp}(b_i/Aa_{i+1})|Aa_{i+1}b$
(iii) $\operatorname{stp}(b_i/Aa)$, $\operatorname{stp}(a_i/Aa)$ are **P**-Moishezon over A.

This is done as follows. Suppose we have already defined $a=a_0,a_1,\dots,a_i$ and $b=b_0,b_1,\dots,b_i$. Note that by the induction hypothesis $a_ib_i\models\operatorname{stp}(ab/A)$. Hence $\operatorname{stp}(a_i/Ab_i)$ is **P**-Moishezon over A. Let $a_{i+1}\models\operatorname{stp}(a_i/Ab_i)|Ab_ia$. We then have that $\operatorname{stp}(a_{i+1}/Ab_i)$ and $\operatorname{stp}(b_i/Aa)$ are **P**-Moishezon over A. So, by Proposition 2.4(g), $\operatorname{stp}(a_{i+1}/Aa)$ is **P**-Moishezon over A. Next let $b_{i+1}\models\operatorname{stp}(b_i/Aa_{i+1})|Aa_{i+1}b$. As $\operatorname{stp}(b/Aa)$ is **P**-Moishezon over A, so is $\operatorname{stp}(b_{i+1}/Aa_{i+1})$. We have just shown that $\operatorname{stp}(a_{i+1}/Aa)$ is **P**-Moishezon over A, and so we get, by Proposition 2.4(g) again, that $\operatorname{stp}(b_{i+1}/Aa)$ is **P**-Moishezon over A, as desired.

Having constructed these sequences, note that by (i) and (ii), $\operatorname{acl}(Aa_i)\cap\operatorname{acl}(Ab_i)=\operatorname{acl}(Aa)\cap\operatorname{acl}(Ab)=Ad$ and $a_ib_i\models\operatorname{stp}(ab/Ad)$, for all $i\geq 0$. Moreover, because of finite U-rank, eventually,

$$a\mathop{\smile\!\!\!\!\!\!\mid}\nolimits_{Ad}a_lb_l$$

for sufficiently large $l\geq 0$. Indeed, this is Lemma 2.2 of [5]—see in particular the statement marked by $(*)$ in the proof of that Lemma. So $\operatorname{tp}(b_l/Aa)$ is the nonforking extension of $\operatorname{tp}(b/Ad)$, and it is **P**-Moishezon over A by (iii). □

We are interested in the following consequence for minimal canonical types.

Corollary 3.6. *Suppose the CBP holds for T and $U(a/A)<\omega$. If $U(a/Ab)=1$, $b=\overline{\operatorname{Cb}}(a/Ab)$, and $b\notin\operatorname{acl}(Aa)$, then $\operatorname{tp}(a/\operatorname{acl}(Aa)\cap\operatorname{acl}(Ab))|Aa$ is **P**-Moishezon over A.*

PROOF. Again, let d be such that $Ad=\operatorname{acl}(Aa)\cap\operatorname{acl}(Ab)$.

We first show that $\operatorname{tp}(b/Ad)|Aa$ is **P**-Moishezon over A. Let $a'=\overline{\operatorname{Cb}}(b/Aa)$. So $a'\in\operatorname{acl}(Aa)$. If $a'\in\operatorname{acl}(Ab)$ then $\overline{\operatorname{Cb}}(b/Aa)\subseteq Ad$ and hence $b\mathop{\smile\!\!\!\!\!\!\mid}\nolimits_{Ad}a$. It follows that $\operatorname{tp}(b/Ad)|Aa=\operatorname{tp}(b/Aa)$. But by Theorem 3.4, $\operatorname{tp}(b/Aa)$ is **P**-Moishezon over A and we are done. Hence we may assume that $a'\notin\operatorname{acl}(Ab)$. But then, as $U(a/Ab)=1$, $a\in\operatorname{acl}(Aba')$. Since $b\mathop{\smile\!\!\!\!\!\!\mid}\nolimits_{Aa'}a$, this implies that $a\in\operatorname{acl}(Aa')$. Hence a and a' are interalgebraic over A. So $a=\overline{\operatorname{Cb}}(b/Aa)$ also. Now Lemma 3.5 applies and we have that $\operatorname{tp}(b/Ad)|Aa$ is **P**-Moishezon over A.

To prove the corollary, we let $a_1b_1\models\operatorname{tp}(ab/Ad)|Aa$ and we show that $\operatorname{tp}(a_1/Aa)$ is **P**-Moishezon over A. Note that $b\notin\operatorname{acl}(Aa)\implies b_1\notin\operatorname{acl}(Aa_1)\implies b_1\notin\operatorname{acl}(Aa_1a)$, where the last implication is because $b_1\mathop{\smile\!\!\!\!\!\!\mid}\nolimits_{Aa_1}a$. Hence there exists $b_2\models\operatorname{tp}(b_1/Aa_1a)$ with $b_2\notin\operatorname{acl}(Ab_1)$. Let p_{b_1} and p_{b_2} be the conjugates of $\operatorname{tp}(a/Ab)$

obtained by replacing b by b_1 and b_2 respectively. As $b_1 = \overline{\mathrm{Cb}}(p_{b_1})$, the fact that $b_2 \notin \mathrm{acl}(Ab_1)$ means that $p_{b_1} \cup p_{b_2}$ must be a forking extension of p_{b_1}. The latter being of rank 1 implies that $p_{b_1} \cup p_{b_2}$ is algebraic. Since b_1 and b_2 have the same type over Aa_1 and $a_1 \models p_{b_1}$, we get that $a_1 \models p_{b_2}$ also and so $a_1 \in \mathrm{acl}(Ab_1b_2)$. But $\mathrm{tp}(b_1/Aa) = \mathrm{tp}(b_2/Aa)$ is the nonforking extension of $\mathrm{tp}(b/Ad)$ to Aa, and is thus **P**-Moishezon over A by the preceeding paragraph. Hence, by Proposition 2.4 parts (e) and (f), $\mathrm{tp}(a_1/Aa)$ is **P**-Moishezon over A. □

We now return to the promised criterion for when an almost internal type is in fact Moishezon. For the sake of convenience we restrict ourselves to finite rank theories — though everything can be formulated so as to apply to finite rank types in arbitrary stable theories (with the CBP).

Theorem 3.7. *Suppose T is a finite U-rank theory with the CBP, and* **P** *is the set of all nonmodular minimal types. Suppose p is a stationary type of U-rank at least two that is almost internal to a nonmodular minimal type. If p admits a diagonal section, then p is* **P***-Moishezon over $\varnothing$.*

In particular, every stationary type of rank at least two that is almost internal to a nonmodular minimal type becomes **P***-Moishezon after taking a nonforking extension to a realization.*

PROOF. We begin by proving the "in particular" clause as a separate lemma.

Lemma 3.8. *Suppose T is a finite U-rank theory with the CBP. Suppose $c = \overline{\mathrm{Cb}}(a/Ac)$ and $\mathrm{stp}(a/Ac)$ is of U-rank at least two and almost internal to a nonmodular minimal type. Then $\mathrm{stp}(a/Ac)|Aca$ is* **P***-Moishezon over A.*

PROOF OF LEMMA 3.8. Replacing a by (a, c) we may also assume that $a \in \mathrm{acl}(Ac)$. Since $\mathrm{stp}(a/Ac)$ is almost internal to a nonmodular minimal type, there exists $c' \supseteq c$ such that $\mathrm{stp}(a/Ac')$ is a stationary canonical type of U-rank 1 and $\mathrm{acl}(Aca) \cap \mathrm{acl}(Ac') = \mathrm{acl}(Ac)$. This is Lemma 5.4 of [5] and expresses the existence of a "rich family of curves" on any U-rank at least two type that is almost internal to a nonmodular minimal type. Note that $c' \notin \mathrm{acl}(Aa)$: indeed, if it were, then we would have $c' \in \mathrm{acl}(Ac)$, contradicting the fact that $U(a/Ac) > 1$ while $U(a/Ac') = 1$. Now, by Corollary 3.6 applied to $\mathrm{stp}(a/Ac')$, we have that $\mathrm{stp}\big(a/\,\mathrm{acl}(Aa) \cap \mathrm{acl}(Ac')\big)|Aa$ is **P**-Moishezon over A. But $\mathrm{acl}(Aa) \cap \mathrm{acl}(Ac') = \mathrm{acl}(Aca) \cap \mathrm{acl}(Ac') = \mathrm{acl}(Ac)$. Hence $\mathrm{stp}(a/Ac)|Aa$ is **P**-Moishezon over A, as desired. □

Now suppose p is a stationary type almost internal to a nonmodular minimal type and admitting a diagonal section. Let $c = \overline{\mathrm{Cb}}(p)$. Recall that p admitting a diagonal section means that there exists $a \in \mathrm{acl}\big(\mathrm{dom}(p)\big)$ such that $\mathrm{tp}(a/c) = p \restriction_c$. Note that p, or rather its unique extension to $\mathrm{acl}\big(\mathrm{dom}(p)\big)$, is an extension of $\mathrm{tp}(a/c)|ca$. Hence it suffices to show that $\mathrm{tp}(a/c)|ca$ is **P**-Moishezon over $\varnothing$. But since p is a nonforking extension of $\mathrm{tp}(a/c)$, the latter is also almost internal to a nonmodular minimal type (the same one). So we can apply Lemma 3.8 with $A = \varnothing$ to conclude that $\mathrm{tp}(a/c)|ca$ is **P**-Moishezon over $\varnothing$, as desired. □

Note that Theorem 3.7 only abstracts a special case of the "in particular" clause of Campana's algebraicity criterion (Theorem 3.1), and that it is not even clear how to formulate a more complete model-theortic counterpart of that result. It may be that the Zariski-type structures of Hrushovski-Zilber, where it is possible to talk about *specialisations*, offer the right context in which to work this out.

4. Coreductions

As was mentioned in the introduction, the original intention for developing a notion of Moishezon-ness in the abstract stable setting was to prove that under certain assumptions on the theory (namely UCBP), for all a and A there exists a minimum algebraically closed set $A \subseteq B \subseteq \operatorname{acl}(Aa)$ such that $\operatorname{stp}(a/B)$ is almost internal to the set of nomodular minimal types. As it turned out, Chatzidakis [2], employing even weaker assumptions (simply CBP), found a much more direct proof of this intended application.[4] In this final section we discuss Chatzidakis' result from the point of view of known results in complex geometry. Even though it was not in the end necessary, we also explain how Moishezon-ness is related to these issues.

Given an irreducible compact complex space X, one of the first constructions in bimeromorphic geometry is to consider its *algebraic reduction*, a holomorphic surjection $f\colon X \to V$ where V is a projective algebraic variety and such that f induces an isomorphism between the rational function field of V and the meromorphic function field of X. The algebraic reduction map satisfies a certain universal property expressing that V is the maximum algebraic image of X. The model-theoretic content of this is expressed in the following easy fact about finite rank types in arbitrary stable theories: Suppose $\mathbf{P}$ is any $\operatorname{Aut}_A(\overline{M})$-invariant set of partial types and $U(a/A) < \omega$. Then there exists $b \in \operatorname{acl}(Aa)$ such that $\operatorname{stp}(b/A)$ is almost $\mathbf{P}$-internal and whenever $c \in \operatorname{acl}(Aa)$ with $\operatorname{stp}(c/A)$ almost $\mathbf{P}$-internal, then $c \in \operatorname{acl}(Ab)$. Clearly this b is uniquely determined upto interalgebraicity over A, and we could call it the $\mathbf{P}$-*reduction of a over A*. Note that finding the $\mathbf{P}$-reduction is the first step in a $\mathbf{P}$-analysis of $\operatorname{stp}(a/A)$. Up to bimeromorphic equivalence, algebraic reductions in complex geometry are just $\mathbb{P}$-reductions in the theory CCM.

Now, in [1], Campana considered a dual notion to algebraic reductions: maximum algebraic fibrations (or quotients). Somewhat loosely speaking, a meromorphic surjection $f : X \to Y$ is an *algebraic coreduction of X* if the general fibres of f are Moishezon and if every compact complex-analytic family of Moishezon subspaces of X is finer than the fibration induced by f. Campana proved that algebraic coreductions of Kähler-type spaces exist. Campana's proof used (and it seems inspired) his "first algebraicity criterion." This is what lead us to study Moishezon types and to prove Theorem 3.7 above. In any case, the model-theoretic content of algebraic coreductions is captured by the following definition:

Definition 4.1. Suppose $\mathbf{P}$ is a set of $\operatorname{Aut}_A(\overline{M})$-invariant partial types. A $\mathbf{P}$-*coreduction of a over A* is a tuple $b \in \operatorname{acl}(Aa)$ such that $\operatorname{stp}(a/Ab)$ is almost $\mathbf{P}$-internal and whenever $c \in \operatorname{acl}(Aa)$ with $\operatorname{stp}(a/Ac)$ almost $\mathbf{P}$-internal, then $b \in \operatorname{acl}(Ac)$.

But do coreductions exist? Unlike reductions, they are not an immediate consequence of finite rank. Nevertheless, Chatzidakis proves — this is Theorem 2.8 of [2] — that *if T has the CBP, $\mathbf{P}$ is an $\operatorname{Aut}_A(\overline{M})$-invariant set of minimal types, and $U(a/A) < \omega$, then a $\mathbf{P}$-coreduction of a over A exists.* In particular this leads to an attractive transfer of results from complex geometry to differential-algebraic geometry: finite rank differential-algebraic varieties admit maximum fibrations by subvarieties that are almost internal to the field of constants. This is a new and potentially useful tool in the study of finite-rank differential algebraic varieties.

---

[4]She then used this result to prove Theorem 3.4 above; that the CBP implies the UCBP.

In order to prove the existence of coreductions for finite rank types it suffices (and is necessary) to show that if $b_1, b_2 \in \operatorname{acl}(Aa)$ are such that $\operatorname{stp}(a/Ab_1)$ and $\operatorname{stp}(a/Ab_2)$ are almost $\mathbf{P}$-internal, then $\operatorname{stp}\big(a/\operatorname{acl}(Ab_1)\cap\operatorname{acl}(Ab_2)\big)$ is almost $\mathbf{P}$-internal. In the following proposition we prove a weaker version of this fact, but unconditionally (without the CBP). It serves as a good illustration of the power of Moishezon-ness over internality.

Proposition 4.2. *We do not assume the CBP. Suppose* $\mathbf{P}$ *is any* $\operatorname{Aut}_A(\overline{M})$*-invariant set of types and* $U(a/A) < \omega$*. If* $b_1, b_2 \in \operatorname{acl}(Aa)$ *with* $\operatorname{stp}(a/Ab_i)|Aa$ $\mathbf{P}$*-Moishezon over* $\operatorname{acl}(Ab_1)\cap\operatorname{acl}(Ab_2)$ *for* $i = 1, 2$*, then* $\operatorname{tp}\big(a/\operatorname{acl}(Ab_1)\cap\operatorname{acl}(Ab_2)\big)$ *is almost* $\mathbf{P}$*-internal.*

Remark 4.3. Note that in the case when the CBP holds and $\mathbf{P}$ is the set of all nonmodular minimals, the hypothesis that $\operatorname{tp}(a/Ab_i)|Aa$ is $\mathbf{P}$-Moishezon over $\operatorname{acl}(Ab_1)\cap\operatorname{acl}(Ab_2)$ would follow by Theorem 3.7 from $\operatorname{tp}(a/Ab_i)$ being simply almost internal to some nonmodular minimal type. But this does not give a new proof of the existence of coreductions under CBP because Theorem 3.7 itself relied on the existence of coreductions (to get UCBP from CBP).

Proof of Proposition 4.2. For the sake of notational convenience, let us rename $b := b_1$ and $c := b_2$. Passing to algebraic closures we assume that $\operatorname{tp}(a/Ab)$ and $\operatorname{tp}(a/Ac)$ are stationary. Replacing A by $\operatorname{acl}(Ab)\cap\operatorname{acl}(Ac)$, we also assume that $A = \operatorname{acl}(Ab)\cap\operatorname{acl}(Ac)$. We wish to show that $\operatorname{tp}(a/A)$ is almost $\mathbf{P}$-internal.

Recursively define $a = a_0, a_1, \dots$, $b = b_0, b_1, \dots$, and $c = c_0, c_1, \dots$ so that

(i) $a_ib_ic_i \models \operatorname{tp}(abc/A)$ for all i,

(ii) for i even, $a_{i+1} \models \operatorname{tp}(a_i/Ac_i)|Ac_ib$ and $c_{i+1} = c_i$,

(iii) for i odd, $a_{i+1} \models \operatorname{tp}(a_i/Ab_i)|Ab_ib$ and $b_{i+1} = b_i$, and

(iv) $\operatorname{stp}(a_i/Ab)$ is almost $\mathbf{P}$-internal for all i.

This is done as follows. Suppose that for even i we have defined $a = a_0, a_1, \dots, a_i$, $b = b_0, b_1, \dots, b_i$, and $c = c_0, c_1, \dots, c_i$ as desired. We first show how to define $a_{i+1}, b_{i+1}, c_{i+1}$. Let $a_{i+1} \models \operatorname{tp}(a_i/Ac_i)|Aa_ib$. (Note that $c_i \in \operatorname{acl}(Aa_i)$.) Since $\operatorname{tp}(a/Ac)|Aa$ is $\mathbf{P}$-Moishezon over A by assumption, we get that $\operatorname{tp}(a_{i+1}/Aa_i) = \operatorname{tp}(a_i/Ac_i)|Aa_i$ is $\mathbf{P}$-Moishezon over A. By (iv) then, $\operatorname{stp}(a_{i+1}/Ab)$ is almost $\mathbf{P}$-internal. Set $c_{i+1} := c_i$. Since $a_{i+1}c_{i+1} \models \operatorname{tp}(a_ic_i/A) = \operatorname{tp}(ac/A)$, there exists b_{i+1} such that $a_{i+1}b_{i+1}c_{i+1} \models \operatorname{tp}(abc/A)$.

We now construct $a_{i+2}b_{i+2}c_{i+2}$ analogously. Let $a_{i+2} \models \operatorname{tp}(a_{i+1}/Ab_{i+1})Aa_{i+1}b$. Since $\operatorname{tp}(a/Ab)|Aa$ is $\mathbf{P}$-Moishezon over A, $\operatorname{tp}(a_{i+2}/Aa_{i+1}) = \operatorname{tp}(a_{i+1}/Ab_{i+1})|Aa_{i+1}$ is $\mathbf{P}$-Moishezon over A. But we have just shown that $\operatorname{stp}(a_{i+1}/Ab)$ is almost $\mathbf{P}$-internal. So $\operatorname{stp}(a_{i+2}/Ab)$ is almost $\mathbf{P}$-internal. Setting $b_{i+2} := b_{i+1}$ we can find c_{i+2} such that $a_{i+2}b_{i+2}c_{i+2} \models \operatorname{tp}(abc/A)$. This completes the recursive construction.

We claim that for all i, $U(a_{i+1}/Ab) \geq U(a_i/Ab)$. Indeed, for i even,

$$\begin{aligned}
U(a_{i+1}/Ab) &= U(a_{i+1}c_{i+1}/Ab) && \text{as } c_{i+1} \in \operatorname{acl}(Aa_{i+1})\\
&= U(a_{i+1}c_i/Ab) && \text{as } c_{i+1} = c_i\\
&= U(a_{i+1}/Abc_i) + U(c_i/Ab)\\
&= U(a_{i+1}/Ac_i) + U(c_i/Ab) && \text{as } a_{i+1} \mathop{\smile\!\!\!\!\!\!|}\nolimits_{Ac_i} b\\
&= U(a_i/Ac_i) + U(c_i/Ab) && \text{as } a_{i+1} \models \operatorname{tp}(a_i/Ac_i)\\
&\geq U(a_i/Abc_i) + U(c_i/Ab)\\
&= U(a_ic_i/Ab)
\end{aligned}$$

$= U(a_i/Ab)$

The argument for i odd is analogous.

Hence, eventually for some $l > 0$, $U(a_j/Ab) = U(a_l/Ab)$ for all $j \geq l$. Let us take l to be even. Now,

$$\begin{aligned} U(a_{l+1}b_{l+1}/Ab) &= U(a_{l+1}/Ab) && \text{as } b_{l+1} \in \operatorname{acl}(Aa_{l+1}) \\ &= U(a_{l+2}/Ab) && \text{by choice of } l \\ &= U(a_{l+2}b_{l+1}/Ab) && \text{as } b_{l+1} = b_{l+2} \in \operatorname{acl}(Aa_{l+2}) \\ &= U(a_{l+2}/Abb_{l+1}) + U(b_{l+1}/Ab) \\ &= U(a_{l+2}/Ab_{l+1}) + U(b_{l+1}/Ab) && \text{as } a_{l+2} \perp_{Ab_{l+1}} b \\ &= U(a_{l+1}/Ab_{l+1}) + U(b_{l+1}/Ab) && \text{as } a_{l+2} \models \operatorname{tp}(a_{l+1}/Ab_{l+1}). \end{aligned}$$

On the other hand, $U(a_{l+1}b_{l+1}/Ab) = U(a_{l+1}/Abb_{l+1}) + U(b_{l+1}/Ab)$. It follows that $U(a_{l+1}/Abb_{l+1}) = U(a_{l+1}/Ab_{l+1})$, which means that $b \perp_{Ab_{l+1}} a_{l+1}$. But $b \perp_{Ac_{l+1}} a_{l+1}$ by (ii). Hence, $\overline{\operatorname{Cb}}(b/Aa_{l+1}) \subseteq \operatorname{acl}(Ab_{l+1}) \cap \operatorname{acl}(Ac_{l+1}) = A$ and we have that $b \perp_A a_{l+1}$. So $\operatorname{tp}(a/A) = \operatorname{tp}(a_{l+1}/A)$ has a nonforking extension—namely $\operatorname{stp}(a_{l+1}/Ab)$—that is almost $\mathbf{P}$-internal. Hence $\operatorname{tp}(a/A)$ is itself almost $\mathbf{P}$-internal. □

4.1. Generating fibrations. In conclusion, let us relate some of these ideas to the notion of "generating families" from [5]. By a *fibration* of a stationary type $p(x) = \operatorname{tp}(a/A)$ we mean simply a stationary type $q(x, y) = \operatorname{tp}(a, b/A)$ such that $s(y) = \operatorname{tp}(b/A)$ and $q_b := \operatorname{tp}(a/Ab)$ are also stationary. The *fibres* of this fibration are the types $q_{b'} := q(x, b')$ where $b' \models s(y)$. Given another realization $a' \models p(x)$, we say that a and a' are *q-connected* if there exists a finite sequence $a = a_0, a_1, \ldots, a_l = a'$ of realizations of $p(x)$ and a sequence $b = b_0, b_1, \ldots, b_{l-1}$ of realizations of $s(y)$ such that a_i and a_{i+1} both realize $q_{b_i}(x)$ for all $i = 0, \ldots, l-1$. We say that $q(x, y)$ *generates* $p(x)$ if every (equivalently some) A-independent pair of realizations of $p(x)$ are q-connected. It turns out that in the finite U-rank context q generates p if and only if $\operatorname{acl}(Aa) \cap \operatorname{acl}(Ab) = \operatorname{acl}(A)$ (see Lemma 2.2 of [5]). The author and Pillay obtained the following criterion for internality to a nomodular minimal type (Theorem 1.3(b) of [5]): If T has the CBP and a finite rank type $p(x)$ is generated by a fibration whose fibres are almost internal to a nonmodular minimal type, then $p(x)$ is itself almost internal to that type.

We can extend these notions to pairs (or indeed finite collections) of fibrations. Let us say that *$p(x)$ is generated by the pair of fibrations $q_1(x, y) = (a, b/A)$ and $q_2(x, z) = \operatorname{tp}(a, c/A)$* if for some (equivalently any) pair of A-independent realizations of $p(x)$, a and a', there exists a finite sequence of realizations of $p(x)$, $a = a_0, a_1, \ldots, a_l = a'$ such that each pair a_i, a_{i+1} is either q_1-connected or q_2-connected. Chatzidakis' proof of the existence of coreductions more or less gives the following internality criterion:

Theorem 4.4. *Suppose T has the CBP, $p(x) \in S(A)$ is a stationary finite rank type, and $\mathbf{P}$ is any set of $\operatorname{Aut}_A(\overline{M})$-invariant minimal types. If p is generated by a pair of fibration whose fibres are almost $\mathbf{P}$-internal, then p is itself almost $\mathbf{P}$-internal.*

Sketch of proof. Suppose $p(x) = \operatorname{tp}(a/A)$ is generated by the fibrations $q_1(x, y) = \operatorname{tp}(a, b/A)$ and $q_2(x, z) = \operatorname{tp}(a, c/A)$. Since $\operatorname{tp}(a/Aa), \operatorname{tp}(a/Ab)$, and

$\text{tp}(a/Ac)$ are all almost **P**-internal, the existence of **P**-coreductions ([2, Theorem 2.8]) implies that $\text{tp}\big(a/\operatorname{acl}(Aa) \cap \operatorname{acl}(Ab) \cap \operatorname{acl}(Ac)\big)$ is almost **P**-internal. But the fact that q_1 and q_2 generate $p(x)$ implies that

$$\operatorname{acl}(Aa) \cap \operatorname{acl}(Ab) \cap \operatorname{acl}(Ac) = A.$$

Indeed, arguing as in Lemma 2.2 of [5], it is not hard to show that there exists a finite sequence $abc = a_0b_0c_0, a_1b_1c_1, \ldots, a_nb_nc_n$ such that $\text{stp}(a_ib_ic_i/A) = \text{stp}(a_{i+1}b_{i+1}c_{i+1}/A)$ and either $b_i = b_{i+1}$ or $c_i = c_{i+1}$, for all $i = 0, \ldots, n-1$, and a_n is independent of a over A. Hence $\operatorname{acl}(Aa) \cap \operatorname{acl}(Ab) \cap \operatorname{acl}(Ac) = \operatorname{acl}(Aa_n) \cap \operatorname{acl}(Ab_n) \cap \operatorname{acl}(Ac_n)$, and so $\operatorname{acl}(Aa) \cap \operatorname{acl}(Ab) \cap \operatorname{acl}(Ac) \subseteq \operatorname{acl}(Aa) \cap \operatorname{acl}(Aa_n) = A$. Thus, $p(x) = \text{tp}(a/A)$ is almost **P**-internal. □

Proposition 4.2 can also be understood in these terms. Loosely speaking, it says that if a finite rank stationary type over A is generated by a pair of fibrations whose fibres satisfy a certain strengthening of almost **P**-internality, where **P** here is any $\text{Aut}_A(\overline{M})$-invariant set of partial types, then the type itself is almost **P**-internal. The precise strengthening of internality that we require of the fibres is that a nonforking extension to a realization is **P**-Moishezon over A.

References

1. F. Campana, *Coréduction algébrique d'un espace analytique faiblement kählérien compact*, Invent. Math. **63** (1981), no. 2, 187–223.
2. Z. Chatzidakis, *A note on canonical bases and one-based types in supersimple theories* (2010), preprint.
3. R. Moosa, *A nonstandard Riemann existence theorem*, Trans. Amer. Math. Soc. **356** (2004), no. 5, 1781–1797.
4. ———, *On saturation and the model theory of compact Kähler manifolds*, J. Reine Angew. Math. **586** (2005), 1–20.
5. R. Moosa and A. Pillay, *On canonical bases and internality criteria*, Illinois J. Math. **52** (2008), no. 3, 901–917.
6. A. Pillay, *Model-theoretic consequences of a theorem of Campana and Fujiki*, Fund. Math. **174** (2002), no. 2, 187–192.
7. A. Pillay and M. Ziegler, *Jet spaces of varieties over differential and difference fields*, Selecta Math. (N.S.) **9** (2003), no. 4, 579–599.

Department of Pure Mathematics, University of Waterloo, 200 University Avenue West, Waterloo, ON N2L 3G1, Canada

E-mail address: rmoosa@math.uwaterloo.ca

Centre de Recherches Mathématiques
CRM Proceedings and Lecture Notes
Volume **53**, 2011

Generic Stability, Regularity, and Quasiminimality

Anand Pillay and Predrag Tanović

Dedicated to Professor Michael Makkai on the occasion of his seventieth birthday.

ABSTRACT. We study the notions generic stability, regularity, homogeneous pregeometries, quasiminimality, and their mutual relations, in arbitrary first order theories. We prove that "infinite-dimensional homogeneous pregeometries" coincide with generically stable strongly regular types $(p(x), x = x)$. We prove that quasiminimal structures of cardinality at least $\aleph_2$ are "homogeneous pregeometries." We prove that the "generic type" of an arbitrary quasiminimal structure is "locally strongly regular." Some of the results depend on a general dichotomy for "regular-like" types: generic stability, or the existence of a suitable definable partial ordering.

1. Introduction

The first author was motivated partly by hearing Wilkie's talks on his program for proving Zilber's conjecture that the complex exponential field is quasiminimal (definable subsets are countable or co-countable), and wondering about the first order (rather than infinitary) consequences of the approach. The second author was partly motivated by his interest in adapting his study of minimal structures (definable subsets are finite or cofinite) and his dichotomy theorems ([6]), to the quasiminimal context.

Zilber's conjecture (and the approach to it outlined by Wilkie) is closely related to the existence and properties of a canonical pregeometry on the complex exponential field. See the end of Section 4 for a more detailed discussion. Also in Section 5 of this paper we discuss to what extent a pregeometry can be recovered just from quasiminimality, sometimes assuming the presence of a definable group structure. This continues in a sense an earlier study of the general model theory of quasiminimality by Itai, Tsuboi, and Wakai [2].

A pregeometry is a closure relation on subsets of a not necessarily saturated structure M satisfying usual properties (including exchange). We will also assume "homogeneity" ($\mathrm{tp}(b/A)$ is unique for $b \notin \mathrm{cl}(A)$) and "infinite-dimensionality"

2010 *Mathematics Subject Classification.* 03C45.
The first author was supported by EPSRC grant EP/F009712/1.
The second author was supported by the Ministry of Science and Technology of Serbia.
This is the final form of the paper.

(dim(M) is infinite). One of the points of this paper then is that the canonical "generic type" p of the pregeometry is "generically stable" and regular. This includes the statement that on realizations of p, the closure operation is precisely forking in the sense of Shelah (see Theorem 4.1). Another main point of the paper is that a quasiminimal structure of cardinality at least $\aleph_2$ carries, in a canonical fashion, a homogeneous (in the above sense), infinite-dimensional pregeometry, improving on results in [2].

Generic stability, the stable-like behavior of a given complete type vis-a-vis forking, was studied in several papers including [5] and [1], but mainly in the context of theories with NIP (i.e., without the independence property). Here we take the opportunity, in Section 2, to give appropriate definitions in an arbitrary ambient theory T, as well as discussing generically stable (strongly) regular types.

The notion of a regular type is central in stability theory and classification theory, where the counting of models of superstable theories is related to dimensions of regular types. Here (Section 3) we give appropriate generalizations of (strong) regularity for an arbitrary theory T (although it does not agree with the established definitions for simple theories). In Section 3 a basic dichotomy theorem (Theorem 3.1) is proved for global regular types p; roughly speaking, either p is generically stable, or there is certain definable partial ordering on the set of realizations of p. A generalization of this theorem is given in Section 6 (Theorem 5.3), which lies behind our result on quasiminimal structures of cardinality $\geq \aleph_2$.

In Section 7 a local version of regularity is given and applied to the analysis of quasiminimal structures (see Corollaries 7.1 and 7.2).

The current paper is a revised and expanded version of a preprint "Remarks on generic stability, pregeometries, and quasiminimality" by the first author, which was written and circulated in June 2009.

The first author would like to thank Clifton Ealy, Krzysztof Krupinski, and Alex Usvyatsov for various helpful conversations and comments. After seeing the first author's preliminary results (and talk at a meeting in Lyon in July 2009), Ealy pointed out that the commutativity of regular groups is problematic (see our Question at the end of Section 3) and suggested possible directions towards a counterexample. Krupinski independently came up with examples such as Example 5.1 of the current paper. And after a talk by the first author on the same subject in Bedlewo in August 2009, Usvyatsov pointed out examples such as Example 3.1 and 3.2 of the current paper. Both authors would like to thank Jonathan Kirby for helpful comments and questions, and for clarifying the connection to exponential algebraicity (see the end of Section 4).

We now give our conventions and give a few basic definitions relevant to the paper. As the referee kindly mentioned this paper maybe of interest to readers who are not so familiar with stability-style model theory, and so following his/her suggestion, we give more details than we usually would concerning some standard constructions.

T denotes an arbitrary complete 1-sorted theory in a language L and $\overline{M}$ denotes a saturated (monster) model of T. As a rule $a, b, c, \ldots$, denote elements of $\overline{M}$, and $\bar{a}, \bar{b}, \bar{c}$ denote finite tuples of elements. (But in some situations $a, b, \ldots$ may denote elements of $\overline{M}^{\mathrm{eq}}$.) A, B, C denote small subsets, and $M, M_0, \ldots$ denote small elementary submodels of $\overline{M}$. By a "global type" we mean morally a complete type over a sufficiently saturated model. In practise we will mean a complete type

$p(\bar{x}) \in S(\overline{M})$ over the "monster model." Such a type p is said to be A-invariant if p is $\mathrm{Aut}(\overline{M}/A)$-invariant; and p is said to be invariant if it is A-invariant for some small A. Notice that by saturation/homogeneity of $\overline{M}$, the A-invariance of p is equivalent to p *not splitting* over A, in the sense that for any L-formula $\phi(\bar{x}, \bar{y})$ and $\bar{b}$ from $\overline{M}$, whether or not $\phi(\bar{x}, \bar{b}) \in p$ depends on $\mathrm{tp}(\bar{b}/A)$. So an A-invariant type $p(\bar{x})$ comes with a kind of "infinitary" defining schema d_p over A. Namely for a given L-formula $\phi(\bar{x}, \bar{y})$, $d_p(\phi(\bar{x}, \bar{y}))$ is the set of $q(\bar{y}) \in S(A)$ such that for some (any) $\bar{b}$ realizing q, $\phi(\bar{x}, \bar{b}) \in p$.

We now explain how to build the "nonforking iterates" $p^{(n)}(\bar{x}_1, \ldots, \bar{x}_n) \in S(\overline{M})$ of p. Let $\phi(\bar{x}_1, \bar{x}_2, \bar{c})$ be a formula over $\overline{M}$ (with witnessed parameters $\bar{c}$). We will put such a formula in $p^{(2)}(\bar{x}_1, \bar{x}_2)$ if for some (any) $\bar{a}_1$ realizing $p|(A, \bar{c})$ (the restriction of p to $A, \bar{c}$), the formula $\phi(\bar{a}_1, \bar{x}_2, \bar{c})$ is in $p(\bar{x}_2)$. Having defined $p^{(n)}$, we put a formula $\phi(\bar{x}_1, \ldots, \bar{x}_n, \bar{x}_{n+1}, \bar{c})$ in $p^{(n+1}$ if for for some (any) $(\bar{a}_1, \ldots, \bar{a}_n)$ realizing $p^{(n)}|(A, \bar{c})$, the formula $\phi(\bar{a}_1, \ldots, \bar{a}_n, \bar{x}_{n+1}, \bar{c})$ is in $p(\bar{x}_{n+1})$. This construction depends only on the defining schema d of p. It is clear that $p^{(n)}(\bar{x}_1, \ldots, \bar{x}_n) \subseteq p^{(n+1)}(\bar{x}_1, \ldots, \bar{x}_{n+1})$, and we let $p^{(\omega)}$ be the increasing union of the $p^{(n)}$'s. It is not hard to see that any realization $(\bar{a}_i : i = 1, 2, \ldots.)$ of $p^{(\omega)}$ (in an elementary extension of the monster model!) is an indiscernible sequence over $\overline{M}$.

Assuming the global type p to be A-invariant, by a *Morley sequence in p over A* we mean a realization, in $\overline{M}$, $(\bar{a}_i : i = 1, 2, \ldots.)$ of $p^{(\omega)}|A$. Clearly this can also be obtained by choosing $\bar{a}_1$ to realize $p|A$, choosing $\bar{a}_2$ to realize $p|(A, \bar{a}_1)$ etc. So a Morley sequence in p over A is among other things an A-indiscernible sequence, and can be stretched to an A-indiscernible sequence of any ordinal length α (a Morley sequence in p over A of length α). In any case clearly the type over A of any Morley sequence of length α in p over A depends only on p and A.

An invariant global type $p(\bar{x})$ is said to be *symmetric* if for any n, formula $\phi(\bar{x}_1, \ldots, \bar{x}_n)$ over $\overline{M}$ and permutation π of $\{1, \ldots, n\}$,

$$\phi(\bar{x}_1, \ldots, \bar{x}_n) \in p^{(n)}(\bar{x}_1, \ldots, \bar{x}_n) \iff \phi(\bar{x}_{\pi(1)}, \ldots, \bar{x}_{\pi(n)}) \in p^{(n)}(\bar{x}_1, \ldots, \bar{x}_n).$$

It is not hard to see that the condition for $n = 2$ implies it for all n. Also note that p is symmetric if and only if for any small A such that p is A-invariant, any Morley sequence in p over A is *totally indiscernible.*

In some parts of this paper we will be considering analogous notions for complete types over "large" but not necessarily saturated or homogeneous (in the sense of model theory) structures M, where we are no longer able to use the expression "invariant" type.

Some other basic notions used in this paper are *definable type*, *heir*, *coheir*, *almost finitely satisfiability.* A *definable type* (over A) refers usually to a complete type $p(x)$ say over a model M such that for any $\phi(x, \bar{y}) \in L$, $\{\bar{b} \in M : \phi(x, \bar{b}) \in p\}$ is a definable set (over A) in M. If $p(x)$ is a complete type over a model M and N is a larger model (elementary extension) then an *heir* of p over N is an extension $q(x) \in S(N)$ of p such that for any formula $\phi(x, \bar{y})$ with parameters from M, if $\phi(x, \bar{c}) \in q(x)$ for some $\bar{c}$ from N, then $\phi(x, \bar{b}) \in p$ for some $\bar{b} \in M$. A basic and easy fact is that if $p(x) \in S(M)$ is a definable type, then it has a unique heir over any larger model (which is precisely given by applying the defining schema for p to the larger model).

In the same context, with $p(x) \in S(M)$, and $q(x) \in S(N)$ an extension of p, q is said to be a *coheir* of p if q is finitely satisfiable in M (any formula $\phi(x)$ in $q(x)$

is realized by some element of M, even though the formula may have parameters outside M). Finally if $q(x)$ is a global complete type and A a small set of parameters, q is said to be *almost finitely satisfiable* in A if q is finitely satisfiable in any model (elementary substructure) of $\overline{M}$ which contains A.

Of course pervasive notions are *dividing* and *forking* in the sense of Shelah. A formula $\phi(x, b)$ (with witnessed parameters b) is said to *divide over A* if for some A-indiscernible sequence $(b_i : i < \omega)$ of realizations of $\mathrm{tp}(b/A)$, $\{\phi(x, b_i : i < \omega\}$ is inconsistent. The formula is said to *fork over A* if it implies a finite disjunction of formulas $\psi_j(x, c_j)$ each of which divides over A. The notions extend naturally to (partial, or complete) types in place of formulas.

In stable theories, the above notions cohere in the following senses:

(1) Given a type $p(x) \in S(M)$ over a model and $N \supseteq M$, p has a unique nonforking extension over N which coincides with its unique heir over N as well as with its unique coheir over N.
(2) Any complete type over a model is definable (in particular any global type is invariant).
(3) A global type $p(x)$ does not fork over A iff it is definable over $\mathrm{acl}^{\mathrm{eq}}(A)$ iff it is almost finitely satisfiable in A.
(4) Moreover any global type is definable in a specific way over some (any) Morley sequence. (See Proposition 2.1(i) below.)

Although stable theories were originally the main environment for studying forking, we saw subsequently the fundamental role of forking in simple theories, and more recently its importance for NIP theories, including o-minimal theories and valued fields. In the current paper we are concerned to a large extent with the role of forking in "pregeometries" in general first order theories.

Recall that (P, cl), where cl is an operation on subsets of P, is a *pregeometry* (or cl is a pregeometry on P) if for all $A, B \subseteq P$ and $a, b \in P$ the following hold:

Monotonicity: $A \subseteq B$ implies $A \subseteq \mathrm{cl}(A) \subseteq \mathrm{cl}(B)$;
Finite character: $\mathrm{cl}(A) = \bigcup\{\mathrm{cl}(A_0) \mid A_0 \subseteq A \text{ finite}\}$;
Transitivity: $\mathrm{cl}(A) = \mathrm{cl}(\mathrm{cl}(A))$;
Exchange (symmetry): $a \in \mathrm{cl}(A \cup \{b\}) \setminus \mathrm{cl}(A)$ implies $b \in \mathrm{cl}(A \cup \{a\})$.

cl is called a *closure operator* if it satisfies the first three conditions.

2. Generic stability

Definition 2.1. A nonalgebraic global type $p(\bar{x}) \in S(\overline{M})$ is generically stable if, for some small A, it is A-invariant and:

if $(\bar{a}_i : i < \alpha)$ (any α, not only ω) is a Morley sequence in p over A then for any formula $\phi(\bar{x})$ (with parameters from $\overline{M}$) $\{i : \models \phi(\bar{a}_i)\}$ is either finite or co-finite.

Remark 2.1. If p is generically stable then as a witness-set A in the definition we can take any small A such that P is A-invariant.

Proposition 2.1. *Let $p(\bar{x}) \in S(\overline{M})$ be generically stable and A-invariant. Then*:

(i) *For any formula $\phi(\bar{x}, \bar{y}) \in L$ there is n_ϕ such that for any Morley sequence $(\bar{a}_i : i < \omega)$ of p over A, and any $\bar{b}$:*

$$\phi(\bar{x}, \bar{b}) \in p \quad \textit{iff} \quad \models \bigvee_{\substack{w \subset \{0,1,\ldots,2n_\phi\} \\ |w| = n_\phi + 1}} \bigwedge_{i \in w} \phi(\bar{a}_i, \bar{b}).$$

(i) *p is definable over A and almost finitely satisfiable in A.*
(ii) *Any Morley sequence of p over A is totally indiscernible.*
(iii) *p is the unique global nonforking extension of $p \mid A$.*

PROOF. This is a slight elaboration of the proof of Proposition 3.2 from [1].

(i) First note that for any $\phi(\bar{x}, \bar{y}) \in L$ there is n_ϕ such that for any Morley sequence $(\bar{a}_i : i < \omega)$ in p over A, and any $\bar{b}$ either at most n_ϕ many $\bar{a}_i$'s satisfy $\phi(\bar{x}, \bar{b})$ or at most n_ϕ many $\bar{a}_i$'s satisfy $\neg\phi(\bar{x}, \bar{b})$. For if not, then by compactness we can find a Morley sequence in p over A of length $\omega + \omega$ which violates Definition 2.1. Note that if at most n_ϕ $\bar{a}_i$'s satisfy $\neg\phi(\bar{x}, \bar{b})$ then $\phi(\bar{x}, \bar{b}) \in p$: for otherwise we could let $(\bar{a}_{\omega+j} : j < \omega)$ realize $p^{(\omega)}$ restricted to $A \cup \{\bar{a}_i : i < \omega\} \cup \{\bar{b}\}$, in which case $(\bar{a}_i : i < \omega + \omega)$ is a Morley sequence in p over A such that all but finitely many $\bar{a}_i$ for $i < \omega$ satisfy $\phi(\bar{x}, \bar{b})$ and all $\bar{a}_{\omega+j}$ satisfy $\neg\phi(\bar{x}, \bar{b})$, again a contradiction.

(ii) By (i) p is definable (over a Morley sequence), so by A-invariance p is definable over A. (If $\psi(\bar{y}, \bar{d})$ is the $\phi(\bar{x}, \bar{y})$ definition of p, then by A-invariance of p, this formula is preserved, up to equivalence, by automorphisms which fix A pointwise, hence is equivalent to a formula over A.) Let M be a model containing A, $\phi(\bar{x}, \bar{c}) \in p$ and let $I = (\bar{a}_i : i < \omega)$ be a Morley sequence in p over A such that $\mathrm{tp}(I/M\bar{c})$ is finitely satisfiable in M. Then, by (i), $\phi(\bar{x}, \bar{c})$ is satisfied by some $\bar{a}_i$ hence also by an element of M.

(iii) This follows from (i) exactly as in the proof of the Proposition 3.2 of [1] (where NIP was not used).

(iv) Let $q(\bar{x})$ be a global nonforking extension of $p \mid A$. We will prove that $q = p$.

Claim. *If $(\bar{a}_0, \ldots, \bar{a}_n, \bar{b})$ are such that $\bar{a}_0 \models q \mid A$, $\bar{a}_{i+1} \models q \mid (A, \bar{a}_0, \ldots, \bar{a}_i)$ and $b \models q \mid (A, \bar{a}_0, \ldots, \bar{a}_n)$ then $(\bar{a}_0, \ldots, \bar{a}_n, \bar{b})$ is a Morley sequence in p over A.*

PROOF. We prove it by induction. Suppose we have chosen $\bar{a}_0, \ldots, \bar{a}_n$ as in the claim and we know (induction hypothesis) that $(\bar{a}_0, \ldots, \bar{a}_n)$ begins a Morley sequence $I = (\bar{a}_i \mid i < \omega)$ in p over A. Suppose that $\phi(\bar{a}_0, \ldots, \bar{a}_n, \bar{x}) \in q$. Then we claim that $\phi(\bar{a}_0, \ldots, \bar{a}_{n-1}, \bar{a}_i, x) \in q$ for all $i > n$. For otherwise, without loss of generality $\neg\phi(\bar{a}_0, \ldots, \bar{a}_{n-1}, \bar{a}_{n+1}, \bar{x}) \in q$. But then, by indiscernibility of $\{\bar{a}_i\bar{a}_{i+1} : i = n, n+2, n+4, \ldots\}$ over $(A, \bar{a}_0, \ldots, \bar{a}_{n-1})$, and the nondividing of q over A, we have that

$$\{\phi(\bar{a}_0, \ldots, \bar{a}_{n-1}, \bar{a}_i, x) \wedge \neg\phi(\bar{a}_0, \ldots, \bar{a}_{n-1}, \bar{a}_{i+1}, \bar{x}) : i = n, n+2, n+4, \ldots\}$$

is consistent, which contradicts Definition 2.1. Hence $\models \phi(\bar{a}_0, \ldots, \bar{a}_{n-1}, \bar{a}_i, \bar{b})$ for all $i > n$. So by part (i), $\phi(\bar{a}_0, \ldots, \bar{a}_{n-1}, \bar{x}, \bar{b}) \in p(\bar{x})$. The inductive assumption gives that $\mathrm{tp}(\bar{a}_0, \ldots, \bar{a}_{n-1}, \bar{a}_n/A) = \mathrm{tp}(\bar{a}_0, \ldots, \bar{a}_{n-1}, \bar{b}/A)$, so by A-invariance of p, $\phi(\bar{a}_0, \ldots, \bar{a}_{n-1}, \bar{x}, \bar{a}_n) \in p(\bar{x})$. Thus $\models \phi(\bar{a}_0, \ldots, \bar{a}_{n-1}, \bar{a}_{n+1}, \bar{a}_n)$ and, by total indiscernibility of I over A (part (iii)), $\models \phi(\bar{a}_0, \ldots, \bar{a}_{n-1}, \bar{a}_n, \bar{a}_{n+1})$. We have shown that $q \mid (A, \bar{a}_0, \ldots, \bar{a}_n) = p \mid (A, \bar{a}_0, \ldots, \bar{a}_n)$, which allows the induction process to continue. The claim is proved. □

Now suppose that $\phi(\bar{x}, \bar{c}) \in q(\bar{x})$. Let $\bar{a}_i$ for $i < \omega$ be such that $\bar{a}_i$ realizes $q \mid (A, \bar{c}, \bar{a}_0, \ldots, \bar{a}_{i-1})$ for all i. By the claim $(\bar{a}_i : i < \omega)$ is a Morley sequence of p over A. But $\phi(\bar{a}_i, \bar{c})$ for all i, hence by the proof of (i), $\phi(\bar{x}, \bar{c}) \in p$. So $q = p$. □

Let us note for the record that for a global type p, generically stable implies definable implies invariant, and these are strict implications.

Now we restate the notion of generic stability for groups from [1] in the context of connected groups. Recall that a definable (or even type-definable) group G is connected if it has no relatively definable subgroup of finite index. A type $p(x) \in S_G(\overline{M})$ is left G-invariant iff for all $g \in G$:

$$(g \cdot p)(x) \stackrel{\text{def}}{=} \{\phi(g^{-1} \cdot x) : \phi(x) \in p(x)\} = p(x);$$

likewise for right G-invariant.

Definition 2.2. Let G be a definable (or even type-definable) connected group in $\overline{M}$. G is *generically stable* if there is a global complete 1-type $p(x)$ extending '$x \in G$' such that p is generically stable and left G-invariant.

As in the NIP context, we show that a generically stable left invariant type is also right invariant and is unique such (and we will call it the generic type):

Lemma 2.1. *Suppose that G is generically stable, witnessed by $p(x)$. Then $p(x)$ is the unique left-invariant and also the unique right-invariant type.*

PROOF. First we prove that p is right invariant. Let (a, b) be a Morley sequence in p over (any small) A. By left invariance $g = a^{-1} \cdot b \models p \mid A$. By total indiscernibility we have $\mathrm{tp}(a, b) = \mathrm{tp}(b, a)$, so $g^{-1} = b^{-1} \cdot a \models p \mid A$. This proves that $p = p^{-1}$. Now, for any $g \in G$ we have

$$p = g^{-1} \cdot p = (g^{-1} \cdot p)^{-1} = p^{-1} \cdot g = p \cdot g,$$

so p is also right-invariant.

For the uniqueness, suppose that q is a left invariant global type and we prove that $p = q$. Let $\phi(x) \in q$ be over A, let $I = (a_i : i \in \omega)$ be a Morley sequence in p over A, and let $b \models q \models (A, I)$. Then, by left invariance of q, for all $i \in \omega$ we have $\models \phi(a_i \cdot b)$. By Proposition 2.1(i) we get $\phi(x \cdot b) \in p(x)$ and, by right invariance, $\phi(x) \in p(x)$. Thus $p = q$. □

3. Global regularity

For stable T, a stationary type $p(x) \in S(A)$ is said to be regular if for any $B \supseteq A$, and b realizing a forking extension of p over B, $p \mid B$ (the unique nonforking extension of p over B) has a unique complete extension over (B, b). Appropriate versions of regularity (where we do not have stationarity) have been given for simple theories for example. Here we give somewhat different versions for global "invariant" types in arbitrary theories. So the reader should be aware that our definitions below are not consistent with usual terminology for *simple theories*, although they are of course consistent with the stable case.

Definition 3.1. Let $p(\bar{x})$ be a global nonalgebraic type.

(i) $p(\bar{x})$ is said to be *regular* if, for some small A, it is A-invariant and for any $B \supseteq A$ and $\bar{a} \models p \mid A$: either $\bar{a} \models p \mid B$ or $p \mid B \vdash p \mid B\bar{a}$.

(ii) Suppose $\phi(\bar{x}) \in p$. We say that $(p(\bar{x}), \phi(\bar{x}))$ is *strongly regular* if, for some small A over which ϕ is defined, p is A-invariant and for all $B \supseteq A$ and $\bar{a}$ satisfying $\phi(\bar{x})$, either $\bar{a} \models p \mid B$ or $p \mid B \vdash p \mid B\bar{a}$.

Remark 3.1. If p is regular then as a witness-set A in the definition we can take any small A such that p is A-invariant. Similarly for strongly regular types. Also note that strongly regular implies regular.

Definition 3.2. Let N be any submodel of $\overline{M}$ (possibly $N = \overline{M}$), and let $p(x) \in S_1(N)$. The operator cl_p is defined on (all) subsets of N by:

$$\mathrm{cl}_p(X) = \{a \in N \mid a \not\models p \mid X\}.$$

Also define $\mathrm{cl}_p^A(X) = \mathrm{cl}_p(X \cup A)$ for any $A \subset N$ (and $X \subseteq N$).

Intuitively, having fixed $p(x) \in S(N)$ and $B \subset N$ then realizations of $p|B$ are considered as 'generic over B' so formulas in $p(x)$ should be considered as defining 'large' subsets of N and their negations as defining 'small' subsets. In this way, $\mathrm{cl}_p(B)$ is the union of all 'small' definable subsets. In the rest of this section we will be interested in the case where $N = \overline{M}$. But later in the paper we will consider other N.

Remark 3.2. Let $p(x) \in S_1(\overline{M})$ be A-invariant, where A is small.

(i) Strong regularity of $(p(x), x = x)$ translates into a simpler expression using cl_p:

$$(p(x), x = x) \text{ is strongly regular} \quad \text{iff} \quad p \mid B \vdash p \mid \mathrm{cl}_p(B) \text{ for any small } B \supseteq A.$$

Since $a \models p \mid B$ is the same as $a \notin \mathrm{cl}_p(B)$, another equivalent way of expressing strong regularity of $(p(x), x = x)$ is:

$$a \models p \mid B \quad \text{iff} \quad a \models p \mid \mathrm{cl}_p(B) \qquad (\text{for all } a \in \overline{M} \text{ and small } B \supseteq A).$$

(ii) The corresponding expression is not so concise when $(p(x), \phi(x))$ is strongly regular or when p is just regular. For example:

$$p \text{ is regular} \quad \text{iff} \quad p \mid AB \vdash p \mid A \cup B \cup \left(\mathrm{cl}_p^A(B) \cap (p \mid A)(\overline{M})\right) \text{ for any small } B.$$

We can also consider the restriction of cl_p^A to $A \cup (p \mid A)(\overline{M})$. Formally we define (for B a small subset of $A \cup (p \mid A)(\overline{M})$) $\mathrm{cl}_{p_A}(B) = A \cup (\mathrm{cl}_p^A(B) \cap (p|A)(\overline{M}))$. Then we have a simple consequence of regularity; p is regular *implies*

$$p \mid AB \vdash p \mid \mathrm{cl}_{p_A}(B) \quad \text{for any small } B \subset (p|A)(\overline{M}).$$

As in (i) this is translated to: if p is regular then

$$a \models p \mid AB \quad \text{iff} \quad a \models p \mid \mathrm{cl}_{p_A}(B) \qquad (\text{for any } a \in \overline{M} \text{ and small } B \subset (p \mid A)(\overline{M})).$$

(iii) Note that in (i) and (ii) we took into account only small B's, while Definition 3.1 mentions all B's. But this does not matter, since in all of them the statements with and without 'small' are easily seen to be equivalent.

Lemma 3.1. *Suppose that A is small and $p \in S_1(\overline{M})$ is A-invariant.*

(i) *$(p(x), x = x)$ is strongly regular iff cl_p^A is a closure operator on $\overline{M}$.*

(ii) *Suppose that $(p(x), x = x)$ is strongly regular. Then cl_p^A is a pregeometry operator on $\overline{M}$ iff every Morley sequence in p over A is totally indiscernible.*

(iii) *Suppose that p is regular. Then cl_{p_A} is a closure operator on $A \cup (p \mid A)(\overline{M})$; it is a pregeometry operator iff every Morley sequence in p over A is totally indiscernible.*

PROOF. (i) Assume that $(p(x), x = x)$ is strongly regular. Then for any small $B \supseteq A$ we have:

$$a \notin \mathrm{cl}_p(B) \quad \text{iff} \quad a \models p \mid B \quad \text{iff} a \models p \mid \mathrm{cl}_p(B) \quad \text{iff} \quad a \notin \mathrm{cl}_p(\mathrm{cl}_p(B)).$$

The first and the last equivalence follow from the definition of cl_p and the middle one is by Remark 3.2(i). Thus $\mathrm{cl}_p(B) = \mathrm{cl}_p(\mathrm{cl}_p(B)$ so cl_p^A is a closure operator on $\overline{M}$. The other direction is similar.

(ii) Suppose that $(p(x), x = x)$ is strongly regular and that every Morley sequence in p over A is totally indiscernible. To show that cl_p^A is a pregeometry operator, by part (i), it suffices to verify the exchange property over finite extensions $B \supset A$. Let $(a_1, \ldots, a_n) \in B^n$ be a maximal Morley sequence in p over A; note that it is finite since it is in $B \setminus A$, which is finite. Then $B \subseteq \mathrm{cl}_p^A(a_1, \ldots, a_n)$ (otherwise any element in the difference would contradict the maximality) and, since cl_p^A is a closure operator, we get $\mathrm{cl}_p^A(B) \subseteq \mathrm{cl}_p^A(a_1, \ldots, a_n)$ and thus $\mathrm{cl}_p^A(B) = \mathrm{cl}_p^A(a_1, \ldots, a_n)$.

To verify exchange, let $a \models p \mid B$ (so $a \notin \mathrm{cl}_p^A(B)$) and let $b \in \overline{M}$. Note that $(a_1, \ldots, a_n, a)$ is a Morley sequence in p over A and $\mathrm{cl}_p^A(Ba) = \mathrm{cl}_p^A(a_1, \ldots, a_n, a)$. We have:

$$\begin{aligned} b \notin \mathrm{cl}_p^A(Ba) \quad & \text{iff} \quad b \notin \mathrm{cl}_p^A(a_1, \ldots, a_n, a) \\ & \text{iff} \quad (a_1, \ldots, a_n, a, b) \text{ is a Morley sequence} \\ & \text{iff} \quad (a_1, \ldots, a_n, b, a) \text{ is a Morley sequence} \\ & \text{iff} \quad (b \notin \mathrm{cl}_p^A(B) \text{ and } a \notin \mathrm{cl}_p^A(Bb)). \end{aligned}$$

In particular $b \notin \mathrm{cl}_p^A(Ba)$ implies $a \notin \mathrm{cl}_p^A(Bb)$. This proves exchange.

(iii) This is similar to (i) and (ii). Suppose that p is regular, let $B \subset (p \mid A)(\overline{M})$ be small and let $a \models p \mid A$. Then:

$$a \notin \mathrm{cl}_{p_A}(B) \quad \text{iff} \quad a \models p \mid AB \quad \text{iff} \quad a \models p \mid \mathrm{cl}_{p_A}(B) \quad \text{iff} \quad a \notin \mathrm{cl}_{p_A}(\mathrm{cl}_{p_A}(B));$$

the first equivalence and the last equivalence follow from the definition of cl_{p_A} and the middle one is by Remark 3.2(ii). Thus $\mathrm{cl}_p(B) = \mathrm{cl}_p(\mathrm{cl}_p(B))$. The other clause is proved as in part (ii). □

Theorem 3.1. *Suppose that A is small and $p(x) \in S_1(\overline{M})$ is A-invariant and regular.*

(i) *If p is symmetric, then cl_{p_A} is a pregeometry operator on $A \cup (p \mid A)(\overline{M})$, In the case where a $(p(x), x = x)$ is strongly regular, cl_p^A is a pregeometry operator on $\overline{M}$.*

(ii) *If p is not symmetric, then there exists a finite extension A_0 of A and a A_0-definable partial order $\leq$ such that every Morley sequence in p over A_0 is strictly increasing.*

PROOF. If p is symmetric then every Morley sequence in p over A is totally indiscernible and (i) follows from Lemma 3.1.

Now suppose that p is not symmetric. Then, as mentioned in the introduction, there is a finite extension $A_0 \supset A$ and a Morley sequence (a, b) of length 2 in p over A_0 such that $\mathrm{tp}(a, b/A_0) \neq \mathrm{tp}(b, a/A_0)$. Choose $\phi(x, y) \in \mathrm{tp}(a, b/A_0)$ witnessing the "asymmetry": $\models \phi(x, y) \to \neg\phi(y, x)$. Then $\phi(a, x) \wedge \neg\phi(x, a) \in \mathrm{tp}(b/aA_0) \subset p(x))$ so $\phi(x, a) \notin p(x)$. We claim that

$$(p \mid A_0)(t) \cup \{\phi(t, a)\} \cup \{\neg\phi(t, b)\}$$

is inconsistent. Otherwise, there is d realizing $p \mid A_0$ such that $\models \phi(d, a) \wedge \neg\phi(d, b)$. Then d does not realize $p \mid (A_0, a)$ (witnessed by $\phi(x, a)$) so, by regularity, $p \mid (A_0, a) \vdash p \mid (A_0, a, d)$ and thus $b \models p \mid (A_0, a, d)$. In particular $b \models p \mid (A_0, d)$ and since, by invariance, $\phi(d, x) \in p(x)$ we conclude $\models \phi(d, b)$. A contradiction.

By this claim, and compactness, we find $\theta(t) \in p \mid A_0$ such that

$$\models (\forall t)(\phi(t, a) \wedge \theta(t) \to \phi(t, b)).$$

Let $x \preccurlyeq y$ be $(\forall t)(\phi(t, x) \wedge \theta(t) \to \phi(t, y))$. Clearly, $\preccurlyeq$ defines a quasi order and $a \preccurlyeq b$. Also:

$$\models \phi(a, b) \wedge \theta(a) \wedge \neg\phi(a, a);$$

The first conjunct follows by our choice of ϕ, the second from $a \models p \mid A_0$, and the third from the asymmetry of ϕ. Altogether they imply $b \not\preccurlyeq a$. Thus if $x < y$ is $x \preccurlyeq y \wedge y \not\preccurlyeq x$ we have $a < b$. □

The next examples concern issues of whether symmetric regular types are definable or even generically stable. But we first give a case where this is true (although it depends formally on Theorem 4.1 of the next section).

Corollary 3.1. *Suppose that $(p(x), x = x)$ is strongly regular and symmetric. Then $p(x)$ is generically stable.*

PROOF. If $(p(x), x = x)$ is invariant-strongly regular then, by Theorem 3.1(i) cl_p is a pregeometry operator on $\overline{M}$, and then $p(x)$ is generically stable by Theorem 4.1(ii). □

Example 3.1. A symmetric, definable, strongly regular type which is not generically stable: Let $L = \{U, V, E\}$ where U, V are unary and E is a binary predicate. Consider the bipartite graph (M, U^M, V^M, E^M) where $U^M = \omega$, V^M is the set of all finite subsets of ω, $M = U^M \cup V^M$, and $E^M = \{(u, v) : u \in U^M, v \in V^M, \text{and } u \in v\}$. Let $A \subset M$ be finite. Then:

If $(c_1, \dots, c_n), (d_1, \dots, d_n) \in (U^M)^n$ have the same quantifier-free type over A then $\mathrm{tp}(c_1, \dots, c_n/A) = \mathrm{tp}(d_1, \dots, d_n/A)$,

since the involution of ω mapping c_i's to d_i's respectively, and fixing all the other elements of ω is an A-automorphism of M. Note that this is expressible by a set of first-order sentences, so is true in the monster.

Further, if $e_1, \dots, e_n \in U^M$ are distinct and have the same type over A then, since every permutation of ω which permutes $\{e_1, \dots, e_n\}$ and fixes all the other elements of ω is an A-automorphism of M, $(e_1, \dots, e_n)$ is totally indiscernible over A. This is also expressible by a set of first-order sentences.

Let $p(x) \in S_1(M)$ be the type of a "new" element of U which does not belong to any element of V^M. Then, by the above, p is definable, its global heir $\bar{p}$ is symmetric, and $(\bar{p}(x), U(x))$ is strongly regular.

Example 3.2. A symmetric, strongly regular type which is not definable: Consider the bipartite graph (M, U^M, V^M, E^M) where $U^M = \omega$, V^M consists of all finite and co-finite subsets of ω, $M = U^M \cup V^M$, and E^M is $\in$. Let $\overline{M}$ be the monster and let $p(x) \in S_1(\overline{M})$ be the type of a new element of $U^{\overline{M}}$, which belongs to all co-finite members of $V^{\overline{M}}$ (and no others). Arguing as in the previous example $(p(x), U(x))$ is strongly regular and symmetric. Since "being a co-finite subset of U^M" is not definable, p is not definable.

Definition 3.3. Let G be a definable group in $\overline{M}$. G is called a *regular group* if for some global type $p(x) \in S_G(\overline{M})$, $(p(x), \text{"}x \in G\text{"})$ is strongly regular (in particular invariant over some small set).

We will see in Example 5.1 that nonsymmetric regular groups, and even fields, exist.

Theorem 3.2. *Suppose that G is a group definable over $\varnothing$, which is regular, witnessed by $p(x) \in S_G(\overline{M})$. Then:*

(i) *$p(x)$ is both left and right translation invariant (and in fact invariant under definable bijections).*
(ii) *A formula $\phi(x)$ is in $p(x)$ iff two left (right) translates of $\phi(x)$ cover G iff finitely many left (right) translates of $\phi(x)$ cover G. (Hence $p(x)$ is the unique "generic type" of G.)*
(iii) *$p(x)$ is definable over $\varnothing$.*
(iv) *$G = G^0$ (i.e., G is connected).*

PROOF. (i) Suppose that $f : G \to G$ is a B-definable bijection and $a \models p \mid B$. Since $p \mid B \vdash p \mid (B, f(a))$ is not possible, by strong regularity, we get $f(a) \models p \mid B$. Thus p is invariant under f.

(ii) Suppose that $D \subseteq G$ is defined by $\phi(x) \in p(x)$ which is over A. Let $g \models p \mid A$ and we show $G = D \cup g \cdot D$. If $b \in G \setminus D$ then b does not realize $p \mid A$ so, by strong regularity, $g \models p \mid (A, b)$. By (i) $g^{-1} \models p \mid (A, b)$, thus $g^{-1} \in D \cdot b^{-1}$ and $b \in g \cdot D$. This proves $G = D \cup g \cdot D$.

For the other direction, if finitely many translates of $\psi(x)$ cover G then at least one of them belongs to $p(x)$ and, by (i), $\psi(x) \in p(x)$.

(iii) and (iv) follow immediately from (ii). □

Question. Is every regular group commutative?

4. Homogeneous pregeometries

If (M, cl) is a pregeometry then, as usual, we obtain notions of independence and dimension: for $A, B \subset M$ we say that A is independent over B if $a \notin \text{cl}(A \setminus \{a\} \cup B)$ for all $a \in A$. Given A and B, all subsets of A which are independent over B and maximal such, have the same cardinality, called $\dim(A/B)$. (M, cl) is infinite-dimensional if $\dim(M/\varnothing) \geq \aleph_0$.

Remark 4.1. (i) If $\bar{c}$ is a tuple of length n then $\dim(\bar{c}/B) \leq n$ for any B.
(ii) If $l(\bar{c}) = n$, $|A| \geq n + 1$ and A is independent over B then there is $a \in A$ such that $a \notin \text{cl}(B \cup \bar{c})$.

Definition 4.1. We call an infinite-dimensional pregeometry (M, cl) *homogeneous* if for any finite $B \subset M$, the set of all $a \in M$ such that $a \notin \text{cl}(B)$ is the set of realizations in M of a complete type $p_B(x)$ over B.

Note that Definition 4.1 relates in some way the closure operation to the first-order structure. But it does not say anything about automorphisms, and nothing is being claimed about the homogeneity or strong homogeneity of M as a first-order structure. In particular we do not want to assume M to be a saturated structure.

Lemma 4.1. *Suppose (M, cl) is a homogeneous pregeometry.*

(i) $p_{\mathrm{cl}}(x) = \bigcup\{p_B(x) : B$ *finite subset of* $M\}$ *is a complete 1-type over M, which we call the generic type of the pregeometry (M, cl).*
(ii) $a \notin \mathrm{cl}(B)$ *iff* $a \models p_{\mathrm{cl}}(x) \mid B$. *In particular* $\mathrm{cl} = \mathrm{cl}_{p_{\mathrm{cl}}}$.
(iii) $I = (a_i : i < \omega)$ *is independent over B iff $a_i \models p_{\mathrm{cl}} \mid (B, a_0, \dots, a_{i-1})$ for all i. In particular, if $M = \overline{M}$ and p_{cl} is B-invariant then I is independent over B iff it is a Morley sequence in p_{cl} over B.*

PROOF. (i) Consistency is by compactness: given $A_1, \dots, A_n$ finite subsets of M and $B = A_1 \cup \dots \cup A_n$, clearly $p_{A_1}(x) \cup \dots \cup p_{A_n}(x) \subseteq p_B(x)$ and the later is consistent. Completeness is clear.

(ii) and (iii) are easy. □

Lemma 4.2. *Suppose (M, cl) is a homogeneous pregeometry. Let $(a_i : i \in \omega)$ be an $\varnothing$-independent subset of M. Then for any L-formula $\phi(x, \bar{y})$ with $l(\bar{y}) = n$, and n-tuple $\bar{b}$ from M:*

$$\phi(x, \bar{b}) \in p_{\mathrm{cl}}(x) \quad \textit{iff} \quad \models \bigwedge_{i \in w} \phi(a_i, b) \ \textit{for some}\ w \subset \{1, \dots 2n\},\ |w| = n+1.$$

In particular $p_{\mathrm{cl}}(x)$ is definable.

PROOF. If $\phi(x, \bar{b})$ is large (namely in p_{cl}), then its negation is small, and thus if $M \models \neg\phi(a, \bar{b})$ then $a \in \mathrm{cl}(\bar{b})$. By Remark 4.1(ii), at most n many a_i's can satisfy $\neg\phi(x, \bar{b})$, hence at least $n+1$ among the first $2n+1$ a_i's satisfy $\phi(x, \bar{b})$. Conversely, if at least $n+1$ a_i's satisfy $\phi(x, \bar{b})$ then, again by Remark 4.1(ii), $\phi(x, \bar{b})$ can not be small, so it is large. □

In the next proposition we make use of Definition 3.2 from the previous section.

Proposition 4.1. *Suppose (M, cl) is a homogeneous pregeometry. Let $p(x)$ be the generic type and let $\bar{p}(x)$ be its (unique by definability) global heir.*

(i) $(\overline{M}, \mathrm{cl}_{\bar{p}})$ *is a homogeneous pregeometry and* cl *is the restriction of* $\mathrm{cl}_{\bar{p}}$ *to M.*
(ii) *If $(a_1, \dots, a_n)$ (from $\overline{M}$) is independent over A then:*

$$\mathrm{tp}(b_1, \dots, b_n/A) = \mathrm{tp}(a_1, \dots, a_n/A) \quad \textit{iff} \quad (b_1, \dots, b_n) \textit{ is independent over } A.$$

(iii) $\bar{p}(x)$ *is $\varnothing$-invariant and generically stable.*
(iv) $(\bar{p}(x), x = x)$ *is strongly regular.*

PROOF. (i) is a reasonably routine exercise, using the fact that $\bar{p}$ is defined by the same schema which defines p, and is left to the reader. But we will briefly point out why exchange holds: Note first that $\bar{p}^{(2)}(x_1, x_2)$ is also definable, over the same parameters as p. Let $\phi(x_1, x_2, \bar{b})$ be an arbitrary formula over $\overline{M}$, with parameters $\bar{b}$ witnessed. To prove exchange it suffices to see that $\phi(x_1, x_2, \bar{b}) \in \bar{p}^{(2)}(x_1, x_2)$ iff $\phi(x_2, x_1, \bar{b}) \in \bar{p}^{(2)}(x_1, x_2)$. By definability of $\bar{p}^{(2)}$, there are formulas $\delta_1(\bar{z})$ and $\delta_2(\bar{z})$ over M such that for any $\bar{b}'$ from $\overline{M}$, we have

$$\phi(x_1, x_2, \bar{b}') \in \bar{p}^{(2)}(x_1, x_2) \quad \text{iff} \quad \models \delta_1(\bar{b}'),$$

and

$$\phi(x_2, x_1, \bar{b}') \in \bar{p}^{(2)}(x_1, x_2) \quad \text{iff} \quad \models \delta_2(\bar{b}').$$

As exchange holds inside M, we have $M \models \forall \bar{z}(\delta_1(\bar{z}) \leftrightarrow \delta_2(\bar{z}))$. Hence this also holds in $\overline{M}$.

(ii) We prove it by induction on n. For $n = 1$, by definition, we have $a_1, b_1 \models \bar{p} \mid A$. Now assume true for n and prove for $n + 1$. Without loss of generality $A = \varnothing$. Suppose first that $\mathrm{tp}(b_1, \ldots, b_{n+1}) = \mathrm{tp}(a_1, \ldots, a_{n+1})$. Let a' realize $p \mid (a_1, \ldots, a_{n+1}, b_1, \ldots, b_{n+1})$. So $\mathrm{tp}(a_1, \ldots, a_n, a_{n+1}) = \mathrm{tp}(a_1, \ldots, a_n, a')$. On the other hand, by the induction assumption (over $\varnothing$), $(b_1, \ldots, b_n)$ is independent, so independent over a' (by symmetry). By induction assumption applied over a', $\mathrm{tp}(b_1, \ldots, b_n, a') = \mathrm{tp}(a_1, \ldots, a_n, a')$. Hence $\mathrm{tp}(b_1, \ldots, b_n, a') = \mathrm{tp}(b_1, \ldots, b_n, b_{n+1})$. As $a' \notin \mathrm{cl}_{\bar{p}}(b_1, \ldots, b_n)$, also $b_{n+1} \notin \mathrm{cl}_{\bar{p}}(b_1, \ldots, b_n)$. Thus $(b_1, \ldots b_n, b_{n+1})$ is independent.

The converse (if $(b_1, \ldots, b_{n+1})$ is independent then it realizes $\mathrm{tp}(a_1, \ldots, a_{n+1})$) is proved in a similar fashion and left to the reader.

(iii) By part (i), Lemma 4.2 also applies to the pregeometry $\mathrm{cl}_{\bar{p}}$. Let $(a_i : i \in \omega)$ be $\mathrm{cl}_{\bar{p}}$-independent. Then $\bar{p}(x)$ is defined over $(a_i : i \in \omega)$ as in Lemma 4.2. But if $(b_i : i \in \omega)$ has the same type as $(a_i : i \in \omega)$ then, by (ii), it is also $\mathrm{cl}_{\bar{p}}$-independent, hence $\bar{p}(x)$ is defined over $(b_i : i \in \omega)$ in the same way. This implies that $\bar{p}$ is $\varnothing$-invariant. Thus, by Lemma 4.1(iii) a Morley sequence in $\bar{p}$ is the same thing as an infinite $\mathrm{cl}_{\bar{p}}$-independent set. By Lemma 4.2 and Definition 2.1, $\bar{p}$ is generically stable.

(iv) By part (i) $(\overline{M}, \mathrm{cl}_{\bar{p}})$ is a pregeometry so, by Lemma 3.1(i), $(\bar{p}(x), x = x)$ is strongly regular. □

We now drop (for a moment) all earlier assumptions and summarize the situation:

Theorem 4.1. *Let T be an arbitrary theory.*

(i) *Let $p(x)$ be a global $\varnothing$-invariant type such that $(p(x), x = x)$ is strongly regular. Then $(\overline{M}, \mathrm{cl}_p)$ is a homogeneous pregeometry.*

(ii) *On the other hand, suppose $M \models T$ and (M, cl) is a homogeneous pregeometry. Then there is a unique global $\varnothing$-invariant generically stable type $p(x)$ such that $(p(x), x = x)$ is strongly regular, and such that the restriction of* cl_p *to M is precisely* cl.

We end this section by pointing out the connection to exponential fields, as mentioned to us by Kirby. In [3], Jonathan Kirby proved that ecl(−), "exponential algebraic closure," as originally defined by Macintyre, gives a pregeometry on *any* exponential field, and this result extends those of Wilkie [7] for the complex exponential field. It is an open question whether for the complex exponential field, ecl(−) is *homogeneous* in the sense of Definition 4.1 above. A positive answer would yield quasiminimality for the complex exponential field as well as generic stability and strong regularity of its (unique) exponentially transcendental type. On the other hand, a positive answer does exist for Zilber's pseudoexponentiation and some other exponential fields.

5. Quasiminimal structures

Recall that a 1-sorted structure M in a countable language is called quasiminimal if M is uncountable and every definable (with parameters) subset of M

is countable or co-countable; the definition was given by Zilber in [8]. Here we investigate the general model theory of quasiminimality, continuing an earlier work by Itai, Tsuboi and Wakai [2].

Throughout this section fix a quasiminimal structure M and let $\overline{M}$ be a saturated elementary extension. The set of all formulas (with parameters) defining a co-countable subset of M forms a complete 1-type $p(x) \in S_1(M)$; we will call it the generic type of M. If $p(x)$ happens to be definable we will denote its (unique) global heir by $\bar{p}(x)$.

Remark 5.1. (i) In quasiminimal structures Zilber's countable closure operator ccl is defined via cl_p:

$$\mathrm{cl}_p^0(A) = \mathrm{cl}_p(A), \quad \mathrm{cl}_p^{n+1}(A) = \mathrm{cl}_p(\mathrm{cl}_p^n(A)) \quad \text{and} \quad \mathrm{ccl}(A) = \bigcup_{n\in\omega} \mathrm{cl}_p^n(A).$$

(ii) If A is countable then $\mathrm{ccl}(A)$ is countable, too. ccl is a closure operator on M.

(iii) cl_p is a closure operator iff $\mathrm{cl}_p = \mathrm{ccl}$ (which is in general not the case, see Example 5.2).

In Section 2 of [2] Itai, Tsuboi and Wakai, studied the case when M is strongly $\aleph_1$-homogeneous, in the model theoretic sense, namely, any partial elementary map between countable subsets extends to an automorphism of M. They proved in Proposition 2.10 there that if ccl is *not* a pregeometry operator then some uncountable subset of M is totally ordered by a formula, and they also remark that Maesono has strengthened in one direction the conclusion to: $\mathrm{Th}(M)$ has the strict order property (Remark 2.16 there). They also proved (assuming model-theoretic strong $\aleph_1$-homogeneity)that ccl is a pregeometry operator whenever $|M| \geq \aleph_2$.

In Theorem 5.1 below we will use a weaker assumption than strong $\aleph_1$-homogeneity, which we call countable basedness, and derive a stronger dichotomy similar to the one for strongly regular types in Theorem 3.1. Then, in Corollary 5.1 we prove that any quasiminimal structure of size at least $\aleph_2$ must be of symmetric type; in particular, (M, ccl) is a homogeneous pregeometry (in the sense of the previous section) after possibly countably many parameters from M into the language. In Theorem 5.2 we will prove that the failure of countable basedness implies the strict order property.

Recall from the introduction that p does not split over A, if $(\phi(x,\bar{b}_1) \leftrightarrow \phi(x,\bar{b}_2)) \in p(x)$ for all $\phi(x,\bar{y})$ over A and all $\bar{b}_1, \bar{b}_2 \in M$ realizing the same type over A. We talk explicitly about splitting rather than invariance, because M need not be saturated/homogeneous. Remember that in this section we are taking $p(x) \in S_1(M)$ to be the unique "generic" type of M consisting of all formulas defining co-countable sets.

Definition 5.1. (i) $p(x)$ is *based on* A if:

$$p \text{ does not split over } A \quad \text{and} \quad \mathrm{ccl}(AC) \subsetneq M \text{ for all finite } C \subset M;$$

(ii) $p(x)$ (or M) is *countably based* if there is a countable $A \subset M$ such that it is based on A.

The technical condition $\mathrm{ccl}(AC) \subsetneq M$ is satisfied by any countable A, so countable baseness is equivalent to the existence of a countable subset A over which p

does not split. In the proof of Corollary 5.1 we will use an uncountable base set, and this is why it is added: it describes a relative smallness of A in M.

If p does not split over A then we say that $(a_1, \dots, a_n)$ is a weak Morley sequence in p over A if a_k realizes $p \mid (A, a_1, \dots, a_{k-1})$ for all relevant k. As in the case of global invariant types weak Morley sequences are indiscernible over A. Now we can state one of our main theorems, whose proof will be presented in a somewhat more general context in the next section, and then draw an important corollary.

Theorem 5.1. *Suppose that $p(x)$ is based on $A \subset M$. Then cl_p^A is a closure operator and exactly one of the following two holds:*

(1) *Every (some) weak Morley sequence in p over A is totally indiscernible; in this case (M, cl_p^A) is a homogeneous pregeometry, p is definable over A, $\bar{p}$ (its unique global heir) is generically stable and $(\bar{p}(x), x = x)$ is strongly regular.*

(2) *There exists a weak Morley sequence in p over A which is not totally indiscernible; in this case there is a finite extension A_0 of A and an A_0-definable partial order $\leq$ such that every weak Morley sequence in p over A_0 is strictly increasing.*

Corollary 5.1. *If $|M| \geq \aleph_2$ then p is definable (hence countably based) and the first case of Theorem 5.1 holds.*

PROOF. Suppose $|M| \geq \aleph_2$ and let $M_0 \subset M$ be ccl-closed of size $\aleph_1$. First we show that $p(x)$ is finitely satisfiable in M_0 (actually it is finitely satisfiable in any uncountable subset of M): for any $\phi(x) \in p(x)$ $\phi(M)$ is co-countable, so $\phi(M)$ intersects M_0. Thus p does not split over M_0 and, since $|\mathrm{ccl}(M_0 C)| = \aleph_1$ for any countable C, p is based on M_0. Theorem 5.1 applies; we will show that case (1) holds. Otherwise, there is a definable $<$ and a strictly increasing sequence $\{a_i \mid i < \omega_2\} \subset M$. Then $x < a_{\omega_1}$ and $a_{\omega_1} < x$ define uncountable mutually disjoint subsets of M. A contradiction. Thus case (1) holds and the conclusion follows. $\square$

Theorem 5.2. *If p is not countably based then there exists a definable partial order (on some M^n) which has uncountable strictly increasing chains.*

PROOF. Suppose that p is not countably based. Then p is finitely satisfiable in no countable subset of M and, inductively, we can find a sequence $(M_i \mid i < \omega_1)$ of countable ccl-closed submodels and a sequence $(\bar{a}_i \mid i < \omega_1)$ of tuples such that for all $i < \omega_1$:

(1) $M_i \prec M_{i+1}$ and $M_\alpha = \bigcup_{j<\alpha} M_j$ for $\alpha < \omega_1$ a limit ordinal;
(2) $\bar{a}_i \in M_{i+1}$;
(3) $p \mid M_i \bar{a}_i$ is not finitely satisfiable in M_i.

For each $i < \omega_1$ witness (3) by an L-formula $\phi_i(x, \bar{y}, \bar{z})$ and $\overline{m}_i \in M_i$ such that $\phi_i(x, \overline{m}_i, \bar{a}_i) \in p(x)$ is not satisfied in M_i. Thus $\neg\phi_i(M, \overline{m}_i, \bar{a}_i)$ is countable and contains M_i. Since $M_{i+1} \supset M_i \bar{a}_i$ is ccl-closed we have:

$$M_i \subseteq \neg\phi_i(M, \overline{m}_i, \bar{a}_i) \subseteq M_{i+1}.$$

We will now find uncountable $S \subset \omega_1$, and $\phi(x, \bar{y}, \bar{z}) \in L$ and $\overline{m}$ from M, such that for all $i \in S$ such that for all $i \in S$ $\phi_i(x, \bar{y}, \bar{z}) = \phi(x, \bar{y}, \bar{z})$ and $\overline{m}_i = \overline{m}$: Without loss of generality assume that the universe of M is ω_1. Then $C = \{\alpha \in \omega_1 \mid M_\alpha = \alpha\}$ is a club subset of ω_1. By the Pressing Down Lemma the function $\alpha \mapsto \overline{m}_\alpha$ is constant on a stationary $S_1 \subset C$, so the $\overline{m}_i$'s are the same for all $i \in S_1$ (say $\overline{m}$). Since there are only countably many possibilities for the ϕ_i there is uncountable $S \subset S_1$ such that ϕ_i's are the same for all $i \in S$ (say ϕ). The family $(\neg\phi(M, \overline{m}, \bar{a}_i) \mid i \in S)$ is a strictly increasing chain of definable subsets of M, yielding the theorem. □

As we see now the existence of a definable group operation on a quasiminimal structure has strong (but easy) consequences.

Theorem 5.3. *Suppose that M is a quasiminimal group. Then $p(x)$ is definable over $\varnothing$, both left and right translation invariant, and $\overline{M}$ is a regular group, in the sense of Definition* 3.3, *witnessed by $\bar{p}(x)$, where $\bar{p}$ is the unique global heir of p.*

PROOF. Let $X \subseteq M$ be definable. First we claim that X is uncountable iff $X \cdot X = M$. If X is uncountable, then X is co-countable, as is X^{-1}. So for any $a \in M$, $a \cdot X^{-1}$ is co-countable, so has nonempty intersection with X. If $d \in X \cap a \cdot X^{-1}$ then $a \in X \cdot X$, proving the claim.

It follows that $p(x)$ is definable over $\varnothing$. In particular it is based on $\varnothing$ and, by Theorem 5.1, cl_p is a closure operator on M. The definability of p implies that $\mathrm{cl}_{\bar{p}}$ is also a closure operator and $(\bar{p}(x), x = x)$ is strongly regular by Lemma 3.1(i). The rest follows from Theorem 3.2. □

Several examples of quasiminimal structures with "bad" properties are given in [2]. For example $\omega_1 \times \mathbb{Q}$ equipped with the lexicographic order is quasiminimal, its generic type is definable over $\varnothing$, but which is non symmetric in the sense that case (2) of Theorem 5.1 holds. We give here some other examples, including algebraic ones.

Example 5.1. A quasiminimal field with countably based "generic type," which is "asymmetric": In fact, every strongly minimal structure M of size $\aleph_1$ can be expanded to such a quasiminimal structure. Let again $I = \omega_1 \times \mathbb{Q}$ and let $\lhd$ be the strict lexicographic order on I. Further, let $B = \{b_i \mid i \in I\}$ be a basis of the strongly minimal structure M. For each $a \in M$ let $i \in I$ be $\lhd$-maximal for which there are $i_1, \dots, i_n \in I$ such that $a \in \mathrm{acl}(b_{i_1}, \dots, b_{i_n}, b_i) \setminus \mathrm{acl}(b_{i_1}, \dots, b_{i_n})$; Clearly, $i = i(a)$ is uniquely determined. Now, expand $(M, \dots)$ to $(M, <, \dots)$ where $b < c$ iff $i(b) \lhd i(c)$. We will prove that $(M, <, \dots)$ is quasiminimal. Suppose that $M_0 \prec M$ is a countable, $<$-initial segment of M and that $B \setminus M_0$ does not have $\leq$-minimal elements, and let $a, a' \in M \setminus M_0$. Then there is an automorphism of $(B, \lhd)$ fixing $B \cap M_0$ pointwise and moving $b_{i(a)}$ to $b_{i(a')}$. It easily extends to an M_0-automorphism of $(M, <, \dots)$, so $b_{i(a)} \equiv b_{i(a')}(M_0)$ (in the expanded structure). Note that replacing $b_{i(a)}$ by a in B (in the definition of $<$) does not affect $<$, so $a \equiv a'(M_0)$ and there is a single 1-type over M_0 realized in $M \setminus M_0$. Since every countable set is contained in an M_0 as above, $(M, <, \dots)$ is quasiminimal. The remaining details are left to the reader.

Question. Is every quasiminimal field algebraically closed?

The following is an example of a quasiminimal structure, which is very far from being regular: $\mathrm{cl}_p(A) \neq \mathrm{cl}_p(\mathrm{cl}_p(A))$ for arbitrarily large countable A's. In particular the generic type is not countably based.

Example 5.2. A quasiminimal structure where $\mathrm{cl}_p \neq \mathrm{ccl}$: Peretyatkin in [4] constructed an $\aleph_0$-categorical theory of a 2-branching tree. Our quasiminimal structure will a model of this theory. The language consists of a single binary function symbol $L = \{\inf\}$. Let Σ be the class of all finite L-structures $(A, \inf)$ satisfying:

(1) $(A, \inf)$ is a semilattice;
(2) $(A, <)$ is a tree (where $x < y$ iff $\inf(x, y) = x \neq y$);
(3) (2-branching) No three distinct, pairwise $<$-incompatible elements satisfy: $\inf(x, y) = \inf(x, z) = \inf(y, z)$.

Then the Fraissé limit of Σ exists and its theory, call it T_2, is $\aleph_0$-categorical and has unique 1-type. If we extend the language to $\{\inf, <, \perp\}$, where $x \perp y$ stands for $x \not\leq y \wedge y \not\leq x$, then T_2 has elimination of quantifiers.

Let $(\overline{M}, <)$ be the monster model of T_2, let $\lhd$ be a lexicographic order on $I = \omega_1 \times Q$, and let $C = \{c_i \mid i \in \omega_1 \times Q\}$ be $<$-increasing. Then we can find a sequence of countable models $\{M_i \mid i \in \omega_1 \times Q\}$ satisfying:

(a) $M_i \prec M_j$ for all $i \lhd j$;
(b) $M_i \cap C = \{c_j \in C \mid j \unlhd i\}$ for all i;
(c) $M_i \cap C < M_j \setminus M_i$ for all $i \lhd j$.

Finally, let $M = \bigcup\{M_i \mid i \in I\}$. Clearly, C is an uncountable branch in M. Moreover, by (c), any other branch is completely contained in some M_i, and is so countable. This suffices to conclude that M is quasiminimal and that the generic type is determined by $C < x$.

Fix $c_i \in C$ and $a \in M \backslash C$ with $c_i < a$. Note that $x \not\leq c_i$ is small, so $M_j \subset \mathrm{cl}_p(c_i)$ for all $j \lhd i$. Also, $x \perp a$ is large so $\mathrm{cl}_p(a)$ is the union of branches containing a. Since $c_i \in \mathrm{cl}_p(a)$ we have $M_j \subseteq \mathrm{cl}_p(c_i) \subset \mathrm{cl}_p^2(a)$; since $M_j \not\subseteq \mathrm{cl}_p(a)$ we conclude that $\mathrm{cl}_p(a) \neq \mathrm{cl}_p^2(a)$ and cl_p is not a closure operator. Similarly, for any countable $A \subset M$ we can find a, c_i as above much bigger than A, and thus both realizing $p \mid A$. Then $x \perp a \wedge \neg(x \perp c_i) \in p(x)$ witness that $p(x)$ splits over A.

6. A general dichotomy theorem

In this section we outline a dichotomy theorem, which yields Theorem 5.1 (the countably based quasiminimal case), and also subsumes the global strongly regular case (Theorem 3.1). Our set up will also apply to κ-quasiminimal structures, for κ regular (defined in the obvious way). We fix an arbitrary model N (allowing the possibility that $N = \overline{M}$), and a complete 1-type $p(x)$ over N. The operator cl_p defined on subsets of N may not be a closure operator, but generates one:

Definition 6.1. For $X \subset N$ define: $\mathrm{Cl}_p(X) = \bigcup\{\mathrm{cl}_p^n(X) \mid n \in \omega\}$ where $\mathrm{cl}_p^0(X) = X$, $\mathrm{cl}_p^{n+1}(X) = \mathrm{cl}_p(\mathrm{cl}_p^n(X))$.

Note that Cl_p is a closure operator and that cl_p is a closure operator if and only if $\mathrm{cl}_p = \mathrm{Cl}_p$. If N is quasiminimal and p is the generic type, then Cl_p agrees with ccl. However, in the general case we can easily have $\mathrm{Cl}_p(\varnothing) = N$ which is not interesting at all; the interesting case is when N is 'infinite-dimensional.'

Definition 6.2. $A \subseteq N$ is *finitely* Cl_p*-generated over* $B \subset N$ if there is a finite $\bar{a} \subset A$ such that $A \subseteq \mathrm{Cl}_p(B\bar{a})$; if $B = \varnothing$ then we simply say that A is finitely Cl_p-generated.

The interesting case for us is when N itself is not finitely Cl_p-generated, which *will be assumed from now on.* This is already a weak regularity assumption on p, as we will see in the next section where it is proved that p is 'locally strongly regular.' The 'relative smallness' of $A \subset N$ is expressed by: N is not finitely Cl_p-generated over A.

Definition 6.3. A sequence $\{a_i \mid i \in \alpha\}$ is cl_p*-free over* $B \subset N$ if for all $i \leq \alpha$ $a_i \notin \mathrm{cl}_p(B \cup \{a_i \mid j < i\})$; cl_p-free means cl_p-free over $\varnothing$. Similarly Cl_p*-free* sequences are defined.

Definition 6.4. p is *based on* $A \subset N$ if p does not split over A, and N is not finitely Cl_p-generated over A.

Note that Cl_p-free sequences are also cl_p-free. Moreover, if $A \subset N$ and p does not split over A then every cl_p-free sequence over (any domain containing) A is indiscernible, by the standard argument.

Lemma 6.1. *Suppose that $p(x)$ is based on A. Then*

(i) cl_p^A *is a closure operator on N.*

(ii) (N, cl_p^A) *is a pregeometry iff every cl_p-free sequence over A is totally indiscernible.*

PROOF. Without loss of generality assume $A = \varnothing$.

(i) Assuming that cl_p is not a closure operator we will find a nonindiscernible cl_p-free sequence over C for some finite $C \subset N$, which is in contradiction with nonsplitting over $\varnothing$. So suppose that cl_p is not a closure operator. Then there are a finite $C \subset N$ and $a \in N$ such that $a \in \mathrm{cl}_p^2(C) \setminus \mathrm{cl}_p(C)$. Since $a \notin \mathrm{cl}_p(C)$ we have $a \models p \mid C$ so, since N is not finitely Cl_p-generated, there are $a_1, a_2 \in N$ such that (a_1, a_2) is a Cl_p-free sequence over Ca. We will prove that (a, a_1, a_2) is not indiscernible over C; since it is cl_p-free over C we have a contradiction as p does not split over C.

Witness $a \in \mathrm{cl}_p^2(C)$ by a formula $\varphi(x, \bar{b}) \in \mathrm{tp}(a/C\bar{b})$ which is not in $p(x)$, where $\varphi(x, y_1, \ldots, y_n)$ is over C and $(b_1, \ldots, b_n) = \bar{b} \in \mathrm{cl}_p(C)^n$. Choose $\theta_i(y_i) \in \mathrm{tp}(b_i/CA)$ witnessing $b_i \in \mathrm{cl}_p(C)$ (i.e., $\theta_i(y_i) \notin p(y_i)$) and let $x_1 \equiv_\varphi x_2$ denote the formula

$$(\forall \bar{y})\Big(\bigwedge_{1 \leq i \leq n} \theta_i(y_i) \to \Big(\varphi(x_1, \bar{y}) \leftrightarrow \varphi(x_2, \bar{y})\Big)\Big).$$

It is, clearly, over C and we show $a \not\equiv_\varphi a_2$: from $\models \varphi(a, \bar{b})$ (witnessing $a \in \mathrm{cl}_p^2(C)$) and $a_2 \notin \mathrm{Cl}_p(C)$ we derive $\models \neg\varphi(a_2, \bar{b})$ and thus $\bar{b}$ witnesses $a \not\equiv_\varphi a_2$. On the other hand, since all realizations of θ_i's are in $\mathrm{cl}_p(C)$, and since $\mathrm{tp}\big(a_1/\mathrm{cl}_p(C)\big) = \mathrm{tp}\big(a_2/\mathrm{cl}_p(C)\big)$, we have $a_1 \equiv_\varphi a_2$. Therefore (a, a_1, a_2) is not indiscernible over C.

(ii) Having proved (i), the proof of Lemma 3.1(ii) goes through: Let A_0 be finite, let $(a_1, \ldots, a_n) \in A_0^n$ be cl_p-free over $\varnothing$ such that $\mathrm{cl}_p(A_0) = \mathrm{cl}_p(a_1, \ldots, a_n)$, let $a \models p \mid A_0$ and let $b \in N$. Then $(a_1, \ldots, a_n, a)$ is cl_p-free over $\varnothing$ and $\mathrm{cl}_p(aA_0) = \mathrm{cl}_p(a_1, \ldots, a_n, a)$. We have:

$$\begin{aligned} b \notin \mathrm{cl}_p(aA_0) \quad &\text{iff} \quad b \notin \mathrm{cl}_p(a_1, \ldots, a_n, a) \\ &\text{iff} \quad (a_1, \ldots, a_n, a, b) \text{ is } \mathrm{cl}_p\text{-free over } \varnothing \\ &\text{iff} \quad (a_1, \ldots, a_n, b, a) \text{ is } \mathrm{cl}_p\text{-free over } \varnothing \\ &\text{iff} \quad (b \notin \mathrm{cl}_p(A_0) \text{ and } a \notin \mathrm{cl}_p(A_0 b)). \end{aligned}$$

□

Theorem 6.1. *Suppose that $p(x)$ is based on $A \subset N$. Then cl_p^A is a closure operator and exactly one of the following two holds:*

(1) *Every cl_p-free sequence over A is totally indiscernible; in this case (N, cl_p^A) is a homogeneous pregeometry, p is definable over A, $\bar{p}$ (its unique global heir) is generically stable and $(\bar{p}(x), x = x)$ is strongly regular.*

(2) *Otherwise. In which case there is a finite extension A_0 of A and an A_0-definable partial order $\leq$ such that every* cl_p*-free sequence over A_0 is strictly increasing.*

PROOF. To simplify the notation assume $A = \varnothing$, it will not affect the generality. First suppose that every cl_p-free sequence over $\varnothing$ is symmetric. Then, by Lemma 6.1(i), cl_p is a closure operator and, by Lemma 6.1(ii), it is a pregeometry operator. Since N is not finitely cl_p-generated it is infinite-dimensional so (N, cl_p) is a homogeneous pregeometry and the conclusion follows from Proposition 4.1.

Now suppose otherwise. Then over some finite A_0 there is a cl_p-free sequence (a, b) over A_0 such that $\mathrm{tp}(a, b/A_0) \neq \mathrm{tp}(b, a)/A_0$. So for some $\phi(x, y)$ over A_0, we have:

(1) $a \models p \mid \mathrm{cl}_p(A_0)$, $b \models p \mid \mathrm{cl}_p(A_0, a)$ and $\models \phi(a, b)$;

(2) $\models \phi(x, y) \to \neg\phi(y, x)$.

We *claim* $\phi(N, a) \subsetneq \phi(N, b)$. To prove it, first note that $\models \phi(a, b) \wedge \neg\phi(b, a)$ and $b \notin \mathrm{cl}_p(Aa)$ imply $\neg\phi(x, a) \in p(x)$, so $\phi(N, a) \subseteq \mathrm{cl}_p(A_0 a)$. Now suppose $d \in \phi(N, a)$ and we will show $d \in \phi(N, b)$. By the above $d \in \mathrm{cl}_p(A_0 a)$. We have two possibilities for d. The first is $d \in \mathrm{cl}_p(A_0)$, where $a \equiv b\,(\mathrm{cl}_p(A_0))$ and $\models \phi(d, a)$ imply $\models \phi(d, b)$ and we're done. The second is $d \notin \mathrm{cl}_p(A_0)$. Then a and d realize $p \mid \mathrm{cl}_p(A_0)$ and, since p does not split over A_0, we have $(\phi(a, x) \leftrightarrow \phi(d, x)) \in p(x)$; since $d \in \mathrm{cl}_p(A_0 a)$ we have $(\phi(a, x) \leftrightarrow \phi(d, x)) \in p \mid \mathrm{cl}_p(aA_0)$ and, since $b \models p \mid \mathrm{cl}_p(A_0 a)$, we get $\models \phi(a, b) \leftrightarrow \phi(d, b)$. Thus $\models \phi(d, b)$. This proves $\phi(N, a) \subseteq \phi(N, b)$.

Finally, the asymmetry of $\phi(x, y)$ implies $\models \neg\phi(a, a)$ so $a \notin \phi(N, a)$ and $a \in \phi(N, b) \setminus \phi(N, a)$; this proves that the inclusion is proper.

'$\phi(N, x) \subsetneq \phi(N, y)$' is an A_0-definable strict-ordering relation $x < y$ on N and we have $a < b$. By nonsplitting, $a' < b'$ is true whenever (a', b') is cl_p-free over A_0, so every cl_p-free sequence over A_0 is increasing. □

7. Local regularity

Here we introduce and study "locally strongly regular types" and give applications to quasiminimal structures (see Corollary 7.2).

Given an arbitrary model M, $p(x) \in S_1(M)$ and $\phi(x) \in p$ we can ask when p extends to a global M-invariant $\bar{p}$, such that $(\bar{p}, \phi)$ is strongly regular. When T is stable the situation is well understood, but we are interested in the general case. Our definition of local strong regularity is actually a necessary condition for such an extension to exist. It is convenient to work with types over arbitrary sets (not just models).

Definition 7.1. A nonisolated type $p(x) \in S_1(A)$ is *locally strongly regular via* $\phi(x) \in p(x)$ if $p(x)$ has a unique extension over $A\bar{b}$ whenever $\bar{b} \in \overline{M}$ is a finite tuple of realizations of $\phi(x)$ no element of which realizes p.

Proposition 7.1. *Suppose that $p(x) \in S_1(M)$ is definable and locally strongly regular via $\phi(x) \in p(x)$, and let $\bar{p}(x)$ be its global heir. Then $\big(\bar{p}(x), \phi(x)\big)$ is strongly regular (and of course definable).*

PROOF. Suppose that $\big(\bar{p}(x), \phi(x)\big)$ is not strongly regular. Then there are $B = M\bar{b}$ and $a \in \mathrm{cl}_{\bar{p}}(B) \cap \phi(\overline{M})$ and $c \models \bar{p} \mid B$ such that c does not realize $\bar{p} \mid Ba$. Witness $a \in \mathrm{cl}_{\bar{p}}(B)$ by $\theta(y, \bar{z})$ which is over M, implies $\phi(y)$, and $\models$

$\theta(a,\bar{b}) \wedge \neg(d_p\theta)(\bar{b})$ (where d_p is the defining schema of p). Similarly, find $\varphi(x,y,\bar{z})$ over M such that $\models \varphi(c,a,\bar{b}) \wedge \neg(d_p\varphi)(a,\bar{b})$.

$$\models (\exists y)\big(\theta(y,\bar{b}) \wedge \neg(d_p\theta)(\bar{b}) \wedge \varphi(c,y,\bar{b}) \wedge \neg(d_p\varphi)(y,\bar{b})\big).$$

Since $\mathrm{tp}(c/M\bar{b})$ is an heir of $p(x)$ there is $\overline{m} \in M$ and a' such that

$$\models \theta(a',\overline{m}) \wedge \neg(d_p\theta)(\overline{m}) \wedge \varphi(c,a',\overline{m}) \wedge \neg(d_p\varphi)(a',\overline{m}).$$

The first two conjuncts witness $a' \in \phi(\overline{M}) \setminus p(\overline{M})$ while the last two witness that c is not a realization of $\bar{p} \mid Ma'$. A contradiction. □

For the sake of this section we will call a sequence $(a_i \mid i \in \alpha)$ a *coheir sequence over C* if $\mathrm{tp}(a_j/M(a_i \mid i < j))$ is finitely satisfiable in C for any $j < \alpha$. In particular $\mathrm{tp}(a_0/C)$ is finitely satisfiable in C.

Proposition 7.2. *Suppose that $p(x) \in S_1(C)$ is locally strongly regular via $x = x$ and that there exists an infinite, totally indiscernible (over C) sequence of realizations of p which is a coheir sequence over C. Then p has a global C-invariant extension $\bar{p}$ such that $(\bar{p}(x), x = x)$ is strongly regular and generically stable.*

PROOF. Let $I = \{a_i : i \in \omega\}$ be a totally indiscernible (over C) sequence of realizations of p which is a coheir sequence in C and let $p_n(x) = \mathrm{tp}(a_{n+1}/Ca_1 \ldots a_n)$. We will first prove that each $p_n(x)$ is locally strongly regular via $x = x$. Suppose, on the contrary, that $p_n(x)$ is not locally strongly regular. Then there are $b_1 \ldots b_k = \bar{b}$, with none realizing $p_n(x)$, such that p_n has at least two extensions in $S_1(C\bar{a}\bar{b})$ (here $\bar{a} = a_1 \ldots a_n$). Let $\varphi(x,\bar{z},\bar{y})$ be over C and such that both $\varphi(x,\bar{a},\bar{b})$ and $\neg\varphi(x,\bar{a},\bar{b})$ are consistent with $p_n(x)$.

Choose $\theta_i(y_i,\bar{a}) \in \mathrm{tp}(b_i/C\bar{a})$ witnessing that b_i does not realize $p_n(y_i)$ and let $\phi(x_1,x_2,\bar{a})$ be

$$(\exists \bar{y})\Big(\bigwedge_{1\le i\le n} \big(\theta_i(y_i,\bar{a}) \wedge \neg\theta_i(x_2,\bar{a})\big) \wedge \neg\big(\varphi(x_1,\bar{a},\bar{y}) \leftrightarrow \varphi(x_2,\bar{a},\bar{y})\big)\Big).$$

It is, clearly, over C and we show $\models \phi(a_{n+2},a_{n+1},\bar{a})$. By our assumptions on φ and $\bar{b}$, there is $\bar{b}' \equiv \bar{b}(C\bar{a})$ such that $\models \neg(\varphi(a_{n+1},\bar{a},\bar{b}) \leftrightarrow \varphi(a_{n+1},\bar{a},\bar{b}'))$. Also $\models \bigwedge_{1\le i\le n}(\theta_i(b_i,\bar{a}) \wedge \theta_i(b_i',\bar{a}) \wedge \neg\theta_i(a_{n+1},\bar{a}))$. Thus for any $c \in C$ either $\bar{b}$ or $\bar{b}'$ in place of $\bar{y}$ witnesses $\models \phi(c,a_{n+1},\bar{a})$ and, since $\mathrm{tp}(a_{n+2}/C\bar{a}a_{n+1})$ is finitely satisfiable in C, we conclude $\models \phi(a_{n+2},a_{n+1},\bar{a})$.

By total indiscernibility, $\mathrm{tp}(\bar{a}/Ca_{n+1}a_{n+2})$ is finitely satisfiable in C, so there are $\bar{c} \in C$ and $\bar{d}$ such that

$$\bigwedge_{1\le i\le n} \big(\theta_i(d_i,\bar{c}) \wedge \neg\theta_i(a_{n+1},\bar{c})\big) \wedge \neg\big(\varphi(a_{n+2},\bar{c},\bar{d}) \leftrightarrow \varphi(a_{n+1},\bar{c},\bar{d})\big).$$

$\bigwedge_{1\le i\le n}(\theta_i(d_i,\bar{c}) \wedge \neg\theta_i(a_{n+1},\bar{c}))$ witnesses that no d_i realizes p, and $\neg(\varphi(a_{n+2},\bar{c},\bar{d}) \leftrightarrow \varphi(a_{n+1},\bar{c},\bar{d}))$ witnesses that p does not have a unique extension over $C\bar{d}$; a contradiction.

Therefore each $p_n(x)$ is locally strongly regular via $x = x$. It easily follows that $p_I(x) = \bigcup_{n\in\omega} p_n(x) \in S_1(CI)$ is locally strongly regular via $x = x$ as well. Moreover, the same is true whenever $I' \subset \overline{M}$ is an indiscernible (over C) extension of I; then I' is also totally indiscernible over C and $p_{I'}(x) \in S(CI')$, defined by:

$$\phi(x, \bar{a}') \in p_{I'} \text{ (where } \phi(x, \bar{y}) \text{ is over } C \text{ and } \bar{a}' \in I') \\ \text{iff} \quad \phi(x, \bar{a}) \in p_I(x) \text{ for some } \bar{a} \in I \text{ with } \bar{a} \equiv \bar{a}' \quad (C)$$

is locally strongly regular via $x = x$ and does not split over C.

Now let $J \subset \overline{M}$ be a maximal indiscernible (over C) extension of I. By total indiscernibility and maximality of J no element of $\overline{M} \setminus CJ$ realizes $p_J(x)$, so the local strong regularity implies that p_J has a unique global extension $\bar{p}$. Since $p_J(x)$ does not split over C, it is easy to conclude that $\bar{p}$ is C-invariant. To show that $(\bar{p}(x), x = x)$ is strongly regular let $A \supseteq CI$ be small and we will prove that $\bar{p} \mid A \vdash \bar{p} \mid Ab$ for any b which does not realize $\bar{p} \mid A$. So fix such a b and let $A_0 \subset A$ be maximal such that $I_0 = I \cup A_0$ is a Morley sequence in $\bar{p}$ over C. Then $\bar{p} \mid CI_0(x)$ is $p_{I_0}(x)$ (as defined above for $I' = I_0$), so is locally strongly regular via $x = x$. The maximality of A_0 implies that no $a \in A$ realizes $\bar{p} \mid CI_0$ so, by local strong regularity, $\bar{p} \mid CI_0 \vdash \bar{p} \mid A$. In particular, since b does not realize $\bar{p} \mid A$, we have that b does not realize $\bar{p} \mid CI_0$ either. Thus no element of Ab realizes $\bar{p} \mid CI_0(x)$ so $\bar{p} \mid CI_0(x) \vdash \bar{p} \mid Ab$; in particular $\bar{p} \mid A \vdash p \mid Ab$ and $(\bar{p}(x), x = x)$ is strongly regular.

Having proved that $(\bar{p}(x), x = x)$ is strongly regular we can apply Corollary 3.1 to $\overline{M}$ and $\bar{p}$. Note that I is a Morley sequence in $\bar{p}$ over C, and is totally indiscernible, so $\bar{p}$ is symmetric, and hence generically stable. □

Our next goal is to prove that the generic type of a quasiminimal structure is locally strongly regular via $x = x$. This we will do in a more general situation, for any M and $p \in S_1(M)$ for which M is not finitely Cl_p-generated (with notation as in Section 6).

Proposition 7.3. *Suppose $p \in S_1(M)$ and M is not finitely Cl_p-generated. Then:*

(i) *Whenever $I \subseteq M$ is a maximal Cl_p-free sequence then $(p \mid I)(x)$ is locally strongly regular via $x = x$, and $(p \mid I)(x) \vdash p(x)$.*

(ii) *$p(x)$ is locally strongly regular via $x = x$.*

Proof. (i) Let $I \subseteq M$ be a maximal Cl_p-free sequence. We will prove that $(p \mid I)(x)$ is locally strongly regular via $x = x$. After passing to a subset and rearranging I if necessary we may assume that $I = \{a_i \mid i < \alpha\}$ where α is a limit ordinal.

Suppose, on the contrary, that there are $d_1, d_2 \in \overline{M}$ realizing $p \mid I$, a formula $\phi(x, \bar{y})$, and a tuple $\bar{b} = b_1 b_2 \dots b_n \in \overline{M}^n$ such that none of b_i's realize $p \mid I$ and:

$$\models \neg\phi(d_1, \bar{b}) \wedge \phi(d_2, \bar{b})).$$

Choose $\theta_i(y_i) \in \mathrm{tp}(b_i/I)$ such that $\theta_i(x) \notin p|I(x)$. To simplify notation we will assume that $\phi(x, \bar{y})$, as well as each $\theta_i(x)$, are over $\varnothing$ (absorbing a few parameters from I into the language won't hurt the generality). At least one of

$$\{i < \alpha \mid \models \phi(a_i, \bar{b})\} \quad \text{and} \quad \{i < \alpha \mid \models \neg\phi(a_i, \bar{b})\}$$

is cofinal in α. Assume the first one is cofinal and let $I_0 = \{a_i \mid \models \phi(a_i, \bar{b})\}$. The cofinality implies: first that $p|I$ is finitely satisfiable in I_0 (so $\mathrm{tp}(d_1/I)$ is finitely satisfiable in I_0); second, that $p(x) \cup \{\phi(x, \bar{b}\}$ is finitely satisfiable in I_0, so there is a type containing it in $S_1(Id_1\bar{b})$ which is finitely satisfiable in I_0; wlog, let d_2 realizes it. Thus, both $\mathrm{tp}(d_1/I)$ and $\mathrm{tp}(d_2/Id_1\bar{b})$ are finitely satisfiable in I_0.

$$\models (\exists \bar{y})\Big(\bigwedge_{1\le i\le n} \theta_i(y_i) \wedge \neg\phi(d_1,\bar{y}) \wedge \phi(d_2,\bar{y})\Big).$$

Since $\mathrm{tp}(d_2/Id_1)$ is finitely satisfiable in I_0, there is $a_i \in I_0$ such that:

$$\models (\exists \bar{y})\Big(\bigwedge_{1\le i\le n} \theta_i(y_i) \wedge \neg\phi(d_1,\bar{y}) \wedge \phi(a_i,\bar{y})\Big).$$

Since $\mathrm{tp}(d_1/M)$ is finitely satisfiable in I_0, there is $a_j \in I_0$ such that:

$$\models (\exists \bar{y})\Big(\bigwedge_{1\le i\le n} \theta_i(y_i) \wedge \neg\phi(a_j,\bar{y}) \wedge \phi(a_i,\bar{y})\Big).$$

Finally, since $a_i, a_j \in M$ there is $\bar{b}' = b_1'b_2'\dots b_n' \in M^n$ satisfying:

$$\models \bigwedge_{1\le i\le n} (\theta_i(b_i') \wedge \neg\phi(a_j,\bar{b}') \wedge \phi(a_i,\bar{b}')).$$

But $\bigwedge_{1\le i\le n}\theta_i(b_i')$ implies $\bar{b}' \subset \mathrm{cl}_p(\varnothing)$ and thus $\mathrm{tp}(a_i/\mathrm{cl}_p(\varnothing)) \ne \mathrm{tp}(a_j/\mathrm{cl}_p(\varnothing))$. A contradiction. Therefore $(p \mid I)(x)$ is locally strongly regular via $x = x$. The maximality of I implies $M = \mathrm{cl}_p(I)$ so, by local strong regularity of $p \mid I$, we have $(p \mid I)(x) \vdash p(x)$.

(ii) $p \mid I$ is locally strongly regular and $(p \mid I)(x) \vdash p(x)$ implies that $p(x)$ is locally strongly regular via $x = x$. □

Corollary 7.1. *The generic type of a quasiminimal structure is locally strongly regular via $x = x$.*

Theorem 7.1. *Suppose that $p \in S_1(M)$ and that (M, Cl_p) is an infinite-dimensional pregeometry. Then p is definable and $(\bar{p}(x), x = x)$ is strongly regular and generically stable (where $\bar{p}$ is the unique global heir of p).*

PROOF. First note that, by Proposition 7.3(ii), p is locally strongly regular via $x = x$. Now we will find an infinite, totally indiscernible sequence of realizations of p which is a coheir sequence over M. Towards this aim we first prove:

Claim. *If $A \subset M$ and $a, b \in M$ are Cl_p-independent over A then* $\mathrm{tp}(a,b/A) = \mathrm{tp}(b,a/A)$.

PROOF. It suffices to prove the claim for A finite. Suppose, on the contrary, that $\phi(x,y)$ is over A and $\models \phi(a,b) \wedge \neg\phi(b,a)$. Since $b \notin \mathrm{Cl}_p(Aa)$ we have $\phi(a,x) \in p(x)$ and, since $a \notin \mathrm{Cl}_p(Ab)$, we have $\neg\phi(b,x) \in p(x)$. By infinite-dimensionality and since Aab is finite, there is $e \in M \setminus \mathrm{Cl}_p(Aab)$. Then e realizes $p \mid Aab$ so $\models \phi(a,e) \wedge \neg\phi(b,e)$. But $a \notin \mathrm{Cl}_p(Ae)$ implies $\phi(x,e) \in p(x)$ and $b \notin \mathrm{Cl}_p(Ae)$ implies $\neg\phi(x,e) \in p(x)$. A contradiction. The claim is proved. □

Now, let $I \subseteq M$ be a maximal Cl_p-free sequence. We can find an infinite M-indiscernible sequence $J = (a_i : i < \omega)$ such that for all i any formula satisfied by a_i over $M \cup \{a_j : j < i\}$ is satisfied by some element of I. Using the claim, we conclude that J is totally indiscernible over M. Now apply Proposition 7.2. □

Corollary 7.2. *Suppose that M is quasiminimal and let $p(x) \in S_1(M)$ be the "generic type" of M (consisting of formulas defining uncountable sets). Suppose that (M, ccl) is a pregeometry. Then for some countable* ccl*-free sequence $A \subset M$, p is based on A, and after adding constants for elements of A, is a homogeneous. Then p is countably based and symmetric. Moreover, as a base set we can choose (some) infinite, countable,* ccl*-free sequence.*

PROOF. Clearly, (M, ccl) is infinite-dimensional so, by Theorem 7.1, the generic type p is definable and symmetric. Countable baseness and symmetry follow. To prove the 'moreover' part, let $M_0 \prec M$ be countable, of infinite ccl-dimension, such that p is definable over M_0 and such that p is the heir of $p \mid M_0$. Further, let $I \subset M_0$ be a maximal ccl-free subset of M_0. Then we can apply Proposition 7.3(i) to M_0 and I: $p \mid I \vdash p \mid M_0$. In particular, p is invariant over I, so I is a base for p. □

Thus if M is quasiminimal and (M, ccl) is a pregeometry we know that the generic type is definable, but defining schema may involve parameters. In general parameters are needed as the following example shows.

Example 7.1. Quasiminimal structure where ccl is a pregeometry operator but the generic type is not $\varnothing$-invariant: This is a slight variation of an example from [2]. We have two unary predicates U, V and binary relation symbols E and f. The domain M is the disjoint union $U(M) \cup V(M)$. $U(M)$ is uncountable and E is an equivalence relation on it having $\aleph_0$ many classes with all of them but one of size $\aleph_0$. $V(M)$ is countable and contains 'names' for E-classes, while $f\colon U(M) \to V(M)$ is an onto projection (and $f(a) = f(a')$ iff $E(a, a')$). Let $p(x)$ be the generic type: $p(x)$ says that x is in the uncountable class. Then $\mathrm{cl}_p(\varnothing) = V(M)$, while $\mathrm{cl}_p^2(\varnothing) = \mathrm{Cl}_p(\varnothing)$ contains also all the countable E-classes. Also $\mathrm{Cl}_p(X) = \mathrm{Cl}_p(\varnothing) \cup X$ so (M, Cl_p) is an infinite-dimensional pregeometry. But p is not $\varnothing$-invariant: all the elements of $U(M)$ have the same type while $E(x, a) \wedge \neg E(x, b) \in p(x)$ for a in the uncountable class and b in a countable one. Therefore p is not definable over $\varnothing$.

Theorem 7.2. *Suppose that $G \subseteq M$ is a definable group and $p(x) \in S_G(M)$ is locally strongly regular via "$x \in G$". Then:*

(i) *$p(x)$ is both left and right translation invariant (and in fact invariant under definable bijections).*
(ii) *A formula $\phi(x)$ is in $p(x)$ iff two left (right) translates of $\phi(x)$ cover G iff finitely many left (right) translates of $\phi(x)$ cover G. (Hence $p(x)$ is the unique generic type of G.)*
(iii) *$p(x)$ is definable over $\varnothing$ and G is connected.*
(iv) *$(\bar{p}(x), \text{“}x \in G\text{”})$ is strongly regular and $\overline{G}$ is a definable-regular group. (Here $\bar{p}$ is the unique heir of $p(x)$ and $\overline{G} \subseteq \overline{M}$ is defined by "$x \in G$").*

PROOF. (i) Suppose that $f\colon \overline{G} \to \overline{G}$ is an M-definable bijection and $a \models p$. Since $p \vdash p \mid (M, f(a))$ is not possible, by local strong regularity we get $f(a) \models p$. Thus p is invariant under f.

(ii) The local strong regularity of $p(x)$ implies that whenever $g, g' \in \overline{G}$ do not realize p then $g \cdot g'$ does not realize p either. It follows that $a \cdot g \models p$ whenever $a \models p$ and $g \in \overline{G}$ does not realize p. Thus:

$$\phi(x) \in p(x) \quad \text{iff} \quad (\forall y \in \overline{G})\big(\neg\phi(y) \to \phi(y \cdot x)\big) \in p(x),$$

and $\phi(x) \in p(x)$ iff $\phi(\overline{G}) \cup a^{-1} \cdot \phi(\overline{G}) = \overline{G}$.

(iii) follows immediately from (ii), and then (iv) follows from Proposition 7.1. $\square$

References

1. E. Hrushovski and A. Pillay, *On NIP and invariant measures*, available at arXiv:0710.2330.
2. M. Itai, A. Tsuboi, and K. Wakai, *Construction of saturated quasi-minimal structure*, J. Symbolic Logic **69** (2004), no. 1, 9–22.
3. J. Kirby, *Exponential algebraicity in exponential fields*, Bull. Lond. Math. Soc. **42** (2010), no. 5, 879–890.
4. M. G. Peretjat′kin, *Complete theories with a finite number of countable models*, Algebra i Logika **12** (1973), 550–576 (Russian); English transl., Algebra and Logic **12** (1973), 310–326.
5. S. Shelah, *Dependent first order theories, continued*, Israel J. Math. **173** (2009), 1–60.
6. P. Tanović, *Minimal first order structures*, Ann. Pure Appl. Logic, to appear.
7. A. J. Wilkie, *Some local definability theory for holomorphic functions*, Model Theory with Applications to Algebra and Analysis. Vol. 1 (Z. Chatzidakis, D. Macpherson, A. Pillay, and A. Wilkie, eds.), London Math. Soc. Lecture Note Ser., vol. 349, Cambridge Univ. Press, Cambridge, 2008, pp. 197–213.
8. B. Zilber, *Dimensions and homogeneity in mathematical structures*, Connections between Model Theory and Algebraic and Analytic Geometry (A. Macintyre, ed.), Quad. Mat., vol. 6, Dept. Math., Seconda Univ. Napoli, Caserta, 2000, pp. 131–148.

University of Leeds, Leeds, UK
E-mail address: A.Pillay@leeds.ac.uk

Mathematical Institute SANU, Belgrade, Serbia
E-mail address: tane@mi.sanu.ac.rs

Centre de Recherches Mathématiques
CRM Proceedings and Lecture Notes
Volume **53**, 2011

Indexed Lawvere Theories for Local State

John Power

ABSTRACT. Monads for global state and local state have been used to provide semantics of programming languages for many years. There is a computationally natural presentation of an ordinary Lawvere theory that corresponds to the monad on Set for global state, inevitably called the Lawvere theory for global state. Here, we introduce a notion of indexed Lawvere theory and use it to give a Lawvere-style account of local state, extending the theorem for global state to local state. En route, we develop the notion of comodel of a Lawvere theory and exploit a universal characterization of the category of worlds for local state. Ultimately, we give both syntactic and semantic characterizations of the operation *block* that allows one to move between worlds and use them to characterize the monad for local state.

1. Introduction

Monads have been used to model aspects of computational effects since the late 1980s [15,16]. The equivalence between finitary monads and Lawvere theories was elaborated in the 1960s. So one might reasonably expect that it would have been clear from the outset that computational effects could be captured in either way. But, as a matter of historical fact, that is not true, as explained in [6].

The relationship between monads and Lawvere theories was an observation of no practical value in regard to computational effects until it was recognized, in 2001, that there is a computationally natural presentation of the Lawvere theory for global state [20]: we recall the details in Section 2. That recognition, followed quickly by corresponding recognition for all the other leading examples except for continuations, opened the way to the modeling of deeper aspects of computational effects [3–5, 19, 22, 23, 26]. This paper may be seen as part of the development. Specifically, we extend the notion of Lawvere theory to one of indexed Lawvere theory in order to characterize the monad for local state in a way that directly and naturally extends the Lawvere theory for global state: it is not as simple as asking

---

2010 *Mathematics Subject Classification.* 18C10, 18C15, 18C20, 18C50, 18D30.

This document in an output from the PMI2 Project funded by the UK Department for Innovation, Universities and Skills (DIUS) for the benefit of the Japanese Higher Education Sector and the UK Higher Education Sector. The views expressed are not necessarily those of DIUS, nor British Council.

This is the final form of the paper.

for all models of an indexed Lawvere theory, but the notion of indexed Lawvere theory is fundamental.

The monad on Set for global state, which was suggested to us by Peter O'Hearn, is $T_S - = (- \times S)^S$, where S is a set of states, typically given by V^{Loc} for a countable set V of values and a finite set Loc of locations. Corresponding to that, the countable Lawvere theory L_S for global state is freely generated by operations lookup: $V \to \mathrm{Loc}$ and update: $1 \to \mathrm{Loc} \times V$ subject to computationally natural equations that we recall in Section 2. In order to extend from global to local state, one replaces Set by the functor category [Inj, Set], where Inj is the category of finite sets and injections [17, 20, 21], or equivalently, the category of natural numbers and monomorphisms. The monad for local state, on [Inj, Set], is

$$(T_{LS}X)_n = \Big(\int^{m \in (n/\mathrm{Inj})} (Sm \times X_m)\Big)^{Sn}$$

where $Sn = V^n$ for a given set of values V; we describe its behavior on maps in Section 6. The idea of the monad is that in a state with n locations, a computation can create $m - n$ new locations and return a value, e.g., a function, that depends on them, cf. [11, 12]. In the case $V = 1$ this reduces to the monad for local names in [28], which was introduced with an eye to the π-calculus.

The monad T_{LS} enriches over both the Cartesian closed structure of [Inj, Set] and the convolution symmetric monoidal structure on [Inj, Set] generated by the sum of natural numbers. But with neither enrichment does the monad correspond to an enriched Lawvere theory for which we have been able to provide a computationally natural presentation. In [20], with an oversight corrected in [21], we did give an algebraic-like presentation of the monad T_{LS} extending that for global state, but it used an uncomfortable and uninvestigated hybrid of the two enrichments and it did not separate syntax from semantics. We managed somewhat better in [25], which at least includes a definition of indexed Lawvere theory, but even that paper had no presentation that separated syntax from semantics. That is what we provide here, gradually moving towards it through the course of the paper.

In order to extend global structure, such as that for global state, to local structure, one must first extend a Lawvere theory L to an indexed version of the same theory, i.e., if L has an operation $f\colon a \to b$, then a local version also needs to have such an operation, but it must have it at each world, and it must be natural in the size of the world. For instance, for state, if one has n locations, there are n possible places at which one may either lookup or update. So we first seek a general construct that extends the operations of a global computational effect to become operations of an induced local effect. The canonical symmetric monoidal structure of Law allows us to do that: Inj is the free symmetric monoidal category for which the unit is the initial object (sometimes called a *coaffine* category [13, 18]) on 1; Law is another symmetric monoidal category with unit the initial object; and so the universal property allows us to extend any Lawvere theory L to an Inj-indexed such theory $L_{\otimes}$, uniformly allowing us to extend global operations to local operations. The details are in Section 3.

This example of an Inj-indexed Lawvere theory motivates an axiomatic definition of indexed Lawvere theory together with one of a model of an indexed Lawvere theory. Those definitions in turn motivate a subtle analysis of the relationship between D-indexed Lawvere theories for an arbitrary small category D and monads on $[D, \mathrm{Set}]$, upon which we check what our abstract theory tells us in the example of state. This constitutes Section 4. Although we focus on state in this paper, the example of π-calculus may well also merit study in this light, cf. [28]. Dependent algebra seems likely to involve a notion of indexed Lawvere theory too, but it may be more complex than that we introduce here.

We next add an operation block: the central fact of locality is the capacity to create a new world. Semantically, block is modeled by a map of the form

$$M(n+1) \to M(n)^V$$

uniform in n and subject to natural coherence conditions, where $M(n)$ is a model of global state with n locations. So we first describe a general construct whose semantics extends a model $M(n)$ at world n to a model of the form $M(n)^V$ at world $n+1$. To do that, we observe that V is the final comodel of the ordinary Lawvere theory L_V for global state with value set V and one location. We then observe that exponentiating by the comodel V, the set $M(n)^V$ has $n+1$ *lookup* and *update* structures on it, n of them because $M(n)$ has n such structures, with an additional one determined by the comodel structure of V. So we develop the idea of exponentiation by a comodel of a Lawvere theory in Section 5.

Having put $M(n+1)$ and $M(n)^V$ into the same category, we are in a position to analyze block. Some of the coherence conditions on *block* are equivalent to naturality. The remaining two conditions can be expressed as a unit condition and a symmetry condition. We describe the conditions, give semantics for *block*, and use it to characterize the monad T_{LS} for local state in Section 6.

Having modeled the semantics of *block*, we seek corresponding syntax. Specifically, we seek a construct that extends a Lawvere theory L for a global effect to include a syntactic correlate of *block*, subject to natural equations, for which a model is as above. We describe such syntax in Section 7 and end the paper by characterizing the monad T_{LS} for local state in terms of the models of this extended notion of Lawvere theory.

As mentioned above, this paper completes the goal enunciated but not achieved in the conference paper [25]. The originality of this paper lies primarily in its development of syntax rather than semantics.

2. Global state

There are a few mildly different albeit equivalent definitions of Lawvere theory in the literature, so we spell out one definition for definiteness of the paper. Let Nat be the category of natural numbers and all maps between them. It has strictly associative finite coproducts given by the sum of natural numbers. Consequently the category $\mathrm{Nat}^{\mathrm{op}}$ has strictly associative finite products.

Definition 2.1. A *Lawvere theory* consists of a small category L with strictly associative finite products and an identity-on-objects strict finite-product preserving functor $J\colon \mathrm{Nat}^{\mathrm{op}} \to L$. The Lawvere theory is typically denoted by L. A *map* of Lawvere theories from L to L' is a functor from L to L' that commutes with J and J': it necessarily strictly preserves finite products. A *model* of L in any category C with finite products is a finite product preserving functor $M\colon L \to C$.

This definition forces the objects of a Lawvere theory L to be exactly the natural numbers. The definition of model specifically does not require strict preservation of finite products. For any Lawvere theory L, the models of L together with all natural transformations form a category $\mathrm{Mod}(L, C)$. If C is locally presentable, the functor $\mathrm{ev}_1\colon \mathrm{Mod}(L, C) \to C$ has a left adjoint, yielding a monad T_L on C. The category $\mathrm{Mod}(L, C)$ is then coherently equivalent to the category T_L-Alg. Taking C to be Set, the monads thus arising are exactly the finitary monads on Set. We denote the category of Lawvere theories by Law.

It is routine to generalize the definition of Lawvere theory to allow for countable arities rather than finite ones: one systematically replaces finiteness by countability, replacing Nat by the category $\aleph_1$, a skeleton of the category of countable (including the possibility of finite) sets and all functions between them. We denote the category of countable Lawvere theories by Law_c.

It is also routine to generalize the notion of Lawvere theory to that of enriched Lawvere theory, the case of primary interest in computing being enrichment in the Cartesian closed category ωCpo, as explained in detail in [4].

The key example that motivated refinement of Moggi's modeling of computational effects by monads was given by global state [20]. Moggi modeled global state by use of the monad $T_S - = (- \times S)^S$ on Set for a countable set S of states. This monad has countable rank, so necessarily corresponds to a countable Lawvere theory. But the Lawvere theory only became interesting when we could see a computationally natural presentation of it as follows [20]:

Example 2.2. Let V be a countable set of values, and let Loc be a finite set of locations. Put $S = V^{\mathrm{Loc}}$, so S is the set of states with Loc locations taking values in V. The countable Lawvere theory L_S for state is freely generated by operations $l\colon V \to \mathrm{Loc}$ and $u\colon 1 \to \mathrm{Loc} \times V$ subject to equations expressed in terms of models in any category C with countable products as follows: a model in C consists of

- an object A of C
- a lookup map $l\colon A^V \to A^{\mathrm{Loc}}$, and
- an update map $u\colon A \to A^{\mathrm{Loc} \times V}$

subject to commutativity of two classes of diagrams. First, we have four interaction diagrams (NB: Paul-André Melliès has just shown that the second axiom is redundant!) as follows:

$$\begin{array}{ccccc}
A & \xrightarrow{u} & A^{\mathrm{Loc}\times V} & \xrightarrow{\cong} & (A^V)^{\mathrm{Loc}} \\
{\scriptstyle A^t}\downarrow & & & & \downarrow{\scriptstyle l^{\mathrm{Loc}}} \\
A^{\mathrm{Loc}} & \xleftarrow{A^\delta} & A^{\mathrm{Loc}\times\mathrm{Loc}} & \xleftarrow{\cong} & (A^{\mathrm{Loc}})^{\mathrm{Loc}}
\end{array}$$

where $\delta\colon \mathrm{Loc} \to \mathrm{Loc} \times \mathrm{Loc}$ and $t\colon \mathrm{Loc} \to 1$ are the diagonal and terminal maps, and the lower unlabeled isomorphism matches the outer Loc of $(A^{\mathrm{Loc}})^{\mathrm{Loc}}$ with the first Loc of $A^{\mathrm{Loc}\times\mathrm{Loc}}$,

$$\begin{array}{ccccc}
(A^V)^V & \xrightarrow{l^V} & (A^{\mathrm{Loc}})^V & \xrightarrow{\cong} & (A^V)^{\mathrm{Loc}} \\
{\scriptstyle \cong}\downarrow & & & & \downarrow{\scriptstyle l^{\mathrm{Loc}}} \\
A^{V\times V} & & & & (A^{\mathrm{Loc}})^{\mathrm{Loc}} \\
{\scriptstyle A^\delta}\downarrow & & & & \downarrow{\scriptstyle \cong} \\
A^V & \xrightarrow[l]{} & A^{\mathrm{Loc}} & \xleftarrow{A^\delta} & A^{\mathrm{Loc}\times\mathrm{Loc}}
\end{array}$$

where the unlabeled isomorphisms match the outer V of $(A^V)^V$ with the first V of $A^{V\times V}$ and similarly for Loc, cf. [9],

$$
\begin{array}{ccccc}
A & \xrightarrow{\;u\;} & A^{\mathrm{Loc}\times V} & \xrightarrow{\;u^{\mathrm{Loc}\times V}\;} & (A^{\mathrm{Loc}\times V})^{\mathrm{Loc}\times V} \\
{\scriptstyle u}\big\downarrow & & & & \big\downarrow{\scriptstyle \cong} \\
A^{\mathrm{Loc}\times V} & \xrightarrow[A^{\mathrm{Loc}\times\pi_1}]{} & A^{\mathrm{Loc}\times V\times V} & \xleftarrow{\;A^{\delta\times V\times V}\;} & A^{\mathrm{Loc}\times\mathrm{Loc}\times V\times V}
\end{array}
$$

where the unlabeled isomorphism matches the outside Loc with the first Loc and similarly for V, and

$$
\begin{array}{ccccc}
A^V & \xrightarrow{\;l\;} & A^L & \xrightarrow{\;u^{\mathrm{Loc}}\;} & (A^{\mathrm{Loc}\times V})^{\mathrm{Loc}} \\
{\scriptstyle u^V}\big\downarrow & & & & \big\downarrow{\scriptstyle A^{\delta\times V}} \\
(A^{\mathrm{Loc}\times V})^V & & \xrightarrow[A^{\mathrm{Loc}\times\delta}]{} & & A^{\mathrm{Loc}\times V}
\end{array}
$$

suppressing two isomorphisms. We also have three commutation diagrams as follows:

$$
\begin{array}{ccccccc}
(A^V)^V & \xrightarrow{\;l^V\;} & (A^L)^V & \xrightarrow{\;\cong\;} & (A^V)^{\mathrm{Loc}} & \xrightarrow{\;l^{\mathrm{Loc}}\;} & (A^{\mathrm{Loc}})^{\mathrm{Loc}} \\
{\scriptstyle s}\big\downarrow & & & & & & \big\downarrow{\scriptstyle s} \\
(A^V)^V & & & & & & (A^{\mathrm{Loc}})^{\mathrm{Loc}} \\
{\scriptstyle l^V}\big\downarrow & & & & & & \big\downarrow \\
(A^{\mathrm{Loc}})^V & \xrightarrow[\cong]{} & (A^V)^{\mathrm{Loc}} & \xrightarrow[l^{\mathrm{Loc}}]{} & (A^{\mathrm{Loc}})^{\mathrm{Loc}} & \xrightarrow{\quad} & A^{\mathrm{Loc}_2}
\end{array}
$$

where s signifies 'swap' maps and Loc_2 denotes the set of ordered pairs of distinct elements of Loc, with the unlabeled maps both given by the same canonical map,

$$
\begin{array}{ccccccc}
A & \xrightarrow{\;u\;} & A^{\mathrm{Loc}\times V} & \xrightarrow{\;u^{\mathrm{Loc}\times V}\;} & (A^{\mathrm{Loc}\times V})^{\mathrm{Loc}\times V} & \xrightarrow{\;s\;} & (A^{\mathrm{Loc}\times V})^{\mathrm{Loc}\times V} \\
{\scriptstyle u}\big\downarrow & & & & & & \big\downarrow \\
A^{\mathrm{Loc}\times V} & & \xrightarrow{\;u^{\mathrm{Loc}\times V}\;} & & (A^{\mathrm{Loc}\times V})^{\mathrm{Loc}\times V} & \xrightarrow{\quad} & A^{\mathrm{Loc}_2\times V\times V}
\end{array}
$$

where s again signifies a swap map and with the unlabeled maps again given by the same canonical map, and

$$
\begin{array}{ccccccc}
A^V & \xrightarrow{\;l\;} & A^{\mathrm{Loc}} & \xrightarrow{\;u^{\mathrm{Loc}}\;} & (A^{\mathrm{Loc}\times V})^{\mathrm{Loc}} & \xrightarrow{\;\cong\;} & (A^{\mathrm{Loc}})^{\mathrm{Loc}\times V} \\
{\scriptstyle u^V}\big\downarrow & & & & & & \big\downarrow \\
(A^{\mathrm{Loc}\times V})^V & \xrightarrow[\cong]{} & (A^V)^{\mathrm{Loc}\times V} & \xrightarrow[l^{\mathrm{Loc}\times V}]{} & (A^{\mathrm{Loc}})^{\mathrm{Loc}\times V} & \xrightarrow{\quad} & A^{\mathrm{Loc}_2\times V}
\end{array}
$$

where, again, the unlabeled maps are given by the same canonical map.

The forgetful functor from $\mathrm{Mod}(L_S, \mathrm{Set})$ to Set induces the monad given by $T_S- = (- \times S)^S$ for global state on Set. So, if one models lookup and update subject to natural computational equations, one derives the monad for global state, rather than needing to take the latter as a primitive [20]. This approach has the advantage of having the algebraic constructs that generate the computational effect, i.e., lookup and update, built into it [4].

3. Tensor powers

The *tensor* of Lawvere theories, cf. [2], takes a pair of Lawvere theories L and L', takes all the operations and equations of each, and insists that each operation of L commutes with each operation of L'. More formally, it is defined as follows:

Definition 3.1. Given Lawvere theories L and L', the Lawvere theory $L \otimes L'$ is defined by the universal property of having maps of Lawvere theories from L and L' to $L \otimes L'$, with commutativity of all operations of L with respect to all operations of L', i.e., given $f\colon n \to m$ in L and $f'\colon n' \to m'$ in L', we demand commutativity of the diagram

$$\begin{array}{ccc} n \times n' & \xrightarrow{n \times f'} & n \times m' \\ {\scriptstyle f \times n'}\downarrow & & \downarrow{\scriptstyle f \times m'} \\ m \times n' & \xrightarrow[m \times f']{} & m \times m'. \end{array}$$

Tensor is characterized as follows [2, 4]:

Theorem 3.2. *Given Lawvere theories L and L', the tensor $L \otimes L'$ induces an equivalence of categories*

$$\mathrm{Mod}(L \otimes L', \mathrm{Set}) \simeq \mathrm{Mod}\big(L, \mathrm{Mod}(L', \mathrm{Set})\big)$$

The tensor forms a symmetric monoidal structure on Law, with unit given by the initial object of Law, which is $\mathrm{Nat}^{\mathrm{op}}$ together with the identity functor. Our primary use of this symmetric monoidal structure on Law is as a way to extend a Lawvere theory L to an n-fold version of it, modeling the idea of having n possible locations. It extends without fuss to countable Lawvere theories.

Theorem 3.3 ([4]). *Let L_S be the countable Lawvere theory for global state set S as in Example 2.2, and let $L_{S'}$ be the countable Lawvere theory for global state set S'. Then the tensor product $L_S \otimes L_{S'}$ is the countable Lawvere theory for global state set $S \times S'$.*

Corollary 3.4. Let L_V be the countable Lawvere theory for global state V, understood to be a set of values V with one location. Then the n-fold tensor product $L_V^{(n)}$ is the countable Lawvere theory for global state V^n, i.e., the Lawvere theory for global state with value set V and n locations.

So, if one has n locations, the n-fold tensor product $L_V^{(n)}$ of L_V describes those operations generated by the operations of global state, namely lookup and update, subject to their equations.

Observe that, although the category Inj of natural numbers and monomorphisms does not have finite coproducts, it has an initial object 0 that, together with the sum of natural numbers, equips Inj with a symmetric monoidal structure for which the unit is the initial object, aka it is *coaffine* [13, 18]. It is elementary to verify that it is the free symmetric monoidal category on 1 with unit the initial object. So the universal property of Inj allows us, starting with an arbitrary Lawvere theory L to generate a functor $\mathrm{Inj} \to \mathrm{Law}$.

4. Indexed Lawvere theories

Motivated by Section 3, we now introduce an axiomatic notion of indexed Lawvere theory. We shall make more sophisticated use of indexed Lawvere theories in analyzing *block* in Section 7. We do not assert definitiveness of this definition: we really only have one substantial example of its use here, and although we could speak a little about π-calculus, cf. [28], and we could write vague remarks about dependent algebra or about structures such as differential graded R-modules, cf. [9], we have not substantially investigated them. When further such examples of indexed structures are studied from this perspective, the definition given here may well need to be refined. But it seems natural and it suits the purpose at hand, hence our introducing it.

Definition 4.1. For any small category D, a *D-indexed Lawvere theory* is a functor $L\colon D \to \mathrm{Law}$. A *model* M of a D-indexed Lawvere theory L in a category C with finite products consists of, for every object d of D, a model

$$M_d\colon L_d \to C$$

of L_d in C, together with, for each map $f\colon d \to d'$ in D, a natural transformation

$$\begin{array}{ccc} L_d & \xrightarrow{M_d} & C \\ {\scriptstyle L_f}\big\downarrow & \Downarrow M_f & \big\downarrow{\scriptstyle \mathrm{Id}} \\ L_{d'} & \xrightarrow[M_{d'}]{} & C \end{array}$$

respecting the composition and identities of D, i.e., M_{id} is the identity and, given maps $f\colon d \to d'$ and $f'\colon d' \to d''$, the natural transformation $M_{f'f}$ is given by pasting $M_{f'}$ on to M_f.

This duly extends to give a category $\mathrm{Mod}(L, C)$: the arrows are D-indexed families of natural transformations that are coherent with respect to the M_f's. For any D-indexed Lawvere theory L and category C with finite products, there is a forgetful functor

$$U\colon \mathrm{Mod}(L, C) \to [D, C]$$

taking a model M to the functor sending d to $M_d(1)$. For each d, recalling that L_d is an ordinary Lawvere theory, the functor U has a d-component of the form $U_d\colon \mathrm{Mod}(L_d, C) \to C$. For any locally presentable category C, each U_d has a left adjoint F_d, exhibiting $\mathrm{Mod}(L_d, C)$ as monadic over C. These left adjoints and this monadicity result can be combined as follows:

Proposition 4.2. For any D-indexed Lawvere theory L and any locally presentable category C, the forgetful functor $U\colon \mathrm{Mod}(L, C) \to [D, C]$ has a left adjoint F given pointwise, i.e., for any functor $X\colon D \to C$,

$$(FX)_d = F_d X_d$$

Moreover, the adjunction exhibits $\mathrm{Mod}(L, C)$ as monadic over $[D, C]$.

Proof. It follows from the study of ordinary Lawvere theories that an adjoint F_d exists for each d. One can routinely verify that such adjoints collectively satisfy the adjointness property for the forgetful functor U. Monadicity can be proved in several ways, one such proof being by Beck's monadicity theorem. □

Our leading class of examples of D-indexed Lawvere theories for this section has $D = \text{Inj}$ and arises as follows:

Definition 4.3. Given any ordinary Lawvere theory L, we denote by $L_\otimes$ the Inj-indexed Lawvere theory freely generated by the tensor structure of Law and the fact of Inj being the free symmetric monoidal category on 1 with unit the initial object. So $(L_\otimes)n = L^{(n)}$, the n-fold tensor product of L.

Applying Proposition 4.2 to the Inj-indexed Lawvere theories that thus arise allows us to conclude the following:

Corollary 4.4. Given any ordinary Lawvere theory L and any locally presentable category C, let T be the monad on C induced by the forgetful functor

$$\text{Mod}(L, C) \to C$$

Then, the forgetful functor $\text{Mod}(L_\otimes, C) \to [\text{Inj}, C]$ exhibits $\text{Mod}(L_\otimes, C)$ as monadic over $[\text{Inj}, C]$ with monad $T_\otimes$ given by

$$(T_\otimes X)_n = T^{(n)} X_n$$

where $T^{(n)}$ is the monad on C induced by $L^{(n)}$.

All the above duly extends to countability and enrichment without fuss. So Corollary 3.4 yields our leading example of an Inj-indexed Lawvere theory and its models.

Example 4.5. By Corollary 3.4, if we let L_V be the countable Lawvere theory for global state with value set V and one location, the n-fold tensor $L_V^{(n)}$ is the countable Lawvere theory for global state with value set V and n locations. There are $n+1$ canonical coprojections from $L_V^{(n)}$ to $L_V^{(n+1)}$ determined by the universal property of tensor as determined by Definition 3.1. Putting $D = \text{Inj}$, the universal property of Inj extends L_V to the functor

$$(L_V)_\otimes \colon \text{Inj} \to \text{Law}$$

sending n to $L_V^{(n)}$, with behavior of $(L_V)_\otimes$ on maps determined by the canonical coprojections. A model M of $(L_V)_\otimes$ in a category C with countable products consists of, for each n, a model M_n of $L_V^{(n)}$, i.e., a model of the countable Lawvere theory for state with value set V and n locations, in C, respecting the structural inclusions of L_V into each $L_V^{(n)}$.

Consider the significance of Corollary 4.4 to the example of global state. Applied to Example 4.5, it allows us to extend and reformulate Example 2.2 as follows:

Example 4.6. The category $[\text{Inj}, \text{Set}]$ is a Cartesian closed category. Given a set V of values, let V also denote the constant functor from Inj to Set at the object V of Set. Let Loc denote the object of $[\text{Inj}, \text{Set}]$ given by the inclusion of Inj into Set. Using these conventions, Example 2.2 can be read as enriched in the Cartesian closed category $[\text{Inj}, \text{Set}]$ [20, 24]. The category $\text{Mod}(L_S, [\text{Inj}, \text{Set}])$ of models of L_S qua $[\text{Inj}, \text{Set}]$-enriched countable Lawvere theory is equivalent to the category $\text{Mod}((L_V)_\otimes, \text{Set})$. By Corollary 4.4, the induced monad T on $[\text{Inj}, \text{Set}]$ is given by

$$(TX)_n = T_{V^n} X_n$$

i..e, the value of TX at n is given by applying the monad for global state with value set V and n locations to the set X_n.

5. Exponentiation by a comodel

The technical development of the paper so far allows us to extend the operations and equations of any global effect to a local effect. But in order to account for locality, we also need to introduce a *block* operation. In order to do that, we introduce a general construction that extends an arbitrary model of the n-fold tensor product $L^{(n)}$ of a Lawvere theory L to a model of $L^{(n+1)}$: for block makes an assignment of structure to a new variable. Our construction is based on the observation that exponentiation by a comodel of a Lawvere theory yields a model of the theory.

Definition 5.1. A *comodel* of a Lawvere theory L in a category C with finite coproducts is a model of L in C^{op}.

Comodels and natural transformations form a category $\mathrm{Comod}(L, C)$, with a canonical forgetful functor to C determined by evaluation at 1.

Theorem 5.2 ([27]). *For any Lawvere theory L, the forgetful functor*

$$U\colon \mathrm{Comod}(L, \mathrm{Set}) \to \mathrm{Set}$$

has a right adjoint, exhibiting $\mathrm{Comod}(L, \mathrm{Set})$ *as comonadic over* Set.

PROOF. A proof and a construction are given by taking a dual of a standard construction for algebra in an axiomatic setting and simplifying it a little: replacing Set by Set^{op} requires care, but we only really need its limits and colimits, which we have [8].

The assertion of the existence of a right adjoint is equivalent to the assertion that the inclusion of $\mathrm{Comod}(L, \mathrm{Set})$ into $[L^{\mathrm{op}}, \mathrm{Set}]$ has a right adjoint. For that, the key point is that for any natural number n, the functor $n \times -\colon \mathrm{Set} \to \mathrm{Set}$ sends a set X to the n-fold coproduct of copies of X, and it preserves limits of ω-cochains.

Given a functor $H\colon L^{\mathrm{op}} \to \mathrm{Set}$, there is, for each n, a canonical map of the form $\phi_n\colon n \times H1 \to H(n \times 1)$. A map in L of the form $f\colon n \to 1$ is sent by H to a function of the form $Hf\colon H1 \to H(n \times 1)$. So we need to pullback Hf along ϕ_n. We need to do that simultaneously for all maps in L, yielding a limit $H_1 1$ together with a function $\sigma_1\colon H_1 1 \to H1$. In general, H_1 will not extend to a finite-coproduct preserving functor because of nontriviality of

$$n \times \sigma_1\colon n \times H_1 1 \to n \times H1$$

But one can continue inductively to build a cochain of length ω: in defining $H_2 1$, one must also account for equalities in L, i.e., given $f\colon n \to 1$ and $g_i\colon n_i \to 1$ for $1 \leq i \leq n$, one must account for $f(g_1, \ldots, g_n)$. But after defining $H_2 1$, it is routine to extend to all n. In the limit, H_ω is functorial as one can define $H_\omega f$ by using the limiting property of the cochain given by $n \times \sigma_k$. It is routine to verify that H_ω is the cofree comodel on H.

For comonadicity, it is routine to verify the conditions for the dual of Beck's monadicity theorem, i.e., that U reflects isomorphisms and that split equalisers lift [1]. □

Enrichment of Theorem 5.2 in a Cartesian closed category is routine: we used the fact that the tensor of n with X had the universal property of a product. And the result generalizes routinely to a countable version. So Theorem 5.2 generalizes to categories such as Poset and ωCpo.

Corollary 5.3. For any Lawvere theory L, the category $\mathrm{Comod}(L, \mathrm{Set})$ has a terminal object, the final comodel.

Our leading example, which is also the leading example of [27], is given by global state:

Example 5.4. Let L_S be the countable Lawvere theory for global state, with $S = V^{\mathrm{Loc}}$ and where Loc is finite. The induced comonad on Set is given by $(-)^S \times S$, and the final comodel is therefore given by $S = V^{\mathrm{Loc}}$. A fortiori, the final comodel of L_V in Set is given by V, with structural maps given by

$$\mathrm{loc} = \delta \colon V \to V \times V$$

and

$$\mathrm{upd} = \pi_2 = t \times V \colon V \times V \to 1 \times V \cong V$$

The category of comodels for the Lawvere theory L_S is equivalent to a category of arrays [27].

Given a comodel $A \colon L^{\mathrm{op}} \to \mathrm{Set}$ of an arbitrary Lawvere theory L in Set and given an arbitrary object X of a category C with products, letting X^{-} denote a product of copies of X in C, the composite

$$L \xrightarrow{A} \mathrm{Set}^{\mathrm{op}} \xrightarrow{X^{-}} C$$

preserves finite products. Thus X^{A1}, the $A1$-fold product in C of copies of X, inherits an L-model structure in C from the L-comodel A in Set. More generally, for any Lawvere theory L' and any model $M \colon L' \to C$ in a category C with products, the composite

$$L \xrightarrow{A} \mathrm{Set}^{\mathrm{op}} \xrightarrow{M^{-}} \mathrm{Mod}(L', C)$$

preserves finite products. By Theorem 3.2, to give such a finite product preserving functor is equivalent to giving a model of $L' \otimes L$. This is natural in M, yielding the following:

Proposition 5.5. For any Lawvere theory L and any comodel A of L in Set, and for any Lawvere theory L' and any category C with products, exponentiation $(-)^A$ induces a functor

$$(-)^A \colon \mathrm{Mod}(L', C) \to \mathrm{Mod}(L' \otimes L, C)$$

Corollary 5.6. For any Lawvere theory L, let F be the final comodel of L in Set. Then, for any category C with products, exponentiation $(-)^F$ induces a functor

$$(-)^F \colon \mathrm{Mod}(L_{\otimes} n, C) \to \mathrm{Mod}(L_{\otimes}(n+1), C)$$

Moreover, this is natural in n as an object of Inj.

This duly generalizes to the countable setting and also to the enriched setting. Our leading example, Example 4.6, extends as follows:

Example 5.7. Consider a model M of the Inj-indexed countable Lawvere theory $(L_V)_{\otimes}$ in Set. Each $(M_n 1)^V$ canonically possesses the structure of a model of $L_V^{(n+1)}$: it inherits n L_V-structures from those of $M_n 1$, while the final L_V-comodel structure of V given by Example 5.4 induces a further L_V-structure on $(M_n 1)^V$.

The significance of this for us is that the semantics of $block_n$ will be a map of $L_V^{(n+1)}$-models from M_{n+1} to M_n^V.

Observe that the above discussion is inherently semantic, i.e., it is about constructs on models. But the spirit of the idea of Lawvere theories is that we seek constructs on theories, and then consider a notion of model of that in any well-behaved semantic category C such as Set. In order to provide a construct at the level of theories, observe that, for any comodel A of a Lawvere theory L, the functor

$$(-)^A\colon \mathrm{Mod}(L', C) \to \mathrm{Mod}(L' \otimes L, C)$$

of Proposition 5.5 makes the following diagram commute:

$$\begin{array}{ccc} \mathrm{Mod}(L', C) & \xrightarrow{(-)^A} & \mathrm{Mod}(L' \otimes L, C) \\ {\scriptstyle \mathrm{ev}_1}\downarrow & & \downarrow{\scriptstyle \mathrm{ev}_1} \\ C & \xrightarrow[(-)^{A1}]{} & C. \end{array}$$

Letting $C = \mathrm{Set}$ for example, the functor $(-)^{A1}\colon C \to C$ is monadic [14,23], with left adjoint given by $A1 \times -$, so with monad given by $(A1 \times -)^{A1}$, the monad for state with state set $A1$. So, letting L_{A1} denote the countable Lawvere theory for state with state set $A1$, it follows that $\mathrm{Mod}(L', C)$ is equivalent to $\mathrm{Mod}(L', \mathrm{Mod}(L_{A1}, C))$, which is in turn equivalent to $\mathrm{Mod}(L' \otimes L_{A1}, C)$, and the diagram

$$\begin{array}{ccc} \mathrm{Mod}(L', C) & \longrightarrow & \mathrm{Mod}(L' \otimes L, C) \\ {\scriptstyle \simeq}\uparrow & & \downarrow{\scriptstyle \mathrm{ev}_1} \\ \mathrm{Mod}(L' \otimes L_{A1}, C) & \xrightarrow[\mathrm{ev}_1]{} & C \end{array}$$

commutes. We can therefore conclude [1]:

Theorem 5.8. *For any Lawvere theory L and any comodel A of L in* Set, *and for any Lawvere theory L' and any locally presentable category C, there is a map of Lawvere theories $f\colon L' \otimes L \to L' \otimes L_{A1}$ such that the functor*

$$\mathrm{Mod}(L' \otimes L_{A1}, C) \xrightarrow{\simeq} \mathrm{Mod}(L', C) \xrightarrow{(-)^A} \mathrm{Mod}(L' \otimes L, C)$$

is of the form $\mathrm{Mod}(f, C)$.

This can duly be iterated, yielding, for any n, a map of Lawvere theories of the form $L' \otimes L^{(n)} \to L' \otimes L_{A1^n}$. We shall use this construct to elucidate *block* at the level of theories in Section 7.

We remark as an aside that, putting $L' = \mathrm{Nat}^{\mathrm{op}}$ and $C = \mathrm{Set}$, the functor

$$(-)^A\colon \mathrm{Set} \to \mathrm{Mod}(L, \mathrm{Set})$$

of Proposition 5.5 has a left adjoint: that left adjoint agrees with the central construct used in [23] to analyze structural operational semantics for computational effects.

6. Semantics for block

We have foreshadowed the semantics of *block* in Section 5. By Corollary 5.3, any Lawvere theory L has a final comodel F. By Corollary 5.6, for any category C with products, exponentiation $(-)^F$ induces a functor

$$(-)^F\colon \mathrm{Mod}(L_{\otimes}n, C) \to \mathrm{Mod}(L_{\otimes}(n+1), C)$$

natural in n.

Definition 6.1. Given an arbitrary Lawvere theory L and a category C with products, a block-*algebra* in C consists of a model M of the Inj-indexed Lawvere theory $L_{\otimes}$ in C, together with an indexed family of maps

$$\mathrm{block}_n\colon M_{n+1} \to M_n^F$$

in $\mathrm{Mod}(L_{\otimes}(n+1), C)$, natural with respect to Inj, satisfying the following two equations:

$$\begin{array}{ccc} M_n(1) & \xrightarrow{(M_{inc})_1} & M_{n+1}(1) \\ & \searrow^{(M_n 1)^t} & \downarrow{\scriptstyle (\mathrm{block}_n)_1} \\ & & (M_n 1)^F \end{array}$$

$$\begin{array}{ccc} M_{n+2} & \xrightarrow{M_{n+s}} & M_{n+2} \\ {\scriptstyle \mathrm{block}_{n+1}}\downarrow & & \downarrow{\scriptstyle \mathrm{block}_{n+1}} \\ M_{n+1}^F & & M_{n+1}^F \\ {\scriptstyle \mathrm{block}_n^F}\downarrow & & \downarrow{\scriptstyle \mathrm{block}_n^F} \\ (M_n^F)^F & \xrightarrow[s]{} & (M_n^F)^F \end{array}$$

where the second occurrence of s means swapping the two copies of F respectively.

The first axiom asserts that *block* only affects the newly created location, while the second asserts that in creating two new locations and assigning structure to them, one can do so in either order providing one remembers which location is being assigned which structure.

One can routinely form a category of block-algebras, the maps being maps of models of $L_{\otimes}$ that respect the block structure. We denote the category by block-Alg.

Theorem 6.2. *If C is locally presentable, the composite forgetful functor*

$$\text{block-Alg} \to [\mathrm{Inj}, C]$$

is monadic.

Proof. There are several proofs of this: a *block*-algebra is, by construction, given by adding operations subject to equations to a model of $L_{\otimes}$, and such models are in turn given by operations and equations on an object of $[\mathrm{Inj}, C]$. Combining these facts, a *block*-algebra consists of an object of $[\mathrm{Inj}, C]$ together with operations subject to universally defined equations, and so the category of such is monadic over $[\mathrm{Inj}, C]$ as an instance of the general theory of [10]. □

This routinely extends to countability and enrichment. We now check that, if L is the countable Lawvere theory for global state with value set V and one location, it yields the usual monad T_{LS} for local state on [Inj, Set], following O'Hearn et al. [11, 12, 17, 28] and agreeing with the monad of [20, 21].

The monad for local state, on [Inj, Set], as suggested to us by Peter O'Hearn, is

$$(T_{LS}X)_n = \Big(\int^{m\epsilon(n/I)} (Sm \times X_m)\Big)^{Sn}$$

where $Sn = V^n$ for a given set of values V, cf. [11, 12]. If $V = 1$ it reduces to the monad for local names in [28].

The behavior of T_{LS} on injective maps $f : n \to n'$ is as follows: decompose n' as the sum $n + n''$, note that $S(p + n'') = Sp \times Sn''$, and use covariance of X. So the map

$$\Big(\int^{m\epsilon(n/I)} (Sm \times X_m)\Big)^{Sn} \times Sn \times Sn'' \to \int^{m''\epsilon((n+n'')/I)} (Sm'' \times X_{m''})$$

evaluates at Sn, then maps the mth component of the first of the two coends into the $(m + n'')$th component of the second, using the above isomorphism for S and functoriality of X.

In [20, 21], an algebraic-like signature generating T_{LS} was given by defining the corresponding category $LS([\text{Inj}, \text{Set}])$ of algebras as follows, subject to some not entirely trivial rewriting: an algebra consists of

- an object A of [Inj, Set]
- a lookup map with components $l_n \colon n \times A(n)^V \to A(n)$
- an update map with components $u_n \colon n \times A(n) \to A(n)^V$
- a block map with components $b_n \colon A(n+1) \to A(n)^V$

subject to commutativity of seven interaction diagrams and six commutativity diagrams. The interaction diagrams consist of the four interaction diagrams for global state listed in Example 2.2 (but with the second now proved redundant by Paul-André Melliès), together with

$$\begin{array}{ccc} A(n+1)^V & \xrightarrow{l_{n+1}\cdot(\mathrm{ev}_{n+1}\times \mathrm{id})} & A(n+1) \\ {\scriptstyle b_n^V}\downarrow & & \downarrow{\scriptstyle b_n} \\ (A(n)^V)^V & \xrightarrow[A(n)^\delta]{} & A(n)^V \end{array}$$

$$\begin{array}{ccc} A(n+1) & \xrightarrow{u_{n+1}\cdot(\mathrm{ev}_{n+1}\times \mathrm{id})} & A(n+1)^V \\ {\scriptstyle b_n}\downarrow & & \downarrow{\scriptstyle b_n^V} \\ A(n)^V & \xrightarrow[(A(n)^t)^V]{} & (A(n)^V)^V \end{array}$$

$$\begin{array}{ccc} A(n) & \xrightarrow{A(\mathrm{inc})} & A(n+1) \\ & {\scriptstyle A(n)^t}\searrow & \downarrow{\scriptstyle b_n} \\ & & A(n)^V. \end{array}$$

The commutation diagrams are those for global state together with

$$\begin{array}{ccc}
A(n+2) & \xrightarrow{A(n+s)} & A(n+2) \\
{\scriptstyle b_{n+1}}\downarrow & & \downarrow{\scriptstyle b_{n+1}} \\
A(n+1)^V & & A(n+1)^V \\
{\scriptstyle b_n^V}\downarrow & & \downarrow{\scriptstyle b_n^V} \\
(A(n)^V)^V & \xrightarrow[s]{} & (A(n)^V)^V
\end{array}$$

and, for k less than or equal to n,

$$\begin{array}{ccc}
A(n+1)^V & \xrightarrow{l_{n+1}\cdot(\mathrm{ev}_k \times \mathrm{id})} & A(n+1) \\
{\scriptstyle b_n^V}\downarrow & & \downarrow{\scriptstyle b_n} \\
(A(n)^V)^V & \xrightarrow[l_n\cdot(\mathrm{ev}_k \times \mathrm{id})]{} & A(n)^V
\end{array}$$

$$\begin{array}{ccc}
A(n+1) & \xrightarrow{u_{n+1}\cdot(\mathrm{ev}_k \times \mathrm{id})} & A(n+1)^V \\
{\scriptstyle b_n}\downarrow & & \downarrow{\scriptstyle b_n^V} \\
A(n)^V & \xrightarrow[u_n\cdot(\mathrm{ev}_k \times \mathrm{id})]{} & (A(n)^V)^V.
\end{array}$$

One can observe directly the correspondence between these axioms and those we have developed: recalling that we refer to n copies of each of *lookup* and *update* rather than one n-parametrized version of each, we have already accounted for precisely the diagrams arising from global state in Example 4.6. So to give A together with l and u is equivalent to giving a model M of $(L_V)_\otimes$ in Set. Now recall, from Example 5.4, that V is the final comodel of L_V. So, using Example 5.7, the first two interaction axioms together with the last two commutation axioms are equivalent to the assertion that $b_n\colon M_{n+1} \to M_n^V$ is a map in $\mathrm{Mod}(L^{(n+1)}, \mathrm{Set})$. The remaining two axioms are exactly the axioms for block in Definition 6.1. This, together with the characterization of T_{LS} in [20, 21], allows us to conclude the following:

Theorem 6.3. *Taking L to be the countable Lawvere theory L_V for global state with value set V and one location, the category* block-Alg *is equivalent to* $LS([\mathrm{Inj}, \mathrm{Set}])$ *and is thus monadic over* $[\mathrm{Inj}, \mathrm{Set}]$ *with monad T_{LS} that for local state.*

7. Syntax for block

We end the paper by applying the notion of indexed Lawvere theory to give a syntactic description of *block*, its category of models characterizing the monad T_{LS} on $[\mathrm{Inj}, \mathrm{Set}]$ for local state. We first need to consider free strict symmetric monoidal categories with initial unit. These are an example of *clubs* and are thus among the easiest of structures for which to describe the free such: see [7] and the long succession of papers about clubs that followed it. We shall describe the one example we need after our definition.

Definition 7.1. Let Inj_2 denote the free strict symmetric monoidal category on the arrow category $\cdot \to \cdot$ with unit the initial object.

For concreteness, assume that the generating arrow for Inj_2 is

$$x \xrightarrow{\gamma} y$$

An arbitrary object of Inj_2 is a list of x's and y's, with the empty list acting as the unit. There are fully faithful functors from Inj into Inj_2, one of them, ι_1, sending n to a list of n copies of x, the other, ι_2, sending n to a list of n copies of y.

Observe that the map $\gamma\colon x \to y$ induces a monoidal natural transformation $\mathrm{Inj}(\gamma)\colon \iota_1 \Rightarrow \iota_2$. This, together with the fact that 0 is initial, makes the following diagrams commute:

$$\begin{array}{ccc} \iota_1(n) & \xrightarrow{\iota_1(n+!)} & \iota_1(n+1) \\ & \searrow^{\iota_1(n)\otimes !} & \downarrow{\scriptstyle \iota_1(n)\otimes\gamma} \\ & & \iota_1(n)\otimes\iota_2(1) \end{array} \tag{1}$$

$$\begin{array}{ccc} \iota_1(n+2) & \xrightarrow{\iota_1(n+s)} & \iota_2(n+2) \\ \downarrow{\scriptstyle \iota_1(n+1)\otimes\gamma} & & \downarrow{\scriptstyle \iota_1(n+1)\otimes\gamma} \\ \iota_1(n+1)\otimes\iota_2(1) & & \iota_1(n+1)\otimes\iota_2(1) \\ \downarrow{\scriptstyle \iota_1(n)\otimes\gamma\otimes\iota_2(1)} & & \downarrow{\scriptstyle \iota_1(n)\otimes\gamma\otimes\iota_2(1)} \\ \iota_2(n)\otimes\iota_1(1)\otimes\iota_1(1) & \xrightarrow[\iota_2(n)\otimes s]{} & \iota_2(n)\otimes\iota_1(1)\otimes\iota_1(1) \end{array} \tag{2}$$

where the two occurrences of s mean the evident twist maps.

By Theorem 5, every Lawvere theory L together with a comodel A in Set generates a map of Lawvere theories $A^*\colon L \to L_{A1}$ and hence an Inj_2-indexed Lawvere theory $L_{\otimes A}$. Note that $L_{\otimes A}\big(\iota_2(n)\big) = L_{A1^n}$, the Lawvere theory for state with state set $A1^n$. We need to consider some, but not all, of the models of the Inj_2-indexed Lawvere theory $L_{\otimes A}$.

I do not currently see a way to avoid a condition on the models like this, even in the case of A being the final comodel. I suspect that that reflects the fact that this is a first paper introducing indexed Lawvere theories, indicating a lack of sophistication in the idea to date: I would not be surprised if there is an interesting, general construct that underlies this that is yet to be discovered; a detailed analysis of further examples, arising, for instance, from dependent algebra, may well uncover it.

Definition 7.2. Given a Lawvere theory L together with a comodel A of L in Set, an *A-model* of the free Inj_2-indexed Lawvere theory $L_{\otimes A}$ on $A^*\colon L \to L_{A1}$ in a category C with products is a model M of $L_{\otimes A}$ in C such that

$$M_{z\otimes\iota_2(n)}- = (M_z-)^{A1^n}\colon L_{\otimes A}\big(z\otimes\iota_2(n)\big) \to C$$

naturally in $z\epsilon\mathrm{Inj}_2$ and strong symmetric monoidal naturally in $n\epsilon\mathrm{Inj}$.

The definition of A-model extends routinely to yield a category $A\,\mathrm{Mod}(L, C)$, the arrows being the maps of models of $L_{\otimes A}$ that respect the A-structure.

Theorem 7.3. *If C is locally presentable, and F is the final comodel of a Lawvere theory L, the categories $F\,\mathrm{Mod}(L, C)$ and* block-Alg *are coherently equivalent.*

PROOF. The proof is given by tedious checking of a mass of detail. The heart of it is the fact that the coherence conditions in Definition 7.2 mean that an A-model of the free Inj_2-indexed Lawvere theory $L_{\otimes A}$ on $A^* \colon L \to L_{A1}$ in C is given by

- a model M of $L_{\otimes}$ in C
- for each n, a map of $L_{\otimes}(n+1)$-models

$$M_{n+1} \to M_n^F$$

- subject to naturality in $n\epsilon\mathrm{Inj}$
- subject to commutativity of the images of (1) and (2).

Considerable calculation shows that preservation of the commutativity of (1) and (2), together with the naturality conditions, yields preservation of all commutative diagrams in Inj_2, while agreeing with the two commutativity conditions in the definition of *block*-algebra. □

The above duly extends to countability and enriches. Recall that if L_V is the countable Lawvere theory for global state with one location and with value set V, the final comodel of L_V is given by V.

Corollary 7.4. Let L_V be the countable Lawvere theory for global state with one location and value set V. Then $V\,\mathrm{Mod}(L_V, \mathrm{Set})$ is monadic over $[\mathrm{Inj}, \mathrm{Set}]$ with monad T_{LS}, the monad for local state.

References

1. M. Barr and C. Wells, *Toposes, triples and theories*, Grundlehren Math. Wiss., vol. 278, Springer, New York, 1985.
2. P. Freyd, *Algebra valued functors in general and tensor products in particular*, Colloq. Math. **14** (1966), 89–106.
3. M. Hyland, P. B. Levy, G. Plotkin, and J. Power, *Combining algebraic effects with continuations*, Theoret. Comput. Sci. **375** (2007), no. 1-3, 20–40.
4. M. Hyland, G. Plotkin, and J. Power, *Combining effects: sum and tensor*, Theoret. Comput. Sci. **357** (2006), no. 1-3, 70–99.
5. M. Hyland and J. Power, *Discrete Lawvere theories and computational effects*, Theoret. Comput. Sci. **366** (2006), no. 1-2, 144–162.
6. ———, *The category theoretic understanding of universal algebra: Lawvere theories and monads*, Computation, Meaning, and Logic: Articles Dedicated to Gordon Plotkin (L. Cardelli, M. Fiore, and G. Winskel, eds.), Electron. Notes Theor. Comput. Sci., vol. 172, Elsevier, Amsterdam, 2007, pp. 437–458.
7. G. M. Kelly, *An abstract approach to coherence*, Coherence in categories (S. Mac Lane, ed.), Lecture Notes in Math., vol. 281, Springer, Berlin, 1972, pp. 106–147.
8. ———, *A unified treatment of transfinite constructions for free algebras, free monoids, colimits, associated sheaves, and so on*, Bull. Austral. Math. Soc. **22** (1980), no. 1, 1–83.
9. ———, *Basic concepts of enriched category theory*, London Math. Soc. Lecture Note Ser., vol. 64, Cambridge Univ. Press, Cambridge, 1982.
10. G. M. Kelly and A. J. Power, *Adjunctions whose counits are coequalizers, and presentations of finitary enriched monads*, J. Pure Appl. Algebra **89** (1993), no. 1-2, 163–179.
11. P. B. Levy, *Call-by-push-value: a subsuming paradigm* (extended abstract), Typed Lambda Calculi and Applications (L'Aquila, 1999) (J.-Y. Girard, ed.), Lecture Notes in Comput. Sci., vol. 1581, Springer, Berlin, 1999, pp. 228–242.

12. ———, *Call-by-push-value: A functional/imperative synthesis*, Semant. Struct. Comput., vol. 2, Kluwer, Dordrecht, 2003.
13. ———, *Global state considered helpful*, Proceedings of the 24th Conference on the Mathematical Foundations of Programming Semantics (Philadelphia, PA, 2008) (A. Bauer and M. Mislove, eds.), Electron. Notes Theor. Comput. Sci., vol. 218, Elsevier, Amsterdam, 2008, pp. 241–259.
14. F. Métayer, *State monads and their algebras*, available at `arXiv:math/0407251`.
15. E. Moggi, *Computational lambda-calculus and monads*, Proceedings of the Fourth Annual Symposium on Logic in Computer Science (Pacific Grove, CA, 1989), IEEE Press, Piscataway, NJ, 1898, pp. 14–23.
16. ———, *Notions of computation and monads*, Inform. and Comput. **93** (1991), no. 1, 55–92.
17. P. W. O'Hearn and R. D. Tennent (eds.), *ALGOL-like languages*, Progr. Theoret. Comput. Sci., Birkhäuser, Boston, MA, 1997.
18. Z. Petrić, *Coherence in substructural categories*, Studia Logica **70** (2002), no. 2, 271–296.
19. G. Plotkin and J. Power, *Adequacy for algebraic effects*, Foundations of Software Science and Computation Structures (Genova, 2001) (F. Honsell and M. Miculan, eds.), Lecture Notes in Comput. Sci., vol. 2030, Springer, Berlin, 2001, pp. 1–24.
20. ———, *Notions of computation determine monads*, Foundations of Software Science and Computation Structures (Grenoble, 2002) (M. Nielsen and U. Engberg, eds.), Lecture Notes in Comput. Sci., vol. 2303, Springer, Berlin, 2002, pp. 342–356.
21. ———, *Erratum*, available at `http://homepages.inf.ed.ac.uk/gdp/publications/`.
22. ———, *Logic for computational effects: work in progress*, 6th International Workshop on Formal Methods (Dublin, 2003).
23. ———, *Tensors of comodels and models for operational semantics*, Proceedings of the 24th Conference on the Mathematical Foundations of Programming Semantics (Philadelphia, PA, 2008) (A. Bauer and M. Mislove, eds.), Electron. Notes Theor. Comput. Sci., vol. 218, Elsevier, Amsterdam, 2008, pp. 295–311.
24. J. Power, *Enriched Lawvere theories*, Theory Appl. Categ. **6** (1999), 83–93.
25. ———, *Semantics for local computational effects*, Proceedings of the 22nd Annual Conference on Mathematical Foundations of Programming Semantics (Genova, 2006) (S. Brookes and M. Mislove, eds.), Electron. Notes Theor. Comput. Sci., vol. 158, Elsevier, Amsterdam, 2006, pp. 355–371.
26. ———, *Generic models for computational effects*, Theoret. Comput. Sci. **364** (2006), no. 2, 254–269.
27. J. Power and O. Shkaravska, *From comodels to coalgebras: State and arrays*, PProceedings of the Workshop on Coalgebraic Methods in Computer Science (Barcelona, 2004) (J. Adámek and S. Milius, eds.), Electron. Notes Theor. Comput. Sci., vol. 106, Elsevier, Amsterdam, 2004, pp. 297–314.
28. I. Stark, *Categorical models for local names*, Lisp and Symbolic Computation **9** (1996), no. 1, 77–107.

Department of Computer Science, University of Bath, Bath, BA2 7AY, United Kingdom

E-mail address: `A.J.Power@bath.ac.uk`

Centre de Recherches Mathématiques
CRM Proceedings and Lecture Notes
Volume **53**, 2011

Model Theory in Additive Categories

Mike Prest

ABSTRACT. A sequence of views of the model theory of modules and other additive structures is presented; starting from the usual set-theoretic perspective we move to a topos-theoretic one via accessible categories.

1. Introduction

In this expository paper I use the model theory of modules to link "classical" set-based model theory with category-theoretic model theory through a short sequence of changes of view of what is a module. We start with the usual view of a module as a set-with-structure. In fact, even there it is useful to take an alternative, functorial, view of modules since it allows us to generalize so as to encompass many other kinds of additive structure. At the next stage we think of modules as taking their structure purely from the category (of modules) in which they lie; this category has a generating set of finitely presented objects, and modules may be regarded as functors on those, hence as structures for a multi-sorted language based on the category of finitely presented modules. At the last stage we introduce the (abelian) category of pp-sorts and we regard modules as functors on that category. At each stage we enrich the language that we use for the model theory of modules, introducing new sorts and function symbols, but all these languages are conservative extensions of the basic language. What we gain is more "structure" in the language; more precisely it becomes clear that underlying each language there is a category of sorts and these categories have progressively stronger properties. To begin with, that underlying category is a ring, more generally any (skeletally) small preadditive category; at the intermediate stage it is an additive category with, in particular, cokernels; finally it is any small abelian category. And modules are, correspondingly, arbitrary additive functors, left exact functors, exact functors. The three viewpoints could be characterized, respectively, as being based on classical model theory, accessible category theory and, finally, topos theory.

2010 *Mathematics Subject Classification.* Primary 03C60; Secondary 14A22, 16D90, 18B25, 18C35, 18E05.

This is the final form of the paper.

2. Three views of modules

2.1. The classical view. We fix a ring R and then define a right (or left) R-module in the usual way — as an abelian group with R acting as a ring of endomorphisms of that group. We use the notation Mod-R for the category whose objects are the right R-modules and whose arrows are the R-linear maps; by R-Mod we denote the category of left R-modules and by mod-R the full subcategory of Mod-R whose objects are the finitely presented (i.e. finitely generated and finitely related) modules. For general background on the model theory of modules the reader can consult [21] and, for a more algebraic/category-theoretic approach, [25] (or [24] for a summary).

Let's set up a running example: vector spaces over a field are just too simple so we take R to be the ring, $\mathrm{UT}_2(K) = \left(\begin{smallmatrix} K & K \\ 0 & K \end{smallmatrix}\right)$ of 2×2 upper-triangular matrices over a field K. Consider the elements $e_1 = \left(\begin{smallmatrix} 1 & 0 \\ 0 & 0 \end{smallmatrix}\right)$, $e_2 = \left(\begin{smallmatrix} 0 & 0 \\ 0 & 1 \end{smallmatrix}\right)$, $\alpha = \left(\begin{smallmatrix} 0 & 1 \\ 0 & 0 \end{smallmatrix}\right)$ of R. We have $e_1^2 = e_1$, $e_2^2 = e_2$, $e_1e_2 = 0 = e_2e_1$, $e_1 + e_2 = 1$, $e_1\alpha = \alpha = \alpha e_2$. If M is any right R-module, then, as a K-vector space, M decomposes as $M = Me_1 \oplus Me_2$. Furthermore, α acts on M by sending Me_1 to Me_2 and sending Me_2 to 0. This gives another way of representing any right R-module, namely as a representation of the quiver $\bullet \to \bullet$. A *quiver* is just a directed graph and a *representation* over K of a quiver is obtained by replacing each vertex by a K-vector space and each arrow by a K-linear map between the corresponding vector spaces. We have just seen how to obtain a K-representation of $\bullet \to \bullet$ from a right R-module and it is equally easy to build a right R-module canonically from a K-representation of $\bullet \to \bullet$. In fact, these processes give an equivalence between Mod-R and the category of K-representations of the quiver $\bullet \to \bullet$.

This example is just a small step up from taking R to be a field; in that case the modules are vector spaces, hence are direct sums of copies of the unique indecomposable (1-dimensional) module and are completely specified by a single dimension. In the case of $R = \mathrm{UT}_2(K)$ there are three indecomposable (i.e. direct-sum indecomposable) modules, every module M is a direct sum of copies of these three and the number of copies of each indecomposable in such a decomposition depends only on M. Thus M is completely specified by three dimensions. We say that R is a ring of *finite representation type*, meaning that every module is a direct sum of indecomposable modules and there are only finitely many indecomposable modules up to isomorphism. This does mean that our chosen ring R has a far simpler representation theory than the "typical" ring. Nevertheless, it is complicated enough to illustrate some points which cannot be seen in the category of vector spaces over a field. We will introduce notation for the three indecomposables, which are:

P_1, which is determined by $P_1e_1 = K$, $P_1e_2 = K$, α acts as an isomorphism (say, $\lambda \mapsto \lambda$) from the first to the second copy of K;
P_2, given by $P_2e_1 = 0$, $P_2e_2 = K$;
S_1, given by $S_1e_1 = K$, $S_1e_2 = 0$

(P_1 and P_2 are projective modules; S_1 and P_2 are simple modules). As representations of the quiver $\bullet \to \bullet$, these modules are $P_1 = K \xrightarrow{1} K$, $P_2 = 0 \xrightarrow{0} K$, $S_1 = K \xrightarrow{0} 0$.

To do model theory on R-modules, we introduce the language $\mathcal{L}_R$ which has symbols $+$ and 0 for the underlying group structure and, for each element $r \in R$, a symbol to express the corresponding action (multiplication by r). The fundamental result is pp-elimination of quantifiers.

A formula $\phi(\bar{x})$ is *pp* (*positive primitive*) if it is equivalent to one of the form $\exists \bar{y}\, \theta(\bar{x}, \bar{y})$ where θ is a conjunction of atomic formulas, that is, linear equations. If $\phi(\bar{x})$ is pp and M is a module then the solution set, $\phi(M)$, of ϕ in M is a subgroup of $M^{l(\bar{x})}$ — a *pp-definable subgroup*; more generally, if we allow parameters in ϕ, then the solution set, if nonempty, is a coset of the subgroup defined by the formula which is obtained by replacing all parameters by 0.

By a pp-pair we will mean a pair ϕ, ψ of pp formulas in the same free variables with $\psi \to \phi$, that is, with $\psi(M) \leq \phi(M)$ for each R-module M; the associated quotient group $\phi(M)/\psi(M)$ plays an important role, indeed we often use the notation ϕ/ψ for a pp-pair.

Theorem 2.1 (pp-elimination of quantifiers). *If R is any ring and $\chi(\bar{x})$ is a formula of $\mathcal{L}_R$ then χ is equivalent, modulo the theory of R-modules, to the conjunction of a sentence and a finite Boolean combination of pp formulas. The sentence, moreover, may be taken to be a finite Boolean combination of statements of the form $|\phi(-)/\psi(-)| \geq k$ where ϕ/ψ is a pp-pair and k is a positive integer.*

Corollary 2.2. *Modules M, N are elementarily equivalent iff for each pp-pair ϕ, ψ the cardinalities $|\phi(M)/\psi(M)|$, $|\phi(N)/\psi(N)|$ are equal if they are finite. Furthermore, an embedding $M \to N$ of elementarily equivalent modules is elementary iff it is* pure *that is, if for every pp formula $\phi(\bar{x})$ we have $\phi(M) = M^{l(\bar{x})} \cap \phi(N)$.*

Another corollary is that, modulo a complete theory of modules, every type (with parameters) is determined by the pp formulas in it; then, typically, notions from stability theory — forking, various ranks, etc. — can be described in terms of pp-types (sets of pp formulas closed under conjunction and implication), see, e.g., [15, 21]. In particular every module is stable.

Let M be a module over $R = \mathrm{UT}_2(K)$; as stated earlier, $M \simeq M_1 \oplus M_2 \oplus M_3$ where M_i is a direct sum of copies of, respectively, P_1, P_2, S_1. If ϕ is a pp formula then $\phi(M) = \phi(M_1) \oplus \phi(M_2) \oplus \phi(M_3)$ and, if, say $M_1 = P_1^{(\kappa)}$ then $\phi(M) = \phi(P_1)^{(\kappa)}$ so, to describe the pp-definable subgroups of M, it suffices to describe those of the three indecomposable modules. Observing that these must be K-vector spaces simplifies such description and, in particular, there is just a short list of pp-definable subgroups of M which have the form $\phi(M)$ where ϕ is pp with a single free variable. In particular we easily see (by [21, Section 5.2 also Section 11.4]) that every R-module is totally transcendental and of Morley rank ≤ 4.

A different kind of corollary is that the algebra and model theory fit together well in the sense that pp formulas are exactly those preserved by homomorphisms and so the algebraic category Mod-R fits better with the model-theoretic one (which has the same objects but just the elementary embeddings for morphisms) than would be typical without pp-elimination of quantifiers.

Lemma 2.3. *If $f \colon M \to N$ is a homomorphism of modules and if ϕ is a pp formula then $f\phi(M) \leq \phi(N)$.*

Another consequence, which we say imprecisely but which has various precise renditions (as we saw in the example above), is that if a module splits as a direct sum then its model theory decomposes correspondingly. One precise expression is the existence of a *largest complete theory of modules*: this is defined to be the theory of any module of the form $\bigoplus\{M_T : T\}$ where T runs over complete theories of modules and M_T is any chosen model of T. A rather more intrinsic definition of T is that for each *proper* pp-pair ϕ/ψ (i.e., such that $\phi(M) > \psi(M)$ for *some* module M) we include the axioms $|\phi(-)/\psi(-)| > k$ for each $k \geq 1$. This largest theory T^* of R-modules has the defining property that if M is any module then M is a direct summand of some model of T^*. This leads to a more subtle effect of pp-elimination of quantifiers in that it moves attention away from complete theories of modules to those theories whose category of models is closed under direct sums and direct summands (equivalently those theories axiomatized by sentences declaring that two pp formulas are equivalent). A useful consequence of this is that many results may be proved for the incomplete version of the largest theory of modules (which is just the theory of modules) and then specialized (or "localized") to arbitrary such theories.

In our example, the largest theory of modules is that of $M^{\aleph_0}$, or just of M if the base field K is infinite, where $M = P_1 \oplus P_2 \oplus S_1$. This theory has Morley rank 4, being built up in an obvious sense from the theories of the three indecomposable modules (which have Morley ranks respectively 2, 1, 1). Let me emphasize that this is much simpler than the typical category of modules, for instance the structure even of $\mathbb{Z}$-modules is impossibly complicated once one moves away from nice (finitely generated, pure-injective, ...) kinds of module. Nevertheless, it is a result of Ziegler [33, 4.10] that every module is elementarily equivalent to a direct sum of pure-injective modules (these are the modules, also termed algebraically compact, which are injective over pure embeddings) (Prior to the proof of pp-elimination of quantifiers, it was already a result, [32, Corollary 4 to Theorem 4], of Sabbagh that every module is elementarily equivalent to a pure-injective module, indeed to its pure-injective hull.)

This has further effects. For example, if T is a superstable theory of modules and if the class of models of T is closed under direct sums then T must be totally transcendental so this condition (closure under direct sums) on T erases some model-theoretic subtleties; on the other hand the condition of total transcendence is the base of a tower of finiteness conditions ("m-dimension," equivalently "Krull–Gabriel dimension," see, e.g., [25]) with totally transcendental = dimension 0 (and with, for example, the theory of abelian groups = $\mathbb{Z}$-modules having dimension 2).

2.2. A slightly category-theoretic view. Everything we have said about modules works for much more general additive structures, including additive functors, chain complexes (of modules), sheaves of modules over sufficiently nice spaces, comodules, "abelian structures" in the sense of Fisher. To reach and precisely describe that level of generality, we first note that modules, even at this simplest level of description, are functors. For a ring is just a preadditive category with a single object, where we say that a category is *preadditive* if each of its hom sets (A, B) (our notation for the set of arrows from the object A to the object B) has an abelian group structure and if composition on either side is linear. If such a category has a single object $*$ then the set $(*, *)$ of endomorphisms of that object is therefore a ring, R say. An additive (i.e., preserving the enriched-in-abelian-groups structure on hom sets) functor from this single-object category to the category **Ab** of abelian groups is given by: an abelian group M (the image of $*$) together with, for each $r \in R = (*, *)$, an endomorphism of M such that, writing composition to the left (so fg means g then f), the result is a left R-module. A right R-module is, therefore, a contravariant functor from this single-object category to **Ab** (usually we regard contravariant functors from a category $\mathcal{C}$ as covariant functors on the opposite category $\mathcal{C}^{\mathrm{op}}$). Furthermore, the R-linear maps between modules are easily checked to be just the natural transformations between the corresponding functors. This view of a module is essentially that expressed by the natural isomorphism $M \simeq \mathrm{Hom}_R(R_R, M)$ (where the free module R_R is, one may check, the contravariant representable functor $(-, *)$).

The point here is that the ring R, a single-object preadditive category, can be replaced by any skeletally small (i.e., up to isomorphism there is just a set of objects) preadditive category $\mathcal{R}$, so right R-modules are replaced by additive functors from $\mathcal{R}^{\mathrm{op}}$ to **Ab**—we will still refer to such functors as $\mathcal{R}$-modules. By choosing the preadditive category $\mathcal{R}$ appropriately, all the above-mentioned kinds of examples, and more, can be realized as $\mathcal{R}$-modules (in particular, [25, Example 10.1.2] should suggest how to cover chain complexes; over sufficiently nice spaces (in particular over (locally) noetherian ones) sheaves of modules form a locally finitely presented category ([29], also [27], see [25, Section 16.3]) so are covered by the next section; ditto for comodules ([6,30], see [25, Section 16.2]); [14, Section 1.9] describes how to incorporate relations into a functional language and hence how to deal with abelian structures in the sense of Fisher). Of course, often the kinds of structures that we want will not be *all* the modules/additive functors, rather those which satisfy some additional axioms, almost invariably of the kind which say that certain pp-pairs are closed (we say that the pp-pair ϕ/ψ is *closed* on a structure, or class of structures, if $|\phi(-)/\psi(-)| = 1$ on that structure/class of structures). That is, typically the kinds of objects we want form a definable subcategory (as defined in Section 2.3) of the category of all $\mathcal{R}$-modules.

In the lines above we referred implicitly to the language $\mathcal{L}_{\mathcal{R}}$ of $\mathcal{R}$-modules: this is set up by introducing a sort, s_A, for each object of $\mathcal{R}$ and, for each arrow $f \in (A, B)$ in $\mathcal{R}$, introducing a function symbol (let us use f also for this) from sort s_B to s_A. On each sort we introduce a symbol for addition and a constant symbol for the 0. Then right $\mathcal{R}$-modules, that is, contravariant functors from $\mathcal{R}$ to **Ab**, are axiomatized as those $\mathcal{L}_{\mathcal{R}}$-structures in which each sort is an abelian group and such that the interpretations of the function symbols satisfy the "addition and multiplication tables" for $\mathcal{R}$. If R is a ring then this reduces precisely to the usual language, and axioms, for R-modules. More detail on these languages, more generally on those for the finitely accessible categories discussed below, may be found in [25, Section 10.2.4, Appendix B].

There is an obvious alternative presentation of our running example, where the category $\mathcal{R}$ has two objects $*_1, *_2$ (corresponding to the vertices of the quiver $\bullet \to \bullet$) and where $(*_1, *_1) = K$, $(*_2, *_2) = K$, $(*_1, *_2) = K$, $(*_2, *_1) = 0$. The K-representations of the quiver can, in a very obvious way, be viewed as functors from this small category to **Ab** (equivalently to Mod-K) and hence as 2-sorted structures.

2.3. The accessible category view. A category $\mathcal{C}$ is *finitely accessible* if it has direct limits (= directed colimits in the more logical terminology), if there is, up to isomorphism, just a set of finitely presented objects and if every object of $\mathcal{C}$ is a direct limit of finitely presented objects. In a general category $\mathcal{C}$ an object $C \in \mathcal{C}$ is said to be *finitely presented* if the functor $(C, -)$ commutes with direct limits; in all the examples we will consider, this will reduce to "finitely generated and finitely related." See [1, 8, 18] for background on these categories. For example, the category Mod-R is finitely accessible since every module is a direct limit of finitely presented modules; indeed it is (as is Mod-$\mathcal{R}$) an example of a *locally finitely presented* category, meaning that it is finitely accessible and has all limits and colimits (this will always mean over (co)limits over *small* diagrams). We denote by $\mathcal{C}^{\mathrm{fp}}$ the full subcategory of finitely presented objects — since $\mathcal{C}$ is finitely accessible this is skeletally small and sometimes, indeed in the next paragraph, we will replace it by a small category to which it is equivalent.

To a finitely accessible category $\mathcal{C}$ we associate a language $\mathcal{L}(\mathcal{C})$: for each object A of (a small version of) $\mathcal{C}^{\mathrm{fp}}$ we have a sort s_A and for each arrow $f\colon A \to B$ in $\mathcal{C}$ we introduce a function symbol (also denoted f) with domain s_B and codomain s_A. Each object C of $\mathcal{C}$ becomes an $\mathcal{L}(\mathcal{C})$-structure by: $s_A(C) = (A, C)$ — that is, the elements of C of sort A are just the morphisms from A to C — and f^C is defined to be "composition with f," that is the function $(f, C)\colon (B, C) \to (A, C)$ which takes an arrow $g\colon B \to C$ to the composition $fg\colon A \to C$. Of course we saw essentially this construction just above and the link becomes explicit when we realize that this way of making objects of $\mathcal{C}$ into $\mathcal{L}(\mathcal{C})$-structures is to regard them as contravariant functors from $\mathcal{C}^{\mathrm{fp}}$ to the category **Set** of sets. That is, our language $\mathcal{L}(\mathcal{C})$ really is the one appropriate for functors from $\mathcal{C}^{\mathrm{fp}}$ to **Set** and what we have done to $\mathcal{C}$ is to apply the restricted Yoneda embedding $\mathcal{C} \to \big((\mathcal{C}^{\mathrm{fp}})^{\mathrm{op}}, \mathbf{Set}\big)$, which takes $C \in \mathcal{C}$ to the representable functor $(-, C)$ and has the obvious effect on morphisms (that this is a full embedding follows from $\mathcal{C}$ being finitely accessible); thus we regard objects of $\mathcal{C}$ as $\mathcal{L}(\mathcal{C})$-structures. As will be illustrated in the next example, it is sufficient to use, in place of $\mathcal{C}^{\mathrm{fp}}$, some full subcategory which is, in a suitable sense, generating.

Consider the case that $\mathcal{C}$ is the category Mod-R of modules over some ring (or small preadditive category) R. Applying the construction of the above paragraph we obtain the view of an R-module M as a structure with a sort for each (isomorphism type of) finitely presented module A, with the elements of M of sort A being exactly the R-linear maps from A to M. We may, however, use smaller languages without losing any expressive power: we could use, as a basis for the language, just the generator R_R in place of the whole of mod-R = (Mod-R)^{fp} then, via the natural identification of M with $\mathrm{Hom}_R(R_R, M)$, we would have the usual language for R-modules. If, instead, we were to use the somewhat larger generating set $\{R_R^n : n \geq 1\}$ then we would have a language where n-tuples of elements of a module are literally elements (of an appropriate sort) of that module. Similarly we can, by choosing generators and relations for $A \in$ mod-R, view the elements of a module M of sort A as certain finite tuples of elements of M which satisfy certain linear relations.

As explained above, the functorial view of this is that we apply the restricted Yoneda embedding Mod-$R \to \big((\text{mod-}R)^{\mathrm{op}}, \mathbf{Ab}\big)$ and thus replace a module M by the representable functor, $(-, M)$ restricted to the finitely presented modules. We view the result using the appropriate language $\mathcal{L}$(mod-R) which has functors from mod-R to **Ab** as natural structures. Also as remarked above we may restrict the functor $(-, M)$ even more, for instance to its value at R_R — in which case we obtain the usual, single-sorted, language for R-modules.

(One might note that the functor category $\big((\text{mod-}R)^{\mathrm{op}}, \mathbf{Ab}\big)$ is itself finitely accessible, indeed is the category of right "modules" over $\mathcal{R}$ = mod-R. The natural language, in the sense defined above, for this category has a sort for each finitely presented functor; note that those functors/sorts which arise from finitely presented *modules*, that is the representable functors, form a proper subset.)

Of course not all additive functors on mod-R are R-modules but the next result identifies those which are.

Theorem 2.4 (see [5, Sections 1.4, 2.2]**).** *Let $\mathcal{C}$ be a finitely accessible additive category. Then the restricted Yoneda embedding $\mathcal{C} \to ((\mathcal{C}^{\mathrm{fp}})^{\mathrm{op}}, \mathbf{Ab})$ identifies $\mathcal{C}$ with the flat objects of $((\mathcal{C}^{\mathrm{fp}})^{\mathrm{op}}, \mathbf{Ab})$, equivalently, if $\mathcal{C}$ has cokernels, with the left exact functors from $(\mathcal{C}^{\mathrm{fp}})^{\mathrm{op}}$ to* **Ab**.

If $\mathcal{R}$ is a small preadditive category then a right $\mathcal{R}$-module F is *flat* if the functor $F \otimes_{\mathcal{R}} -\colon \mathcal{R}\text{-Mod} \to \mathbf{Ab}$, which takes a left $\mathcal{R}$-module L to the abelian group $F \otimes_{\mathcal{R}} L$ is *exact* (i.e., takes exact sequences in $\mathcal{R}$-Mod to exact sequences of abelian groups). The tensor product used here generalizes the usual one over rings and is characterized by being *right exact* (takes right exact sequences to right exact sequences) and, for each object A of $\mathcal{R}$, taking the value FA on the left $\mathcal{R}$-module $(A,-)$. Such exactness properties can be axiomatized by declaring certain pp-pairs to be closed and so the image of $\mathcal{C}$ in $((\mathcal{C}^{\mathrm{fp}})^{\mathrm{op}}, \mathbf{Ab})$ is a definable subcategory of $((\mathcal{C}^{\mathrm{fp}})^{\mathrm{op}}, \mathbf{Ab})$ where, for a skeletally small preadditive category $\mathcal{R}$, we say that a subcategory of Mod-$\mathcal{R}$ is *definable* if its objects form the class of models of a theory which is axiomatized by declaring certain pp-pairs to be closed. Equivalent conditions are given in the next result.

Theorem 2.5 (see [25, 3.4.7]**).** *Suppose that $\mathcal{R}$ is a skeletally small preadditive category and let $\mathcal{D}$ be a full subcategory of* Mod-$\mathcal{R}$. *Then*

(i) *$\mathcal{D}$ is a definable subcategory iff*
(ii) *$\mathcal{D}$ is an axiomatisable subcategory of* Mod-$\mathcal{R}$ *closed under (finite, hence arbitrary) direct sums and direct summands iff*
(iii) *$\mathcal{D}$ is closed under direct products, direct limits and pure subobjects.*

Indeed, the categories which arise as definable subcategories of such functor = "module" categories seem to be the context within which the model theory of modules generalises in a natural way; see [25, Part III], also [26]. An alternative characterization of these categories will be given in Section 3.

Probably it has not escaped the reader's notice that the module context, where we consider additive functors to **Ab**, though not a specialization of, is parallel to (alternatively, is an enrichment of), the general context where we consider arbitrary functors to **Set**. We will say more about this in Section 3.

We return to our running example. In that case mod-R consists just of finite direct sums of the three indecomposable R-modules P_1, P_2, S_1. The value of an additive functor on such direct sums is entirely determined by its values on the three indecomposables and on the morphisms between them; it follows easily that the functor category $((\text{mod-}R)^{\mathrm{op}}, \mathbf{Ab})$ is equivalent to the category of contravariant additive functors on the full subcategory of mod-R with objects these three modules. But any category of the form $(\mathcal{A}^{\mathrm{op}}, \mathbf{Ab})$ where $\mathcal{A}$ has just finitely many objects is equivalent to the category of left modules over S where S is the endomorphism ring of the direct sum of all the objects in $\mathcal{A}$. When R is a ring of finite representation type, the endomorphism ring, S, of the direct sum of one copy of each indecomposable module is referred to as the *Auslander algebra* of R. Thus, the functor category $(\text{mod-}R, \mathbf{Ab})$ is equivalent to S-Mod where $S = \mathrm{End}(S_1 \oplus P_1 \oplus P_2)$.

It is not difficult to calculate this ring explicitly, obtaining $S = \begin{pmatrix} K & K & 0 \\ 0 & K & K \\ 0 & 0 & K \end{pmatrix}$. It happens, in this example, that S also is of finite representation type and so a complete description may be given of its modules and hence the functor category — we will give this when we have obtained another view of this category.

(The switch to $(\text{mod-}R, \mathbf{Ab})$ from $\big((\text{mod-}R)^{\text{op}}, \mathbf{Ab}\big)$ is for use below; in any case, for any finite-dimensional algebra R there is a duality $(\text{mod-}R)^{\text{op}} \simeq R\text{-mod}$, so the method allows us to compute both functor categories.)

2.4. The topos-like view. In moving to the accessible category view of R-modules, we added new sorts and function symbols to the language $\mathcal{L}_R$, corresponding exactly to the finitely presented objects of Mod-R; the next step is to add new sorts and function symbols corresponding to finitely presented functors. We will explain this but will take a different route to the underlying category.

We define, following [14] and [9], a pp variant of Shelah's imaginaries — the category $\mathbb{L}_R^{\text{eq}+}$ of *pp imaginaries* for R-modules. The objects are pp-pairs ϕ/ψ and the morphisms from ϕ/ψ to ϕ'/ψ' are the pp-definable functions from the first sort to the second. That is, if ϕ (so ψ) is in free variables $\bar{x}$ and if ϕ' is in free variables $\bar{y}$ then an arrow from ϕ/ψ to ϕ'/ψ' is given by a pp formula $\rho(\bar{x}, \bar{y})$ such that in every R-module $\phi(\bar{x}) \wedge \rho(\bar{x}, \bar{y}) \to \phi'(\bar{y})$ and $\psi(\bar{x}) \wedge \rho(\bar{x}, \bar{y}) \to \psi'(\bar{y})$ (so ρ well-defines a function from $\phi(M)/\psi(M)$ to $\phi'(M)/\psi'(M)$ for every M). One may check directly, but Theorem 2.7 gives an indirect proof, that this is an abelian category. If T is a theory of modules which is axiomatized by declaring certain pp-pairs to be closed then one may define a similar category relative to T, the only change being that the requirements on ρ above are that it well-define a function in every model of T. The result is again an abelian category which is, in fact, a quotient category (in the sense of categories of fractions) of the first-defined one. This is part of a general theory of localization from the theory of all R-modules to theories of the above form, equivalently moving from Mod-R to the definable subcategory consisting of all models of T.

We can arrive at the category $\mathbb{L}_R^{\text{eq}+}$ by another route.

Observe that if ϕ is a pp formula in $\mathcal{L}_R$ then its application to R-modules defines a functor F_ϕ from Mod-R to $\mathbf{Ab}$, namely $F_\phi(M) = \phi(M)$ and, since morphisms preserve pp formulas (Lemma 2.3), we can define $F_\phi f$, where $f\colon M \to N$, to be f restricted, respectively, corestricted to $\phi(M)$, resp. $\phi(N)$. In fact F_ϕ is determined by its action on finitely presented modules because every module is a direct limit of finitely presented modules and because pp formulas commute with direct limits in the following sense.

Proposition 2.6. *Suppose that $\mathcal{C}$ is a locally finitely presentable additive category and that ϕ is a pp formula in a language based on a set $\mathcal{G}$ of finitely presented objects (i.e., the sorts and function symbols are based in the sense we have seen already on the full subcategory $\mathcal{G}$ of $\mathcal{C}^{\text{fp}}$). Suppose that $(C_\lambda)_\lambda$ is a directed system of objects of $\mathcal{C}$ with direct limit C. Then $\phi(C) = \varinjlim \phi(C_\lambda)$.*

PROOF (sketch). Because solution sets of pp formulas are preserved by morphisms in $\mathcal{C}$ the $\phi(C_\lambda)$ do form a directed system. It is enough to show that every element of $\phi(C)$ is the image of some element of some $\phi(C_\lambda)$ under the canonical map from C_λ to C. Because solutions of ϕ and witnesses to quantifiers are morphisms from finitely presented objects, these lift through the system (being the defining property of finitely presented) and yield a λ and a solution of the sort required. $\square$

So we will regard F_ϕ as an additive (covariant) functor from mod-R to **Ab** but also identify it with its unique extension to a functor on Mod-R which commutes with direct limits. If ϕ/ψ is a pp-pair then F_ψ is a subfunctor of F_ϕ and so we may form the quotient functor F_ϕ/F_ψ, which we will denote $F_{\phi/\psi}$, which takes M to the quotient group $\phi(M)/\psi(M)$. (As explained earlier it is reasonable to think of these additive functors as modules — indeed they are "modules" over the preadditive category mod-R — and concepts, such as sub- and quotient objects, which make sense for modules make equal sense for functors.) It is not difficult to show that the pp-defined functions between sorts of the form ϕ/ψ give natural transformations between the corresponding functors $F_{\phi/\psi}$; with some more effort one can show that this gives an equivalence ([3]).

Theorem 2.7. *The category $\mathbb{L}_R^{\mathrm{eq}+}$ of pp imaginaries for R-modules is equivalent to the category* (mod-R, **Ab**)^{fp} *of finitely presented* (*covariant*) *functors from finitely presented modules to abelian groups.*

Because the functor category (mod-R, **Ab**) is locally coherent ([2]) the full subcategory of finitely presented objects is abelian, hence $\mathbb{L}_R^{\mathrm{eq}+}$ is an abelian category. (There are strong relations between the various functor categories $\big((\text{mod-}R)^{\mathrm{op}}, \mathbf{Ab}\big)$, (mod-$R$, **Ab**), ($R$-mod, **Ab**); in particular, there is an anti-equivalence between the categories (mod-R, **Ab**)^{fp} and (R-mod, **Ab**)^{fp}. This is part of "elementary duality." For all this see [9, 10, 23, 25].)

We return to our running example. As said above, for a ring R of finite representation type, the functor category (mod-R, **Ab**) is equivalent to the category of left modules over the Auslander algebra of R. In our example this algebra is $S = \mathrm{End}(S_1 \oplus P_1 \oplus P_2) = \left(\begin{smallmatrix} K & K & 0 \\ 0 & K & K \\ 0 & 0 & K \end{smallmatrix}\right)$. This is the path algebra of the quiver $\bullet \xrightarrow{\beta} \bullet \xrightarrow{\gamma} \bullet$ with the relation $\beta\gamma = 0$. It is easy to compute (the Auslander – Reiten quiver and hence) the indecomposable modules over this ring — it turns out that there are 5 of them. (If you want to do the computation you should bear in mind that left S-modules are representations of the same quiver but with arrows reversed.) Three of these correspond to the representable functors $(S_1, -), (P_1, -)$ and $(P_2, -)$; there is also the simple quotient of the second by the first of these, and then the simple quotient of $(P_2, -)$ by that simple. These indecomposables are, respectively, the functors which take a module M to: $Me_1; Me_1 \oplus M\alpha; M\alpha; Me_2; Me_2/M\alpha$ (with notation as when the example was first introduced). Clearly one can, for each of these, write down a pp-pair which defines the corresponding sort.

Thus, in this example, every pp-sort is a direct sum of copies of the five indecomposable sorts listed above. With this approach one can also compute the pp-defined maps between sorts and hence obtain a complete description of the category $\mathbb{L}_R^{\mathrm{eq}+}$. As remarked earlier, this example is, however, far from typical and, in general, one can expect to obtain only partial information about the category of functors/pp-sorts.

The theorem above holds just as well if the ring R is replaced by a skeletally small preadditive category $\mathcal{R}$. If $\mathcal{D}$ is a definable subcategory of Mod-$\mathcal{R}$ then the category $\mathbb{L}^{\mathrm{eq}+}(\mathcal{D})$ of pp-imaginaries for $\mathcal{D}$ is in fact a localization of

$\mathbb{L}_{\mathcal{R}}^{\mathrm{eq}+} \simeq (\text{mod-}\mathcal{R}, \mathbf{Ab})^{\mathrm{fp}}$, namely that obtained by forming the quotient by the (Serre) subcategory consisting of all pp-pairs which are closed on $\mathcal{D}$ (equivalently, those finitely presented functors which are zero on $\mathcal{D}$). If $\mathcal{D}$ is axiomatised by T then we also write $\mathbb{L}^{\mathrm{eq}+}(T)$ for this category of pp-imaginaries.

The connection with the topos view is that the category of pp-imaginaries is a variant (the version for regular formulas, in the terminology of [12]) of the syntactic category of a geometric theory, see [12, Vol. 2, Section D1.4; 16, Section 10.5; 19]. It is not quite the same since, if T' is a conservative extension of a geometric (or coherent, or regular,...) theory T constructed using only formulas of the kind allowed (geometric, coherent, regular,...) then the syntactic categories of T and T' are not necessarily equivalent categories (though they will lead to the same category of sheaves — the classifying topos), whereas our construction of the category $\mathbb{L}^{\mathrm{eq}+}$ from T is invariant under pp-defined conservative extensions of T.

We recall the shape of things here. A Grothendieck site consists of a small category $\mathcal{C}$ together with a Grothendieck topology J — a notion of "covering" — on $\mathcal{C}$; one way of giving the latter is to choose, for each object $C \in \mathcal{C}$ a set of subfunctors of the representable functor $(-, C)\colon \mathcal{C} \to \mathbf{Set}$ — of course the sets of subfunctors so chosen must together satisfy certain axioms (see [12, 16]). This generalizes topological spaces with the usual notion of "covering": in that case the small category is the collection of open sets, regarded as a poset and hence as a category, and a set of objects (open sets) of that category is said to cover another object if the latter is their poset-join (i.e., union). One defines a *presheaf* over $\mathcal{C}$ to be simply a contravariant functor from $\mathcal{C}$ to $\mathbf{Set}$. The category $(\mathcal{C}^{\mathrm{op}}, \mathbf{Set})$ of presheaves depends only on $\mathcal{C}$ but one may then define what it is for a presheaf to be a sheaf for the chosen Grothendieck topology J. There is a corresponding process of sheafification, which is a functor from the category, $(\mathcal{C}^{\mathrm{op}}, \mathbf{Set})$, of presheaves to the category, $\mathrm{Sh}_J(\mathcal{C})$, of J-sheaves on $\mathcal{C}$. This includes and generalizes the usual notion of sheafification for sheaves of sets over topological spaces. A category which arises in this way, as a category of sheaves, is said to be a *Grothendieck topos.*

The category $\mathrm{Sh}_J(\mathcal{C})$ of sheaves turns out to be the classifying topos for the models of a geometric theory: the language for that theory is defined in terms of the small category $\mathcal{C}$ and, very roughly, the axioms say when a set of objects covers another (see [4] for an exposition). In general that theory will be "geometric," meaning that it is axiomatized by formulas of the form $\phi \to \psi$ where ϕ and ψ are geometric formulas which, in turn, means built from atomic formulas (including "false") using finite conjunction, existential quantification and infinite disjunctions. If $\mathcal{E}$ is a topos then one may makes sense of what it is for an object E of $\mathcal{E}$ to be a model in $\mathcal{E}$ of a geometric theory T; in particular one may consider the category (with homomorphisms, not elementary embeddings), $\mathrm{Mod}_{\mathbf{Set}}(T)$, of models of T in the topos $\mathbf{Set}$.

If T is defined from the site $(\mathcal{C}, J)$ in the way referred to above, more generally if T is any geometric theory, then one builds from T its syntactic category $\mathbf{B}(T)$ together with a Grothendieck topology J on $\mathbf{B}(T)$ and proves that the category $\mathcal{B}(T) = \mathrm{Sh}_J(\mathbf{B}(T))$ of sheaves for that site is the classifying topos for models of T in the sense that, for any (Grothendieck) topos $\mathcal{E}$, in particular for $\mathcal{E} = \mathbf{Set}$, the $\mathcal{E}$-models of T are in natural bijection to the geometric morphisms from $\mathcal{E}$ to $\mathcal{B}(T)$ (a geometric morphism is a generalization of the direct and inverse image functors on sheaves induced by a continuous map of topological spaces); more precisely, there is a "universal" model U for T in $\mathcal{B}(T)$ which has the property that for any (Grothendieck) topos $\mathcal{E}$ and model E of T in $\mathcal{E}$, there is a unique geometric morphism from $\mathcal{E}$ to $\mathcal{B}(T)$ such that E is the pullback along this geometric morphism of the universal model U: $\mathrm{Mod}_{\mathcal{E}}(T) \simeq \mathrm{Geom}(\mathcal{E}, \mathcal{B}(T))$.

We compare with what we have in the additive context. The additive analogue of the syntactic category $\mathbf{B}(T)$, at least if one adds enough sorts to the language, is the category $\mathbb{L}^{\mathrm{eq}+}(T)$ of pp-imaginaries. The equivalence of the latter, at least in the case where T is the theory of $\mathcal{R}$-modules for some small preadditive category $\mathcal{R}$, with the category of finitely presented functors in $(\text{mod-}\mathcal{R}, \mathbf{Ab})$ also is essentially there in the topos context, with the category of coherent objects of the classifying topos $\mathcal{B}(T)$ replacing the category of finitely presented objects ([19], also [31] for a proof like that in the additive context). This comparison suggests that the functor category $(\text{mod-}\mathcal{R}, \mathbf{Ab})$ is the additive analogue of the classifying topos for (the theory of) the category of right $\mathcal{R}$-modules. The extension to more general additive definable categories also follows the topos picture in that the general case can be obtained by localizing from such (well-behaved) "presheaf" categories — this modus operandi is often used in the additive case in [25] and in the topos context in [4].

3. Additive geometry

If we recall that topos theory stemmed from Grothendieck's scheme-theoretic development of algebraic geometry then we naturally ask whether some of that development has parallels in the additive context and, indeed, this is the case, as indicated next.

One starting point is the observation of Ivo Herzog that if R is a ring and $\mathbb{L}_R^{\mathrm{eq}+}$ is the category of pp-imaginaries defined earlier ($\text{Mod-}R^{\mathrm{eq}+}$ in the notation of [9]) then the R-modules can be identified with the exact functors from $\mathbb{L}_R^{\mathrm{eq}+}$ to $\mathbf{Ab}$. The general result (stated in slightly different terms) is due to Henning Krause.

Theorem 3.1 ([13, 2.9]). *If $\mathcal{D}$ is a definable additive category and if $\mathbb{L}^{\mathrm{eq}+}(\mathcal{D})$ is its category of pp-imaginaries then $\mathcal{D}$ is equivalent to the category $\mathrm{Ex}(\mathbb{L}^{\mathrm{eq}+}(\mathcal{D}), \mathbf{Ab})$ of exact functors from $\mathbb{L}^{\mathrm{eq}+}(\mathcal{D})$ to $\mathbf{Ab}$.*

The equivalence is given by evaluation: to an object D of $\mathcal{D}$ we associate the functor which takes $F \in \mathbb{L}^{\mathrm{eq}+}(\mathcal{D})$ to the value, FD, of F at D; this is an exact functor on $\mathbb{L}^{\mathrm{eq}+}(\mathcal{D})$ and it turns out that every exact functor from $\mathbb{L}^{\mathrm{eq}+}(\mathcal{D})$ to $\mathbf{Ab}$ arises in this way from an object of $\mathcal{D}$.

Krause [13, 7.2] also showed that if $\mathcal{C}$ is a finitely accessible additive category and if $\mathcal{D}$ is a definable additive category then the exact functors from $\mathbb{L}^{\mathrm{eq}+}(\mathcal{D})$ to $\mathbb{L}^{\mathrm{eq}+}(\mathcal{C})$ are in natural bijection with the functors from $\mathcal{C}$ to $\mathcal{D}$ which commute with direct products and direct limits (recall, Theorem 2.5, that every definable category has direct products and direct limits).

Now, the functors from a definable category $\mathcal{C}$ to a definable category $\mathcal{D}$ which commute with direct products and direct limits are exactly the "interpretation functors" — the functors which are induced by a pp-model-theoretic interpretation of models of the theory of $\mathcal{D}$ (i.e., objects of $\mathcal{D}$) in some theory extending the theory of $\mathcal{C}$ (the prefix "pp" means that the formulas used to define new sorts and function symbols all should be pp). For more on these functors see [23] or [26]. The general result, removing the restriction that $\mathcal{C}$ be finitely accessible, is [26, 12.10] and can be expressed as an equivalence of 2-categories, as follows.

Let $\mathbb{DEF}$ be the 2-category whose objects are the definable additive categories, whose (1-)arrows are the functors which commute with direct products and direct limits, equivalently the interpretation functors, and whose 2-arrows are the natural transformations between such functors. Let $\mathbb{ABEX}$ be the 2-category whose objects are the skeletally small abelian categories, whose arrows are the exact functors and whose 2-arrows are the natural transformations.

Theorem 3.2 ([28, 2.3]). *There is an anti-equivalence of 2-categories* $\mathbb{DEF} \simeq (\mathbb{ABEX})^{\mathrm{op}}$.

Thus pp-interpretation functors between definable additive categories correspond naturally (and contravariantly) to exact functors between their categories of pp-imaginaries. In fact, much of this picture, in the **Set**-context (but see [17, Section 6] for the additive context), is already in the work [11, 17] of Makkai and Hu.

Theorem 3.2 above can be seen as an extension of the anti-equivalence between the category of commutative rings and the category of affine schemes, as follows. First we recall Freyd's notion, and proof of existence ([7, 4.1]) of the free abelian category of a ring R, more generally of any small preadditive category $\mathcal{R}$. This is an additive functor $\mathcal{R} \to \mathrm{Ab}(\mathcal{R})$ with $\mathrm{Ab}(\mathcal{R})$ abelian and such that every additive functor from $\mathcal{R}$ to an abelian category $\mathcal{B}$ factors through $\mathcal{R} \to \mathrm{Ab}(\mathcal{R})$ via an exact functor, unique up to natural equivalence, from $\mathrm{Ab}(\mathcal{R})$ to $\mathcal{B}$. In particular a homomorphism $R \to S$ of rings induces an essentially unique exact functor $\mathrm{Ab}(R) \to \mathrm{Ab}(S)$. The relevance of this is explained by the next result.

Theorem 3.3. *For any small preadditive category* $\mathcal{R}$, *the free abelian category* $\mathrm{Ab}(\mathcal{R})$ *is naturally equivalent to the category* $(\mathcal{R}\text{-mod}, \mathbf{Ab})^{\mathrm{fp}}$ *hence, by Theorem* 2.7, *to* $\left((\text{mod-}\mathcal{R}, \mathbf{Ab})^{\mathrm{fp}}\right)^{\mathrm{op}}$, *that is, to* $(\mathbb{L}_{\mathcal{R}}^{\mathrm{eq}+})^{\mathrm{op}}$.

Thus any homomorphism of rings $R \to S$ induces an exact functor between the associated abelian categories $\mathbb{L}_R^{\mathrm{eq}+} \to \mathbb{L}_S^{\mathrm{eq}+}$ (which, in terms of formulas, is just replacement of (function symbols corresponding to) elements of R by their images in S) and this gives us a (non-full) embedding of the category of rings and ring homomorphisms into $\mathbb{ABEX}$. If R is a commutative ring then, [28, 5.2], it may be recovered as the *center* of the category $\mathrm{Ab}(R)$ (that is, the ring of endomorphisms of the identity functor on $\mathrm{Ab}(R)$), so this is the sense in which $\mathbb{ABEX}$ generalizes the category of commutative rings.

On the other side of the algebraic-geometric picture we have the category of schemes — sheaves of rings obtained by glueing together affine schemes, where an affine scheme is the prime spectrum of a commutative ring equipped with the Zariski topology and the sheaf of localizations of that ring. In our case we have, at the moment, just $\mathbb{DEF}$. But to each definable additive category $\mathcal{D} \in \mathbb{DEF}$ we may associate a topological space, the rep-Zariski spectrum $\mathrm{Zar}(\mathcal{D})$. This has, for its underlying set, the set $\mathrm{pinj}(\mathcal{D})$ of isomorphism classes of indecomposable pure-injective objects of $\mathcal{D}$. A basis for the topology consists of the sets of the form $[\phi/\psi] = \{N \in \mathrm{pinj}(\mathcal{D}) : \phi(N) = \psi(N)\}$ where ϕ/ψ ranges over pp-pairs in the language for $\mathcal{D}$ (this is the "Hochster dual" of the Ziegler topology on $\mathrm{pinj}(\mathcal{D})$). See [26, 28] for details of this and what follows. In the case that $\mathcal{D}$ is the definable category of injective modules over a commutative noetherian ring, this space is exactly the prime spectrum of the ring with the Zariski topology. Over the space $\mathrm{Zar}(\mathcal{D})$ there is a sheaf of small abelian categories — it is the sheafification of the "presheaf of definable scalars." That presheaf is defined on a basis of open sets as follows. To each extension T' of the theory of $\mathcal{D}$ obtained by declaring finitely many pp-pairs to be closed, we have the corresponding category $\mathcal{D}'$ of models — a definable subcategory of $\mathcal{D}$; the corresponding open set is just $\mathcal{D}' \cap \mathrm{pinj}(\mathcal{D})$ and to this open set we assign the category $\mathbb{L}^{\mathrm{eq}+}(\mathcal{D}')$ of pp-definable imaginaries of $\mathcal{D}'$. It is the sheafification of this presheaf — a "categoried space" — which we associate to $\mathcal{D}$ and which we regard as our generalization of the notion of affine scheme (taking the view that a scheme is a ringed space — I remark that the functor-of-points view also makes sense in the general context we are outlining). In the paper [28] we show in a precise sense how this does extend the usual ringed space associated to a commutative ring. In the case that $\mathcal{D}$ is Mod-R, or any definable subcategory of Mod-R, we can look just at the definable maps from the home sort to itself and thus obtain a (pre)sheaf of rings which are in some sense localizations of R — this is the origin of the term "definable scalars." Again, in the case that $\mathcal{D}$ is the category of injective modules over a commutative Noetherian ring, this is the usual structure sheaf of localizations of the ring.

References

1. J. Adámek and J. Rosický, *Locally presentable and accessible categories*, London Math. Soc. Lecture Note Ser., vol. 189, Cambridge Univ. Press, Cambridge, 1994.
2. M. Auslander, *Coherent functors*, Proceedings of the Conference on Categorical Algebra (La Jolla, CA, 1965) (S. Eilenberg, D. K. Harrison, S. Mac Lane, and H Röhrl, eds.), Springer, New York, 1966, pp. 189–231.
3. K. Burke, *Some model-theoretic properties of functor categories for modules*, Doctoral Thesis, University of Manchester, 1994.
4. O. Caramello, *The duality between Grothendieck toposes and geometric theories*, Doctoral Thesis, University of Cambridge, 2009.
5. W. Crawley-Boevey, *Locally finitely presented additive categories*, Comm. Algebra **22** (1994), no. 5, 1641–1674.
6. S. Crivei, M. Prest, and G. Reynders, *Model theory of comodules*, J. Symbolic Logic **69** (2004), no. 1, 137–142.
7. P. Freyd, *Representations in abelian categories*, Proceedings of the Conference on Categorical Algebra (La Jolla, CA, 1965) (S. Eilenberg, D. K. Harrison, S. Mac Lane, and H Röhrl, eds.), Springer, New York, 1966, pp. 95–120.
8. P. Gabriel and F. Ulmer, *Lokal präsentierbare Kategorien*, Lecture Notes in Math., vol. 221, Springer, Berlin, 1971.
9. I. Herzog, *Elementary duality of modules*, Trans. Amer. Math. Soc. **340** (1993), no. 1, 37–69.
10. ———, *Contravariant functors on the category of finitely presented modules*, Israel J. Math. **167** (2008), 347–410.
11. H. Hu, *Dualities for accessible categories*, Category Theory 1991 (Montréal, QC, 1991) (R. A. G. Seely, ed.), CMS Conf. Proc., vol. 13, Amer. Math. Soc., Providence, RI, 1992, pp. 211–242.
12. P. T. Johnstone, *Sketches of an elephant: a topos theory compendium.* Vol. 1, Oxford Logic Guides, vol. 43, Oxford Univ. Press, Oxford, 2002; Vol. 2, vol. 44.
13. H. Krause, *Exactly definable categories*, J. Algebra **201** (1998), no. 2, 456–492.
14. T. G. Kucera and M. Prest, *Imaginary modules*, J. Symbolic Logic **57** (1992), no. 2, 698–723.
15. ———, *Four concepts from "geometrical" stability theory in modules*, J. Symbolic Logic **57** (1992), no. 2, 724–740.
16. S. Mac Lane and I. Moerdijk, *Sheaves in geometry and logic*, Universitext, Springer, New York, 1992.
17. M. Makkai, *A theorem on Barr-exact categories, with an infinitary generalization*, Ann. Pure Appl. Logic **47** (1990), no. 3, 225–268.
18. M. Makkai and R. Paré, *Accessible categories: the foundations of categorical model theory*, Contemp. Math., vol. 104, Amer. Math. Soc., Providence, RI, 1989.

19. M. Makkai and G. E. Reyes, *First order categorical logic: Model-theoretical methods in the theory of topoi and related categories*, Lecture Notes in Math., vol. 611, Springer, Berlin, 1977.
20. M. Prest, *Elementary torsion theories and locally finitely presented categories*, J. Pure Appl. Algebra **18** (1980), no. 2, 205–212.
21. ———, *Model theory and modules*, London Math. Soc. Lecture Note Ser., vol. 130, Cambridge University Press, Cambridge, 1988.
22. ———, *Remarks on elementary duality*, Ann. Pure Appl. Logic **62** (1993), no. 2, 185–205.
23. ———, *Interpreting modules in modules*, Ann. Pure Appl. Logic **88** (1997), no. 2-3, 193–215.
24. ———, *Model theory and modules*, Handbook of Algebra. Vol. 3 (M. Hazewinkel, ed.), North-Holland, Amsterdam, 2003, pp. 227–253.
25. ———, *Purity, spectra and localisation*, Encyclopedia Math. Appl., vol. 121, Cambridge Univ. Press, Cambridge, 2009.
26. ———, *Definable additive categories: purity and model theory*, Mem. Amer. Math. Soc. **210** (2011).
27. M. Prest, V. Puninskaya, and A. Ralph, *Some model theory of sheaves of modules*, J. Symbolic Logic **69** (2004), no. 4, 1187–1199.
28. M. Prest and R. Rajani, *Structure sheaves of definable additive categories*, J. Pure Appl. Algebra **214** (2010), no. 8, 1370–1383.
29. M. Prest and A. Ralph, *Locally finitely presented categories of sheaves of modules*, preprint, available at `http://www.maths.manchester.ac.uk/~mprest/publications.html`.
30. M. Prest and R. Wisbauer, *Finite presentation and purity in categories $\sigma[M]$*, Colloq. Math. **99** (2004), no. 2, 189–202.
31. R. Rajani and M. Prest, *Model-theoretic imaginaries and coherent sheaves*, Appl. Categ. Structures **17** (2009), no. 6, 517–559.
32. G. Sabbagh, *Aspects logiques de la pureté dans les modules*, C. R. Acad. Sci. Paris Sér. A-B **271** (1970), A909–A912.
33. M. Ziegler, *Model theory of modules*, Ann. Pure Appl. Logic **26** (1984), no. 2, 149–213.

School of Mathematics, University of Manchester, Oxford Road, Manchester M13 9PL, UK

E-mail address: `mprest@manchester.ac.uk`

Centre de Recherches Mathématiques
CRM Proceedings and Lecture Notes
Volume **53**, 2011

A Derivation of Einstein's Vacuum Field Equations

Gonzalo E. Reyes

For Mihály: teacher, student, collaborator, friend

ABSTRACT. The aim of this paper is to present a new derivation of Einstein's vacuum field equations which we believe is more physically motivated that the original one. It uses only the notion of connection rather than metric. The starting point is the same as that of Sachs/Wu [13], although the development is quite different. We use infinitesimals extensively as available in Synthetic Differential Geometry to simplify formulation of notions and proofs.

1. Introduction

In his work on General Relativity ([2]), Einstein started from the field formulation of Newton's gravitational theory due to Poisson, i.e., the equation

$$\nabla^2\phi = 4\pi\kappa\rho$$

where ϕ is the gravitational potential, $\nabla^2\phi = \sum_\alpha \partial^2\phi/\partial(x^\alpha)^2$ the Laplacian operator, κ a constant and ρ is the density of matter.

He proceeded to find the analogue of the Laplacian operator in the relativistic 4-dimensional space-time manifold that replaced Newton's 3-dimensional space. For this, he had first to look for the analogue of gravitational potentials. To achieve this, he postulated the existence of a metric that he took as the analogue of the potential. Then he looked for the analogue of the Laplacian operator. He did this somewhat formally: only second order derivatives of the metric should appear, etc. Finally, he sought the analogue of density of matter.

Since we are interested only in the vacuum equations, we start rather from Laplace equation

$$\nabla^2\phi = 0$$

where ϕ is the gravitational potential. Thus, we leave out the question of the analogue of the density of matter.

This limitation has three motivations

2010 *Mathematics Subject Classification.* Primary 83CO5; Secondary 03G30.
Key words and phrases. Einstein vacuum field equations, Synthetic Differential Geometry.
This is the final form of the paper.

(1) According to Einstein himself ([5]), "... [the vacuum equations are the only equations which follow without ambiguity from the fundamental hypothesis of General Relativity...." On the analogue of the density of matter of Poisson's equation equations he says ([4]): "The right side of the [field] equations [the tensor $-kT_{ik}$] is a formal condensation of all things whose comprehension in the sense of a field theory is still problematic." Further, he says "... this formulation was merely a makeshift..."

(2) The vacuum field equations suffice to derive the Schwarzschild solution and deduce the testable consequences of the theory that Einstein mentioned in his work, i.e., the gravitational red shift of spectral lines, the deflection of light by the sun and the precession of the perihelia of the orbits of the inner planets.

(3) The whole derivation can be carried out from the assumption of a space-time with a symmetric connection, rather than a metric.

Returning to Laplace equation, a little familiarity with Linear Algebra brings to light the trivial observation that $\nabla^2\phi$ is the *trace* of the linear transformation Ψ given by the matrix $(\partial^2\phi/\partial x^\alpha \partial x^\beta)_{\alpha\beta}$. This suggests to concentrate on Ψ instead and find a relativistic analogue of this linear transformation.

This is what Sachs and Wu do in their approach to derive Einstein's field equations in [13]. To achieve this aim, they provide a geometrical interpretation of Ψ in terms of acceleration of a neighbor (a vector field along a curve, to be defined below) whose analogue is readily available in the relativistic theory, namely covariant acceleration of the corresponding relativistic neighbor. The map Ψ is in fact defined from Newton's law of motion of freely falling particles (in a gravitational field). The corresponding map in Relativity is likewise defined in terms of the Einstein's Law of Motion of free falling particles.

Einstein's vacuum field equation, in analogy with Laplace's vacuum field formulation of Newtonian theory can be formulated as the statement that the trace of the relativistic analogue of Ψ is zero.

Thus, we derive the vacuum field equations from the law of motion in both cases.

Our approach to derive Einstein's vacuum field equations follows that of Sachs and Wu. However, and in spite of the same starting point, our derivation differs considerably from theirs as anyone browsing their book can see. Among other differences, we do not require a metric but the simpler notion of parallel transport, as mentioned before.

Another difference from both Einstein's and Sachs and Wu's approaches is the extensive use of infinitesimals. In fact, we place ourselves in the context of Synthetical Differential Geometry (SDG). SDG is a mathematical theory whose aim is to provide a rigorous foundation for infinitesimals and infinitesimal structures as used by geometers such as Sophus Lie, Élie Cartan as well as some physicists and engineers. (See [7, 10, 11]).

In SDG, the reals R have the structure of a (nontrivial) ring, not a field, and the main axiom states one of the basic insights of the creators of calculus, namely, that "in the infinitely small, every curve is a line in a unique way." To formulate this axiom, let $D = \{x \in R \mid x^2 = 0\}$. An infinitesimal curve is a function $f \in R^D$.

Axiom of Kock–Lawvere (KL). The map $R \times R \to R^D$ that sends (a, b) into the curve $[d \mapsto a + db]$ is a bijection.

As an application, let $f \in R^R$ and $x_0 \in R$. To define $f'(x_0)$, the derivative of f at the point x_0, consider the infinitesimal curve $\gamma\colon D \to R$ defined by $\gamma(d) = f(x_0 + d)$. By the axiom, $\gamma(d) = a + db$ for a unique couple (a, b). We define $f'(x_0) = b =$ the slope of the line. Thus, $f(x_0 + d) = f(x_0) + df'(x_0)$, since clearly $a = f(x_0)$.

The usual properties of derivatives can be proved, by algebraic manipulations but we will not go into these matters. (See books cited above).

An important property of D that holds in practically all models is that the functor $(-)^D$, itself a right adjoint of $(-) \times D$, has a right adjoint $(-)^{1/D}$. Thus $(-) \times D \dashv (-)^D \dashv (-)^{1/D}$. In particular (and this is the only property of D we will use) $(-)^D$ preserves arbitrary unions.

Besides D, the first order infinitesimals, there are higher order ones such as $D_n = \{x \in R \mid x^{n+1} = 0\}$ and $D_\infty = \{x \in R \mid \exists n\, x^n = 0\}$, the nilpotent reals. Infinitesimals make the main notions of Differential Geometry such as connexion, parallel transport, covariant derivative, etc., quite intuitive and (several) proofs simple algebraic computations. As an example, the tangent bundle of a space M is simply the exponential space M^D and the structure map $\pi_M\colon M^D \to M$ is just "evaluation at 0," i.e., $\pi_M(\gamma) = \gamma(0)$. For "good" spaces, the so-called *microlinear spaces* that play a role in SDG similar to smooth manifolds in classical Differential Geometry, the fiber $M_m = \pi_M^{-1}(m)$ is an R-module.

Before launching ourselves head on into our main concern, let us notice a circumstance that will simplify enormously our task: the notions of velocity and acceleration of particles in Newtonian theory are defined *at a single point.* So is the Laplacian. These notions, for their definition, require to take derivatives which according to SDG require functions to be defined only on an *infinitesimal neighborhood* of the point (rather that an ordinary neighborhood as in classical DG). From now on everything will be infinitesimal: manifolds, geodesics, vector fields, etc.

In more details, open covers to define manifolds will be replaced by infinitesimal covers to obtain D_∞-manifolds.

A *D_∞-manifold of dimension n* is a space having the property that each point has a neighborhood isomorphic to D^n_∞. Clearly, D_∞-manifolds are closed under product and the dimension of the product is the sum of the dimension of the factors. Furthermore, embedded (classical) manifolds in smooth toposes are D_∞-manifolds. These manifolds are microlinear spaces.

Notice that in D_∞-manifolds we can introduce coordinates and make calculations with them, just as in classical manifolds. Although our definitions are invariant and hence independent of coordinates, these will allow to compare ours with the classical ones in the literature.

Trajectories will be *infinitesimal* curves $\gamma\colon D_\infty \to M$. If γ is such a curve, we define its *derivative* to be the map $\gamma^\bullet\colon D_\infty \to M^D$ by $\gamma^\bullet(t)(d) = \gamma(t + d)$. Notice that $\pi_M \circ \gamma^\bullet = \gamma$. If $M = R^n$, then γ may be written in terms of its components, i.e., $\gamma = (\gamma^1, \ldots, \gamma^n)$ and a simple application of KL shows that $\gamma^\bullet(t)(d) = \gamma(t) + d\big((\gamma^1)'(t), \ldots, (\gamma^n)'(t)\big)$. By writing $\gamma'(t) = \big((\gamma^1)'(t), \ldots, (\gamma^n)'(t)\big)$, we have $\gamma^\bullet(t)(d) = \gamma(t) + d\gamma'(t)$.

Furthermore, vector fields on a D_∞ manifold will be maps $Q\colon U \to U^D$ with U an infinitesimal neighborhood of M.

Working with these infinitesimal objects, we are able to carry out our whole derivation completely in the realm of the infinitesimal. This simplifies things considerably. Roughly speaking algebra rather than analysis is required to carry out our proof and in fact, the whole derivation of Einstein's equations from the Sachs and Wu reformulation of Laplace's equation can be carried out synthetically (or axiomatically). As an example, existence and uniqueness of "infinitesimal" solutions of differential equations can be proved (as done, e.g., in [8]). Other examples will appear below.

From a classical point of view we need to deal only with "formal analysis" (i.e., analysis with elements of formal power series, rather than ordinary functions) to carry out our derivation.

The reader is referred to [10, 12] for all that we need from SDG.

Acknowledgements. Among those who have helped and encouraged me, I must mention first and foremost Mihály Makkai who followed a very long one person seminar at McGill with contagious enthusiasm. He wasn't put off by the high density of mistakes per phrase along the way and corrected errors, made criticisms, suggested better alternatives and kept himself and the speaker cheerful during the long process. I cannot thank him enough. Bill Lawvere was responsible for making these ideas "public" by obtaining an invitation for me to participate in the Boston Meeting on "Trends in the Mathematical Representation of Space: Philosophical and Historical Perspectives" on 2007 (November 31 and December 1). His support has given me confidence to continue this work.

2. Newton's theory of Gravitation

Newton's law of motion states that the trajectory of particles of freely falling particles in the gravitational field obey the second-order differential equation

$$\gamma''(t) + \nabla(\phi(\gamma(t)) = 0 \quad \text{for all } t \in D_\infty.$$

In other words, the trajectory are solutions of the autonomous second order differential equation G:

$$y'' + \nabla(\phi(y)) = 0, \quad \text{where } \phi \text{ is the gravitational potential.}$$

written in vector form. Equivalently, the system

$$(y^\alpha)'' + \nabla^\alpha(\phi(y^1, y^2, y^3)) = 0 \quad \text{for } \alpha = 1, 2, 3$$

written in terms of scalar components.

In invariant terms, this differential equation may be considered as a symmetric vector field $G\colon E^D \to (E^D)^D \approx E^{D\times D}$ satisfying $s \circ G = G$ where $s\colon E^{D\times D} \to E^{D\times D}$ is the map induced by the twisting map $\tau\colon D \times D \to D \times D$ given by $\tau(d_1, d_2) = (d_2, d_1)$ (cf. [12, Second order DE's and geodesics.pdf]). The solutions of G are the trajectories of freely falling particles.

The fact that G is not an ordinary vector field on E (since it is defined on E^D rather than E) creates problems for further developments.

Fortunately, G is reducible to a family $\{G^u\}_u$ of ordinary vector fields indexed by non zero vectors in the sense given by the following

Theorem 2.1. *Let $\gamma\colon D_\infty \to E$ be a curve with $\gamma(0) = x$ and $\gamma^\bullet(0) = u$, a nonzero vector in E_x. If γ is a solution of G, then γ is an integral curve of G^u. Furthermore, integral curves of any G^u are solutions of G.*

PROOF. See Section 4 (Theorem 4.3). □

The key notion of the Sachs and Wu approach is that of (the trajectories of) a swarm of infinitely close particles freely falling in the gravitational field.

Let $\{\gamma_h\}_{h\in D}$ be the trajectories of the swarm. We may view this family as a single map

$$W\colon D_\infty \to E^D$$

by defining $W(t)(h) = \gamma_h(t)$.

In fact, it is clear that we have the following one-to-one correspondence for every $v \in M_x = \pi_M^{-1}(x)$.

$$\frac{\text{Families } \{\gamma_h\colon D_\infty \to E\}_{h\in D} \text{ with } \gamma_h(0) = v(h) \text{ for all } h \in D}{\text{Maps } W\colon D_\infty \to E^D \text{ with } W(t)(0) = \gamma_0(t) \text{ and } W(0) = v.}$$

Letting $\gamma = \gamma_0$, we have $W(t)(0) = \gamma_0(t) = \gamma(t)$.

Then W is *a vector field along* γ in the sense that the following diagram

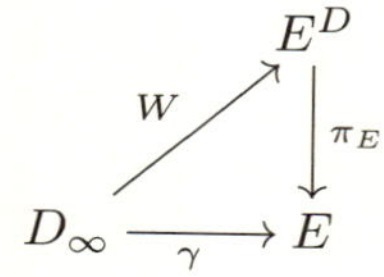

is commutative.

As an example, $\gamma^\bullet$ is a vector field along γ.

Let $\gamma\colon D_\infty \to E$ be a curve and $x = \gamma(0)$, $u = \gamma^\bullet(0)$ a nonzero vector. A *neighbor of* γ *in a vector field* Q is a vector field along γ, $W\colon D_\infty \to E^D$ such that for every $h \in D$, the curve γ_h defined by $\gamma_h(t) = W(t)(h)$ is an integral curve of Q.

Theorem 2.2. *Let γ be a trajectory of a free falling particle, $\gamma(0) = x$, $\gamma^\bullet(0) = u$ (a nonzero vector) and let $v \in E_x$. Then there is a unique neighbor W^v of γ in G^u such that $W^v(0) = v$.*

PROOF. Define $W^v(t)(h) = \gamma_h(t)$, where γ_h is the integral curve of G^u with $\gamma_h(0) = v(h)$. Uniqueness is a consequence of uniqueness of solutions of DEs (cf. [8]). □

By an application of KL, $W^v(t)(h) = \gamma_0(t) + h\delta(t) = \gamma(t) + h\delta(t)$ for a unique function δ.

Let $u \in E_x$ be a non zero vector. By existence and uniqueness of second-order DEs there is a (unique) solution $\gamma\colon D_\infty \to E$ of G with $\gamma(0) = x$ and $\gamma^\bullet(0) = u$. Thus γ is an integral curve of G^u. Define a map

$$\Psi^u\colon E_x \to E_x$$

as follows. Let $v \in E_x$ be arbitrary. Pick the unique neighbor W^v of γ in G^u with $W^v(0) = v$. Define a vector field $(W^v)^{\bullet\bullet}$ along γ by

$$(W^v)^{\bullet\bullet}(t)(h) = \gamma(t) + h\delta''(t)$$

for every t.

Finally, define

$$\Psi^u(v) = (W^v)^{\bullet\bullet}(0).$$

This vector field along γ is already well-known in differential geometry. In fact, leaving out the superscript v from W,

Theorem 2.3. *The following holds*

$$W^{\bullet\bullet} = \nabla^2_{\gamma^\bullet} W$$

where the connection is the Euclidean (or flat) one.

PROOF. Recall that $\nabla_{\gamma^\bullet} W$ is uniquely defined by the property $h\nabla_{\gamma^\bullet} W(t) = \nabla(\gamma(t+h), W(t+h))(-h) - W(t)$ for every $h \in D$. In words, we transport $W(t+h)$ back along γ to $\gamma(t)$ and subtract $W(t)$ (cf. [12, covariant-derivation.pdf, p. 7]).

Since the connection is Euclidean, vectors are preserved and the result is just $h\nabla_{\gamma^\bullet} W(t) = \lambda d[\gamma(t) + d\delta(t+h)] - \lambda d[\gamma(t) + d\delta(t)] = \lambda d(\gamma(t) + d\delta'(t))$ with $\delta(t)$ given by $\gamma_d(t) = \gamma(t) + d\delta(t)$ (KL).

Iterating,

$$(\nabla^2 W)(t) = \lambda d\big(\gamma(t) + d\delta''(t)\big).$$

finishing the proof of the proposition. □

We now prove the main theorem of this section. Notice first that E_x is canonically isomorphic to R^3 by the isomorphism $v \in E_x \mapsto v_0 \in R^3$ where $v(d) = x + dv_0$ (KL). In classical notation, $v = (x, v_0)$ with $x = (x^1, x^2, x^3)$ and $v_0 = (v_0^1, v_0^2, v_0^3)$.

Thus, we may view Ψ^u as a map $\Psi^u : R^3 \to R^3$ and $\Psi^u(v) = \delta''(0)$.

Theorem 2.4 (cf. [13]). *The map $\Psi^u\colon R^3 \to R^3$ is an R-linear transformation. In fact $(-\Psi^u)$ is the linear transformation defined by the matrix $(\partial^2\phi/\partial x^\alpha \partial x^\beta)_{\alpha\beta}$. In particular, Ψ^u is independent of u.*

PROOF. Since G^u is a G-field, the trajectory γ_h is a solution of G for each $h \in D$, i.e.,

$$(\gamma_h)''(t) = -\nabla\phi\big(\gamma_h(t)\big) \quad \text{for each } t$$

where ϕ is the gravitational potential. Thus, at time t,

$$-h\delta''(t) = -[(\gamma_h'') - (\gamma)''](t) = \nabla\phi\big(\gamma_h(t)\big) - \nabla\phi\big(\gamma(t)\big).$$

In particular, at time $t = 0$,

$$-h\delta''(0) = \nabla\phi\big(\gamma_h(0)\big) - \nabla\phi\big(\gamma(0)\big).$$

By developing the RHS in Taylor series, we have that its αth component is

$$h\sum_\beta(\partial^2\phi/\partial x^\alpha \partial x^\beta)(x)v_0^\beta.$$

Then $\Phi(v) = -\delta''(0) = -\Psi^u(v)$ where

$$\Phi\colon R^3 \to R^3$$

is the linear transformation defined by the matrix $\big((\partial^2\phi/\partial x^\alpha \partial x^\beta)(x)\big)_{\alpha\beta}$. □

Writing Ψ for Ψ^u, we may formulate

Newton's vacuum field equation.

$$\operatorname{trace}(\Psi) = 0.$$

In fact, this is just the Laplace equation.

N.B. We may view the preceding as the construction of a map

$$\begin{cases} E_x^* \to \mathrm{Lin}_R(E_x, E_x) \\ u \mapsto \Psi^u \end{cases}$$

where E_x^* the set of non zero tangent vectors in E_x, from the law of motion G (the Newtonian law).

Indeed, let $u \in E_x^*$. By existence and uniqueness of solutions of (autonomous) second order equations there is a unique solution $\gamma\colon D_\infty \to E$ of G with $\gamma(0) = x$ and $\gamma^\bullet(0) = u$. Now, proceed as before starting with Theorem 2.2.

The vacuum field equation is just the statement $\mathrm{trace}(\Psi^u) = 0$ for every u. In this particular case, Ψ^u is independent of u.

The same scheme will be used to derive the vacuum field equations in Einstein's relativistic theory of gravitation that we now discuss.

3. Einstein's Relativistic Theory of Gravitation

The General Theory of Relativity or rather the Relativistic Theory of Gravitation (as it should be properly called according to the Russian physicist V. Fock [6]) rests on the following fundamental assumptions.

(1) Space-time is a 4-dimensional "curved" manifold.
(2) Free falling particles describe geodesics ("straightest lines") in this manifold.
(3) There is a reciprocal action of matter on curvature and curvature on matter described by Einstein's field equations. In the words of John Wheeler: "Matter tells space-time how to curve, and curved space tells matter how to move."

Some comments: one can say that the basic idea is to change the geometry so that will incorporate gravitation. More precisely, to abandon "flat" 3 dimensional Euclidean space for 4-dimensional "curved" space.

This has the following consequence: in classical physics, the nature of gravitation had been a problem since Newton's time. In the words of A. Koyré [9, p. 16] "...in spite of the rational plausibility and mathematical simplicity of the Newtonian law...there was in it something that baffled the mind. Bodies attract each other, act upon each other (or at least behave as they did). But how do they manage to perform this action, to overcome the chasm of the void that so radically separates them and isolates them from each other? We must confess that nobody, not even Newton, could (or can) explain, or understand this *how*. Newton himself, as we well know, never admitted gravitation as a "physical" force."

These problems were swept by the incorporation of gravitation into the new geometry: there is no mysterious "gravitational" force acting on a particle: the particle just follows a geodesic! It is fair to say, nevertheless, that the extent to which Einstein's theory has solved this problem has been debated up to now.

The first two assumptions can be formulated in a straightforward way. The mathematical formulation of the third, i.e., the field equations, is the real subject matter of this paper. A lot of work has to be done to carry out the program of the introduction. We do this for the vacuum only.

We reformulate Einstein's physical postulates as follows:

Postulate 1 (Structure of space-time). Space-time is a D_∞-manifold M of dimension n with a symmetric connection ∇.

We leave the dimension of the space-time M indeterminate (rather than $n = 4$), since our derivation does not depend on the value of n.

Postulate 2 (Law of motion). Freely falling particles near an attracting body describe geodesics. In other words, trajectories of particles under inertia and gravitation are solutions of the second order differential equations defining geodesics:

$$\gamma^{\bullet\bullet} = \nabla(\gamma^{\bullet}, \gamma^{\bullet})$$

or equivalently

$$\nabla_{\gamma^{\bullet}}\gamma^{\bullet} = 0$$

(cf. [12, affineconnections.pdf, p. 12]).

From here one can derive the second order differential equation for geodesics in terms of coordinates that can be found in texts of Differential Geometry (e.g. [14])

$$\frac{\mathrm{d}^2 x^k}{\mathrm{d}t^2} + \sum_{ij} \Gamma^k_{ij} \frac{\mathrm{d}x^i}{\mathrm{d}t}\frac{\mathrm{d}x^j}{\mathrm{d}t} = 0$$

(cf. [12, affineconnections.pdf, p. 14]).

Imitating what we did for the Newtonian gravitational theory, we assume that space-time is a D_∞-manifold, that vector fields are maps $Q\colon U \to U^D$ with U an infinitesimal neighborhood of M and trajectories, in particular geodesics, curves with domain D_∞.

Since a D_∞-manifold M is microlinear, each tangent space $\pi_M^{-1}(m)$ is a R-modules ($m \in M$). (Here $\pi_M\colon M^D \to M$ is the evaluation at 0). A microlinear space *has dimension* n if all the tangent spaces are free R-modules of dimension n.

In the first section, Laplace field formulation of Newton's law of gravitation was stated in terms of the linear transformation $\Psi\colon E_x \to E_x$ as trace$(\Psi) = 0$.

Trying to reproduce the computations of Section 1 in our relativistic context, we take for G the second order DE of geodesics. Furthermore, we keep the same notion of neighbor. For the record:

A *neighbor of an integral curve* γ *in a vector field* X is a vector field W along γ such that the trajectories of W, namely the γ_h's defined by $\gamma_h(t) = W(t)(h)$ are integral curves of X.

The existence of neighbors need not be postulated, but can be proved

Lemma 3.1 (There are enough neighbors). *Let X be a vector field, γ an integral curve of X and $v \in M_x$. Then there is a unique neighbor W of γ in X with $W(0) = v$.*

PROOF. Cf. Lemma 4.1 □

Thus, a G-field Q is a vector field whose integral curves are solutions of G, i.e., geodesics. The following will be used in the sequel

Lemma 3.2. *The integral curves of Q are geodesics iff $\nabla_Q Q = 0$.*

PROOF. Cf. Lemma 4.2. □

We call such fields *geodesic fields*. Thus, a *geodesic field* is a vector field whose integral curves are trajectories of free falling particles, i.e., geodesics according to Postulate 2. (Recall that in the Newtonian framework, the integral curves of the gravitational field G are the trajectories of free falling particles.)

As before, we do not postulate the existence of geodesic fields either, but prove their existence:

Lemma 3.3 (There are enough geodesic fields). *For every $u \in M_x$ not zero, there is a geodesic field G^u such that $G^u(x) = u$.*

PROOF. In fact, G is "reducible" to a family of vector fields G^u indexed by nonzero vectors $u \in M_x$. Then G^u is the sought geodesic field. Cf. Lemma 4.3. □

Following the analogy with Newtonian gravitation we define a map

$$\Psi^u \colon M_x \to M_x$$

for each non zero vector $u \in M_x$ as follows: let $\gamma\colon D_\infty \to M$ be the integral curve of G^u with initial condition $\gamma(0) = x$.

Notice that $\gamma^\bullet(0) = G^u(\gamma(0)) = G^u(x) = u$.

Let now $v \in M_x$ be arbitrary. Then there is a unique neighbor W of γ in G^u with $W(0) = v$. Define

$$\Psi^u(v) = \nabla^2_{\gamma^\bullet} W(0).$$

We would like to formulate Einstein's vacuum field equation equation as

$$\operatorname{trace}(\Psi^u) = 0$$

for every nonzero vector $u \in M_x$.

However, there are several problems with this formulation: first we have to prove that Ψ^u is R-linear. Furthermore, Ψ^u seems to depend not only on u but on G^u and, worst of all, it is not informative, in fact, we would like to tie Ψ^u to the curvature of the space.

The following takes care of all of these problems

Theorem 3.4 (Main Theorem, cf. [13]).

$$(\nabla^2_{\gamma^\bullet} W)(t) = R_{\gamma^\bullet(t) W(t)} \gamma^\bullet(t)$$

where R is the Riemann – Christoffel tensor.

N.B. The definition of this tensor, as well as all the properties that we require in our proofs can be found in [12, the-riemann-christoffel-tensor.pdf].

Corollary 3.5.

$$\Psi^u(v) = R_{uv} u.$$

To prove the theorem we need the following

Lemma 3.6. *Let $\gamma\colon D_\infty \to M$ be an infinitesimal curve in M, $p = \gamma(0)$ and Q a vector field defined on an infinitesimal neighborhood of p with $Q(p) \neq 0$. Then for every neighbor W of γ in Q there is a vector field $\widetilde{W}$ of M such that $[Q, \widetilde{W}] = 0$ and $\widetilde{W} \circ \gamma = W$.*

N.B. Definitions and properties of Lie derivatives $L_Q W$ and Lie brackets $[X, Y]$ can be found in [12, lie-derivatives.pdf].

PROOF. Cf. Lemma 4.4 □

PROOF (OF THE MAIN THEOREM). We recall the following results (cf. [10])

(1) $\nabla_X Y = \nabla_Y X + [X, Y]$, provided that ∇ is symmetric

(2) $R_{XY} Z = \nabla_X(\nabla_Y Z) - \nabla_Y(\nabla_X Z) - \nabla_{[X,Y]}(Z)$

The rest of the proof is now a simple computation: letting $Q = G^u$ we have

$$\begin{aligned}\nabla_Q^2 \widetilde{W} &= \nabla_Q \nabla_Q \widetilde{W}\\ &= \nabla_Q(\nabla_{\widetilde{W}} Q + [Q, \widetilde{W}])\\ &= \nabla_Q \nabla_{\widetilde{W}} Q\\ &= R_{Q,\widetilde{W}} Q + \nabla_{\widetilde{W}} \nabla_Q Q + \nabla_{[Q,\widetilde{W}]} Q\\ &= R_{Q\widetilde{W}} Q.\end{aligned}$$

Post-composing with γ, we obtain the desired result. In more details:

Claim 1. *$\tilde{Q}_{\gamma(t)}(\delta) = \gamma(t+\delta)$ for every $\delta \in D_2$.*

Since M is microlinear, it is enough to show that $(Q_{d_1} \circ Q_{d_2})(\gamma(t)) = \gamma(t + d_1 + d_2)$. This is clear. In fact, since γ is an integral curve of Q,

$$Q_{d_1} \circ Q_{d_2}(\gamma(t)) = Q_{d_1}(Q_{d_2}(\gamma(t))) = Q_{d_1}(\gamma(t+d_2)) = \gamma(t + d_2 + d_1)$$

Claim 2. $(\nabla_Q \widetilde{W})(\gamma(t)) = (\nabla_{\gamma^\bullet} W)(t)$.

This is proved by starting from Claim 1 and the definition of $\nabla_X Y$ (cf. [11, p. 200; 12]).

$$\begin{aligned}h(\nabla_Q \widetilde{W})(\gamma(t)) &= \nabla([d \mapsto \tilde{Q}_{\gamma(t)}(d+h)], \widetilde{W}_{Q_{\gamma(t)}(h)})(-h) - \widetilde{W}_{\gamma(t)}\\ &= \nabla([d \mapsto \gamma(t+h+d)], \widetilde{W}_{\gamma(t+h)})(-h) - W(t)\\ &= \nabla(\gamma^\bullet(t+h), W(t+h))(-h) - W(t)\\ &= h(\nabla_{\gamma^\bullet} W)(t).\end{aligned}$$

The claim results from canceling the universally quantified h.

Claim 3. $(R_{Q\widetilde{W}} Q)_{\gamma(t)} = R_{\gamma^\bullet(t) W(t)} \gamma^\bullet(t)$.

This follows from the definition of the Riemann – Christoffel tensor, by noticing that $Q(\gamma(t)) = \gamma^\bullet(t) \in M_{\gamma(t)}$ and $\widetilde{W}(\gamma(t)) = W(t) \in M_{\gamma(t)}$. □

By analogy with Newton's gravitational theory, we postulate

Einstein's vacuum field equation (First formulation).

$$\operatorname{trace}(\Psi^u) = 0 \quad \text{for all non zero } u \in M_x.$$

This is not the usual formulation of this equation, but it is equivalent.

First, we define the *Ricci tensor* as the contraction of the Riemann – Christoffel curvature tensor, i.e., in classical notation $R_{ki} = \sum_l R^l_{kil}$. From a more conceptual or invariant point of view, recall that the Riemann – Christoffel curvature tensor is given by a map $R\colon M^D \times_M M^D \times_M M^D \to M^D$.

The contraction is a new map $\mathrm{Ric}\colon M^D \times_M M^D \to M^D$ defined as follows: if $u, v \in M_x$, let $\Psi(u,v)\colon M_x \to M_x$ be the R-linear map defined by $\Psi(u,v)(w) = R_{uw}v$. Then $\mathrm{Ric}(u,v) = \operatorname{trace}(\Psi(u,v))$.

To connect this invariant definition of Ric with the classical expression of the Ricci tensor we first take any basis of M_x to represent $\Psi(u,v)$ as a matrix. By definition, $\mathrm{Ric}(u,v)$ is the trace of this matrix. Take, for instance the basis $\{\partial/\partial x^i|_x\}_i$. The matrix of $\Psi(u,v)$ relative to this basis is $(\omega^i(\Psi(u,v)(\partial/\partial x^j|_x)))_{ij}$, where $\{\omega^i\}_i$ is the dual basis, and its trace is $\mathrm{Ric}(u,v) = \sum_i \omega^i(R_{u(\partial/\partial x^i|_x)}v)$.

Next, we start with the following notations and identifications

$$\begin{cases} R^k_{ij} = R_{\partial_i \partial_j} \partial_k \text{ (where } \partial_i = \partial/\partial x_i) \\ (R^k_{ij})_l \text{ in our notation } = R^l_{kij} \text{ in Einstein's notation} \end{cases}$$

Thus $R_{ki} = \sum_l R^l_{kil} = \sum_l (R^k_{il})_l = \sum_l \omega^l(R_{\partial_i \partial_l} \partial_k) = \operatorname{trace}\big(\Psi(\partial_i, \partial_k)\big) = \operatorname{Ric}(\partial_i, \partial_k)$.

Lemma 3.7. *If ∇ is symmetric, then*

$$\operatorname{Ric}(u, v) = \operatorname{Ric}(v, u).$$

PROOF. Cf. [12, metrics.pdf, p. 8]. □

This allows us to rewrite the first formulation of Einstein vacuum field equation as $\operatorname{Ric}(u, u) = 0$ for every nonzero $u \in M_x$, since $\Psi^u(v) = R_{uv}$ and hence $\operatorname{trace}(\Psi^u) = \operatorname{Ric}(u, u)$. But this implies that $\operatorname{Ric} = 0$.

Indeed, let $v, w \in M_x$ be *arbitrary* vectors. Write $v = \sum v_i e_i$ and $w = \sum w_i e_i$. Then $\operatorname{Ric}(u, v) = \sum_{ij} v_i w_j \operatorname{Ric}(e_i, e_j)$. But $\operatorname{Ric}(e_i + e_j, e_i + e_j) = \operatorname{Ric}(e_i, e_i) + 2\operatorname{Ric}(e_i, e_j) + Ric(e_j, e_j)$. By hypothesis, the LHS and the extreme terms of the RHS are 0, so the middle term must be zero, i.e., $\operatorname{Ric}(e_i, e_j) = 0$. Thus, $\operatorname{Ric} = 0$. In other words, we may reformulate the above equation as

Einstein's vacuum field equation (Final formulation).

$$\operatorname{Ric} = 0$$

which is the usual form of this equation.

4. Missing Proofs

In this final section we provide proofs for the lemma and propositions required to prove the main theorem

Lemma 4.1 (There are enough neighbors). *Let X be a vector field, γ an integral curve of X and $v \in M_x$ (with $\gamma(0) = x$). Then there is a unique neighbor V of γ in X with $V(0) = v$.*

PROOF. Let $\gamma \colon D_\infty \to M$ an integral curve of X and $u \in M_x$. Define $V \colon D_\infty \to M^D$ by $V(t)(h) = \gamma_h(t)$, where γ_h is the unique integral curve of X with initial condition $\gamma_h(0) = v(h)$.

Thus, $V(t)(0) = \gamma_0(t) = \gamma(t)$, since both γ and γ_0 are integral curves of Q with the same initial condition. In other words, V is a vector field along γ. Furthermore, $L_X V = 0$, by construction. Finally, $V(0)(h) = \gamma_h(0) = u(h)$, and this shows that $V(0) = u$.

To show uniqueness, assume that V^1 and V^2 are two such. Then V^1_h and V^2_h satisfy the same (first-order) differential equation, namely X, with the same initial conditions since $V^1_h(0) = V^1(0)(h) = v(h) = V^2(0)(h) = V^2_h(0)$. Thus, they coincide. □

Lemma 4.2. *The integral curves of Q are geodesics iff $\nabla_Q Q = 0$.*

PROOF. By definition of ∇,

$$h\nabla_Q Q\big(\gamma(t)\big) = \nabla([d \mapsto \tilde{Q}\big(\gamma(t)\big)(h + d)], Q(Q(\gamma(t)(h)))(-h) - Q\big(\gamma(t)\big)$$

where $\tilde{Q}$ is the canonical "extension" of Q, i.e, $\tilde{Q} \colon D_2 \to M^M$ with $\tilde{Q}(d) = Q(d)$ for $d \in D$ (cf. [8]).

We notice that $Q(Q(\gamma(t))(h) = Q(\gamma(t+h)) = \gamma^\bullet(t+h)$. Furthermore,

$$\begin{aligned}\tilde{Q}(\gamma(t))(h+d) &= (Q_h \circ Q_d)(\gamma(t)) \\ &= Q_h(Q_d(\gamma(t)) \\ &= Q_h(\gamma(t+d)) \\ &= \gamma(t+h+d) \\ &= \gamma^\bullet(t+h)(d)\end{aligned}$$

By replacing in the definition

$$h\nabla_Q Q(\gamma(t)) = \nabla(\gamma^\bullet(t+h), \gamma^\bullet(t+h))(-h) - \gamma^\bullet(t) = h\nabla_{\gamma^\bullet}\gamma^\bullet(t)$$

Since this holds for every $h \in D$,

$$\nabla_Q Q(\gamma(t)) = \nabla_{\gamma^\bullet}\gamma^\bullet(t).$$

This shows the implication from right to left. For the other direction, assume that all integral curves of Q are geodesics and let $x \in M$. If γ is the unique integral curve of Q with $\gamma(0) = x$, then $\nabla_{\gamma^\bullet}\gamma^\bullet(0) = 0$ and hence $\nabla_Q Q(x) = 0$. □

The following generalizes a result of [13]:

Theorem 4.3 (Reduction theorem). *Let Y be an autonomous second order equation, i.e., a symmetric vector field $Y: M^D \to (M^D)^D$. Then there is a family Y^u of Y-vector fields indexed by nonzero vectors $u \in M_x$ such that Y is* reducible *to the family $(Y^u)_u$ in the sense that if $\gamma: D_\infty \to M$ is a solution of Y with $\gamma(0) = x$ and $\gamma^\bullet(0) = u$, then γ is an integral curve of Y^u.*

N.B. We say that a vector field Q is a *Y-field* iff all integral curves of Q are solutions of Y.

PROOF. By choosing appropriate coordinates we may assume that $x = (0, \dots, 0)$ and $u = (0, \dots, 0, (0, \dots 0, u_0))$ with u_0 invertible.

Define a map $\phi: D^n_\infty \to D^n_\infty$ by the formula

$$\phi(\delta^1, \dots, \delta^{n-1}, t) = \gamma_{(\delta^1, \dots, \delta^{n-1}, 0)}(t)$$

where $\gamma_{(\delta^1, \dots, \delta^{n-1}, 0)}$ is the solution of the equation Y starting at $(\delta^1, \dots, \delta^{n-1}, 0)$ with initial "velocity" u, i.e.,

$$\begin{cases}\gamma_{(\delta^1, \dots, \delta^{n-1}, 0)}(0) = (\delta^1, \dots, \delta^{n-1}, 0) \\ \gamma^\bullet_{(\delta^1, \dots, \delta^{n-1}, 0)}(0) = (\delta^1, \dots, \delta^{n-1}, 0, (0, \dots, 0, u_0)).\end{cases}$$

(For the existence and uniqueness of these solutions, see [8; 12, Second order DE and geodesics.pdf]).

We claim that ϕ is invertible:

Write $\phi = (\phi^1, \dots, \phi^n)$. Spelling the initial conditions of $\gamma_{(\delta^1, \dots, \delta^{n-1}, t)}$ in terms of the components of ϕ we get

$$\begin{cases}\phi^i(\delta^1, \dots, \delta^{n-1}, 0) = \delta^i \quad (i < n) \\ \phi^n(\delta^1, \dots, \delta^{n-1}, 0) = 0\end{cases}$$

for the first and

$$\begin{cases}\partial\phi^i/\partial t(\delta^1, \dots, \delta^{n-1}, 0) = 0 \qquad (i < n) \\ \partial\phi^n/\partial t(\delta^1, \dots, \delta^{n-1}, 0) = u_0\end{cases}$$

for the second.

By developing in Taylor series,

$$\phi^i(\delta^1,\ldots,\delta^j+d,\ldots,\delta^{n-1},0)=\phi^i(\delta^1,\ldots,\delta^j,\ldots,\delta^{n-1},0)$$
$$+\,d(\partial\phi^i/\partial y^j)(\delta^1,\ldots,\delta^j,\ldots,\delta^{n-1},0).$$

By identifying corresponding terms in this equation,

$$\begin{cases}(\partial\phi^i/\partial y^j)(\delta^1,\ldots,\delta^j,\ldots,\delta^{n-1},0)=0 & (i\neq j)\\ (\partial\phi^i/\partial y^i)(\delta^1,\ldots,\delta^j,\ldots,\delta^{n-1},0)=1\end{cases}$$

Therefore $\phi(0,\ldots,0,0)=(0,\ldots,0,0)$ and the Jacobian of ϕ at $(0,\ldots,0,0)$ is $u_0\neq 0$. The claim is then a consequence of the [12, Infinitesimal Inverse function theorem.pdf].

We let ψ be the inverse of ϕ.

Define the vector field Y^u or simply Q by the formula

$$Q(x^1,\ldots,x^n)=\gamma^\bullet_{(\delta^1,\ldots,\delta^{n-1},0)}(t)$$

with $\phi(\delta^1,\ldots,\delta^{n-1},t)=(x^1,\ldots,x^n)$.

Claim 1. *Q is a vector field and $Q(x)=u$.*

The fact that Q is a vector field is obvious:

$$\begin{aligned}Q(x^1,\ldots,x^n)(0)&=\gamma^\bullet_{(\delta^1,\ldots,\delta^{n-1},0)}(t)(0)\\&=\gamma_{(\delta^1,\ldots,\delta^{n-1},0)}(t)\\&=\phi(\delta^1,\ldots,\delta^{n-1},t)\\&=(x^1,\ldots,x^n).\end{aligned}$$

On the other hand,

$$Q(x)=Q(0,\ldots,0)=\gamma^\bullet_{(0,\ldots,0,0)}(0)=(0,\ldots,0,(0,\ldots,0,u_0))=u$$

by the second initial condition that γ satisfies.

Claim 2. *The $\gamma_{(\delta^1,\ldots,\delta^{n-1},0)}(-)$'s are integral curves of Q, i.e.,*

$$\gamma^\bullet_{(\delta^1,\ldots,\delta^{n-1},0)}(t)=Q(\gamma_{(\delta^1,\ldots,\delta^{n-1},0)}(t)).$$

In fact, let $\gamma_{(\delta^1,\ldots,\delta^{n-1},0)}(t)=(x^1,\ldots,x^n)$. We saw that, $\gamma_{(\delta^1,\ldots,\delta^{n-1},0)}(t)=\phi(\delta^1,\ldots,\delta^{n-1},t)$. On the other and, by definition of Q,

$$Q(\gamma_{(\delta^1,\ldots,\delta^{n-1},0)}(t))=Q(x^1,\ldots,x^n)=\gamma^\bullet_{(\delta^{1\prime},\ldots,\delta^{n-1\prime},0)}(t')$$

where $\phi(\delta^{1\prime},\ldots,\delta^{n-1\prime},t')=(x^1,\ldots,x^n)$.

Since $\phi(\delta^1,\ldots,\delta^{n-1},t)=\gamma_{(\delta^1,\ldots,\delta^{n-1},0)}(t)=(x^1,\ldots,x^n)$ and ϕ is a bijection, $\delta^i=\delta^{i\prime}$ and $t=t'$, completing the proof.

Claim 3. *The integral curves of Q are precisely those of the form $\delta(t)=\gamma_{(x^1,\ldots,x^{n-1},0)}(x^n+t)$, for $(x^1,\ldots,x^n)\in D^n_\infty$.*

In fact, all curves of this form are integral curves of Q : let $u(t)=x^n+t$ and $\gamma(t)=\gamma_{(x^1,\ldots,x^{n-1},0)}(t)$. Then $\delta=\gamma\circ u$.

We claim that

$$\begin{cases}\gamma^\bullet\circ u=(\gamma\circ u)^\bullet=\delta^\bullet\\ \gamma^{\bullet\bullet}\circ u=(\gamma\circ u)^{\bullet\bullet}=\delta^{\bullet\bullet}.\end{cases}$$

Indeed, computing the first term of the first equation at (t,d) with $d \in D$, we have

$$\begin{aligned}\gamma^\bullet(u(t))(d) &= \gamma(u(t)+d)\\ &= \gamma((x^n+t)+d)\\ &= \gamma(x^n+(t+d))\\ &= \gamma(u(t+d))\\ &= (\gamma\circ u)(t+d)\\ &= (\gamma\circ u)^\bullet(t)(d)\end{aligned}$$

The second equation is proved by iteration.

As a consequence, we have the two equations

$$\begin{cases}\delta^\bullet = (\gamma\circ u)^\bullet = \gamma^\bullet\circ u = (Q\circ\gamma)\circ u = Q\circ(\gamma\circ u) = Q\circ\delta\\ \delta^{\bullet\bullet} = \gamma^{\bullet\bullet}\circ u = (Y\circ\gamma^\bullet)\circ u = Y\circ(\gamma^\bullet\circ u) = Y\circ\delta^\bullet\end{cases}$$

The first shows that curves of the stated form are integral curves of Q. Furthermore, the second equation shows that Q is an Y-field, i.e., any integral curve of Q is a solution of Y. In more detail: let α be an integral curve of Q with initial condition $\alpha(0) = (x_0^1,\dots,x_0^n)$. Since ϕ is bijective, there is a unique n-tuple $(\delta^1,\dots,\delta^n)$ such that $\phi(\delta^1,\dots,\delta^n) = (x_0^1,\dots,x_0^n)$. Define $\delta(t) = \gamma_{(\delta^1,\dots,\delta^{n-1},0)}(\delta^n+t)$

Then, $\delta=\alpha$ since both are integral curves of Q with the same initial condition: $\delta(0) = \gamma_{(\delta^1,\dots,\delta^{n-1},0)}(\delta^n) = \phi(\delta^1,\dots,\delta^n) = (x_0^1,\dots,x_0^n) = \alpha(0)$. □

Lemma 4.4. *Let $\gamma\colon D_\infty\to M$ be an infinitesimal curve in M, $p=\gamma(0)$ and Q a vector field defined on an infinitesimal neighborhood of p with $Q(p)\neq 0$. Then for every neighbor W of γ in Q there is a vector field $\widetilde{W}$ of M such that $[Q,\widetilde{W}]=0$ and $\widetilde{W}\circ\gamma = W$.*

N.B. Definitions and properties of Lie Derivatives (L_QW) and Lie Brackets ($[Q,\widetilde{W}]$) used in this proof can be found in '[12, lie-derivatives.pdf].

PROOF. Since $Q(p)\neq 0$, by a change of coordinates we may assume that, locally around p, $Q = \partial/\partial x^1$ (see [12, Straightening out Theorem.pdf]). Write $W(t) = \sum_i w^i(t)\partial/\partial x^i|_{\gamma(t)}$. Then

$$0 = L_QW(t) = \sum_i w^i(t)L_Q\Big(\frac{\partial}{\partial x^i}\Big|_{\gamma(t)}\Big) + \sum_i (w^i)'(t)\frac{\partial}{\partial x^i}\Big|_{\gamma(t)}.$$

Now, $L_Q((\partial/\partial x^i)\circ\gamma) = L_{\partial/\partial x^1}((\partial/\partial x^i)\circ\gamma) = [\partial/\partial x^1, \partial/\partial x^i]\circ\gamma$. But $[\partial/\partial x^i,\partial/\partial x^j]$ $=0$. This follows from the definition of Lie bracket

$$\Big[\frac{\partial}{\partial x^i},\frac{\partial}{\partial x^j}\Big]_x(d_1d_2) = \Big(\frac{\partial}{\partial x^j}\Big)_{-d_2}\circ\Big(\frac{\partial}{\partial x^i}\Big)_{-d_1}\circ\Big(\frac{\partial}{\partial x^j}\Big)_{d_2}\circ\Big(\frac{\partial}{\partial x^i}\Big)_{d_1}(x)$$

and the definition of the vector field $(\partial/\partial x^i)$, namely $(\partial/\partial x^i)_d(x) = x + de_i$.

Since $\{\partial/\partial x^i|_{\gamma(t)}\}_i$ is a basis, $(w^i)'(t)=0$, i.e., $w^i(t)=a^i$ is a constant. Define

$$\widetilde{W}_x = \sum_i a^i\frac{\partial}{\partial x^i}\Big|_x.$$

This vector field has the required properties. Indeed, obviously $\widetilde{W} \circ \gamma = W$. The second property is a consequence of bi-linearity of $[-,-]$

$$[Q, \widetilde{W}] = \left[\frac{\partial}{\partial x^1}, \widetilde{W}\right] = \sum_i a_i \left[\frac{\partial}{\partial x^1}, \frac{\partial}{\partial x^i}\right] = 0.$$

Appendix A

A.1. A reformulation of Laplace equation. From what we did in the first section, we have an intuitive reformulation of Laplace equation due to Sachs/Wu that we now present. This follows from the following

Lemma A.1. *Let $\Phi\colon R^3 \to R^3$ be any linear transformation. Then*

$$\frac{1}{4\pi}\int_{u\in S^2} \Phi(u)\cdot u\sigma = \frac{1}{3}\,\mathrm{Trace}(\Phi)$$

where σ is the volume form of S^2.

PROOF. Compute the integral using spherical coordinates:

$$\begin{cases} x_1 = \cos\phi_1 \\ x_2 = \sin\phi_1\cos\phi_2 \\ x_3 = \sin\phi_1\sin\phi_2 \end{cases}$$

$(0 \leq \phi_1 \leq \pi,\ 0 \leq \phi_2 < 2\pi)$

In these coordinates, the volume form of S^2, namely $\sigma = u_1\,\mathrm{d}u_2 \wedge \mathrm{d}u_3 + u_2\,\mathrm{d}u_3 \wedge \mathrm{d}u_1 + u_3\,\mathrm{d}u_1 \wedge \mathrm{d}u_2$ (see, e.g., [1, p. 96]) becomes $\sigma = \sin\phi_1\,\mathrm{d}\phi_1\,\mathrm{d}\phi_2$

Let $(a_{ij})_{ij}$ be a matrix representation of Φ. Then $\Phi(u)\cdot u = \sum_{i,j} a_{ij}u_iu_j$ and a simple computation shows that

$$\int_{u\in S^2} u_iu_j\sigma = \begin{cases} 0 & \text{if } i \neq j \\ 4\pi/3 & \text{otherwise} \end{cases}$$

Thus, $\int_{u\in S^2}\Phi(u)\cdot u\sigma = (4\pi/3)\sum_i a_{ii} = (4\pi/3)\,\mathrm{Trace}(\Phi)$ □

In particular, this proposition applies to the linear transformation Φ given by the matrix $(\partial^2\phi/\partial x^\alpha x^\beta)_{\alpha\beta}$. In this case, there is a simple intuitive interpretation of this equation given in [13]: "Assume that you are in the middle of a cabin freely falling near the earth surface with lots of apples around you. Look at those apples situated at the same distance (which we assume small) from you and measure their acceleration towards (or away from) you. Their average is 0."

In our context, we may formulate this intuitive interpretation as a precise mathematical statement by defining "small" as being of square 0, i.e., an element of D.

Recall from the section on Newton's theory of gravitation that the difference between the acceleration of the apple whose trajectory is γ_h and yours (trajectory γ) at time 0 is $h\Psi(u)$.

We define the average or mean acceleration of all the apples at distance $h \in D$ from 0 to be

$$\frac{1}{4\pi}\int_{u\in S^2} h\Psi(u)\cdot u\sigma$$

where $\sigma = x_1\,\mathrm{d}x_2 \wedge \mathrm{d}x_3 - x_2\,\mathrm{d}x_1 \wedge x_3 + x_3\,\mathrm{d}x_1 \wedge \mathrm{d}x_2$ is the volume form of S^2. Notice that since $\int_{S^2}\sigma = 4\pi$, the term "mean" is justified.

Applying the previous proposition to Ψ we have the following

Corollary A.2. *The mean relative acceleration of all particles situated at distance h from 0 is*

$$\frac{1}{4\pi}\int_{u\in S^2} h\Psi(u)\cdot u\sigma = \frac{h}{3}\,\mathrm{trace}(\Psi)$$

where $\sigma = x_1\,\mathrm{d}x_2\wedge \mathrm{d}x_3 - x_2\,\mathrm{d}x_1\wedge \mathrm{d}x_3 + x_3\,\mathrm{d}x_1\wedge \mathrm{d}x_2$ *is the volume form of* S^2.

The Sachs and Wu interpretation is the statement that the integral in the corollary is 0 for every $h \in D$. But this is equivalent to the statement that $\nabla^2\phi = 0$, since $\Psi = -\Phi$ and $trace(\Phi) = \nabla^2\phi$. □

1.2. The first two postulates of Einstein. The aim of this section is to show how to derive postulates 1 and 2 from other principles discussed in the literature.

Following Einstein, space-time is assumed to be a 4-dimensional semi-Riemannian manifold M with metric g. Recall that a *semi-Riemannian manifold* is a manifold with a metric g having the property that around each point p there are local coordinates ξ with

$$(g_\xi)_{\alpha\beta}(p) = \eta_{\alpha\beta}$$

where the η's are the entries of the matrix

$$\begin{pmatrix} -1 & 0 & 0 & 0\\ 0 & 1 & 0 & 0\\ 0 & 0 & 1 & 0\\ 0 & 0 & 0 & 1\end{pmatrix}.$$

In particular, $(\Gamma_\xi)^\gamma_{\alpha\beta}(p) = 0$ (see in [12, metrics.pdf, Corollary 0.7, p. 6]).

Ad Postulate 1. From g we can manufacture a symmetric connexion ∇. (Details in [12, metrics.pdf].)

The unique symmetric connection compatible with a given metric is called, improperly according to Spivak ([14, Vol. II, p. 256]), the *Levi-Civita connection* for the metric.

Ad Postulate 2. This axiom is introduced as such in General Relativity under the name "Equation of motion" and it is deduced/motivated by the so-called *equivalence principle* (see Einstein [2] and Weinberg [15]).

I will not go into the questions of validity of this principle (see V. Fock [6]), or the shortcomings of Weinberg's "proof," but point out that Postulate 2 can be deduced from a very weak "point-wise" form of the equivalence principle that seems un-objectionable and may be formulated as follows:

Let $\gamma\colon D_\infty \to M = R^n$ be the trajectory of a free falling particle and p a point of γ. Whenever ξ are local geodesic coordinates around p (i.e., with $\Gamma^\delta_{\alpha\beta}(p) = 0$), the trajectory satisfies

$$\frac{\mathrm{d}^2\xi^k}{\mathrm{d}t^2}(p) = 0$$

at the point p.

From this principle we deduce that γ is indeed a geodesic with equation

$$\frac{\mathrm{d}^2x^k}{\mathrm{d}t^2} + \sum_{\alpha,\beta}\Gamma^k_{\alpha\beta}\frac{\mathrm{d}x^\alpha}{\mathrm{d}t}\frac{\mathrm{d}x^\beta}{\mathrm{d}t} = 0.$$

An indication of the proof: since ξ^k is a function of the x^α, a simple application of the chain rule gives (at p)

$$0 = \sum_\alpha \frac{\partial \xi^k}{\partial x^\alpha} \frac{\mathrm{d}^2 x^\alpha}{\mathrm{d}t^2} + \sum_{\alpha,\beta} \frac{\partial^2 \xi^k}{\partial x^\alpha \partial x^\beta} \frac{\mathrm{d}x^\alpha}{\mathrm{d}t} \frac{\mathrm{d}x^\beta}{\mathrm{d}t}.$$

By multiplying this equation by $\partial x^\alpha / \partial \xi^k$ and adding with respect to α, we obtain

$$\frac{\mathrm{d}^2 x^k}{\mathrm{d}t^2} + \sum_{\alpha\beta} (\Gamma^k_{\alpha\beta})_{x\xi} \frac{\mathrm{d}x^\alpha}{dt} \frac{\mathrm{d}x^\beta}{dt} = 0.$$

since ξ and x are inverse of each other.

By applying the transformation rule for Γ at point p, $(\Gamma^k_{\alpha\beta})_{x\xi} = (\Gamma^k_{\alpha\beta})_x$. Since p was arbitrary, this concludes the proof.

N.B. Recall that going from coordinates x to x', Γ is transformed as follows

$$(\Gamma^\gamma_{\alpha\beta})_{x'} = \sum_{ijk} (\Gamma^k_{ij})_x \frac{\partial x^i}{\partial x'^\alpha} \frac{\partial x^j}{\partial x'^\beta} \frac{\partial x'^\gamma}{\partial x^k} + \sum_i \frac{\partial^2 x^i}{\partial x'^\alpha \partial x'^\beta} \frac{\partial x'^\gamma}{\partial x^i}$$

(cf. [12, affineconnections.pdf]).

References

1. H. Cartan, *Formes différentielles. Applications élémentaires au calcul des variations et à la théorie des courbes et des surfaces*, Hermann, Paris, 1967.
2. A. Einstein, *The foundations of the general theory of relativity*, The Principle of Relativity, Dover Books on Physics, Dover, New York, 1952, pp. 109–164.
3. ———, *The meaning of relativity*, 5th ed., Princeton Univ. Press, Princeton, NJ, 1955.
4. ———, *Autobiographical notes*, Albert Einstein: Philosopher-Scientist, Vol. 1 (P. A. Schilpp, ed.), Library of Living Philosophers, vol. 7, Harper & Row, New York, 1959, pp. 3–95.
5. A. Einstein, L. Infeld, and B. Hoffmann, *The gravitational equations and the problem of motion*, Ann. of Math. (2) **39** (1938), no. 1, 65–100.
6. V. Fock, *The theory of space time and gravitation*, Pergamon Press, New York, 1959.
7. A. Kock, *Synthetic differential geometry*, London Math. Soc. Lecture Note Ser., vol. 51, Cambridge Univ. Press, Cambridge, 1981.
8. A. Kock and G. E. Reyes, *Ordinary differential equations and their exponentials*, Cent. Eur. J. Math. **4** (2006), no. 1, 64–81.
9. A. Koyré, *Newtonian Studies*, Univ. of Chicago Press, Chicago, IL, 1968.
10. R. Lavendhomme, *Basic concepts of synthetic differential geometry*, Kluwer Texts Math. Sci., vol. 13, Kluwer, Dordrecht, 1996.
11. I. Moerdijk and G. E. Reyes, *Models for smooth infinitesimal analysis*, Springer, New York, 1991.
12. G. E. Reyes. series of pamphlets on Synthetic Differential Geometry in the author's webpage `reyes-reyes.com`.
13. R. K. Sachs and H. H. Wu, *General relativity for mathematicians*, Grad. Texts in Math., vol. 48, Springer, New York, 1977.
14. M. Spivak, *A comprehensive introduction to differential geometry*. Vol. I–V, 2nd ed., Publish or Perish, Wilmington, DE, 1979.
15. S. Weinberg, *Gravitation and cosmology: Principles and applications of the general theory of relativity*, Wiley, New York, 1972.

DÉPARTEMENT DE MATHÉMATIQUES ET STATISTIQUE, UNIVERSITÉ DE MONTRÉAL, CP 6128, SUCC CENTRE-VILLE, MONTRÉAL, QC H3C 3J7, CANADA

E-mail address: `reyes@dms.umontreal.ca`

Centre de Recherches Mathématiques
CRM Proceedings and Lecture Notes
Volume **53**, 2011

An Adventure into Hungarian Word Order with Cyclic Pregroups

Mehrnoosh Sadrzadeh

Abstract. Pregroups are partially ordered monoids where each element has a left and a right adjoint. They have been developed by Lambek and applied to the analysis of syntax of natural languages, e.g. from English and French to Japanese, Arabic, and Persian. In recent joint work with Casadio, we introduced the concept of a cyclic pregroup to formalize clitic movement in French, Italian, and Persian. Our cyclic meta-rules formalize limited kinds of movements of words within a sentence. This paper is an encounter between cyclic pregroups and a fragment of Hungarian grammar dealing with a representative group of word order changes. These allow for communicating different kinds of emphasis on the constituents of a sentence. We develop a more general version of our previous rules and use them to derive the new type of the verb from its original type in each case of movement.

1. Introduction

Pregroups are mathematical structures developed by Lambek for the syntactic analysis of language, improving on his prior Syntactic Calculus (also known as the Lambek Calculus). They formalize regularities of numbers and have applications to linguistics [12]. In their categorized version, they are compact closed categories and have been applied to physics [2, 14]. In the linguistic world, pregroups have earned rapid success in describing syntax of natural languages, they have been applied to English, French, Italian, but also to Japanese, Arabic, and Persian, and many more, for references see [8]. Hungarian is missing from the list! We do not claim to attempt to add it in this paper, nonetheless we offer a touristic encounter with a small fragment of its syntax. This is the section that deals with change of word order, due to change of focus or emphasis on constituents of the sentence. We attempt to explain and formalize these forms of movement by using a pregroup grammar that is equipped with special *cyclic rules*.

In previous work [9], we have argued that certain *cyclic rules*, once limited to particular parts of speech such as verbs, formalize movement in the analysis of syntax of natural languages. We have shown how, when applied to the syntactic

2010 *Mathematics Subject Classification.* Primary 06F99; Secondary 68T99.
Key words and phrases. Mathematical Linguistics, Pregroups, Hungarian, Syntax, Cyclicity.
Thanks go to J. Lambek, C. Casadio, the referee, and EPSRC PDRF Grant EP/E035698/1.
This is the final form of the paper.

type of the verb, these rules formalize the main patterns of clitic movement, in a particularly appealing and uniform way, in languages Persian, French and Italian. In this paper we develop more general versions of our previous rules and show how they formalize the change of word order in Hungarian. We use the rules to derive the type of the verb after each movement from the original type of the verb in a sentence with basic word order and no emphasis. Our approach to explaining and formalizing these kinds of movement differs from that suggested in [1, 15, 18]. In our analysis we do not make a reference to the concept of a *focus phrase*. These may be empty even in the presence of a movement, hence cause ambiguity in the meaning of the sentence. Instead, we introduce the new notion of a *mobilizer*, this is a constituent that causes the movement because the speaker wants to convey extra information about it. In our approach, whenever there is movement, there is a mobilizer, hence the ambiguity gets discarded.

In this paper and in previous work, we have added the cyclic rules as what Lambek calls meta-rules to the pregroup dictionaries of natural languages, so that they are only applicable to the type of the verb. It is tempting to try to abstract away and instead add them (in their more general form of this paper or the more restricted form of previous work) as axioms to a pregroup, thus make them applicable to all the types rather than just that of the verb. This temptation should be resisted, since as we shall show, such a pregroup will reduce to a partially ordered group.

To analyze movement, e.g., in the wh-questions of English, Lambek refers to the *trace* terminology of Chomsky. He argues that whenever there is such a movement present, e.g., in a whom-question when the object moves to the beginning of the sentence and becomes the question word 'whom,' then there should be an iterated adjoint present in the types involved, e.g. in the type of 'whom.' Our pregroup treatment of movement is not in conflict with Lambek's approach. As shown in Section 2.3, our rules are derivable from the cyclic rules of classical noncommutative Linear Logic. These have double adjoints in them and as demonstrated in section 2.3, one can apply Lambek's idea to the equations of (2.1) to explain the movement in a comprehensive manner. We have chosen to work with these rules and not with types with double adjoints, since they express the movements that we are interested in, in a more direct and transparent manner.

2. Pregroups equipped with clitic rules

A pregroup is a partially ordered monoid where each element has a left and a right adjoint. It is denoted by a tuple $(P, \leq, 1, \cdot, (-)^{\mathrm{l}}, (-)^{\mathrm{r}})$ satisfying the following axioms, where the monoid operation $\cdot$ is (as usual) omitted for shorthand purposes:

$$\begin{aligned} p \leq q \implies pr \leq qr \quad &\text{and} \qquad rp \leq rq \\ p^{\mathrm{l}}p \leq 1 \leq pp^{\mathrm{l}} \quad &\text{and} \quad pp^{\mathrm{r}} \leq 1 \leq p^{\mathrm{r}}p \\ p1 = 1 \quad &\text{and} \qquad 1p = 1 \end{aligned}$$

It follows that adjoints are unique and that the unit is self adjoint, i.e.

$$1^{\mathrm{l}} = 1 = 1^{\mathrm{r}}$$

The adjoint of multiplication is multiplication of adjoints but in the reverse order, i.e.

$$(pq)^{\mathrm{l}} = q^{\mathrm{l}}p^{\mathrm{l}} \quad \text{and} \quad (pq)^{\mathrm{r}} = q^{\mathrm{r}}p^{\mathrm{r}}$$

An example of a pregroup from arithmetic is the set of all unbounded monotone maps on integers. The free pregroup generated over a partially ordered set exists, for more details on these see [4, 11, 12].

Since their introduction by Lambek in [11], pregroups have been applied to the analysis of syntax of a wide range of natural languages. The natural language applications are based on the following two main definitions.

Definition 2.1. A pregroup dictionary is a relation, given by

$$\mathcal{D} \subseteq \Sigma \times T(\mathcal{B})$$

where Σ is the set of all words of a natural language, $\mathcal{B}$ is a partially ordered set of grammatical types, and $T(\mathcal{B})$ is the free pregroup generated over $\mathcal{B}$.

A pregroup dictionary assigns to each word of a language, a pregroup type, based on the grammatical roles it may take in the language. These types are used to decide whether a string of words constitutes a grammatical phrase or not, via the following definition.

Definition 2.2. Given a set α of designated grammatical types $\alpha \subseteq T(\mathcal{B})$, a string of words $w_1, \ldots, w_n \in \Sigma$ is grammatical iff $w_1 \cdots w_n \leq \sigma$ for $\sigma \in \alpha$.

Examples of elements of α are types standing for declarative sentence and questions.

As a rough and inclusive example of English grammar, consider the set of types $\{\pi, o, s, q, \bar{q}, i\}$, where π stands for subject, o for object, s for declarative sentence, q for yes-no question, $\bar{q}$ for wh-question, and i for the infinitive of a verb. Consider the following dictionary

$$\{(\text{I}, \pi), (\text{him}, o), (\text{saw}, \pi^{\text{r}} s o^{\text{l}}), (\text{did}, q i^{\text{l}} \pi^{\text{l}}), (\text{whom}, \bar{q} o^{\text{ll}} q^{\text{l}}), (\text{see}, i o^{\text{l}})\}.$$

According to this dictionary and taking α to include $\{s, q, \bar{q}\}$, the following strings of words are grammatical in English

$$\begin{array}{cccc} \text{I} & \text{saw} & \text{him.} & \\ \pi & (\pi^{\text{r}} s o^{\text{l}}) & o & \leq s \end{array}$$

$$\begin{array}{ccccc} \text{Did} & \text{I} & \text{see} & \text{him?} & \\ (q i^{\text{l}} \pi^{\text{l}}) & \pi & i o^{\text{l}} & o & \leq q \end{array}$$

$$\begin{array}{ccccc} \text{Whom} & \text{did} & \text{I} & \text{see?} & \\ (\bar{q} o^{\text{ll}} q^{\text{l}}) & q i^{\text{l}} \pi^{\text{l}} & \pi & i o^{\text{l}} & \leq \bar{q} \end{array}$$

According to Lambek [13] the double adjoint o^{ll} in the type of whom marks the movement of the object from the end of the sentence to its beginning.

To each pregroup inequality, one can assign a diagram. In the case of inequalities corresponding to grammatical strings of words, these diagrams are referred to as *reduction diagrams*. They depict the grammatical structure of a string of words and are one dimensional alternatives to parse-trees. For example the reduction diagrams of the above sentences are as follows:

$$\begin{array}{ccc} \text{I} & \text{saw} & \text{him.} \\ \pi & (\pi^{\text{r}} s o^{\text{l}}) & o \end{array}$$

$$\begin{array}{cccc} \text{Did} & \text{I} & \text{see} & \text{him?} \\ (qi^{l}\pi^{l}) & \pi & io^{l} & o \end{array}$$

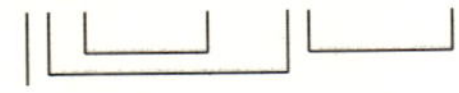

$$\begin{array}{cccc} \text{Whom} & \text{did} & \text{I} & \text{see?} \\ (\bar{q}o^{ll}q^{l}) & qi^{l}\pi^{l} & \pi & io^{l} \end{array}$$

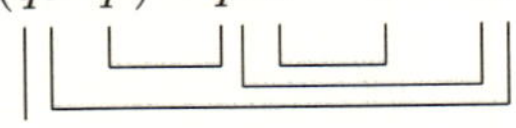

A similar approach is used to formalize grammar of different languages. Reduction diagrams are useful tools in comparing the grammatical structures of different languages. For instance, consider the following two diagrams.

They both reduce the sentence "He bought a book from the bookshop," the left-hand side applies to Persian and Hindi, and the right-hand side to English and French. Comparing these diagrams, without having any knowledge of the corresponding grammars, one can prescribe that English and French are verb-in-the-middle, whereas Persian and Hindi are verb-at-the-end. For more examples, including a poem by Omar Khayam, and details of typing see [16].

2.1. Clitic movement in Persian. When formalizing the Persian grammar using pregroups in [16], I noticed the following phenomenon. The basic word order for a Persian sentence with transitive verb is SOV (Subject-Object-Verb), e.g. in the following sentence.

$$\begin{array}{ccc} \text{I} & \text{him} & \text{saw} \\ \text{man} & \text{u-ra} & \text{didam.} \\ \pi & o & (o^{r}\pi^{r}s) \end{array}$$

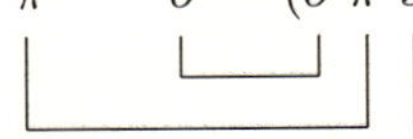

But there is also another possibility: the subject and object, especially when they are pronouns, may become clitic pronouns and attach themselves to the end of the verb. In so doing, their order of occurrence with regard to each other does not change: the subjective clitic is before the objective one in the clitic sentence, exactly like in the original sentence where the subject is before the object. However, their order does change with regard to the verb. In the clitic sentence the the subject is the closer of the two to the verb, whereas in the normal sentence the object is closer to the verb. For examples, the clitic version of the above sentence is as follows

$$\begin{array}{ccc} \text{saw} & \text{I} & \text{him} \\ \text{did} & \text{am} & \text{ash.} \\ (s\bar{o}^{l}\bar{\pi}^{l}) & \bar{\pi} & \bar{o} \end{array}$$

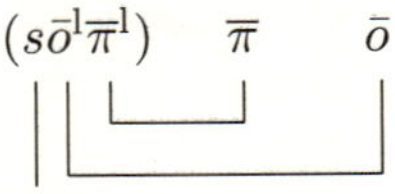

The overlined types $\bar{\pi}, \bar{o}$, stand for the clitic versions of the subject and object pronouns. According to this analysis, the verb takes two different unrelated types: one in its original form and another in its clitic form (i.e., with the clitic pronouns attached to it). Whereas the clitics seem to have a pattern to their movement, implying that, based on this pattern the type of the verb in its clitic form must be derivable from the original type of the verb. One solution would be to add a rule to the pregroup dictionary of Persian to formalize the pattern and as a result to be able to derive the clitic type of the verb from its original type. This rule is as follows

Clitic Rule 1. If $p^{\mathrm{r}}q$ is *in* the *original* type of the verb, so is $q\bar{p}^{\mathrm{l}}$.

The overlined types $\bar{p}^{\mathrm{l}}$, $\bar{p}^{\mathrm{r}}$ are introduced as a notational convenience to distinguish the clitic pronouns from the nonclitic pronouns or arguments. For any clitic pronoun $\bar{p}$, we postulate the partial order $\bar{p} \leq p$ to express the fact that a clitic pronoun is also a kind of pronoun. We assume that for all $p, q \in P$, we have $\overline{pq} = \bar{p}\,\bar{q}$. The calculation to derive the clitic type of the verb from its original type in the above example is as follows

To derive the new type of the verb, we apply rule 1 to its original type $o^{\mathrm{r}}\pi^{\mathrm{r}}s$ by taking q to be s and p to be πo, and noting that $o^{\mathrm{r}}\pi^{\mathrm{r}} = (\pi o)^{\mathrm{r}}$ and $(\overline{\pi o})^{\mathrm{l}} = (\bar{\pi}\,\bar{o})^{\mathrm{l}} = \bar{o}^{\mathrm{l}}\bar{\pi}^{\mathrm{l}}$. Denoting the application of rule 1 by the squiggle arrow, this derivation can be abbreviated to

$$o^{\mathrm{r}}\pi^{\mathrm{r}}s = (\pi o)^{\mathrm{r}}s \rightsquigarrow^{1} s(\overline{\pi o})^{\mathrm{l}} = s(\bar{\pi}\,\bar{o})^{\mathrm{l}} = s\bar{o}^{\mathrm{l}}\bar{\pi}^{\mathrm{l}}$$

Mention of the word *original* in the rule is to prevent the repetition of its application, so that one cannot apply the rule over and over to the types resulting from the application of the rule itself. If this was allowed, then nongrammatical types would be generated. The bars are inserted to prevent the formation of the verb type allowing also nonclitic pronouns to appear after the verb, again resulting in ungrammatical sentences. The mention of the word *in* is to formalize partial clitic movements, when only one of the subject or object becomes a clitic. These partial movements are not very common in Persian, so we will not discuss them here.

2.2. Clitic movement in French. After discussions with Casadio, I came to realize that a similar phenomenon happens in many other languages, including French and Italian. We started to work on a similar formalization for French and Italian, leading to joint work [9]. Interestingly, it turned out that the converse (up to the bars) of the above rule is needed for the clitic movement in French and in Italian needs both rules.

Consider a bitransitive sentence in French

Jean	donne	une pomme	à Marie.
π	$(\pi^{\mathrm{r}}s\,w^{\mathrm{l}}o^{\mathrm{l}})$	o	w

The clitic movement for this sentence has the following pattern

Jean	la	lui	donne.
π	$\bar{o}$	$\bar{w}$	$(\bar{w}^{\mathrm{r}}\bar{o}^{\mathrm{r}}\pi^{\mathrm{r}}s)$

To be able to derive the clitic type of the verb from its original type, we add the following rule to the pregroup dictionary of French.

Clitic Rule 2. If qp^{l} is *in* the *original* type of the verb, so is $\bar{p}^{\mathrm{r}}q$.

Note that, up to a bar, this is the converse of the clitic rule for Persian. The calculation for the derivation of the above example is as follows

To derive the new type of the verb, we apply rule 2 to its original type $(\pi^{\mathrm{r}} s\, w^{\mathrm{l}} o^{\mathrm{l}})$ by taking q to be $\pi^{\mathrm{r}} s$ and p to be ow, and noting that $w^{\mathrm{l}} o^{\mathrm{l}} = (ow)^{\mathrm{l}}$ and $(\overline{ow})^{\mathrm{r}} = (\bar{o}\,\overline{w})^{\mathrm{r}} = \overline{w}^{\mathrm{r}}\, \bar{o}^{\mathrm{r}}$. This derivation can be abbreviated to

$$\pi^{\mathrm{r}} s\, w^{\mathrm{l}} o^{\mathrm{l}} = \pi^{\mathrm{r}} s (ow)^{\mathrm{l}} \rightsquigarrow^{2} (\overline{ow})^{\mathrm{r}} \pi^{\mathrm{r}} s = (\bar{o}\, \overline{w})^{\mathrm{r}} \pi^{\mathrm{r}} s = \overline{w}^{\mathrm{r}}\, \bar{o}^{\mathrm{r}} \pi^{\mathrm{r}} s$$

Clitics can have a partial movement, i.e., it is possible that only one of the objects moves to before the verb where as the other one stays at its original position after the verb. An example of a sentence with a partial clitic movement is "Jean lui donne une pomme.", which we type as follows

Jean	lui	donne	une pomme.
π	$\overline{w}$	$(\overline{w}^{\mathrm{r}} \pi^{\mathrm{r}} s o^{\mathrm{l}})$	o

To derive the type of the verb, the same clitic rule is applied, but this time to a subtype of the type of the original verb, via the usage of the word *in*, as follows

Apply rule 2 the subtype $\pi^{\mathrm{r}} s w^{\mathrm{l}}$ of the original type of the verb and turn it into $\overline{w}^{\mathrm{r}} \pi^{\mathrm{r}} s$. Then attach it to the left over o^{l} and obtain the desired type $\overline{w}^{\mathrm{r}} \pi^{\mathrm{r}} s o^{\mathrm{l}}$.

For more examples in Persian, French, and also Italian, including some exceptions, see [9].

2.3. Cyclic pregroups and groups. To formalize clitic movement, we have added rules to the pregroup dictionaries of languages. It is tempting to add instead the following axioms to the pregroup

(1) $p^{\mathrm{r}} q \leq q p^{\mathrm{l}}$ (2) $q p^{\mathrm{l}} \leq p^{\mathrm{r}} q$.

In so doing, these become applicable to all the types within the pregroup, rather than just to the type of the verb as we have been doing so far.

One can then define the notions of a *left cyclic pregroup*, as a pregroup with axiom (1), a *right cyclic pregroup*, as a pregroup with axiom (2), and a *cyclic pregroup*, as one with both. The free models in each case would be generated over the already existing partial order $\mathcal{B}$ plus the new type $\bar{p}$ and the new order $p \leq \bar{p}$. However, addition of either of these axioms makes the pregroup reduce to a partially ordered group, which is an example of a pregroup where for all $p \in P$ we have $p^{\mathrm{l}} = p^{\mathrm{r}}$. To see this, consider axiom (1), take $q = 1$, we obtain $p^{\mathrm{r}} \leq p^{\mathrm{l}}$ for all $p \in P$, apply left adjoints to both sides and obtain $p^{\mathrm{ll}} \leq p$. Here take $p = w^{\mathrm{r}}$ for some $w \in P$ and obtain $w^{\mathrm{l}} \leq w^{\mathrm{r}}$. Now since we have $p^{\mathrm{r}} \leq p^{\mathrm{l}}$ for all p, we obtain $w^{\mathrm{l}} = w^{\mathrm{r}}$. The argument for axiom (2) is similar.

Interestingly, it turns out that the above axioms are derivable from the translation of Abrusci and Yetter's cyclic rules for noncommutative Linear Logic [3, 17]. These are as follows

(2.1) (left) $ab \leq ba^{\mathrm{ll}}$ (right) $ab \leq b^{rr} a$.

To see this, consider axiom (1), it is derivable from the left axiom by taking a to be p^{r} and b to be q, since $(p^{\mathrm{r}})^{\mathrm{ll}} = p^{\mathrm{l}}$. Axiom (2) is derivable from the right axiom by taking a to be q, and b to be p^{l}, since $(p^{\mathrm{l}})^{rr} = p^{\mathrm{r}}$.

Lambek argues that whenever there is a movement in the formation of the sentence, then iterated adjoints appear in the involved types. This idea exposes itself in a beautiful way in the above equations. These basically say that the juxtaposition of types ab can also be given as ba, but in so doing, the moved type is marked with a double adjoint. Hence, the left equation $ab \leq ba^{ll}$ describes the case when the type a moves from the left to the right of the type b, that is why it is annotated with the double adjoint ll. The right equation $ab \leq b^{rr}a$, describes the other possible movement; that it is b that moves from the right side of a to its left, hence it is annotated with the double adjoint rr.

2.4. An example from Number Theory. In [12] Lambek has shown that the set of all monotone maps on integers, i.e., $\mathbb{Z}^{\mathbb{Z}}$ is a pregroup under function composition and the natural order. This example has been referred to as the Lambek pregroup. He discusses in detail the example $f(n) = 2n$ by computing the following

$$f^{r}(n) = [n/2] \quad \text{and} \quad f^{l}(n) = [(n+1)/2]$$

where $[x]$ is the biggest integer smaller than x. We use this to build an example for a left cyclic pregroup. For all $n \in \mathbb{Z}$, we have that

$$[n/2] \leq [(n+1)/2]$$

From this and for $k \in \mathbb{Z}^{+}$ it follows that

$$k[n/2] \leq k[(n+1)]/2 \quad \text{and} \quad [(kn)/2] \leq k[n/2]$$

By transitivity from the above two it follows that

$$[(kn)/2] \leq k[(n+1)/2]$$

So we have shown that

$$f^{r} \circ g \leq g \circ f^{l}$$

for

$$g(n) = kn \quad \text{and} \quad f(n) = 2n$$

In other words we have shown that for the above choice of functions from the Lambek pregroup, the left cyclic axiom holds. Finding a direct example where the right cyclic axiom holds has been more challenging so far.

3. Movement in a Hungarian sentence

Hungarian belongs to the family of Finno-Ugric languages. This is the largest subgroup of Uralic, spoken by people in northern Eurasia, with Hungary serving as its western limits. Uralic includes more than 30 languages that date as far as the third millennium B.C. It has no relatives in other language families. However (to my delight and surprise), proto-Uralic seems to have been influenced by Indo-Iranian. This similarity is established in [5] and is witnessed by the identity of words, thought to be loan words. The loan dates back to $2-3$ millennium B.C. and its assumed place is the banks of river Volga[1]. Here are some examples of the loan

[1]In second thoughts, this similarity should not be surprising, as it was about the same time that the ancient Aryans (i.e., proto Indo-Iranians) descended from Siberia and on their way south to the plateau of Iran (must have) crossed the Ural mountains and the river Volga.

words

Word	Indo-Iranian	Hungarian
hundred	sata	száz
seven	hapt	hét
horn	sarva	szarv

The words for seven and hundred are still in use in modern Persian.

My references in Hungarian grammar have been [1, 5, 10, 15, 18]. According to the more aged of these references, i.e., [5, 10], word order in a Hungarian sentence is either SOV (Subject, Object, Verb) for predicative sentences or SVO (Subject-Verb-Object) for verbal sentences. They argue that since Hungarian is an agglutinative language, i.e., grammatical relations are expressed by means of affixes, word order does not need much discussion. [10] refers to it as the "analytic" approach to Hungarian grammar and moves to choose a "synthetic" approach which he finds to be more "conspicuous and decisive." Benkö and Imre [5] discuss it under the section on "Clause Structure," where the closest to the kind of movement that we have in mind is called the "Emphatic Order." The cases discussed therein are rather limited. They only deal with instances where the emphasized constituent moves to the "head of the clause," hence only covering our *multi moves* of Section 3.2.

More recent references, e.g., [1, 15, 18] present more material on word order. They suggest that exactly due to the agglutinative nature of the language, one should not look at word order in Hungarian from the usual SVO or SOV perspective, but rather as a focus-oriented structure. Abondiolo [1] suggests to approach it (Hungarian as well as the whole family of Uralic languages) as a TFV for Topic-Focus-Verb. The other two follow him and work with the "topic+focus+verb+rest" order. The approach of this paper is closer to these more recent ones, but, as we shall see, it is not exactly the same. Our example sentence is from [18], as it had the most comprehensive examples of the three.

The basic word order for a sentence with a transitive verb in Hungarian is as follows

János	tegnap	elvitt	két könyvet	Péternek.
János	yesterday	took	two books	to Péternek.

Taking π to stand for the type of subject, o for first object, w for second object, and λ for adverb, we assign the following types[2] to the constituents of the above sentence

János	tegnap	elvitt	két könyvet	Péternek.
π	λ	$(\lambda^r \pi^r s\, w^l o^l)$	o	w

The basic order changes according to the information structure of the context in which the sentence is uttered. This is to enable the speaker to communicate extra information about a constituent, e.g. emphasize it or make it inclusive. For instance, the speaker might want to emphasize that it was indeed János who took the books, or that it was only two books and nothing more that were taken from one person to another. In any case, we call the constituent about which some extra information is conveyed, a *mobilizer*. This is different from the *focus* phrases of [1] and [15], where the focus can be empty but a movement is still possible. This causes

[2]These types are the most primitive options that serve the purpose of this paper, it is likely that a more sophisticated typing is needed for the pregroup typing of the full Hungarian grammar.

an ambiguity on the existence of a focus and hence on interpreting the meaning of the sentence. From a pure syntactic point of view, if the verb has a prefix, e.g., in *elvitt* and a focus is present, then the prefix becomes a suffix, e.g., *vitt el.* But if there is no prefix then it would be ambiguous as whether there is a focus present or not. Even if not, the word order that corresponds to the existence of the focus is still possible, but will convey a different meaning. Hence, the same sentence may have different possible information structures. Our approach discards this ambiguity: whenever there is a movement, we have a mobilizer. It is based on these mobilizers that we classify and discuss each possible movement.

Our aim is to derive the type of the verb after each possible movement from the type of the verb in the basic sentence, i.e., from $(\lambda^{\mathrm{r}}\pi^{\mathrm{r}} s\, w^{\mathrm{l}} o^{\mathrm{l}})$. As the constituents that move are not necessarily clitics and the movement is fairly free, we generalize our previous *clitic rules* in two ways: (1) we remove the bars from the moved types, so that they are not limited to clitic pronouns, and (2) we remove the word *original* from the rule definition, so that the rules may be repetitively applied to intermediate types. This will not produce ungrammatical types anymore, exactly because the movement is not restricted to clitic pronouns. The new rules are as follows:

Movement Rule 1. If $p^{\mathrm{r}}q$ is $\boxed{in}$ the type of the verb, so is qp^{l}.

Movement Rule 2. If qp^{l} is $\boxed{in}$ the type of the verb, so is $p^{\mathrm{r}}q$.

Rule 1 describes the movement of a word with type p from a position before the verb to one after it. In the original sentence, this type will cancel out with a type p^{r} that occurs in the type of the verb. After the movement, it will cancel out with a type p^{l} in the new type of the verb. Rule 2 describes an opposite movement, i.e., when a word with type p moves from a position after the verb to one before it, hence originally it cancels out with a type p^{l} within the type of the verb, but after the movement with a p^{r}.

In the definition of the rules, we have included the word *in.* This enables a *local* application of the rules, i.e., to subtypes within the type of the verb, but might cause over-generation. To avoid this, we limit the rules as follows:

Rule 1 should not turn a right adjoint x^{r} that is originally positioned before s into its left adjoint x^{l} still located before s.

Rule 2 should not turn a left adjoint x^{l} that is originally positioned after s into its right adjoint x^{r} still located after s.

These limitations will, for instance, prevent formation of types such as $(\pi^{\mathrm{r}}\lambda^{\mathrm{l}}sw^{\mathrm{l}}o^{\mathrm{l}})$, which is obtained by applying rule 1 to subtype $\lambda^{\mathrm{r}}\pi^{\mathrm{r}}$ of the original type of verb, to turn it into $\pi^{\mathrm{r}}\lambda^{\mathrm{l}}$. Our first limitation prevents this, since π^{r}, which is a left adjoint type before s has now become a left adjoint type π^{l} and is still before s. Similarly, the second limitation prevents formation of $(\lambda^{\mathrm{r}}\pi^{\mathrm{r}}so^{\mathrm{r}}w^{\mathrm{l}})$, which is obtained by applying rule 2 to subtype $w^{\mathrm{l}}o^{\mathrm{l}}$ to turn it into $o^{\mathrm{r}}w^{\mathrm{l}}$. In any case (and as observed by the referee), these limitations are coherent with the constituent structure of the sentence, i.e., that in all of these moves, as we shall see below, the whole argument of the verb (and not part of it) is moved.

We distinguish between two types of movement: single and multi. Taking the basic word order as the reference frame, the former is when *only one word* changes its place of occurrence in the sentence, e.g., by moving from a position after the verb to one before it. The latter is when *two or more words* do single moves. The second type of movement and respectively its formalization is more complex. There is a large body of literature on this matter in generative linguistics, e.g. see [6]. Our approach provides some preliminary insights into the question: we deal with them by invoking a local and limited but repetitive application of the rules.

3.1. Single moves. The mobilizer of this type can be the subject or either of the objects. In each case, they will appear right before the verb after the movement. If the mobilizer is the subject, i.e., *János*, then the temporal adverb *yesterday* moves to after the verb, as follows

János	vitt el	tegnap	két könyvet	Péternek.
János	took	yesterday	two books	to Péternek.
π	$(\pi^{\mathrm{r}} s w^{\mathrm{l}} o^{\mathrm{l}} \lambda^{\mathrm{l}})$	λ	o	w

To derive the new type of the verb, we apply rule 1 to the original type $(\lambda^{\mathrm{r}}\pi^{\mathrm{r}} s w^{\mathrm{l}} o^{\mathrm{l}})$, by taking q to be $\pi^{\mathrm{r}} s w^{\mathrm{l}} o^{\mathrm{l}}$ and p to be λ.

If the mobilizer is the first object, i.e., *two books*, then it moves to before the verb. The above sentence and its typing change as follows

János	tegnap	**két könyvet**	vitt el	Péternek.
János	yesterday	**two books**	took	to Péternek.
π	λ	o	$(o^{\mathrm{r}}\lambda^{\mathrm{r}}\pi^{\mathrm{r}} s w^{\mathrm{l}})$	w

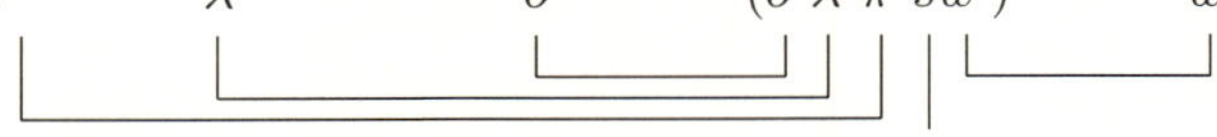

To derive the new type of the verb, we apply rule (2) to the original type $(\lambda^{\mathrm{r}}\pi^{\mathrm{r}} s w^{\mathrm{l}} o^{\mathrm{l}})$, by taking q to be $\lambda^{\mathrm{r}}\pi^{\mathrm{r}} s w^{\mathrm{l}}$, and p to be o.

If the mobilizer is the second object *Péternek*, then it moves to before the verb, as follows

János	tegnap	**Péternek**	vitt el	két könyvet.
János	yesterday	**to Péternek**	took	two books.
π	λ	w	$(w^{\mathrm{r}}\lambda^{\mathrm{r}}\pi^{\mathrm{r}} s o^{\mathrm{l}})$	o

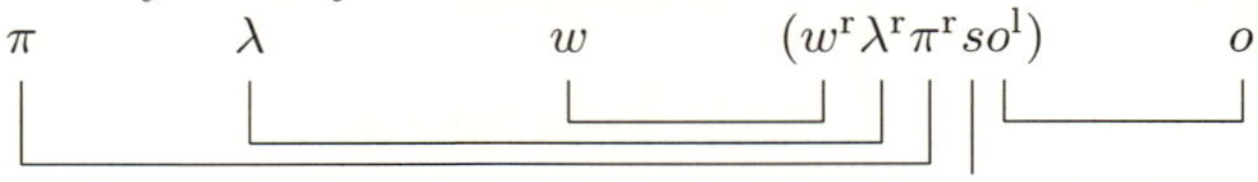

To derive the new type of the verb, we apply rule 2 to the subtype $\lambda^{\mathrm{r}}\pi^{\mathrm{r}} s w^{\mathrm{l}}$ of the original type, by taking q to be $\lambda^{\mathrm{r}}\pi^{\mathrm{r}} s$ and p to be w.

The verb can also be a mobilizer, e.g., when extra information is conveyed to mark the completion of its corresponding task, as in the task of *taking*. In this case the verb moves to the beginning of the sentence, as follows

Elvitt	János	tegnap	két könyvet	Péternek.
Took	János	yesterday	two books	to Péternek.
$(s w^{\mathrm{l}} o^{\mathrm{l}} \lambda^{\mathrm{l}} \pi^{\mathrm{l}})$	π	λ	o	w

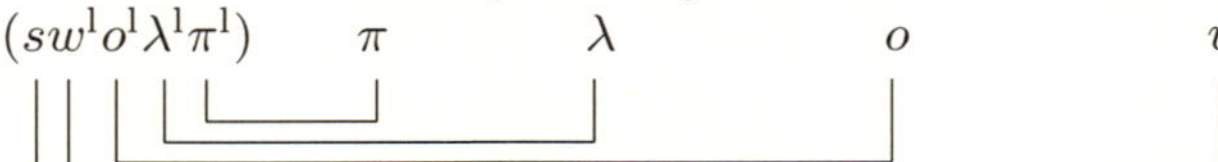

To derive the new type of the verb, we apply rule (1) to the original type $(\lambda^r\pi^r sw^l o^l)$, taking p to be $(\pi\lambda)$, and q to be $sw^l o^l$. The calculation is abbreviated to

$$\lambda^r\pi^r sw^l o^l = (\pi\lambda)^r sw^l o^l \rightsquigarrow^1 sw^l o^l(\pi\lambda)^l = sw^l o^l \lambda^l \pi^l$$

3.2. Multi moves. Here a different kind of information, rather than just emphasis, is conveyed about some of the constituents. For instance, the information that the object is the sole or not the sole entity on which the verb has taken place. In this case, not only the mobilizer moves to before the verb, but also, together with the verb, they move to the beginning of the sentence.

If the mobilizer is the first object, e.g., we want to say that it was to Péter and no one else that the two books were taken, the sentence becomes as follows

Péternek	vitt el	tegnap	János	két könyvet.
To Péternek	took	yesterday	János	two books.
w	$(w^r so^l\pi^l\lambda^l)$	λ	π	o

One way to see this as a multi move is: not only Péternek has moved to the beginning of the sentence, but also first tegnap and then János have moved from before the verb to after it, and in so doing have changed their order with regard to each other. The calculation for deriving the new type of the verb reflects the above complications and needs repetitive applications of the rules. It is as follows

Start from the original type of the verb $(\lambda^r\pi^r sw^l o^l)$, according to our above description, first Péternek moves to the front, for this we apply rule 2 to the subtype $\lambda^r\pi^r sw^l$, take p to be w, and obtain $w^r\lambda^r\pi^r so^l$. Then tegnap moves to after the verb, for this we apply rule 1 to the subtype $\lambda^r\pi^r so^l$, take p to be λ, and obtain $w^r\pi^r so^l\lambda^l$. Finally János moves to after tegnap, for this we apply rule 1 to the subtype $\pi^r so^l$, take p to be π, and obtain $w^r so^l\pi^l\lambda^l$.

If the mobilizer is the second object, e.g., if we want to say that two books and nothing else were taken to Péternek, then the sentence becomes as follows

Két könyvet	vitt el	János	tegnap	Péternek.
Two books	took	János	yesterday	to Péternek.
o	$(o^r sw^l\lambda^l\pi^l)$	π	λ	w

One way to see this as a multi move is: first két könyvet moves to the beginning of the sentence, then the two words János and tagnap move to after the verb. The corresponding calculation is as follows.

First apply rule 2 to $(\lambda^r\pi^r sw^l o^l)$, take p to be o, and obtain $o^r\lambda^r\pi^r sw^l$, then apply rule 1 to the subtype $\lambda^r\pi^r sw^l$, take p to be $\pi\lambda$, note $\lambda^r\pi^r = (\pi\lambda)^r$, similarly $(\pi\lambda)^l = \lambda^l\pi^l$, and obtain $o^r sw^l\lambda^l\pi^l$.

The following is also a possibility, it is uttered when one wants to insist that two books where taken to Péternek but at the same time not exclude the possibility that may be also other things were taken.

Két könyvet	tegnap	elvitt	János	Péternek.
Two books	yesterday	took	János	to Péternek.
o	λ	$(\lambda^{r} o^{r} s w^{l} \pi^{l})$	π	w

One way to see this as a multi move is: first János moves to after the verb then két könyvet moves to the beginning of the sentence. The calculation is as follows

> Starting from the original type of the verb $(\lambda^{r}\pi^{r}sw^{l}o^{l})$, the calculation to derive these series of moves is as follows: apply rule 1 to the subtype $\pi^{r}sw^{l}$, take p to be π, and obtain $\lambda^{r}sw^{l}\pi^{l}o^{l}$, then apply rule 2 to the subtype $sw^{l}\pi^{l}o^{l}$, take p to be o, and obtain $\lambda^{r}o^{r}sw^{l}\pi^{l}$.

4. Summary

We have added two meta rules to a preliminary pregroup dictionary of Hungarian. These rules are generalizations of our previous *cyclic rules* [9]. They are more general since they allow for repetitive applications and can also be applied to any word that does not necessarily have a clitic pronoun assigned to it. We have used these rules to formalize and reason about change of word order in Hungarian. We have done so by deriving the new type of the verb after the movement, from its original type before the movement.

Change of word order in Hungarain is due to change of emphasis in a sentence, it occurs when an extra piece of information needs to be communicated about a constituent of the sentence. Some references on Hungarian grammar, e.g., [1, 15, 18] explain this movement by using the notion of *focus*; we introduce the new concept of a *mobulizer*. In some cases there is a movement in the sentence which changes the meaining, but the focus is empty. This causes ambiguity in interpretting the meaning of the sentence. We discrad this ambiguity by assigning a mobilizer to each possible movement. This is the constituent that causes the movement and about which extra information is communicated. It is according to the position of the mobilzer after the movement that a different meaning is assigned to a sentence.

We have shown that the axiomatic version of our rules are derivable from Abrusci and Yetter's cyclic axioms for noncommutative linear logic [3, 17]. We have also reviewed how adding the meta rules as axioms to pregroups, hence developing the new notion of a *cyclic pregroups*, reduces them to ordered groups. We have found an example of our first rule (left cyclic) within the arithmetics model of a Lambek pregroup.

References

1. D. Abondolo (ed.), *The Uralic languages*, Routledge Language Family Descriptions, Routledge, London–New York, 1998.
2. S. Abramsky and B. Coecke, *A categorical semantics of quantum protocols*, 19th Annual IEEE Symposium on Logic in Computer Science (Turku, 2004), IEEE Computer Society, Los Alamitos, CA, 2004, pp. 415–425.
3. V. M. Abrusci, *Classical conservative extensions of Lambek calculus: The Lambek calculus in logic and linguistics*, Studia Logica **71** (2002), no. 3, 277–314.
4. M. Barr, *On subgroups of the Lambek pregroup*, Theory Appl. Categ. **12** (2004), no. 8, 262–269.
5. L. Benkö and S. Imre, *The Hungarian language*, Mouton, The Hague–Paris, 1972.

6. M. Brody and A. Szabolcsi, *Overt scope in Hungarian*, Syntax **6** (2003), no. 1, 1951.
7. W. Buszkowski, *Lambek grammars based on pregroups*, Logical Aspects of Computational Linguistics (Le Croisic, 2001) (P. de Groote, G. Morrill, and C. Retoré, eds.), Lecture Notes in Comput. Sci., vol. 2099, Springer, Berlin, 2001, pp. 95–109.
8. C. Casadio and J. Lambek (eds.), *Recent computational algebraic approaches to morphology and syntax*, Polimetrica, Milan, 2008.
9. C. Casadio and M. Sadrzadeh, *In search for a uniform pattern of clitic movement*, Eighth International Tbilisi Symposium on Language, Logic and Computation (Bakuriani, 2009), Lecture Notes in Comput. Sci., Springer, Berlin, to appear.
10. Jr. Hall Robert Anderson, *Hungarian grammar*, Language Monographs, vol. 21, Linguistic Society of America, Baltimore, MD, 1944.
11. J. Lambek, *Type grammar revisited*, Logical Aspects of Computational Linguistics (Nancy, 1997) (A. Lecomte, F. Lamarche, and G. Perrier, eds.), Lecture Notes in Comput. Sci., vol. 1582, Springer, Berlin, 1999, pp. 1–27.
12. ———, *Iterated Galois connections in arithmetic and linguistics*, Galois Connections and Applications (K. Denecke, M. Erné, and S. L. Wismath, eds.), Math. Appl., vol. 565, Kluwer, Dordrecht, 2004, pp. 389–397.
13. ———, *Pregroups and natural language processing*, Math. Intelligencer **28** (2006), no. 2, 41–48.
14. ———, *Compact monoidal categories from linguistics to physics*, New Structures for Physics (B. Coecke, ed.), Lecture Notes in Phys., vol. 813, Springer, New York, 2009, pp. 467–487.
15. B. Megyesi, *The Hungarian language: A short descriptive grammar*, Department of Linguistics, Stockholm University, 1998, `http://stp.ling.uu.se/~bea/publ/megyesi-hungarian.pdf`.
16. M. Sadrzadeh, *Pregroup analysis of Persian sentences*, Recent Computational Algebraic Approaches to Morphology and Syntax (C. Casadio and J. Lambek, eds.), Polimetrica, Milan, 2008, pp. 121–144.
17. D. N. Yetter, *Quantales and (noncommutative) linear logic*, J. Symbolic Logic **55** (1990), no. 1, 41–64.
18. Wikipedia, *Hungarian grammar — Wikipedia, the free encyclopedia*, `http://en.wikipedia.org/wiki/Hungarian_grammar`.

Oxford University Computing Laboratory, Wolfson Building, Parks Road, Oxford, OX1 3QD, UK

E-mail address: `mehrs@comlab.ox.ac.uk`

Centre de Recherches Mathématiques
CRM Proceedings and Lecture Notes
Volume **53**, 2011

No Limit Model in Inaccessibles

Saharon Shelah

Dedicated to Michael Makkai

ABSTRACT. Our aim is to improve the negative results, i.e., non-existence of limit models, and the failure of the generic pair property from [6] to inaccessible λ as promised there. In [6], the negative results were obtained only for non-strong limit cardinals.

0. Introduction

Let $\lambda = \lambda^{<\lambda} > \kappa$ be regular cardinals. A complete first order theory T may have (some variant of) (λ, κ)-limit model, which, if exists, is unique, see history in [6] and Definition 0.9. There we prove existence for the theory of linear order and non-existence for first order theories which are strongly independent and then just independent and even the parallel for $\kappa = 2$ (one direction of the so-called generic pair conjecture). Those non-existence results in [6] were for $\lambda = 2^{\kappa}$, here we deal with strongly inaccessible λ. In [7] there are existence results but for λ measurable, and we promise there the non-existence results for λ strongly inaccessible as complimentary results.

Let λ be strongly inaccessible ($> |T|$) such that $\lambda^+ = 2^{\lambda}$; this for transparency only.

Here in Section 1 we prove that for strongly independent T (see Definition 0.2), a strong version of the generic pair conjecture (see Definition 0.7(2)) holds. We also prove the non-existence of (λ, κ)-limit models, a related property (for all versions of "limit model").

In Section 2, we prove this even for independent T. The use of $\lambda^+ = 2^{\lambda}$ is just to have a more transparent formulation of the conjecture. See more on the generic pair conjecture for dependent T in [8].

We thank Itay Kaplan for much helpful criticism.

Notation 0.1. (1) $\mathscr{D}_{\lambda}$ is the club filter on λ for λ regular uncountable.

2010 *Mathematics Subject Classification.* Primary 03C45; Secondary 03C55.

The author would like to thank the Israel Science Foundation for partial support of this research (Grant No. 242/03). Publication 906.

The author thanks Alice Leonhardt for the beautiful typing.

This is the final form of the paper.

(2) $S^\lambda_\kappa = \{\delta < \lambda : \mathrm{cf}(\delta) = \kappa\}$.
(3) For a limit ordinal δ let $\mathscr{P}^{\mathrm{ub}}(\delta) = \{\mathscr{U} : \mathscr{U}$ is an unbounded subset of $\delta\}$.
(4) T denotes a complete first order theory.
(5) For a model M, $\varphi(\bar{x}, \bar{y}) \in \mathbb{L}(\tau_M)$ and $\bar{d} \in {}^{lg(\bar{y})}M$, let $\varphi(M, \bar{d}) = \{\bar{c} \in {}^{lg(\bar{x})}M : M \models \varphi[\bar{c}, \bar{d}]\}$.
(6) $\mathbf{S}^n(A, M) = \{\mathrm{tp}(\bar{b}, A, N) : M \prec N$ and $\bar{b} \in {}^nN\}$ where $\mathrm{tp}(\bar{c}, A, N) = \{\varphi(\bar{x}, \bar{a}) : \varphi(\bar{x}, \bar{y}) \in \mathbb{L}(\tau_M), \bar{a} \in {}^{lg(\bar{y})}A$ and $M \models \varphi[\bar{c}, \bar{a}]\}$.
(7) $\mathbf{S}^n(M) = \mathbf{S}^n(M, M)$ and $\mathbf{S}^{<\omega}(M) = \bigcup\{\mathbf{S}^n(M) : n \in \omega\}$.
(8) $\beth_\alpha(\lambda) = \lambda + \Sigma\{2^{\beth_\beta(\lambda)} : \beta < \alpha\}$ and $\beth_\alpha = \beth_\alpha(\aleph_0)$.

Recall (as in [6, 2.3])

Definition 0.2. (1) T has the strong independence property (or is strongly independent) *when*: some $\varphi(\bar{x}, \bar{y}) \in \mathbb{L}(\tau_T)$ has it, where:

(2) $\varphi(\bar{x}, \bar{y}) \in \mathbb{L}(\tau_T)$ has the strong independence property for T *when* for every model M of T and pairwise disjoint finite $\mathbf{I}_1, \mathbf{I}_2 \subseteq {}^{lg(\bar{y})}(M)$ for some $\bar{a} \in {}^{lg(\bar{x})}M$ we have $l \in \{1,2\} \wedge \bar{b} \in \mathbf{I}_l \Rightarrow M \models \varphi[\bar{a}, \bar{b}_l]^{\mathrm{if}(l=2)}$.

Remark 0.3. (1) Elsewhere we use $\varphi(x, y)$, i.e., the x and y are singletons, but the proofs are not affected.

(2) Also we may restrict ourselves to $\mathbf{I}_1, \mathbf{I}_2 \subseteq \psi(M, \bar{d})$ where $\psi \in \mathbb{L}(\tau_T)$ such that $\psi(M, \bar{d})$ is infinite, and we may restrict ourselves to $\mathbf{I}_1$, $\mathbf{I}_2$ such that every $\bar{b} \in \mathbf{I}_1 \cup \mathbf{I}_1$ realizes a fixed nonalgebraic type $p \in \mathbf{S}^m(A, M)$ with M being $(|A|^+ + \aleph_0)$-saturated. The results are not really affected.

Question 0.4. (1) Assume $\lambda_2 = \lambda_2^{<\lambda_1} \geq \lambda_1 > |T|$, T a complete first order dependent theory. Is the theory $T^*_{\lambda_1,\lambda_2}$ a dependent theory or at least when is $T^*_{\lambda_1,\lambda_2}$ a dependent theory? where
(a) $T^*_{\lambda_1,\lambda_2} = \mathrm{Th}(K^+_{\lambda_1,\lambda_2})$ where
(b) $K^+_{\lambda_1,\lambda_2} = \{(N, M) : M$ is a λ_1-saturated model of T of cardinality λ_2, N a λ_2^+-saturated elementary extension of $M\}$.

(2) Similarly for other properties of $T^*_{\lambda_1,\lambda_2}$; note[1] that this theory is complete if $\lambda_1 = \lambda_2$.

(3) When can we prove that $T^*_{\lambda_1,\lambda_2}$ does not depend on the cardinals at least for many pairs?

Remark 0.5. (1) Concerning failure of Question 0.4(1) see Kaplan – Shelah [3].

(2) Any solution of the generic pair conjecture answers positively Question 0.4(3) for dependent T in the relevant cases.

(3) It is known that in Question 0.4(1) if T extends PA or ZFC then in $T^* = \mathrm{Th}(N, M)$ we can interpret the second order theory of λ_2.

But may well be that as in Baldwin – Shelah [1]

Question 0.6. Assume $|T| < \kappa \leq \lambda_1 \leq \lambda_2 = \lambda_2^{<\lambda_1}$, T a complete first order theory. For which T's can we interpret in $M \in K^+_{\lambda_1,\lambda_2}$ a model of PA of

[1] let $(N_l, M_l) \in K^+_{\lambda_1,\lambda_2}$ for $l = 1, 2$ and let f_* be an isomorphism from M_1 onto M_2 and let $\mathscr{F} = \{f : f$ is a (N_1, N_2)-elementary mapping extending f_* of cardinality $\leq \lambda_1\}$. Now we can prove that any $f \in \mathscr{F}$ preserve satisfaction for first order formulas.

cardinality $\geq \lambda_1$ by first order formula *or* just an $\mathbb{L}_{\infty,\kappa}(\tau_T)$-formulas with parameters, the intention is that we assume λ_2 is enough larger than λ_1 which is large enough than $|T|$; if $2^\kappa \geq \lambda_1$ this is trivial.

Recall (from [6, 0.2])

Definition 0.7. (0) Let $\mathrm{EC}_\lambda(T)$ be the class of models M of (the first order) T of cardinality λ. Let $\mathrm{EC}_{\lambda,\kappa}(T)$ be the class of κ-saturated models $M \in \mathrm{EC}_\lambda(T)$.

(1) Assume $\lambda > |T|$, (we usually assume $\lambda = \lambda^{<\lambda}$) and $2^\lambda = \lambda^+$, $M_\alpha \in \mathrm{EC}_\lambda(T)$ is $\prec$-increasing continuous for $\alpha < \lambda^+$ with $M = \bigcup\{M_\alpha : \alpha < \lambda^+\} \in \mathrm{EC}_{\lambda^+}(T)$, and M is saturated. The generic pair property (for T, λ) says that for some club E of λ^+ for all pairs $\alpha < \beta$ of ordinals from E of cofinality λ, (M_β, M_α) has the same isomorphism type (we denote this property of T by $\mathrm{Pr}^2_{\lambda,\lambda}(T)$).

(2) The generic pair conjecture for $\lambda > \aleph_0$, (usually $\lambda = \lambda^{<\lambda}$) such that $2^\lambda = \lambda^+$ says that for any complete first order T of cardinality $< \lambda$, T is dependent iff it has the generic pair property for λ.

(3) Let $\mathbf{n}_{\lambda,\kappa}(T)$ be $\min\{|\{M_\delta/\cong : \delta \in E \text{ has cofinality } \kappa\}| : E \text{ a club of } \lambda^+\}$ for λ and $\overline{M} = \langle M_\alpha : \alpha < \lambda^+\rangle$ as above and $\kappa = \mathrm{cf}(\kappa) \leq \lambda$; clearly the choice of $\overline{M}$ is immaterial.

Remark 0.8. (1) Note that to say $\mathbf{n}_{\lambda,\kappa}(T) = 1$ is a way to say that T has (some variant of) a (λ,κ)-limit model, see Definition 0.9 below. There are other variants of the definition of limit.

(2) Recall that we conjecture that for $\lambda = \lambda^{<\lambda} > \kappa = \mathrm{cf}(\kappa) > |T|$, $2^\lambda = \lambda^+$ we have $\mathbf{n}_{\lambda,\kappa}(T) = 1 \Leftrightarrow \mathbf{n}_{\lambda,\kappa}(T) < 2^\lambda \Leftrightarrow T$ is dependent. The use of "$\lambda^+ = 2^\lambda$" is just for clarity. See more in [6–8].

(3) Recall that if $\lambda = \kappa = \lambda^{<\lambda}$, then $\mathbf{n}_{\lambda,\lambda}(T) = 1$ means T has a unique saturated model; (and parallely if $\lambda > \mathrm{cf}(\lambda) = \kappa$, λ strong limit). So we concentrate on the case $\mathbf{n}_{\lambda,\kappa}(T) = 1$ where $\kappa < \lambda$.

Definition 0.9. We define when M_* is a (λ,κ)-limit model of T where $\lambda \geq \kappa = \mathrm{cf}(\kappa)$ and $\lambda \geq |T|$. In general it means that: letting $K_\lambda = \{M : M \text{ is a model of } T \text{ with universe an ordinal} \in [\lambda,\lambda^+)\}$, for some function $\mathbf{F}$ with domain K and satisfying $M \prec \mathbf{F}(M) \in K$ we have:

$\oplus$ if $M_\alpha \in K$ for $\alpha < \lambda^+$ is $\prec$-increasing continuous and $\alpha < \lambda \Rightarrow \mathbf{F}(M_{\alpha+1}) \prec M_{\alpha+2}$ then for some club E of λ^+ we have:

$$\delta \in E \wedge \mathrm{cf}(\delta) = \kappa \implies M_\delta \cong M_*.$$

Remark 0.10. If $2^\lambda = \lambda^+$ we have: $\mathbf{n}_{\lambda,\kappa}(T) = 1$ iff T has a (λ,κ)-limit model.

1. Strongly independent T

Context 1.1. (1) T is a fixed first order complete theory and $\mathfrak{C} = \mathfrak{C}_T$ a monster for it; for notational simplicity τ_T is relational.

(2) We let λ be a regular uncountable cardinal $> |T|$; we deal mainly with strongly inaccessible λ.

Here for λ strongly inaccessible and (complete first order) T with the strong independence property (of cardinality $< \lambda$) we prove the non-existence of (λ,κ)-limit models for $\kappa = \mathrm{cf}(\kappa) < \lambda$ (in Theorem 1.9) and the generic pair conjecture for

λ and T, in Theorem 1.10 (which shows nonisomorphism). Recall that the generic pair property speaks on the isomorphism type of pairs of models.

Definition 1.2 gives us a more constructive invariant of $(N, M)/\cong$. Unfortunately it seemed opaque how to manipulate it so we shall use a related but different version, the one from Definition 1.4. Naturally it concentrates on types in one formula $\varphi(\bar{y}, \bar{x})$ witnessing the strong independence property. But mainly gives the pair (N, M) an invariant $\langle \mathscr{P}_\delta : \delta < \lambda \rangle / \mathscr{D}_\lambda$ where $\mathscr{P}_\delta \subseteq \mathscr{P}(\mathscr{P}(\delta))$. Now always $|\mathscr{P}_\delta| \leq 2^{|\delta|}$ and it is easily computable from one $\mathscr{P} \subseteq \mathscr{P}(\delta)$, in fact from the invariant $\mathrm{inv}_4(M, N)$ from Definition 1.2, but in our proofs its use is more transparent. It has monotonicity property and we can increase it.

We need different but similar version for the proof of non-existence of (λ, κ)-limit models.

Definition 1.2. (1) Let $\mathscr{E}^*_T$ be the following two-place relation on $\{(M, \mathbf{P}) : M \models T \text{ and } \mathbf{P} \subseteq \mathbf{S}^{<\omega}(M)\}$; let $(M_1, \mathbf{P}_1)\mathscr{E}^*_T(M_2, \mathbf{P}_2)$ iff there is an isomorphism h from M_1 onto M_2 mapping $\mathbf{P}_1$ onto $\mathbf{P}_2$.

(2) For models $M \subseteq N$ we define (the important case is $M \prec N \models T$):

(a) $\mathrm{inv}_1(M, N) = \{p \in \mathbf{S}^{<\omega}(M) : p \text{ is realized in } N\}$

(b) $\mathrm{inv}_2(M, N) = (M, \mathrm{inv}_1(M, N))/\mathscr{E}^*_T$.

(3) If $M \prec N$ are models of T such that the universe of N is $\subseteq \lambda$, recalling $\mathscr{D}_\lambda$ is the club filter on λ, let:

(a) for any ordinal $\delta < \lambda$

$$\begin{aligned} &\mathrm{inv}_3(\delta, M, N) \\ &\quad = (M \restriction \delta, \{p \in \mathbf{S}^{<\omega}(M \restriction \delta) : p \text{ is realized by some sequence from } N \restriction \delta\})/\mathscr{E}^*_T) \end{aligned}$$

(b) $\mathrm{inv}_4(M, N) = \langle \mathrm{inv}_3(\delta, M, N) : \delta < \lambda \rangle / \mathscr{D}_\lambda$.

(4) If $M \prec N$ are models of T of cardinality λ then $\mathrm{inv}_4(M, N)$ is equal to $\mathrm{inv}_4(f(M), f(N))$ for every one-to-one function f from N into λ (equivalently some f, see Observation 1.3(1), (2) below).

Observation 1.3. (0) *In Definition* 1.2(3) *for a club of* δ*'s below* λ *we have* $M \restriction \delta \prec M$ *and* $N \restriction \delta \prec N$ *and so* $M \restriction \delta \prec N \restriction \delta \models T$.

(1) *Concerning Definition* 1.2(3), *if* $M \prec N$ *are models of* T *of cardinality* λ *and* f_1, f_2 *are one-to-one functions from* N *into* λ then $\mathrm{inv}_4(f_1(M), f_1(N)) = \mathrm{inv}_4(f_2(M), f_2(N))$ *using Definition* 1.2(3)(b).

(2) *Definitions* 1.2(3), 1.2(4) *are compatible and in Definition* 1.2(4), *"some* f *such that* f *is a one-to-one function from* N *to* λ*" is equivalent to "every* f *such that...."*

PROOF (OBSERVATION 1.3). Straight, e.g., (this argument will be used several times).

(1) Let $E = \{\delta < \lambda : \delta \text{ is a limit ordinal such that } M \restriction \delta \prec M,\ N \restriction \delta \prec N$ and $\mathrm{Rang}(f_l \restriction \delta) = \mathrm{Rang}(f_l) \cap \delta$ for $l = 1, 2\}$. So E is a club of λ and $\delta \in E \Rightarrow f_2 \circ f_1^{-1}$ is an isomorphism from $f_1(N \restriction \delta)$ onto $f_2(N \restriction \delta)$, mapping $f_1(M \restriction \delta)$ onto $f_2(M \restriction \delta)$. □

Definition 1.4. Assume $\varphi = \varphi(\bar{x}, \bar{y}) \in \mathbb{L}(\tau_T)$ and $N_1 \prec N_2$ are models of T of cardinality λ.

(1) For one-to-one mapping f from N_2 to λ and $\delta < \lambda$ we define

$\mathrm{inv}_5^\varphi(\delta, f, N_1, N_2) = \{\mathscr{P} \subseteq \mathscr{P}(\delta)$: there are $\bar{a}_\gamma \in {}^{lg(\bar{x})}N_2$ satisfying $f(\bar{a}_\gamma) \in {}^{lg(\bar{x})}\delta$ for $\gamma < \delta$ such that for every $\mathscr{U} \subseteq \delta$ the following are equivalent:

(i) $\mathscr{U} \in \mathscr{P}$

(ii) for some $\bar{b} \in {}^{lg(\bar{y})}N_1$ we have $\gamma < \delta \Rightarrow N_2 \models \varphi[\bar{a}_\gamma, \bar{b}]^{\mathrm{if}(\gamma \in \mathscr{U})}\}$.

(2) We let $\mathrm{inv}_6^\varphi(N_1, N_2)$ be $\langle \mathrm{inv}_5^\varphi(\delta, f, N_1, N_2) : \delta < \lambda\rangle/\mathscr{D}_\lambda$ for some (equivalently every) f as above.

Claim 1.5. (1) *In Definition* 1.4(2) *we have* $\mathrm{inv}_6^\varphi(N_1, N_2)$ *is well defined.*

(2) *In Definition* 1.4, *for* δ, λ, N_1, N_2, $\varphi(\bar{x}, \bar{y})$ *as there*

(a) *the set* $\mathrm{inv}_5^\varphi(\delta, f, N_1, N_2)$ *has cardinality at most* $2^{|\delta|}$

(b) *if* π *is a one-to-one function from* $f(N_2)$ *into* λ *mapping* $f(N_2) \cap \delta$ *onto* $\pi(f(N_2)) \cap \delta$ *then* $\mathrm{inv}_5^\varphi(\delta, \pi \circ f, N_1, N_2) = \mathrm{inv}_5^\varphi(\delta, f, N_1, N_2)$.

PROOF (CLAIM 1.5). Easy. □

Definition 1.6. (1) For $\varphi = \varphi(\bar{x}, \bar{y}) \in \mathbb{L}(\tau_T)$, a model N of T with universe λ, δ a limit ordinal $< \lambda$ and $\kappa < \lambda$ let

$\mathrm{inv}_{7,\kappa}^\varphi(\delta, N) = \{\mathscr{P} \subseteq \mathscr{P}(\delta)$: we can find $\bar{a}_\gamma^i \in {}^{lg(\bar{x})}\delta$ for $\gamma < \delta$, $i < \kappa$ such that the following conditions on $\mathscr{U} \subseteq \delta$ are equivalent:

(i) $\mathscr{U} \in \mathscr{P}$

(ii) for some $\bar{b} \in {}^{lg(\bar{y})}N$ we have: for every $i < \kappa$ large enough for every $\gamma < \delta$ we have $N \models \varphi[\bar{a}_\gamma^i, \bar{b}]^{\mathrm{if}(\gamma \in \mathscr{U})}\}$.

(2) For $\varphi = \varphi(\bar{y}, \bar{x}) \in \mathbb{L}(\tau_T)$ and a model N of T of cardinality λ let $\mathrm{inv}_{8,\kappa}^\varphi(N) = \langle \mathrm{inv}_7^\varphi(\delta, N') : \delta < \lambda\rangle/\mathscr{D}_\lambda$ for every, equivalently some model N' isomorphic to N with universe λ.

Observation 1.7. (1) $\mathrm{inv}_{8,\kappa}^\varphi(N)$ *is well defined for* $N \in \mathrm{EC}_\lambda(T)$ *when* $|T| + \kappa < \lambda$.

(2) *In Definition* 1.6(1) *we have* $|\mathrm{inv}_{7,\kappa}^\varphi(\delta, N)| \leq 2^{|\delta|+\kappa}$.

PROOF (OBSERVATION 1.7). Easy. □

Claim 1.8. *Assume* $\lambda > |T|$ *is regular,* $S \subseteq \lambda$ *is stationary,* $\varphi = \varphi(\bar{x}, \bar{y})$ *and*

(a) $\langle N_i : i < \kappa\rangle$ *is a* $\prec$*-increasing sequence*

(b) $N_i \in \mathrm{EC}_\lambda(T)$

(c) $N = \bigcup\{N_i : i < \kappa\}$

(d) $\overline{\mathscr{P}} = \langle \mathscr{P}_\alpha : \alpha < \lambda\rangle$ *where* $\mathscr{P}_\alpha \subseteq \mathscr{P}(\alpha)$

(e) f *is a one-to-one function from* N *onto* λ

(f) *for a club of* δ*'s below* λ *there are* $\bar{a}_\gamma^j \in N_{j+1} \cap f^{-1}(\delta)$ *for* $\gamma < \delta$, $j < \kappa$ *satisfying*

(α) *for every* $\bar{c} \in {}^{lg(\bar{x})}(N_j)$ *there is* $\mathscr{U} \in \mathscr{P}_\delta$ *such that* $\gamma < \delta \Rightarrow N \models \varphi[\bar{a}_\gamma^j, \bar{c}]^{\mathrm{if}(\gamma \in \mathscr{U})}$

(β) *for every* $\mathscr{U} \in \mathscr{P}_\delta$ *for some* $\bar{b} \in {}^{lg(\bar{y})}(N_\delta)$ *we have* $\gamma < \delta \Rightarrow N \models \varphi[\bar{a}_\gamma^j, \bar{b}]^{\mathrm{if}(\gamma \in \mathscr{U})}$.

Then $\{\delta \in S : \mathscr{P}_\delta \in \mathrm{inv}^{\varphi}_{7,\kappa}(\delta, f(N))\} \in \mathscr{D}_\lambda + S$.

PROOF (CLAIM 1.8). Straight. □

Now we come to the main two results of this section.

Theorem 1.9. *For some club E of λ^+, if $\delta_1 \neq \delta_2$ belong to $E \cap S^{\lambda^+}_\kappa$* then *$M_{\delta_1}, M_{\delta_2}$ are not isomorphic, moreover $\mathrm{inv}^{\varphi}_{8,\kappa}(M_{\delta_1}) \neq \mathrm{inv}^{\varphi}_{8,\kappa}(M_{\delta_2})$* when :

⊠ (a) *T has the strong independence property (see Definition 0.2)*
(b) *$\lambda = \lambda^{<\lambda}$ is regular uncountable, $\lambda > |T|$, $\lambda > \kappa = \mathrm{cf}(\kappa)$ and $\lambda^+ = 2^\lambda$*
(c) *M is a saturated model of T of cardinality λ^+*
(d) *$\langle M_\alpha : \alpha < \lambda^+\rangle$ is $\prec$-increasing continuous sequence with union M, each of cardinality λ.*

Theorem 1.10. *Assume ⊠ of Theorem 1.9.*

(1) *For some club E of λ^+, if $\delta_1 < \delta_2 < \delta_3$ are from E and $\delta_l \in S^{\lambda^+}_\lambda$ for $l =$* 1, 2, 3 then *$(M_{\delta_2}, M_{\delta_1}) \ncong (M_{\delta_3}, M_{\delta_1})$, moreover $\mathrm{inv}^{\varphi}_6(M_{\delta_2}, M_{\delta_1}) \neq \mathrm{inv}^{\varphi}_6(M_{\delta_3}, M_{\delta_1})$ for some φ.*

(2) *If $M \prec N_0$ are models of T of cardinality λ*, then *for some elementary extension $N_1 \in \mathrm{EC}_\lambda(T)$ of N_0 we have $N_1 \prec N_2 \in \mathrm{EC}_\lambda(T) \Rightarrow (N_0, M) \ncong (N_2, M)$.*

Discussion 1.11. We shall below start with $M \in \mathrm{EC}_\lambda(T)$ and a sequence $\langle b_i : i < \lambda\rangle$ of distinct members such that $\langle \varphi(\bar{b}_i, \bar{y}) : i < \lambda\rangle$ are independent, and like to find N, $\langle \bar{a}_i : i < \lambda\rangle$ such that $M \prec N \in \mathrm{EC}_\lambda(T)$ and the $\langle \bar{b}_i : i < \lambda\rangle$ has a real affect on the relevant φ-invariant, in the case of Theorem 1.10(1) this is $\mathrm{inv}^{\varphi}_5(M, N)$: for a stationary set of δ's below λ it adds something to the δth component in a specific representation, i.e., assuming $f \colon N \to \lambda$ is a one-to-one function and we deal with $\langle \mathrm{inv}^{\varphi}_5(\delta, f, M, N) : \delta < \lambda\rangle$; we have freedom about $\varphi(\bar{a}_\alpha, \bar{b}_i)$ and we can assume $\bar{b} \in {}^{lg(\bar{y})}M \backslash \{\bar{b}_i : i < \lambda\} \Rightarrow N \models \neg\varphi[\bar{a}_\alpha, \bar{b}]$.

But the relevant $\mathscr{P}_\delta$ is influenced not just by say $\langle \bar{b}_i : i \in [\delta, 2^{|\delta|})\rangle$ but also by later $\bar{b}_i$'s (and earlier $\bar{b}_i$). To control this we use below $\langle \bar{a}_\alpha : \alpha < \lambda\rangle$, S, E such that we deal with different $\delta \in S$ in an independent way to large extent; this is the reason for choosing the C^*_α's.

PROOF (THEOREM 1.9). By the proof of [6, Section 2] without loss of generality λ is strongly inaccessible. Choose $\theta \in \mathrm{Reg} \cap \lambda \backslash \{\aleph_0\}$, will be needed when we generalize the proof in Section 2.

Let $\langle \mathscr{U}_i : i < \kappa\rangle$ be a $\subseteq$-increasing sequence of subsets of λ such that the set $\mathscr{U}^-_i = \mathscr{U}_i \backslash \bigcup\{\mathscr{U}_j : j < i\}$ has cardinality λ for each $i < \kappa$ and let $\mathscr{U}_\kappa = \bigcup\{\mathscr{U}_i : i < \kappa\}$. Let $\varphi(\bar{x}, \bar{y}) \in \mathbb{L}(\tau_T)$ have the strong independence property, see Definition 0.2. We can choose $\langle C^*_\alpha : \alpha < \lambda^+\rangle$ such that $C^*_\alpha \subseteq \mathrm{nacc}(\alpha)$, $\mathrm{otp}(C^*_\alpha) \leq \kappa$, $\beta \in C^*_\alpha \Rightarrow C^*_\beta = C^*_\alpha \cap \beta$ and $\lambda | \alpha \wedge \mathrm{cf}(\alpha) = \kappa \Rightarrow \alpha = \sup(C^*_\alpha)$ and $\mathrm{cf}(\alpha) \neq \kappa \Rightarrow \mathrm{otp}(C^*_\alpha) < \kappa$.

[See [4] but for completeness we show this; by induction on $\alpha < \lambda^+$ we choose $\langle C^*_\varepsilon : \varepsilon < \lambda\alpha\rangle$ such that:

(a) the relevant demand holds
(b) if $\alpha = \beta + 1$, $C \subseteq \lambda\beta$, $(\forall i \in C)(C^*_i = C \cap i)$ and $\mathrm{otp}(C) < \kappa$ then for some $i \in (\lambda\beta, \lambda\alpha)$ we have $C^*_{i+1} = C$.

As $\lambda = \lambda^{<\kappa}$ this is easy but we elaborate. For $\alpha = 0$ trivial for α limit obvious. Assume $\alpha = \beta + 1$ let $\alpha_* = \lambda\alpha$, $\beta_* = \lambda\beta$ and $\langle C^*_i : i < \beta_*\rangle$ has been defined.

First, we choose $C^*_{\beta_i}$. If $\mathrm{cf}(\beta_*) \neq \kappa$ let $C^*_{\beta_*} = \varnothing$, so assume $\mathrm{cf}(\beta_*) = \kappa$ then necessarily $\mathrm{cf}(\beta) = \kappa$.

Let $\langle \alpha_\varepsilon : \varepsilon < \kappa \rangle$ be increasing with limit β, and choose $\beta_\varepsilon \in [\lambda\alpha_\varepsilon, \lambda\alpha_\varepsilon + \lambda)$ by induction on $\varepsilon < \kappa$ such that $C^*_{\beta_\varepsilon} = \{\beta_\zeta : \zeta < \varepsilon\}$.

Lastly, let $C^*_{\lambda\alpha} := \{\beta_\varepsilon : \varepsilon < \kappa\}$. So $C^*_{\beta_*}$ has been defined in any case.

Now let $\mathscr{C}_\alpha = \{C \subseteq \lambda\beta : \mathrm{otp}(C) < \kappa \text{ and } \gamma \in C \Rightarrow C^*_\gamma = C \cap \gamma\}$, so $|\mathscr{C}| \leq \lambda$, also $\varnothing \in \mathscr{C}$, so let $\langle C^*_{\lambda\beta+i} : i \in (0,\lambda)\rangle$ list $\mathscr{C}_\alpha$ possibly with repetitions. So we have defined $\langle C^*_\varepsilon : \varepsilon < \lambda\alpha\rangle$, so have carried the induction.]

Let $S_* = \{\mu : \mu = \beth_{\alpha+\omega} \text{ for some } \alpha < \lambda\}$. Let E_*, $\langle C_\alpha : \alpha < \lambda\rangle$ be such that:

$\circledast_1$ (a) $C_\alpha \subseteq \alpha \cap S_*$
- (b) $\beta \in C_\alpha \Rightarrow C_\beta = C_\alpha \cap \beta$
- (c) $\mathrm{otp}(C_\alpha) \leq \theta$
- (d) E_* is the club of λ included in$\{\delta < \lambda : \theta < \delta = \beth_\delta\}$
- (e) $\mathrm{otp}(C_\alpha) = \theta$ iff $\alpha \in E_* \cap S^\lambda_\theta$
- (f) if $\alpha \in S := E_* \cap S^\lambda_\theta$ then $\alpha = \sup(C_\alpha)$
- (g) if $\alpha \in E_*$ and $i < \kappa$ then $|\alpha \cap \mathscr{U}^-_i| = |\alpha|$.

[Why can we choose? By induction on the cardinal $\chi \in [\aleph_0, \lambda)$ we choose $\langle C_\alpha : \alpha < \beth_\chi\rangle$ and $E_\chi = E_* \cap \beth_\chi$ such that the relevant demands hold and: if $\chi = 2^{\chi_1}$ and C is a subset of $S_* \cap \beth_{\chi_1}$ of order type $< \theta$ satisfying $\alpha \in C \Rightarrow C_\alpha = C \cap \alpha$ *then* for some $\alpha \in S_* \cap (\beth_{\chi_1}, \beth_\chi)$ we have $C_\alpha = C$. Why this extra induction hypothesis help? As arriving to $\alpha \in S$ so $\alpha = \beth_\alpha$ let $\langle \chi_i : i < \theta\rangle$ be an increasing sequence of cardinals with limit $\beth_\alpha = \alpha$ and we choose $\alpha_i \in (\beth_{\chi_i}, \beth_{2^{\chi_i}}) \cap S_*$ by induction on $i < \theta$ such that $C_{\alpha,i} = \{\alpha_j : j < i\}$ and the let $C_\chi = \{\alpha_i : i < \theta\}$.]

We shall prove that

$\circledast_2$ if $\boxdot_2$ below holds, then there is a β such that $\odot_2$ holds *where*:

$\boxdot_2$ (a) $\alpha < \lambda^+$, $i < \kappa$
- (b) f is a one-to-one function from M_α into $\mathscr{U}'_i = \bigcup\{\mathscr{U}_j : j < i\}$
- (c) $E \subseteq E_*$ is a club of λ such that $\delta \in E \Rightarrow f(M_\alpha) \restriction \delta \prec f(M_\alpha)$
- (d) $\overline{\mathscr{P}} = \langle \mathscr{P}_\delta : \delta \in S\rangle$
- (e) $\mathscr{P}_\delta \subseteq \mathscr{P}(\delta)$ and $\varnothing \in \mathscr{P}_\delta$ and $\mathscr{P}_\delta \subseteq \bigcup_{l\leq 2} \mathscr{P}^{*,l}_\delta$ where[2]
  - (α) $\mathscr{P}^{*,0}_\delta = \{A \subseteq \delta : \sup(A) = \delta \text{ and } A \subseteq \bigcup\{[\mu, 2^\mu) : \mu \in C_\delta\}\}$,
  - (β) $\mathscr{P}^{*,1}_\delta = \bigcup\{\mathscr{P}^{*,0}_{\delta_1} : \delta_1 \in S \cap \delta\}$,
  - (γ) $\mathscr{P}^{*,2}_\delta = \{A \subseteq \delta : \text{for some } \delta_1 \in \lambda\backslash(\delta+1) \text{ we have } A \subseteq \bigcup\{[\partial, 2^\partial) : \partial \in C_{\delta_1} \cap \delta\}\}$
- (f) if $\delta_1 < \delta_2$ are from S *then*
  - (α) $A \in \mathscr{P}_{\delta_1} \Rightarrow A \in \mathscr{P}_{\delta_2}$
  - (β) $A \in \mathscr{P}_{\delta_2} \Rightarrow A \cap \delta_1 \in \mathscr{P}_{\delta_1}$,
  - (γ) for any $\delta \in S$ the family $\mathscr{P}^{*,1}_\delta \cup \mathscr{P}^{*,2}_\delta$ is a set of bounded subsets of δ; (this follows)
- (g) $\bar{b}_{\delta,\mathscr{U}} \in M_\alpha$ for $\delta \in E \cap S$, $\mathscr{U} \in \mathscr{P}_\delta$ are such that $\bar{b}_{\delta_1,\mathscr{U}_1} = \bar{b}_{\delta_2,\mathscr{U}_2} \wedge \mathscr{U}_1 \in \mathscr{P}_{\delta_1} \wedge \mathscr{U}_2 \in \mathscr{P}_{\delta_2} \Rightarrow \delta_1 = \delta_2 \wedge \mathscr{U}_1 = \mathscr{U}_2$

$\odot_2$ (α) $\beta \in (\alpha, \lambda^+)$
- (β) there are $\bar{a}_\gamma \subseteq M_\beta$ for $\gamma < \lambda$ such that for a club of $\delta \in E$, if $\delta \in S$ then the following conditions on $\mathscr{U} \subseteq \delta$ are equivalent:
  - (i) $\mathscr{U} \in \mathscr{P}_\delta$

---

[2]Note that $\mathscr{P}^{*,1}_{\delta,i}$, $\mathscr{P}^{*,2}_{\delta,i}$ are the families of sets we like to ignore as they are influenced by our choices for $\delta_1 \in S\backslash\{\delta\}$, so we work to have them families of bounded subsets of δ.

(ii) for some $\bar{b} \in {}^{lg(\bar{y})}M_\alpha$ we have: for every $\gamma < \delta$, $M_\beta \models \varphi[\bar{a}_\gamma, \bar{b}]$ iff $\gamma \in \mathscr{U}$

(iii) clause (ii) holds for $\bar{b} = \bar{b}_{\delta,\mathscr{U}}$ and $\mathscr{U} \in \mathscr{P}_\delta$

[Why? For each $\delta \in E \cap S$ let $\langle \mathscr{U}_{\delta,\varepsilon} : \varepsilon < |\mathscr{P}_\delta| \leq 2^{|\delta|}\rangle$ list $\mathscr{P}_\delta$ and let $\bar{b}_{\delta,\varepsilon} := \bar{b}_{\delta,\mathscr{U}_{\delta,\varepsilon}}$.

Let

$$\Gamma = \{\varphi(\bar{x}_\gamma, \bar{b}_{\delta,\varepsilon})^{\text{if}(\gamma\in\mathscr{U}_{\delta,\varepsilon})} : \gamma < \lambda, \delta \in E \text{ and } \varepsilon < |\mathscr{P}_{\delta,i}|\} \cup \{\neg\varphi(\bar{x}_\gamma, \bar{b}) : \gamma < \lambda,\ \bar{b} \in {}^{lg(\bar{y})}(M_\alpha) \text{ and for no } \delta \in E,\ \varepsilon < |\mathscr{P}_\delta| \text{ do we have } \bar{b} = \bar{b}_{\delta,\varepsilon}\}.$$

As $\varphi(\bar{x}, \bar{y})$ has the strong independence property, recalling that by clause (g) of $\boxdot_2$ the sequence $\langle \bar{b}_{\delta,\varepsilon} : \delta \in E \cap S \text{ and } \varepsilon < |\mathscr{P}_\delta|\rangle$ is with no repetitions, clearly Γ is finitely satisfiable in M_α, but M is λ^+-saturated, $M_\alpha \prec M$ and $|\Gamma| = \lambda$ hence we can find $\bar{a}_\gamma \in {}^{lg(\bar{x})}M$ for $\gamma < \lambda$ such that the assignment $\bar{x}_\gamma \mapsto \bar{a}_\gamma (\gamma < \lambda)$ satisfies Γ in M. Lastly, choose $\beta \in (\alpha, \lambda^+)$ such that $\{\bar{a}_\gamma : \gamma < \lambda\} \subseteq M_\beta$.

Now check recalling $\emptyset \in \mathscr{P}_\delta$ for $\delta \in S$.]

Note

$\odot_3$ in $\odot_2$ if h is a one-to-one mapping from M_β into $\mathscr{U}_i$ extending f *then* for some club E of λ if for every $\delta \in S \cap E$ we have $(\forall \gamma < \lambda)(\gamma < \delta \to h(\bar{a}_\gamma) \in {}^{lg(\bar{x})}\delta)$ and so for every $\mathscr{U} \subseteq \delta$ the conditions (i), (ii), (iii) from $\odot_2$ are equivalent.

Next we can choose $\bar{f}$ such that

$\circledast_3$ (a) $\bar{f} = \langle f_\alpha : \alpha < \lambda^+\rangle$

(b) f_α is a one-to-one function from M_α into $\mathscr{U}_{\text{otp}(C^*_\alpha)}$

(c) if $\alpha \in C^*_\beta$ then $f_\alpha \subseteq f_\beta$.

Now

$\circledast_4$ for every $\alpha < \lambda^+$ there is $\overline{\mathscr{P}}^\alpha = \langle \mathscr{P}^\alpha_\varepsilon : \varepsilon \in S\rangle$ such that

(i) $\mathscr{P}^\alpha_\varepsilon \subseteq \mathscr{P}(\varepsilon)$ are as in $\boxdot_2$(e), (f) above

(ii) for every $\beta \leq \alpha$, for a club of δ's from S we have

$$\mathscr{P}^\alpha_\delta \notin \text{inv}^\varphi_{7,\kappa}(\delta, f_\beta(M_\beta)).$$

[Why? For every $\beta \leq \alpha$ and $\delta \in (\kappa, \lambda)$ we have $\text{inv}^\varphi_{7,\kappa}(\delta, f_\beta(M_\beta))$ is a subset of $\mathscr{P}(\mathscr{P}(\delta))$ of cardinality $\leq 2^{|\delta|}$. As the number of β's is $\leq \lambda$, by diagonalization we can do this: let $\alpha+1 = \bigcup_{\varepsilon<\lambda} u_\varepsilon$ and $u_\varepsilon \in [\alpha+1]^{<\lambda}$ increasing continuous for $\varepsilon < \lambda$; moreover, $\alpha < \lambda \Rightarrow u_\varepsilon = \alpha$ and $\alpha \geq \lambda \Rightarrow u_\varepsilon \cap \lambda \subseteq \varepsilon$ and $|u_\varepsilon| \leq |\varepsilon|$. By induction on $\varepsilon \in (\kappa, \lambda) \cap S$ choose $\mathscr{P}^\alpha_\varepsilon \subseteq \bigcup_{l<3} \mathscr{P}^{*,l}_{\alpha_\varepsilon}$ which includes $\bigcup\{\mathscr{P}^\alpha_\zeta : \zeta \in u_\varepsilon \cap S\} \cup \mathscr{P}^{*,2}_\varepsilon$ and satisfies $\mathscr{P}^{*,0}_\varepsilon \cap \mathscr{P}^\alpha_\varepsilon \in \mathscr{P}(\mathscr{P}^{*,0}_{\varepsilon,i})) \backslash \{\mathscr{P} \cap \mathscr{P}^{*,0}_\varepsilon : \mathscr{P} \in \text{inv}^\varphi_{7,\kappa}(\varepsilon, f_\beta(M_\beta)), \beta \in u_\varepsilon\}$. Note that for each $\beta \leq \alpha$ the set $\{\varepsilon < \lambda : \beta \in u_\varepsilon\}$ contains an end-segment of λ hence a club of λ as required.]

Now choose pairwise distinct $\bar{b}_{\delta,\mathscr{U}} \in {}^{lg(\bar{y})}(M_0)$ for $\delta \in E_*$, $\mathscr{U} \in \mathscr{P}^{*,0}_\delta$

$\circledast_5$ for every $\alpha_* \leq \alpha < \lambda^+$ for some $\beta \in (\alpha, \lambda^+)$ and $\bar{a}_\gamma \in {}^{lg(\bar{x})}M_\beta$ for $\gamma < \lambda$ the condition in clause (γ) of $\odot_2$ holds with $\overline{\mathscr{P}}^{\alpha_*}$ here standing for $\overline{\mathscr{P}}$ there and the $\bar{b}_{\delta,\mathscr{U}}$ chosen above.

[Why? By $\circledast_2$.]

$\circledast_6$ let $E = \{\delta < \lambda^+ : \delta$ is a limit ordinal such that for every $\alpha_* \leq \alpha < \delta$ there is $\beta < \delta$ as in $\circledast_5\}$.

Clearly E is a club of λ^+.

$\circledast_7$ if $\delta_1 < \delta_2$ are from $E \cap S^{\lambda^+}_\kappa$ then $M_{\delta_1}, M_{\delta_2}$ are not isomorphic.

[Why? Let $\alpha_* = \min(C^*_{\delta_2} \backslash \delta_1)$. We consider $\overline{\mathscr{P}}^{\alpha_*}$ which is from $\circledast_4$. On the one hand $\{\varepsilon < \lambda : \mathscr{P}^{\alpha_*}_\varepsilon \notin \mathrm{inv}^{\varphi}_{7,\kappa}(\varepsilon, f_{\delta_1}(M_{\delta_1}))\}$ contains a club by $\circledast_4$(ii). Note that $\langle f_\alpha : \alpha \in C^*_{\delta_2} \backslash \delta_1\rangle$ is $\subseteq$-increasing sequence of functions with union f_{δ_2}.

On the other hand choose an increasing $\langle \alpha_i : i < \kappa\rangle$ with limit δ_2 satisfying $\alpha_0 = 0$, $\alpha_1 = \delta_1$ such that $(\alpha_*, \alpha_{1+i}, \alpha_{1+i+1})$ are like $(\alpha_*, \alpha, \beta)$ in $\circledast_5$ for each $i < \kappa$ and $i \in (1, \kappa) \Rightarrow \alpha_i \in C^*_{\delta_2}$. Now by Claim 1.8, $\{\varepsilon < \lambda : \mathscr{P}^{\alpha_*}_\varepsilon \in \mathrm{inv}^{\varphi}_{7,\kappa}(\varepsilon, f_{\delta_2}(M_{\delta_2}))\}$ contains a club. Hence by the last sentence and the end of the previous paragraph $M_{\delta_1} \ncong M_{\delta_2}$ as required.]

So we are done. $\square$

Remark 1.12. We can avoid using C^*_δ and also C_δ (e.g., using $A \in \mathscr{P}^{*,0}_\delta \Rightarrow \mathrm{otp}(A) = \delta$) but this seems less transparent.

PROOF (THEOREM 1.10). Similar but easier (for λ regular not strong limit (but $2^\lambda > 2^{<\lambda}$) also easy), or see the proof of Theorem 2.9. $\square$

2. Independent T

We would like to do something similar to Section 1, but our control on the relevant family of subsets of M is less tight. We control it to some extent by using the completion of a free Boolean algebra.

Context 2.1. T a complete first order theory, $\varphi(x, \bar{y})$ has the independence property (of course the existence of such φ follows from the strong independence property but is weaker).

We continue [6, 2.1–2.12], but we do not rely on it.

Definition 2.2. For a set I let

(a) $\mathbb{B} = \mathbb{B}_I$ be the Boolean algebra generated by $\langle e_t : t \in I\rangle$ freely,
(b) $\mathbb{B}^{\mathrm{c}}_I$ is the completion of $\mathbb{B}$
(c) for $J \subseteq I$ let $\mathbb{B}^{\mathrm{c}}_{I,J}$ be the complete subalgebra of $\mathbb{B}^{\mathrm{c}}_I$ generated by $\{e_s : s \in J\}$
(d) let $\mathrm{uf}(\mathbb{B}^{\mathrm{c}}_I)$ be the set of ultrafilters on I.

Claim 2.3. *Assume*

$\circledast$ (a) $M \models T$
(b) $\bar{b}_t \in {}^{lg(\bar{y})}M$ *for* $t \in I$
(c) $\langle \varphi(x, \bar{b}_t) : t \in I\rangle$ *is an independent sequence of formulas.*

Then *there is a function F from ${}^{lg(\bar{y})}M$ to $\mathbb{B} = \mathbb{B}^{\mathrm{c}}_I$ such that*

(α) $F(\bar{b}_t) = e_t$
(β) *for every ultrafilter D of $\mathbb{B}$ there is $p = p_D = p_{F,D} \in \mathbf{S}_\varphi(M)$, in fact, a unique one, such that for every $\bar{b} \in {}^{lg(\bar{y})}M$ we have $\varphi(x, \bar{b}) \in p \Leftrightarrow F(\bar{b}) \in D$.*

Remark 2.4. (1) Note that the mapping $D \mapsto p_D$ is not necessarily one to one, but $D_1 \cap \{e_t : t \in I\} \neq D_2 \cap \{e_t : t \in I\} \Rightarrow p_{D_1} \neq p_{D_2}$.

(2) If $I = I_1 \cup I_2, I_1 \cap I_2 = \varnothing$ and $|I_2| = |I_1|^{\aleph_0}$ *then* we can find a mapping F from ${}^{lg(\bar{y})}M$ onto (not just into) $\mathbb{B} = \mathbb{B}^c_{I_1}$ such that clause (α), (β) are satisfied.

Proof (Claim 2.3). Clearly $\mathscr{P}(M)$ is a Boolean algebra and $\{\varphi(M, \bar{b}_t) : t \in M\}$ generates freely a subalgebra of $\mathscr{P}(M)$ which we call $\mathbb{B}'$. So there is a homomorphism h from $\mathbb{B}'$ into $\mathbb{B}$ mapping $\varphi(M, \bar{b}_t)$ to e_t (moreover h is unique and is an isomorphism from $\mathbb{B}'$ onto $\mathbb{B}_I \subseteq \mathbb{B}^c_I$). So h is a homomorphism from $\mathbb{B}' \subseteq \mathscr{P}(M)$ into $\mathbb{B}^c_I$, which is a complete Boolean algebra hence there is a homomorphism h^+ from the Boolean algebra $\mathscr{P}(M)$ into $\mathbb{B}^c$ extending h.

Lastly, define $F\colon {}^{lg(\bar{y})}M \to \mathbb{B}^c$ by $F(\bar{b}) = h^+(\varphi(M, \bar{b}))$. Now check. □

Conclusion 2.5. *Assume* ⊛ *from Claim* 2.3 *and*

⊡ (a) $I = \lambda$ *is regular uncountable*
(b) $|M| \subseteq \mathscr{U} \subseteq \lambda$
(c) D_α *is an ultrafilter of* $\mathbb{B}^c_I$ *for* $\alpha < \lambda$
(d) $\mathscr{U} \backslash |M|$ *is unbounded in* λ.

Then *we can find* $\langle a_\alpha : \alpha < \lambda \rangle$ *and* N *such that*

(α) $M \prec N$
(β) $|N| \subseteq \mathscr{U}$
(γ) $a_\alpha \in N$ *for* $\alpha < \lambda$
(δ) a_α *realizes* $p_{D_\alpha} \in \mathbf{S}_\varphi(M)$.

Remark 2.6. Conclusion 2.5 is easy but intended to clarify how we shall use the ultrafilters, so is quoted toward the end of the section.

Proof (Conclusion 2.5). Should be clear. □

Discussion 2.7. Note that compared to Section 1 instead $\bar{x}$, $\bar{y}$, $\bar{a}_\alpha$, $\bar{b}_\beta$ we have x, $\bar{y}$, a_α, $\bar{b}_\beta$. Compared to Section 1, we have less control over $\{\text{tp}(a, M, N) : a \in N\}$. There, for the sequences $\bar{b}$ of M which are not among $\{\bar{b}_\gamma : \gamma < \lambda\}$, we can demand $N \models \neg\varphi[\bar{a}_\gamma, \bar{b}]$ for $\gamma < \lambda$ so $\text{tp}_\varphi(\bar{a}_\gamma, M, N)$ can be clearly read. Here the complete Boolean algebra $\mathbb{B}^c_I$ is helping, a small price is that we need $\theta > \aleph_0$.

In order to try to keep track of what is going on we shall use only $\text{tp}(a_\gamma, M, N)$ of the form p_D for ultrafilter D on $\mathbb{B}^c_I$. Further, we better have, e.g., a nice function π from ${}^\lambda 2$ to $\text{uf}(\mathbb{B}^c_I)$ such that $(e_\alpha \in \pi(\eta)) \Leftrightarrow \eta(\alpha) = 1$.

A possible approach is: we define $\langle M_{\eta,u} : \eta \in \mathscr{T} \subseteq \text{des}(\lambda), u \in \mathscr{P}(n_\eta)\rangle$ as in [5, Section 3] and we define $D_\eta \in \text{uf}(\mathbb{B}^c \cap M)$ such that $\alpha \in M_\eta \cap \lambda \Rightarrow [e^{\eta(\alpha)}_\eta \in D_\eta]$ and $\bigcup_\eta \mathbb{D}_\eta \in \text{uf}(\mathbb{B}^c)$.

We need some continuity so each "$e \in D_\eta$" ($e \in \mathbb{B}^c$) depends on $\eta \restriction u_e$ for some "small" $u_e \subseteq \lambda$.

Theorem 2.8. *In Theorem* 1.9 *it suffices to assume* ⊠′ *which means clauses* (b), (c), (d) *of* ⊠ *and*

(a)′ *T has the independence property.*

Theorem 2.9. *In Theorem* 1.10 *it suffices to assume* ⊠′ *of Theorem* 2.8.

Proof (Theorem 2.8). Just combine the proofs of Theorem 1.9 from Section 1 and Theorem 2.9 below. □

Proof (Theorem 2.9). As in the proof of Theorem 1.9 we can assume λ is strongly inaccessible though the proof is just easier otherwise. We let

$\circledast_1$ (a) $E_* = \{\delta < \lambda : \delta = \beth_\delta\}$, a club of λ
(b) $S_* = \{\beth_{\alpha+\omega} : \alpha < \lambda\}$
(c) choose a regular uncountable $\theta < \lambda$

and let

$\circledast_2$ (a) $S = \{\delta \in E_* : \mathrm{cf}(\delta) = \theta\} = S^\lambda_\theta \cap E_*$
(b) let $\overline{C}$ be as in $\circledast_1$ of the proof of Theorem 1.9, in particular $\overline{C} = \langle C_\alpha : \alpha < \lambda\rangle$, $C_\alpha \subseteq S_*$, $\mathrm{otp}(C_\alpha) \leq \theta$, $\alpha \in C_\beta \Rightarrow C_\alpha = C_\beta \cap \alpha$ and $\alpha \in S \Leftrightarrow \alpha = \sup(C_\alpha) \Leftrightarrow \mathrm{otp}(C_\alpha) = \theta$ and $\alpha \in S \Rightarrow \sup(C_\alpha) = \alpha$
(c) for $\mu \in S$ let $A^*_\mu = \bigcup\{[\chi, 2^\chi] : \chi \in C_\mu\}$.

Let $\mathbb{D}_*$ be an ultrafilter of $\mathbb{B}^{\mathrm{c}}_\lambda$ such that $e_\alpha \notin \mathbb{D}_*$ for $\alpha < \lambda$.

Now for $\eta \in {}^\lambda 2$ we choose $\mathbb{D}_\eta$ such that

$\circledast_3$ (a) $\mathbb{D}_\eta$ is an ultrafilter of $\mathbb{B}^{\mathrm{c}}_\lambda$
(b) if $e \in \mathbb{D}_* \subseteq \mathbb{B}^{\mathrm{c}}_\lambda$ belongs to $\mathbb{B}^{\mathrm{c}}_{\lambda,\eta^{-1}\{0\}}$ (see Definition 2.2, the completion of the subalgebra of $\mathbb{B}^{\mathrm{c}}_\lambda$ generated by $\{e_\alpha : \eta(\alpha) = 0\}$) then $e \in \mathbb{D}_\eta$.
(c) if $\alpha < \lambda$ and $\eta(\alpha) = 1$ then $e_\alpha \in \mathbb{D}_\eta$.

So

$\circledast_4$ (a) if $\eta \in {}^\lambda 2$ is constantly zero then $\mathbb{D}_\eta = \mathbb{D}_*$
(b) $e_\alpha \in \mathbb{D}_\eta \Leftrightarrow \eta(\alpha) = 1$ for $\alpha < \lambda$, $\eta \in {}^\lambda 2$.

Now let $\bar\eta = \langle \eta_\varepsilon : \varepsilon < \lambda\rangle$ be a sequence of members of ${}^\lambda 2$ and below we shall be interested mainly in the case $\alpha = \mu \in S$.

Define

$\circledast_5$ for $e \in \mathbb{B}^{\mathrm{c}}_\lambda$ and $\alpha \leq \lambda$ we let $Y^\alpha_{\bar\eta,e} := \{\varepsilon < \alpha : e \in \mathbb{D}_{\eta_\varepsilon}\}$
$\circledast_6$ $\mathscr{P}_{\bar\eta,\alpha} := \{Y^\alpha_{\bar\eta,e} : e \in \mathbb{B}^{\mathrm{c}}_\lambda\}$.

Now what can we say on $\mathscr{P}_{\bar\eta,\mu}$ for $\mu \in S$? As we can consider $e \in \{e_\alpha : \alpha \in [\mu, 2^\mu)\}$, clearly

$\circledast_7$ $\{\{\varepsilon < \mu : \eta_\varepsilon(\alpha) = 1\} : \alpha \in [\mu, 2^\mu)\} \subseteq \mathscr{P}_{\bar\eta,\mu} \subseteq \mathscr{P}(\mu)$.

This may be looked at as a "lower bound" of $\mathscr{P}_{\bar\eta,\mu}$. Naturally we try to get also an "upper bound" to $\mathscr{P}_{\bar\eta,\mu}$; now note

$\circledast_8$ if $e \in \mathbb{B}^{\mathrm{c}}_\lambda$ then $Y^\mu_{\bar\eta,-e} = \mu \backslash Y^\mu_{\bar\eta,e}$.

Now define (recalling A^*_μ is from $\circledast_2$(c))

$\circledast_9$ Ξ is the set of $\bar\eta$ of the form $\langle \eta_\varepsilon : \varepsilon < \lambda\rangle$ such that:
(a) $\eta_\varepsilon \in {}^\lambda 2$ for every $\varepsilon < \lambda$,
(b) if $\eta_\varepsilon(\alpha) = 1$ *then* $(\exists \mu \in S)[\mu \leq \alpha < 2^\mu \wedge \varepsilon \in A^*_\mu]$,
(c) if $\mu \in S$ and $u \subseteq [\mu, 2^\mu)$ is countable *then* $\{\varepsilon \in A^*_\mu$: if $\alpha \in u$ then $\eta_\varepsilon(\alpha) = 0\}$ is of cardinality μ.

Also (by our knowledge of the completion of a free Boolean algebra, $\mathbb{B}^{\mathrm{c}}_\lambda$ satisfies the c.c.c.) for every $e \in \mathbb{B}^{\mathrm{c}}_\lambda$ we can choose u_e such that:

$\boxplus_1$ (a) $u_e \subseteq \lambda$ is countable
(b) $e \in \mathbb{B}^{\mathrm{c}}_{\lambda,u_e}$.

So by clause (b) of $\circledast_3$ clearly

$\boxplus_2$ if $\bar\eta \in \Xi$, $e \in \mathbb{B}^{\mathrm{c}}_\lambda$, $\varepsilon < \mu \in S$ and $u_e \subseteq \eta_\varepsilon^{-1}\{0\}$ *then* $e \in \mathbb{D}_{\eta_\varepsilon} \Leftrightarrow e \in \mathbb{D}_*$

hence

$\boxplus_3$ if $\bar\eta \in \Xi$, $e \in \mathbb{B}^{\mathrm{c}}_\lambda \cap \mathbb{D}_*$ and $\mu \in S$ then $Y^\mu_{\bar\eta,e} \supseteq \{\varepsilon < \mu : u_e \subseteq \eta_\varepsilon^{-1}\{0\}\}$.

Next

$\boxplus_4$ for $\bar\eta \in \Xi$

(a) let $\mathscr{D}_{\bar\eta,\mu}$ be the filter on μ generated by $\{A^*_\mu\} \cup \{\{\varepsilon < \mu : u \subseteq \eta_\varepsilon^{-1}\{0\}$ and $\varepsilon > \zeta\} : \zeta < \mu$ and $u \subseteq \lambda$ is countable$\}$

(b) let $\mathscr{I}_{\bar\eta,\mu}$ be the dual ideal.

Clearly

$\boxplus_5$ (a) if $\bar\eta \in \Xi$, $\mu \in S$ and $\alpha \in \lambda\backslash[\mu, 2^\mu)$ then $\{\varepsilon < \mu : \eta_\varepsilon(\alpha) \neq 0\}$ is a bounded subset of μ,

(b) if $\bar\eta \in \Xi$ and $\mu \in S$ then $\mathscr{D}_{\bar\eta,\mu}$ is a uniform $\aleph_1$-complete filter on μ (recalling $\mathrm{cf}(\mu) = \theta > \aleph_0$ as $\mu \in S$) and $\varnothing \notin \mathscr{D}_{\bar\eta,\mu}$.

[Why? See $\circledast_9$.]

Now by $\boxplus_3$ we have $\bar\eta \in \Xi \wedge e \in \mathbb{B}^{\mathrm{c}}_\lambda \cap \mathbb{D}_* \Rightarrow Y^\mu_{\bar\eta,e} \in \mathscr{D}_{\bar\eta,\mu}$ so recalling $\circledast_8$ we have $e \in \mathbb{B}^{\mathrm{c}}_\lambda \backslash \mathbb{D}_* \Rightarrow Y^\mu_{\bar\eta,e} = \varnothing \bmod \mathscr{D}_{\bar\eta,\mu}$ hence

$\boxplus_6$ $\mathscr{P}_{\bar\eta,\mu} \subseteq \{X \subseteq \mu : X \in \mathscr{D}_{\bar\eta,\mu}$ or $\mu\backslash X \in \mathscr{D}_{\bar\eta,\mu}\}$.

Now

$\odot_1$ if $\mu \in S$ then we can find $\bar A^\xi_\mu$ for $\xi < 2^{2^\mu}$ such that:

(a) $\bar A^\xi_\mu = \langle A^\xi_\gamma : \gamma \in [\mu, 2^\mu)\rangle$

(b) A^ξ_γ is an unbounded subset of A^*_μ

(c) $\mathscr{D}^\xi_\mu, \mathscr{I}^\xi_\mu$ are well defined, i.e. $\varnothing \in \mathscr{D}^\xi_\mu$ when we let

(α) $\mathscr{D}^\xi_\mu$ be the $\aleph_1$-complete filter of subsets of μ generated by $\{A^\xi_\gamma\backslash\beta : \gamma \in [\mu, 2^\mu)$ and $\beta < \mu\}$ so $A^*_\mu\backslash\beta \in \mathscr{D}^\xi_\mu$ for $\beta < \mu$

(β) $\mathscr{I}^\xi_\mu = \{\mu\backslash B : B \in \mathscr{D}^\xi_\mu\}$, i.e., the dual ideal

(d) moreover if $\xi^1 \neq \xi^2$ are $< 2^{2^\mu}$, then

$$\{A^*_\mu\backslash A^{\xi^1}_\gamma : \gamma \in [\mu, 2^\mu)\} \nsubseteq \mathscr{D}^{\xi^2}_\mu \cup \mathscr{I}^{\xi^2}_\mu.$$

[Why $\odot_1$ holds? As $|A^*_\mu| = |\mu|$ is a strong limit cardinal of cofinality $\theta > \aleph_0$ clearly $\mu = |A^*_\mu| = |A^*_\mu|^{\aleph_0}$ hence by [2] there is a sequence $\langle B_\gamma : \gamma \in [\mu, 2^\mu)\rangle$ of subsets of A^*_μ such that any nontrivial Boolean combination of countably many of them has cardinality μ. Let $\langle U_\xi : \xi < 2^{2^\mu}\rangle$ be a sequence of pairwise distinct subsets of $[\mu, 2^\mu)$ each of cardinality $2^{|\mu|}$ no one included in another and let $\langle A^\xi_\gamma : \gamma \in [\mu, 2^{|\mu|})\rangle$ list $\{B_\gamma : \gamma \in U_\xi\}$.

Now check.]

$\odot_2$ in $\odot_1$ it follows that

(e) for every $\mathscr{P} \subseteq \mathscr{P}(\mu)$ for at most one $\xi < 2^{2^\mu}$ we have

$$\{A^*_\mu\backslash A^\xi_\gamma : \gamma \in [\mu, 2^\mu)\} \subseteq \mathscr{P} \subseteq \mathscr{D}^\xi_\mu \cup \mathscr{I}^\xi_\mu.$$

$\odot_3$ for every $\bar\xi = \langle \xi(\mu) : \mu \in S\rangle \in \Pi\{2^{2^\mu} : \mu \in S\}$ there is $\bar\eta = \bar\eta_{\bar\xi}$ such that:

(a) $\bar\eta_{\bar\xi} \in \Xi$ so $\bar\eta_{\bar\xi} = \langle \eta_{\bar\xi,\varepsilon} : \varepsilon < \lambda\rangle$

(b) if $\mu \in S$, $\gamma \in [\mu, 2^\mu)$ then $\{\varepsilon \in \lambda : \eta_{\bar\xi,\varepsilon}(\gamma) = 1\} = A^*_\mu\backslash A^{\xi(\mu)}_\gamma$.

[Why? Just read the definition of Ξ in $\circledast_9$ and $\bar A^\xi_\mu$ in $\odot_1$.]

$\odot_4$ if $\mu \in S$ then $\mathscr{D}_{\bar\eta_{\bar\xi},\mu} \cup \mathscr{I}_{\bar\eta_{\bar\xi},\mu} = \mathscr{D}^{\xi(\mu)}_\mu \cup \mathscr{I}^{\xi(\mu)}_\mu$.

[Why? Easy, recalling $\boxplus_5$(a).]

$\odot_5$ if $\gamma(*) < \lambda^+$ and $\bar{\mathbf{P}}^\gamma = \langle \mathbf{P}^\gamma_\mu : \mu \in S\rangle$, for $\gamma < \gamma(*)$ where $\mathbf{P}^\gamma_\mu \subseteq \mathscr{P}(\mathscr{P}(\mu))$ has cardinality $\leq 2^\mu$ for $\mu \in S$, $\gamma < \gamma(*)$ *then* we can find $\bar\xi = \langle \xi(\mu) :$

$\mu \in S\rangle \in \Pi\{2^{2^\mu} : \mu \in S\}$ such that for every $\gamma < \gamma(*)$ the following set is not stationary: $S_{\bar\eta,\gamma} = \{\mu \in S$: for some $\mathscr{P} \in \mathbf{P}^\gamma_\mu$ we have $\{A^{\xi(\mu)}_\gamma : \gamma \in [\mu, 2^\mu)\} \subseteq \mathscr{P} \subseteq \mathscr{D}_{\bar\eta,\mu} \cup \mathscr{I}_{\bar\eta,\mu}\}$.

[Why? Let $\langle u_\alpha : \alpha < \lambda\rangle$ be an increasing continuous sequence of subsets of $\gamma(*)$ with union $\gamma(*)$ such that $|u_\alpha| \leq |\alpha|$ for $\alpha < \lambda$. Now for each $\mu \in S$, the family $\bigcup\{\mathbf{P}^\gamma_\mu : \gamma \in u_\mu\}$ is a family of $\leq |u_\mu| \times 2^\mu$ subsets of $\mathscr{P}(\mu)$.

Now by clause (e) of $\odot_1$ for each $\mu \in S$, $\gamma \in u_\mu$, $\mathscr{P} \in \mathbf{P}^\gamma_\mu$ let $\xi_{\mu,\gamma,\mathscr{P}} < 2^{2^\mu}$ be such that: if for some $\xi < 2^{2^\mu}$ we have $\{A^\xi_\gamma : \gamma \in [\mu, 2^\mu)\} \subseteq \mathscr{P} \subseteq \mathscr{D}^\xi_\mu \cup \mathscr{I}^\xi_\mu$ then $\xi_{\mu,\gamma,\mathscr{P}}$ is the first such ξ. Choose $\xi(\mu) < 2^{2^\mu}$ which does not belong to $\{\xi_{\mu,\gamma,\mathscr{P}} : \gamma \in u_\mu \text{ and } \mathscr{P} \in \mathbf{P}^\gamma_\mu\}$.

So let $\bar\eta = \bar\eta_{\langle \xi(\mu):\mu\in S\rangle} \in \Xi$ be as in $\odot_3$ now $\bar\eta$ is as required by $\odot_2, \odot_3, \odot_4$.

Let us elaborate, why is it as required in $\odot_5$?

First, clearly $\eta_\varepsilon \in {}^\lambda 2$ for $\varepsilon < \lambda$. Second, fix $\gamma < \gamma(*)$, then there is $\alpha < \lambda$ such that $\gamma \in u_\alpha$, so it suffices to show that, for any $\mu \in S\backslash\alpha$, we have $\mu \notin S_{\bar\eta,\gamma}$. So assume $\mathscr{P} \in \mathbf{P}^\gamma_\mu$ satisfies clause (e) of $\odot_1$, and we should prove that $\neg[\{A^{\xi(\mu)}_\gamma : \gamma \in [\mu, 2^\mu)\} \subseteq \mathscr{P} \subseteq \mathscr{D}_{\bar\eta,\mu} \cup \mathscr{I}_{\bar\eta,\mu}]$; but if for some $\xi < 2^{2^\mu}$ we have $\{A^\xi_\gamma : \gamma \in [\mu, 2^\mu)\} \subseteq \mathscr{P} \subseteq \mathscr{D}_{\bar\eta,\mu} \cup \mathscr{I}_{\bar\eta,\mu}$ then necessarily $\xi = \xi_{\mu,\gamma,\mathscr{P}} \neq \xi(\mu)$, contradiction to $\odot_1$.]

$\odot_6$ if $\langle M_\gamma : \gamma \leq \gamma(*)\rangle$ is a $\prec$-increasing continuous and $M_\gamma \in \mathrm{EC}_\lambda(T)$ and $\bar b_\alpha \in {}^{lg(\bar y)}(M_0)$ for $\alpha < \lambda$ are such that $\langle \varphi(x, \bar b_\alpha) : \alpha < \lambda\rangle$ is independent, *then* we can find N such that

(α) $M_{\gamma(*)} \prec N \in \mathrm{EC}_\lambda(T)$

(β) if $N \prec N' \in \mathrm{EC}_\lambda(T)$ and $\gamma < \gamma(*)$ *then*[3] we have

$$\mathrm{inv}^\varphi_6(M_\gamma, N') \notin \{\mathrm{inv}^\varphi_6(M_{\gamma_1}, M_{\gamma_2}) : \gamma \leq \gamma(*) \text{ and } \gamma_1 < \gamma_2 \leq \gamma(*)\}.$$

[Why? Without loss of generality the universe of $M_{\gamma(*)}$ is $\mathscr{U}_1 \in [\lambda]^\lambda$ such that $\lambda\backslash\mathscr{U}_1$ has cardinality λ. Let $\langle u_\alpha : \alpha < \lambda\rangle$ be as in the proof of $\odot_5$.

For $\delta \leq \lambda$ and $\gamma(1) < \gamma(2) \leq \gamma(*)$ let $\mathbf{P}^{\gamma(1),\gamma(2)}_\delta = \mathrm{inv}^5_\varphi(\delta, \mathrm{id}_{N_{\gamma(2)}}, M_{\gamma(1)}, M_{\gamma(2)})$, see Definition 1.4, clearly $\mathrm{inv}^\varphi_6(M_{\gamma(1)}, M_{\gamma(2)}) = \langle \mathbf{P}^{\gamma(1),\gamma(2)}_\delta : \delta < \lambda\rangle/\mathscr{D}_\lambda$. So it is enough[4] to find N and sequence $\langle a_\alpha : \alpha < \lambda\rangle$ of elements of N such that $M_{\gamma(*)} \prec N$, $|N| = \lambda$ and for each $\gamma(0) \leq \gamma(*)$, for every $\mu \in S$ except nonstationarily many, the family

$$\{\{\gamma < \mu : N \models \varphi[a_\gamma, \bar b]\} : \bar b \in {}^{lg(\bar y)}(M_{\gamma(0)})\}$$

is not in $\mathbf{P}_\mu := \bigcup\{\mathbf{P}^{\gamma(1),\gamma(2)}_\mu : \gamma(1) < \gamma(2) \leq \gamma(*) \text{ are from } u_\mu\}$.

We choose $\bar\xi = \langle \xi(\mu) : \mu \in S\rangle$ as in $\odot_5$; let $\bar\eta = \bar\eta_{\bar\xi}$, see $\odot_3$, so recalling $\circledast_3$ clearly $\langle \mathbb{D}_{\eta_\varepsilon} : \varepsilon < \lambda\rangle$ is well defined. Now for each $\varepsilon < \alpha$ letting F be from Claim 2.3 for the model $M_{\gamma(*)}$ and the sequence $\langle \varphi(x, \bar b_\alpha) : \alpha < \lambda\rangle$, let $p_\varepsilon \in \mathbf{S}_\varphi(M_{\gamma(*)})$ be such that for every $\bar b \in {}^{lg(\bar y)}(M_{\gamma(*)})$ we have $\varphi(x, \bar b) \in p_\varepsilon \Leftrightarrow F(\bar b) \in \mathbb{D}_{\eta_\varepsilon}$ so $\neg\varphi(x, \bar b) \in p_\varepsilon \Leftrightarrow F(\bar b) \notin \mathbb{D}_{\eta_\varepsilon}$.

So by Conclusion 2.5 we can find an elementary extension N of $M_{\gamma(*)}$ and $a_\alpha \in N$ for $\alpha < \lambda$ such that a_α realizes p_α, and without loss of generality N has universe $\subseteq \lambda$ such that $\lambda\backslash|N|$ has cardinality λ. Concerning inv^6_φ our demand concerns what occurs for a club of $\delta < \lambda$ for this. Let $E \subseteq E_*$ be a club of λ such

---

[3]Really any pregiven set of $\leq \lambda$ "forbidden" inv^φ_6 is O.K. and can make it work for $\mathrm{inv}^\varphi_6(N_\gamma, N')$ for every $\gamma < \gamma(*)$.

[4]Can demand $\alpha < \lambda \Rightarrow {}^{\omega>}(\omega(\alpha+1))$ if $(N\backslash M_{\gamma(*)}) \cap [\omega\alpha, \omega\alpha+\omega)$ is infinite for every $\alpha < \lambda$.

that $\gamma < \delta \in E \Rightarrow a_\gamma \in N \cap \delta$. Now in $\odot_6(\beta)$ we promise something (given N) on "every N' such that...," so let $N \prec N' \in \mathrm{EC}_\lambda(T)$, and without loss of generality the universe of N' is $\subseteq \lambda$ and let $\delta \in S \cap E$. For any $\gamma \leq \gamma(*)$ by $\odot_5$, i.e., by the choice of $\bar{\xi}, \bar{\eta}_\xi$ above there is a club $E_\gamma \subseteq E$ of λ such that for any $\mu \in S \cap E_\gamma$, the set $S_{\bar{\eta},\mu}$ from $\odot_5$ is disjoint to E_γ, hence the set $\mathscr{P}_{\mu,\gamma} := \{\{\gamma < \mu : N' \models \varphi[a_\gamma, \bar{b}]\} : \bar{b} \in {}^{lg(\bar{y})}(M_\gamma)\}$ does not belong to $\bigcup\{\mathbf{P}_\mu^{\gamma(1),\gamma(n)} : \gamma(1) < \gamma(2) \leq \gamma(*)$ are from $u_\mu\}$ so we are done.]

$\odot_7$ if $\langle M_\alpha : \alpha < \lambda^+\rangle$ is as in $\boxtimes'$ from Theorem 2.8 then for some club E of λ^+, we have: if $\alpha_1 < \alpha_2$, $\beta_1 < \beta_2$ are from E and $\alpha_2 \neq \beta_2$ then

$$(M_{\alpha_2}, M_{\alpha_1}) \not\cong (M_{\beta_2}, M_{\beta_1}).$$

[Why? For every $\beta < \lambda^+$ we apply $\odot_6$ to $\langle M_\alpha : \alpha \leq \beta\rangle$ and get N_β as there so $M_\beta \prec N_\beta \in \mathrm{EC}_\lambda(T)$. As $M = \bigcup\{M_\gamma : \gamma < \lambda^+\}$ is saturated, without loss of generality $N_\beta \prec M$ hence for some $\xi_\beta < \lambda^+$ we have $N_\beta \prec M_{\xi_\beta}$.

Let $E = \{\delta < \lambda^+ : \delta$ a limit ordinal such that $\beta < \delta \Rightarrow \xi_\beta < \delta\}$.

Let $\alpha_1 < \alpha_2$, $\beta_1 < \beta_2$ be from E such that $\alpha_2 \neq \beta_2$ and we shall prove that $(M_{\alpha_2}, M_{\alpha_1})$ is not isomorphic to $(M_{\beta_2}, M_{\beta_1})$. By symmetry without loss of generality $\alpha_2 < \beta_2$ and let $\gamma(*) = \max\{\alpha_2, \beta_1\}$ so $\gamma(*) < \beta_2$. Now we apply $\odot_6$ with $\langle M_\gamma : \gamma \leq \gamma(*)\rangle$, N, N', γ, β_1, α_2, α_2 here standing for $\langle M_\gamma : \gamma \leq \gamma(*)\rangle$, N, N', $\gamma(0)$, $\gamma(1)$, $\gamma(2)$ there so we are clearly done. □

References

1. J. T. Baldwin and S. Shelah, *Second-order quantifiers and the complexity of theories*, Notre Dame J. Formal Logic **26** (1985), no. 3, 229–303.
2. R. Engelking and M. Karłowicz, *Some theorems of set theory and their topological consequences*, Fund. Math. **57** (1965), 275–285.
3. I. Kaplan and S. Shelah, *A counterexample to existence of indiscernibles and other examples*, preprint.
4. S. Shelah, *Remarks on squares*, Around Classification Theory of Models, Lecture Notes in Math., vol. 1182, Springer, Berlin, 1986, pp. 276–279.
5. ———, *Anti-homogeneous partitions of a topological space*, Sci. Math. Jpn. **59** (2004), no. 2, 203–255.
6. ———, *Dependent T and existence of limit models*, Tbil. Math. J., submitted, available at arXiv:math/0609636.
7. ———, *Dependent theories and the generic pair conjecture*, Commun. Contemp. Math., submitted, available at arXiv:math/0702292.
8. ———, *A dependent dream and recounting types*, Publ. 950 in List of Shelah's papers.

Einstein Institute of Mathematics, Edmond J. Safra Campus, Givat Ram, The Hebrew University of Jerusalem, Jerusalem, 91904, Israel and Department of Mathematics, Rutgers University, Hill Center — Busch Campus, 110 Frelinghuysen Road, Piscataway, NJ 08854-8019, USA

E-mail address: shelah@math.huji.ac.il

URL: http://shelah.logic.at

Centre de Recherches Mathématiques
CRM Proceedings and Lecture Notes
Volume **53**, 2011

The Strict ω-Groupoid Interpretation of Type Theory

Michael A. Warren

ABSTRACT. Hofmann and Streicher showed that there is a model of the intensional form of Martin-Löf's type theory obtained by interpreting closed types as groupoids. We show that there is also a model when closed types are interpreted as strict ω-groupoids. The nonderivability of various truncation and uniqueness principles in intensional type theory is then an immediate consequence. In the process of constructing the interpretation we develop some ω-categorical machinery including a version of the Grothendieck construction for strict ω-categories.

1. Introduction

In [9], Hofmann and Streicher established the existence of a model of the intensional form of Martin-Löf's type theory [16] using the category **Gpd** of small groupoids. This seminal result — and the related observation that the identity types in this theory form a kind of weak groupoid — is the starting point for much of the subsequent research relating category theory, homotopy theory and type theory. Type theoretically, the importance of the groupoid model is that it establishes the nonderivability of certain type theoretic principles relating to the identity types.

Recall that in Martin-Löf's type theory if A is a type and a, b are terms of type A, then there exists a type $\underline{A}(a, b)$ called the *identity types of* A (*from* a *to* b). Under the Curry–Howard correspondence $\underline{A}(a, b)$ may be thought of as the proposition which asserts that a and b are equal. The distinguishing feature of the *intensional* form of the theory from the *extensional* form is the absence in the former of the *reflection rule* which states that if there exists a term of type $\underline{A}(a, b)$, then $a = b$, where this latter equality is the primitive equality relation of the theory. Thus, identity types are trivial in the extensional theory. Prior to [9] it was unknown whether certain principles of a similar character to the reflection rule were in fact derivable. For example, it was unknown whether there could be more than one distinct "identity proof." I.e., whether $f = g$ when $f, g : \underline{A}(a, b)$. This principle is referred to as *uniqueness of identity proofs* in the literature and it is refuted in the groupoid model. (Note that in ibid. the principle of uniqueness of identity proofs really only refers to uniqueness up to propositional equality, but the groupoid model refutes both of these principles.) For in the groupoid model, a

2010 *Mathematics Subject Classification.* 03B15, 18D05.
This is the final form of the paper.

(closed) type A is interpreted as a groupoid $[\![A]\!]$, a term $a : A$ is interpreted as an object of $[\![A]\!]$ and the identity type $\underline{A}(a,b)$ is interpreted as the (discrete) groupoid consisting of all arrows $[\![a]\!] \to [\![b]\!]$ in $[\![A]\!]$. Thus, any groupoid G containing a pair of distinct parallel arrows refutes the uniqueness of identity proofs.

Although the groupoid model is useful for refuting certain type theoretic principles such as uniqueness of identity proofs, it fails to adequately exhibit higher-dimensional structure of the identity types. E.g., although the reflection rule is refuted the corresponding rule which asserts that all identity types of identity types are trivial is valid in the model since identity types are interpreted as discrete groupoids. I.e., the rule

$$\frac{p : \underline{\underline{A}(a,b)}(f,g)}{f = g : \underline{A}(a,b)}$$

is valid. At the time Hofmann and Streicher conjectured that it should be possible to obtain models which would refute such "truncation principles" by interpreting types as some kind of higher-dimensional groupoids. It is the aim of this paper to show that this is indeed the case. In particular, we construct a model of intensional type theory which extends the original groupoid model by using strict ω-groupoids. As such, this is the first known model of intensional type theory (aside from the syntactic model) which is truly higher-dimensional. In the terminology of [1] the model refutes *all* of the truncation rules TR_n.

The interpretation of type theory using ω-groupoids directly extends the original groupoid interpretation. Contexts Γ are interpreted as small ω-groupoids $[\![\Gamma]\!]$ and types $\Gamma \vdash A$ in context are interpreted as morphisms (of ω-categories) $[\![\Gamma]\!] \to \omega$-**Gpd**. A term $\Gamma \vdash a : A$ is interpreted as a section of the projection $\int [\![A]\!] \to [\![\Gamma]\!]$ where $\int [\![A]\!]$ denotes the (ω-categorical version of the) Grothendieck construction of $[\![A]\!]$. For a closed type A, the identity type $\underline{A}(a,b)$ is then interpreted as the ω-groupoid which has as objects arrows $[\![a]\!] \to [\![b]\!]$ in $[\![A]\!]$. As in the groupoid model, the dependent sums and products are given by left- and right-adjoints, respectively, to suitable reindexing functors.

Relation to other work. There has been much recent work on relating Martin-Löf's type theory to homotopy theory and higher-dimensional category theory and we will now briefly summarize the relation of the current paper to other such work. First, following a suggestion by Moerdijk, it was shown in [2] that intensional type theory (at least the fragment containing identity types and dependent sums) can be interpreted in weak factorization systems and Quillen's model categories which satisfy certain coherence conditions. It was then shown by Gambino and Garner [5] that there is a weak factorization system in the category of contexts. It was also shown, by van den Berg and Garner [25] and independently by Lumsdaine [15], that the identity type construction gives rise to weak ω-groupoids in the sense of Batanin and Leinster [3, 14]. Garner [6] has also introduced a "two-dimensional" type theory which is a form of intensional type theory and he has proved that it is sound and complete with respect to a semantics valued in the 2-categories (as such, these theories necessarily satisfy the corresponding truncation rule). In [1] the homotopy theory of the category of groupoids and a fortiori homotopy 1-types is related to the homotopy theory of the 1-truncated form of intensional type theory.

We would also like to mention the recent work of Lafont, Métayer and Worytkiewicz [12] in which they describe a model structure on the category of strict

ω-categories which corresponds to the "natural" or "folk" model structure on the category of small categories. Ultimately we would like to be able to relate their work to the model constructed here. In particular, one should be able to use exponentiation by the ω-groupoid interval $\mathbf{I}$ — which is the co-ω-groupoid object in ω-**Cat** with exactly two 0-cells, two invertible 1-cells between them, two invertable 2-cells between those, and so forth — to interpret identity types. (In the setting of ordinary 1-dimensional groupoids this kind of construction has been studied in [27] and related results on cocategory objects were obtained in [28].) Exponentiation by $\mathbf{I}$ has been studied by Métayer [17] and is used in [12] for the construction of path objects in the model structure. In order to develop such an approach using $\mathbf{I}$ it appears to be necessary to generalize also the notion of split fibration to higher-dimensions. The notion of Grothendieck 2-fibration has been studied by Hermida [7] and, for 2-groupoids, by Moerdijk and Svennson [18]. However, we are unaware of similar work in higher dimensions.

Summary of the paper. In Section 2 we recall some basic facts and definitions concerning ω-categories. In Section 3 we extend the familiar Grothendieck construction to the setting of ω-categories and functors $A\colon \mathcal{C} \to \omega\text{-}\mathbf{Cat}$. This construction is crucial since it allows us to interpret context extension and it provides the data necessary for the interpretation of open terms. In Section 4 we introduce the notion of ω-groupoid with which we will be concerned. We also describe a "dual" version of the Grothendieck construction. For ω-groupoid valued functors there is an induced functor $\neg$ relating the Grothendieck construction and its dual which we will require in order to interpret identity types and dependent products. In Section 5 we finally describe the interpretation of identity types. For closed types A and terms a and b of type A the identity type $\underline{A}(a,b)$ is easily described as the ω-groupoid which has as objects arrows $a \to b$ and with $(n+1)$-cells given by $(n+2)$-cells of A bounded in dimension 0 by a and b. However, in order to most efficiently prove the soundness of the elimination rule for identity types it is necessary to give an interpretation of the corresponding open identity type. This is essentially a fibred version of the foregoing type, but is considerably more involved in the case of ω-groupoids than it was for groupoids. Section 6 is concerned with the proof of the soundness of the introduction and elimination rules for identity types. Finally, the main results of the paper (the soundness theorem and its consequences) are collected in Section 7. Appendix A contains a summary of the rules governing identity types, dependent products, and dependent sums in the theory considered in this paper. Appendix B contains a brief description of the comma ω-categories featuring in the description of the universal property of the Grothendieck construction and the proof that the Grothendieck construction does indeed possess the correct universal property.

It is worth mentioning that we assume familiarity on the part of the reader with Martin-Löf's type theory and we refer the reader to [4, 8, 10, 16, 19, 23, 24] for further details. There is also a brief survey of intensional type theory contained in [1] which includes a summary of the kinds of truncation principles which are refuted by the model.

Acknowledgements. The results of the present paper originally occur in the author's Ph.D. thesis [27] and I thank my thesis supervisor Steve Awodey for numerous insightful comments and discussions of this work. I also thank the other members of my thesis committee — Nicola Gambino, Alex Simpson and Thomas Streicher — for discussion of and comments on this work. The anonymous referee also provided a number of useful comments and suggestions. Finally, I thank Pieter Hofstra, Peter LeFanu Lumsdaine and Phil Scott for useful conversations regarding this material.

2. Generalities on ω-categories

In this section we recall the definition of strict ω-categories and ω-groupoids, fix notation and recall several useful facts about these structures. We refer the reader to [14, 21, 22, 26] for further details regarding the theory of strict ω-categories.

2.1. ω-categories, functors and transformations. Just as categories are (directed) reflexive graphs equipped with the additional structure given by identities and composition, so too n-categories are reflexive n-globular sets with additional structure and ω-categories are reflexive globular sets with additional structure. Here we recall that *reflexive globular sets* A are the (unbounded) higher-dimensional analogues of reflexive graphs and, as such, consists of sets A_n graded by nonnegative integers n together with *source*, *target* and *identity* functions

$$A_0 \underset{t}{\overset{s}{\leftrightarrows}} \xrightarrow{i} A_1 \underset{t}{\overset{s}{\leftrightarrows}} \xrightarrow{i} A_2 \underset{t}{\overset{s}{\leftrightarrows}} \xrightarrow{i} \cdots$$

such that the globular identities

$$s \circ s = s \circ t, \quad t \circ t = t \circ s,$$

and the equations

$$s \circ i = 1_{A_n} = t \circ i$$

are satisfied. A (*strict*) ω-*category* consists of a reflexive globular set A and, for all $p \geq 0$ and $n > 0$, *composition* functions

$$A_n \times_{A_p} A_n \xrightarrow{*_p} A_n,$$

where $A_n \times_{A_p} A_n$ is the pullback

$$\begin{array}{ccc} A_n \times_{A_p} A_n & \longrightarrow & A_n \\ \downarrow & & \downarrow t^{(n-p)} \\ A_n & \xrightarrow[s^{(n-p)}]{} & A_p \end{array}$$

with

$$l^k := \underbrace{l \circ l \circ \cdots \circ l}_{k\text{-times}},$$

for any $k \geq 0$ and $l = s, t$. This data is required to satisfy the following conditions:

Domain and Codomain Laws.

$$l(g *_p f) = \begin{cases} l(g) *_p l(f) & \text{if } p < (n-1) \\ = \begin{cases} s(f) & \text{if } l = s \\ t(g) & \text{if } l = t \end{cases} & \text{if } p = (n-1). \end{cases}$$

for $l = s, t$.

Associativity Laws. Each operation $*_p$ is associative.

Unit Laws. Given f in A_n,

$$i^{(n-p)}\big(t^{(n-p)}(f)\big) *_p f = f = f *_p i^{(n-p)}\big(s^{(n-p)}(f)\big).$$

Interchange Laws. Given $q < p < n$ and f, g, h, k in A_n such that the composites $(y *_q x)$, $(k *_q h)$, $(h *_p f)$ and $(k *_p g)$ are defined,

$$(k *_p g) *_q (h *_p f) = (k *_q h) *_p (g *_q f),$$

and

$$i(g) *_q i(f) = i(g *_q f).$$

Sometimes we refer to composition $*_0$ along 0-cells as *horizontal composition*. Given two n-cells f and g, if $s(f) = s(g)$ and $t(f) = t(g)$, then we say that f *and* g *are parallel (to one another)*. For example, if f and g are n-cells such that $(g *_p f)$ is defined, then $s^{(n-p)}(f)$, $s^{(n-p)}(g)$, $t^{(n-p)}(f)$ and $t^{(n-p)}(g)$ are all parallel. When no confusion will result we often omit mention of identity maps. E.g., if $f\colon x \to y$ is a 1-cell and $\alpha\colon g \Rightarrow h$ is a 2-cell with $s(g) = y$, we denote $\alpha *_0 i(f)$ by $\alpha *_0 f$.

A (*strict* ω-)*functor* $F\colon \mathcal{C} \to \mathcal{B}$ between ω-categories is a map of globular sets which preserves all composition and identities. We often refer to ω-functors simply as *functors* when it is understood that we are dealing with ω-categories. The category of small ω-categories and functors between them is denoted by ω-**Cat**. Just as the category **Cat** of small categories and functors between them is monadic over the category of directed graphs, so too ω-**Cat** is monadic over the category of globular sets (cf. [14] for an explicit description of the monad). Indeed, ω-**Cat** is a Cartesian closed category with all (small) limits and colimits. Henceforth we often denote ω-categories by $\mathcal{C}, \mathcal{B}, \ldots$. Clearly every ω-category is also an n-category, for $1 \leq n$ and similarly for ω-functors.

Given functors $F, G\colon \mathcal{C} \rightrightarrows \mathcal{B}$ between ω-categories $\mathcal{C}$ and $\mathcal{B}$, a *natural transformation* $\alpha\colon F \Rightarrow G$ consists of an assignment of 1-cells $\alpha_x\colon Fx \to Gx$ for objects x of $\mathcal{C}$ such that the following (somewhat schematic) diagram commutes:

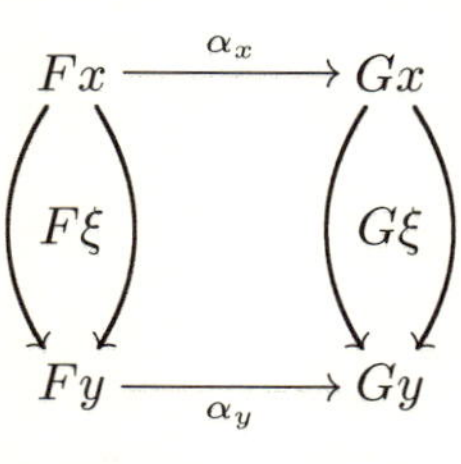

for every k-cell ξ bounded by 0-cells x and y (see Section 2.2 for an explanation of the notation employed here). I.e., if ξ is any k-cell, for $k \geq 1$, such that $s^k\xi = x$ and $t^k\xi = y$, then

$$\alpha_y *_0 F\xi = G\xi *_0 \alpha_x. \tag{2.1}$$

Passing up one dimension, suppose we are given functors F and G as above together with natural transformations α and β from F to G. Then, a *modification* or *2-transformation* $\varphi\colon \alpha \Rightarrow \beta$ consists of an assignment of 2-cells $\varphi_x\colon \alpha_x \to \beta_x$ of $\mathcal{B}$ parameterized by objects x of $\mathcal{C}$ subject to the condition that, for any arrow

$f\colon x \to y$ of $\mathcal{C}$, the following diagram commutes:

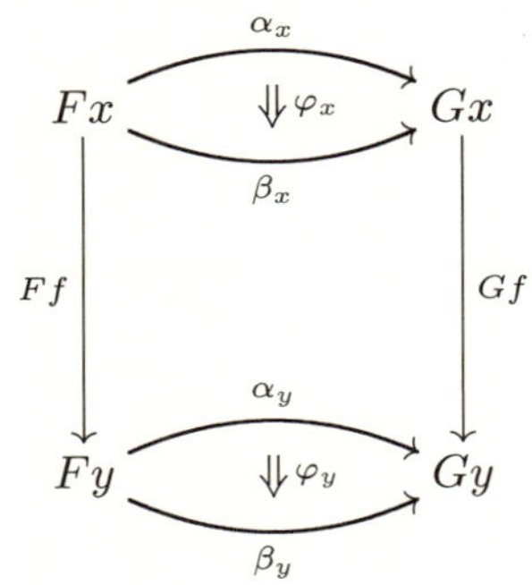

I.e.,

$$\varphi_y *_0 Ff = Gf *_0 \varphi_x \tag{2.2}$$

for $f\colon x \to y$ an arrow of $\mathcal{C}$.

It is possible to generalize inductively to higher-dimensional transformations. In particular, assuming we have defined n-transformations, for $n \geq 2$, in such a way that the obvious boundary conditions are satisfied a $(n+1)$-*transformation* ψ from an n-transformation γ to a n-transformation δ consists of a family of n-cells $\psi_x\colon \gamma_x \Rightarrow \delta_x$ in $\mathcal{B}$ parameterized by objects x of $\mathcal{C}$ such that, whenever $f\colon x \to y$ is an arrow in $\mathcal{C}$, the naturality condition

$$\psi_y *_0 Ff = Gf *_0 \psi_x \tag{2.3}$$

is satisfied. With these definitions it is straightforward to verify that the following more general naturality conditions are also satisfied:

Lemma 2.1. *If ξ is a k-cell of $\mathcal{C}$ bounded by 1-cells $f, g\colon x \rightrightarrows y$ and φ is a $(n+1)$-transformation bounded by functors $F, G\colon \mathcal{C} \to \mathcal{B}$, then*

$$\varphi_y *_0 F\xi = G\xi *_0 \varphi_x.$$

With ω-functors and these higher-dimensional transformations ω-**Cat** itself exhibits the combinatorial structure of an ω-category.

Proposition 2.2. *The category ω-**Cat** is a large ω-category with $(n+1)$-cells given by n-transformations.*

2.2. Notation and conventions. It will be convenient to introduce some conventions governing diagrams in higher dimension. In particular, we will often want to describe the various boundaries of m-cells φ in ω-categories. In particular, we may wish to indicate diagrammatically the n-cells bounding such a φ even when $m > n+2$ so that drawing the details of the diagram would be cumbersome. As such, we will instead often include diagrams such as the following:

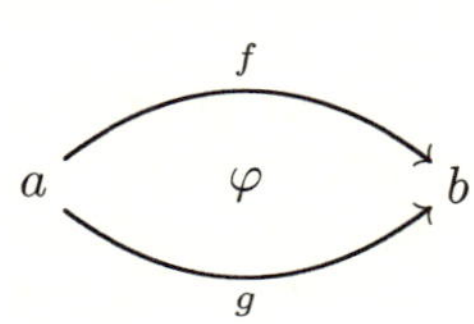

where a and b are n-cells and φ is an m-cell for $m > n+2$. Such diagrams are oriented from "top-to-bottom" unless otherwise stated. I.e., the diagram indicates that

$$s^{m-n-1}(\varphi) = f \quad \text{and} \quad t^{m-n-1}(\varphi) = g.$$

In the few cases where there is no "top-to-bottom" option available, the cells should be oriented "left-to-right." In this section, and throughout this chapter, we will often be dealing with "hom-ω-categories" of the form $\mathcal{C}(a,b)$ where $\mathcal{C}$ is an ω-category and a, b are parallel cells of $\mathcal{C}$. In this setting, or similar ones, the index of a composition $(\gamma *_n \delta)$ always refers to the dimension in $\mathcal{C}$ and not in $\mathcal{C}(a,b)$.

There exists an endofunctor $(-)^+$ on ω-**Cat** called the *dimension shift functor* which shifts the dimension of an ω-category. Specifically, given an ω-category $\mathcal{C}$, $\mathcal{C}^+$ has as objects 1-cells of $\mathcal{C}$ and, in general, n-cells of $\mathcal{C}^+$ are $(n+1)$-cells of $\mathcal{C}$. Similarly, given ω-category $\mathcal{C}$ and objects x and y of $\mathcal{C}$, the hom set $\mathcal{C}(x,y)$ can be made into an ω-category — which we sometimes denote by $\mathcal{C}_1(x,y)$ to emphasize the dimension — by defining 0-cells to be arrows $f\colon x \to y$ and $(n+1)$-cells to be n-cells in the obvious way. Similarly, given parallel $(n+1)$-cells f, g of $\mathcal{C}$, there is an ω-category $\mathcal{C}_{n+2}(f,g)$ which has 0-cells $(n+2)$-cells $\alpha\colon f \Rightarrow g$ and so forth. For parallel n-cells f and g with $n \geq 1$, there exists an inclusion functor

$$\mathcal{C}_{(n+1)}(f,g) \to \big(\mathcal{C}_n(sf,tg)\big)^+$$

which sends a $(n+1)$-cell $\alpha\colon f \to g$ to itself, and similarly for higher-dimensional cells. We will occasionally make use of such functors without giving them an explicit name.

It will also be convenient to fix notation for dealing with functors $A\colon \mathcal{C} \to \omega$-**Cat**. Given a small ω-category $\mathcal{C}$ and a functor $A\colon \mathcal{C} \to \omega$-**Cat** we denote by A_x the category obtained by applying A to an object x of $\mathcal{C}$ and, when $f\colon x \to y$ is an arrow in $\mathcal{C}$, $A_f\colon A_x \to A_y$ denotes the induced functor. We employ similar notation for higher-dimensional cells. This convention will later allow us to avoid excessive use of parentheses. When z is any 0-cell of A_x we denote $A_f(z)$ by $(z.f)$. Similarly, by definition A_γ, for γ a $(n+1)$-cell with $n \geq 1$, is given by a family of n-cells parameterized by 0-cells of its domain category (say) A_x and we denote by $(z.\gamma)$ the n-cell $\big(A_\gamma\big)_z$. In many of the constructions given below we will deal with ω-categories having as their cells tuples of cells of other ω-categories. Such tuples will often be abbreviated using "vector" notation $\vec{f}$.

3. The ω-categorical Grothendieck construction

When dealing with categories, there is, for each $\mathcal{C}$, a monadic adjunction

$$[\mathcal{C}, \mathbf{Cat}] \underset{R}{\overset{L}{\underset{\longrightarrow}{\overset{\longleftarrow}{\bot}}}} \mathbf{Cat}/\mathcal{C}$$

of 2-categories where the left-adjoint is given by the comma category

$$L(F) = (F \downarrow -),$$

for $F\colon \mathcal{D} \to \mathcal{C}$, and the right-adjoint is given by taking the projection from the Grothendieck construction

$$R(G) = \int G \xrightarrow{\pi} \mathcal{C},$$

for $G\colon \mathcal{C} \to \mathbf{Cat}$. An analogous situation occurs in the case of ω-categories and, in particular, the 2-category $[\mathcal{C}, \omega\text{-}\mathbf{Cat}]$ is also monadic over $\omega\text{-}\mathbf{Cat}/\mathcal{C}$. The right-adjoint is again given by an ω-categorical version of the Grothendieck construction. This construction will feature prominently in the interpretation of type theory since it allows us to interpret the extension of contexts. As in the case of $\mathbf{Cat}$, the *Grothendieck construction* $\int A$ *of a functor* $A\colon \mathcal{C} \to \omega\text{-}\mathbf{Cat}$ (or *category of elements of* A) can be described (in terms of its universal property) as the coend

$$\int^{x} (x \downarrow \mathcal{C}) \times A_x, \tag{3.1}$$

which exists since $\omega\text{-}\mathbf{Cat}$ is bicomplete. Here $(x \downarrow \mathcal{C})$, like the comma construction which gives the left-adjoint L, is a *lax* version of the usual comma category. I.e., instead of having as 1-cells commutative triangles, an arrow from an object $g\colon x \to y$ to $h\colon x \to z$ is a pair (f, f') such that $f\colon y \to z$ and

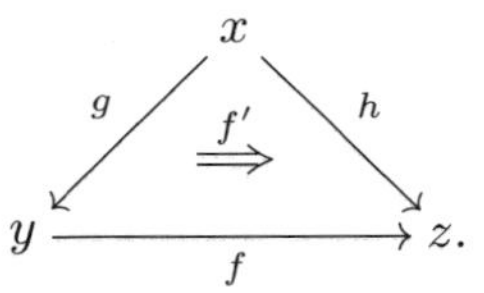

The higher-dimensional cells of this ω-category will not be necessary for our purposes and are described in Appendix B. Although (3.1) is a convenient description of the Grothendieck construction of A, it will be useful for our purposes to have a direct (combinatorial) description of this ω-category. It is to this matter which we now turn. (The verification that the combinatorial description given below of the Grothendieck construction satisfies the universal property of the coend (3.1) is contained in Appendix B.)

3.1. Combinatorial description of the Grothendieck construction. We now give a direct description of the ω-category $\int A$. We leave the proof that the data given below constitutes an ω-category to the reader (the proof can also be found in [27]). In the first two dimensions $\int A$ is the familiar Grothendieck construction of A.

0-Cells. The 0-cells of $\int A$ are pairs (x, x_-) such that x is an object of $\mathcal{C}$ and x_- is an object of A_x.

1-Cells. The 1-cells $(x, x_-) \to (y, y_-)$ are pairs (f, f_-) consisting of a 1-cell $f\colon x \to y$ in $\mathcal{C}$ and a 1-cell $f_- : (x_-.f) \to y_-$ in A_y.

Already at this low dimension there are several features of the definition which should be emphasized. First, in order to define the component f_- of arrows in $\int A$ we have made a choice of "weight" or "orientation." Namely, we have determined that the codomain of f_- should be y_- where we could have as easily determined that its domain should be this object of A_y. Secondly, fixing objects $\vec{x}$ and $\vec{y}$ of $\int A$, there exists a functor

$$\mathcal{C}_1(x, y) \xrightarrow{\mathfrak{d}^1_{\vec{x},\vec{y}}} A_y$$

defined by

$$\mathfrak{d}^1_{\vec{x},\vec{y}}(\gamma) := (x_-.\gamma),$$

for any m-cell γ of $\mathcal{C}(x, y)$. Although this functor depends on $\vec{x}$ and $\vec{y}$ we often write $\mathfrak{d}^1$ when no confusion will result. With this definition we observe that an arrow $\vec{f}\colon \vec{x} \to \vec{y}$ has

$$s(f_-) = \mathfrak{d}^1(f).$$

In this situation, the object $\mathfrak{d}^1(f)$ is said to be the *weighted face of* f_-. We will see that the higher-dimensional cells resulting from the construction of $\int A$ also possess suitably "weighted" faces.

2-Cells. Given 1-cells $\vec{f}$ and $\vec{g}$ with common source and target $\vec{x}$ and $\vec{y}$, respectively, a 2-cell $\vec{f} \to \vec{g}$ consists of $\alpha\colon f \Rightarrow g$ in $\mathcal{C}$ together with a 2-cell α_- of A_y as indicated in the following diagram:

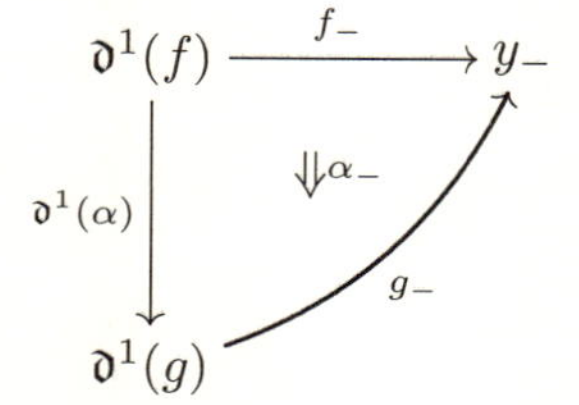

Now, holding $\vec{f}$ and $\vec{g}$ fixed, there exists a functor

$$\mathcal{C}_2(f,g) \xrightarrow{\mathfrak{d}^2_{\vec{f},\vec{g}}} (A_y)_1\big(\mathfrak{d}^1_{\vec{x},\vec{y}}(f), y_-\big)$$

defined by

$$\mathfrak{d}^2_{\vec{f},\vec{g}}(\gamma) := g_- *_0 \mathfrak{d}^1_{\vec{x},\vec{y}}(\gamma)$$

where γ is a n-cell of $\mathcal{C}_2(f,g)$. Note that in this case $\mathfrak{d}^1(\gamma)$ is a $(n+1)$-cell of A_y so that this definition makes sense. It is straightforward to verify that this is functorial. Note that an arrow $\vec{\alpha}\colon \vec{f} \to \vec{g}$ has

$$t(\alpha_-) = \mathfrak{d}^2_{\vec{f},\vec{g}}(\alpha).$$

As above, $\mathfrak{d}^2(\alpha)$ is the *weighted face of* α_-. In general, we will see that $(n+1)$-cells of $\int A$ are given by pairs (φ, φ_-) and that each component φ_- comes equipped with a weighted face. Namely, the weighted face of φ_- is $s(\varphi_-)$ if $(n+1)$ is even and it is $t(\varphi_-)$ if $(n+1)$ is odd. At each stage n we will construct, along with the definition of the n-cells, a functor $\mathfrak{d}^n(-)$ which gives an explicit description of the weighted faces of n-cells.

In general, $(\int A)_n$ is defined by induction on n alternating between even and odd steps in such a way that the following conditions are satisfied:

(1) At each stage n an element of $(\int A)_n$ is a pair (f, f_-) such that f is an n-cell of $\mathcal{C}$ and f_- is an n-cell of A_y.

(2) At each stage $(n+1)$, for $n \geq 1$, there is also constructed, for each pair $\vec{\alpha}, \vec{\beta}$ of parallel n-cells with source $\vec{f}$ and target $\vec{g}$, a functor $\mathfrak{d}^{n+1}_{\vec{\alpha},\vec{\beta}}$ such that

$$\mathcal{C}_{n+1}(\alpha,\beta) \xrightarrow{\mathfrak{d}^{n+1}_{\vec{\alpha},\vec{\beta}}} (A_y)_n\big(\mathfrak{d}^n_{\vec{f},\vec{g}}(\alpha), g_-\big)$$

if $(n+1)$ is even and

$$\mathcal{C}_{n+1}(\alpha,\beta) \xrightarrow{\mathfrak{d}^{n+1}_{\vec{\alpha},\vec{\beta}}} (A_y)_n\big(f_-, \mathfrak{d}^n_{\vec{f},\vec{g}}(\beta)\big)$$

if $(n+1)$ is odd. The functor $\mathfrak{d}^{n+1}$ is called the *weighted face functor* (*in dimension* $(n+1)$ *determined by* $\vec{\alpha}$ *and* $\vec{\beta}$).

(3) At each stage $(n+1)$, for $n \geq 0$, it is required that the following *weighted face conditions* are satisfied:

$$s(\varphi_-) = \begin{cases} \alpha_- & \text{if } (n+1) \text{ is even,} \\ \eth^{n+1}_{\vec{\alpha},\vec{\beta}}(\varphi) & \text{if } (n+1) \text{ is odd;} \end{cases}$$

and

$$t(\varphi_-) = \begin{cases} \eth^{n+1}_{\vec{\alpha},\vec{\beta}}(\varphi) & \text{if } (n+1) \text{is even,} \\ \beta_- & \text{if } (n+1) \text{ is odd,} \end{cases}$$

when $\vec{\varphi}$ is an $(n+1)$-cell $\vec{\varphi}\colon \vec{\alpha} \to \vec{\beta}$.

By the discussion above, the base cases $n = 0, 1, 2$ satisfy these conditions. We now consider the induction steps.

$(n+1)$ is odd. Fix parallel n-cells $\vec{f}$ and $\vec{g}$ of $\int A$. A $(n+1)$-cell $\vec{f} \Rightarrow \vec{g}$ consists of a pair (α, α_-) with $\alpha\colon f \Rightarrow g$ an $(n+1)$-cell in $\mathcal{C}$ and α_- is a $(n+1)$-cell of A_y subject to conditions which we will now describe. Let $\vec{v} = s(\vec{f})$ and $\vec{w} = t(\vec{f})$. Then $\vec{v}$ and $\vec{w}$ are $(n-1)$-cells of $\int A$ and therefore, by the induction hypothesis, there exists a weighted face functor $\eth^n_{\vec{v},\vec{w}}\colon \mathcal{C}_n(v,w) \to (A_y)_{n-1}(\eth^{n-1}(v), w_-)$. As such, α_- is required to be a $(n+1)$-cell of A_y as indicated in the following diagram:

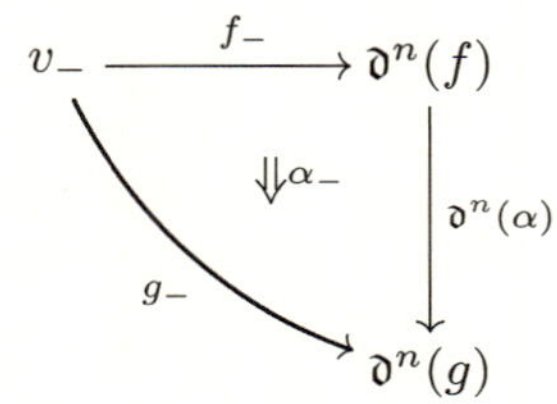

For the weighted functor, we hold $\vec{f}$ and $\vec{g}$ fixed and define

$$\eth^{(n+1)}_{\vec{f},\vec{g}}(\gamma) := \eth^n(\gamma) *_{(n-1)} f_-,$$

for γ a m-cell of $\mathcal{C}_{(n+1)}(f,g)$. The weighted face conditions are then trivially satisfied.

$(n+1)$ is even. Given parallel n-cells $\vec{f}$ and $\vec{g}$ of $\int A$, a $(n+1)$-cell $\vec{f} \Rightarrow \vec{g}$ consists, as above, of a pair (α, α_-) with $\alpha\colon f \Rightarrow g$ in $\mathcal{C}$ and α_- a $(n+1)$-cell of A_y as indicated in the following diagram:

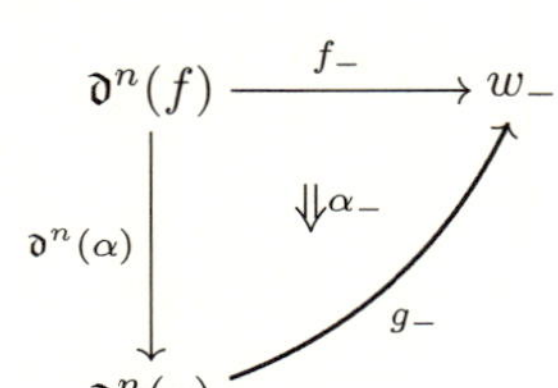

Finally, we define:

$$\eth^{(n+1)}_{\vec{f},\vec{g}}(\gamma) := g_- *_{(n-1)} \eth^n(\gamma),$$

for γ a m-cell of $\mathcal{C}_{(n+1)}(f,g)$. The weighted face conditions are then trivial.

For horizontal composition, suppose we are given a pair of composable arrows $\vec{f}\colon \vec{x} \to \vec{y}$ and $\vec{h}\colon \vec{y} \to \vec{z}$ in $\int A$. Then we obtain

$$A_h\big(\mathfrak{d}^1_{\vec{x},\vec{y}}(f)\big) \xrightarrow{A_h(f_-)} A_h(y_-) = \mathfrak{d}^1_{\vec{y},\vec{z}}(h) \xrightarrow{h_-} z_-$$

in A_z. Moreover,

$$A_h(\mathfrak{d}^1_{\vec{x},\vec{y}}f) = A_h A_f(x_-) = \mathfrak{d}^1_{\vec{x},\vec{z}}(h \circ f),$$

and therefore we define

$$(\vec{h} *_0 \vec{f})_- := h_- *_0 A_h(f_-).$$

This is the familiar definition of composition in the (1-dimensional) category of elements. Now, suppose that we are given m-cells $\vec{\varphi}$ and $\vec{\psi}$ of $\int A$, for $m > 1$, which are bounded by 0- and 1-cells as indicated in the following diagram:

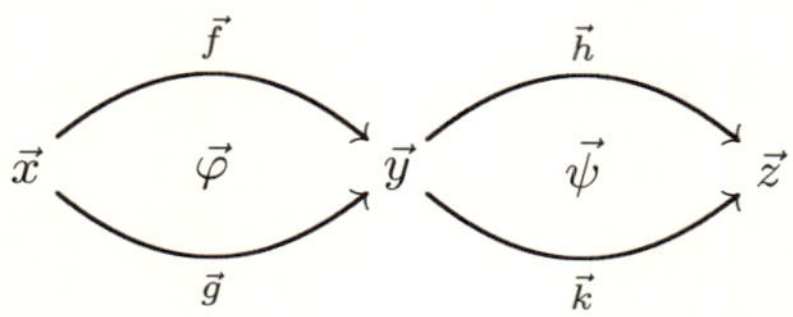

Then,

$$t^m A_h(\varphi_-) = A_h(tg_-) = A_h(y_-) = \mathfrak{d}^1_{\vec{y},\vec{z}}(h),$$

and we set

$$(\vec{\psi} *_0 \vec{\varphi})_- := \psi_- *_0 A_h(\varphi_-).$$

For general composition, assume $n > 0$ and suppose we are given m-cells which are composable along an n-cell as indicated in the following diagram:

(3.2) 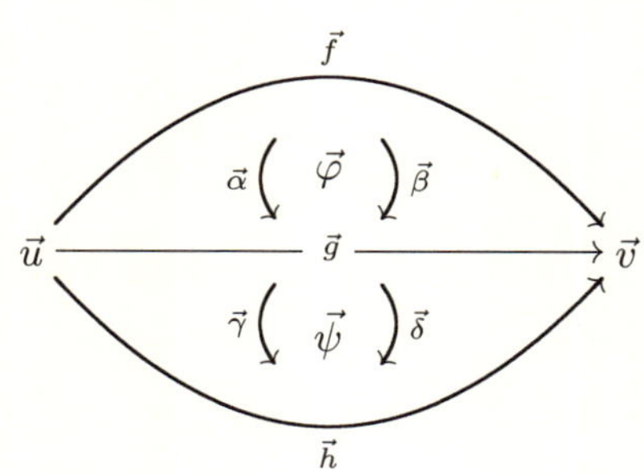

where $\vec{f}, \vec{g}$ and $\vec{h}$ are n-cells in $\int A$. Here, $\vec{\alpha}$ and $\vec{\beta}$ are the $(n+1)$-cells bounding $\vec{\varphi}$. I.e., $\vec{\beta} = t^{(m-n-1)}\vec{\varphi}$, etc. As such, when $m = n+1$ we have $\vec{\beta} = \vec{\alpha} = \vec{\varphi}$ and similarly for $\vec{\psi}$. We would like to define the composite $(\vec{\psi} *_n \vec{\varphi})$. Since the first component will be the composite $(\psi *_n \varphi)$ taken in $\mathcal{C}$, it remains only to define the second component $(\vec{\psi} *_n \vec{\varphi})_-$. The definition will alternate between those cases where $(n+1)$ is even and those where it is odd. First, assume $(n+1)$ is even. Then we obtain the following diagram in A_y:

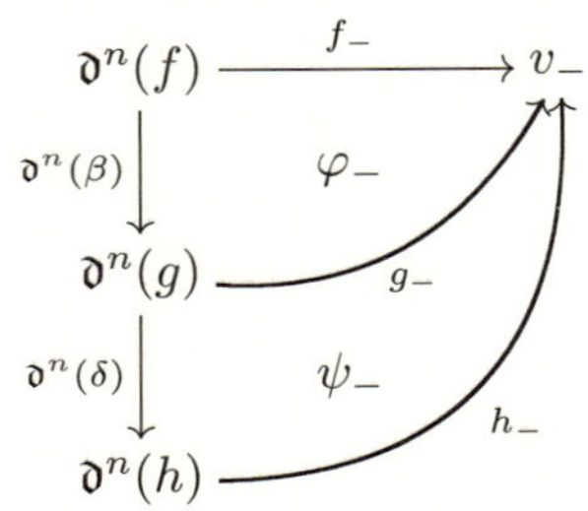

where $\vec{x}$ and $\vec{y}$ are the 0-cells bounding all of the cells in question. To see that we have correctly identified the n-cells of A_y bounding φ_- and ψ_- note that when $m = n+1$ this is trivially the case. When $m > n+1$,

$$s^{(m-n)}(\varphi_-) = s(\alpha_-) = f_-.$$

Similarly, $t^{(m-n)}\varphi_- = t\beta_-$ which, since $(n+1)$ is even, is equal to $\partial^{n+1}_{\vec{f},\vec{g}}(\beta)$, as required. Similar calculations show that ψ_- is as indicated in the diagram. Note also that

$$\partial^n_{\vec{u},\vec{v}}(\delta) *_{(n-1)} \partial^n_{\vec{u},\vec{v}}(\beta) = \partial^n(\delta *_n \beta),$$

by functoriality of $\partial^n_{\vec{u},\vec{v}}(-)$. These observations suggest that, when $(n+1)$ is even, we define $(\vec{\psi} *_n \vec{\varphi})_-$ to be the composite

$$\left(\psi_- *_{(n-1)} \partial^n_{\vec{u},\vec{v}}(\beta)\right) *_n \varphi_-.$$

Similarly, when $(n+1)$ is odd, we obtain

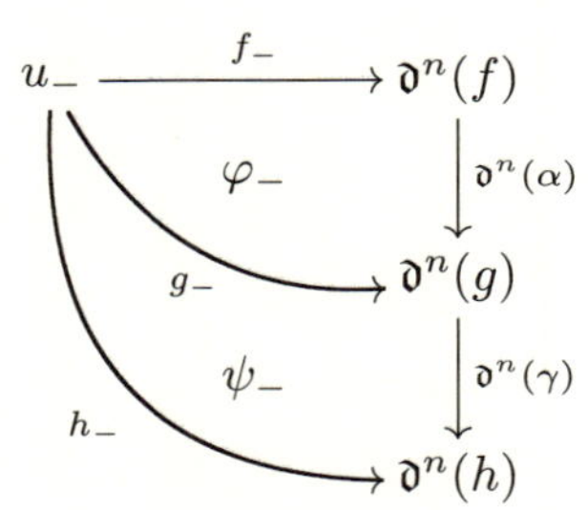

in A_y and we may define $(\vec{\psi} *_n \vec{\varphi})_-$ to be the analogous composite. Explicitly, given $\vec{\varphi}$ and $\vec{\psi}$ as above, we define

$$(\vec{\psi} *_n \vec{\varphi})_- := \begin{cases} \left(\psi_- *_{(n-1)} \partial^n_{\vec{u},\vec{v}}(\beta)\right) *_n \varphi_- & \text{if } (n+1) \text{ is even,} \\ \psi_- *_n \left(\partial^n_{\vec{u},\vec{v}}(\gamma) *_{(n-1)} \varphi_-\right) & \text{if } (n+1) \text{ is odd.} \end{cases}$$

where $\vec{\beta}$ and $\vec{\gamma}$ are the bounding $(n+1)$-cells as indicated in (3.2)).

With these definitions we obtain the following proposition, the detailed proof of which can be found in [27]:

Proposition 3.1. *Given a (small) ω-category $\mathcal{C}$ and a functor $A\colon \mathcal{C} \to \omega$-**Cat**, the Grothendieck construction $\int A$ of A is a (small) ω-category.*

4. ω-groupoids

The purpose of this section is to introduce the particular notion of *ω-groupoid*, which we will be using in order to interpret type theory, and to present several results and constructions on such ω-groupoids. The definition adopted here is the following:

Definition 4.1. A strict ω-category $\mathcal{C}$ is a *(strict) ω-groupoid* if every $(n+1)$-cell $f\colon a \to b$ possesses a *strict* $*_n$-inverse $f^{-1}\colon b \to a$. I.e.,

$$(f^{-1} *_n f) = a, \tag{4.1}$$

and

$$(f *_n f^{-1}) = b. \tag{4.2}$$

This definition generalizes both the usual definition of (1-)groupoid as well as the definition of (strict) 2-groupoid occurring in the work of Moerdijk and Svensson [18]. It should be contrasted with the *weaker* notions of ω-groupoid, also defined in the general setting of strict ω-categories, due to Street [21], and Kapranov and Voevodsky [11]. The essential difference with the definition from [21] is that there the notion of inverse is weakened so that, instead of equations, it is required that there exist (systems of) higher-dimensional cells $(f^{-1} *_n f) \Rightarrow a$, etc. In [11] it is further required that the higher-dimensional cells witnessing invertibility of f satisfy additional coherence conditions.

With Definition 4.1 at hand we obtain the following corollary to Proposition 3.1:

Corollary 4.2. *If $\mathcal{C}$ is a (small) ω-groupoid and $A\colon \mathcal{C} \to \omega$-**Gpd**, then $\int A$ is a (small) ω-groupoid.*

In particular, the inverse $(\vec{f})^{-1}$ of $\vec{f}\colon \vec{x} \to \vec{y}$ is the pair $\big(f^{-1}, A_{f^{-1}}(f_-)^{-1}\big)$ and, for $n > 0$, the inverse of a $(n+1)$-cell $\vec{\varphi}\colon \vec{\alpha} \Rightarrow \vec{\beta}$ is given by

$$(\vec{\varphi})^{-1} := \big(\varphi^{-1}, \varphi_-^{-1} *_{(n-1)} \eth^n_{s\vec{\alpha},t\vec{\alpha}}(\varphi^{-1})\big)$$

when $(n+1)$ is even, and

$$(\vec{\varphi})^{-1} := \big(\varphi^{-1}, \eth^n_{s\vec{\alpha},t\vec{\alpha}}(\varphi^{-1}) *_{(n-1)} \varphi_-^{-1}\big)$$

when $(n+1)$ is odd.

As mentioned above, when describing the Grothendieck construction $\int A$ we could well have chosen to orient our cells (in terms of the weighted faces) differently. For example, we could define an arrow $\vec{x} \to \vec{y}$ in $\int A$ to consist of $f\colon x \to y$ and $f_-\colon y_- \to x_-.f$. By choosing this orientation instead we obtain an ω-category which we call the *dual Grothendieck construction of* A and denote by $\int^* A$. We will give a quick sketch of this construction. When $A\colon \mathcal{C} \to \omega$-**Gpd** and $\mathcal{C}$ is an ω-groupoid, there is a functor $\neg\colon \int A \to \int^* A$ which will be required for the interpretation of type theory. This functor is obtained from the isomorphism $\sigma\colon \mathcal{C} \to \mathcal{C}^{\mathrm{op}}$ which extends the usual isomorphism which is the identity on objects and sends an arrow $f\colon a \to b$ in $\mathcal{C}$ to $f^{-1}\colon a \to b$ in $\mathcal{C}^{\mathrm{op}}$ to the setting of ω-groupoids. Informally, $\neg$ simply turns the various triangles from the Grothendieck construction "inside out."

4.1. The dual Grothendieck construction. When a category $\mathcal{C}$ is an ordinary 1-dimensional groupoid, then there exists an isomorphism $\sigma\colon \mathcal{C} \to \mathcal{C}^{\mathrm{op}}$ which is the identity on objects and sends an arrow $f\colon x \to y$ to its inverse $f^{-1} : y \to x$. Now, when $\mathcal{C}$ is an ω-groupoid there is a corresponding functor σ which we now consider. We write $\mathcal{C}^{\mathrm{op}}$ for the ω-category obtained by reversing *only* the 1-cells of $\mathcal{C}$. For example, given a 2-cell $\alpha : f \to g$ in $\mathcal{C}$, $\sigma\alpha : f^{-1} \to g^{-1}$ is defined to be $(g^{-1} *_0 \alpha^{-1} *_0 f^{-1})$. As a diagram:

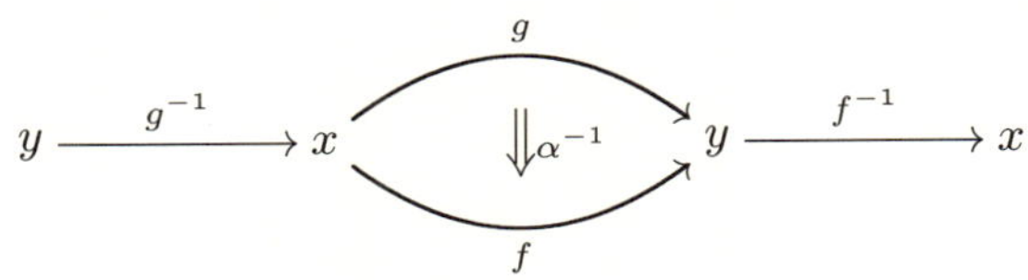

Then, where $\varphi\colon \alpha \to \beta$ is a 3-cell,

$$\sigma(\varphi) := g^{-1} *_0 (\beta^{-1} *_1 \varphi^{-1} *_1 \alpha^{-1}) *_0 f^{-1}.$$

At higher-dimensions the construction is the same, adding a new "inner" block obtained by composing φ^{-1} with the inverses of its boundary maps. The *dual Grothendieck construction* $\int^* A$ of a functor $A\colon \mathcal{C} \to \omega\text{-}\mathbf{Cat}$ can be described as the coend

$$\textstyle\int^* A = \displaystyle\int^x (\mathcal{C} \downarrow \sigma(x)) \times A_x,$$

for $\sigma\colon \mathcal{C}^{\mathrm{op}} \to \mathcal{C}$ as above. In concrete terms, $\int^* A$ has the same 0-cells as $\int A$, but 1-cells $\vec{f}\colon \vec{x} \to \vec{y}$ are given by $f\colon x \to y$ in $\mathcal{C}$ and $f_-\colon y_- \to A_f(x_-)$ in A_x. As with $\int A$, the dual weighted face of such an arrow f_- is $A_f(x_-)$ and the dual weighted face functor

$$\mathcal{C}_1(x, y) \xrightarrow{\check{\eth}^1_{\vec{x},\vec{y}}} A_y$$

is obtained by defining $\check{\eth}^1_{\vec{x},\vec{y}}(\gamma)$ to be $x_-.\gamma$. Inductively, we then have the following:

$(n+1)$ is even. A $(n{+}1)$-cell $\vec{\alpha}\colon \vec{f} \Rightarrow \vec{g}$, with $\vec{f}, \vec{g}\colon \vec{v} \rightrightarrows \vec{w}$, is a pair consisting of a $(n{+}1)$-cell $\alpha\colon f \rightrightarrows g$ in $\mathcal{C}$ and a $(n{+}1)$-cell α_- in A_y as indicated in the following diagram:

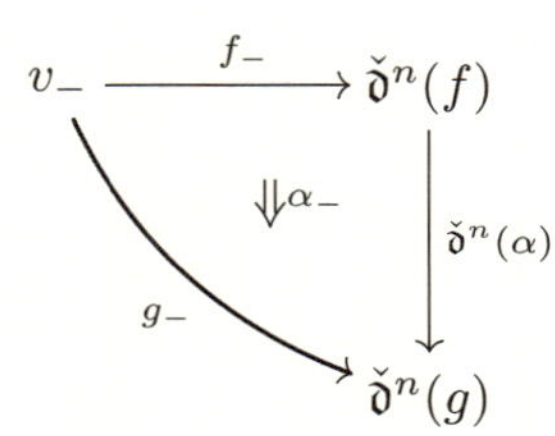

The dual weighted face functor

$$\mathcal{C}_{n+1}(f, g) \xrightarrow{\check{\eth}^{n+1}_{\vec{f},\vec{g}}} (A_y)_n(v_-, \check{\eth}^n(g))$$

is given by defining $\check{\eth}^{n+1}_{\vec{f},\vec{g}}(\gamma)$ to be $\check{\eth}^n(\gamma) *_{(n-1)} f_-$.

$(n+1)$ is odd. On the other hand, when $(n{+}1)$ is odd a $(n{+}1)$-cell $\vec{\alpha}\colon \vec{f} \Rightarrow \vec{g}$ is given by $\alpha\colon f \Rightarrow g$ as above together with a $(n+1)$-cell of A_y as indicated in the following diagram:

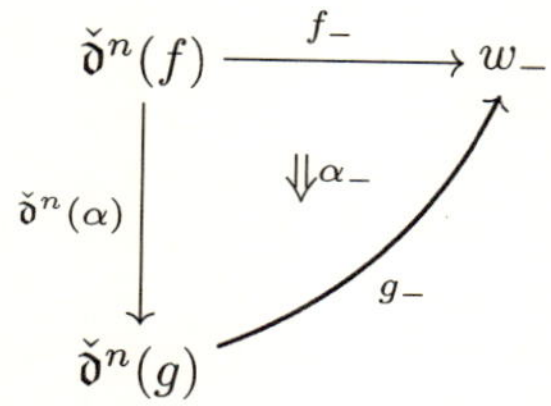

Here the dual weighted face functor

$$\mathcal{C}_{n+1}(f,g) \xrightarrow{\breve{\mathfrak{d}}^{n+1}_{\vec{f},\vec{g}}} (A_y)_n\big(\breve{\mathfrak{d}}^n(f), w_-\big)$$

is obtained by defining $\breve{\mathfrak{d}}^{n+1}_{\vec{f},\vec{g}}(\gamma)$ to be $g_- *_{(n-1)} \breve{\mathfrak{d}}^n(\gamma)$.

In the case where $\mathcal{C}$ is an ω-groupoid there are functors $\bar{\sigma}_x \colon (x \downarrow \mathcal{C}) \to (\mathcal{C}^{\mathrm{op}} \downarrow x)$, for each object x of $\mathcal{C}$, corresponding to σ which send an object $x \to y$ of $(x \downarrow \mathcal{C})$ to its inverse and which act on 1-cells by sending

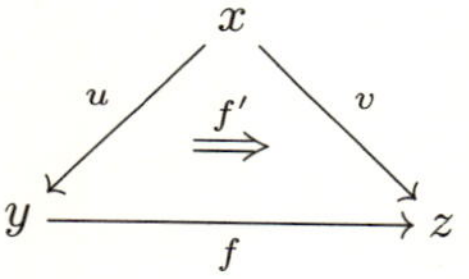

to

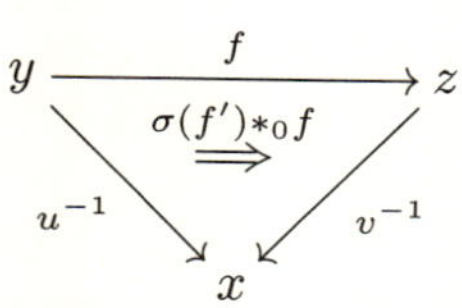

For higher-dimensional cells matters are even more straightforward since a n-cell (α, α'), with $n > 1$, is sent to $(\alpha, \sigma(\alpha') *_0 \alpha)$. The maps $\bar{\sigma}_x$ induce, by the universal property of $\int A$, a canonical functor $\neg\colon \int A \to \int^* A$ such that the squares

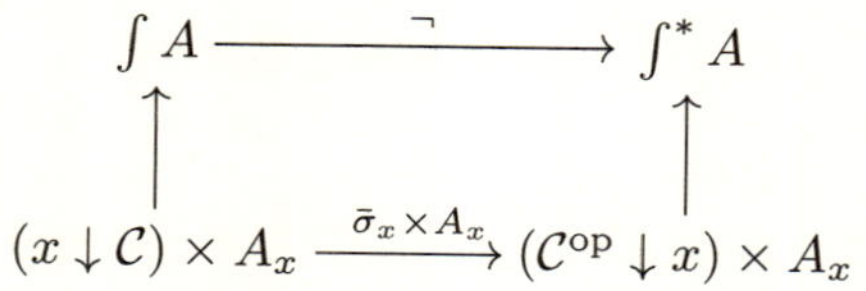

commute, where the vertical maps are the coend inclusions. Consequently, $\neg$ commutes with projections in the sense that the triangle

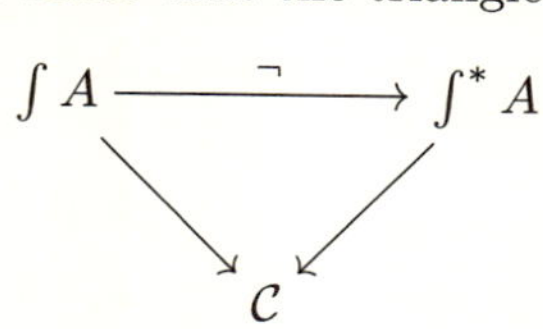

commutes. (This can also be seen from the direct, combinatorial construction of $\neg$ given in [27].) We refer to $\neg$ as the *duality functor* and we will often denote the second component of $\neg(\vec{\varphi})$ by $\neg\varphi_-$. Because the first component of $\neg(\vec{\varphi})$ is φ in all dimensions this should lead to no confusion.

5. The interpretation of identity types

We will now describe the interpretation of type theory using ω-groupoids. Roughly, the interpretation, which generalizes directly the Hofmann – Streicher [9] interpretation using regular 1-dimensional groupoids, is obtained by interpreting dependent types as "indexed ω-categories." However, before going into the details several remarks are in order.

First, whereas in ibid. the entire logical framework is interpreted, we here only interpret the basic form of intensional type theory theory (called $\mathbb{T}_\omega$ in [1,27]). We note, however, that we could just as well have interpreted the entire logical framework in this setting. Secondly, the interpretation we give below can be organized into a (large) comprehension category, or a category with attributes, or a category with families. In this case we believe that the model most naturally can be described as a category with attributes or a category with families [4]. We assume that the reader is familiar with these forms of semantics (all of which are essentially inspired by Lawvere's notion of hyperdoctrine [13]). Because the ideas behind the basic interpretation are not new, we do not go into the full details of the interpretation of the basic syntax. Regarding dependent products and sums we note that, because terms are interpreted (here and in the original groupoid model) as sections of projections $\int A \to \mathcal{C}$ and not as morphisms of split fibrations they correspond, via an equivalence of categories, to 1-cells in the category $[\mathcal{C}, \omega\text{-}\mathbf{Gpd}]_{\text{ps}}$ which has the same objects as $[\mathcal{C}, \omega\text{-}\mathbf{Gpd}]$ but which has as 1-cells (ω-categorical) pseudo natural transformations. Explicitly, the notion of pseudo natural transformation employed here can be extracted from the fact that $[\mathcal{C}, \omega\text{-}\mathbf{Gpd}]_{\text{ps}}$ is equivalent to be the category of algebras and pseudo morphisms of algebras for the 2-monad on $\omega\text{-}\mathbf{Gpd}/\mathcal{C}$ which was mentioned in Section 3. We omit the direct combinatorial description of pseudo natural transformations as it will not be required here and merely mention these facts as they serve to provide some motivation for the constructions given below. As such, where $\pi\colon \int A \to \mathcal{C}$ is a "dependent projection," dependent sums give the left-adjoint to the reindexing functor

$$[\mathcal{C}, \omega\text{-}\mathbf{Gpd}]_{\text{ps}} \xrightarrow{\pi^*} [\textstyle\int A, \omega\text{-}\mathbf{Gpd}]_{\text{ps}}$$

and dependent products give the right-adjoint. For the right-adjoint we emphasize that it is crucial that we are dealing with ω-groupoids since the dual functors from Section 4 are necessary (essentially this is because dependent products are homs and therefore exhibit a certain contravariance). Both of these adjoints are described explicitly in Section 7.1 below. The verification that these give the required adjoints is by a lengthy, but routine, combinatorial argument which we omit. Finally, we mention that the type of natural numbers is to be interpreted as the discrete ω-groupoid of natural numbers as in the original groupoid model.

5.1. Contexts, types and terms. The idea of the interpretation, which should be familiar in light of the discussions in the foregoing chapters, is to regard closed types as ω-groupoids. Explicitly, contexts Γ are interpreted as small ω-groupoids. To begin with, the empty context () is interpreted as the terminal ω-groupoid:

$$[\![()]\!] := \mathbf{1}.$$

Now, given a context Γ together with its interpretation $[\![\Gamma]\!]$ as an ω-groupoid, judgements of the form $\Gamma \vdash A$: type are interpreted as functors

$$[\![\Gamma]\!] \xrightarrow{[\![\Gamma \vdash A:\text{type}]\!]} \omega\text{-}\mathbf{Gpd}.$$

The extended context $(\Gamma, x : A)$ is then interpreted as the ω-groupoid given by applying the Grothendieck construction from Section 3 to the functor in question:

$$[\![\Gamma, x : A]\!] := \int [\![\Gamma \vdash A : \text{type}]\!].$$

A judgement of the form $\Gamma \vdash a : A$ is then interpreted as a section

$$\begin{array}{ccc} [\![\Gamma]\!] & \longrightarrow & \int [\![\Gamma \vdash A : \text{type}]\!] \\ & {\scriptstyle 1_{[\![\Gamma]\!]}}\searrow \quad \swarrow{\scriptstyle \pi} & \\ & [\![\Gamma]\!] & \end{array}$$

of the projection functor.

5.2. Sections of projection functors. Because terms are interpreted as sections of projection functors it will be useful to give a detailed analysis of such sections. A section F of a projection $\pi\colon \int A \to \mathcal{C}$ as indicated in the following diagram:

$$\begin{array}{ccc} \mathcal{C} & \xrightarrow{\quad F \quad} & \int A \\ & {\scriptstyle 1_{\mathcal{C}}}\searrow \quad \swarrow{\scriptstyle \pi} & \\ & \mathcal{C} & \end{array}$$

consists exactly of the following data:

Objects. To each object x of $\mathcal{C}$ there is assigned an object x_F of A_x. I.e., $(x, x_F) = F(x)$.

1-Cells. To an arrow $f\colon x \to y$ in $\mathcal{C}$ there is assigned an arrow $f_F\colon \mathfrak{d}^1_{F(x),F(y)}(f) \to y_F$ of A_y.

$(n+1)$-Cells. When $(n+1)$ is even, there is assigned to an $(n+1)$-cell $\alpha\colon f \Rightarrow g$ an $(n+1)$-cell $\alpha_F\colon f_F \Rightarrow \mathfrak{d}^{n+1}_{F(f),F(g)}(\alpha)$ of A_y. When $(n+1)$ is odd, $\alpha_F\colon \mathfrak{d}^{n+1}_{F(f),F(g)}(\alpha) \Rightarrow g_F$.

Note that such an assignment is made into a map of globular sets by defining $F(\varphi) := (\varphi, \varphi_F)$. These assignments are required to be functorial in the sense of preserving identities and composition. Preservation of composition amounts to the following. Given m-cells, for $m > 0$, ψ and φ in $\mathcal{C}$ such that $(\psi *_0 \varphi)$ is defined, it is required in order for the assignment $(-)_F$ to constitute a section that

$$(\psi *_0 \varphi)_F = \psi_F *_0 A_h(\varphi_F).$$

Assume that the composite $(\psi *_n \varphi)$ is defined and that $t^{m-n-1}\varphi = \beta$ and $s^{m-n-1}\psi = \gamma$. Furthermore, let u and v be the $(n-1)$-cells bounding both φ and ψ. Then it is required that

$$(\psi *_n \varphi)_F = \begin{cases} \big(\psi_F *_{(n-1)} \mathfrak{d}^n_{Fu,Fv}(\beta)\big) *_n \varphi_F & \text{if } (n+1) \text{ is even,} \\ \psi_F *_n \big(\mathfrak{d}^n_{Fu,Fv}(\gamma) *_{(n-1)} \varphi_F\big) & \text{if } (n+1) \text{ is odd.} \end{cases}$$

We also note that given any functor $\sigma\colon \mathcal{D} \to \mathcal{C}$, there exists a functor $\{\sigma\}_A\colon \int(A \circ \sigma) \to \int A$ such that the following diagram is a pullback in ω-**Cat**:

$$\begin{array}{ccc} \int(A\circ\sigma) & \xrightarrow{\{\sigma\}_A} & \int A \\ \downarrow & & \downarrow \\ \mathcal{D} & \xrightarrow[\sigma]{} & \mathcal{C} \end{array}$$

Namely, $\{\sigma\}_A$ sends $\vec{x}$ in $\int A\sigma$ to $(\sigma(x), a)$ and similarly for cells in all dimensions. Consequently, there corresponds to any section F of the projection $\int A \to \mathcal{C}$ a canonical section $F[\sigma]$ of $\int A\sigma \to \mathcal{D}$ for which

$$F \circ \sigma = \{\sigma\}_A \circ F[\sigma].$$

Finally, note that, by taking $\mathcal{D}$ to be $\int A$ itself and σ to be π, we obtain $\int A\pi$ as the pullback of π along itself and there exists a canonical map $\delta_A\colon \int A \to \int A\pi$ induced by the identity functor $1_{\int A}$.

5.3. Substitution and weakening. Suppose we are given an ω-groupoid $\mathcal{C}$ interpreting a context Γ together with $A\colon \mathcal{C} \to \omega$-**Gpd**, $B\colon \int A \to \omega$-**Gpd** and $C\colon \int B \to \omega$-**Gpd** interpreting judgements

$$\Gamma \vdash A : \text{type}, \quad \Gamma, x : A \vdash B(x) : \text{type}, \quad \text{and} \quad \Gamma, x : A, y : B(x) \vdash C(x,y) : \text{type},$$

respectively. Moreover, let a section a interpreting the judgement $\Gamma \vdash a : A$ be given. Then the judgement $\Gamma \vdash B[a/x]$ is interpreted as the composite functor

$$\mathcal{C} \xrightarrow{a} \int A \xrightarrow{B} \omega\text{-}\mathbf{Gpd}.$$

Similarly,

$$[\![\Gamma, y : B(a) \vdash C(a,y) : \text{type}]\!] := C \circ \{a\}_B,$$

in the notation of Section 5.2. Finally, if c is a section of $\int C \to \int B$ interpreting the judgement $\Gamma, x : A, y : B(x) \vdash c(x,y) : C(x,y)$ we define

$$[\![\Gamma, y : B(a) \vdash c(a,y) : C(a,y)]\!] := c[a].$$

Finally, for weakening, we note that when functors $A, B\colon \mathcal{C} \to \omega$-**Gpd** interpret the judgements $\Gamma \vdash A$: type and $\Gamma \vdash B$: type, the "weakened" judgement $\Gamma, x : A \vdash B$: type is interpreted by the composite

$$\int A \xrightarrow{\pi} \mathcal{C} \xrightarrow{B} \omega\text{-}\mathbf{Gpd}.$$

5.4. Identity types. When $A\colon \mathcal{C} \to \omega$-**Gpd** has as its domain an ω-groupoid $\mathcal{C}$, the *identity type* (*for* A) is a functor $I_A\colon \int A\pi \to \omega$-**Gpd** where $\pi\colon \int A \to \mathcal{C}$ is the projection. By definition, $\int A \circ \pi$ is has as objects tuples $\vec{x} = (x, x_-, x_-)$ where x_- and x_+ are themselves objects of A_x. Similarly, n-cells $\vec{f}$ in $\int A\pi$ are tuples (f, f_-, f_+) such that both (f, f_-) and (f, f_+) are n-cells in $\int A$. I.e., the following diagram is a pullback

$$\begin{array}{ccc} \int A\pi & \longrightarrow & \int A \\ \downarrow & & \downarrow{\scriptstyle \pi} \\ \int A & \xrightarrow[\pi]{} & \mathcal{C} \end{array}$$

with the nameless functors the obvious projections. With this in mind, it is straightforward to describe the action of I_A on objects. Namely, $I_A(x, x_-, x_+)$ is defined to be the ω-groupoid $A_x(x_-, x_+)$. Perhaps though, in light of the discussion of the combinatorics of the Grothendieck construction from the previous sections, matters are more complicated in higher dimensions. It is to this task which we now turn.

Remark 5.1. Because, when $\int A\pi$ is involved, we are dealing with two instances of the Grothendieck construction $\int A$ it will be convenient to introduce some notation to describe the various weighted face functors. In particular, because we adopt the convention of notating cells $\vec{f}$ of $\int A\pi$ by (f, f_-, f_+) we will also notate the corresponding weighted face functors accordingly. I.e., we write $\partial^n_-(f)$ and $\partial^n_+(f)$ for the instances of these functors corresponding to the appropriate "negative" and "positive" projections $\int A\pi \to \int A$. When subscripts are necessary we write, e.g., $\partial^n_{\vec{\alpha}\vec{\beta};\xi}$ with $\xi = +, -$. We adopt also a corresponding convention for the "co-weighted face" functors.

5.5. Identity types in dimensions 1 and 2. Given an arrow $\vec{f}\colon \vec{x} \to \vec{y}$ in $\int A\pi$, $I_A(\vec{f})$ is the functor

$$A_x(x_-, x_+) \to A_y(x_-, x_+)$$

which sends any cell γ of $A_x(x_-, x_+)$ to the following composite:

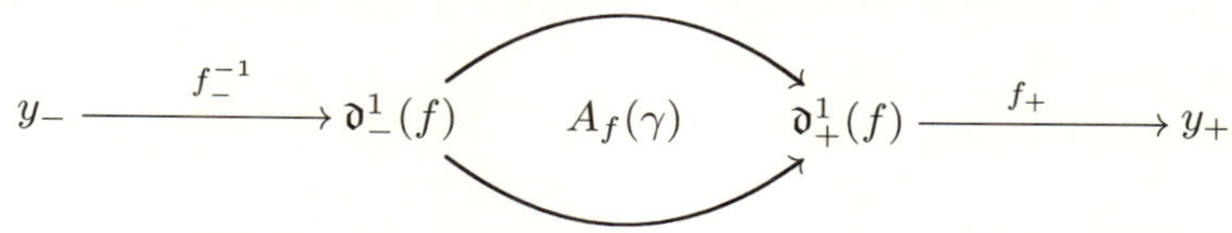

I.e., $I_A(\vec{f})(\gamma)$ is defined to be $(f_+ *_0 A_f(\gamma) *_0 f_-^{-1})$. Already at this stage we have tacitly made use of the dual functor $\neg$ since $\neg f_-$ is f_-^{-1}.

Now, given a 2-cell $\vec{\alpha}\colon \vec{f} \Rightarrow \vec{g}$ we must provide a natural transformation $I_A(\vec{\alpha})$ as indicated in the following diagram:

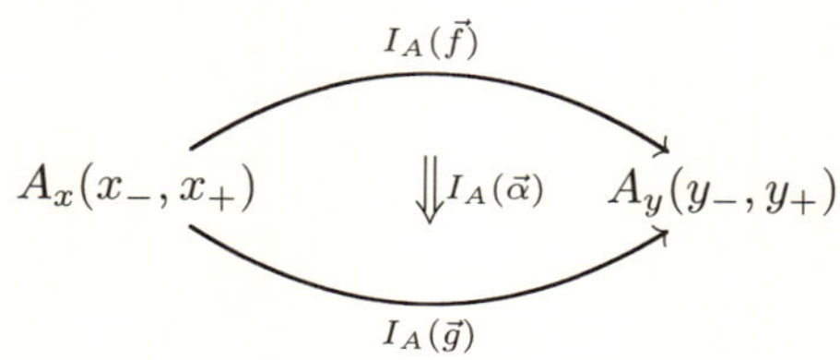

Fixing an object $h\colon x_- \to x_+$ of $A_x(x_-, x_+)$, the component $I_A(\vec{\alpha})_h$ of this transformation at h is described by the composite of the following diagram in A_y:

(5.1) 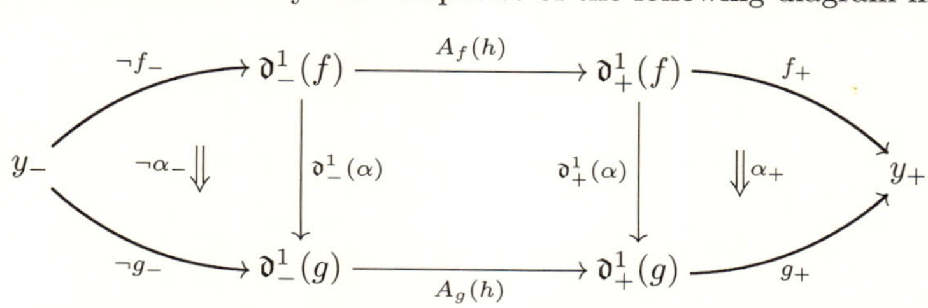

where the middle square commutes (on the nose) by naturality of A_α. Explicitly,

$$I_A(\vec{\alpha})_h := (f_+ *_0 A_g(h) *_0 \neg\alpha_-) *_1 (\alpha_+ *_0 A_f(h) *_0 \neg f_-).$$

With this definition in mind, we now turn to the introduction of some auxiliary functors which will allow us to describe the identity types in higher dimensions.

5.6. Auxiliary functors. Holding an arrow $\vec{f}\colon \vec{x} \to \vec{y}$ of $\int A\pi$ fixed together with an object h of $A_x(x_-, x_+)$ we define functors

$$(A_y)_1(\mathfrak{d}^1_+(f), y_+) \xrightarrow{\Psi^1_{\vec{f},h}} (A_y)_1(y_-, y_+),$$
$$(A_y)_1(y_-, \mathfrak{d}^1_-(f)) \xrightarrow{\check{\Psi}^1_{\vec{f},h}} (A_y)_1(y_-, y_+)$$

by setting

$$\Psi^1_{\vec{f},h}(-) := (- *_0 A_f h *_0 \neg f_-),$$
$$\check{\Psi}^1_{\vec{f},h}(-) := (f_+ *_0 A_f h *_0 -).$$

As usual, we omit either one or both of the subscripts when no confusion will result. The first thing we observe about these functors is that

$$\Psi^1_{\vec{f}}(f_+) = \check{\Psi}^1_{\vec{f}}(\neg f_-). \tag{5.2}$$

The next feature which should be emphasized is that these functors interact in an important way with the usual weighted face functors. In particular, the following diagram (of ω-categories) commutes:

$$\begin{array}{ccc} \mathcal{C}_2(f,g) & \xrightarrow{\check{\mathfrak{d}}^2_{\neg\vec{f},\neg\vec{g};-}} & (A_y)_1(y_-, \check{\mathfrak{d}}^1_-(g)) \\ {\scriptstyle \mathfrak{d}^2_{\vec{f},\vec{g};+}}\downarrow & & \downarrow{\scriptstyle \check{\Psi}^1_{\vec{g},h}} \\ (A_y)_1(\mathfrak{d}^1_+(f), y_+) & \xrightarrow[\Psi^1_{\vec{f},h}]{} & (A_y)_1(y_-, y_+) \end{array} \tag{5.3}$$

To see this, we note that

$$\begin{aligned} \check{\Psi}^1_{\vec{g}}(\check{\mathfrak{d}}^2_-(\gamma)) &= g_+ *_0 A_g h *_0 \mathfrak{d}^1_-(\gamma) *_0 \neg f_- \\ &= g_+ *_0 \mathfrak{d}^1_+(\gamma) *_0 A_f h *_0 \neg f_- \\ &= \Psi^1_{\vec{f}}(\mathfrak{d}^2_+(\gamma)), \end{aligned}$$

where the second equation is by naturality of A_γ. We now observe that, when $\vec{\alpha}$ is as above, the component $I_A(\vec{\alpha})$ at h can be described using these functors as follows:

$$I_A(\vec{\alpha})_h = \check{\Psi}^1_{\vec{g},h}(\neg\alpha_-) *_1 \Psi^1_{\vec{f},h}(\alpha_+)$$

In particular, $I_A(\vec{\alpha})_h$ is obtained by composing

$$\Psi^1_{\vec{f}}(f_+) \xrightarrow{\Psi^1_{\vec{f}}(\alpha_+)} \Psi^1_{\vec{f}}(\mathfrak{d}^2_+(\alpha)) = \check{\Psi}^1_{\vec{g}}(\check{\mathfrak{d}}^2_-(\alpha)) \xrightarrow{\check{\Psi}^1_{\vec{g}}(\neg\alpha_-)} \check{\Psi}^1_{\vec{g}}(\neg g_-) = \Psi^1_{\vec{g}}(g_+). \tag{5.4}$$

As such, we have employed both (5.2) and (5.3) in order to show that the composite defining $I_A(\vec{\alpha})_h$ makes sense. We emphasize this point because it provides the first look at what will be required in higher dimensions.

At the next stage, holding a 2-cell $\vec{\alpha}\colon \vec{f} \to \vec{g}$ and an arrow $h\colon x_- \to x_+$ as above fixed, we define functors

$$(A_y)_2\big(f_+, \mathfrak{d}^2_{\vec{f},\vec{g};+}(\alpha)\big) \xrightarrow{\Psi^2_{\vec{\alpha},h}} (A_y)_2\big(\Psi^1_{\vec{f},h}(f_+), \Psi^1_{\vec{g},h}(g_+)\big),$$
$$(A_y)_2\big(\check{\mathfrak{d}}^2_{\neg\vec{f},\neg\vec{g};-}(\alpha), \neg g_-\big) \xrightarrow{\check{\Psi}^2_{\vec{\alpha},h}} (A_y)_2\big(\Psi^1_{\vec{f},h}(f_+), \Psi^1_{\vec{g},h}(g_+)\big)$$

as follows

$$\Psi^2_{\vec{\alpha},h}(-) := \check{\Psi}^1_{\vec{g},h}(\neg\alpha_-) *_1 \Psi^1_{\vec{f},h}(-),$$
$$\check{\Psi}^2_{\vec{\alpha},h}(-) := \check{\Psi}^1_{\vec{g},h}(-) *_1 \Psi^1_{\vec{f},h}(\alpha_+).$$

The motivation for these definitions can perhaps best be seen in consultation with (5.1). It follows, using the same reasoning from (5.4), that these functors are well-defined and possess the appropriate boundaries. An immediate consequence of the definition is that

$$\Psi^2_{\vec{\alpha},h}(\alpha_+) = \check{\Psi}^2_{\vec{\alpha},h}(\neg\alpha_-).$$

Moreover, (5.3) also generalizes to dimension 2 to give:

$$\begin{array}{ccc}
\mathcal{C}_3(\alpha,\beta) & \xrightarrow{\check{\mathfrak{d}}^3_{\neg\vec{\alpha},\neg\vec{\beta};-}} & (A_y)_2\big(\check{\mathfrak{d}}^2_-(\alpha), \neg g_-\big) \\
{\scriptstyle \mathfrak{d}^3_{\vec{\alpha},\vec{\beta};+}}\downarrow & & \downarrow{\scriptstyle \check{\Psi}^2_{\vec{\alpha},h}} \\
(A_y)_2\big(f_+, \mathfrak{d}^2_+(\beta)\big) & \xrightarrow[\Psi^2_{\vec{\beta},h}]{} & (A_y)_2\big(\Psi^1_{\vec{f},h}(f_+), \Psi^1_{\vec{g},h}(g_+)\big)
\end{array}$$

when $\vec{\alpha}, \vec{\beta}\colon \vec{f} \rightrightarrows \vec{g}$ are fixed 2-cells. To see that the equation holds we reason as follows:

$$\begin{aligned}
\Psi^2_{\vec{\beta},h}\big(\mathfrak{d}^3_+(\gamma)\big) &= \check{\Psi}^1_{\vec{g},h}(\neg\beta_-) *_1 \Psi^1_{\vec{f},h}\big(\mathfrak{d}^3_+(\gamma)\big) \\
&= \check{\Psi}^1_{\vec{g},h}(\neg\beta_-) *_1 \Psi^1_{\vec{f},h}\big(\mathfrak{d}^2_+(\gamma)\big) *_1 \Psi^1_{\vec{f},h}(\alpha_+) \\
&= \check{\Psi}^1_{\vec{g},h}(\neg\beta_-) *_1 \check{\Psi}^1_{\vec{g},h}\big(\check{\mathfrak{d}}^2_-(\gamma)\big) *_1 \Psi^1_{\vec{f},h}(\alpha_+) \\
&= \check{\Psi}^1_{\vec{g},h}\big(\neg\beta_- *_1 \check{\mathfrak{d}}^2_-(\gamma)\big) *_1 \Psi^1_{\vec{f},h}(\alpha_+) \\
&= \check{\Psi}^1_{\vec{g},h}\big(\check{\mathfrak{d}}^3_-(\gamma)\big) *_1 \Psi^1_{\vec{f},h}(\alpha_+) \\
&= \check{\Psi}^2_{\vec{\alpha},h}\big(\check{\mathfrak{d}}^3_-(\gamma)\big),
\end{aligned}$$

where the third equation is by (5.3). We will now show that this construction can be generalized to all dimensions $(n+1)$ with $n \geq 2$. In particular we will prove that at each stage $(n+1)$, for every $(n+1)$-cell $\vec{\varphi}\colon \vec{\alpha} \to \vec{\beta}$ and arrow $h\colon x_- \to x_+$, there exist functors $\Psi^{n+1}_{\vec{\varphi},h}$ and $\check{\Psi}^{n+1}_{\vec{\varphi},h}$ satisfying the following conditions:

(1) When $(n+1)$ is odd,

$$(A_y)_{n+1}\big(\mathfrak{d}^{n+1}_{\vec{\alpha},\vec{\beta};+}(\varphi), \beta_+\big) \xrightarrow{\Psi^{n+1}_{\vec{\varphi},h}} (A_y)_{n+1}\big(\Psi^n_{\vec{\alpha},h}(\alpha_+), \Psi^n_{\vec{\beta},h}(\beta_+)\big),$$
$$(A_y)_{n+1}\big(\neg\alpha_-, \check{\mathfrak{d}}^{n+1}_{\neg\vec{\alpha},\neg\vec{\beta};-}(\varphi)\big) \xrightarrow{\check{\Psi}^{n+1}_{\vec{\varphi},h}} (A_y)_{n+1}\big(\Psi^n_{\vec{\alpha},h}(\alpha_+), \Psi^n_{\vec{\beta},h}(\beta_+)\big).$$

Similarly, when $(n+1)$ is even,

$$(A_y)_{n+1}\big(\alpha_+, \eth^{n+1}_{\vec{\alpha},\vec{\beta};+}(\varphi)\big) \xrightarrow{\Psi^{n+1}_{\vec{\varphi},h}} (A_y)_{n+1}\big(\Psi^n_{\vec{\alpha},h}(\alpha_+), \Psi^n_{\vec{\beta},h}(\beta_+)\big),$$

$$(A_y)_{n+1}\big(\check{\eth}^{n+1}_{\neg\vec{\alpha},\neg\vec{\beta};-}(\varphi), \neg\beta_-\big) \xrightarrow{\check{\Psi}^{n+1}_{\vec{\varphi},h}} (A_y)_{n+1}\big(\Psi^n_{\vec{\alpha},h}(\alpha_+), \Psi^n_{\vec{\beta},h}(\beta_+)\big)$$

(2) When $\vec{\varphi}$ is an $(n+1)$-cell as above,

$$\Psi^{n+1}_{\vec{\varphi},h}(\varphi_+) = \check{\Psi}^{n+1}_{\vec{\varphi},h}(\neg\varphi_-). \tag{5.5}$$

(3) Let parallel $(n+1)$-cells $\vec{\varphi}, \vec{\psi} \colon \vec{\alpha} \rightrightarrows \vec{\beta}$ be given. When $(n+1)$ is odd, the following diagram commutes:

$$\begin{array}{ccc} \mathcal{C}_{n+2}(\varphi,\psi) & \xrightarrow{\check{\eth}^{n+2}_{\neg\vec{\varphi},\neg\vec{\psi};-}} & (A_y)_{n+1}\big(\neg\alpha_-, \check{\eth}^{n+1}_-(\psi)\big) \\ {\scriptstyle \eth^{n+2}_{\vec{\varphi},\vec{\psi};+}}\big\downarrow & & \big\downarrow{\scriptstyle \check{\Psi}^{n+1}_{\vec{\psi},h}} \\ (A_y)_{n+1}\big(\eth^{n+1}_+(\varphi), \beta_+\big) & \xrightarrow[\Psi^{n+1}_{\vec{\varphi},h}]{} & (A_y)_{n+1}\big(\Psi^{n+1}_{\vec{\varphi},h}(\alpha_+), \Psi^{n+1}_{\vec{\psi},h}(\beta_+)\big) \end{array} \tag{5.6}$$

And, when $(n+1)$ is even,

$$\begin{array}{ccc} \mathcal{C}_{n+1}(\varphi,\psi) & \xrightarrow{\check{\eth}^{n+2}_{\neg\vec{\varphi},\neg\vec{\psi};-}} & (A_y)_{n+1}\big(\check{\eth}^{n+1}_-(\varphi), \neg\beta_-\big) \\ {\scriptstyle \eth^{n+2}_{\vec{\varphi},\vec{\psi};+}}\big\downarrow & & \big\downarrow{\scriptstyle \check{\Psi}^{n+1}_{\vec{\varphi},h}} \\ (A_y)_{n+1}\big(\alpha_+, \eth^{n+1}_+(\psi)\big) & \xrightarrow[\Psi^{n+1}_{\vec{\psi},h}]{} & (A_y)_{n+1}\big(\Psi^n_{\vec{\alpha},h}(\alpha_+), \Psi^n_{\vec{\beta},h}(\beta_+)\big) \end{array} \tag{5.7}$$

commutes.

Lemma 5.2. *The conditions* (1) – (3) *described above are satisfied when, for* $\vec{\varphi} \colon \vec{\alpha} \to \vec{\beta}$ *an* $(n+1)$*-cell of* $\int A\pi$ *and* $h \colon x_- \to x_+$ *as above, the functors* $\Psi^{n+1}_{\vec{\varphi},h}$ *and* $\check{\Psi}^{n+1}_{\vec{\varphi},h}$ *are defined as follows*:

$$\Psi^{n+1}_{\vec{\varphi},h}(-) := \begin{cases} \check{\Psi}^n_{\vec{\beta},h}(\neg\varphi_-) *_n \Psi^n_{\vec{\alpha},h}(-) & \text{if } (n+1) \text{ is even,} \\ \Psi^n_{\vec{\beta},h}(-) *_n \check{\Psi}^n_{\vec{\alpha},h}(\neg\varphi_-) & \text{if } (n+1) \text{ is odd;} \end{cases}$$

and

$$\check{\Psi}^{n+1}_{\vec{\varphi},h}(-) := \begin{cases} \check{\Psi}^n_{\vec{\beta},h}(-) *_n \Psi^n_{\vec{\alpha},h}(\varphi_+) & \text{if } (n+1) \text{ is even,} \\ \Psi^n_{\vec{\beta},h}(\varphi_+) *_n \check{\Psi}^n_{\vec{\alpha},h}(-) & \text{if } (n+1) \text{ is odd.} \end{cases}$$

PROOF. We give the proof in the case where $(n+1)$ is odd as the case where it is even is essentially dual. First, to see that $\Psi^{n+1}_{\vec{\varphi}}$ is well defined and possesses the source and target as stated in condition (1) above, suppose we are given a m-cell ζ of $(A_y)_{n+1}(\eth^{n+1}_+\varphi, \beta_+)$. Then, ζ is a $(m+1)$-cell of $(A_y)_n(f_+, \eth^n_+\beta)$ where

$\vec{\alpha}, \vec{\beta}\colon \vec{f} \rightrightarrows \vec{g}$. As such, we may apply $\Psi^n_{\vec{\beta}}$ to obtain

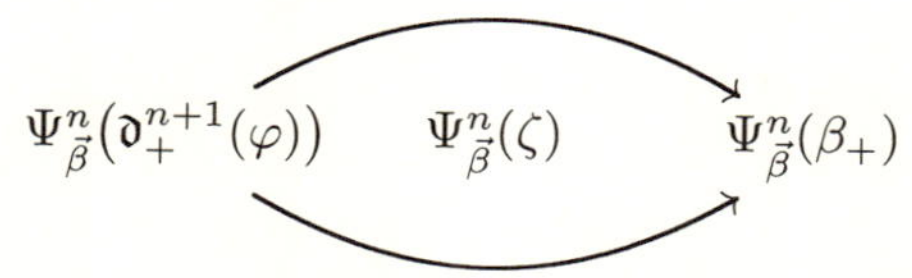

By definition of $\neg\vec{\varphi}$, we have also $\neg\varphi_-\colon \neg\alpha_- \to \check{\mathfrak{d}}^{n+1}_-(\varphi)$. By the induction hypothesis,

$$\check{\Psi}^n_{\vec{\alpha}}(\check{\mathfrak{d}}^{n+1}_-(\varphi)) = \Psi^n_{\vec{\beta}}(\mathfrak{d}^{n+1}_+(\varphi)),$$

and therefore applying $\check{\Psi}^n_{\vec{\alpha}}$ to $\neg\varphi_-$ yields

$$\Psi^n_{\vec{\alpha}}(\alpha_+) = \check{\Psi}^n_{\vec{\alpha}}(\neg\alpha_-) \xrightarrow{\check{\Psi}^n_{\vec{\alpha}}(\neg\varphi_-)} \Psi^n_{\vec{\beta}}(\mathfrak{d}^{n+1}_+(\varphi)).$$

Here the equation is by the induction hypothesis. As such, the composite

$$\Psi^{n+1}_{\vec{\varphi}}(\zeta) := \Psi^n_{\vec{\beta}}(\zeta) *_n \check{\Psi}^n_{\vec{\alpha}}(\neg\varphi_-)$$

is defined and possesses the correct boundary. A similar argument shows that $\check{\Psi}^{n+1}_{\vec{\varphi}}$ is well-defined with the appropriate boundary. Note also that, with these definitions, condition (2) from above is trivially satisfied.

Finally, to see that (3) is satisfied we note that, when $\vec{\varphi}$ and $\vec{\psi}$ are parallel $(n+1)$-cells as above and γ is a cell of $\mathcal{C}_{n+2}(\varphi, \psi)$,

$$\begin{aligned}
\Psi^{n+1}_{\vec{\varphi}}(\mathfrak{d}^{n+2}_+(\gamma)) &= \Psi^n_{\vec{\beta}}(\mathfrak{d}^{n+2}_+(\gamma)) *_n \check{\Psi}^n_{\vec{\alpha}}(\neg\varphi_-) \\
&= \Psi^n_{\vec{\beta}}(\psi_+ *_n \mathfrak{d}^{n+1}_+(\gamma)) *_n \check{\Psi}^n_{\vec{\alpha}}(\neg\varphi_-) \\
&= \Psi^n_{\vec{\beta}}(\psi_+) *_n \Psi^n_{\vec{\beta}}(\mathfrak{d}^{n+1}_+(\gamma)) *_n \check{\Psi}^n_{\vec{\alpha}}(\neg\varphi_-) \\
&= \Psi^n_{\vec{\beta}}(\psi_+) *_n \check{\Psi}^n_{\vec{\alpha}}(\check{\mathfrak{d}}^{n+1}_-(\gamma)) *_n \check{\Psi}^n_{\vec{\alpha}}(\neg\varphi_-) \\
&= \Psi^n_{\vec{\beta}}(\psi_+) *_n \check{\Psi}^n_{\vec{\alpha}}(\check{\mathfrak{d}}^{n+2}_-(\gamma)) \\
&= \check{\Psi}^{n+1}_{\vec{\psi}}(\check{\mathfrak{d}}^{n+2}_-(\gamma)),
\end{aligned}$$

where the fourth equation is by the induction hypothesis. □

5.7. Definition of the identity types. With the functors Ψ^n and $\check{\Psi}^n$ at our disposal it is possible to give a very efficient description of the "identity type" functor $I_A\colon \int A\pi \to \omega$-**Gpd**. In particular, the official definition of I_A in all dimensions is as follows:

Objects. $I_A(x, x_-, x_+)$ is given by the ω-groupoid $A_x(x_-, x_+)$.

1-Cells. Given $\vec{f}\colon \vec{x} \to \vec{y}$, the functor $I_A(\vec{f})\colon I(\vec{x}) \to I(\vec{y})$ is defined by setting

$$I_A(\vec{f})(\gamma) := f_+ *_0 A_f(\gamma) *_0 \neg f_-,$$

for any m-cell γ of $A_x(x_-, x_+)$.

2-Cells. A 2-cell $\vec{\alpha}\colon \vec{f} \Rightarrow \vec{g}$ is sent to the natural transformation $I_A(\vec{\alpha})$ which is defined, for an object $h\colon x_- \to x_+$ of $A_x(x_-, x_+)$, as follows:

$$I_A(\vec{\alpha})_h := \Psi^2_{\vec{\alpha},h}(\alpha_+) = \check{\Psi}^2_{\vec{\alpha},h}(\neg\alpha_-).$$

That $I_A(\vec{\alpha})_h$ possesses the correct domain and codomain is an immediate consequence of the results of Section 5.6.

$(n+1)$-Cells. In general, given an $(n+1)$-cell $\vec{\varphi}\colon \vec{\alpha} \Rightarrow \vec{\beta}$ in $\int A\pi$ and $h\colon x_- \to x_+$, we define

$$I_A(\vec{\varphi})_h := \Psi^{(n+1)}_{\vec{\varphi},h}(\varphi_+) = \check{\Psi}^{(n+1)}_{\vec{\varphi},h}(\neg\varphi_-).$$

Again, that $I_A(\vec{\varphi})_h$ possesses the correct domain and codomain follows directly from the definition of I_A at lower dimensional cells together with the results of Section 5.6.

It remains only to verify that I_A is functorial. To this end, we first prove that the data given in the definition are of the appropriate kinds. E.g., that $I(\vec{\alpha})$ is a natural transformation, etc.

Lemma 5.3. *As defined above, when $\vec{\alpha}\colon \vec{f} \Rightarrow \vec{g}$ is a 2-cell of $\int A\pi$, $I_A(\vec{\alpha})$ is an ω-natural transformation.*

Proof. Explicitly, we must show that, for any m-cell γ of $A_x(x_-, x_+)$ with $m > 0$ such that $s^m(\gamma) = h$ and $t^m(\gamma) = k$, the following "schematic" diagram commutes:

$$\begin{array}{ccc} I_A(\vec{f})(h) & \xrightarrow{I_A(\vec{\alpha})_h} & I_A(\vec{g})(h) \\ \Big\downarrow I_A(\vec{f})(\gamma) \Big\downarrow & & \Big\downarrow I_A(\vec{g})(\gamma) \Big\downarrow \\ I_A(\vec{f})(k) & \xrightarrow[I_A(\vec{\alpha})_k]{} & I_A(\vec{g})(k) \end{array}$$

in $A_y(y_-, y_+)$. I.e., we must prove that

$$I_A(\vec{\alpha})_k *_1 I_A(\vec{f})(\gamma) = I_A(\vec{g})(\gamma) *_1 I_A(\vec{\alpha})_h. \tag{5.8}$$

By definition of $\Psi^2_{\vec{\alpha},h}(\alpha_+)$ and interchange it follows that the right-hand side of (5.8) is equal to

$$(g_+ *_0 A_g(\gamma) *_0 \neg\alpha_-) *_1 \Psi^1_{\vec{f},h}(\alpha_+).$$

Because A_α is itself a transformation $A_f \Rightarrow A_g$ we obtain

$$\mathfrak{d}^1_+(\alpha) *_0 A_f(\gamma) = A_g(\gamma) *_0 \mathfrak{d}^1_-(\alpha).$$

Thus,

$$\begin{aligned} A_g(\gamma) *_0 \neg\alpha_- &= \big(A_g(k) *_1 A_g(\gamma)\big) *_0 \big(\neg\alpha_- *_1 \check{\mathfrak{d}}^2_-(\alpha)\big) \\ &= \big(A_g(k) *_0 \neg\alpha_-\big) *_1 \big(A_g(\gamma) *_0 \check{\mathfrak{d}}^2_-(\alpha)\big) \\ &= \big(A_g(k) *_0 \neg\alpha_-\big) *_1 \big(\mathfrak{d}^1_+(\alpha) *_0 A_f(\gamma) *_0 f_-^{-1}\big). \end{aligned}$$

Thus, the right-hand side of (5.8) is equal to

$$\check{\Psi}^1_{\vec{g},k}(\neg\alpha_-) *_1 \big(\mathfrak{d}^2_+(\alpha) *_0 A_f(\gamma) *_0 f_-^{-1}\big) *_1 \Psi^1_{\vec{f},h}(\alpha_+). \tag{5.9}$$

Moreover, the interchange and unit laws yield

$$\big(\mathfrak{d}^2_+(\alpha) *_0 A_f(\gamma) *_0 f_-^{-1}\big) *_1 \Psi^1_{\vec{\alpha},h}(\alpha_+) = \alpha_+ *_0 A_f(\gamma) *_0 f_-^{-1}.$$

Thus, the right-hand side of (5.8) is equal to

$$\check{\Psi}^1_{\vec{\alpha},k}(\neg\alpha_-) *_1 \big(\alpha_+ *_0 A_f(\gamma) *_0 f_-^{-1}\big) = I_A(\vec{\alpha})_k *_1 I_A(\vec{f})(\gamma),$$

as required. □

A similar argument yields the following fact:

Lemma 5.4. *Given parallel n-cells $\vec{\alpha}$ and $\vec{\beta}$ in $\int A\pi$ bounded by 1-cells $\vec{f}, \vec{g} \colon \vec{x} \rightrightarrows \vec{y}$ together with a $(n+1)$-cell $\vec{\varphi} \colon \vec{\alpha} \Rightarrow \vec{\beta}$, $I_A(\vec{\varphi})$, as defined above, is a modification $I_A(\vec{\alpha}) \Rightarrow I_A(\vec{\beta})$.*

Proposition 5.5. *As defined, I_A is a functor $\int(A \circ \pi) \to \omega$-**Gpd**.*

Proof. First we consider the case of vertical composition. Let p-cells $\vec{\varphi}$ and $\vec{\psi}$ be given, for $p \geq m \geq n+1 > 1$, which are bounded by 0-cells $\vec{x}$ and $\vec{y}$ and by n-cells $\vec{f}$, $\vec{g}$ and $\vec{h}$ as indicated in the following diagram:

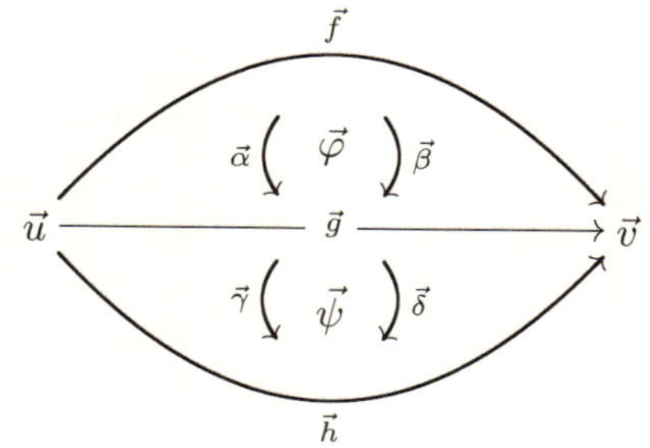

Then, for any object $k \colon x_- \to x_+$ of $A_x(x_-, x_+)$, we will prove by induction on m the stronger fact that when m is odd

$$\Psi^m_{s^{p-m}(\vec{\psi} *_n \vec{\varphi}),k}\big((\vec{\psi} *_n \vec{\varphi})_+\big) = \Psi^m_{s^{(p-m)}\vec{\psi},k}(\psi_+) *_n \Psi^m_{s^{(p-m)}\vec{\varphi},k}(\varphi_+),$$

$$\check{\Psi}^m_{t^{p-m}(\vec{\psi} *_n \vec{\varphi}),k}\big(\neg(\vec{\psi} *_n \vec{\varphi})_-\big) = \check{\Psi}^m_{t^{(p-m)}\vec{\psi},k}(\neg\psi_-) *_n \check{\Psi}^m_{t^{(p-m)}\vec{\varphi},k}(\neg\varphi_-);$$

and

$$\Psi^m_{t^{p-m}(\vec{\psi} *_n \vec{\varphi}),k}\big((\vec{\psi} *_n \vec{\varphi})_+\big) = \Psi^m_{t^{(p-m)}\vec{\psi},k}(\psi_+) *_n \Psi^m_{t^{(p-m)}\vec{\varphi},k}(\varphi_+),$$

$$\check{\Psi}^m_{s^{p-m}(\vec{\psi} *_n \vec{\varphi}),k}\big(\neg(\vec{\psi} *_n \vec{\varphi})_-\big) = \check{\Psi}^m_{s^{(p-m)}\vec{\psi},k}(\neg\psi_-) *_n \check{\Psi}^m_{s^{(p-m)}\vec{\varphi},k}(\neg\varphi_-);$$

when m is even.

First, assume $m = n+1$ is even. We also assume that $n+1 > 2$ since the case where $n+1 = 2$ is a straightforward calculation (using ideas essentially the same as those used here). Then

$$\Psi^{n+1}_{\vec{\delta} *_n \vec{\beta}}\big((\vec{\psi} *_n \vec{\varphi})_+\big) = \check{\Psi}^n_{\vec{h}}\big(\neg(\vec{\delta} *_n \vec{\beta})_-\big) *_n \Psi^n_{\vec{f}}\big((\vec{\psi} *_n \vec{\varphi})_+\big)$$

And this is equal, by definition of composition, to

$$\check{\Psi}^n_{\vec{h}}\big(\neg\delta_- *_n (\check{\eth}^n_-(\delta) *_{(n-1)} \neg\beta_-)\big) *_n \Psi^n_{\vec{f}}\big((\psi_+ *_{(n-1)} \eth^n_+(\beta)) *_n \varphi_+\big) \tag{5.10}$$

Now we will investigate in more detail each of the larger terms in this composite. First:

$$\check{\Psi}^n_{\vec{h}}\big(\neg\delta_- *_n (\check{\eth}^n_-(\delta) *_{(n-1)} \neg\beta_-)\big) = \check{\Psi}^n_{\vec{h}}(\neg\delta_-) *_n \check{\Psi}^n_{\vec{h}}\big(\check{\eth}^n_-(\delta) *_{(n-1)} \neg\beta_-\big).$$

By definition of $\check{\Psi}^n$ and functoriality this is equal to

$$\check{\Psi}^n_{\vec{h}}(\neg\delta_-) *_n \big(\Psi^{n-1}_{\vec{v}}(h_+) *_{(n-1)} \check{\Psi}^{n-1}_{\vec{u}}(\check{\eth}^n_-(\delta)) *_{(n-1)} \check{\Psi}^{n-1}_{\vec{u}}(\neg\beta_-)\big),$$

which by (5.7) is equal to:

$$\begin{aligned}\check{\Psi}^n_{\vec{h}}(\neg\delta_-) *_n \big(\Psi^{n-1}_{\vec{v}}(h_+) *_{(n-1)} \Psi^{n-1}_{\vec{v}}(\eth^n_+(\delta)) *_{(n-1)} \check{\Psi}^{n-1}_{\vec{u}}(\neg\beta_-)\big)\\ = \check{\Psi}^n_{\vec{h}}(\neg\delta_-) *_n \big(\Psi^{n-1}_{\vec{v}}(\eth^{n+1}_+(\delta)) *_{(n-1)} \check{\Psi}^{n-1}_{\vec{u}}(\neg\beta_-)\big)\end{aligned}$$

Similarly, the other half of (5.10) is equal to

$$\big(\Psi^{n-1}_{\vec{v}}(\psi_+) *_{(n-1)} \check{\Psi}^{n-1}_{\vec{u}}(\check{\eth}^{n+1}_{-}(\beta))\big) *_n \Psi^{n}_{\vec{f}}(\varphi_+).$$

By these observations and a routine calculation it follows that (5.10) is equal to

$$\check{\Psi}^{n}_{\vec{h}}(\neg\delta_-) *_n \big(\Psi^{n-1}_{\vec{v}}(\psi_+) *_{(n-1)} \check{\Psi}^{n-1}_{\vec{u}}(\neg\beta_-)\big) *_n \Psi^{n}_{\vec{f}}(\varphi_+).$$

Finally, using the unit and interchange laws this is seen to be the same as $\Psi^{n+1}_{\vec{\delta}}(\psi_+)$ $*_n \Psi^{n+1}_{\vec{\beta}}(\varphi_+)$. The base cases where $n+1$ is odd are dual and the induction steps are trivial. Thus, I_A is a functor. $\square$

6. Reflexivity and elimination terms

In this section we define the functors which will interpret reflexivity and elimination terms. As in [9] we will interpret terms as sections of the projection map $\int A \to \mathcal{C}$ associated to the functor A which interprets their type. We begin by summarizing some of the basic facts about such sections and the related structures resulting from the Grothendieck construction.

6.1. Reflexivity terms. We end this section by describing briefly the "reflexivity term" associated to a functor $A\colon \mathcal{C} \to \omega$-**Gpd**. By definition, the reflexivity term should be a section r_A:

(6.1)
$$\begin{array}{ccc} \int A & \xrightarrow{\;r_A\;} & \int (I_A \circ \delta_A) \\ & {\scriptstyle 1_{\int A}}\searrow \quad \swarrow & \\ & \int A & \end{array}$$

where δ_A is as in Section 5.2. However, we will require the explicit description of this map and we will also introduce some notation related to this map. For objects, given an object $\vec{x}$ of $\int A$, we must provide an object $\vec{x}_r$ of $I_A(x, x_-, x_-)$. I.e., $\vec{x}_r$ should be an arrow $x_- \to x_-$ in A_x. We define $\vec{x}_r$ to be the identity x_-. (Note that here and throughout we omit the identity maps and write x_- instead of $i(x_-)$.) Next, given an arrow $\vec{f}\colon \vec{x} \to \vec{y}$, we need to provide an arrow

$$(\partial I)^1_{r(\vec{x}),r(\vec{y})}(\vec{f}) \xrightarrow{\vec{f}_r} \vec{y}_r = y_-$$

in A_y, where $(\partial I)^n$ denotes the weighted face functor for $\int I_A$ (and so, in this case, also $\int I_A\delta_A$). But, by definition, $(\partial I)^1(\vec{f})$ is just y_- and we therefore define $\vec{f}_r$ to be y_-. Indeed, at *every* dimension $n \geq 1$, when $\vec{\varphi}$ is a n-cell of $\int A$ bounded by objects $\vec{x}$ and $\vec{y}$, we define $\vec{\varphi}_r$ to be y_-. Alternatively, the map r_A may be constructed, as the anonymous referee was kind enough to mention to us, as the canonical section induced by the fact that

$$\begin{array}{ccc} \int (I_A \circ \delta_A) & \longrightarrow & \int I_A \\ \big\downarrow & & \big\downarrow \\ \int A & \xrightarrow[\delta_A]{} & \int (A \circ \pi) \end{array}$$

is a pullback. Note that r is constant y_- in all dimensions $n \geq 1$ since $(\partial I)^1_{r(\vec{x}),r(\vec{y})}(\vec{\varphi})$ is equal to y_- for any $\vec{\varphi}$ in $\int A$ bounded by $\vec{x}$ and $\vec{y}$. As such, we define $\vec{\varphi}_r$ to be

y_- in all dimensions $n \geq 1$. With this definition functoriality of r is trivial and we have proved:

Lemma 6.1. *Given $A\colon \mathcal{C} \to \omega$-**Gpd** as above, the assignment r defined above induces a section r_A as indicated in* (6.1).

6.2. Setting up the construction of elimination terms. Turning now to elimination terms, suppose we are given a functor $D\colon \int I_A \to \omega$-**Gpd** together with a section $d\colon \int A \to \int D\{\delta_A\}_{I_A} r_A$ of the projection $\int D\{\delta_A\}_{I_A} r_A \to \int A$. We would like to prove that this extends to a section J of $\int D \to \int I_A$. We begin by fixing notation.

As we are dealing with multiple cases of the Grothendieck construction it will be convenient to introduce some notation to deal with the different weighted face functions which occur. First, we denote by $\Theta^n(-)$ the weighted face functor for $\int D$. As usual, we denote by $\mathfrak{d}^n_-$ and $\mathfrak{d}^n_+$ the functors for the two projections of $\int I_A$. Finally, we denote by $\widetilde{\Theta}^n$ the weighted face functor for $\int D \circ \{\delta_A\}_{I_A} \circ r_A$.

Next, we observe that there is an endofunctor $\downarrow(-)\colon \int I_A \to \int I_A$ defined as the following composite:

$$\int I_A \xrightarrow{\pi_0} \int A \xrightarrow{r_A} \int I_A \circ \delta_A \to \int I_A.$$

I.e., $\downarrow$ sends an object $\vec{x} = (x, x_-, x_+, x_\to)$ to (x, x_-, x_-, x_-) and similarly for higher-dimensional cells. Here, as throughout, we omit mention of identity arrows. I.e., writing out identities we have that $\downarrow \vec{x}$ is $(x, x_-, x_-, i(x_-))$ or $(x, x_-, x_-, 1_{x_-})$.

We will often be concerned with the situation where we consider, given $\vec{x}$ an object of $\int I_A$, the restriction (x, x_-) of $\vec{x}$ to $\int A$. Rather than write $\pi_0(\vec{x})$ every time for this object we instead denote this pair by $\check{x}$. Similarly, $\check{\gamma}$ denotes $\pi_0(\vec{\gamma})$ for general n-cells $\vec{\gamma}$ of $\int I_A$.

6.3. Naturality cells. The construction of the elimination terms is rather technical and proceeds in several stages. First, we describe "naturality cells" which exhibit what amounts to a (suitable notion of) pseudo natural transformation ε_- from $\downarrow(-)$ to the identity $1_{\int I_A}$. The construction proceeds by induction on dimension as usual.

Dimension 0. First, in dimension 0, given $\vec{x}$ in $\int I_A$, we define an arrow $\varepsilon_{\vec{x}}\colon \downarrow(\vec{x}) \to \vec{x}$ as follows:

$$\varepsilon_{\vec{x}} := (x, x_-, x_\to, x_\to),$$

where $\vec{x} = (x, x_-, x_+, x_\to)$. Next, holding 0-cells $\vec{x}$ and $\vec{y}$ of $\int I_A$ fixed we define functors

$$\left(\int I_A\right)_1 (\vec{x}, \vec{y}) \xrightarrow{\nabla^1_{\vec{x},\vec{y};\xi}} \left(\int I_A\right)_1 (\downarrow \vec{x}, \vec{y})$$

for $\xi = -, +$ as follows:

$$\nabla^1_{\vec{x},\vec{y};\xi} := \begin{cases} (- *_0 \varepsilon_{\vec{x}}) & \text{if } \xi = -, \\ (\varepsilon_{\vec{y}} *_0 \downarrow (-)) & \text{if } \xi = +. \end{cases}$$

As usual, we omit the subscripts $\vec{x}$ and $\vec{y}$ when these are understood.

Note that with these definitions, when $\vec{\varphi}$ is any m-cell, with $m \geq 1$,

$$\nabla^1_-(\vec{\varphi}) = (\varphi, \varphi_-, \varphi_+ *_0 A_f(x_\to), \varphi_\to), \tag{6.2}$$

$$\nabla^1_+(\vec{\varphi}) = (\varphi, \varphi_-, y_\to *_0 \varphi_-, y_\to), \tag{6.3}$$

where $s^{m-1}\vec{\varphi} = \vec{f}$ and $\vec{f}\colon \vec{x} \to \vec{y}$ is as above.

Remark 6.2. Because we will sometimes want to refer to the different elements of such a pair $\nabla^1_\xi(\vec{\varphi})$ we denote by $[\nabla^1_\xi(\vec{\varphi})]_k$ the kth component, for $k = 0, 1, 2, 3$. E.g., $[\nabla^1_+(\vec{\varphi})]_2$ is $(y_\to *_0 \varphi_-)$,.

Dimension 1. Next, we define, given an arrow $\vec{f}\colon \vec{x} \to \vec{y}$ in $\int I_A$, a 2-cell

$$\nabla^1_-(\vec{f}) \xrightarrow{\varepsilon_{\vec{f}}} \nabla^1_+(\vec{f}).$$

I.e., $\varepsilon_{\vec{f}}$ is as indicated in the following "naturality" diagram:

$$\begin{array}{ccc} \downarrow(\vec{x}) & \xrightarrow{\ \varepsilon_{\vec{x}}\ } & \vec{x} \\ {\scriptstyle \downarrow(\vec{f})}\big\downarrow & \varepsilon_{\vec{f}}\Downarrow & \big\downarrow{\scriptstyle \vec{f}} \\ \downarrow(\vec{y}) & \xrightarrow[\ \varepsilon_{\vec{y}}\ ]{} & \vec{y} \end{array}$$

In particular, $\varepsilon_{\vec{f}}$ is defined to be $(f, f_-, f_\to *_0 f_-, f_\to)$. This definition is easily seen to make sense using (6.2) and (6.3). Now, holding parallel arrows $\vec{f}$ and $\vec{g}$ fixed, we define functors

$$\left(\int I_A\right)_2(\vec{f}, \vec{g}) \xrightarrow{\nabla^2_\xi} \left(\int I_A\right)_2(\nabla^1_-(\vec{f}), \nabla^1_+(\vec{g}))$$

by

$$\nabla^2_\xi(\vec{\gamma}) := \begin{cases} \nabla^1_+(\vec{\gamma}) *_1 \varepsilon_{\vec{f}} & \text{if } \xi = -, \\ \varepsilon_{\vec{g}} *_1 \nabla^1_-(\vec{\gamma}) & \text{if } \xi = +. \end{cases}$$

With these definitions it is straightforward to verify that

$$\nabla^2_-(\vec{\varphi}) = (\varphi, \varphi_-, f_\to *_0 \varphi_-, f_\to), \tag{6.4}$$

when $\vec{\varphi}$ is any cell of $(\int I_A)_2(\vec{f}, \vec{g})$. In order to obtain a similar analysis of $\nabla^2_+(\vec{\varphi})$ we require a further fact about the duality functor $\neg$.

Lemma 6.3. *Given any m-cell, for $m \geq 1$, $\vec{\varphi}$ of $\int A$,*

$$\neg\varphi_- *_0 \varphi_- = \mathfrak{d}^1_{\vec{x},\vec{y}}(\varphi),$$

$$\varphi_- *_0 \neg\varphi_- = y_-$$

where $\vec{x}$ and $\vec{y}$ are the 0-cells bounding $\vec{\varphi}$.

PROOF. This is a direct consequence of the easily proved fact that, for $0 \leq n \leq m-2$

$$\rho_{\partial\vec{\varphi}}^{m-n-2}(\varphi_-^{-1}) *_n \varphi_- = t^{m-n-1}(\vec{\varphi})_-^- *_n \left(\varphi_- *_{(n+1)} \rho_{\partial\vec{\varphi}}^{m-n-3}(\varphi_-^{-1})\right)$$

when n is even (or 0), and

$$\varphi_- *_n \rho_{\partial\vec{\varphi}}^{m-n-2}(\varphi_-^{-1}) = \left(\rho_{\partial\vec{\varphi}}^{m-n-3}(\varphi_-^{-1}) *_{(n+1)} \varphi_-\right) *_n s^{m-n-1}(\vec{\varphi})_-^{-1},$$

when n is odd. Iteratively applying these facts and canceling inverses yields the required result. □

Using Lemma 6.3 it follows, by a (lengthy but) straightforward calculation, that, where $\vec{\beta}$

$$\nabla^2_+(\vec{\varphi}) = \Big(\varphi, \varphi_-, (g_\rightarrow *_1 \Psi^2_{\vec{\beta},x_\rightarrow}(\varphi_+)) *_0 \beta_-, \varphi_\rightarrow\Big) \tag{6.5}$$

for any m-cell $\vec{\varphi}$ of $(\int I_A)_2(\vec{f},\vec{g})$ with $\vec{\beta} = t^{m-2}\vec{\varphi}$. Using (6.4) and (6.5) we define $\varepsilon_{\vec{\alpha}}$, for $\vec{\alpha}\colon \vec{f} \rightrightarrows \vec{g}$ a 2-cell of $\int I_A$, as follows:

$$\varepsilon_{\vec{\alpha}} := (\alpha, \alpha_-, \alpha_\rightarrow *_0 \alpha_-, \alpha_\rightarrow).$$

In higher dimensions this procedure is carried out as follows:

Dimension $(n+1)$. Given $\vec{\alpha}$ and $\vec{\beta}$ of dimension n together with the appropriate ∇^n_- we first observe that, using decompositions of $\nabla^n_\xi(\vec{\varphi})$ corresponding to (6.4) and (6.5), and proved by a standard calculation using Lemma 6.3, it follows that, if, $\vec{\varphi}\colon \vec{\alpha} \Rrightarrow \vec{\beta}$ is a $(n+1)$-cell, then

$$\nabla^n_-(\vec{\varphi}) = \begin{cases} (\varphi, \varphi_-, \alpha_\rightarrow *_0 \varphi_-, \alpha_\rightarrow) & \text{if } n \text{ is even,} \\ (\varphi, \varphi_-, (\partial I)^n(\vec{\varphi}) *_0 \varphi_-, \varphi_\rightarrow) & \text{if } n \text{ is odd;} \end{cases}$$

and

$$\nabla^n_+(\vec{\varphi}) = \begin{cases} (\varphi, \varphi_-, (\partial I)^n(\vec{\varphi}) *_0 \varphi_-, \alpha_\rightarrow) & \text{if } n \text{ is even,} \\ (\varphi, \varphi_-, \beta_\rightarrow *_0 \varphi_-, \beta_\rightarrow) & \text{if } n \text{ is odd.} \end{cases}$$

Thus we define

$$\nabla^n_-(\vec{\alpha}) \xrightarrow{\varepsilon_{\vec{\varphi}}} \nabla^n_+(\vec{\beta})$$

by

$$(\varphi, \varphi_-, \varphi_\rightarrow *_0 \varphi_-, \varphi_\rightarrow).$$

Now, holding $\vec{\alpha}$ and $\vec{\beta}$ fixed, we define

$$\left(\int I_A\right)_{n+1}(\vec{\alpha},\vec{\beta}) \xrightarrow{\nabla^{n+1}_\xi} \left(\int I_A\right)_{n+1}(\nabla^n_-(\vec{\alpha}), \nabla^n_+(\vec{\beta}))$$

for $\xi = -, +$ as

$$\nabla^{n+1}_\xi(\vec{\gamma}) := \begin{cases} \nabla^n_+(\vec{\gamma}) *_n \varepsilon_{\vec{\alpha}} & \text{if } \xi = -, \\ \varepsilon_{\vec{\beta}} *_n \nabla^n_-(\vec{\gamma}) & \text{if } \xi = +. \end{cases}$$

6.4. Elimination terms in dimensions 0, 1. Assume we are given an object $\vec{x} = (x, x_-, x_+, x_\rightarrow)$ of $\int I_A$. We would like to provide a corresponding object, which for the sake of notational convenience we simply denote by x_J, of $D(\vec{x})$. This x_J is obtained by a kind of Yoneda style argument. Namely, we observe that, by assumption there is a term $\breve{x}_d$ in $D(\downarrow(\vec{x}))$. Applying the functor $D(\varepsilon_{\vec{x}})$ yields the required x_J in $D(\vec{x})$. I.e.,

$$x_J := D(\varepsilon_{\vec{x}})(\breve{x}_d).$$

Because it will greatly simplify matters in the later stages, we introduce a special notation for $D(\varepsilon_{\vec{x}})$ and its higher-dimensional generalizations $D(\varepsilon_{\vec{\gamma}})$. Namely, we define

$$\langle\vec{\gamma}\rangle := D(\varepsilon_{\vec{\gamma}}).$$

With this notation $x_J = \langle\vec{x}\rangle(\breve{x}_d)$.

In dimension 1, given an arrow $\vec{f}\colon \vec{x} \to \vec{y}$ in $\int I_A$ we have by hypothesis the arrow $\breve{f}_d\colon \widetilde{\Theta}^1_{\breve{x},\breve{y}}(\breve{f}) \to \breve{y}_d$ in $D(\downarrow \vec{y})$ and, applying $\langle \vec{y} \rangle$,

$$\langle \vec{y} \rangle (\widetilde{\Theta}^1 \breve{f}) \xrightarrow{\langle \vec{y} \rangle (\breve{f}_d)} y_J$$

in $D(\vec{y})$. Now,

$$\langle \vec{y} \rangle (\widetilde{\Theta}^1 \breve{f}) = \langle \vec{y} \rangle (D(\downarrow \vec{f})\breve{x}_d) = D(\nabla^1_+ \vec{f})\breve{x}_d.$$

Also,

$$\Theta^1(\vec{f}) = D(\vec{f})(x_J) = D(\nabla^1_- \vec{f})\breve{x}_d$$

and therefore we define f_J to be the composite

$$\Theta^1(\vec{f}) \xrightarrow{\langle \vec{f} \rangle_{\breve{x}_d}} D(\nabla^1_+ \vec{f})\breve{x}_d \xrightarrow{\langle \vec{y} \rangle \breve{f}_d} y_J.$$

Again, it will be useful to introduce some additional notation to clarify the situation in higher-dimensions. First, holding fixed objects $\vec{x}$ and $\vec{y}$ of $\int I_A$, we define a functor

$$\left(\int I_A\right)_1 (\vec{x}, \vec{y}) \xrightarrow{\vartheta^1_{\vec{x},\vec{y}}} D(\vec{y})$$

by

$$\vartheta^1_{\vec{x},\vec{y}}(\vec{\gamma}) := \langle \vec{y} \rangle \big(\widetilde{\Theta}^1_{\breve{x},\breve{y}}(\breve{\gamma})\big).$$

The next ingredient is to define, for $\vec{f}\colon \vec{x} \to \vec{y}$, arrows $\lhd(\vec{f})\colon \Theta^1(\vec{f}) \to \vartheta^1(\vec{f})$ and $\rhd(\vec{f})\colon \vartheta^1(\vec{f}) \to y_J$ as follows:

$$\lhd(\vec{f}) := \langle \vec{f} \rangle_{\breve{x}_d},$$
$$\rhd(\vec{f}) := \langle \vec{y} \rangle \breve{f}_d.$$

Thus, with this notation f_J is just $\rhd(\vec{f}) *_0 \lhd(\vec{f})$. We will see below that, in general, γ_J will always be formed as a composite of the form $\rhd(\vec{\gamma}) *_{k-1} \lhd(\vec{\gamma})$ along a $(k-1)$-cell $\vartheta^k(\vec{\gamma})$.

6.5. Elimination terms in dimensions 2. In dimension 2, let arrows $\vec{f}, \vec{g}\colon \vec{x} \rightrightarrows \vec{y}$ in $\int I_A$ be given together with a 2-cell $\vec{\alpha}\colon \vec{f} \Rightarrow \vec{g}$. In order to define α_J we will describe 2-cells filling both the square and triangle as indicated in the following diagram:

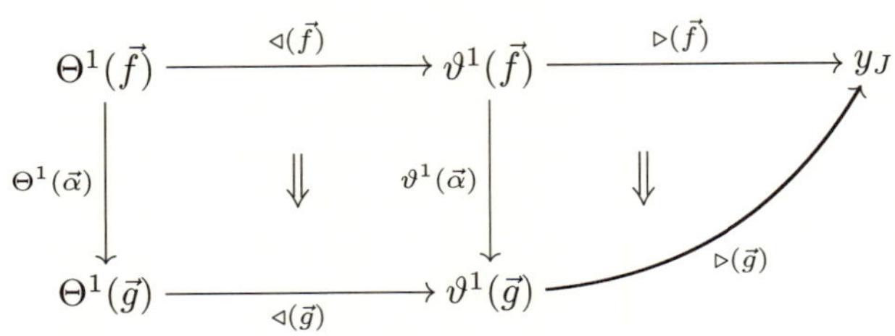

Defining the functor

$$\left(\int I_A\right)_2 (\vec{f}, \vec{g}) \xrightarrow{\vartheta^2_{\vec{f},\vec{g}}} D(\vec{y})_1(\Theta^1(\vec{f}), y_J)$$

by

$$\vartheta^2(-) := \rhd(\vec{g}) *_0 \vartheta^1(-) *_0 \lhd(\vec{f}).$$

we see that our goal is precisely to provide 2-cells

$$f_J \xrightarrow{\triangleleft(\vec{\alpha})} \vartheta^2_{\vec{f},\vec{g}}(\vec{\alpha}) \xrightarrow{\triangleright(\vec{\alpha})} \Theta^2_{\vec{f},\vec{g}}(\vec{\alpha}).$$

The strategy for filling the square and triangle from above is fairly simple. For the triangle, we use $\breve{\alpha}_d$, and for the square we use the naturality cell $\varepsilon_{\vec{\alpha}}$. To begin with, we define

$$\triangleleft(\vec{\alpha}) := \langle\vec{y}\rangle(\breve{\alpha}_d) *_0 \triangleleft(\vec{f}).$$

For the square, we observe that

$$\Theta^1(\vec{\gamma}) = D(\vec{\gamma})(x_J),$$
$$\vartheta^1(\vec{\gamma}) = D(\nabla^1_+\vec{\gamma})(\breve{x}_d)$$

where $\vec{\gamma}$ is any cell of $\int I_A(\vec{x},\vec{y})$. Thus,

$$\vartheta^1(\vec{\gamma}) *_0 \triangleleft(\vec{f}) = D(\nabla^2_-\vec{\gamma})(\breve{x}_d),$$
$$\triangleleft(\vec{g}) *_0 \Theta^1(\vec{\gamma}) = D(\nabla^2_+\vec{\gamma})(\breve{x}_d).$$

As such, we define

$$\triangleright(\vec{\alpha}) := \triangleright(\vec{g}) *_0 \langle\vec{\alpha}\rangle_{\breve{x}_d}.$$

Thus, α_J is $(\triangleright(\vec{\alpha}) *_1 \triangleleft(\vec{\alpha}))$. It will often be convenient to omit parentheses when dealing with the arrows $\triangleright(\vec{\alpha})$ and $\triangleleft(\vec{\alpha})$. In order to avoid confusion, we adopt the convention that $\triangleright$ and $\triangleleft$ bind more tightly than composition. I.e., $\triangleleft\vec{\gamma} *_k \vec{\varphi}$ should be read as $\triangleleft(\vec{\gamma}) *_k \vec{\varphi}$.

Before moving on to dimension 3, we first introduce some additional machinery which is the final technical ingredient required in order to make the induction to higher dimensions possible. Namely, for $\vec{f},\vec{g}\colon \vec{x} \rightrightarrows \vec{y}$ parallel 1-cells, we define functors

$$D(\vec{y})_1(\vartheta^1\vec{f}, y_J) \xrightarrow{\breve{H}^1_{\vec{f},\vec{g}}} D(\vec{y})_1(\Theta^1\vec{f}, y_J) \xrightarrow{H^1_{\vec{f},\vec{g}}} D(\vec{y})_1(\Theta^1\vec{f}, \vartheta^1\vec{g})$$

as follows:

$$\breve{H}^1_{\vec{f},\vec{g}}(-) := (- *_0 \triangleleft\vec{f}),$$
$$H^1_{\vec{f},\vec{g}}(-) := (\triangleright\vec{g} *_0 -).$$

With these functors at our disposal, we are in the position to make several remarks regarding their interaction with the other structures with which we are concerned. To begin with, when $\vec{\alpha}$ is a 2-cell $\vec{f} \Rightarrow \vec{g}$,

$$\triangleright\vec{\alpha} = H^1_{\vec{f},\vec{g}}\big(\langle\vec{\alpha}\rangle_{\breve{x}_d}\big), \tag{6.6}$$

$$\triangleleft\vec{\alpha} = \breve{H}^1_{\vec{f},\vec{g}}\big(\langle\vec{y}\rangle(\breve{\alpha}_d)\big). \tag{6.7}$$

Also, these functors interact with ϑ^2 in the sense that

$$\breve{H}^1_{\vec{f},\vec{g}}\big(\triangleright\vec{g} *_0 \vartheta^1(\vec{\gamma})\big) = \vartheta^2_{\vec{f},\vec{g}}(\vec{\gamma}) = H^1_{\vec{f},\vec{g}}\big(\vartheta^1(\vec{\gamma}) *_0 \triangleleft\vec{f}\big). \tag{6.8}$$

In an informal sense, the problem of providing the elimination maps φ_J will be seen to always amount, as above, to filling both a triangle and a square. In each case, the tactic is essentially the same as above and the functors H^k and $\breve{H}^k$ allow us to express in the most perspicuous way the combinatorics of the situation for the squares and triangles, respectively.

6.6. Elimination terms in dimension 3. In dimension 3, given $\vec{\varphi}\colon \vec{\alpha} \Rrightarrow \vec{\beta}$ a 3-cell of $\int I_A$, we would like to describe the 3-cells indicated in the following diagram:

$$\begin{array}{ccccc} f_J & \xrightarrow{\quad \lhd\vec{\alpha} \quad} & \vartheta^2(\vec{\alpha}) & \xrightarrow{\quad \rhd\vec{\alpha} \quad} & \Theta^2(\vec{\alpha}) \\ & \Downarrow & \Big\downarrow {\scriptstyle \vartheta^2(\vec{\varphi})} & \Downarrow & \Big\downarrow {\scriptstyle \Theta^2(\vec{\varphi})} \\ & \searrow^{\lhd\vec{\beta}} & \vartheta^2(\vec{\beta}) & \xrightarrow[\quad \rhd\vec{\beta} \quad]{} & \Theta^2(\vec{\beta}) \end{array} \tag{6.9}$$

With this picture in mind we begin by defining, for fixed parallel 2-cells $\vec{\alpha}, \vec{\beta}\colon \vec{f} \rightrightarrows \vec{g}$, functors

$$D(\vec{y})_2(f_J, \vartheta^2\vec{\beta}) \xrightarrow{\breve{H}^2_{\vec{\alpha},\vec{\beta}}} D(\vec{y})_2(f_J, \Theta^2\vec{\beta}) \xrightarrow{H^2_{\vec{\alpha},\vec{\beta}}} D(\vec{y})_2(\vartheta^2\vec{\alpha}, \Theta^2\vec{\beta})$$

as follows:

$$\begin{aligned} \breve{H}^2_{\vec{\alpha},\vec{\beta}}(-) &:= \rhd\vec{\beta} *_1 \breve{H}^1_{\vec{f},\vec{g}}(-), \\ H^2_{\vec{\alpha},\vec{\beta}}(-) &:= H^1_{\vec{f},\vec{g}}(-) *_1 \lhd\vec{\alpha}. \end{aligned}$$

Next,

$$\left(\int I_A\right)_3 (\vec{\alpha}, \vec{\beta}) \xrightarrow{\vartheta^3_{\vec{\alpha},\vec{\beta}}} D(\vec{y})_2(f_J, \Theta^2(\vec{\beta}))$$

is defined by

$$\vartheta^3(-) := \rhd(\vec{\beta}) *_1 \vartheta^2(-) *_1 \lhd(\vec{\alpha})$$

Before going any further it is useful to establish several facts. First, we note that by a straightforward calculation:

$$\vartheta^2_{\vec{f},\vec{g}}(\vec{\gamma}) = \breve{H}^1_{\vec{f},\vec{g}}\big(\langle\vec{y}\rangle\widetilde{\Theta}^2\breve{\gamma}\big). \tag{6.10}$$

We call (6.10) the *triangle-law for dimension* 2 and note that together with (6.7) it follows that the triangle from (6.9) may be filled with the 3-cell $\breve{H}^1_{\vec{f},\vec{g}}(\langle\vec{y}\rangle\breve{\varphi}_d)$. Accordingly, we define

$$\rhd\vec{\varphi} := \breve{H}^2_{\vec{\alpha},\vec{\beta}}\big(\langle\vec{y}\rangle\breve{\varphi}_d\big).$$

Turning to the square, observe that

$$\begin{aligned} \Theta^2_{\vec{f},\vec{g}}(\vec{\gamma}) *_1 \rhd\vec{\alpha} &= (g_J *_0 \Theta^1\vec{\gamma}) *_1 (\rhd\vec{g} *_0 \langle\vec{\alpha}\rangle_{\breve{x}_d}) \\ &= (\rhd\vec{g} *_0 \lhd\vec{g} *_0 \Theta^1\vec{\gamma}) *_1 (\rhd\vec{g} *_0 \langle\vec{\alpha}\rangle_{\breve{x}_d}) \\ &= \rhd\vec{g} *_0 \big((\lhd\vec{g} *_0 \Theta^1\vec{\gamma}) *_1 \langle\vec{\alpha}\rangle_{\breve{x}_d}\big) \\ &= \rhd\vec{g} *_0 D\big(\nabla^3_-\vec{\gamma}\big)_{\breve{x}_d}. \end{aligned}$$

Consequently, we obtain the *source square-law for dimension 2*:

$$\Theta^2_{\vec{f},\vec{g}}(\vec{\gamma}) *_1 \rhd\vec{\alpha} = H^1_{\vec{f},\vec{g}}\big(D(\nabla^3_-\vec{\gamma})_{\breve{x}_d}\big). \tag{6.11}$$

Another straightforward calculation yields the *target square-law for dimension* 2:

$$\rhd\vec{\beta} *_1 \vartheta^2_{\vec{f},\vec{g}}(\vec{\gamma}) = H^1_{\vec{f},\vec{g}}\big(D(\nabla^3_+\vec{\gamma})_{\breve{x}_d}\big). \tag{6.12}$$

Thus, the filler of the square in (6.9) is defined to be $H^1_{\vec{f},\vec{g}}(\langle\vec{\varphi}\rangle_{\breve{x}_d})$. Finally, we set

$$\triangleleft\vec{\varphi} := H^2_{\vec{\alpha},\vec{\beta}}(\langle\vec{\varphi}\rangle_{\breve{x}_d}),$$
$$\varphi_J := \triangleright\vec{\varphi} *_2 \triangleleft\vec{\varphi}.$$

6.7. The construction in higher dimensions. Now, at higher dimensions, the construction of the elimination terms is by induction on dimension. In particular, we proceed by induction on $n \geq 2$ in such a way that *at stage* $(n+1)$ — in addition to the existence of the required $(n+1)$-cells φ_J — the following conditions are satisfied:

(1) For all parallel n-cells $\vec{\alpha}, \vec{\beta}\colon \vec{f} \rightrightarrows \vec{g}$, there is a functor $\vartheta^{n+1}_{\vec{\alpha},\vec{\beta}}$ parallel to $\Theta^{n+1}_{\vec{\alpha},\vec{\beta}}$. I.e., .

(2) For any $(n+1)$-cell $\vec{\varphi}\colon \vec{\alpha} \to \vec{\beta}$, there exist corresponding $(n+1)$-cells $\triangleright\vec{\varphi}$ and $\triangleleft\vec{\varphi}$ such that

$$\alpha_J \xrightarrow{\triangleleft\vec{\varphi}} \vartheta^{n+1}_{\vec{\alpha},\vec{\beta}}\vec{\varphi} \xrightarrow{\triangleright\vec{\varphi}} \Theta^{n+1}_{\vec{\alpha},\vec{\beta}}\vec{\varphi}$$

when $(n+1)$ is even, and

$$\Theta^{n+1}_{\vec{\alpha},\vec{\beta}}\vec{\varphi} \xrightarrow{\triangleleft\vec{\varphi}} \vartheta^{n+1}_{\vec{\alpha},\vec{\beta}}\vec{\varphi} \xrightarrow{\triangleright\vec{\varphi}} \beta_J$$

when $(n+1)$ is odd.

(3) There are, for $\vec{\alpha}$ and $\vec{\beta}$ parallel n-cells, functors $H^n_{\vec{\alpha},\vec{\beta}}$ and $\breve{H}^n_{\vec{\alpha},\vec{\beta}}$ such that

$$D(\vec{y})_n(\vartheta^n\vec{\alpha}, g_J) \xrightarrow{\breve{H}^n_{\vec{\alpha},\vec{\beta}}} D(\vec{y})_n(\Theta^n\vec{\alpha}, g_J),$$
$$D(\vec{y})_n(\Theta^n\vec{\alpha}, \vartheta^n\vec{\beta}) \xrightarrow{H^n_{\vec{\alpha},\vec{\beta}}} D(\vec{y})_n(\Theta^n\vec{\alpha}, g_J)$$

if $(n+1)$ is even; and

$$D(\vec{y})_n(f_J, \vartheta^n\vec{\beta}) \xrightarrow{\breve{H}^n_{\vec{\alpha},\vec{\beta}}} D(\vec{y})_n(f_J, \Theta^n\vec{\beta}),$$
$$D(\vec{y})_n(\vartheta^n\vec{\alpha}, \Theta^n\vec{\beta}) \xrightarrow{H^n_{\vec{\alpha},\vec{\beta}}} D(\vec{y})_n(f_J, \Theta^n\vec{\beta})$$

if $(n+1)$ is odd.

(4) The following *triangle-law* is satisfied:

$$\vartheta^{n+1}_{\vec{\alpha},\vec{\beta}}\vec{\gamma} = \breve{H}^n_{\vec{\alpha},\vec{\beta}}(\langle\vec{y}\rangle\widetilde{\Theta}^{n+1}\breve{\gamma}),$$

when $\vec{\gamma}$ is any cell in the domain of ϑ^{n+1}.

(5) If $\vec{\varphi}, \vec{\psi}\colon \vec{\alpha} \rightrightarrows \vec{\beta}$ are parallel $(n+1)$-cells, the following *square-laws* are satisfied:

$$H^n_{\vec{\alpha},\vec{\beta}}(D(\nabla^{n+2}_{-}\vec{\gamma})_{\breve{x}_d}) = \begin{cases} \Theta^{n+1}\vec{\gamma} *_n \triangleright\vec{\varphi} & \text{if } (n+1) \text{ is even,} \\ \vartheta^{n+1}\vec{\gamma} *_n \triangleleft\vec{\varphi} & \text{if } (n+1) \text{ is odd;} \end{cases}$$

and

$$H^n_{\vec{\alpha},\vec{\beta}}(D(\nabla^{n+2}_{+}\vec{\delta})_{\breve{x}_d}) = \begin{cases} \triangleright\vec{\psi} *_n \vartheta^{n+1}\vec{\delta} & \text{if } (n+1) \text{ is even,} \\ \triangleleft\vec{\psi} *_n \Theta^{n+1}\vec{\delta} & \text{if } (n+1) \text{ is odd,} \end{cases}$$

for appropriate cells $\vec{\gamma}$ and $\vec{\delta}$. Note that the ∇^{n+2}_{ξ} here are defined with respect to $\vec{\varphi}$ and $\vec{\psi}$.

Assuming we have carried out the construction up to stage n, we claim that the following definitions at stage $(n+1)$ will satisfy the required conditions:

- For parallel n-cells $\vec{\alpha}$ and $\vec{\beta}$,

$$\vartheta^{n+1}_{\vec{\alpha},\vec{\beta}}(-) := \rhd\vec{\alpha} *_{(n-1)} \vartheta^{n}(-) *_{(n-1)} \lhd\vec{\beta}.$$

- If $\vec{\alpha}$ and $\vec{\beta}$ are parallel n-cells $\vec{f} \rightrightarrows \vec{g}$, then we define

$$H^{n}_{\vec{\alpha},\vec{\beta}}(-) := \begin{cases} \rhd\vec{\beta} *_n H^{n-1}_{\vec{f},\vec{g}}(-) & \text{if } (n+1) \text{ is even,} \\ H^{n-1}_{\vec{f},\vec{g}}(-) *_n \lhd\vec{\alpha} & \text{if } (n+1)\text{is odd;} \end{cases}$$

and

$$\breve{H}^{n}_{\vec{\alpha},\vec{\beta}}(-) := \begin{cases} \breve{H}^{n-1}_{\vec{f},\vec{g}}(-) *_n \lhd\vec{\alpha} & \text{if } (n+1) \text{ is even,} \\ \rhd\vec{\beta} *_n \breve{H}^{n-1}_{\vec{f},\vec{g}}(-) & \text{if } (n+1) \text{ is odd.} \end{cases}$$

- Given $\vec{\varphi}\colon \vec{\alpha} \Rightarrow \vec{\beta}$ a $(n+1)$-cell, we define

$$\lhd\vec{\varphi} := \begin{cases} \breve{H}^{n}_{\vec{\alpha},\vec{\beta}}\big(\langle\vec{y}\rangle\breve{\varphi}_d\big) & \text{if } (n+1) \text{ is even,} \\ H^{n}_{\vec{\alpha},\vec{\beta}}\big(\langle\vec{\varphi}\rangle_{\breve{x}_d}\big) & \text{if } (n+1) \text{ is odd;} \end{cases}$$

and

$$\rhd\vec{\varphi} := \begin{cases} H^{n}_{\vec{\alpha},\vec{\beta}}\big(\langle\vec{\varphi}\rangle_{\breve{x}_d}\big) & \text{if } (n+1) \text{ is even,} \\ \breve{H}^{n}_{\vec{\alpha},\vec{\beta}}\big(\langle\vec{y}\rangle\breve{\varphi}_d\big) & \text{if } (n+1) \text{ is odd.} \end{cases}$$

- In all dimensions,

$$\alpha_J := \rhd\vec{\alpha} *_n \lhd\vec{\alpha}$$

when $\vec{\alpha}$ is a $(n+1)$-cell.

The reader can readily verify that we have already satisfied the conditions of the induction in the base case where $n = 2$. We now turn to the induction step.

Lemma 6.4. *With the definitions given above, the conditions of the construction are satisfied in all dimensions $(n+1)$.*

PROOF. First, assume $(n+1)$ is even with $n > 2$ and let an $(n+1)$-cell $\vec{\varphi}\colon \vec{\alpha} \rightrightarrows \vec{\beta}$ in $\int I_A$ be given. Then, by the induction hypothesis and examination of the following diagram

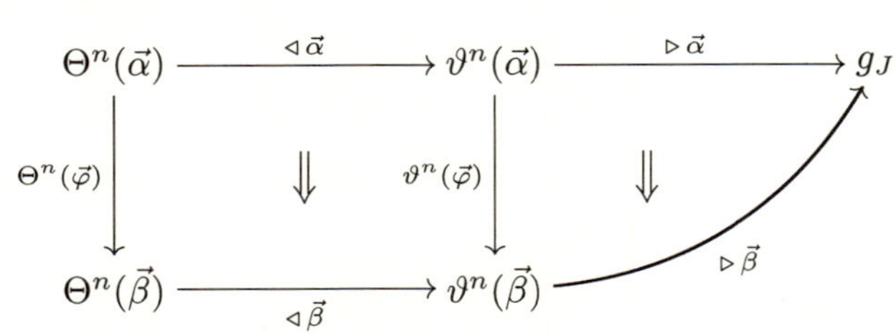

where $\vec{\alpha}, \vec{\beta}\colon \vec{f} \rightrightarrows \vec{g}$, it follows that conditions (1) – (3) are satisfied with the definitions given above.

For the triangle law, we reason as follows:

$$\begin{aligned}\breve{H}^n_{\vec{\alpha},\vec{\beta}}(\langle\vec{y}\rangle\widetilde{\Theta}^{n+1}\breve{\gamma}) &= \breve{H}^{n-1}_{\vec{f},\vec{g}}(\langle\vec{y}\rangle\breve{\beta}_d *_{(n-1)} \langle\vec{y}\rangle\widetilde{\Theta}^n\breve{\gamma}) *_{(n-1)} \lhd\vec{\alpha}\\ &= \breve{H}^{n-1}_{\vec{f},\vec{g}}(\langle\vec{y}\rangle\breve{\beta}_d) *_{(n-1)} \breve{H}^{n-1}_{\vec{f},\vec{g}}(\langle\vec{y}\rangle\widetilde{\Theta}^n\breve{\gamma}) *_{(n-1)} \lhd\vec{\alpha}\\ &= \rhd\vec{\beta} *_{(n-1)} \vartheta^n_{\vec{f},\vec{g}}(\vec{\gamma}) *_{(n-1)} \lhd\vec{\alpha}\\ &= \vartheta^{n+1}_{\vec{\alpha},\vec{\beta}}(\vec{\gamma}),\end{aligned}$$

where the penultimate equation is by definition of $\rhd\vec{\beta}$ and the induction hypothesis.

Next assume given $(n+1)$-cells $\vec{\varphi},\vec{\psi}\colon \vec{\alpha} \rightrightarrows \vec{\beta}$. For the "source" square law, we have

$$\begin{aligned}H^n_{\vec{\alpha},\vec{\beta}}(D(\nabla^{n+2}_{-}\vec{\gamma})_{\breve{x}_d}) &= H^n_{\vec{\alpha},\vec{\beta}}(D(\nabla^{n+1}_{+}\vec{\gamma})_{\breve{x}_d} *_n \langle\vec{\varphi}\rangle_{\breve{x}_d})\\ &= \left(\rhd\vec{\beta} *_{(n-1)} H^{n-1}_{\vec{f},\vec{g}}(D(\nabla^{n+1}_{+}\vec{\gamma}))\right) *_n \rhd\vec{\varphi}\\ &= (\rhd\vec{\beta} *_{(n-1)} \lhd\vec{\beta} *_{(n-1)} \Theta^n\vec{\gamma}) *_n \rhd\vec{\varphi}\\ &= (\beta_J *_{(n-1)} \Theta^n\vec{\gamma}) *_n \rhd\vec{\varphi}\\ &= \Theta^{n+1}_{\vec{\alpha},\vec{\beta}}\vec{\gamma} *_n \rhd\vec{\varphi},\end{aligned}$$

where the third equation is by the induction hypothesis. For the "target" square law, we reason similarly and note that

$$\begin{aligned}H^n_{\vec{\alpha},\vec{\beta}}(\langle\vec{\psi}\rangle_{\breve{x}_d} *_n D(\nabla^{n+1}_{-}\vec{\gamma})_{\breve{x}_d}) &= \rhd\vec{\psi} *_n \left(\rhd\vec{\beta} *_{(n-1)} H^{n-1}_{\vec{f},\vec{g}}(D(\nabla^{n+1}_{-}\vec{\gamma})_{\breve{x}_d})\right)\\ &= \rhd\vec{\psi} *_n (\rhd\vec{\beta} *_{(n-1)} \vartheta^n\vec{\gamma} *_{(n-1)} \lhd\vec{\alpha})\\ &= \rhd\vec{\psi} *_n \vartheta^{n+1}_{\vec{\alpha},\vec{\beta}}\vec{\gamma},\end{aligned}$$

as required.

The induction step where $(n+1)$ is odd is essentially dual. □

Using the lemma, we now have the following fundamental result.

Proposition 6.5. *The cells of the form φ_J constitute a section $J\colon \int I_A \to \int D$ of the projection map $\int D \to \int I_A$.*

PROOF. In light of Lemma 6.4 the only thing which remains is to verify that the assignment $(-)_J$ is functorial. This however is a consequence of the functoriality of d and the construction of the "terms" φ_J using the functors H^k and $\breve{H}^k$. □

7. The main results

With the machinery from the preceding sections at our disposal it is now possible to establish our main results.

7.1. Dependent sums and products. Before defining the dependent products and sums, we begin by describing some basic features of general setup. First, given a functor $A\colon \mathcal{C} \to \omega$-**Gpd** we note that, by the basic properties of ω-**Gpd**, there exists an ω-groupoid denoted by $\Gamma(A)$ of sections of the projection $\int A \to \mathcal{C}$.

I.e., the objects of $\Gamma(A)$ are sections

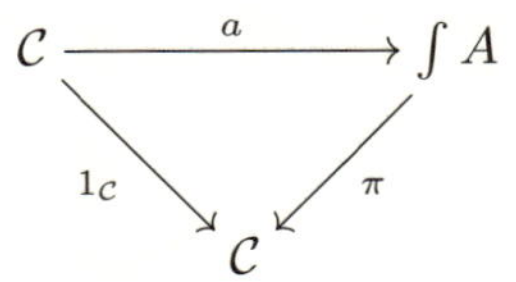

and arrows are 2-cells $\alpha : a \Rightarrow b$ for which $\pi \circ \alpha = 1_{\mathcal{C}}$ and so forth at higher dimensions. Now, given a further functor $B\colon \int A \to \omega$-**Gpd** and an object x of $\mathcal{C}$, we define a functor $\ulcorner B_x \urcorner\colon A_x \to \omega$-**Gpd** as follows:

Objects. Given an object x_- of A_x, $\ulcorner B_x \urcorner(x_-)$ is the ω-groupoid $B(x, x_-)$.

1-Cells. Given an arrow $h\colon x_- \to x_+$ in A_x, we define the functor $\ulcorner B_x \urcorner(h)\colon B(x, x_-) \to B(x, x_+)$ as follows:

$$\ulcorner B_x \urcorner(h)(-) := B(x, h)(-).$$

Here we note that this is possible since $(x, h)\colon (x, x_-) \to (x, x_+)$, where we have written (x, h) instead of $(1_x, h)$.

$(\boldsymbol{n+1})$**-Cells** Given a $(n+1)$-cell $\alpha\colon f \Rightarrow g$ in A_x bounded by 0-cells x_- and x_+, together with an object y of $B(x, x_-)$, the transformation $\ulcorner B_x \urcorner(\alpha)$ is defined at y by

$$\ulcorner B_x \urcorner(\alpha)_y := B(x, \alpha)_y.$$

Now, we would like to describe, in this same setting, the dependent sum $\Sigma_{A,B} : \mathcal{C} \to \omega$-**Gpd**. Note that, when A and B are apparent, we will often omit the subscripts.

Objects. Given an object x of $\mathcal{C}$, we define

$$\Sigma_{A,B}(x) := \int \ulcorner B_x \urcorner.$$

Given $f\colon x \to y$ in $\mathcal{C}$ we would like to define $\Sigma_{A,B}(f)\colon \int \ulcorner B_x \urcorner \to \int \ulcorner B_y \urcorner$, which we will write, for the sake of avoiding too many parentheses, as Σ_f. An object $\vec{v}$ of $\int \ulcorner B_x \urcorner$ is a pair $(v_-, v_\sharp)$ such that v_- is an object of A_x and $v_\sharp$ is an object of $B(x, v_-)$. We adopt a similar notation in higher dimensions. We also write $\partial(x)^n$ for the weighted face functor of $\int \ulcorner B_x \urcorner$ and similarly for $\partial(y)^n$, etc. With this in mind we adopt the following definition (which we will discuss below):

1-Cells. Given $f\colon x \to y$ in $\mathcal{C}$, Σ_f is defined on objects $\vec{v}$ of $\int \ulcorner B_x \urcorner$ by

$$\Sigma_f(\vec{v}) := \bigl(v_-.f, B(f, v_-.f)(v_\sharp)\bigr),$$

and on $(n+1)$-cells $\vec{\varphi}$ of $\int \ulcorner B_x \urcorner$ by

$$\Sigma_f(\vec{\varphi}) := \bigl(\varphi_-.f, B(f, w_-.f)(\varphi_\sharp)\bigr),$$

where $\vec{v}$ and $\vec{w}$ are the objects bounding $\vec{\varphi}$ at source and target position, respectively.

Note that this is a correct definition since, by functoriality of B, it is proved (simultaneously with the verification of the definitions) at each stage that

$$B(f, w_-.f)\bigl(\partial(x)^{n+1}_{\vec{\alpha},\vec{\beta}}(\varphi_-)\bigr) = \partial(y)^{n+1}_{\Sigma_f\vec{\alpha},\Sigma_f\vec{\beta}}(\varphi_-.f),$$

where $\vec{\alpha}$ and $\vec{\beta}$ are the n-cells bounding $\vec{\varphi}$.

In dimension $(n+1)$, $\Sigma_{A,B}$ is given as follows:

$(n+1)$-Cells. Given a $(n+1)$-cell $\varphi\colon \alpha \to \beta$ in $\mathcal{C}$, $\Sigma_\varphi\colon \Sigma_\alpha \to \Sigma_\beta$ has as its component at the object $\vec{v}$ of $\int\ulcorner B_x \urcorner$ the n-cell

$$(\Sigma_\varphi)_{\vec{v}} := \left(v_-.\varphi, B(\varphi, v_-.\varphi)(v_\sharp)\right)$$

of $\int\ulcorner B_y \urcorner$.

To see that this makes sense, let n-cells $\lambda, \mu\colon \alpha \to \beta$ be given. Then it follows, by induction on n that,

$$\partial(y)^{(n+1)}_{(\Sigma_\lambda)_{\vec{v}},(\Sigma_\mu)_{\vec{v}}}(v_-.\gamma) = \begin{cases} B(\mu, v_-.\gamma)_{v_\sharp} & \text{if } (n+1) \text{ is even,} \\ B(\lambda, v_-.\gamma)_{v_\sharp} & \text{if } (n+1) \text{ is odd,} \end{cases}$$

where γ is any m-cell with $m \geq n+1$ bounded by λ and μ (i.e., $s^{m-n}\gamma = \lambda$ and $t^{m-n}\gamma = \mu$). This is a straightforward induction using functoriality of B. This completes the description of the interpretation $\Sigma_{A,B}$ of dependent sums in this setting. With these definitions, the verification of functoriality is a routine calculation.

Next, we define the dependent product $\Pi_{A,B}\colon \mathcal{C} \to \omega$-**Gpd** to be the functor which sends an object x to the ω-groupoid $\Gamma(\ulcorner B_x \urcorner)$ of sections of the projection $\int\ulcorner B_x \urcorner \to A_x$. Explicitly, $\Pi_{A,B}$ is as follows:

Objects. $\Pi_{A,B}(x)$ is defined to be $\Gamma(\ulcorner B_x \urcorner)$.

1-Cells. Given an arrow $f\colon x \to y$ in $\mathcal{C}$, $\Pi_{A,B}(f)$ is the functor sending a section a in $\Gamma(\ulcorner B_x \urcorner)$ to the section $\Sigma_f \circ a \circ A_{f^{-1}}$. In general, $\Pi_{A,B}(f)$ sends an arbitrary cell φ of $\Gamma(\ulcorner B_x \urcorner)$ to the composite $\Sigma_f \circ \varphi \circ A_{f^{-1}}$ as indicated in the following diagram:

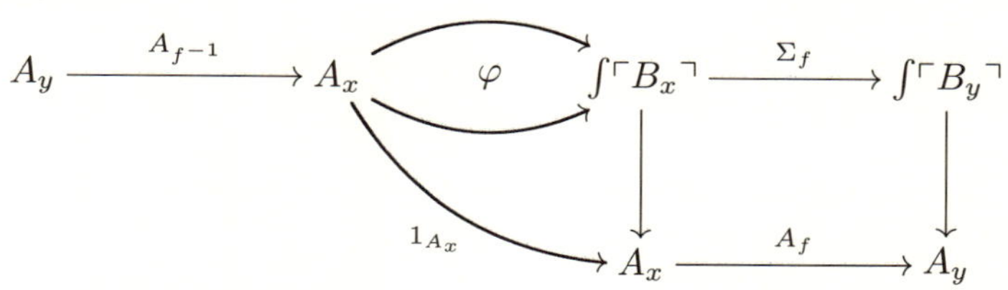

$(n+1)$-Cells. In general, when $\gamma\colon \alpha \to \beta$ is a $(n+1)$-cell in $\mathcal{C}$ and a is an object of $\Pi_{A,B}(x)$, the dependent product is defined by setting

$$\Pi_{A,B}(\gamma)_a := \Sigma_\gamma \circ a \circ A_{\sigma(\gamma)},$$

where $\sigma\colon \mathcal{C} \to \mathcal{C}^{\mathrm{op}}$ is the dual functor for $\mathcal{C}$ as described in Section 4.

This completes the description of the dependent sums and products. The introduction and elimination terms are then obtained as generalizations of those given in [9].

7.2. Results. With the interpretations given we obtain the following result extending the original groupoids model from ibid. to the setting of ω-groupoids:

Theorem 7.1. *The interpretation given above soundly interprets the basic form of intensional type theory with dependent sums, dependent products and identity types (i.e., the theory $\mathbb{T}_\omega$ from [1]). Moreover, the η-rules for dependent sums and products are also satisfied.*

PROOF. Using the interpretation given above it is straightforward to verify that all of the required laws, including all of the Beck – Chevalley laws, are satisfied. $\square$

Recall from [1, 27] that we employ the following notation for iterated identity types:

$$\underline{A}^{n+1}(a_1, b_1; \ldots; a_{n+1}, b_{n+1}) := \underline{\underline{A}^n(a_1, b_1; \ldots; a_n, b_n)}(a_{n+1}, b_{n+1}).$$

In this notation, the principle of *n-dimensional uniqueness of identity proofs* is

$$\frac{\vdash a_{n+1}, b_{n+1} : \underline{A}^n(a_1, b_1; \ldots; a_n, b_n)}{\vdash a_{n+1} = b_{n+1} : \underline{A}^n(a_1, b_1; \ldots; a_n, b_n)}$$

It is a trivial consequence of the interpretation that, for all $n \geq 1$, the ω-groupoid model refutes UIP_n. To see this, form the globular set G which is described as follows:

0-Cells. Two 0-cells $\bot^0$ and $\top^0$.
$(n+1)$-Cells. Two parallel $(n+1)$-cells $\bot^{n+1}, \top^{n+1} \colon \bot^n \to \top^n$.

Then the free ω-groupoid on G refutes uniqueness of identity proofs at all dimensions (in both the forms which deal with definitional and propositional equality). This is the case because the terms of identity types in any type interpreted as this free groupoid are simply the cells of this groupoid. I.e., we have:

Corollary 7.2. *For all $n \geq 1$, the principle* UIP_n *is not derivable in* $\mathbb{T}_\omega$. *Similar, the principle of uniqueness of identity proofs with respect to propositional equality is also not derivable.*

Recall from [1] that the *n-truncation rule* is as follows:

$$\frac{\vdash a_{n+1}, b_{n+1} : \underline{A}^n(a_1, b_1; \ldots; a_n, b_n) \qquad \vdash p : \underline{A}^{n+1}(a_1, b_1; \ldots; a_{n+1}, b_{n+1})}{\vdash a_{n+1} = b_{n+1} : \underline{A}^n(a_1, b_1; \ldots; a_n, b_n)}\ \mathrm{TR}_n$$

Our main corollary of Theorem 7.1 then states that the identity types in intensional type theory are nontrivial in all dimensions in the sense that all of the truncation rules are refuted by the ω-groupoid model. The same counterexample from above suffices to give this result as well. I.e., we have:

Corollary 7.3. *For $n \geq 0$,* TR_n *is not derivable in* $\mathbb{T}_\omega$.

We note that all of these results also apply in the case of type theory with the added η-rules for dependent products and sums, since these rules are valid in the model.

Appendix A. Rules governing identity types, dependent products and dependent sums

In this appendix we describe the syntax of the system considered in this paper (called $\mathbb{T}_\omega$ in [1]). All rules below are stated in an ambient context which is omitted for ease of presentation.

A.1. Identity types.

$$\frac{a, b : A}{\vdash \underline{A}(a, b) : \mathsf{type}}\ \text{formation}$$

$$\frac{a : A}{r(a) : \underline{A}(a, a)}\ \text{introduction}$$

$$\frac{\begin{array}{c} x : A, y : A, z : \underline{A}(x,y) \vdash B(x,y,z) : \text{type} \\ x : A \vdash \varphi(x) : B\big(x,x,r(x)\big) \\ f : \underline{A}(a,b) \end{array}}{\mathrm{J}\big(\varphi(x), a, b, f\big) : B(a,b,f)} \text{ elimination}$$

$$\frac{a : A}{\mathrm{J}\big(\varphi(x), a, a, r(a)\big) = \varphi(a) : B\big(a,a,r(a)\big)} \text{ conversion}$$

A.2. Dependent products.

$$\frac{x : A \vdash B(x) : \text{type}}{\prod_{x:A} B(x) : \text{type}} \textstyle\prod \text{ formation}$$

$$\frac{x : A \vdash f(x) : B(x)}{\lambda_{x:A} f(x) : \prod_{x:A} B(x)} \textstyle\prod \text{ introduction}$$

$$\frac{f : \prod_{x:A} B(x) \qquad a : A}{\mathrm{app}(f,a) : B(a).} \textstyle\prod \text{ elimination}$$

$$\frac{\lambda_{x:A} f(x) : \prod_{x:A} B(x) \qquad a : A}{\mathrm{app}\big(\lambda_{x:A} f(x), a\big) = f(a) : B(a)} \textstyle\prod \text{ conversion}$$

A.3. Dependent sums.

$$\frac{x : A \vdash B(x) : \text{type}}{\sum_{x:A} B(x) : \text{type}} \textstyle\sum \text{ formation}$$

$$\frac{a : A \qquad b : B(a)}{\mathrm{pair}(a,b) : \sum_{x:A} B(x)} \textstyle\sum \text{ introduction}$$

$$\frac{p : \sum_{x:A} B(x) \qquad x : A, y : B(x) \vdash \psi(x,y) : C\big(\mathrm{pair}(x,y)\big)}{\mathrm{R}\big(\psi(x,y), p\big) : C(p)} \textstyle\sum \text{ elimination}$$

$$\frac{a : A \qquad b : B(a) \qquad x : A, y : B(x) \vdash \psi(x,y) : C\big(\mathrm{pair}(x,y)\big)}{\mathrm{R}\big(\psi(x,y), \mathrm{pair}(a,b)\big) = \psi(a,b) : C\big(\mathrm{pair}(a,b)\big)} \textstyle\sum \text{ conversion}$$

Appendix B. The universal property of the Grothendieck construction

In this appendix we provide a detailed proof that the Grothendieck construction $\int A$ for strict ω-categories introduced in Section 3 can be described as the coend $\int^x (x \downarrow \mathcal{C}) \times A_x$. First we turn to a more detailed description of the comma ω-category $(x \downarrow \mathcal{C})$ when x is a 0-cell of an ω-category $\mathcal{C}$.

B.1. The comma ω-category. Given an ω-category $\mathcal{C}$ and an object x of $\mathcal{C}$ the *comma ω-category* $(x \downarrow \mathcal{C})$ will be constructed inductively in such a way that at each stage a n-cell of $(x \downarrow \mathcal{C})$ is a pair (f, f_-) such that f is a n-cell of $\mathcal{C}$ and f_- is a $(n+1)$-cell of $\mathcal{C}$. At each stage it will also be necessary to construct auxiliary functors which describe the boundaries of the cells f_- in terms of f and lower dimensional cells. These are like the weighted face functors discussed in the construction of the Grothendieck construction given above and we denote them also using the notation $\mathfrak{d}$. Because we will not require the combinatorial details of the comma category $(x \downarrow \mathcal{C})$ we will merely describe the different cells and leave the proof that this yields an ω-category to the reader. We mention that slicing constructions in the higher-dimensional setting have been studied in recent work by Palm [20].

0-Cells. A 0-cell of $(x \downarrow \mathcal{C})$ is a pair $\vec{y} = (y, y_-)$ such that y is an object of $\mathcal{C}$ and y_- is an arrow $x \to y$ in $\mathcal{C}$. For fixed 0-cells $\vec{y}, \vec{z}$ define $\partial_{\vec{y},\vec{z}}(-) := - *_0 y_-$. When $\vec{y}, \vec{z}$ are understood the subscript will be omitted.

1-Cells. A 1-cell $\vec{y} \to \vec{z}$ of $(x \downarrow \mathcal{C})$ is a tuple $\vec{f} = (f, f_-)$ with $f : y \to z$ a 1-cell in $\mathcal{C}$ and $f_- : \partial(f) \to z_-$ a 2-cell in $\mathcal{C}$. Fixing parallel 1-cells $\vec{f}, \vec{g} \colon \vec{y} \to \vec{z}$, we define $\partial_{\vec{f},\vec{g}}(-) := g_- *_1 \partial(-)$.

$(n+1)$-Cells for $(n+1)$ even. A $(n+1)$-cell $\vec{\varphi} \colon \vec{\gamma} \to \vec{\delta}$ is given by a $(n+1)$-cell $\varphi \colon \gamma \to \delta$ in $\mathcal{C}$ and a $(n+2)$-cell $\varphi_- : \gamma_- \to \partial(\varphi)$ in $\mathcal{C}$. Given parallel $(n+1)$-cells $\vec{\varphi}$ and $\vec{\psi}$ we define $\partial_{\vec{\varphi},\vec{\psi}}(-) := \partial(-) *_{n+1} \varphi_-$.

$(n+1)$-Cells for $(n+1)$ odd. A $(n+1)$-cell $\vec{\varphi} \colon \vec{\gamma} \to \vec{\delta}$ is given by a $(n+1)$-cell $\varphi \colon \gamma \to \delta$ in $\mathcal{C}$ and a $(n+2)$-cell $\varphi_- : \partial(\varphi) \to \delta_-$ in $\mathcal{C}$. Given parallel $(n+1)$-cells $\vec{\varphi}$ and $\vec{\psi}$ we define $\partial_{\vec{\varphi},\vec{\psi}}(-) := \psi_- *_{n+1} \partial(-)$.

Here we use the notation ∂ instead of $\mathfrak{d}$ to describe the required "weighted boundaries" of the construction since we will be mixing both the Grothendieck construction and its $\mathfrak{d}$ and comma categories in Section B.2 below.

Allowing x to vary we obtain a functor (of ω-categories) $(- \downarrow \mathcal{C}) : \mathcal{C}^{\mathrm{op}} \to \omega\text{-}\mathbf{Cat}$ which we describe as follows.

Dimension 1. Given an arrow $l \colon v \to x$ in $\mathcal{C}^{\mathrm{op}}$ we have a functor $l^* : (x \downarrow \mathcal{C}) \to (v \downarrow \mathcal{C})$ which sends a cell $\vec{\varphi} = (\varphi, \varphi_-)$ of any dimension of $(x \downarrow \mathcal{C})$ to

$$l^*(\vec{\varphi}) := (\varphi, \varphi_- *_0 l).$$

This is trivially functorial.

Dimension 2. Given parallel arrows $l, m \colon v \to x$ in $\mathcal{C}$ and a 2-cell $\lambda \colon l \to m$ the induced transformation $\lambda^* : l^* \to m^*$ has as its component at an object $\vec{y} = (y, y_-)$ of $(x \downarrow \mathcal{C})$ the cell

$$(\lambda^*)_{\vec{y}} := (1_y, y_- *_0 \lambda).$$

With this definition the naturality condition for natural transformations of ω-categories is straightforward.

Dimension $n > 2$. Given parallel $(n-1)$-cells α and β in $\mathcal{C}$ (bounded by v and x, respectively, at dimension 0), and a n-cell $\varphi \colon \alpha \to \beta$, the component of the induced transformation $\varphi^* : \alpha^* \to \beta^*$ at an object $\vec{y}$ of $(x \downarrow \mathcal{C})$ is given by $(1_y, y_- * \varphi)$. This clearly satisfies the naturality condition.

This process is clearly functorial (since it is just given by precomposition).

B.2. Description of the coend inclusions. We now turn to the description of the coend inclusion functors $\iota_x\colon (x\downarrow \mathcal{C})\times A_x \to \int A$, for $A\colon \mathcal{C}\to \omega\text{-}\mathbf{Cat}$. Again, ι_x is given by an inductive construction where we must verify certain equations at each stage. We will begin with the first few dimensions and then we will explain precisely the conditions of the induction.

0-Cells. A 0-cell of $(x\downarrow \mathcal{C})\times A_x$ is a tuple $\vec{y}=(y,y_-,y_o)$ such that y is a 0-cell of $\mathcal{C}$, y_- is a 1-cell $y_-: x\to y$ in $\mathcal{C}$, and y_o is an object of A_x. We define

$$\iota_x(\vec{y}) := (y, y_o\cdot y_-).$$

So, in particular, $\iota_x(\vec{y})_- = y_o\cdot y_-$.

Now, given two 0-cells $\vec{y}$ and $\vec{z}$ of $(x\downarrow \mathcal{C})\times A_x$ we have the weighted boundary functors $\partial_{\vec{y},\vec{z}}(-)$ (for $(x\downarrow \mathcal{C})$) and $\mathfrak{d}_{\iota_x(\vec{y}),\iota_x(\vec{z})}(-)$ (for $\int A$). These two functors are moreover related by the law

$$\mathfrak{d}_{\iota_x(\vec{y}),\iota_x(\vec{z})}(-) = y_o\cdot \partial_{\vec{y},\vec{z}}(-) \tag{B.1}$$

since

$$\mathfrak{d}_{\iota_x(\vec{y}),\iota_x(\vec{z})}(-) = y_o\cdot y_-\cdot - = y_o\cdot(-*_0 y_-) = y_o\cdot\partial_{\vec{y},\vec{z}}(-).$$

1-Cells. A 1-cell $\vec{f}\colon \vec{y}\to\vec{z}$ of $(x\downarrow \mathcal{C})\times A_x$ is a tuple $\vec{f}=(f,f_-,f_o)$ such that f is a 1-cell $f\colon y\to z$ in $\mathcal{C}$, $f_-: \partial_{\vec{y},\vec{z}}(f)\to z_-$ is a 2-cell in $\mathcal{C}$, and $f_o\colon y_o\to z_o$ is a 1-cell in A_x. Then $\iota_x(\vec{f})$ is defined to be the pair $(f,\iota_x(\vec{f})_-)$ where $\iota_x(\vec{f})_-$ is the 1-cell

$$\mathfrak{d}_{\iota_x(\vec{y}),\iota_x(\vec{z})}(f) = y_o\cdot\partial_{\vec{y},\vec{z}}(f) \xrightarrow{y_o\cdot f_-} y_o\cdot z_- \xrightarrow{\quad f_o\cdot z_-\quad} z_o\cdot z_-,$$

where the equation is by (B.1). I.e.,

$$\iota_x(\vec{f})_- := (f_o\cdot z_-)*_0(y_o\cdot f_-).$$

Now, before moving on to the next stage we observe that, for 1-cells $\vec{f},\vec{g}$ of $(x\downarrow \mathcal{C})\times A_x$, the following equation relating the weighted boundary functors holds:

$$\mathfrak{d}_{\iota_x(\vec{f}),\iota_x(\vec{g})}(-) = (g_o\cdot z_-)*_0(y_o\cdot\partial_{\vec{f},\vec{g}}(-)), \tag{B.2}$$

since

$$\begin{aligned}
\mathfrak{d}_{\iota_x(\vec{f}),\iota_x(\vec{g})}(-) &= \iota_x(\vec{g})_- *_0 \mathfrak{d}_{\iota_x(\vec{y}),\iota_x(\vec{z})}(-)\\
&= ((g_o\cdot z_-)*_0(y_o\cdot g_-))*_0(y_o\cdot\partial_{\vec{y},\vec{z}}(-))\\
&= (g_o\cdot z_-)*_0(y_o\cdot(g_- *_1 \partial_{\vec{y},\vec{z}}(-)))\\
&= (g_o\cdot z_-)*_0(y_o\cdot\partial_{\vec{f},\vec{g}}(-)),
\end{aligned}$$

where the first equation is by definition of $\partial_{\iota_x(\vec{f}),\iota_x(\vec{g})}(-)$, the second equation is by (B.1), the third equation is by associativity of $*_0$ and functoriality of A, and the fourth equation is by definition of $\partial_{\vec{f},\vec{g}}(-)$.

2-Cells. A 2-cell $\vec{\alpha}\colon\vec{f}\to\vec{g}$ in $(x\downarrow \mathcal{C})\times A_x$ consists of a tuple $\vec{\alpha}=(\alpha,\alpha_-,\alpha_o)$ such that $\alpha: f\to g$ is a 2-cell in $\mathcal{C}$, α_- is a 3-cell $\alpha_-: f_-\to\partial_{\vec{f},\vec{g}}(\alpha)$ in $\mathcal{C}$, and α_o is a 2-cell $\alpha_o\colon f_o\to g_o$ in A_x. Then $\iota_x(\vec{\alpha})$ is a pair $(\alpha,\iota_x(\vec{\alpha})_-)$ where $\iota_x(\vec{\alpha})_-$ is the

2-cell

$$\begin{array}{ccc}
\iota_x(\vec{f})_- & & \boldsymbol{\partial}_{\iota_x(\vec{f}),\iota_x(\vec{g})}(\alpha) \\
\Vert & & \Vert \\
(f_o \cdot z_-) *_0 (y_o \cdot f_-) & \xrightarrow{(\alpha_o \cdot z_-) *_0 (y_o \cdot \alpha_-)} & (g_o \cdot z_-) *_0 \big(y_o \cdot \partial_{\vec{f},\vec{g}}(\alpha)\big).
\end{array}$$

I.e.,

$$\iota_x(\vec{\alpha})_- := (\alpha_o \cdot z_-) *_0 (y_o \cdot \alpha_-).$$

Again, for 2-cells $\vec{\alpha}$ and $\vec{\beta}$ of $(x \downarrow \mathcal{C}) \times A_x$, there is an equation

$$\boldsymbol{\partial}_{\iota_x(\vec{\alpha}),\iota_x(\vec{\beta})}(-) = (\alpha_o \cdot z_-) *_0 (y_o \cdot \partial_{\vec{\alpha},\vec{\beta}}(-)) \tag{B.3}$$

relating the weighted boundary functors. This equation is obtained as follows:

$$\begin{aligned}
\boldsymbol{\partial}_{\iota_x(\vec{\alpha}),\iota_x(\vec{\beta})}(-) &= \boldsymbol{\partial}_{\iota_x(\vec{f}),\iota_x(\vec{g})}(-) *_1 \iota_x(\vec{\alpha})_- \\
&= \big((g_o \cdot z_-) *_0 (y_o \cdot \partial_{\vec{f},\vec{g}}(-)\big) *_1 \big((\alpha_o \cdot z_-) *_0 (y_o \cdot \alpha_-)\big) \\
&= (\alpha_o \cdot z_-) *_0 \big(y_o \cdot (\partial_{\vec{f},\vec{g}}(-) *_2 \alpha_-)\big) \\
&= (\alpha_o \cdot z_-) *_0 (y_o \cdot \partial_{\vec{\alpha},\vec{\beta}}(-)),
\end{aligned}$$

where the penultimate equation is by the interchange law and functoriality of A.

3-Cells. A 3-cell $\vec{\varphi}\colon \vec{\alpha} \to \vec{\beta}$ of $(x \downarrow \mathcal{C}) \times A_x$ is a tuple $\vec{\varphi} = (\varphi, \varphi_-, \varphi_o)$ such that $\varphi\colon \alpha \to \beta$ is a 3-cell in $\mathcal{C}$, φ_- is a 4-cell $\varphi_-\colon \partial_{\vec{\alpha},\vec{\beta}}(\varphi) \to \beta_-$ in $\mathcal{C}$, and $\varphi_o\colon \alpha_o \to \beta_o$ is a 3-cell in A_x. Then $\iota_x(\vec{\varphi})$ is the pair $(\varphi, \iota_x(\vec{\varphi})_-)$ where $\iota_x(\vec{\varphi})_-$ is the 3-cell

$$\begin{array}{ccc}
\boldsymbol{\partial}_{\iota_x(\vec{\alpha}),\iota_x(\vec{\beta})}(\varphi) & & \iota_x(\vec{\beta})_- \\
\Vert & & \Vert \\
(\alpha_o \cdot z_-) *_0 (y_o \cdot \partial_{\vec{\alpha},\vec{\beta}}(\varphi)) & \xrightarrow{(\varphi_o \cdot z_-) *_0 (y_o \cdot \varphi_-)} & (\beta_o \cdot z_-) *_0 (y_o \cdot \beta_-).
\end{array}$$

in A_z. I.e.,

$$\iota_x(\vec{\varphi})_- := (\varphi_o \cdot z_-) *_0 (y_o \cdot \varphi_-).$$

Now, given two 3-cells $\vec{\varphi}, \vec{\psi}\colon \vec{\alpha} \to \vec{\beta}$ in $(x \downarrow \mathcal{C}) \times A_x$, we have the equation

$$\boldsymbol{\partial}_{\iota_x(\vec{\varphi}),\iota_x(\vec{\psi})}(-) = (\psi_o \cdot z_-) *_0 (y_o \cdot \partial_{\vec{\varphi},\vec{\psi}}(-)), \tag{B.4}$$

since

$$\begin{aligned}
\boldsymbol{\partial}_{\iota_x(\vec{\varphi}),\iota_x(\vec{\psi})}(-) &= \iota_x(\vec{\psi})_- *_2 \boldsymbol{\partial}_{\iota_x(\vec{\alpha}),\iota_x(\vec{\beta})}(-) \\
&= \big((\psi_o \cdot z_-) *_0 (y_o \cdot \psi_-)\big) *_2 \big((\alpha_o \cdot z_-) *_0 (y_o \cdot \partial_{\vec{\alpha},\vec{\beta}}(-))\big) \\
&= (\psi_o \cdot z_-) *_0 \big(y_o \cdot (\psi_- *_3 \partial_{\vec{\alpha},\vec{\beta}}(-))\big) \\
&= (\psi_o \cdot z_-) *_0 (y_o \cdot \partial_{\vec{\varphi},\vec{\psi}}(-))
\end{aligned}$$

Indeed, as suggested by these cases, we define at all subsequent stages of the construction

$$\iota_x(\vec{\xi})_- := (\xi_o \cdot z_-) *_0 (y_o \cdot \xi_-), \tag{B.5}$$

and that we must verify that the equation

$$\boldsymbol{\partial}_{\iota_x(\vec{\xi}),\iota_x(\vec{\zeta})}(-) = (\zeta_o \cdot z_-) *_0 (y_o \cdot \partial_{\vec{\xi},\vec{\zeta}}(-) \tag{B.6}$$

holds. However, the verification that this equation holds generalizes directly from the cases of dimensions 2 and 3 given above (since the definitions of $\eth$ and ∂ also generalize directly from these cases). That ι_x is functorial is straightforward and is left to the reader.

The functors ι_x, for x an object of $\mathcal{C}$, constitute a dinatural transformation $(- \downarrow \mathcal{C}) \times A_- \to \int A$ in the sense that, for any $l, m : v \to x$ in $\mathcal{C}$ and any n-cell φ bounded by l and m (including the degenerate case where $l = m = \varphi$), the diagram

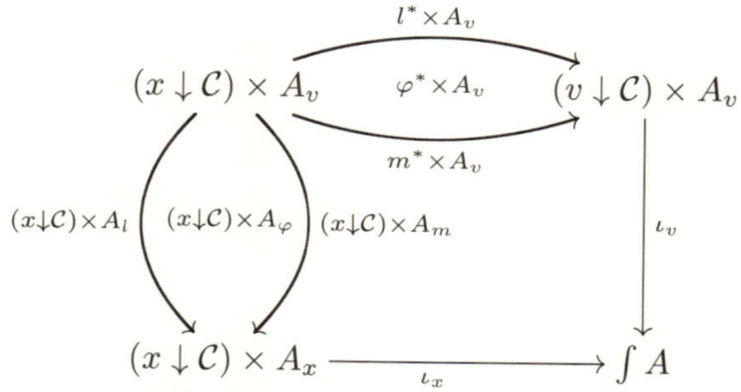

in ω-**Cat** commutes. To see this note that, for 0-cells $\vec{y} = (y, y_-, y_o)$ of $(x \downarrow \mathcal{C}) \times A_v$,

$$\iota_v \circ (l^* \times A_v)(\vec{y}) = (y, y_o \cdot (y_- *_0 l)) = (y, y_o \cdot l \cdot y_-) = \iota_x \circ ((x \downarrow \mathcal{C}) \times A_l)(\vec{y}),$$

and, for $\vec{\varphi}$ any higher dimensional cell bounded by 0-cells $\vec{y}$ and $\vec{z}$, we have

$$\begin{aligned}\iota_v \circ (l^* \times A_v)(\vec{\varphi}) &= \big(\varphi, (\varphi_o \cdot (z_- *_0 l)) *_0 (y_o \cdot (\varphi_- *_0 l))\big)\\ &= \big(\varphi, (\varphi_o \cdot l \cdot z_-) *_0 (y_o \cdot l \cdot \varphi_-)\big)\\ &= \iota_x \circ ((x \downarrow \mathcal{C}) \times A_l)(\vec{\varphi}).\end{aligned}$$

B.3. Verification of the universal property. We will now show that the ι_- constitute the universal dinatural transformation of this form. I.e., we prove that if $\mathcal{A}$ is any ω-category together with a family of dinatural transformations (in the ω-categorical sense of a ω-categorical natural transformation) $\xi\colon (- \downarrow \mathcal{C}) \times A_- \to \mathcal{A}$, then there exists a canonical functor (of ω-categories) $F\colon \int A \to \mathcal{A}$ such that

(B.7)

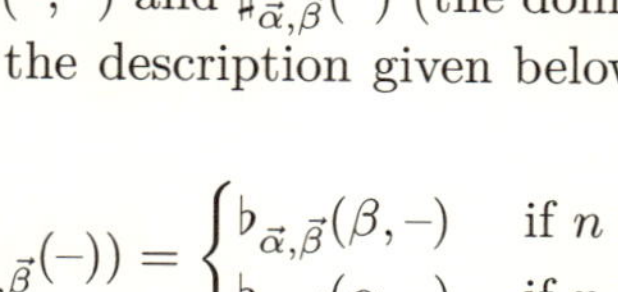

commutes for every object x of $\mathcal{C}$. The construction of F is a fairly involved induction on dimension. In particular, we will need to construct, for each pair of n-cells $\vec{\alpha}, \vec{\beta}$, auxiliary functors $\flat_{\vec{\alpha},\vec{\beta}}(-,-)$ and $\sharp_{\vec{\alpha},\vec{\beta}}(-)$ (the domains and codomains of these functors will be clear from the description given below). Moreover, we will prove at each stage that

$$\sharp_{\vec{\alpha},\vec{\beta}}(\eth_{\vec{\alpha},\vec{\beta}}(-)) = \begin{cases}\flat_{\vec{\alpha},\vec{\beta}}(\beta,-) & \text{if } n \text{ is even},\\ \flat_{\vec{\alpha},\vec{\beta}}(\alpha,-) & \text{if } n \text{ is odd}.\end{cases}$$

Additionally, we will require that at even stages $F(\vec{\alpha}) = \flat_{\vec{\alpha},\vec{\beta}}(\alpha, 1_\alpha)$ and $F(\vec{\beta}) = \sharp_{\vec{\alpha},\vec{\beta}}(\beta_-)$, and that the dual equations hold at odd stages. The final fact we require is that, as arrows in $(y \downarrow \mathcal{C})$, if $\varphi\colon \alpha \to \beta$ is an n-cell in $\mathcal{C}$ (bounded by 0-cells x and y), then $(\varphi, 1_\varphi)\colon (\alpha, 1_\alpha) \to (\beta, \varphi)$ when n is odd. Similarly, when n is even, $(\varphi, 1_\varphi)\colon (\alpha, \varphi) \to (\beta, 1_\beta)$. This last fact is a straightforward induction, using the definition of the ω-category $(y \downarrow \mathcal{C})$. The second condition will be trivially satisfied by definition in each case and therefore we omit explicit mention of it in the construction below. We will also omit the proof of functoriality of the auxiliary functors as this is also straightforward.

We describe F as follows:

0-Cells. Given a 0-cell $\vec{x} = (x, x_-)$ of $\int A$, we define

$$F(\vec{x}) := \xi_x(x, 1_x, x_-).$$

For 0-cells $\vec{x}, \vec{y}$, we define

$$\begin{aligned} \flat_{\vec{x},\vec{y}}(-,-) &:= \xi_x(-,-,x_-), \\ \sharp_{\vec{x},\vec{y}}(-) &:= \xi_y(y, 1_y, -). \end{aligned}$$

Then we have

$$\begin{aligned} \sharp_{\vec{x},\vec{y}}(\eth_{\vec{x},\vec{y}}(-)) &= \xi_y(y, 1_y, x_- \cdot -) \\ &= \xi_y \circ ((y \downarrow \mathcal{C}) \times A_-)(y, 1_y, x_-) \\ &= \xi_x \circ (-^* \times A_x)(y, 1_y, x_-) \\ &= \xi_x(y, -, x_-) \\ &= \flat_{\vec{x},\vec{y}}(y, -), \end{aligned}$$

where the third equation is by dinaturality of ξ.

$(n+1)$-Cells for $(n+1)$ odd. Given a $(n+1)$-cell $\vec{\varphi} = (\varphi, \varphi_-)\colon \vec{\alpha} \to \vec{\beta}$, we define $F(\vec{\varphi})\colon F(\vec{\alpha}) \to F(\vec{\beta})$ as the following composite

$$\begin{array}{ccccccc} F(\vec{\alpha}) & & & & & & F(\vec{\beta}) \\ \| & & & & & & \| \\ \flat_{\vec{\alpha},\vec{\beta}}(\alpha, 1_\alpha) & \xrightarrow{\flat_{\vec{\alpha},\vec{\beta}}(\varphi, 1_\varphi)} & \flat_{\vec{\alpha},\vec{\beta}}(\beta, \varphi) & = & \sharp_{\vec{\alpha},\vec{\beta}}(\eth_{\vec{\alpha},\vec{\beta}}(\varphi)) & \xrightarrow{\sharp_{\vec{\alpha},\vec{\beta}}(\varphi_-)} & \sharp_{\vec{\alpha},\vec{\beta}}(\beta_-) \end{array}$$

I.e.,

$$F(\vec{\varphi}) := \sharp_{\vec{\alpha},\vec{\beta}}(\varphi_-) *_n \flat_{\vec{\alpha},\vec{\beta}}(\varphi, 1_\varphi).$$

Now, given parallel $(n+1)$-cells $\vec{\varphi}, \vec{\psi}\colon \vec{\alpha} \to \vec{\beta}$ we define

$$\flat_{\vec{\varphi},\vec{\psi}}(-,-) := \sharp_{\vec{\alpha},\vec{\beta}}(\psi_-) *_n \flat_{\vec{\alpha},\vec{\beta}}(-,-) \sharp_{\vec{\varphi},\vec{\psi}}(-) := \sharp_{\vec{\alpha},\vec{\beta}}(-) *_n \flat_{\vec{\alpha},\vec{\beta}}(\varphi, 1_\varphi).$$

With this definition, we observe that

$$\begin{aligned} \sharp_{\vec{\varphi},\vec{\psi}}(\eth_{\vec{\varphi},\vec{\psi}}(-)) &= \sharp_{\vec{\varphi},\vec{\psi}}(\psi_- *_n \eth_{\vec{\alpha},\vec{\beta}}(-)) \\ &= \sharp_{\vec{\alpha},\vec{\beta}}(\psi_- *_n \eth_{\vec{\alpha},\vec{\beta}}(-)) *_n \flat_{\vec{\alpha},\vec{\beta}}(\varphi, 1_\varphi) \\ &= \sharp_{\vec{\alpha},\vec{\beta}}(\psi_-) *_n \flat_{\vec{\alpha},\vec{\beta}}(\beta, -) *_n \flat_{\vec{\alpha},\vec{\beta}}(\varphi, 1_\varphi) \\ &= \sharp_{\vec{\alpha},\vec{\beta}}(\psi_-) *_n \flat_{\vec{\alpha},\vec{\beta}}(\varphi, -) \\ &= \flat_{\vec{\alpha},\vec{\beta}}(\varphi, -), \end{aligned}$$

where the third equation is by functoriality and the corresponding equation for $\sharp_{\vec{\alpha},\vec{\beta}}(-)$.

$(n+1)$-Cells for $(n+1)$ even. Given a $(n+1)$-cell $\vec{\varphi}\colon \vec{\alpha} \to \vec{\beta}$ we define $F(\vec{\varphi})$ to be the composite

$$\begin{array}{ccccccc} F(\vec{\alpha}) & & & & & & F(\vec{\beta}) \\ \Vert & & & & & & \Vert \\ \sharp_{\vec{\alpha},\vec{\beta}}(\alpha_-) & \xrightarrow{\sharp_{\vec{\alpha},\vec{\beta}}(\varphi_-)} & \sharp_{\vec{\alpha},\vec{\beta}}(\partial_{\vec{\alpha},\vec{\beta}}(\varphi)) & = & \flat_{\vec{\alpha},\vec{\beta}}(\alpha,\varphi) & \xrightarrow{\flat_{\vec{\alpha},\vec{\beta}}(\varphi,1_\varphi)} & \flat_{\vec{\alpha},\vec{\beta}}(\beta,1_\beta) \end{array}$$

I.e.,

$$F(\vec{\varphi}) := \flat_{\vec{\alpha},\vec{\beta}}(\varphi, 1_\varphi) *_n \sharp_{\vec{\alpha},\vec{\beta}}(\varphi_-).$$

Finally, for parallel $(n+1)$-cells $\vec{\varphi}, \vec{\psi}\colon \vec{\alpha} \to \vec{\beta}$ we define

$$\begin{aligned} \flat_{\vec{\varphi},\vec{\psi}}(-,-) &:= \flat_{\vec{\alpha},\vec{\beta}}(-,-) *_n \sharp_{\vec{\alpha},\vec{\beta}}(\varphi_-), \\ \sharp_{\vec{\varphi},\vec{\psi}}(-) &:= \flat_{\vec{\alpha},\vec{\beta}}(\psi, 1_\psi) *_n \sharp_{\vec{\alpha},\vec{\beta}}(-). \end{aligned}$$

The verification that the corresponding equation holds is by an argument dual to the one for the case of $(n+1)$ odd given above.

Next, we will prove inductively that F makes the diagram (B.7) commute. The key point which will allow us to prove this fact is that the action of the functor F has a simplified description for those cells of $\int A$ in the image of ι_x. (To prove the uniqueness of F we will in fact show that all cells of $\int A$ occur as composites of cells in the images of the ι_x, but more on this later.) In order to explain this it will be convenient to again introduce additional auxiliary operations $\diamondsuit_{\vec{\alpha},\vec{\beta}}(\vec{\varphi})$ and $\clubsuit_{\vec{\alpha},\vec{\beta}}(\vec{\varphi})$ at each stage of the proof which are indexed by parallel n-cells $\vec{\alpha}, \vec{\beta}$ of $(x \downarrow C) \times A_x$ and which act on m-cells $\vec{\varphi}$ of $(x \downarrow C) \times A_x$ for $m \geq n$. We will also prove that these auxiliary functions interact in an appropriate way with each other and with $\sharp_{\iota_x(\vec{\alpha}),\iota_x(\vec{\beta})}(-)$ and $\flat_{\iota_x(\vec{\alpha}),\iota_x(\vec{\beta})}(-,-)$. These matters are described below in the proof.

0-Cells. Given $\vec{y} = (y, y_-, y_o)$ in $(x \downarrow C) \times A_x$, we have

$$F \circ \iota_x(\vec{y}) = F(y, y_o \cdot y_-) = \xi_y(y, y, y_o \cdot y_-) = \xi_x(y, y_-, y_o),$$

where the final equation is by dinaturality of the ξ.

Given two 0-cells $\vec{y}$ and $\vec{z}$, and a n-cell $\vec{\varphi}$ bounded by $\vec{y}$ and $\vec{z}$, respectively, we define

$$\begin{aligned} \clubsuit_{\vec{y},\vec{z}}(\vec{\varphi}) &:= \xi_x(z, \varphi_-, y_o), \\ \diamondsuit_{\vec{y},\vec{z}}(\vec{\varphi}) &:= \xi_x(\alpha, \alpha *_0 y_-, y_o), \end{aligned}$$

and we observe that

$$\sharp_{\iota_x(\vec{y}),\iota_x(\vec{z})}\big(\iota_x(\vec{\varphi})_-\big) = \xi_x(z, z_-, \varphi_o) *_0 \clubsuit_{\vec{y},\vec{z}}(\vec{\varphi}), \tag{B.8}$$

$$\flat_{\iota_x(\vec{y}),\iota_x(\vec{z})}(\varphi, \varphi) = \diamondsuit_{\vec{y},\vec{z}}(\vec{\varphi}). \tag{B.9}$$

To see that these equations hold we reason as follows:

$$\begin{aligned} \sharp_{\iota_x(\vec{y}),\iota_x(\vec{z})}\big(\iota_x(\vec{\varphi})_-\big) &= \xi_z(z, z, (\varphi_o \cdot z_-) *_0 (y_o \cdot \varphi_-)) \\ &= \xi_z(z, z, \varphi_o \cdot z_-) *_0 \xi_z(z, z, y_o \cdot \varphi_-) \\ &= \xi_x(z, z_-, \varphi_o) *_0 \xi_x(z, \varphi_-, y_o), \end{aligned}$$

where the final equation is by dinaturality of ξ, and similarly

$$\begin{aligned}\flat_{\iota_x(\vec{y}),\iota_x(\vec{z})}(\varphi,\varphi) &= \xi_y(\varphi,\varphi,y_o\cdot y_-)\\ &= \xi_x(\varphi,\varphi *_0 y_-,y_o)\end{aligned}$$

where the final equation is by dinaturality of ξ.

We also note that, by definition,

$$\clubsuit_{\vec{y},\vec{z}}(\vec{\varphi}) *_0 \diamondsuit_{\vec{y},\vec{z}}(\vec{\varphi}) = \xi_x(f,f_-,y_o)$$

1-Cells. Given $\vec{f} = (f,f_-,f_o)\colon \vec{y}\to\vec{z}$ in $(x\downarrow\mathcal{C})\times A_x$, we have

$$\begin{aligned}F\circ\iota_x(\vec{f}) &= \sharp_{\iota_x(\vec{y}),\iota_x(\vec{z})}\big(\iota_x(\vec{f})_-\big) *_0 \flat_{\iota_x(\vec{y}),\iota_x(\vec{z})}(f,f)\\ &= \xi_x(z,z_-,f_o) *_0 \clubsuit_{\vec{y},\vec{z}}(\vec{f}) *_0 \diamondsuit_{\vec{y},\vec{z}}(\vec{f})\\ &= \xi_x(z,z_-,f_o) *_0 \xi_x(f,f_-,y_o)\\ &= \xi_x(f,f_-,f_o).\end{aligned}$$

where the second equation is by (B.8) and (B.9).

Now, given parallel 1-cells $\vec{f},\vec{g}\colon\vec{y}\to\vec{z}$, we define

$$\begin{aligned}\clubsuit_{\vec{f},\vec{g}}(\vec{\varphi}) &:= \clubsuit_{\vec{y},\vec{z}}(\vec{g}) *_0 \diamondsuit_{\vec{y},\vec{z}}(\vec{\varphi}),\\ \diamondsuit_{\vec{f},\vec{g}}(\vec{\varphi}) &:= \clubsuit_{\vec{y},\vec{z}}(\vec{\varphi}) *_0 \diamondsuit_{\vec{y},\vec{z}}(\vec{f}),\end{aligned}$$

and we observe that

$$\sharp_{\iota_x(\vec{f}),\iota_x(\vec{g})}\big(\iota_x(\vec{\varphi})_-\big) = \xi_x(z,z_-,\varphi_o) *_0 \diamondsuit_{\vec{f},\vec{g}}(\vec{\varphi}), \tag{B.10}$$

$$\flat_{\iota_x(\vec{f}),\iota_x(\vec{g})}(\varphi,\varphi) = \xi_x(z,z_-,g_o) *_0 \clubsuit_{\vec{f},\vec{g}}(\vec{\varphi}). \tag{B.11}$$

The proof that these equations hold is immediate from the definition together with (B.8) and (B.9). Similarly, we also have that

$$\clubsuit_{\vec{f},\vec{g}}(\vec{\varphi}) *_1 \diamondsuit_{\vec{f},\vec{g}}(\vec{\varphi}) = \xi_x(\varphi,\varphi_-,y_o).$$

2-Cells. Given $\vec{\alpha} = (\alpha,\alpha_-,\alpha_o)\colon\vec{f}\to\vec{g}$ in $(x\downarrow\mathcal{C})\times A_x$ we have

$$\begin{aligned}F\circ\iota_x(\vec{\alpha}) &= \flat_{\iota_x(\vec{f}),\iota_x(\vec{g})}(\alpha,\alpha) *_1 \sharp_{\iota_x(\vec{f}),\iota_x(\vec{g})}\big(\iota_x(\vec{\alpha})_-\big)\\ &= \big(\xi_x(z,z_-,g_o) *_0 \clubsuit_{\vec{f},\vec{g}}(\vec{\alpha})\big) *_1 \big(\xi_x(z,z_-,\alpha_o) *_0 \diamondsuit_{\vec{f},\vec{g}}(\vec{\alpha})\big)\\ &= \xi_x(z,z_-,\alpha_o) *_0 \big(\clubsuit_{\vec{f},\vec{g}}(\vec{\alpha}) *_1 \clubsuit_{\vec{f},\vec{g}}(\vec{\alpha})\big)\\ &= \xi_x(z,z_-,\alpha_o) *_0 \xi_x(\alpha,\alpha_-,y_o)\\ &= \xi_x(\vec{\alpha}),\end{aligned}$$

where the first equation is (B.10) and (B.11).

This calculation can be visualized as follows:

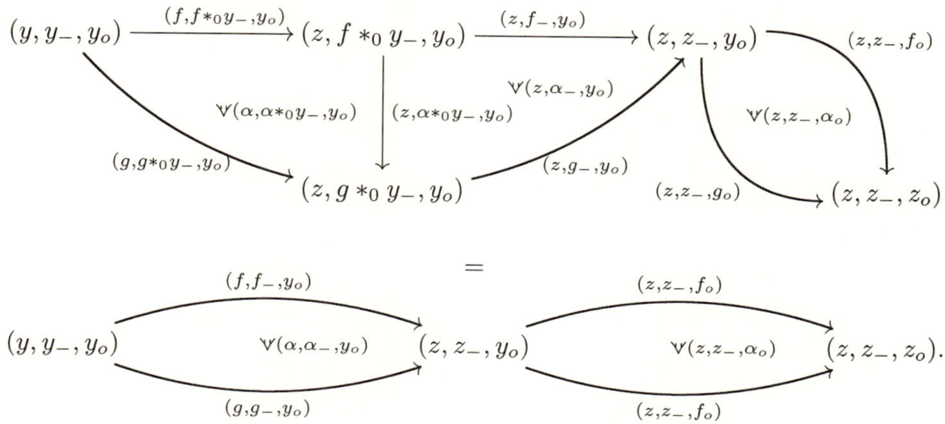

Now, given parallel 2-cells $\vec{\alpha}, \vec{\beta}\colon \vec{f} \to \vec{g}$ we define

$$\clubsuit_{\vec{\alpha},\vec{\beta}}(\vec{\varphi}) := \clubsuit_{\vec{f},\vec{g}}(\vec{\beta}) *_1 \diamondsuit_{\vec{f},\vec{g}}(\vec{\varphi}),$$
$$\diamondsuit_{\vec{\alpha},\vec{\beta}}(\vec{\varphi}) := \clubsuit_{\vec{f},\vec{g}}(\vec{\varphi}) *_1 \diamondsuit_{\vec{f},\vec{g}}(\vec{\alpha}).$$

It then follows that

$$\flat_{\iota_x(\vec{\alpha}),\iota_x(\vec{\beta})}(\varphi,\varphi) = \xi_x(z, z_-, \alpha_o) *_0 \diamondsuit_{\vec{\alpha},\vec{\beta}}(\vec{\varphi}), \tag{B.12}$$

$$\sharp_{\iota_x(\vec{\alpha}),\iota_x(\vec{\beta})}\big(\iota_x(\vec{\varphi})_-\big) = \xi_x(z, z_-, \varphi_o) *_0 \clubsuit_{\vec{\alpha},\vec{\beta}}(\vec{\varphi}).,\tag{B.13}$$

and that

$$\clubsuit_{\vec{\alpha},\vec{\beta}}(\vec{\varphi}) *_2 \diamondsuit_{\vec{\alpha},\vec{\beta}}(\vec{\varphi}) = \xi_x(\varphi, \varphi_-, y_o).$$

All three of these equations are easily verified using the corresponding equations for $\clubsuit_{\vec{f},\vec{g}}$ and $\diamondsuit_{\vec{f},\vec{g}}$.

We now describe the general structure of the inductive proof. At each stage $(n+1)$ we will prove that $F \circ \iota_x(\vec{\varphi}) = \xi_x(\vec{\varphi})$, for $\vec{\varphi}$ a $(n+1)$-cell. We will also define, for parallel $(n+1)$-cells $\vec{\varphi}, \vec{\psi}\colon \vec{\alpha} \to \vec{\beta}$, Next, given parallel $(n+1)$-cells $\vec{\varphi}, \vec{\psi}\colon \vec{\alpha} \to \vec{\beta}$ we define

$$\clubsuit_{\vec{\varphi},\vec{\psi}}(\vec{\gamma}) := \clubsuit_{\vec{\alpha},\vec{\beta}}(\vec{\psi}) *_n \diamondsuit_{\vec{\alpha},\vec{\beta}}(\vec{\gamma})\diamondsuit_{\vec{\varphi},\vec{\psi}}(\vec{\gamma}) := \clubsuit_{\vec{\alpha},\vec{\beta}}(\vec{\gamma}) *_n \diamondsuit_{\vec{\alpha},\vec{\beta}}(\vec{\varphi}),$$

$\clubsuit_{\vec{\varphi},\vec{\psi}}$ and we will prove the corresponding equations:

$$\flat_{\iota_x(\vec{\varphi}),\iota_x(\vec{\psi})}(\gamma,\gamma) = \begin{cases} \xi_x(z, z_-, \varphi_o) *_0 \diamondsuit_{\vec{\varphi},\vec{\psi}}(\vec{\gamma}) & \text{for } (n+1) \text{ even,} \\ \xi_x(z, z_-, \psi_o) *_0 \clubsuit_{\vec{\varphi},\vec{\psi}}(\vec{\gamma}) & \text{for } (n+1) \text{ odd,} \end{cases}$$

and

$$\sharp_{\iota_x(\vec{\varphi}),\iota_x(\vec{\psi})}\big(\iota_x(\vec{\gamma})_-\big) = \begin{cases} \xi_x(z, z_-, \gamma_o) *_0 \clubsuit_{\vec{\varphi},\vec{\psi}}(\vec{\gamma})\text{for } (n+1) \text{ even,} \\ \xi_x(z, z_-, \gamma_o) *_0 \diamondsuit_{\vec{\varphi},\vec{\psi}}(\vec{\gamma})\text{for } (n+1) \text{ odd.} \end{cases}$$

Finally, we will show at each stage that

$$\clubsuit_{\vec{\varphi},\vec{\psi}}(\vec{\gamma}) *_{(n+1)} \diamondsuit_{\vec{\varphi},\vec{\psi}}(\vec{\gamma}) = \xi_x(\gamma, \gamma_-, y_o).$$

We now turn to the proof.

$(n+1)$-Cells for $(n+1)$ odd. Given a $(n+1)$-cell $\vec{\varphi}\colon \vec{\alpha} \to \vec{\beta}$ we have

$$\begin{aligned}
F \circ \iota_x(\vec{\varphi}) &= \sharp_{\iota_x(\vec{\alpha}),\iota_x(\vec{\beta})}\big(\iota_x(\vec{\varphi})_-\big) *_n \flat_{\iota_x(\vec{\alpha}),\iota_x(\vec{\beta})}(\varphi,\varphi)\\
&= \big(\xi_x(z,z_-,\varphi_o) *_0 \clubsuit_{\vec{\alpha},\vec{\beta}}(\vec{\varphi})\big) *_n \big(\xi_x(z,z_-,\alpha_o) *_0 \diamondsuit_{\vec{\alpha},\vec{\beta}}(\vec{\varphi})\big)\\
&= \xi_x(z,z_-,\varphi_o) *_0 \big(\clubsuit_{\vec{\alpha},\vec{\beta}}(\vec{\varphi}) *_n \diamondsuit_{\vec{\alpha},\vec{\beta}}(\vec{\varphi})\big)\\
&= \xi_x(\vec{\varphi}).
\end{aligned}$$

Next, given parallel $(n+1)$-cells $\vec{\varphi},\vec{\psi}\colon \vec{\alpha} \to \vec{\beta}$, we prove that the required equations hold as follows:

$$\begin{aligned}
\flat_{\iota_x(\vec{\varphi}),\iota_x(\vec{\psi})}(\gamma,\gamma) &= \sharp_{\iota_x(\vec{\alpha}),\iota_x(\vec{\beta})}\big(\iota_x(\vec{\psi})_-\big) *_n \flat_{\iota_x(\vec{\alpha}),\iota_x(\vec{\beta})}(\gamma,\gamma)\\
&= \big(\xi_x(z,z_-,\gamma_o) *_0 \clubsuit_{\vec{\alpha},\vec{\beta}}(\vec{\gamma})\big) *_n \big(\xi_x(z,z_-,\varphi_o) *_0 \diamondsuit_{\vec{\alpha},\vec{\beta}}(\vec{\gamma})\big)\\
&= \xi_x(z,z_-,\gamma_o) *_0 \big(\clubsuit_{\vec{\alpha},\vec{\beta}}(\vec{\gamma}) *_n \diamondsuit_{\vec{\alpha},\vec{\beta}}(\vec{\gamma})\big)\\
&= \xi_x(\vec{\gamma}),
\end{aligned}$$

where we have used the equations given by the induction hypothesis. Similarly, we see that

$$\clubsuit_{\vec{\varphi},\vec{\psi}}(\vec{\gamma}) *_{n+1} \diamondsuit_{\vec{\varphi},\vec{\psi}}(\vec{\gamma}) = \xi_x(\gamma,\gamma_-,y_o).$$

$(n+1)$-Cells for $(n+1)$ even. The calculations here are dual (in the appropriate sense) to those for the odd case and are left to the reader.

This completes the proof of the following lemma:

Lemma B.1. *With the definition of F given above,* (B.7) *commutes.*

Finally, in order to show that F is the canonical functor $\int A \to \mathcal{A}$ making (B.7) commute, we will prove that every cell of $\int A$ can be written as a composite of cells in the image of the maps ι_x. The construction of this decomposition of the cells of $\int A$, which is again by induction on dimension, involves constructing auxiliary operations (which mimic $\flat$ and $\sharp$) and then proving that these satisfy suitable equations.

0-Cells. Given a 0-cell $\vec{x} = (x, x_-)$ in $\int A$ we have that

$$\vec{x} = \iota_x(x, x, x_-).$$

Given 0-cells $\vec{x}$ and $\vec{y}$ and a $(n+1)$-cell $\vec{\varphi}$ bounded by $\vec{x}$ and $\vec{y}$ in $\int A$, we define

$$\begin{aligned}
\hat{\flat}_{\vec{x},\vec{y}}(\vec{\varphi}) &:= \iota_x(\varphi,\varphi,x_-),\\
\hat{\sharp}_{\vec{x},\vec{y}}(\vec{\varphi}) &:= \iota_y(y,y,\varphi_-).
\end{aligned}$$

We observe that we have the equations

$$\begin{aligned}
\hat{\flat}_{\vec{x},\vec{y}}(\vec{\varphi}) &= (\varphi, \mathfrak{d}_{\vec{x},\vec{y}}(\varphi)),\\
\hat{\sharp}_{\vec{x},\vec{y}}(\vec{\varphi}) &= (y, \varphi_-),
\end{aligned}$$

since

$$\iota_x(-,-,x_-)_- = (x_- \cdot -) *_0 (x_- \cdot -) = x_- \cdot - = \mathfrak{d}_{\vec{x},\vec{y}}(-),$$

and the first equation is trivial by definition of $\iota_y(y,y,-)$.

1-Cells. Given $\vec{f} = (f, f_-)\colon \vec{x} \to \vec{y}$ we have

$$\iota_y(y,y,f_-) *_0 \iota_x(f,f,x_-) = \vec{f}.$$

by the equations above (since $\mathfrak{d}_{\vec{x},\vec{y}}(f)$ is the domain of f_-).

Next, given parallel $\vec{f}, \vec{g} \colon \vec{x} \to \vec{y}$ in $\int A$, we define

$$\hat{\flat}_{\vec{f},\vec{g}}(\vec{\varphi}) := \hat{\sharp}_{\vec{x},\vec{y}}(\vec{g}) *_0 \hat{\flat}_{\vec{x},\vec{y}}(\vec{\varphi}),$$
$$\hat{\sharp}_{\vec{f},\vec{g}}(\vec{\varphi}) := \hat{\sharp}_{\vec{x},\vec{y}}(\vec{\varphi}) *_0 \hat{\flat}_{\vec{x},\vec{y}}(\vec{f}).$$

Now, we also have the equations

$$\hat{\flat}_{\vec{f},\vec{g}}(\vec{\varphi}) = (\varphi, \mathfrak{d}_{\vec{f},\vec{g}}(\varphi)),$$
$$\hat{\sharp}_{\vec{f},\vec{g}}(\vec{\varphi}) = (f, \varphi_-),$$

as the reader may readily verify.

We may now describe the general inductive construction. Explicitly, for parallel $(n+1)$-cells $\vec{\varphi}, \vec{\psi} \colon \vec{\alpha} \to \vec{\beta}$, we define

$$\hat{\flat}_{\vec{\varphi},\vec{\psi}}(\vec{\gamma}) := \begin{cases} \hat{\sharp}_{\vec{\alpha},\vec{\beta}}(\vec{\psi}) *_n \hat{\flat}_{\vec{\alpha},\vec{\beta}}(\vec{\gamma}) & \text{for } (n+1) \text{ odd}, \\ \hat{\flat}_{\vec{\alpha},\vec{\beta}}(\vec{\gamma}) *_n \hat{\sharp}_{\vec{\alpha},\vec{\beta}}(\vec{\varphi}) & \text{for } (n+1) \text{ even}, \end{cases}$$

and

$$\hat{\sharp}_{\vec{\varphi},\vec{\psi}}(\vec{\gamma}) := \begin{cases} \hat{\sharp}_{\vec{\alpha},\vec{\beta}}(\vec{\gamma}) *_n \hat{\flat}_{\vec{\alpha},\vec{\beta}}(\vec{\varphi}) & \text{for } (n+1) \text{ odd}, \\ \hat{\flat}_{\vec{\alpha},\vec{\beta}}(\vec{\psi}) *_n \hat{\sharp}_{\vec{\alpha},\vec{\beta}}(\vec{\gamma}) & \text{for } (n+1) \text{ even}. \end{cases}$$

We note that with these definitions one easily proves by induction that the equations

$$\hat{\flat}_{\vec{\varphi},\vec{\psi}}(\vec{\gamma}) = (\gamma, \mathfrak{d}_{\vec{\varphi},\vec{\psi}}(\gamma)),$$
$$\hat{\sharp}_{\vec{\varphi},\vec{\psi}}(\vec{\gamma}) = (\varphi, \gamma_-)$$

hold. So, if $\vec{\gamma} \colon \vec{\varphi} \to \vec{\psi}$ is any $(n+1)$-cell of $\int A$, then

$$\vec{\gamma} = \hat{\sharp}_{\vec{\varphi},\vec{\psi}}(\vec{\gamma}) *_n \hat{\flat}_{\vec{\varphi},\vec{\psi}}(\vec{\gamma}).$$

Therefore, since each cell of the form $\hat{\flat}_{\vec{\varphi},\vec{\psi}}(\vec{\gamma})$ or $\hat{\sharp}_{\vec{\varphi},\vec{\psi}}(\vec{\gamma})$ is by construction a composite of cells in the image of the ι it follows that every cell of $\int A$ can be represented as a composite of cells in the image of the ι and therefore F is the canonical map $\int A \to \mathcal{A}$ making (B.7) commute. This completes the proof of the following:

Theorem B.2. *Given a functor $A \colon \mathcal{C} \to \omega$-**Cat** of ω-categories, the Grothendieck construction $\int A$ described in Section 3 is the coend $\int^x (x \downarrow \mathcal{C}) \times A_x$.*

References

1. S. Awodey, P. Hofstra, and M. A. Warren, *Martin-Löf complexes*, submitted, available at `arXiv:0906.4521`.
2. S. Awodey and M. A. Warren, *Homotopy theoretic models of identity types*, Math. Proc. Cambridge Philos. Soc. **146** (2009), no. 1, 45–55.
3. M. A. Batanin, *Monoidal globular categories as a natural environment for the theory of weak n-categories*, Adv. Math. **136** (1998), no. 1, 39–103.
4. P. Dybjer, *Internal type theory*, Types for Proofs and Programs (Torino, 1995) (Berardi Sefano and M. Coppo, eds.), Lecture Notes in Comput. Sci., vol. 1158, Springer, Berlin, 1996, pp. 120–134.
5. N. Gambino and R. Garner, *The identity type weak factorisation system*, Theoret. Comput. Sci. **409** (2008), no. 1, 94–109.
6. R. Garner, *Two-dimensional models of type theory*, Math. Structures Comput. Sci. **19** (2009), no. 4, 687–736.
7. C. Hermida, *Some properties of* **Fib** *as a fibred 2-category*, J. Pure Appl. Algebra **134** (1999), no. 1, 83–109.

8. M. Hofmann, *Syntax and semantics of dependent types*, Semantics and Logics of Computation (Cambridge, 1995) (A. M. Pitts and P. Dybjer, eds.), Publ. Newton Inst., vol. 14, Cambridge Univ. Press, Cambridge, 1997, pp. 79–130.
9. M. Hofmann and T. Streicher, *The groupoid interpretation of type theory*, Twenty-Five Years of Constructive Type Theory (Venice, 1995) (G. Sambin and J. M. Smith, eds.), Oxford Logic Guides, vol. 36, Oxford Univ. Press, New York, 1998, pp. 83–111.
10. B. Jacobs, *Categorical logic and type theory*, Stud. Logic Found. Math., vol. 141, North-Holland, Amsterdam, 1999.
11. M. M. Kapranov and V. A. Voevodsky, *∞-groupoids and homotopy types*, Cahiers Topologie Géom. Différentielle Catég. **32** (1991), no. 1, 29–46.
12. Y. Lafont, F. Métayer, and K. Worytkiewicz, *A folk model structure on omega-cat*, Adv. Math. **224** (2010), no. 3, 1183–1231.
13. F. W. Lawvere, *Adjointness in foundations*, Repr. Theory Appl. Categ. **16** (2006), 1–16. Reprinted from Dialectica **23** (1969).
14. T. Leinster, *Higher operads, higher categories*, London Math. Soc. Lecture Note Ser., vol. 298, Cambridge Univ. Press, Cambridge, 2004.
15. P. L. Lumsdaine, *Weak ω-categories from intensional type theory*, Typed Lambda Calculi and Applications (Brasília, 2009) (P.-L. Curien, ed.), Lecture Notes in Comput. Sci., vol. 5608, Springer, Berlin, 2009, pp. 172–187.
16. P. Martin-Löf, *An intuitionistic theory of types: predicative part*, Logic Colloquium '73 (Bristol, 1973) (H. E. Rose and J. C. Shepherdson, eds.), Stud. Logic Found. Math., vol. 80, North-Holland, Amsterdam, 1975, pp. 73–118.
17. F. Métayer, *Resolutions by polygraphs*, Theory Appl. Categ. **11** (2003), no. 7, 148–184.
18. I. Moerdijk and J.-A. Svensson, *Algebraic classification of equivariant homotopy 2-types*. I, J. Pure Appl. Algebra **89** (1993), no. 1-2, 187–216.
19. B. Nordström, K. Petersson, and J. M. Smith, *Programming in Martin-Löf's type theory: An introduction*, Internat. Ser. Monogr. Comput. Sci., vol. 7, Oxford Univ. Press, New York, 1990.
20. T. Palm, *Categories with slicing*, Theory Appl. Categ. **22** (2009), no. 5, 97–135.
21. R. Street, *The algebra of oriented simplexes*, J. Pure Appl. Algebra **49** (1987), no. 3, 283–335.
22. ———, *The petit topos of globular sets*, J. Pure Appl. Algebra **154** (2000), no. 1-3, 299–315.
23. T. Streicher, *Semantics of type theory*, with a foreword by M. Wirsing, Progr. Theoret. Comput. Sci., Birkhäuser, Boston, MA, 1991.
24. ———, *Investigations into intensional type theory*, Habilitationsschrift, Ludwig-Maximilians-Universität München, 1993.
25. B. van den Berg and R. Garner, *Types are weak ω-groupoids*, submitted, available at `arXiv:0812.0298`.
26. D. Verity, *Complicial sets characterising the simplicial nerves of strict ω-categories*, Mem. Amer. Math. Soc. **193** (2008), no. 905.
27. M. A. Warren, *Homotopy theoretic aspects of constructive type theory*, Ph.D. Thesis, Carnegie Mellon University, 2008.
28. ———, *A characterization of representable intervals*, submitted, available at `arXiv:0903.3743`.

Department of Mathematics and Statistics, University of Ottawa, 585 King Edward Ave., Ottawa, ON K1N 6N5, Canada

E-mail address: `mwarren@uottawa.ca`

Centre de Recherches Mathématiques
CRM Proceedings and Lecture Notes
Volume **53**, 2011

Lax Monoidal Fibrations

Marek Zawadowski

Dedicated to Mihály Makkai on the occasion of his 70th birthday

ABSTRACT. We introduce the notion of a lax monoidal fibration and we show how it can be conveniently used to deal with various algebraic structures that play an important role in some definitions, cf. [3, 12, 24, 25] of the opetopic sets. We present the 'standard' such structures, the exponential fibrations of basic fibrations and three areas of applications. The first area is related to the T-categories of A. Burroni. The monoids in the Burroni lax monoidal fibrations form the fibration of T-categories. The construction of the relative Burroni fibrations and free T-categories in this context, allow us to extend the definition of the set of opetopes given in [19] to the category of opetopic sets (internally to any Grothendieck topos, if needed). We also show that the fibration of (1-level) multicategories considered in [12] is equivalent to the fibration of (finitary, cartesian) polynomial monads. This equivalence is induced by the equivalence of lax monoidal fibrations of amalgamated signatures, polynomial diagrams, and polynomial (finitary, endo) functors. Finally, we develop a similar theory for symmetric signatures, analytic diagrams (a notion introduced here), and (finitary, multivariable) analytic (endo)functors, cf. [15]. Among other things we show that the fibrations of symmetric multicategories is equivalent to the fibration of analytic monads. We also give a characterization (Corollary 7.6) of such a fibration of analytic monads. An object of this fibration is a weakly cartesian monad on a slice of Set whose functor part is a finitary functor weakly preserving wide pullbacks. A morphism of this fibration is a weakly cartesian morphism of monads whose functor part is a pullback functor.

1. Introduction

The notion of a lax monoidal fibration studied here, is designed to help understand connections between various definitions of opetopic sets. More specifically, various algebraic/categorical mechanisms used to define them. The primary goal was to understand the relation between the notion of opetopic set of Baez – Dolan, cf. [3] and multitopic set of Hermida – Makkai – Power, cf. [12]. However, it turned out that many other notions connected with the development of higher categories can be successfully organized into a lax monoidal fibration. This is not a mere encoding for its sake but in the context of lax monoidal fibration many notions can be

2010 *Mathematics Subject Classification.* 18D10, 18D30, 18D50, 18C15.
This is the final form of the paper.

conveniently compared, characterized, and developed beyond what was previously known. This paper provides many examples of such applications of lax monoidal fibrations but the comparison of the mentioned definitions of opetopic sets, as well as a yet another definition of opetopic sets (!), will be presented in the forthcoming paper [25].

The lax monoidal fibrations provide a convenient tool to deal with many-level structures, like categories that have objects and morphisms, multicategories[1] of various kinds (that have objects-types and multiarrows-function symbols) or T-categories of Burroni, cf. [6]. These are examples of structures that have just two levels but by building a tower of fibrations, see Sections 5.7 and 5.8, or iterating a construction inside a single fibration, cf. [24, 25], we can deal with many-level structures like opetopic sets, polygraphs, n-categories, ω-categories and others. To define monoids of interest in this setting we are not doing it in 'one big step' but we divide it into three smaller steps. First we define a fibration, then we define the monoidal structure in this fibration and finally we define a fibration of monoids over the same base as the fibration we started with. In that way if we want, as we do in Section 7, to compare multicategories with nonstandard amalgamation with symmetric multicategories we can compare the fibrations of amalgamated and symmetric signatures first, then compare the tensors (there is more than one possibility) and finally we get a comparison of suitable multicategories.

A lax monoidal fibration is a fibration $p\colon \mathcal{E} \to \mathbf{B}$ equipped with two functors $\otimes\colon \mathcal{E} \times_{\mathbf{B}} \mathcal{E} \to \mathcal{E}$ and $I\colon \mathbf{B} \to \mathcal{E}$ commuting over the base but not required to be morphisms of fibrations[2]. We call such morphism lax morphism of fibrations (or fibred morphisms), as the fact that a morphism of fibrations commute over the base already forces some lax preservation of prone morphisms. There are also coherence morphisms α, λ, ρ satisfying the usual conditions but they are not required to be isomorphisms, as in many examples they are not. The direction of these morphism are so chosen to cover all our examples. The fibres of such fibration are monoidal categories and reindexing functors are monoidal functors. It is in fact often the case, that the fibres are strong monoidal categories but reindexing functors are almost never strong even in the lax monoidal fibration whose monoids are small categories! This makes the whole context unavoidably lax. The morphism of lax monoidal fibrations are lax morphisms of fibrations that are monoidal in the only reasonable sense. The 2-cells are also defined in the only reasonable way. Then in the analogy with the nonfibred situation, cf. [3], a lax monoidal fibration may act on arbitrary fibrations. So we have a 2-category of actions of lax monoidal fibrations, as well. It is quite surprising how many things can be explained in terms of actions and their exponential adjoints. This will be carefully explained in Section 4 and used many times in the following sections.

The paper is organized as follows. In Sections 2 and 3, we introduce the main notions of the paper of a lax monoidal fibration, an action of a lax monoidal fibrations and 2-categories of these structures. The examples presented there are very basic.

[1]We follow mostly the terminology from [19], in particular for us (various) multicategories are the same things as (various) colored set operads.

[2]This means that we do not require that $\otimes$ or I send prone (formerly cartesian) morphisms to prone morphisms.

In Section 4, we discuss the lax monoidal fibrations $\mathcal{E}^{\mathcal{E}} \to \mathbf{B}$ that arises as exponential fibrations of bifibrations. It turns out that the exponentiation in $\mathrm{Cat}_{/\mathbf{B}}$, the slice of Cat over the base, is much more interesting than the exponentiation in $\mathrm{Fib}(\mathbf{B})$, the category of fibrations over $\mathbf{B}$. Among such fibrations there are even more special ones, the exponential fibrations of the basic fibrations $\mathrm{cod}\colon \mathbf{B}^{\to} \to \mathbf{B}$. If $\mathbf{B}$ has pullbacks such a fibration, denoted $\mathrm{Exp}(\mathbf{B}) \to \mathbf{B}$, always exists and if a lax monoidal fibration $\mathcal{E} \to \mathbf{B}$ acts on the basic fibrations $\mathrm{cod}\colon \mathbf{B}^{\to} \to \mathbf{B}$ we have a representation morphism of lax monoidal fibrations

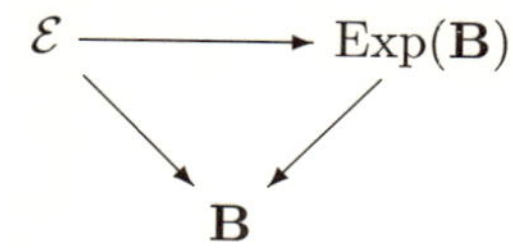

that compares an arbitrary fibration with a standard one.

In Section 5, we show how one can split the definition of a T-category of Burroni, cf. [6, p. 225–227], into three parts. The fibration of T-graphs, denoted $p_T\colon \mathrm{Gph}(T) \to \mathcal{C}$, the monoidal part and finally the monoids in such a lax monoidal fibrations. We call such fibrations Burroni fibrations to honor A. Burroni who was the first to consider them, cf. [6, p. 262]. The monoids in such fibrations are exactly the T-categories of Burroni. Since it is not necessary to have a cartesian monad[3] T to build such a fibration we can recover that way all the T-categories that were considered in [6]. The fibres of such a fibration p_T are not necessarily strong monoidal unless T is cartesian. However, as we already mentioned, the reindexing functors are almost never strong monoidal functors. The Burroni fibrations always acts on basic fibrations and hence they have representation morphisms of lax monoidal fibration into the standard ones

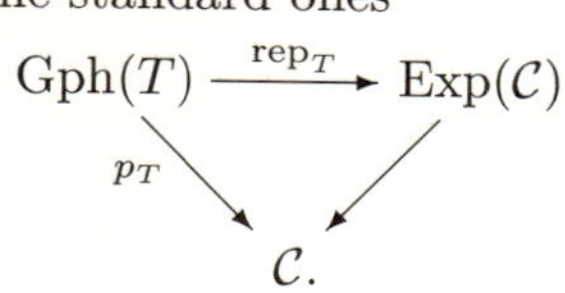

If T is cartesian then this morphism is a morphism of bifibrations, Proposition 5.2. The construction of T-categories can be made relative with respect to a fibration if the monad T is already fibred. Moreover, in this relative context the construction due to M. Kelly, cf. [16, p. 69], see also [4], together with the characterization of T. Leinster, cf. [19, p. 334], gives a characterization of those fibred cartesian monads for which the free relative T-categories exists. This allows us to extend the definition of the set of opetopes given by T. Leinster, cf. [19, p. 179], to the definition of the whole category of opetopic sets, and this category can be build internally in any Grothendieck topos not only in Set. We simply iterate ω times the construction of relative T-graph fibration starting from the identity monad.

In Section 6, we show that two seemingly different languages used to define opetopes and opetopic sets, cf. [12, 18], are in fact equivalent. We show that the lax monoidal fibrations of amalgamated signatures $p_{\mathrm{a}}\colon \mathrm{Sig}_{\mathrm{a}} \to \mathrm{Set}$ and of polynomial diagrams $p_{\mathrm{pd}}\colon \mathcal{PolyDiag} \to \mathrm{Set}$ are equivalent. The difference is rather in style that can be easily explained in the context of lax monoidal fibrations. The amalgamated signatures are 'more concrete' and come naturally equipped with an action on the

[3]The only requirement is that the category have pullbacks.

basic fibration cod: $\mathrm{Set}^{\to} \to \mathrm{Set}$ whereas polynomial diagrams come equipped with a representation into the exponential fibration $\mathrm{Exp}(\mathrm{Set}) \to \mathrm{Set}$. This representation is the exponential adjoint of the action and its essential image is the lax monoidal fibration of (finitary) polynomial (endo)functors $p_{\mathrm{poly}}\colon \mathcal{P}oly \to \mathrm{Set}$. As these three fibrations are equivalent as lax monoidal fibrations we obtain in particular that the fibration of (1-level) multicategories with nonstandard amalgamations is equivalent to the fibration of polynomial monads (i.e., cartesian monads on slices of Set whose functor parts are polynomial functors) and as morphisms cartesian morphism of monads whose functor parts are pullback functors (counted as morphism in the opposite direction), see Corollary 6.13. It is possible to give a natural definition of opetopic sets in this context, see [24, 25]. We end this section by showing how to deal with the so called single tensor and 2-level objects, the original setting for the definition in [12].

In Section 7, we develop a parallel theory to the one from the previous section but this time we start with the lax monoidal fibration of symmetric signatures $p_{\mathrm{s}}\colon \mathrm{Sig}_{\mathrm{s}} \to \mathrm{Set}$ instead of amalgamated signatures, whose monoids form exactly the fibration of symmetric multicategories. This fibration is also naturally equipped with an action on the basic fibration and taking an adjoint we get again representation morphism

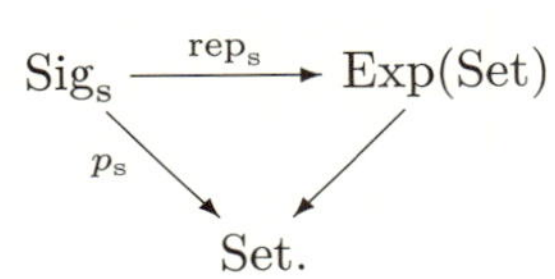

As in the previous case, this morphism is faithful and full on isomorphisms. Its essential image, denoted by $p_{\mathrm{an}}\colon \mathcal{A}n \to \mathrm{Set}$, is the lax monoidal fibration of multivariable analytic (endo)functors, cf. [15], and analytic natural transformations between them. As a consequence, the fibration of symmetric multicategories in Set is equivalent to the fibration of analytic monads. We also provide an abstract characterization of the fibration of analytic functors extending the one from [15]. We show, see Theorem 7.5, that this fibration of (multivariable) analytic (endo)functors, which is a lax monoidal subfibration of the exponential fibration $\mathrm{Exp}(\mathrm{Set}) \to \mathrm{Set}$, consists of finitary functors on slices of Set that preserve weakly wide pullbacks and has as morphisms weakly cartesian natural transformations. The proof of this characterization is based on ideas from [2, 15]. As a consequence, we obtain Corollary 7.6 saying that the notions of a symmetric multicategory and of an analytic monad are equivalent. So analytic functors is yet another tool that could be used to define the category of opetopic sets. In Section 7.4 we introduce an intermediate notion of an analytic diagrams that is related to symmetric signatures and analytic functors as polynomial diagrams are related to amalgamated signatures and polynomial functors. These diagrams are polynomial diagrams of a special kind in the category of symmetric sets $\sigma\mathrm{Set}$, i.e., the category of presheaves on the coproduct (in Cat) of finite symmetric groups. However the representation is given via a composition of five functors not three as in the case of usual polynomial diagrams. In the last subsection of the paper we compare the notions studied in Sections 6 and 7. We describe the following diagram of lax monoidal fibrations and their morphisms

$$\begin{array}{ccccc}
\mathrm{Sig}_{\mathrm{a}} & & \xrightarrow{\;K_{\mathrm{sig}}\;} & & \mathrm{Sig}_{\mathrm{s}} \\
{\scriptstyle \iota_a}\downarrow & & \overset{\Psi}{\Rightarrow} & & \downarrow{\scriptstyle \iota_s} \\
\mathcal{PolyDiag} & & \xrightarrow{\;K_{\mathrm{diag}}\;} & & \mathcal{AnDiag} \\
{\scriptstyle \mathrm{rep}_{\mathrm{pd}}}\downarrow & & \overset{\Phi}{\Rightarrow} & & \downarrow{\scriptstyle \mathrm{rep}_{\mathrm{and}}} \\
\mathcal{Poly} & & \xrightarrow{\;K_{\mathrm{fu}}\;} & & \mathcal{An} \\
\downarrow & & & & \downarrow \\
\mathrm{Cart(Set)} & & \longrightarrow & & \mathrm{wCart(Set)} \\
& \searrow & & \swarrow & \\
& & \mathrm{Exp(Set)}. & &
\end{array}$$

All the arrows are strong morphisms of lax monoidal fibrations. We show among other things that the horizontal morphisms are full and faithful. The three named horizontal arrows are morphisms comparing signatures, diagrams, and functors, respectively. The four named vertical arrows are equivalences of lax monoidal fibrations. The five unnamed arrows are inclusions. Many more interesting connections between these lax monoidal fibrations will be explained in [25]. We finish with an observation that a weakly cartesian natural transformations between polynomial functors are cartesian.

There are two possible notions of an analytic functor on $\mathrm{Set}_{/O}$. The species and the analytic functors of one variable Set $\to$ Set and of many variables $\mathrm{Set}_{/O} \to \mathrm{Set}$, for a finite set O, were introduced by A. Joyal in [14, 15] to study enumerative combinatorics. Clearly, an O-tuple of multivariable analytic functors taken together form an endofunctor $\mathrm{Set}_{/O} \to \mathrm{Set}_{/O}$, that should be considered as analytic, as well. The concept of analytic functor was studied in the category Set, in category of vector spaces Vect, cf. [15] but also in an arbitrary monoidal category, cf. [2]. In that way, we have two kinds of analytic functors on slices of Set (and powers of other monoidal categories). An analytic functor from $\mathrm{Set}_{/O}$ to $\mathrm{Set}_{/O}$ can be defined as the left Kan extension of a functor $f\colon \mathbf{B} \to \mathrm{Set}_{/O}$, where $\mathbf{B}$ is the category of finite sets and bijections, cf. [2], or as an O-tuple of multivariable analytic functors $\mathrm{Set}_{/O} \to \mathrm{Set}$, cf. [15]. The first notion does not allow functors that are not coproducts of functors between fibres. In this paper we consider only the second notion.

The idea of equipping fibrations with some kind of a monoidal structure goes back to N. Saavedra Rivano [21] and M.-F. Gouzou – R. Grunig, cf. [10]. It was taken up later by M. Shulman in [22, p. 698]. These notions '$\mathbf{B}$-$\otimes$-catégorie fibrée' in [21], 'catégorie fibrée sur $\mathbf{B}$ multiplicative' in [10], and 'monoidal fibration' in [22] are different than the notion of a lax monoidal fibrations presented here. Also the motivations in each case are different than ours. The total category of a lax monoidal fibration is not a monoidal category and in this sense the notion is closer to the notions considered in [21] and [10]. On the other hand, we do not require our tensor or unit to be morphisms of fibrations (i.e., preserve the prone morphisms) as it would eliminate most of our examples. This causes that our reindexing morphisms are not necessarily strong monoidal functors. In our applications the actions of lax monoidal fibrations play an important role. This does not have an analog in the other approaches.

I would like to thank George Janelidze and Thomas Streicher for the conversations related to the matters contained in this paper, André Joyal for explaining to me some aspects of his theory of analytic functors. Special thanks are due to Krzysztof Kapulkin, Magdalena Kędziorek, Karol Szumiło and Stanisław Szawiel, the members of an informal Category Seminar held in Spring 2008 at Warsaw University, for giving me an opportunity to present the essential notions introduced and studied in this paper. I would like also to thank the anonymous referee for the very thorough report that helped considerably to improve the presentation of the paper. Last but not least I would like to thank, our jubilee, Mihály Makkai for introducing me to the subject of higher-dimensional categories many years ago and to him and Victor Harnik for countless discussions of the related matters.

The diagrams for this paper were prepared with the help of catmac1 of Michael Barr.

2. Lax monoidal fibrations

2.1. Preliminaries, fibrations. Our standard reference no fibrations (opfibrations and bifibrations) is [23]. However the terminology used here follows more the one used by P. Tayor and P. T. Johnston. We call *prone* and *supine* morphisms what [23] would call *cartesian* and *cocartesian.* The fibre of a (bi)fibration $p\colon \mathcal{E} \to \mathbf{B}$ over $B \in \mathbf{B}$ will be denoted $\mathcal{E}_B$. If p is a fibration, $u\colon B \to B'$ is a morphism in $\mathbf{B}$ then we have (using axiom of choice for classes) a *reindexing functor* $u^*\colon \mathcal{E}_{B'} \to \mathcal{E}_B$ defined with the help of prone morphisms; if p is an opfibration, we have a *coreindexing functor* $u_!\colon \mathcal{E}_B \to \mathcal{E}_{B'}$ defined with the help of supine morphisms. In a bifibration both functors exist and are adjoint $u_! \dashv u^*$. The unit and counit of this adjunction will be denoted by η^u and ε^u, respectively. We call a bifibration $P\colon \mathcal{E} \to \mathbf{B}$ *cartesian* if the fibres of p have pullbacks, $u_!$ preserves them and both η^u and ε^u are cartesian natural transformations, for all morphisms u in $\mathbf{B}$. Note that this notation suppresses the fact that these functors and natural transformations are related to a specific (bi,op)fibration. The fibration we have in mind should be always read from the context. A fibration has fibred (co)limits of type K if and only if each fibre has (co)limits of type K and reindexing functors preserve them. In the paper, we consider (bi)fibrations that are equipped additionally with a monoidal structure. It is not always the case that the morphisms we want to consider between (bi)fibrations preserves all the structure involved (prone morphisms, supine morphisms, and/or tensor). Therefore as the basic morphisms between fibration we consider lax morphisms that only make the square below commute. A *lax morphism* from a fibration p to p' is a pair of functors (G, G') making the square

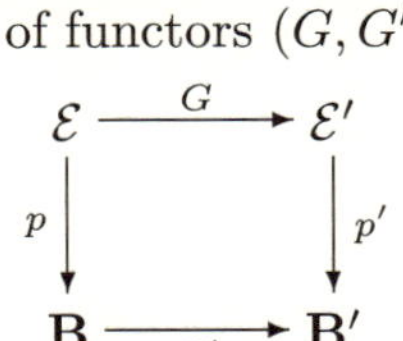

commute. If p and p' are (op)fibrations and G preserves prone (supine) morphisms, (G, G') will be called a *morphism of fibrations* (*opfibrations*). If p and p' are bifibrations and G preserves both prone and supine morphisms, (G, G') will be called a *morphism of bifibrations.* The *fibred natural transformation* $(\tau, \tau')\colon (G, G')$

$\to (H, H')$ is a pair of natural transformations $\tau\colon G \to H$ and $\tau'\colon G' \to H'$ such that $p'(\tau) = \tau'_p$. If $G = G' = \mathrm{id}_{\mathbf{B}}$ and $\tau' = \mathrm{id}_{\mathrm{id}_{\mathbf{B}}}$ then a fibred natural transformation is natural transformation $\tau\colon G \to H$ whose components are vertical morphisms. A *fibred left adjoint* to (G, G') is a fibred morphism (F, F') such that $F \dashv G$ and $F' \dashv G'$ are adjunctions with units and counits (η, ε) and (η', ε'), respectively, so that $p'(\eta) = \eta'$ and $p'(\varepsilon) = \varepsilon'$.

2.2. The basic definition. A *lax monoidal fibration* $(p\colon \mathcal{E} \to \mathbf{B}, I, \otimes, \alpha, \lambda, \varrho)$ is

(1) a fibration $p\colon \mathcal{E} \to \mathbf{B}$,
(2) equipped with two lax morphisms of fibrations $\otimes$ and I

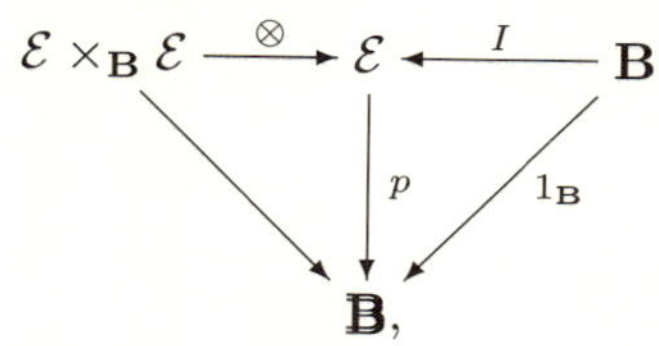

(3) three fibred natural transformations α, λ, ϱ

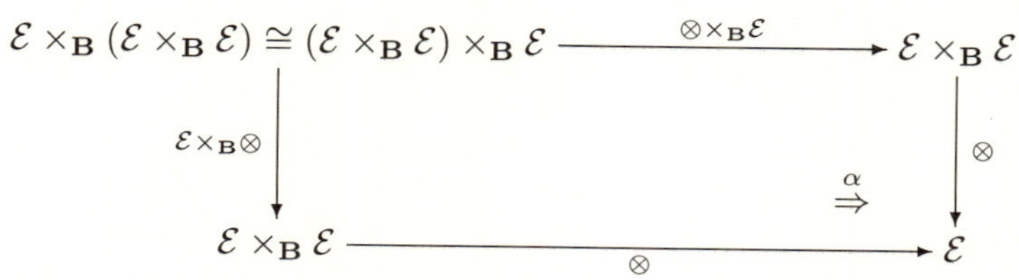

(where the unnamed iso $\cong$ is the canonical one between pullbacks), i.e., there are morphisms

$$A \otimes (B \otimes C) \xrightarrow{\alpha_{A,B,C}} (A \otimes B) \otimes C$$

for any $O \in \mathbf{B}$ and any $A, B, C \in \mathcal{E}_O$ so that $p(\alpha_{A,B,C}) = 1_O$ and these morphisms are natural in A, B, and C, in the obvious sense. Moreover,

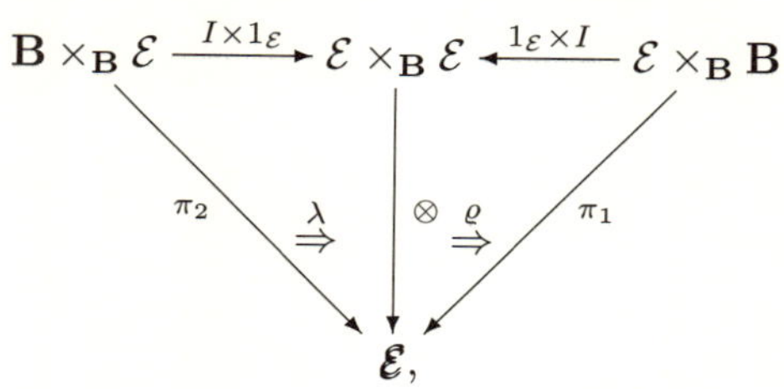

i.e., there are morphisms

$$A \otimes I_O \xrightarrow{\rho_A} A \xrightarrow{\lambda_A} I_O \otimes A$$

for any $O \in \mathbf{B}$ and any $A \in \mathcal{E}_O$ so that $p(\rho_A) = 1_O = p(\lambda_A)$ and these morphisms are natural in A.

(4) The diagrams

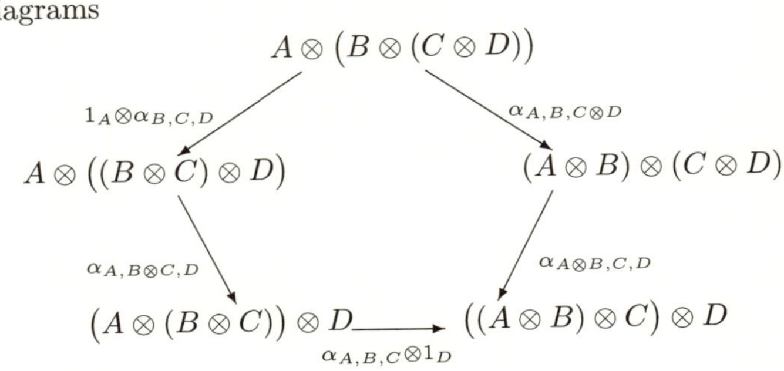

and

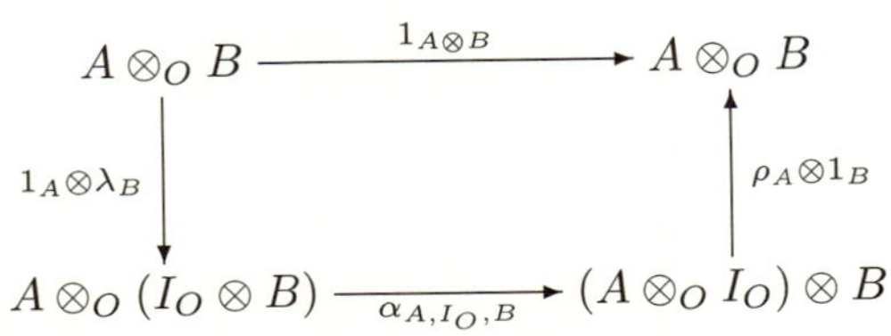

commute, and finally ρ_{I_O} and λ_{I_O} are isomorphisms and

$$\rho_{I_O} = \lambda_{I_O}^{-1} \colon I_O \otimes I_O \to I_O$$

where $O \in \mathbf{B}$ and $A, B, C, D \in \mathcal{E}_O$.

End of the definition of a lax monoidal fibration.

Remarks. (1) The tensor operation can be applied to objects in the same fibre of $p\colon \mathcal{E} \to \mathbf{B}$ only, and to morphisms that lie over the same map in the base. Sometimes we emphasize this by writing $a \otimes_O b$ and $f \otimes_u g$ to indicate that the tensor is in the fibre over O or over a morphism u. So the fibres are (lax) monoidal categories and reindexing 'functors' are lax monoidal. But the total category $\mathcal{E}$ is not monoidal. Both facts are important for the examples we have in mind.

(2) For any $u\colon O \to Q \in \mathbf{B}$ and $A, B \in \mathcal{E}_Q$ we have (unique) morphisms $\psi^0_u\colon I_O \to u^*(I_Q)$ and $\psi^2_{u,A,B}\colon u^*(A) \otimes_O u^*(B) \to u^*(A \otimes_Q B)$ so that the triangles

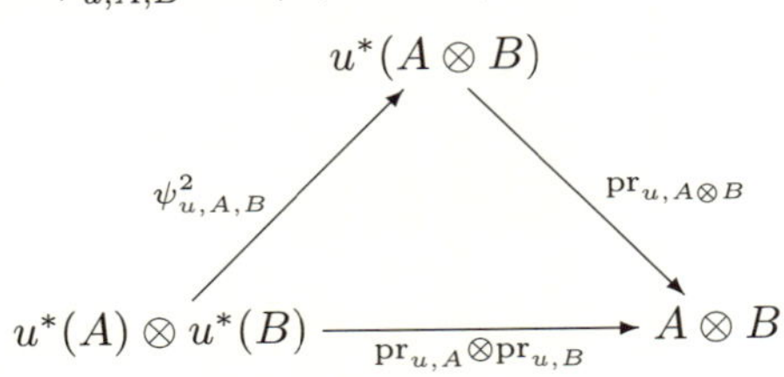

and

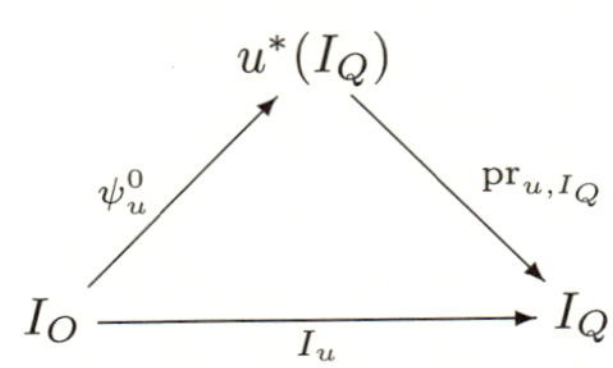

commute, where $\mathrm{pr}_{u,A}$ is a prone morphism over u with codomain A.

Due to the fact that we deal with fibrations, lax morphisms preserve prone morphisms in the lax sense. Thus even if we do not require the tensor and the unit to be morphisms of fibrations, we still have that the 'reindexing' functors are lax monoidal, i.e., they 'respect' the monoidal structure (to some extent).

There are many more diagrams involving ψ's, α's, λ's and ϱ's that commute.

(3) The directions of the natural transformations α's, λ's and ϱ in the definition of a lax monoidal fibration are so chosen to cover all the examples we have in mind. But it is sometimes convenient to consider natural transformations λ or ϱ that go in the other direction.

2.3. Monoids in a lax monoidal fibration. A *monoid* in a fibre over O is a triple $(M, m\colon M \otimes M \to M, e\colon I_O \to M)$ where M is an object in $\mathcal{E}_O$, m, e are morphisms in $\mathcal{E}_O$ making the diagrams

$$\begin{array}{ccccc} M\otimes(M\otimes M) & \xrightarrow{\alpha} & (M\otimes M)\otimes M & \xrightarrow{m\otimes 1_M} & M\otimes M \\ {\scriptstyle 1_M\otimes m}\big\downarrow & & & & \big\downarrow{\scriptstyle m} \\ M\otimes M & & \xrightarrow[m]{\qquad\qquad} & & M \end{array}$$

and

$$\begin{array}{ccccc} I_O\otimes M & \xrightarrow{e\otimes 1_M} & M\otimes M & \xleftarrow{1_M\otimes e} & M\otimes I_O \\ {\scriptstyle \lambda_M}\big\uparrow & & \big\downarrow{\scriptstyle m} & & \big\downarrow{\scriptstyle \rho_M} \\ M & \xrightarrow[1_M]{} & M & \xleftarrow[1_M]{} & M \end{array}$$

commute.

A *morphism of monoids* $f\colon (M, m, e) \to (M', m', e')$ over $u\colon O \to Q \in \mathbf{B}$ is a morphism $f\colon M \to M'$ in $\mathcal{E}$ over u such that the squares

$$\begin{array}{ccc} M\otimes_O M & \xrightarrow{f\otimes_u f} & M'\otimes_Q M' \\ {\scriptstyle m}\big\downarrow & & \big\downarrow{\scriptstyle m'} \\ M & \xrightarrow{f} & M' \\ {\scriptstyle e}\big\uparrow & & \big\uparrow{\scriptstyle e'} \\ I_O & \xrightarrow[I_u]{} & I_Q \end{array}$$

commute.

Then the category of monoids is again fibred over $\mathbf{B}$ and the forgetful functor is a morphism of fibrations

$$\begin{array}{ccc} \mathcal{E} & \xleftarrow{\mathcal{U}} & \mathrm{Mon}(\mathcal{E}, \otimes, I) \\ & {\scriptstyle p}\searrow \quad \swarrow{\scriptstyle q} & \\ & \mathrm{Set}. & \end{array}$$

Remark. As we will see it is not always true that the category of all monoids is of real interest. If the coherence transformations are indeed not isomorphisms it may happen that we may want to consider only those monoids that satisfy some additional conditions, see Section 6.6.

2.4. The 2-category of lax monoidal fibrations. A *morphism of lax monoidal fibrations*

$$(F, K, \varphi_0, \varphi_2) \colon (p \colon \mathcal{E} \to \mathbf{B}, \otimes, I, \alpha, \lambda, \varrho) \to (p' \colon \mathcal{E}' \to \mathbf{B}', \otimes', I', \alpha', \lambda', \varrho')$$

is data (1) – (3) subject to conditions (4) – (6) below ($O \in \mathbf{B}$, $A, B, C \in \mathcal{E}_O$):

(1) $(F, K) \colon (\mathcal{E}, p) \to (\mathcal{E}', p')$ a lax morphism of fibrations,

(2) $\varphi_0 \colon I_K \to F \circ I$ a fibred natural transformation (i.e., for any $O \in \mathbf{B}$ we have a morphisms $(\varphi_0)_O \colon I'_{K(O)} \to F(I_O)$ in $\mathcal{E}'_O$ which is natural in O),

(3) $\varphi_2 \colon F(-) \otimes' F(=) \to F\big((-) \otimes (=)\big)$ a fibred natural transformation (i.e., for $A, B \in \mathcal{E}_O$ we have a morphism $(\varphi_2)_{A,B} \colon F(A) \otimes' F(B) \to F(A \otimes B))$, in $\mathcal{E}'_O$ which is natural in A and B)

(4) the square

$$\begin{array}{ccc}
I'_{K(O)} \otimes F(A) & \xrightarrow{\ \varphi_0 \otimes 1_{F(A)}\ } & F(I_O) \otimes F(A) \\
{\scriptstyle \lambda'_{F(A)}} \big\uparrow & & \big\downarrow {\scriptstyle \varphi_2} \\
F(A) & \xrightarrow[\ F(\lambda_A)\ ]{} & F(I_O \otimes A)
\end{array}$$

commutes, where $O \in \mathbf{B}$ and $A \in \mathcal{E}_O$;

(5) the square

$$\begin{array}{ccc}
F(A) \otimes I'_{K(O)} & \xrightarrow{\ 1_{F(A)} \otimes \varphi_0\ } & F(A) \otimes F(I_O) \\
{\scriptstyle \varrho'_{F(A)}} \big\downarrow & & \big\downarrow {\scriptstyle \varphi_2} \\
F(A) & \xleftarrow[\ F(\varrho_A)\ ]{} & F(A \otimes I_O)
\end{array}$$

commutes, where $O \in \mathbf{B}$ and $A \in \mathcal{E}_O$;

(6) the diagram

$$\begin{array}{ccc}
F(A) \otimes' \big(F(B) \otimes' F(C)\big) & \xrightarrow{\ \alpha'_{F(A),F(B),F(C)}\ } & \big(F(A) \otimes' F(B)\big) \otimes' F(C) \\
{\scriptstyle 1_{F(A)} \otimes' \varphi_2} \big\downarrow & & \big\downarrow {\scriptstyle \varphi_2 \otimes' 1_{F(C)}} \\
F(A) \otimes' F(B \otimes C) & & F(A \otimes B) \otimes' F(C) \\
{\scriptstyle \varphi_2} \big\downarrow & & \big\downarrow {\scriptstyle \varphi_2} \\
F\big(A \otimes (B \otimes C)\big) & \xrightarrow[\ F(\alpha_{A,B,C})\ ]{} & F\big((A \otimes B) \otimes C\big)
\end{array}$$

commutes, where $O \in \mathbf{B}$ and $A, B, C \in \mathcal{E}_O$.

End of the definition of a morphism of lax monoidal fibrations.

A morphism of lax monoidal fibrations is called *strong* if the transition morphisms φ_0, φ_2 are isomorphisms and (F, K) is a morphism of fibrations.

A *transformation* between two morphisms of lax monoidal fibrations is a pair of natural transformations

$$(\tau, \sigma) \colon (F, K, \varphi_0, \varphi_2) \to (F', K', \varphi'_0, \varphi'_2)$$

such that

(1) $\sigma\colon K \to K'$ and $\tau\colon F \to F'$ are natural transformations,
(2) $p'(\tau) = \sigma_p$, i.e., τ is fibred over σ,
(3) the diagrams

$$\begin{array}{ccc} I'_{K(O)} & \xrightarrow{I'_{\sigma_O}} & I'_{K'(O)} \\ {\scriptstyle (\phi_0)_O}\downarrow & & \downarrow{\scriptstyle (\phi'_0)_O} \\ F(I_O) & \xrightarrow[\tau_{I_O}]{} & F'(I_O) \end{array}$$

and

$$\begin{array}{ccc} F(A) \otimes_{K(O)} F(B) & \xrightarrow{\tau_A \otimes_{K(O)} \tau_B} & F'(A) \otimes_{K(O)} F'(B) \\ {\scriptstyle (\phi_2)_{A,B}}\downarrow & & \downarrow{\scriptstyle (\phi'_2)_{A,B}} \\ F(A \otimes_O B) & \xrightarrow[\tau_{A \otimes_O B}]{} & F'(A \otimes_O B) \end{array}$$

commute, for $O \in \mathbf{B}$ and $A, B \in \mathcal{E}_O$.

End of the definition of a transformation between two morphisms of lax monoidal fibrations.

Proposition 2.1. *The morphisms of lax monoidal fibrations induce morphisms between the fibrations of monoids. The transformations between morphisms of lax monoidal fibrations induce natural transformations between the induced functors.*

PROOF. Exercise. □

2.5. Simple examples. (1) *Categories.* Probably the simplest nontrivial example of a lax monoidal fibration (in the above sense) is the fibration of graphs over sets, say $p\colon \mathrm{Gph} \to \mathrm{Set}$, where p sends the parallel pair of arrows to their common codomain. The tensor

$$(A, d^A, c^A\colon A \to O) \otimes_O (B, d^B, c^B\colon B \to O) = (A \times_O B, c^A \circ \pi_1, d^B \circ \pi_2\colon A \times_O B \to O)$$

where $A \times_O B$ denotes the pullback of the following pair of morphisms

$$\begin{array}{ccc} A \times_O B & \xrightarrow{\pi_2} & B \\ {\scriptstyle \pi_1}\downarrow & & \downarrow{\scriptstyle c^B} \\ A & \xrightarrow[d^A]{} & O. \end{array}$$

The unit for the tensor in the fibre over O is a pair of identities on O, $(O, 1_O, 1_O\colon O \to O)$. The total category of the fibration of monoids $q\colon \mathrm{Mon}(\mathrm{Gph}) \to \mathrm{Set}$ in this fibration is the category of small categories and functors. The monoids in a fibre $\mathrm{Mon}(\mathrm{Gph})_O$ are categories with the set of objects O.

(2) *Lambek's multicategories.* Let $\big((-)^*, \eta, \mu\big)$ be the monad for monoids on the category Set. Then, we can define a fibration of multisorted signatures $p_{\mathrm{m}}\colon \mathrm{Sig}_{\mathrm{m}} \to \mathrm{Set}$ as follows. An object of $\mathrm{Sig}_{\mathrm{m}}$ in the fibre over the set O is a triple (A, ∂, O) such that A is a set and $\partial\colon A \to O \times O^*$ function. $(f, u)\colon (A, \partial^A, O) \to (A', \partial^{A'}, O')$ is

a morphism in $\mathrm{Sig_m}$ over a function $u\colon O \to O'$ if $f\colon A \to A'$ is a function making the square

$$\begin{array}{ccc} A & \xrightarrow{\;f\;} & A' \\ {\scriptstyle \partial^A}\downarrow & & \downarrow{\scriptstyle \partial^{A'}} \\ O \times O^* & \xrightarrow[u\times u^*]{} & O' \times O'^* \end{array}$$

commute. The tensor $(A, \partial^A, O) \otimes (B, \partial^B, O) = (A \times_{O^*} B^*, \partial, O)$ of two object in the same fibre is given by the pullback and multiplication μ in the monad $(-)^*$ and the unit for this tensor is $(I_O, \partial, O) = (O, \langle 1_O, \eta\rangle, O)$. The category of monoids in this fibration is the category of Lambek's multicategories. As this construction will be described in Section 5 in a more general case of arbitrary monad over a category with pullbacks we don't go into the details here.

3. Actions of lax monoidal fibrations

3.1. The basic definition. An *action* $(\star, \psi_2, \psi_0)$ of a lax monoidal fibration $(\mathcal{E}, p, I, \otimes, \alpha, \lambda, \varrho)$ on a fibration $\pi\colon \mathcal{X} \to \mathbf{B}$ is

(1) a lax morphism of fibrations

$$\begin{array}{ccc} \mathcal{E} \times_{\mathbf{B}} \mathcal{X} & \xrightarrow{\;\star\;} & \mathcal{X} \\ & {\scriptstyle p}\searrow \quad \swarrow{\scriptstyle \pi} & \\ & \mathbf{B}; & \end{array}$$

(2) a fibred natural transformation ψ_0

$$\begin{array}{ccc} \mathcal{X} \cong \mathbf{B} \times_{\mathbf{B}} \mathcal{X} & \xrightarrow{\;I \times_{\mathbf{B}} 1_{\mathcal{X}}\;} & \mathcal{E} \times_{\mathbf{B}} \mathcal{X} \\ & {\scriptstyle 1_{\mathcal{X}}}\searrow \overset{\psi_0}{\Rightarrow} \swarrow{\scriptstyle \star} & \\ & \mathcal{X}, & \end{array}$$

i.e., for $X \in \mathcal{X}_O$, we have a morphism

$$(\psi_0)_X \colon X \to I_O \star X$$

in the fibre $\mathcal{X}_O$ which is natural in O;

(3) a fibred natural transformation ψ_2

$$\begin{array}{ccc} \mathcal{E} \times_{\mathbf{B}} (\mathcal{E} \times_{\mathbf{B}} \mathcal{X}) \cong (\mathcal{E} \times_{\mathbf{B}} \mathcal{E}) \times_{\mathbf{B}} \mathcal{X} & \xrightarrow{\;\otimes \times_{\mathbf{B}} \mathcal{X}\;} & \mathcal{E} \times_{\mathbf{B}} \mathcal{X} \\ {\scriptstyle \mathcal{E} \times_{\mathbf{B}} \star}\downarrow & \overset{\psi_2}{\Rightarrow} & \downarrow{\scriptstyle \star} \\ \mathcal{E} \times_{\mathbf{B}} \mathcal{X} & \xrightarrow[\star]{} & \mathcal{X}, \end{array}$$

i.e., for $O \in \mathbf{B}$, $A, B \in \mathcal{E}_O$ and $X \in \mathcal{X}_O$ we have a morphism

$$(\psi_2)_{A,B,X} \colon A \star (B \star X) \to (A \otimes B) \star X$$

in the fibre $\mathcal{X}_O$, which is natural in A, B and X; the unnamed iso $\cong$ is the canonical one between pullbacks;

(4) making the pentagon

$$\begin{array}{ccc} & A\star\big(B\star(C\star X)\big) & \\ {\scriptstyle (\psi_2)_{A,B,C\star X}}\swarrow & & \searrow{\scriptstyle 1_A\star(\psi_2)_{B,C,X}} \\ (A\otimes B)\star(C\star X) & & A\star\big((B\otimes C)\star X\big) \\ {\scriptstyle (\psi_2)_{A\otimes B,C,X}}\downarrow & & \downarrow{\scriptstyle (\psi_2)_{A,B\otimes C,X}} \\ \big((A\otimes B)\otimes C\big)\star X & \xleftarrow{\ \alpha_{A,B,C}\star 1_X\ } & \big(A\otimes(B\otimes C)\big)\star X \end{array}$$

(5) and two squares

$$\begin{array}{ccc} A\star X & \xrightarrow{(\psi_0)_{A\star X}} & I\star(A\star X) \\ {\scriptstyle 1_{A\star X}}\downarrow & & \downarrow{\scriptstyle (\psi_2)_{I,A,X}} \\ A\star X & \xrightarrow[\lambda_A\star 1_X]{} & (I\otimes A)\star X \end{array} \qquad \begin{array}{ccc} A\star X & \xrightarrow{1_A\star(\psi_0)_X} & A\star(I\star X) \\ {\scriptstyle 1_{A\star X}}\downarrow & & \downarrow{\scriptstyle (\psi_2)_{A,I,X}} \\ A\star X & \xleftarrow[\varrho_A\star 1_X]{} & (A\otimes I)\star X \end{array}$$

commute, where $O \in \mathbf{B}$, $A, B, C \in \mathcal{E}_O$ and $X \in \mathcal{X}_O$.

End of the definition of an action of a lax monoidal fibration on a fibration.

An action of a lax monoidal fibration is called *strong* if the transition morphisms ψ_0, ψ_2 are isomorphisms and $\star$ is a morphism of fibrations.

3.2. Actions of monoids along an action of a lax monoidal fibration. Let $(\star, \psi_2, \psi_0)$ be an action of a lax monoidal fibration $(p, I, \otimes, \alpha, \lambda, \varrho)$ on a fibration $\pi\colon \mathcal{X} \to \mathbf{B}$, O an object of $\mathbf{B}$, (M, m, e) a monoid in $\mathrm{Mon}(\mathcal{E})_O$, X an object in $\mathcal{X}_O$ and $\nu\colon M \star X \to X$ a morphism in $\mathcal{X}_O$. The pair (X, ν) is an *action* of (M, m, e) on X along the action $(\star, \psi_2, \psi_0)$ (or just $\star$, for short) if the following diagrams

$$\begin{array}{ccccc} M\star(M\star X) & \xrightarrow{(\psi_2)_{M,M,X}} & (M\otimes M)\star X & \xrightarrow{m\star 1_X} & M\star X \\ {\scriptstyle 1_M\star\nu}\downarrow & & & & \downarrow{\scriptstyle \nu} \\ M\star X & & \xrightarrow[\nu]{} & & X \end{array}$$

and

$$\begin{array}{ccc} I\star X & \xrightarrow{e\star 1_X} & M\star X \\ {\scriptstyle (\psi_0)_X}\uparrow & & \downarrow{\scriptstyle \nu} \\ X & \xrightarrow[1_X]{} & X \end{array}$$

commute. A *morphism of actions* $(f, g, u)\colon (M, X, \nu) \to (M', X', \nu')$ is a triple of morphisms $u\colon O \to O'$ in $\mathbf{B}$, and $f\colon M \to M'$ in $\mathrm{Mon}(\mathcal{E})$, $g\colon X \to X'$ in $\mathcal{X}$ both over u so that the square in $\mathcal{X}$

$$\begin{array}{ccc} M\star X & \xrightarrow{f\star g} & M'\star X' \\ {\scriptstyle \nu}\downarrow & & \downarrow{\scriptstyle \nu'} \\ X & \xrightarrow[g]{} & X' \end{array}$$

commutes.

The category of actions $\mathrm{Act}(\mathcal{E}, \mathcal{X}, \star)$ is fibred over $\mathrm{Mon}(\mathcal{E})$. It might happen that monoids in the fibre over O can be interpreted as algebras for a single monoid $\mathcal{M}_O$ in a fibre over $K(O)$. If this association is functorial we have commuting squares

$$\begin{array}{ccc} \mathrm{Mon}(\mathcal{E}) & \xrightarrow{R} & \mathrm{Act}(\mathcal{E}, \mathcal{X}, \star) \\ \downarrow{\scriptstyle q} & & \downarrow{\scriptstyle \pi_\mu} \\ \mathbf{B} & \xrightarrow{\mathcal{M}} & \mathrm{Mon}(\mathcal{E}) \\ \downarrow{\scriptstyle 1_\mathbf{B}} & & \downarrow{\scriptstyle q} \\ \mathbf{B} & \xrightarrow[K]{} & \mathbf{B}. \end{array}$$

The functor R is the representing functor that interprets monoids as algebras. If the upper square is a pullback then we say that $\mathcal{M}$ is the functor of *metamonoid*[4] and the triple $(R, \mathcal{M}, K)$ *strongly represents* monoids in $\mathrm{Mon}(\mathcal{E})$ as algebras. This means in particular that the category of O-monoids is equivalent to the category of $\mathcal{M}_O$-algebras. If the functor R is an embedding (faithful and reflects isomorphisms) on fibres then we say that the triple $(R, \mathcal{M}, K)$ *weakly represents* monoids in $\mathrm{Mon}(\mathcal{E})$.

3.3. The 2-category of actions of lax monoidal fibrations. We define below the morphisms of actions of lax monoidal fibrations and transformations of such morphisms. In that way we shall define the 2-category ACTION of actions of lax monoidal fibrations on fibrations.

A *morphism of actions of lax monoidal fibrations*

$$(F, H, K, \phi_0, \phi_2, \tau)\colon (\mathcal{E}, p, \mathcal{X}, \pi, \star, \psi_0, \psi_2) \to (\mathcal{E}', p', \mathcal{X}', \pi', \star', \psi_0', \psi_2')$$

consists of the data (1) – (4) subject to the conditions (5) – (6) below:

(1) functors

$$F\colon \mathcal{E} \to \mathcal{E}', \quad H\colon \mathcal{X} \to \mathcal{X}', \quad K\colon \mathbf{B} \to \mathbf{B}';$$

(2) a morphism of lax monoidal fibrations

$$(F, K, \phi_0, \phi_2)\colon (\mathcal{E}, p, \otimes, \alpha, \lambda, \varrho) \to (\mathcal{E}', p', \otimes', \alpha', \lambda', \varrho');$$

(3) a lax morphism of fibrations

$$(H, K)\colon (\mathcal{X}, \pi) \to (\mathcal{X}', \pi');$$

(4) a natural transformation

$$\tau\colon \star' \circ (F \times_K H) \to H \circ \star,$$

i.e., we have a morphism

$$\tau_{A,X}\colon F(A) \star' H(X) \to H(A \star X)$$

which is natural in $A \in \mathcal{E}_O$, $X \in \mathcal{X}_O$ and $O \in \mathbf{B}$.

[4]This is what seems to be the intension of the notion of operad for (colored) operads introduced by J. Baez and J. Dolan in [3].

So we have a diagram

$$
\begin{array}{ccccc}
\mathcal{E}\times_{\mathbf{B}}\mathcal{X} & & \xrightarrow{\quad\star\quad} & & \mathcal{X} \\
 & \searrow & & \swarrow{\scriptstyle \pi} & \\
\Big\downarrow{\scriptstyle F\times_K H} & & \mathbf{B} & & \Big\downarrow{\scriptstyle H} \\
 & & \Big\downarrow{\scriptstyle K} & & \\
 & & \mathbf{B}' & & \\
 & \nearrow & & \nwarrow{\scriptstyle \pi'} & \\
\mathcal{E}'\times_{\mathbf{B}}\mathcal{X}' & & \xrightarrow[\quad\star'\quad]{} & & \mathcal{X}'
\end{array}
$$

in which the triangles and internal squares commute, but external square commutes up to the natural transformation τ.

(5) The square

$$
\begin{array}{ccc}
H(X) & \xrightarrow{(\psi'_0)_{H(X)}} & I'_{K(O)}\star' H(X) \\
\Big\downarrow{\scriptstyle H\left((\psi_0)_X\right)} & & \Big\downarrow{\scriptstyle (\phi_0)_O\star' 1_{H(X)}} \\
H(I_O\star X) & \xleftarrow[\tau_{I_O,X}]{} & F(I_O)\star' H(X)
\end{array}
$$

commutes, for $X\in\mathcal{X}_O$, and $O\in\mathbf{B}$.

(6) The hexagon

$$
\begin{array}{ccc}
 & F(A)\star'\big(F(B)\star' H(X)\big) & \\
{\scriptstyle (\psi'_2)_{F(A),F(B),H(X)}}\swarrow & & \searrow{\scriptstyle 1_{F(A)}\star'\tau_{B,X}} \\
\big(F(A)\otimes' F(B)\big)\star' H(X) & & F(A)\star' H(B\star X) \\
\Big\downarrow{\scriptstyle (\phi_2)_{A,B}\star' 1_{H(X)}} & & \Big\downarrow{\scriptstyle \tau_{A,B\star X}} \\
F(A\otimes B)\star' H(X) & & H\big(A\star(B\star X)\big) \\
{\scriptstyle \tau_{A\otimes B,X}}\searrow & & \swarrow{\scriptstyle H\left((\psi_2)_{A,B,X}\right)} \\
 & H\big((A\otimes B)\star X\big) &
\end{array}
$$

commutes, for $A,B\in\mathcal{E}_O$, $X\in\mathcal{X}_O$, and $O\in\mathbf{B}$.

End of the definition of a morphism of actions of lax monoidal fibrations.

Let

$$
\begin{aligned}
&(F,H,K,\phi_0,\phi_2,\tau),(F',H',K',\phi'_0,\phi'_2,\tau)\\
&\qquad\qquad : (\mathcal{E},p,\mathcal{X},\pi,\star,\psi_0,\psi_2)\to(\mathcal{E}',p',\mathcal{X}',\pi',\star',\psi'_0,\psi'_2)
\end{aligned}
$$

be two morphisms of actions of lax monoidal fibrations. A *transformation of morphisms of actions of lax monoidal fibrations*

$$(\zeta_0,\zeta_1,\zeta_2)\colon (F,H,K,\phi_0,\phi_2,\tau)\to(F',H',K',\phi'_0,\phi'_2,\tau)$$

consists of data (1) – (3) subject to the condition (4) below:

(1) natural transformations $\zeta_2\colon F\to F'$, $\zeta_1\colon H\to H'$ and $\zeta_0\colon K\to K'$;

(2) a transformation of lax monoidal fibrations

$$(\zeta_2, \zeta_0)\colon (F, K, \phi_0, \phi_2) \to (F', K', \phi'_0, \phi'_2);$$

(3) a fibred natural transformation of morphisms of fibrations

$$(\zeta_1, \zeta_0)\colon (H, K) \to (H', K')$$

between fibrations $(\mathcal{X}, \pi)$ and $(\mathcal{X}', \pi')$;

(4) so that the square

$$\begin{array}{ccc} F(A) \star' H(X) & \xrightarrow{\tau_{A,X}} & H(A \star X) \\ {\scriptstyle (\zeta_2 \star \zeta_1)_{A,X}} \downarrow & & \downarrow {\scriptstyle (\zeta_1)_{A \star X}} \\ F'(A) \star' H'(X) & \xrightarrow[\tau'_{A \star X}]{} & H'(A \star X) \end{array}$$

commutes, for $A \in \mathcal{E}_O$ and $X \in \mathcal{X}_O$.

Proposition 3.1. *The morphisms of actions of lax monoidal fibrations induce morphisms between fibrations of actions of monoids along actions of monoidal fibrations. The transformations between morphisms of actions of lax monoidal fibrations induce natural transformations between the induced functors.*

PROOF. Exercise. □

3.4. Simple examples. (1) The lax monoidal fibration of graphs $p\colon$ Gph $\to$ Set acts naturally on the basic fibration cod: Set$^{\to}$ $\to$ Set. The action of a graph $(d, c\colon A \to O)$ on a function $\xi\colon X \to O$ is defined as the composition of the horizontal arrows on the top of the following diagram

$$\begin{array}{ccccc} O & \xleftarrow{d} & A & \longleftarrow & A \star X \\ & & {\scriptstyle c} \downarrow & & \downarrow \\ & & O & \xleftarrow[\xi]{} & X \end{array}$$

in which the square is a pullback. Then the action of monoids along this action are all presheaves on all small categories.

(2) If we replace in the previous example Set by any category $\mathcal{C}$ with pullbacks we get all internal presheaves on all internal categories in $\mathcal{C}$.

(3) The fibration of multisorted signatures also acts on the basic fibration. But this example will be described in Section 5.

4. The exponential fibrations

Most of the material of this section belongs to folklore. We present it here as we need it later in this form. **Cat** is the category of large categories, so Set and *Cat* are objects of **Cat**.

If X is an object of a cartesian closed category $\mathcal{C}$ then X^X carries a natural structure of a monoid. If $p_{\mathcal{X}}\colon \mathcal{X} \to \mathbf{B}$ is a fibration then we can form an exponential fibration $p\colon [\mathcal{X} \Rightarrow \mathcal{X}] \to \mathbf{B}$ in Fib($\mathbf{B}$), the category of fibrations over $\mathbf{B}$, which also carries a natural structure of a lax monoidal fibration. Then any strong action of a lax monoidal fibration $p_{\mathcal{E}}\colon \mathcal{E} \to \mathbf{B}$ on $p_{\mathcal{X}}\colon \mathcal{X} \to \mathbf{B}$ gives rise to a representation of $p_{\mathcal{E}}\colon \mathcal{E} \to \mathbf{B}$ in the lax monoidal fibration of the internal endomorphisms of

$p_{\mathcal{X}}\colon \mathcal{X} \to \mathbf{B}$, i.e., a strong morphism of lax monoidal fibrations from $p_{\mathcal{E}}\colon \mathcal{E} \to \mathbf{B}$ to $p\colon [\mathcal{X} \Rightarrow \mathcal{X}] \to \mathbf{B}$ in Fib($\mathbf{B}$). However, the examples of actions we have in mind, are almost never strong. But even in this case we can still find reasonable representations if we will consider the exponentiation of $p_{\mathcal{X}}\colon \mathcal{X} \to \mathbf{B}$ in $\mathbf{Cat}/\mathbf{B}$ instead of Fib($\mathbf{B}$). To distinguish these two kinds of exponentiation we denote the exponential object $p_{\mathcal{X}}\colon \mathcal{X} \to \mathbf{B}$ to $p_{\mathcal{Y}}\colon \mathcal{Y} \to \mathbf{B}$ in $\mathbf{Cat}/\mathbf{B}$ as $p\colon \mathcal{Y}^{\mathcal{X}} \to \mathbf{B}$. It is well known, cf. [9], for $p\colon \mathcal{Y}^{\mathcal{X}} \to \mathbf{B}$ to be a well defined object of $\mathbf{Cat}/\mathbf{B}$ it is necessary and sufficient for $\mathcal{X}$ to be a so-called Conduché fibration. But as we want $p\colon \mathcal{Y}^{\mathcal{X}} \to \mathbf{B}$ to be a fibration we shall assume that $p_{\mathcal{X}}$ is a bifibration, i.e., both fibration and opfibration. In fact, as we are mainly interested in the case where $\mathcal{X} = \mathcal{Y}$, in order to get a better description $p\colon \mathcal{Y}^{\mathcal{X}} \to \mathbf{B}$, it won't be a big restriction when we shall assume that both $\mathcal{X}$ and $\mathcal{Y}$ are bifibrations.

4.1. The exponential bifibrations in Cat/B. For any bifibration $p_{\mathcal{X}}\colon \mathcal{X} \to \mathbf{B}$ the exponential fibration $p_{\exp}\colon \mathcal{X}^{\mathcal{X}} \to \mathbf{B}$ in $\mathbf{Cat}/\mathbf{B}$ is a lax monoidal fibration with tensor being the composition of functors in fibres. The monoids in a fibre $\mathcal{X}^{\mathcal{X}}$ over B are monads on $\mathcal{X}_B$, and a morphism of monoids over u is a usual morphism of monads whose functor part is u^*. The counit of the exponential adjunction, the evaluation $\mathrm{ev}_{\mathcal{X}}\colon \mathcal{X}^{\mathcal{X}} \times_{\mathbf{B}} \mathcal{X} \to \mathcal{X}$, is an action of the lax monoidal fibration $p_{\exp}\colon \mathcal{X}^{\mathcal{X}} \to \mathbf{B}$ on $p_{\mathcal{X}}\colon \mathcal{X} \to \mathbf{B}$. Finally, the algebras for this action are Eilenberg–Moore algebras for all the monads taken together. As we shall need it later, we shall describe all of this below in detail. The case of real interest in this paper is when $p_{\mathcal{X}}\colon \mathcal{X} \to \mathbf{B}$ is a basic bifibration cod: $\mathcal{C}^{\to} \to \mathcal{C}$ of the category $\mathcal{C}$ with pullbacks and very likely being just Set.

Let $\mathbf{1}$, $\mathbf{2}$, $\mathbf{3}$ be the obvious categories generated by the graphs $\{\bullet\}$, $\{\bullet \to \bullet\}$, $\{\bullet \to \bullet \to \bullet\}$, respectively. For an object $B \in \mathbf{B}$, $r_B\colon \mathbf{1} \to \mathbf{B}$ is the functor picking the object B. Similarly, $r_u\colon \mathbf{2} \to \mathbf{B}$ is a functor picking the morphism $u\colon B' \to B$ in $\mathbf{B}$, and $r_{u,v}\colon \mathbf{3} \to \mathbf{B}$ is a morphism picking a composable pair $u \circ v$ in $\mathbf{B}$. Let $p_{\mathcal{X}}\colon \mathcal{X} \to \mathbf{B}$ and $p_{\mathcal{Y}}\colon \mathcal{Y} \to \mathbf{B}$ be two bifibration. We can form pullbacks

$$
\begin{array}{ccc}
\mathcal{X}_B & \longrightarrow & \mathcal{X} \\
\downarrow & & \downarrow{\scriptstyle p_{\mathcal{X}}} \\
\mathbf{1} & \xrightarrow{r_B} & \mathbf{B}
\end{array}
\qquad
\begin{array}{ccc}
\mathcal{X}_u & \longrightarrow & \mathcal{X} \\
\downarrow & & \downarrow{\scriptstyle p_{\mathcal{X}}} \\
\mathbf{2} & \xrightarrow{r_u} & \mathbf{B}
\end{array}
\qquad
\begin{array}{ccc}
\mathcal{X}_{u,v} & \longrightarrow & \mathcal{X} \\
\downarrow & & \downarrow{\scriptstyle p_{\mathcal{X}}} \\
\mathbf{3} & \xrightarrow{r_{u,v}} & \mathbf{B}
\end{array}
$$

in the category $\mathbf{Cat}$, i.e., products in $\mathbf{Cat}/\mathbf{B}$. If the exponential object $p_{\exp}\colon \mathcal{Y}^{\mathcal{X}} \to \mathbf{B}$ exists in $\mathbf{Cat}/\mathbf{B}$ then the objects of $\mathcal{Y}^{\mathcal{X}}_B$ correspond to morphisms from r_B to $p_{\exp}$ in $\mathbf{Cat}/\mathbf{B}$ and we have a sequence of correspondences

$$
\begin{array}{c}
\begin{array}{ccc}
\mathbf{1} & \longrightarrow & \mathcal{Y}^{\mathcal{X}} \\
& {\scriptstyle r_B}\searrow \quad \swarrow{\scriptstyle p_{\exp}} & \\
& \mathbf{B} &
\end{array} \\
\hline
\begin{array}{ccc}
\mathcal{X}_B & \longrightarrow & \mathcal{Y} \\
& \searrow \quad \swarrow{\scriptstyle p_{\mathcal{Y}}} & \\
& \mathbf{B} &
\end{array} \\
\hline
\mathcal{X}_B \longrightarrow \mathcal{Y}_B
\end{array}
$$

showing that we can (and we will) identify objects of $\mathcal{Y}^{\mathcal{X}}_B$ with functors from $\mathcal{X}_B$ to $\mathcal{Y}_B$. Similarly, using r_u and $r_{u,v}$ we see that morphisms in $\mathcal{Y}^{\mathcal{X}}$ over a morphism u are functors from $\mathcal{X}_u$ to $\mathcal{Y}_u$ commuting over $\mathbf{2}$, and the composable pairs are morphisms from $\mathcal{X}_{u,v}$ to $\mathcal{Y}_{u,v}$ commuting over $\mathbf{3}$. The Conduché condition is saying in elementary terms[5] that for any composable pair of morphisms $v\colon B'' \to B'$, $u\colon B' \to B$ in $\mathbf{B}$ the square of the obvious embeddings

$$\begin{array}{ccc} \mathcal{X}_{B'} & \xrightarrow{d_u} & \mathcal{X}_u \\ {\scriptstyle c_v}\downarrow & & \downarrow{\scriptstyle \kappa_u} \\ \mathcal{X}_v & \xrightarrow[\kappa_v]{} & \mathcal{X}_{u,v} \end{array}$$

is a pushout in $\mathbf{Cat}$. Then the composition of morphisms $F\colon \mathcal{X}_u \to \mathcal{Y}$ over u and $G\colon \mathcal{X}_v \to \mathcal{Y}$ over v such that $F \circ d_u = G \circ c_v$ is the unique functor $[F, G]\colon \mathcal{X}_{u,v} \to \mathcal{Y}$ such that $[F, G] \circ \kappa_u = F$ and $[F, G] \circ \kappa_v = G$ composed with the embedding $\mathcal{X}_{u\circ v} \to \mathcal{X}_{u,v}$.

Recall that for $u\colon B' \to B \in \mathbf{B}$ we have the reindexing functor $u^*\colon \mathcal{X}_{B'} \to \mathcal{X}_B$ and the coreindexing functor $u_!\colon \mathcal{X}_B \to \mathcal{X}_{B'}$ defined with the use of prone and supine morphisms in $\mathcal{X}$. We denote such functors in different fibrations by the same symbols. The following lemma describes morphisms in $\mathcal{Y}^{\mathcal{X}}$ more conveniently in five different ways.

Lemma 4.1. *Let $p_{\mathcal{X}}\colon \mathcal{X} \to \mathbf{B}$ and $p_{\mathcal{Y}}\colon \mathcal{Y} \to \mathbf{B}$ be two bifibrations. Let $u\colon B' \to B$ be a morphism in $\mathbf{B}$, Q an object in $\mathcal{Y}^{\mathcal{X}}_{B'}$, i.e., a functor from $\mathcal{X}_{B'}$ to $\mathcal{Y}_{B'}$, P an object in $\mathcal{Y}^{\mathcal{X}}_{B}$, i.e., a functor from $\mathcal{X}_B$ to $\mathcal{Y}_B$. There is a natural correspondence between*

(1) *functors from $F\colon \mathcal{X}_u \to \mathcal{Y}_u$ over $\mathbf{2}$ such that $F \circ d_u = Q$ and $F \circ c_u = P$;*
(2) *natural transformations $\tau\colon Qu^* \to u^*P$ in $\mathbf{Cat}(\mathcal{X}_B, \mathcal{Y}_{B'})$;*
(3) *natural transformations $\sigma\colon u_!Q \to Pu_!$ in $\mathbf{Cat}(\mathcal{X}_{B'}, \mathcal{Y}_B)$;*
(4) *natural transformations $\overline{\tau}\colon u_!Qu^* \to P$ in $\mathbf{Cat}(\mathcal{X}(B), \mathcal{Y}(B))$;*
(5) *natural transformations $\overline{\sigma}\colon Q \to u^*Pu_!$ in $\mathbf{Cat}(\mathcal{X}(B'), \mathcal{Y}(B'))$.*

Moreover, if both $p_{\mathcal{X}}$ and $p_{\mathcal{Y}}$ are cartesian bifibrations and both P and Q (weakly) preserve pullbacks in fibres, then under the above correspondences the (weakly) cartesian natural transformations correspond to the (weakly) cartesian natural transformations.

Note that in the above lemma, the two occurrences of the symbols u^* in (2), and $u_!$ in (3), do NOT denote the same functors! In each of the conditions (2) to (5) one of the functors u^*, $u_!$ refers to the bifibration $p_{\mathcal{X}}\colon \mathcal{X} \to \mathbf{B}$ and one of the functors u^*, $u_!$ to the other bifibration $p_{\mathcal{Y}}\colon \mathcal{Y} \to \mathbf{B}$.

Proof. As the conditions (4) and (5) are easily seen to be equivalent to (2) and (3), respectively, we shall concentrate on equivalence (1), (2), (3).

Fix $u\colon B' \to B$ in $\mathbf{B}$, a functor

$$\begin{array}{ccc} \mathcal{X}_u & \xrightarrow{F} & \mathcal{Y}_u \\ & \searrow \quad \swarrow & \\ & \mathbf{B} & \end{array}$$

[5]This condition will never be used in the explicit form but for the interested reader we recall it here, cf. [9]. The functor $p_{\mathcal{X}}\colon \mathcal{X} \to \mathbf{B}$ is called a *Conduché fibration*, if for any morphism f in $\mathcal{X}$ and a pair of morphisms u, v in $\mathbf{B}$ such that $p_{\mathcal{X}}(f) = u \circ v$, there are morphisms g and h in $\mathcal{X}$ such that $f = g \circ h$, $p_{\mathcal{X}}(g) = u$ and $p_{\mathcal{X}}(h) = v$. Moreover, such a factorization of f is unique up to a zigzag of morphisms in $\mathcal{X}$ that belong to the fibre over the domain of u.

as in (1), and two natural transformations

$$\tau\colon Qu^* \to u^*P, \quad \sigma\colon u_!Q \to Pu_!$$

from the diagram

$$\begin{array}{ccc} \mathcal{X}_{B'} & \xrightarrow{\quad Q \quad} & \mathcal{Y}_{B'} \\ {\scriptstyle u^*}\big\uparrow\big\downarrow{\scriptstyle u_!} & & {\scriptstyle u^*}\big\uparrow\big\downarrow{\scriptstyle u_!} \\ \mathcal{X}_B & \xrightarrow[\quad P \quad]{} & \mathcal{Y}_B \end{array}$$

as in (2) and (3).

Let $f\colon X' \to X$ be a morphism in $\mathcal{X}$ over $u\colon B' \to B$ with a factorization via prone $\mathrm{pr}_{u,X}$ and supine $\mathrm{su}_{u,X'}$ morphisms in $\mathcal{X}$ as follows

$$\begin{array}{ccc} X' & \xrightarrow{\mathrm{su}_{u,X'}} & u_!(X') \\ {\scriptstyle \hat{f}}\big\downarrow & \searrow^{f} & \big\downarrow{\scriptstyle \check{f}} \\ u^*(X) & \xrightarrow[\mathrm{pr}_{u,X}]{} & X. \end{array}$$

The mutual correspondence between the functor F and transformations τ and σ can be read off from the following diagram

$$\begin{array}{ccc} & & u_!\big(F(X')\big) \\ & \nearrow^{\mathrm{su}_{u,F(X')}} & \big\downarrow{\scriptstyle \sigma_{X'}} \\ F(X') & \xrightarrow{F(\mathrm{su}_{u,X'})} & F\big(u_!(X')\big) \\ {\scriptstyle F(\hat{f})}\big\downarrow & \searrow^{F(f)} & \big\downarrow{\scriptstyle F(\check{f})} \\ F\big(u^*(X)\big) & \xrightarrow[F(\mathrm{pr}_{u,X})]{} & F(X) \\ {\scriptstyle \tau_X}\big\downarrow & \nearrow_{\mathrm{pr}_{u,F(X)}} & \\ u^*\big(F(X)\big) & & \end{array}$$

whose part is just F applied to the previous diagram, having in mind that F restricted to the fibre $\mathcal{X}_{B'}$ is Q and to the fibre $\mathcal{X}_B$ is P.

To see the remaining part of the Lemma, we recall the direct relation between τ and σ. From τ we get σ as follows

$$u_!Q \xrightarrow{u_!Q(\eta^u)} u_!Qu^*u_! \xrightarrow{u_!(\tau_{u_!})} u_!u^*Pu_! \xrightarrow{\varepsilon^u_{Pu_!}} Pu_!$$

and we get back τ from σ as follows

$$Qu^* \xrightarrow{\eta_{Qu^*}} u^*u_!Qu^* \xrightarrow{u^*(\sigma_{u^*})} u^*Pu_!u^* \xrightarrow{u^*P(\varepsilon^u)} u^*P$$

where, as usual, η^u and ε^u are the unit and the counit of the adjunction $u_! \dashv u^*$. From this description it is easy to see that with the assumptions of the Lemma, τ is (weakly) cartesian if and only if σ is. □

The second of the above five above descriptions of morphisms in $\mathcal{Y}^{\mathcal{X}}$ seems to be the most convenient for us, and from now on we shall assume that the morphisms in $\mathcal{Y}^{\mathcal{X}}$ are given in that form. The composition in $\mathcal{Y}^{\mathcal{X}}$ is defined as follows. For morphisms $\sigma\colon Q \to P$ and $\tau\colon R \to Q$ in $\mathcal{Y}^{\mathcal{X}}$ over $u\colon B' \to B$ and $v\colon B'' \to B'$, respectively, we have

$$\begin{array}{ccccc} & & v^*Qu^* & & \\ & \overset{\tau_{u^*}}{\nearrow} & & \overset{v^*(\sigma)}{\searrow} & \\ Rv^*u^* & & & & v^*u^*P \\ \cong\uparrow & & & & \downarrow\cong \\ R(u\circ v)^* & & \xrightarrow{\quad\sigma\circ\tau\quad} & & (u\circ v)^*P \end{array}$$

where unnamed isomorphisms come from canonical isomorphisms between functors $(u\circ v)^*$ and $u^*\circ v^*$. The prone morphism over $u\colon B' \to B$ with the codomain P in $\mathcal{Y}^{\mathcal{X}}_B$

$$\mathrm{pr}_{u,P}\colon u^*Pu_! \to P$$

is the natural transformation in $\mathbf{Cat}(\mathcal{X}_B, \mathcal{Y}_{B'})$ defined with the help of the counit ε^u

$$u^*P(\varepsilon^u)\colon u^*Pu_!u^* \to u^*P.$$

Then, for any morphism $v\colon B'' \to B'$ in $\mathbf{B}$ and any morphisms $\tau\colon Q \to P$ in $\mathcal{Y}^{\mathcal{X}}$ over $u\circ v$ we have a (unique!) morphism $\widehat{\tau}\colon Q \to u^*Pu_!$ in $\mathcal{Y}^{\mathcal{X}}$ over v defined as a natural transformation

$$Qv^* \xrightarrow{Qv^*(\eta^u)} Qv^*u^*u_! \xrightarrow{\tau_{u_!}} v^*u^*Pu_!$$

so that $\tau = \mathrm{pr}_{u,P}\circ\widehat{\tau}$ in $\mathcal{Y}^{\mathcal{X}}$, i.e., the triangle of natural transformations

$$\begin{array}{ccc} Qv^*u^* \cong Q(uv)^* & & \\ \widehat{\tau}_{u^*}\downarrow & \overset{\tau}{\searrow} & \\ v^*u^*Pu_!u^* & \xrightarrow[v^*(\mathrm{pr}_{u,P})]{} & v^*u^*P \cong (uv)^*P \end{array}$$

commutes.

Similarly, the supine morphism over $u\colon B' \to B$ with the domain Q in $\mathcal{Y}^{\mathcal{X}}_{B'}$

$$\mathrm{su}_{u,Q}\colon Q \to u_!Qu^*$$

is the natural transformation in $\mathbf{Cat}(\mathcal{X}_B, \mathcal{Y}_{B'})$ defined with the help of the unit η^u

$$\eta^u_{Qu^*}\colon Qu^* \to u^*u_!Qu^*.$$

Then, for any $w\colon B' \to B''$ in $\mathbf{B}$ and any morphism $\sigma\colon Q \to P$ over $w\circ u$ we have a (unique) morphism $\check{\sigma}\colon u_!Qu^* \to P$ in $\mathcal{Y}^{\mathcal{X}}$ over w defined as a natural transformation

$$u_!Qu^*w^* \xrightarrow{u_!(\sigma)} u_!u^*w^*P \xrightarrow{\varepsilon^u_{w^*P}} w^*P$$

so that the triangle

$$\begin{array}{ccc} Q(vu)^* \cong Qu^*w^* & \xrightarrow{(\mathrm{su}_{u,Q})_{w^*}} & u^*u_!Qu^*w^* \\ & {\searrow}^{\sigma} & \downarrow{\scriptstyle u^*(\check{\sigma})} \\ & & (vu)^*P \cong u^*w^*P \end{array}$$

commutes in $\mathcal{Y}^{\mathcal{X}}$.

Proposition 4.2. *For any bifibration $p_{\mathcal{X}}\colon \mathcal{X} \to \mathbf{B}$ the exponential object $p_{\mathrm{exp}}\colon \mathcal{X}^{\mathcal{X}} \to \mathbf{B}$ in $\mathbf{Cat}/\mathbf{B}$ is a bifibration and it has the structure of a lax monoidal fibration, whose fibres are strict monoidal categories.*

PROOF. The fact that p_{exp} is a bifibration we have already seen. We describe the monoidal structure in $p_{\mathrm{exp}}\colon \mathcal{X}^{\mathcal{X}} \to \mathbf{B}$, and leave the reader to verify the axioms. We have an obvious isomorphism of fibrations

$$\mathbf{B} \times_{\mathbf{B}} \mathcal{X} \to \mathcal{X}$$

whose exponential adjoint in $\mathbf{Cat}/\mathbf{B}$

$$I\colon \mathbf{B} \to \mathcal{X}^{\mathcal{X}}$$

is the unit for the tensor. Thus, for $B \in \mathbf{B}$, the unit I_B in fibre $\mathcal{X}^{\mathcal{X}}_B$ is the identity functor on the fibre $\mathcal{X}_B$. The tensor functor

$$\mathcal{X}^{\mathcal{X}} \times_{\mathbf{B}} \mathcal{X}^{\mathcal{X}} \xrightarrow{\otimes} \mathcal{X}^{\mathcal{X}}$$

is the exponential adjoint in $\mathbf{Cat}/\mathbf{B}$ to the morphism

$$\mathcal{X} \times_{\mathbf{B}} \mathcal{X}^{\mathcal{X}} \times_{\mathbf{B}} \mathcal{X}^{\mathcal{X}} \xrightarrow{\mathrm{ev} \times 1_{\mathcal{X}^{\mathcal{X}}}} \mathcal{X} \times_{\mathbf{B}} \mathcal{X}^{\mathcal{X}} \xrightarrow{\mathrm{ev}} \mathcal{X}.$$

The tensor on objects is the composition of functors. As we will use it later, we describe explicitly the action of the tensor on morphisms. Let $\sigma\colon P_1 \to P_0$ and $\tau\colon Q_1 \to Q_0$ be two morphisms in $p_{\mathrm{exp}}\colon \mathcal{X}^{\mathcal{X}} \to \mathbf{B}$ over a morphism $u\colon B_1 \to B_0$, i.e., they are natural transformations $\sigma\colon P_1u^* \to u^*P_0$ and $\tau\colon Q_1u^* \to u^*Q_0$ in $\mathbf{Cat}\big(\mathcal{X}(B_0), \mathcal{X}(B_1)\big)$. Then their tensor $\sigma \otimes_u \tau\colon P_1 \otimes_{B_1} Q_1 = P_1 \circ Q_1 \to P_0 \circ Q_0 = P_0 \otimes_{B_0} Q_0$ is defined from the commutative diagram below

$$\begin{array}{ccc} P_1u^*u_!Q_1u^* & \xrightarrow{\sigma_{u_!} * \tau = \sigma * u_!(\tau)} & u^*P_0u_!u^*Q_0 \\ \uparrow{\scriptstyle P_1(\eta^u_{Q_1u^*})} & & \downarrow{\scriptstyle u^*P_0(\varepsilon^u_{Q_0})} \\ P_1Q_1u^* & \xrightarrow[\sigma \otimes_u \tau]{} & u^*P_0Q_0. \end{array}$$

□

The monoids in $(p_{\mathrm{exp}}\colon \mathcal{X}^{\mathcal{X}} \to \mathbf{B}, \otimes, I)$ are monads over fibres of p. A morphism of monoids $(f, u)\colon (M', m', e') \to (M, m, e)$ over $u\colon B' \to B$ is a morphism of monads from (M, m, e) to (M', m', e') whose functor part is $u^*\colon \mathcal{X}_B \to \mathcal{X}_{B'}$ and $f\colon M'u^* \to u^*M$ is a natural transformation satisfying the usual conditions (see Section 5). The evaluation morphism

$$\mathrm{ev}\colon \mathcal{X}^{\mathcal{X}} \times \mathcal{X} \to \mathcal{X}$$

in $\mathbf{Cat}/\mathbf{B}$ is the action of the lax monoidal fibration p_{exp} on the fibration $p_{\mathcal{X}}$. The algebras for this action are all algebras for all monads in $\mathrm{Mon}(p_{\mathrm{exp}}\colon \mathcal{X}^{\mathcal{X}} \to \mathbf{B}, \otimes, I)$

taken together, i.e., organized into a single category fibred over $\mathbf{B}$ as well as over $\mathrm{Mon}(p_{\mathrm{exp}}\colon \mathcal{X}^{\mathcal{X}} \to \mathbf{B}, \otimes, I)$.

We will apply this construction mainly to the basic fibration $\mathrm{cod}\colon \mathcal{C}^{\to} \to \mathcal{C}$ of a category $\mathcal{C}$ with pullbacks (usually $\mathcal{C} = \mathrm{Set}$). Such a fibration is always a cartesian bifibration. We write $p_{\mathrm{exp}}\colon \mathrm{Exp}(\mathcal{C}) \to \mathcal{C}$ (or $p_{\mathrm{exp},\mathcal{C}}$ if we want to indicate the category $\mathcal{C}$) for $p_{\mathrm{exp}}\colon \mathcal{C}^{\to \mathcal{C}^{\to}} \to \mathcal{C}$. The monoids in the exponential fibration $p_{\mathrm{exp}}\colon \mathrm{Exp}(\mathcal{C}) \to \mathcal{C}$ are monads in slices of $\mathcal{C}$. The morphism between two monoids over $u\colon c \to c' \in \mathcal{C}$ is a morphism of monads (in the opposite direction!) whose functor part is the pullback functor $u^*\colon \mathcal{C}/c' \to \mathcal{C}/c$.

Let $\mathit{Cart}(\mathcal{C})$ denote the subcategory of $\mathrm{Exp}(\mathcal{C})$ whose objects are pullback preserving functors and cartesian natural transformations between them. Moreover, let $\mathit{wCart}(\mathcal{C})$ denote the subcategory of $\mathrm{Exp}(\mathcal{C})$ whose objects are functors weakly preserving pullbacks and weakly cartesian natural transformations between them. Restricting p_{exp} to $\mathit{Cart}(\mathcal{C})$ and $\mathit{wCart}(\mathcal{C})$ we get functors $p_{\mathrm{ca},\mathcal{C}}\colon \mathit{Cart}(\mathcal{C}) \to \mathcal{C}$ and $p_{\mathrm{wca},\mathcal{C}}\colon \mathit{wCart}(\mathcal{C}) \to \mathcal{C}$, respectively. We have

Proposition 4.3. *The functors $p_{\mathrm{ca},\mathcal{C}}$ and $p_{\mathrm{wca},\mathcal{C}}$ described above are lax monoidal bifibrations with all that structure inherited from $p_{\mathrm{exp}}\colon \mathrm{Exp}(\mathcal{C}) \to \mathcal{C}$. In particular, the embeddings*

$$\begin{array}{ccccc} \mathit{Cart}(\mathcal{C}) & \longrightarrow & \mathit{wCart}(\mathcal{C}) & \longrightarrow & \mathrm{Exp}(\mathcal{C}) \\ & \searrow^{p_{\mathrm{ca},\mathcal{C}}} & \downarrow{\scriptstyle p_{\mathrm{wca},\mathcal{C}}} & \swarrow_{p_{\mathrm{exp}}} & \\ & & \mathcal{C} & & \end{array}$$

are morphisms of lax monoidal fibrations and of bifibrations, that are faithful and full on isomorphisms. The monoids in $p_{\mathrm{ca},\mathcal{C}}$ ($p_{\mathrm{wca},\mathcal{C}}$) are (weakly) cartesian monads on slices of $\mathcal{C}$.

PROOF. To see that $p_{\mathrm{ca},\mathcal{C}}$ is a fibration one has to notice that the prone morphism over $u\colon c \to c'$ with the codomain $P\colon \mathcal{C}_{/c'} \to \mathcal{C}_{/c'}$, being a pullback preserving functor, is a cartesian natural transformation. Moreover to see that a factorization via a prone morphism of a morphism τ in $\mathrm{Exp}(\mathcal{C})$, being a cartesian natural transformation between pullback preserving functors, is also a cartesian natural transformation. All this follows directly from the explicit formulas given above for the prone morphisms and the factorization via a prone morphism in $p_{\mathrm{exp}}\colon \mathcal{Y}^{\mathcal{X}} \to \mathbf{B}$ and the fact that both indexing and reindexing functors in p_{exp} preserve pullbacks. All the above remains true if we replace pullbacks by weak pullbacks. Thus both $p_{\mathrm{ca},\mathcal{C}}$ and $p_{\mathrm{wca},\mathcal{C}}$ are fibrations and subfibrations of p_{exp}. The argument that these functors are subopfibrations of p_{exp} is similar.

Clearly, the composition of functors (weakly) preserving pullbacks (weakly) preserves pullbacks. From this it is easy to see that the whole lax monoidal structure of $p_{\mathrm{ca},\mathcal{C}}$ and $p_{\mathrm{wca},\mathcal{C}}$ is inherited from p_{exp}.

The remaining part of the proposition is obvious. □

Remark. There are many more interesting subfibrations of $p_{\mathrm{exp}}\colon \mathrm{Exp}(\mathcal{C}) \to \mathcal{C}$. The fibrations $p_{\mathrm{ca},\mathcal{C}}$ and $p_{\mathrm{wca},\mathcal{C}}$ have also their 'wide pullback versions.' If slices of $\mathcal{C}$ are sufficiently cocomplete (e.g., if $\mathcal{C}$ is Set) then finitary or even accessible functors form full subfibrations of p_{exp}. However, the functors preserving finite limits (or just the terminal object) do not constitute a subfibration of p_{exp}, as the functor $u_!\colon \mathcal{C}_{/c} \to \mathcal{C}_{/c'}$ does not preserve the terminal object, in general.

The following result is the main reason we consider exponential fibrations. It will be used later many times.

Proposition 4.4. *Let $(p_{\mathcal{E}} \colon \mathcal{E} \to \mathbf{B}, I, \otimes, \alpha, \lambda, \varrho)$ be a lax monoidal fibration and $p_{\mathcal{X}} \colon \mathcal{X} \to \mathbf{B}$ be a bifibration. Then the bijective correspondence given by the exponential adjunction in $\mathbf{Cat}/\mathbf{B}$ between morphisms*

$$\begin{array}{ccc} \mathcal{E} \times_{\mathbf{B}} \mathcal{X} & \xrightarrow{\ \star\ } & \mathcal{X} \\ & \searrow \quad \swarrow_{p_{\mathcal{X}}} & \\ & \mathbf{B} & \end{array}$$

and morphisms

$$\begin{array}{ccc} \mathcal{E} & \xrightarrow{\ \check{\star}\ } & \mathcal{X}^{\mathcal{X}} \\ & {}_{p_{\mathcal{E}}}\searrow \quad \swarrow_{p_{\exp}} & \\ & \mathbf{B} & \end{array}$$

induces a bijective correspondence between actions $(\star, \psi_0, \psi_2)$ of $p_{\mathcal{E}}$ on $p_{\mathcal{X}}$ and morphisms of lax monoidal fibrations $(\check{\star}, \phi_0, \phi_2)$ from $p_{\mathcal{E}}$ to $p_{\exp}$. The correspondence relates the coherence natural transformations as follows. The transformation

$$\phi_0 \colon I \to \check{\star} \circ I$$

for $O \in \mathbf{B}$, is a natural transformation of functors $(\phi_0)_O \colon \mathrm{id}_{\mathcal{X}_O} \to I_O \star (-)$, i.e., for $X \in \mathcal{X}_O$ we have

$$\big((\phi_0)_O\big)_X = (\psi_0)_{O,X} \colon X \to I_O \star X.$$

Moreover, for $A, B \in \mathcal{E}_O$, we have

$$\begin{array}{ccc} \check{\star}(A) \circ \check{\star}(B) & \xrightarrow{(\phi_2)_{A,B}} & \check{\star}(A \otimes B) \\ \Big\| & & \Big\| \\ A \star \big(B(-)\big) & \xrightarrow[(\psi_2)_{A,B,-}]{} & (A \otimes B) \star (-). \end{array}$$

i.e., $X \in \mathcal{X}_O$,

$$\big((\phi_2)_{A,B}\big)_X = (\psi_2)_{A,B,X} \colon A \star (B \star X) \to (A \otimes B) \star X.$$

This correspondence is natural in $\mathcal{E}$.

PROOF. Exercise. □

From the above proposition follows that if we have an action of the lax monoidal fibration $p_{\mathcal{E}} \colon \mathcal{E} \to \mathbf{B}$ on a bifibration $p_{\mathcal{X}} \colon \mathcal{X} \to \mathbf{B}$, we get a morphism from the lax monoidal fibration $p_{\mathcal{E}}$ into a lax monoidal fibration whose fibres are strict monoidal categories. If $\mathcal{X}$ is sufficiently concrete (like $\mathrm{Set}^{\to}$) and this morphism is an embedding we can view this kind of phenomena as representation theorems. We represent objects of $\mathcal{E}$ as endofunctors of fibres of $p_{\mathcal{X}} \colon \mathcal{X} \to \mathbf{B}$, and monoids in $(p_{\mathcal{E}} \colon \mathcal{E} \to \mathbf{B}, \otimes, I)$ as monads over fibres of $p_{\mathcal{X}} \colon \mathcal{X} \to \mathbf{B}$. Similar things can be said about morphisms. We will see many examples of such representations later.

4.2. The exponential fibrations in Fib/B. Fib/**B** is a cartesian closed category. For any fibration $p_{\mathcal{X}}\colon \mathcal{X} \to \mathbf{B}$ the exponential fibration $p_{\mathrm{fiexp}}\colon [\mathcal{X} \Rightarrow \mathcal{X}] \to \mathbf{B}$ is lax monoidal with tensor being (again) the internal composition. Monoids in p_{fexp} are compatible families of monads and cartesian morphisms between them.

Having a strong action $\star\colon \mathcal{E} \times_{\mathbf{B}} \mathcal{X} \to \mathcal{X}$ we could also represent $\mathcal{E}$ in $[\mathcal{X} \Rightarrow \mathcal{X}]$. But strong actions are less common, and such (nontrivial) representations are more difficult to achieve in practice. This is why we are not going consider this kind of exponential fibrations in the following.

5. The Burroni fibrations and opetopic sets

5.1. The Burroni fibrations and T-categories. Let $\mathcal{C}$ be a category with pullbacks, $\langle T, \eta, \mu\rangle$ a monad on $\mathcal{C}$. The category $\mathrm{Gph}(T)$ is the category of T-graphs. An object $\langle A, O, \gamma, \delta\rangle$ of $\mathrm{Gph}(T)$ is a span

$$\begin{array}{ccc} & A & \\ {\scriptstyle\gamma}\swarrow & & \searrow{\scriptstyle\delta} \\ O & & T(O) \end{array}$$

in $\mathcal{C}$. The morphisms γ and δ are called *codomains* and *domains* of the T-graph $\langle A, O, \gamma, \delta\rangle$, respectively. Sometimes we write A instead of $\langle A, O, \gamma, \delta\rangle$, for short, when it does not lead to a confusion.

A morphism of T-graphs $\langle f, u\rangle\colon \langle A, O, \gamma, \delta\rangle \to \langle A', O', \gamma', \delta'\rangle$ is a pair of morphisms $f\colon A \to A'$ and $u\colon O \to O'$ in $\mathcal{C}$ making the squares

$$\begin{array}{ccc} A & \xrightarrow{\ f\ } & A' \\ {\scriptstyle\gamma}\downarrow & & \downarrow{\scriptstyle\gamma'} \\ O & \xrightarrow[\ u\ ]{} & O' \end{array} \qquad\qquad \begin{array}{ccc} A & \xrightarrow{\ f\ } & A' \\ {\scriptstyle\delta}\downarrow & & \downarrow{\scriptstyle\delta'} \\ T(O) & \xrightarrow[T(u)]{} & T(O') \end{array}$$

commute. Let $\mathrm{Gph}(T)$ denotes the category of T-graphs and T-graph morphisms. We have a projection functor

$$p_T\colon \mathrm{Gph}(T) \to \mathcal{C}$$

sending the morphism $\langle f, u\rangle\colon \langle A, O, \gamma, \delta\rangle \to \langle A', O', \gamma', \delta'\rangle$ to the morphism $u\colon O \to O'$ which is easily seen to be a fibration, cf. [6, p. 235]. The lax monoidal structure in p_T is defined as follows. Let $\langle A, O, \gamma_A, \delta_A\rangle$ and $\langle B, O, \gamma_B, \delta_B\rangle$ be two objects in the fibre over O, i.e., in $\mathrm{Gph}(T)_O$. Then the tensor

$$\langle A, O, \gamma_A, \delta_A\rangle \otimes_O \langle B, O, \gamma_B, \delta_B\rangle = \langle A \otimes B, O, \gamma_\otimes, \delta_\otimes\rangle$$

is defined from the following diagram

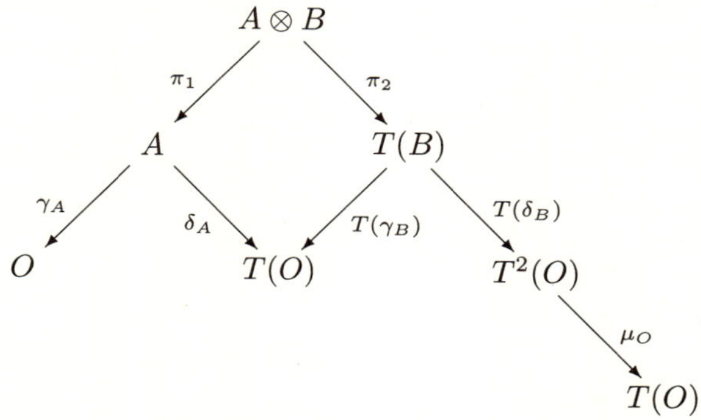

in which the square is a pullback and

$$\gamma_\otimes = \gamma_A \circ \pi_1, \quad \delta_\otimes = \mu_O \circ T(\delta_B) \circ \pi_2.$$

The unit in the fibre over O is

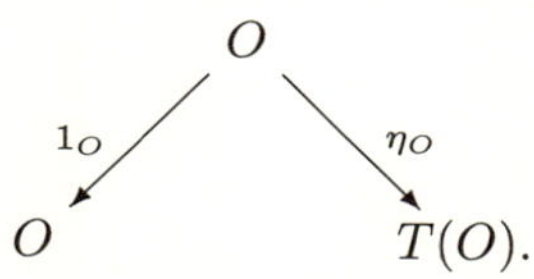

The coherence morphisms are defined using the universal properties of pullbacks. For an object $\langle A, O, \gamma_A, \delta_A \rangle$ on the fibre over O the left unit morphism is

$$\lambda_{\langle A,O,\gamma_A,\delta_A\rangle} = \langle 1_O, \eta_A), 1_O \rangle \colon \langle A, O, \gamma_A, \delta_A \rangle \to \langle O \otimes A, O, \gamma_\otimes, \delta_\otimes \rangle$$

the right unit morphism is

$$\varrho_{\langle A,O,\gamma_A,\delta_A\rangle} = \langle 1_A, 1_O \rangle \colon \langle A \otimes O, O, \gamma_\otimes, \delta_\otimes \rangle \to \langle A, O, \gamma_A, \delta_A \rangle$$

(the right unit morphism is always an isomorphism and in fact we can assume that it is an identity as $A \otimes O$ is a pullback of δ_A along the identity). The associativity morphism

$$\alpha_{A,B,C} \colon A \otimes (B \otimes C) \to (A \otimes B) \otimes C$$

is also defined similarly, using universal properties of pullbacks. We leave the details to the reader.

Proposition 5.1. *Let $\mathcal{C}$ be a category with pullbacks. The functor $p_T \colon \mathrm{Gph}(T) \to \mathcal{C}$ is a bifibration and together with the monoidal structure $(\otimes, I, \alpha, \lambda, \varrho)$ described above is a lax monoidal fibration. The total category of the fibration $q_T \colon \mathrm{Mon}(T) \to \mathcal{C}$ of monoids in $(\mathrm{Gph}(T), p_T, \otimes, I, \alpha, \lambda, \varrho)$ is equivalent to the category of T-categories of Burroni. If moreover, the monad (T, η, μ) is cartesian then the fibres of p_T are (strong) monoidal categories, i.e., the coherence morphisms, λ and α, are isomorphisms.*

PROOF. A simple tedious check. □

Remark. If the monad (T, η, μ) is cartesian then the fibres of $p_T \colon \mathrm{Gph}(T) \to \mathcal{C}$ are strong monoidal categories but the reindexing functors are still only lax monoidal. This is already so for the identity monad $(1_\mathcal{C}, 1_{1_\mathcal{C}}, 1_{1_\mathcal{C}})$ on $\mathcal{C}$. The category $\mathrm{Mon}(1_\mathcal{C})$ is the category of internal categories in $\mathcal{C}$.

5.2. Tautologous actions of Burroni fibrations. If (T, η, μ) is a monad on a category $\mathcal{C}$ with pullbacks then the lax monoidal fibration $p_T\colon \mathrm{Gph}(T) \to \mathcal{C}$ has a natural action on the basic fibration $\mathrm{cod}\colon \mathcal{C}^{\to} \to \mathcal{C}$. The functor part

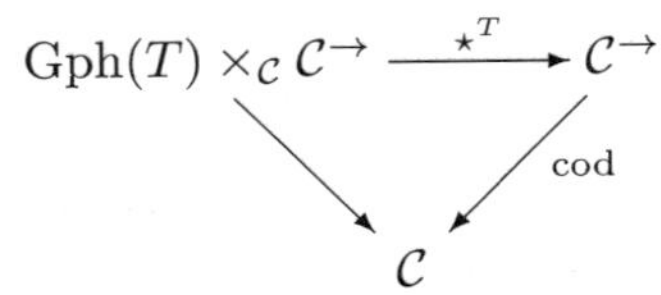

is defined on objects by

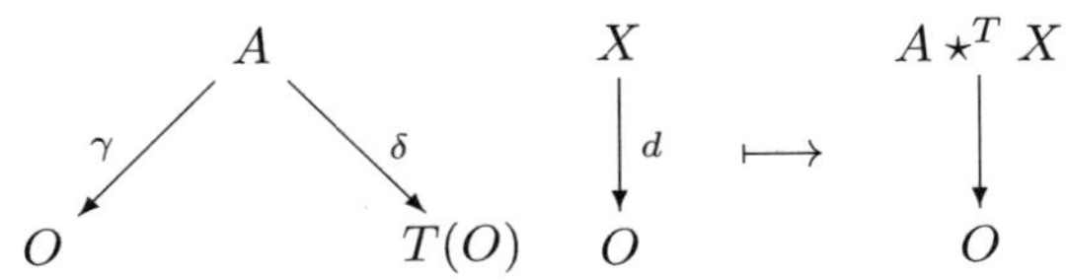

where the right vertical arrow in the above diagram is the composite of the upper horizontal arrows in the following diagram

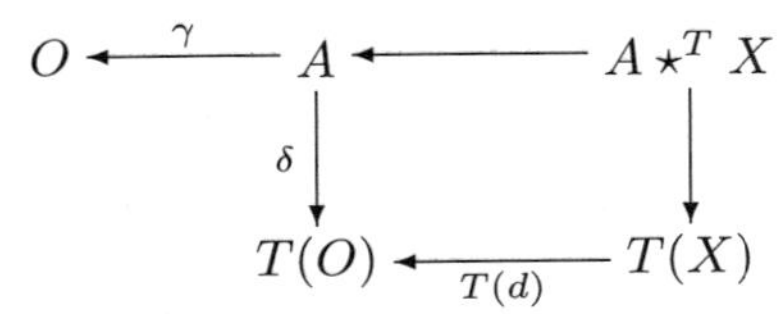

in which the square is a pullback. By adjunction, we get a morphism of lax monoidal fibrations

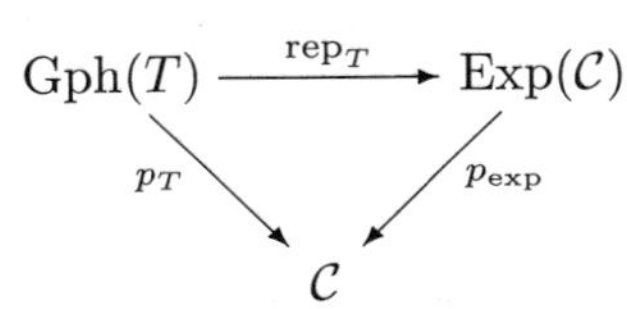

that represents T-graphs as endofunctors on slices of $\mathcal{C}$. Under this representation the T-categories correspond to (some) monads on slices of $\mathcal{C}$.

Example. The actions $\star^T$ of the lax monoidal fibration defined above do not preserve prone morphisms, in general. Even if T is the free monoid monad on Set, the action $\star^T$ is only a lax morphism of fibrations. To see this, consider a morphism $u\colon [1] \to [0]$, from two element set $[1] = \{0,1\}$ to one element set $[0] = \{0\}$. Let (A, γ, δ) be an object in $\mathrm{Gph}(T)$ over $[1]$, such that $A = \{a\}$, and $\partial(a) = 00$, the word of length 2 of zeros, $\gamma(a) = 0$. The identity $1_{[0]}$ on $[0]$ is a morphism in $\mathrm{Set}_{/[0]}$. Then the object $(A, \gamma, \delta) \star ([0], 1_{[0]})$ (in the basic fibration over Set) has one element in the domain, operation a with inputs and outputs in $[0]$, i.e., $\langle a; 0, 00\rangle$. The domain of the prone morphism $u^*\big((A, \gamma, \delta) \star ([0], 1_{[0]})\big) \to (A, \gamma, \delta) \star ([0], 1_{[0]})$ over u has two elements, in the domain, the operation a with the output either 0 or 1 and inputs as before, i.e., $\{\langle a; 0, 00\rangle, \langle a; 1, 00\rangle\}$. On the other hand, the image under $\star^T$ of the prone morphisms over u, whose codomains are (A, γ, δ) and $1_{[0]}$, respectively, $\mathrm{pr}_{u,A} \star^T \mathrm{pr}_{u,1_{[0]}} : u^*(A) \star^T u^*([0]) \to A \star^T [0]$ has in the domain of its domain eight elements, i.e., the operation a with both inputs and outputs either 0 or 1, i.e., $\{\langle a; 0, 00\rangle, \langle a; 1, 00\rangle, \ldots, \langle a; 1, 10\rangle \langle a; 1, 11\rangle\}$. Thus the domains of those morphisms are not isomorphic and hence the prone morphisms are not preserved.

As the morphism rep_T is the exponential transpose of $\star^T$ in $\mathrm{CAT}_{/\mathcal{C}}$ (not in $\mathrm{Fib}(\mathcal{C})$) one of these morphisms can be a morphism of fibrations even if the other one is not. We have

Proposition 5.2. *Let (T, η, μ) be a cartesian monad on a category with pullbacks $\mathcal{C}$. Then the functor* rep_T *defined above is a strong morphism of lax monoidal fibrations and of bifibrations. The image of* rep_T *is in* $p_{\mathrm{ca}}\colon \mathit{Cart}(\mathcal{C}) \to \mathcal{C}$.

PROOF. First, we describe the functor rep_T in details. For an object $A = (A, \gamma, \delta)$ in $\mathrm{Gph}(T)_O$, we have a functor

$$\mathrm{rep}_T(A) = A \star^T (-)\colon \mathcal{C}_{/O} \to \mathcal{C}_{/O}.$$

In the following, we omit the superscript T. For $u\colon O \to Q$ in $\mathcal{C}$ and $(h, u)\colon A \to B$, a morphism in $\mathrm{Gph}(T)$ over u, we have a natural transformation in $\mathbf{Cat}(\mathcal{C}_{/Q}, \mathcal{C}_{/O})$

$$\mathrm{rep}_T(h, u)\colon A \star u^*(-) \to u^*\big(B \star (-)\big)$$

so that for $d^Y : Y \to Q$ in $\mathcal{C}$, the value $\mathrm{rep}_T(h, u)_Y$ is defined from the following diagram

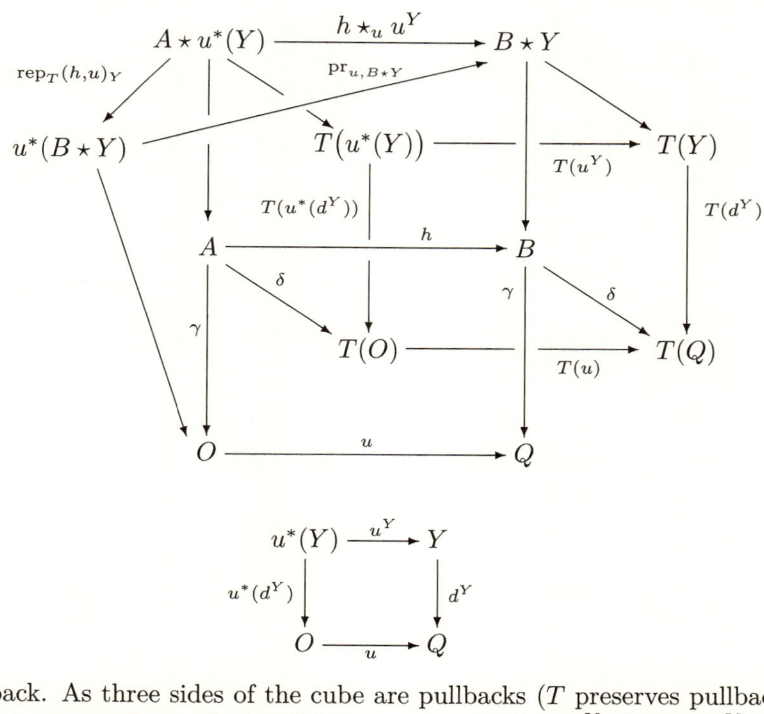

where

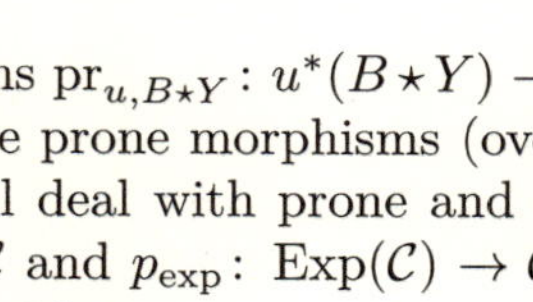

is a pullback. As three sides of the cube are pullbacks (T preserves pullbacks), so is the front square. In particular, if h is an iso, so is $h \star_u u^Y$, for any $d^Y : Y \to Q$ in $\mathcal{C}$.

Note that the morphisms $\mathrm{pr}_{u,B\star Y} : u^*(B \star Y) \to B \star Y$ and $u^Y : u^*(Y) \to Y$, in the above two diagrams, are prone morphisms (over u) in the basic fibration over $\mathcal{C}$. In the following, we will deal with prone and supine morphisms in two other fibrations $p_T\colon \mathrm{Gph}(T) \to \mathcal{C}$ and $p_{\mathrm{exp}}\colon \mathrm{Exp}(\mathcal{C}) \to \mathcal{C}$. Thus in total, we have three different sorts of prone morphisms.

The codomain of a supine morphism over $u\colon O \to Q$ whose domain is $A = (A, \gamma, \delta)$ is $u_!(A) = (A, u\circ\gamma, T(u)\circ\delta)$. The supine morphism is $\mathrm{su}_{u,A} = (1_A, u)\colon A \to$

$u_!(A)$. We have a diagram in Gph(T) over u:

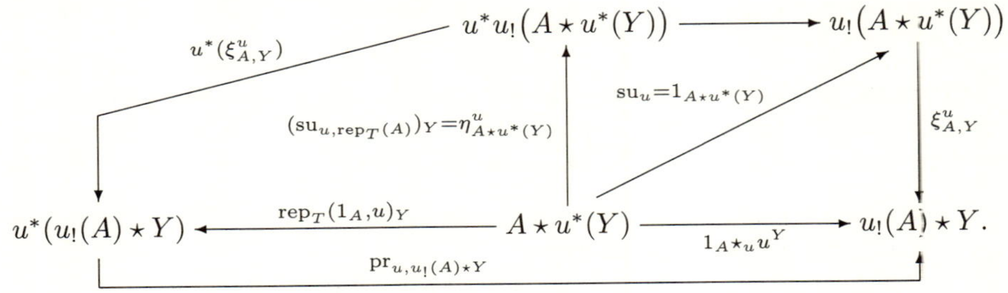

The morphism $\xi^u_{A,Y}$ is the second part of the factorization of $1_A \star_u u^Y$ via a supine morphism. Note that the morphisms $1_A \star_u u^Y$ and $\xi^u_{A,Y}$ considered as morphisms in $\mathcal{C}$ are equal but the first is a part of a morphism in Gph(T) over u, and the second is in the fibre over Q. By a remark below the previous diagram, $1_A \star_u u^Y$ is an isomorphism and hence, so are $\xi^u_{A,Y}$ and $u^*(\xi^u_{A,Y})$, as well. One can verify that the left hand triangle commutes, as it is a triangle in the fibre over O in the basic fibration over $\mathcal{C}$ and commutes when composed with the prone morphism $\mathrm{pr}_{u,u_!(A)\star Y}$.

Thus rep_T preserves the supine morphisms.

One can verify that the prone morphism in Gph(T) over $u\colon O \to Q$ with codomain B is $(u^B, u)\colon u^*(B) \to B$ where

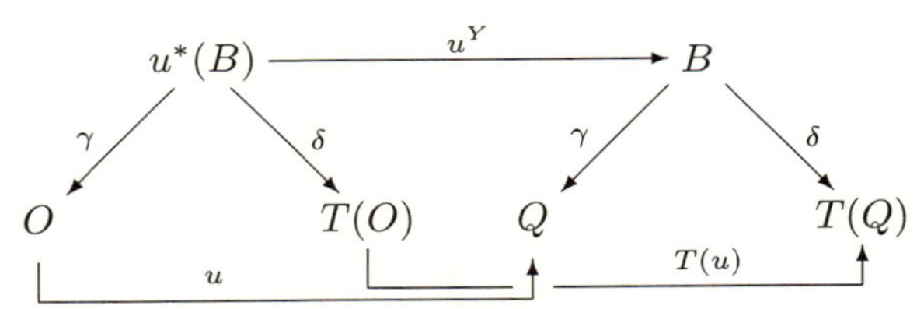

is a limiting cone. For $d^Y : Y \to Q$ in $\mathcal{C}$ with

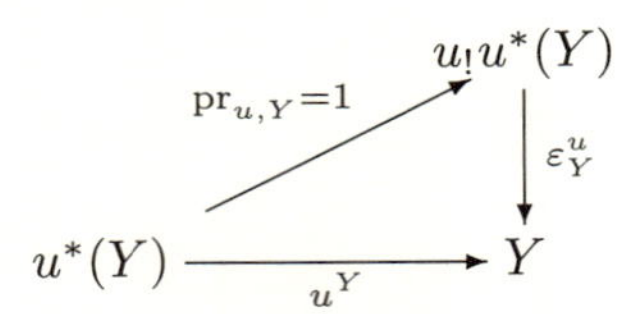

(i.e., $u^Y = \varepsilon^u_Y$) we can form a diagram

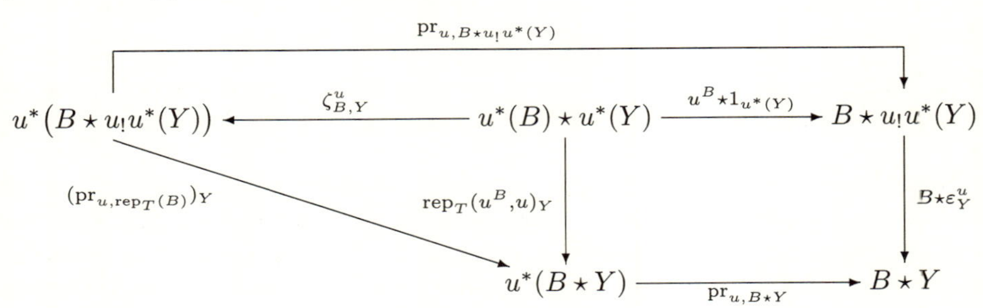

($(\mathrm{pr}_{u,\mathrm{rep}_T(B)})_Y = u^*(B\star\varepsilon^u_Y)$) in which one can verify, using properties of pullbacks, that $u^B \star_u 1_{u^*(Y)}$ is prone in the basic fibration over $\mathcal{C}$. Thus $\zeta^u_{B,Y}$, the first part of the factorization of $u^B \star_u 1_{u^*(Y)}$ via a prone morphism, is an iso. As the left triangle commutes, rep_T preserves prone morphisms, as well. □

Remark. From the above proof, it follows that for any monad T on a category $\mathcal{C}$ with pullbacks, we have morphisms

$$u_!\big(A \star u^*(Y)\big) \xrightarrow{\xi^u_{A,Y}} u_!(A) \star Y$$

$$u^*(B) \star u^*(Y) \xrightarrow{\zeta^u_{B,Y}} u^*\big(B \star u_! u^*(Y)\big)$$

natural in A, B and Y, that are isomorphisms if the monad T is cartesian. Note that these isomorphisms express a kind of Beck – Chevalley condition for actions of lax monoidal fibrations.

Example. If T is the identity monad on a category with pullbacks then rep_T sends the internal category $\mathbf{C} = (C_1, C_0, m, i, d, c)$ in $\mathcal{C}$ to a monad $\mathrm{rep}_T(\mathbf{C})$ on the slice category $\mathcal{C}_{/C_0}$ whose algebras are internal presheaves on $\mathbf{C}$.

5.3. Multisorted signatures vs monotone polynomial diagrams. In this section we shall examine the considerations from the previous section on a specific example of the free monoid monad T on the category Set. Note that $\mathrm{Gph}(T)$ can be thought of as a category of multisorted signatures. An object $\langle A, O, \gamma, \delta \rangle$ of $\mathrm{Gph}(T)$ can be seen as a set of operations A a set of types O, functions γ and δ associating to operations in A their types of codomains in O and their lists of types of their domains in $T(O)$. To emphasize this, we shall denote the fibration $p_T\colon \mathrm{Gph}(T) \to \mathrm{Set}$ for this particular monad T as $p_{\mathrm{m}}\colon \mathrm{Sig}_{\mathrm{m}} \to \mathrm{Set}$. As we already mentioned, cf. [6], the category of monoids in p_{m} is equivalent to the category of Lambek's multicategories. The action of $p_{\mathrm{m}}\colon \mathrm{Sig}_{\mathrm{m}} \to \mathrm{Set}$ on $\mathrm{cod}\colon \mathrm{Set}^{\to} \to \mathrm{Set}$ is as defined above. Thus, by adjunction, we have a representation morphism

$$\begin{array}{ccc} \mathrm{Sig}_{\mathrm{m}} & \xrightarrow{\mathrm{rep}_{\mathrm{m}}} & \mathrm{Exp}(\mathrm{Set}) \\ & {\scriptstyle p_{\mathrm{m}}}\searrow \quad \swarrow{\scriptstyle p_{\mathrm{exp}}} & \\ & \mathrm{Set}. & \end{array}$$

We shall describe the image of this representation in a different way. A *monotone polynomial diagram*[6] over the set O is a diagram of the following form

$$O \xleftarrow{\;s\;} E \xrightarrow{\;p\;} B \xrightarrow{\;t\;} O$$

of sets and functions, moreover the fibres of the morphism p are finite and linearly ordered. We write (t, p, s) to denote such a diagram. A *morphism of monotone diagrams* $(f, g, u)\colon (t, p, s) \to (t', p', s')$ over a function $u\colon O \to O'$ is a triple of functions with $g\colon E \to E'$ and $f\colon B \to B'$ so that the diagram

$$\begin{array}{ccccccc} O & \xleftarrow{\;s\;} & E & \xrightarrow{\;p\;} & B & \xrightarrow{\;t\;} & O \\ {\scriptstyle u}\downarrow & & \downarrow{\scriptstyle g} & & \downarrow{\scriptstyle f} & & \downarrow{\scriptstyle u} \\ O' & \xleftarrow[\;s'\;]{} & E' & \xrightarrow[\;p'\;]{} & B' & \xrightarrow[\;t'\;]{} & O' \end{array}$$

[6]The name is so chosen to indicated the obvious relation with the notion of a polynomial diagram that will be considered in the next section.

commutes, and the middle square is a pullback in the category of posets (i.e., $g_{\lceil p^{-1}(b)}: p^{-1}(b) \to p'^{-1}(f(b))$ is an order isomorphism, for $b \in B$). We compose morphisms of monotone polynomial diagrams in the obvious way, by placing one on top of the other. In this way we defined the category $\mathcal{MPolyDiag}$ of monotone polynomial diagrams. The category $\mathcal{MPolyDiag}$ is fibred over Set, where the projection functor

$$p_{\text{mpd}}: \mathcal{MPolyDiag} \to \text{Set}$$

is given by

$$(f, g, u): (t, p, s) \to (t', p', s') \longmapsto u: O \to O'.$$

This is a lax monoidal fibration[7] which also acts on the basic fibration cod: $\text{Set}^{\to} \to \text{Set}$. The action

$$\begin{array}{ccc} \mathcal{MPolyDiag} \times_{\text{Set}} \text{Set}^{\to} & \xrightarrow{\ \star\ } & \text{Set}^{\to} \\ & \searrow \quad \swarrow_{\text{cod}} & \\ & \text{Set} & \end{array}$$

is given by the well known formula defining polynomial functors (see Section 6), i.e., for (t, p, s) in $\mathcal{MPolyDiag}_O$, and $d^X : X \to O$ a function, we have

$$(t, p, s) \star d^X = t_! p_* s^*(d^X).$$

Thus, by adjointness, we have a morphism of lax monoidal fibrations

$$\begin{array}{ccc} \mathcal{MPolyDiag} & \xrightarrow{\ \text{rep}_{\text{mpd}}\ } & \text{Exp}(\text{Set}) \\ & {}_{p_{\text{mpd}}}\searrow \quad \swarrow_{p_{\text{exp}}} & \\ & \text{Set}. & \end{array}$$

The class of functors in the image rep_{mpd} coincides with the class of finitary polynomial endofunctors. However the linear structure in the fibres of monotone polynomial diagrams restricts the class of natural transformations between them. For every polynomial transformation $\tau: P \to Q$ between polynomial endofunctors on Set^I, there is a monotone morphisms between monotone diagrams $(f, g, u): (t, p, s) \to (t', p', s')$ so that $\text{rep}_{\text{mpd}}(f, g, u)$ is isomorphic to τ (just order fibres in the polynomial diagrams defining P and Q in a compatible way). This observation says that the essential image of rep_{mpd} consists of polynomial functors and polynomial natural transformations, see Section 6. However this is not saying that the monotone polynomial monads on polynomial functors are the same as polynomial monads. For a monad (T, η, μ) on a polynomial functor to be linear means,[8] that we can find one ordering of the fibres of the polynomial diagram defining T so that both morphisms $\eta: 1_C \to T$ and $\mu: T^2 \to T$ are defined by the morphisms of diagrams respecting these orderings (the order of the fibres of the diagram defining T^2 is determined by the order of the diagram defining T). As we shall see later, this might be not possible. We note for the record

[7]The definition of the lax monoidal structure is left to be defined by the reader. It is close to the structure on the fibration of polynomial functors defined in Section 6.

[8]Here by a monotone monad we mean a monad that is an image of a monoid in $p_{\text{mpd}}: \mathcal{MPolyDiag} \to \text{Set}$.

Proposition 5.3. *The representations* $\mathrm{rep}_{\mathrm{m}}$ *and* $\mathrm{rep}_{\mathrm{mpd}}$ *are faithful and they are equivalent as morphisms of lax monoidal fibrations into* $\mathrm{Exp}(\mathrm{Set}) \to \mathrm{Set}$. *As a consequence,* $\mathrm{rep}_{\mathrm{mpd}}$ *is a morphism of bifibrations and the category of Lambek's multicategories is equivalent to the category of monoids in* $\mathcal{MPolyDiag}$. *Moreover, the monads in the image of the morphism of fibrations of monoids*

$$\begin{array}{ccc} \mathrm{Mon}(\mathcal{MPolyDiag}) & \longrightarrow & \mathrm{Mon}(\mathrm{Exp}(\mathrm{Set})) \\ & \searrow \quad \swarrow & \\ & \mathrm{Set} & \end{array}$$

induced by $\mathrm{rep}_{\mathrm{mpd}}$ *are exactly monotone monads on polynomial functors.*

Remark. A bad thing about the representations $\mathrm{rep}_{\mathrm{m}}$ and $\mathrm{rep}_{\mathrm{mpd}}$ is that they are not full, even on isomorphism. As a consequence the monotone monads do not determine the monotone diagrams defining them uniquely (up to isomorphism). The lack of fullness on isomorphisms is due to the fact that fibres in monotone diagrams are linearly ordered. As we shall see in the next two sections, similar representations of both (finitary) polynomial (endo)functors and (finitary multivariable) analytic (endo)functors are full on isomorphisms.

5.4. Morphisms of monads. Morphisms of monads induce morphisms of Burroni fibrations and morphisms of tautologous actions of Burroni fibrations. In details, it looks as follows. Let (S, η^S, μ^S) be a monad on $\mathcal{C}$ and (T, η^T, μ^T) be a monad on $\mathcal{D}$, $F\colon \mathcal{C} \to \mathcal{D}$ a functor preserving pullbacks, and $\xi\colon TF \to FS$ be a natural transformation so that $(F, \xi)\colon (S, \eta^S, \mu^S) \to (T, \eta^T, \mu^T)$ is a monad morphism, i.e., the diagram

$$\begin{array}{ccccc} & \overset{\eta^T_F}{\nearrow} & TF & \xleftarrow{\ \mu^T_F\ } & T^2F \\ F & & \Big\downarrow \xi & & \Big\downarrow T(\xi) \\ & & & & TFS \\ & \underset{F(\eta^S)}{\searrow} & & & \Big\downarrow \xi_S \\ & & FS & \xleftarrow[\ F(\mu^S)\ ]{} & FS^2 \end{array}$$

commutes. Then we can define the functor from S-graphs to T-graphs

$$\begin{array}{ccc} \mathrm{Gph}(S) & \xrightarrow{\ \mathrm{Gph}(F,\xi)\ } & \mathrm{Gph}(T) \\ {\scriptstyle p_S}\Big\downarrow & & \Big\downarrow{\scriptstyle p_T} \\ \mathcal{C} & \xrightarrow[\ F\ ]{} & \mathcal{D} \end{array}$$

as

$$\begin{array}{ccccccccc} & A & & & & F(A) & \longleftarrow & \mathrm{Gph}(F,\xi)(A) & \\ {\scriptstyle\gamma}\swarrow & & \searrow{\scriptstyle\delta} & \longmapsto & {\scriptstyle F(\gamma)}\swarrow & & \searrow{\scriptstyle F(\delta)} & & \searrow \\ O & & S(O) & & F(O) & & FS(O) & \xleftarrow{\ \xi_O\ } & TF(O) \end{array}$$

where the square on the right in a pullback. This functor has an obvious structure (ϕ_0 and ϕ_2) of a morphism of lax monoidal fibrations.

In particular, as any monad (T,η,μ) on a category $\mathcal{C}$ with pullbacks has a monad morphism to the identity monad $1_{\mathcal{C}}$, any Burroni fibration on $\mathcal{C}$ has a morphism into the $1_{\mathcal{C}}$-fibration. This is another way of saying that the category of T-categories has a forgetful functors into the category of internal categories in $\mathcal{C}$.

Now a routine verification will show that such a morphism of lax monoidal fibrations of graphs together with a fibred morphism of basic fibrations

$$\begin{array}{ccc} \mathcal{C}^{\to} & \xrightarrow{F^{\to}} & \mathcal{D}^{\to} \\ {\scriptstyle \mathrm{cod}}\downarrow & & \downarrow{\scriptstyle \mathrm{cod}} \\ \mathcal{C} & \xrightarrow[F]{} & \mathcal{D} \end{array}$$

gives rise to a morphism of tautologous actions.

5.5. Relative Burroni fibrations and relative *T*-categories. The construction of a lax monoidal fibration of T-graphs can be performed even on a fibred monad on a fibration. Suppose $p\colon \mathcal{E} \to \mathbf{B}$ is a fibration such that the fibres of p have pullbacks. Moreover (T,η,μ) is a monad on the category $\mathcal{E}$ so that T is a lax morphism of fibrations

$$\begin{array}{ccc} \mathcal{E} & \xrightarrow{T} & \mathcal{E} \\ & {\scriptstyle p}\searrow \quad \swarrow{\scriptstyle p} & \\ & \mathbf{B} & \end{array}$$

and η, μ are fibred natural transformations (i.e., their components lie in the fibres of p). Having such data we can repeat the construction of the category of T-graphs but restricting the objects to such spans

$$\begin{array}{ccc} & A & \\ {\scriptstyle \gamma}\swarrow & & \searrow{\scriptstyle \delta} \\ O & & T(O) \end{array}$$

that are in fibres of p (i.e., $p(\gamma) = p(\delta) = 1_{p(O)}$). The morphisms are defined as before. In this way, we get a relative Burroni fibration $p_T\colon \mathrm{Gph}(T,p) \to \mathcal{E}$ of T-graphs over p. Clearly, p_T is a lax monoidal fibration with the tensor structure defined as before. Thus we have a fibration of monoids with a forgetful to $\mathrm{Gph}(T,p)$ as in the diagram

$$\begin{array}{ccc} \mathrm{Gph}(T,p) & \xleftarrow{\mathcal{U}_T} & \mathrm{Mon}(T,p) \\ {\scriptstyle p_T}\downarrow & \swarrow{\scriptstyle q_T} & \\ \mathcal{E} & & \\ {\scriptstyle p}\downarrow & & \\ \mathbf{B} & & \end{array}$$

of functors and categories. As for any category $\mathcal{C}$, the functor $!\colon \mathcal{C} \to \mathbf{1}$ into the terminal category $\mathbf{1}$ is a fibration, this construction is a generalization of the previous one.

Remark. We can also define a basic fibration $\mathrm{cod}\colon \mathcal{E}^{\to,p} \to \mathcal{E}$ relative to a fibration $p\colon \mathcal{E} \to \mathbf{B}$, so that the objects of $\mathcal{E}^{\to,p}$ are morphisms of $\mathcal{E}$ in fibres of p and morphisms are commuting squares. Then, as previously for the Burroni fibrations, we have a tautologous action the lax monoidal fibration $p_T\colon \mathrm{Gph}(T,p) \to \mathcal{E}$ on a fibration $\mathrm{cod}\colon \mathcal{E}^{\to,p} \to \mathcal{E}$

$$\begin{array}{ccc} \mathrm{Gph}(T,p) \times_{\mathcal{E}} \mathcal{E}^{\to,p} & \xrightarrow{\star^{(T,p)}} & \mathcal{E}^{\to,p} \\ & \searrow \quad \swarrow_{\mathrm{cod}} & \\ & \mathcal{E}. & \end{array}$$

If we take the exponential adjoint of this morphism, as in Section 5.2, we obtain a (relative) representation of relative T-graphs and relative T-categories.

5.6. Free relative T-categories. The full characterization of those monads T for which the forgetful functor $\mathcal{U}_T$ defined above has a left adjoint seem to be unknown. However there are various reasonable sufficient conditions, cf. [4, 6, 16, 19] in case the monad (T, η, μ) is cartesian. Recall that a monad (T, η, μ) on a category $\mathcal{C}$ with finite products is cartesian if T preserves pullbacks and both η, μ are cartesian natural transformations. A. Burroni in [6, pp. 267 – 269] provided one such characterization and he noticed that if such an adjoint exists $\mathcal{U}_T$ is automatically monadic [6, p. 304]. He also noticed that in certain cases one can iterate the T-category construction [6, p. 269]. However the condition for the iteration in [6] is too strong[9] to be used for our construction below. T. Leinster in [19] used a weaker condition for iteration but he was interested in iteration in particular fibres rather than of the whole fibration. With the help of this kind of iteration he defined the set of opetopes [19, p. 179 and Appendix D]. The construction of the free monoids described in [19] is the same as the earlier and more detailed, yet compact, construction described in [4, Appendix B]. The inductive formula defining the free monoids given in both [4] and [19] seem to appear first in [1, p. 591] to describe free algebras for a functor and then in a long comprehensive study [16, p. 69] that extends and unifies some earlier developments of this and related subjects. The prerequisites for the construction of the free monoids as well as the final goals differ in [4] and [19]. In [4] the prerequisites are given directly in terms of the properties of the category and the tensor involved to get a left adjoint to the forgetful functor from the monoids to the monoidal category. In [19] the prerequisites are given also in terms of the properties of the category however the property of the tensor is not specified directly but through the property of the monad the tensor is coming from. Moreover, in [19] the aim is not only to get a left adjoint but also to make sure that a monad (and a category it is defined on) deduced from the new adjunction satisfies the same properties, so that one can iterate the construction, as in [6].

Below we give a characterization of those fibrations p and fibred monads T on them for which one can iterate the process of taking T-graphs over a fibration p. In the exposition we use ideas from all the mentioned papers. The notions of a suitable fibrations and a fibrewise suitable monad are very much inspired by the notions of a suitable category and a suitable monad, respectively, cf. [19, Appendix D]. The main difference of our approach with respect to [19] is that we iterate whole fibrations over fibrations and get as a final result the category of opetopic sets, whereas in [19] the construction is done fibre by fibre and gives the set of opetopes as a result.

[9] One of the requirement is that the monad T commutes with coproducts.

From the perspective of our construction this set of opetopes is the set of cells in the terminal opetopic set.

We say that a fibration $p: \mathcal{E} \to \mathbf{B}$ is *suitable* if and only if

(1) p has fibred pullbacks, finite coproducts, and filtered colimits,
(2) finite coproducts and filtered colimits are universal in fibres of p,
(3) filtered colimits commutes with pullbacks in fibres of p.

Let $p: \mathcal{E} \to \mathbf{B}$ be a fibration with fibred pullbacks. A monad (T, η, μ) on $\mathcal{E}$ is *cartesian relative to* p if and only if (T, η, μ) is a fibred monad over p (i.e., $p \circ T = p$, $p(\eta) = 1_p = p(\mu)$) and the restriction of the monad (T, η, μ) to every fibre of p is a cartesian monad on this fibre.

Let $p: \mathcal{E} \to \mathbf{B}$ be a suitable fibration. We say that a monad (T, η, μ) on $\mathcal{E}$ is *suitable relative to* p if and only if (T, η, μ) is cartesian relative to p and T preserves filtered colimits in the fibres of p.

The following theorem is the key to the definition of the tower of fibrations that defines the category of opetopic sets.

Theorem 5.4. *Let (T, η, μ) be a suitable monad relative to a suitable fibration $p: \mathcal{E} \to \mathbf{B}$. Then*

(1) *the fibration p_T over p is again suitable;*
(2) *the forgetful functor $\mathcal{U}_T$ is monadic;*
(3) *the monad $(\widetilde{T}, \widetilde{\eta}, \widetilde{\mu})$ induced by the adjunction $\mathcal{F}_T \dashv \mathcal{U}_T$ is suitable relative to p_T.*

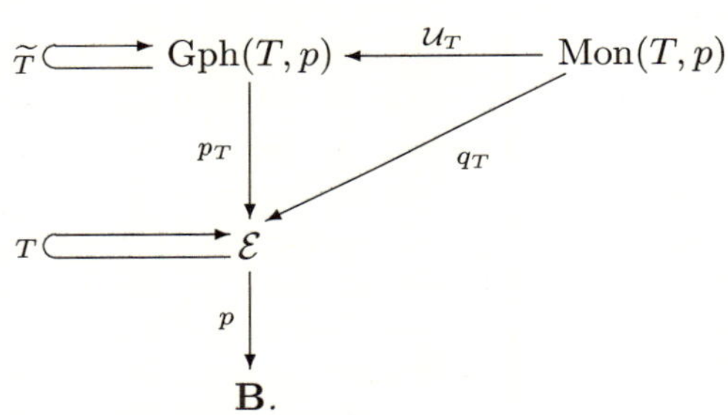

In the proof of this theorem we shall use the following easy lemma.

Lemma 5.5. *Suppose p and q are fibrations and we have two lax morphisms of fibrations $\mathcal{U}$ and $\mathcal{F}$, as in the diagram*

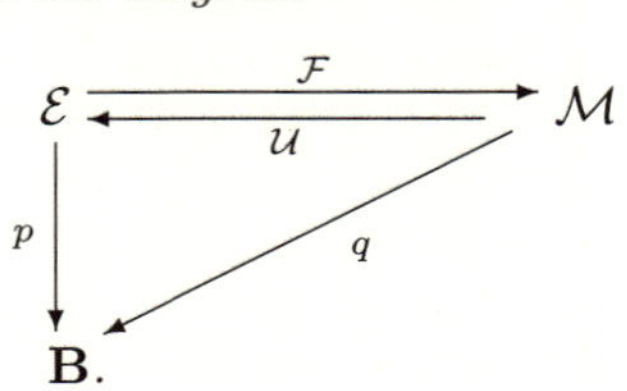

If $\mathcal{U}$ is a morphism of fibrations and $\mathcal{F}$ is a left adjoint to $\mathcal{U}$ when restricted to each fibre then $\mathcal{F}$ is a left adjoint to $\mathcal{U}$.

Remark. This lemma could be compared with Lemma 1.8.9 of [13]. However we don't require the Beck–Chevalley condition as we don't expect $\mathcal{F}$ to be a morphism of fibrations, as in our application it won't be.

PROOF OF THEOREM 5.4. The functor $\mathrm{Gph}(T,p)\to\mathcal{E}$ sending T-graph (A,γ,δ) to A creates pullbacks, finite coproducts and filtered colimits. Thus those limits and colimits have the same exactness properties in the fibres of p_T as they had in fibres of p. The fact that they are fibred in p_T follows from the fact that they are fibred in p and that finite coproducts and filtered colimits are universal. Thus, that p_T is a suitable fibration.

Recall the construction of the free monoid from [3 [10] ,4,16,19]. For an object (A,γ,δ) in a fibre $\mathcal{E}_O$ we construct a filtered diagram. We write A for (A,γ,δ) and O for the unit of the tensor $(O,1_O,\eta_O)$, for short.

$$
\begin{array}{c}
A^0 = O \\
\Big\downarrow{\scriptstyle e_0} \\
A^1 = O + A\otimes O \\
\Big\downarrow{\scriptstyle e_1=1+(1\otimes e_0)} \\
A^2 = O + A\otimes(O + A\otimes O) \\
\Big\downarrow{\scriptstyle e_2=1+(1\otimes e_1)} \\
A^3 = O + A\otimes A^2 \\
\Big\downarrow{\scriptstyle e_3=1+(1\otimes e_2)} \\
\cdots
\end{array}
$$

with the help of binary coproducts and tensors. The colimit of this diagram in $\mathcal{E}_O$ is the universe of $\mathcal{F}_T(A,\gamma,\delta)$. To see the definition of multiplication for the monoid $\mathcal{F}_T(A,\gamma,\delta)$ and unit see [4]. As all the operations involved are functorial in the whole fibration, $\mathcal{F}_T$ is functorial, as well. Thus, by Lemma 5.5, to show that $\mathcal{F}_T$ is a left adjoint to $\mathcal{U}_T$ we need to verify that they are adjoint when restricted to each fibre. But this is clear from [4, 19]. As T preserves filtered colimits in fibres of p so does $\mathcal{U}_T$ in the fibres of p_T and hence $\widetilde{T}$ preserves them, as well.

The monadicity of $\mathcal{U}_T$ follows from Lemme 1 of [6, p. 304] or can be proved directly using the above explicit construction of the free functor $\mathcal{F}_T$. If (M,m,e) is a monoid in $\mathrm{Gph}(T,p)\to\mathcal{E}$ then using m and e we can construct inductively an algebra $(M,\alpha\colon \widetilde{T}(M)\to M)$ and having a $\widetilde{T}$-algebra (M,α) we define a monoid by putting e equal to $I_O\to\widetilde{T}(M)\xrightarrow{\alpha} M$ and m equal to $M{\otimes}M\to I{+}M{\otimes}(I{+}M{\otimes}I)\to \widetilde{T}(M)\xrightarrow{\alpha} M$. The remaining details are left for the readers.

The fact that the induced monad $(\widetilde{T},\widetilde{\eta},\widetilde{\mu})$ is cartesian relative to p_T is also easy. □

Remark. The monadicity of $\mathcal{U}_T$ was already noticed in [6, Proposition II.1.19] for a monad T satisfying slightly stronger conditions.

[10]The assumptions that we have on the monad (T,η,μ) obviously sufficient to for this construction to work.

5.7. A tower of fibrations for opetopic sets. Using the above Theorem 5.4, and starting with any fibrewise suitable monad T_0 on a fibrewise suitable fibration $p\colon \mathcal{E}_0 \to \mathbf{B}$, we can build a tower of (fibrewise suitable) lax monoidal fibrations and fibrewise suitable monads as in the diagram below:

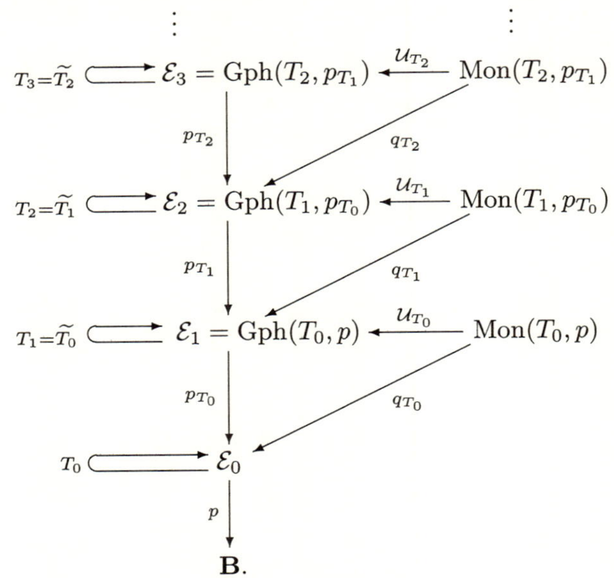

So as p and T_0 are fibrewise suitable, we have the monad $T_1 = \widetilde{T}$ on p_{T_0}. By Theorem 5.4 p_{T_0} and T_1 are again suitable and hence we can repeat the construction again. The identity monad 1_{Set} on Set is of course a fibrewise suitable on the fibrewise suitable fibration $!\colon \mathrm{Set} \to \mathbf{1}$, where $\mathbf{1}$ is the terminal category. Thus we can build a tower of fibrations, as above, starting form this fibration. We obtain

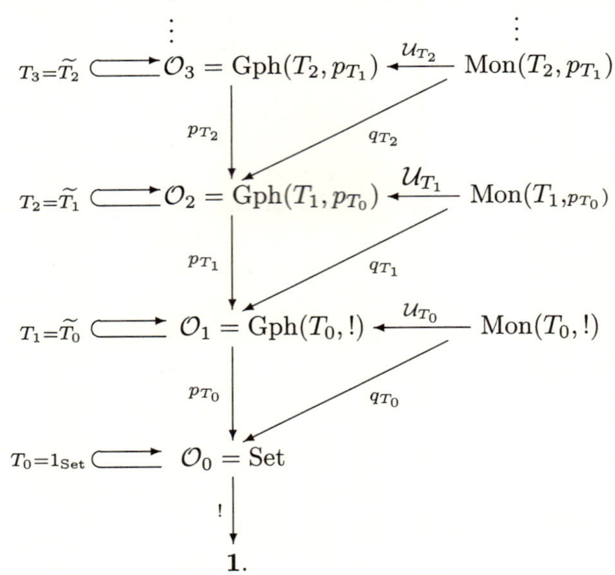

An *opetopic set* is an infinite sequence of objects $\{A_n\}_{n\in\omega}$ such that

(1) A_n is an object in $\mathcal{O}_n$,
(2) A_{n+1} lies in the fibre over A_n, i.e., $p_{T_n}(A_{n+1}) = A_n$,

for $n \in \omega$. A *morphism of opetopic sets* $\{f_n\}_{n\in\omega}\colon \{A_n\}_{n\in\omega} \to \{B_n\}_{n\in\omega}$ is a family of morphisms such that

(1) $f_n\colon A_n \to B_n$ is a morphism in $\mathcal{O}_n$
(2) f_{n+1} lies in the fibre over f_n, i.e., $p_{T_n}(f_{n+1}) = f_n$,

for $n \in \omega$.

Unraveling this definition, we see that an opetopic set (in the above sense) is an ∞-span as the diagram below:

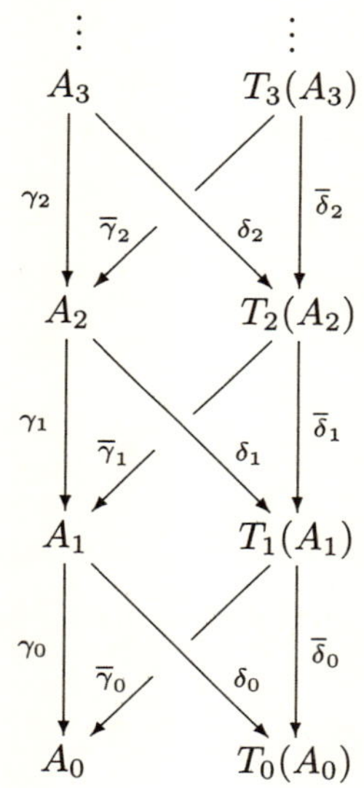

with

$$\gamma_n \circ \gamma_{n+1} = \overline{\gamma}_n \circ \delta_{n+1}, \qquad \delta_n \circ \gamma_{n+1} = \overline{\delta}_n \circ \delta_{n+1}$$
$$\gamma_n \circ \overline{\gamma}_{n+1} = \overline{\gamma}_n \circ \overline{\delta}_{n+1}, \qquad \delta_n \circ \overline{\gamma}_{n+1} = \overline{\delta}_n \circ \overline{\delta}_{n+1}$$

for $n \in \omega$. To describe the terminal opetopic set A, we need to start with $A_0 = 1$ the terminal object in Set. And then choose A_{n+1} as the terminal object in the fibre of p_{T_n} over A_n. Thus A_1 is 1 and A_{n+1} for $n > 0$ can be taken as the limit in the following diagram:

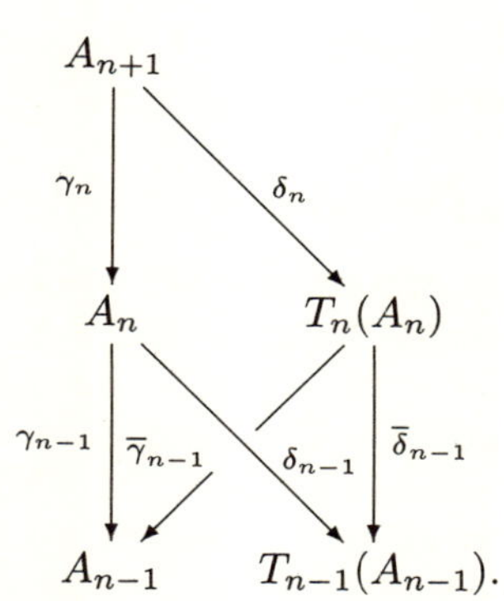

The disjoint union of the sets $\{A_n\}_{n\in\omega}$ is the set of opetopes in the sense of T. Leinster.

The proof of the following theorem uses ordered face structures, cf. [27], and will not be given here.

Theorem 5.6. *The category of opetopic sets so defined is equivalent to the category of multitopic sets.*

Remark (Internal opetopic sets). Clearly the fibration $!\colon \mathrm{Set} \to \mathbf{1}$ is not the only interesting suitable one to start the process of iteration. For example, we can start with $!\colon \mathcal{E} \to \mathbf{1}$ where $\mathcal{E}$ is a sufficiently cocomplete topos. Thus, we have the category of internal opetopic sets in any Grothendieck topos, even in the category of opetopic sets itself!

5.8. A tower of fibrations for n-categories. If we start with the (fibred) identity monad $1_{\mathcal{E}}$ on a fibration $p\colon \mathcal{E} \to \mathbf{B}$ whose fibres have pullbacks then the fibration of monoids over p, $q_{1_{\mathcal{E}}}\colon \mathrm{Mon}(1_{\mathcal{E}}, p) \to \mathcal{E}$ again has pullbacks in the fibres. Thus we can iterate this process and get another tower of fibrations based on monoids, this time:

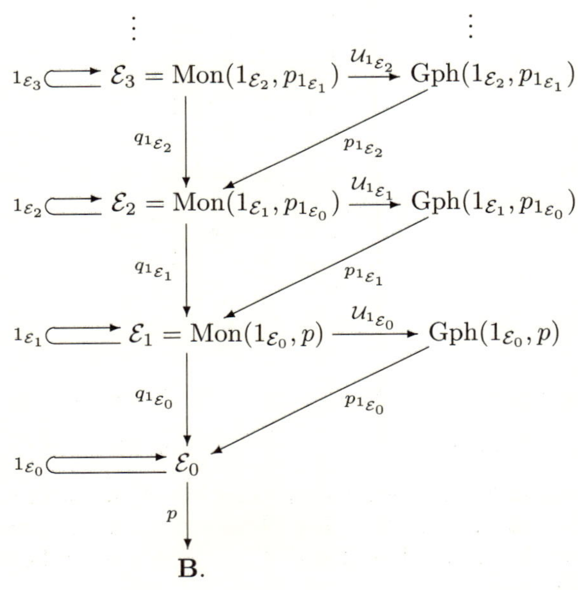

If we start with a suitable fibration $p\colon \mathcal{C} \to \mathbf{1}$, then after nth iteration we recover the D. Bourn [5] construction of internal n-categories in $\mathcal{C}$.

6. Amalgamated signatures vs. polynomial functors

6.1. The amalgamated signatures fibration $p_{\mathrm{a}}\colon \mathbf{Sig_a} \to \mathbf{Set}$. This example is one of the main reasons for considering lax monoidal fibrations in the context of higher category theory at all. The monoids in this fibration are precisely the (1-level) multicategories with nonstandard amalgamation. They are like the multicategories considered by C. Hermida, M. Makkai, and J. Power in [12] to define the multitopic sets, except that there the 2-level version is used by. This modification will be explained at the end of the section.

Notation. Let $[n] = \{0, \ldots, n\}$, $(n] = \{1, \ldots, n\}$, for $n \in \omega$. In particular $[n] = [0] \cup (n]$ and $(0] = \varnothing$. For a set O, we put $O^\dagger_n = O^{[n]}$, $O^*_n = O^{(n]}$ and $O^\dagger = \bigcup_{n\in\omega} O^{[n]}$, $O^* = \bigcup_{n\in\omega} O^{(n]}$. S_n acts on both $O^\dagger_n$ and O^*_n on the right by composition (i.e., we leave 0 fixed in the domain of the elements of $O^\dagger_n$). If $d\colon [n] \to O$ is a function, then its restriction to the positive numbers is denoted by $d^+\colon (n] \to O$ and to $[0]$ by $d^-\colon [0] \to O$. This restrictions establish a bijection $\langle (-)^-, (-)^+ \rangle\colon O^\dagger \to O \times O^*$. Clearly $(-)^\dagger\colon \mathrm{Set} \to \mathrm{Set}$ is a functor.

The base category of our fibration is Set. The total category $\mathrm{Sig_a}$ of our fibration has as objects triples, (A, ∂, O) such that A and O are sets and $\partial\colon A \to O^\dagger$ is a function. We write $\partial_a\colon [n] \to O$ for the effect of ∂ on $a \in A$, and n in this case will be referred to as $|a|$. A morphism $(f, \sigma, u)\colon (A, \partial, O) \to (B, \partial, Q)$ in $\mathrm{Sig_a}$ is a pair of functions $f\colon A \to B$ and $u\colon O \to Q$, and for any $a \in A$ with $n = |a|$ a permutation $\sigma_a\colon [n] \to [n] \in \mathrm{S}_n$ (with $\sigma_a(0) = 0$) making the square

$$\begin{array}{ccc}
[n] & \xleftarrow{\ \sigma_a\ } & [n] \\
{\scriptstyle \partial_a}\downarrow & & \downarrow{\scriptstyle \partial_{f(a)}} \\
O & \xrightarrow[\ u\ ]{} & Q
\end{array}$$

commute. A morphism (f, σ, u) is called *strict* if σ_a is an identity, for $a \in A$. The projection functor $p\colon \mathrm{Sig_a} \to \mathrm{Set}$ sends the morphism $(f, \sigma, u)\colon (A, \partial, O) \to (B, \partial, Q)$ to $u\colon O \to Q$.

Remarks. (1) The category $\mathrm{Sig_m}$ is isomorphic to the full subcategory of $\mathrm{Sig_a}$ whose morphisms are strict.

(2) We think of an object (A, ∂, O) of $\mathrm{Sig_a}$ as a signature with O as the set of its types, A the set of its operation symbols, and ∂ the typing function associating arities to function symbols $a\colon \partial_a(1), \ldots, \partial_a(|a|) \to \partial_a(0)$, i.e., is $\partial_a(1), \ldots, \partial_a(|a|)$ are types of the arguments (inputs) of a and $\partial_a(0)$ is the type of values (outputs) of a.

The lax monoidal structure on p_a. We have two lax morphisms of fibrations

$$\begin{array}{ccccc}
\mathrm{Sig_a} \times_{\mathrm{Set}} \mathrm{Sig_a} & \xrightarrow{\ \otimes\ } & \mathrm{Sig_a} & \xleftarrow{\ I\ } & \mathrm{Set} \\
& {\scriptstyle p'_\mathrm{a}}\searrow & \downarrow{\scriptstyle p_\mathrm{a}} & \swarrow{\scriptstyle 1_{\mathrm{Set}}} & \\
& & \mathrm{Set}. & &
\end{array}$$

Let (A, ∂, O) and (B, ∂, O) be two object in the fibre over O. Their tensor $(A \otimes_O B, \partial^\otimes, O)$ is defined as follows

$$A \otimes_O B = \{\langle a, b_i \rangle_{i \in (|a|]} : a \in A,\ b_i \in B,\ \partial_a(i) = \partial_{b_i}(0), \text{ for } i \in (|a|]\}$$

and for $\langle a, b_i \rangle_{i\in(|a|]} \in A \otimes_O B$,

$$\partial^\otimes_{\langle a, b_i \rangle_i} = [\partial^-_a, \partial^+_{b_i}]_i\colon [|\langle a, b_i \rangle_i|] = \left[\sum_{i=1}^{|a|} |b_i|\right] \to O.$$

Note that just saying that we have a coproduct determines the function $[\partial^-_a, \partial^+_{b_i}]_i$ only up to a permutation. In principle we don't need more than that for as far as

$[\partial_a^-, \partial_{b_i}^+]_i(0) = \partial_a^-(0)$. But to be on the safe side, we will always tacitly assume that the domains of $\partial_{b_i}^+$ are placed one after the other.

For a pair of maps in $\mathrm{Sig_a}$

$$\bar{f} = (f, \sigma, u)\colon (A, \partial, O) \to (A', \partial, Q), \quad \bar{g} = (g, \tau, u)\colon (B, \partial, O) \to (B', \partial, Q)$$

over the same map $u\colon O \to Q$ we define the map

$$\bar{f} \otimes_u \bar{g} = (f \otimes_u g, \sigma \otimes_u \tau, u)\colon (A \otimes_O B, \partial^{\otimes}, O) \to (A' \otimes_Q B', \partial^{\otimes}, Q)$$

so that, for $\langle a, b_i\rangle_{i\in(|a|]} \in A \otimes_O B$,

$$f \otimes_u g(\langle a, b_i\rangle_{i\in(|a|]}) = \langle f(a), g(b_{\sigma_a(j)})\rangle_{j\in(|f(a)|]}.$$

Clearly, $|a| = |f(a)|$ and

$$n = |\langle a, b_i\rangle_{i\in(|a|]}| = \sum_{i\in|a|} |b_i| = \sum_{i\in|f(a)|} |g(b_i)| = |f \otimes_u g(\langle a, b_i\rangle_{i\in(|a|]})|.$$

Moreover, we put

$$(\sigma \otimes_u \tau)_{\langle a,b_i\rangle_i} = [\tau^+_{b_{\sigma_a(i)}}]_i$$

making the square

$$\begin{array}{ccc} [n] & \xleftarrow{(\sigma\otimes_u\tau)_{\langle a,b_i\rangle_i}} & [n] \\ {\scriptstyle \partial^{\otimes}_{\langle a,b_i\rangle_i}}\downarrow & & \downarrow{\scriptstyle \partial^{\otimes}_{\langle f(a),g(b_i)\rangle_i}} \\ O & \xrightarrow{\quad u \quad} & Q \end{array}$$

commute. This ends the definition of the tensor $\otimes$.

The unit I_O in the fibre over O, is $I_O = (O, \partial^{I_O}, O)$ such that for $x \in O$, $\partial_x^{I_O}\colon [1] \to O$ is a constant function equal to x. We note for the record

Lemma 6.1. *The fibration $p_a\colon \mathrm{Sig_a} \to \mathrm{Set}$ with the structure described above is a lax monoidal fibration whose fibres are strong monoidal categories.*

Pulling back the monoidal structure. We shall describe how reindexing functors interact with the monoidal structure in the fibration p_a.

Any object B in the fibre over Q of $p_a\colon \mathrm{Sig_a} \to \mathrm{Set}$ can be pulled back along a function $u\colon O \to Q$:

$$\begin{array}{ccc} u^*(B) & \xrightarrow{\pi_B} & B \\ {\scriptstyle \partial}\downarrow & & \downarrow{\scriptstyle \partial} \\ O^{\dagger} & \xrightarrow{u^{\dagger}} & Q^{\dagger} \\ O & \xrightarrow{u} & Q \end{array}$$

thus

$$u^*(B) = \{\langle b, d\rangle : b \in B,\ d\colon [|b|] \to O \text{ such that } u^{\dagger}(d) = \partial_b\}$$

$(u^{\dagger}(d) = u \circ d)$ and

$$\partial_{\langle b,d\rangle} = d.$$

We have

$$u^*(I_Q) = \{\langle x, x'\rangle \in O^2 : u(x) = u(x')\}$$

and

$$\phi_0 \colon I_O \to u^*(I_Q)$$
$$x \mapsto \langle x, x \rangle.$$

Moreover, for objects A and B over Q we have

$$u^*(A \otimes B) = \left\{ \langle \langle a, b_i \rangle_{i \in (|a|]}, d \rangle : \langle a, b_i \rangle_{i \in (|a|]} \in A \otimes B, u^\dagger(d) = \partial^{\otimes}_{\langle a, b_i \rangle_{i \in (|a|]}} \right\}$$

and

$$u^*(A) \otimes u^*(B) = \{ \langle \langle a, d \rangle, \langle b_i, d_i \rangle \rangle_{i \in (|a|]} : a \in A,\ b_i \in B,\ u^\dagger(d) = \partial_a,\ u^\dagger(d_i) = \partial_{b_i}, \\ d(i) = d_i(0) \text{ for } i \in (|a|] \}.$$

Thus we have a transformation

$$\phi_{2,A,B} \colon u^*(A) \otimes u^*(B) \to u^*(A \otimes B)$$

such that

$$\langle \langle a, d \rangle, \langle b_i, d_i \rangle \rangle_{i \in (|a|]} \mapsto \langle \langle a, b_i \rangle_{i \in (|a|]}, [d^-, d_i^+]_{i \in (|a|]} \rangle.$$

All the morphisms defined above π_B, ϕ_0, and $\phi_{2,A,B}$ are strict, i.e., with amalgamations being identities.

Lemma 6.2. *The data u^*, ϕ_0, ϕ_2 above, make the usual (three) diagrams for coherence of monoidal functor (not necessarily strong) commute.*

PROOF. Exercise. □

Moreover we have

Proposition 6.3. *The total category of the fibration $q_{\mathrm{a}} \colon \mathrm{Mon}(\mathrm{Sig_a}) \to \mathrm{Set}$ of monoids in $p_{\mathrm{a}} \colon \mathrm{Sig_a} \to \mathrm{Set}$ is equivalent to the category of (1-level) multicategories with nonstandard amalgamations. The fibred forgetful functor from the fibration of monoids to the fibration of amalgamated signatures $\mathcal{U} \colon \mathrm{Mon}(\mathrm{Sig_a}) \to \mathrm{Sig_a}$*

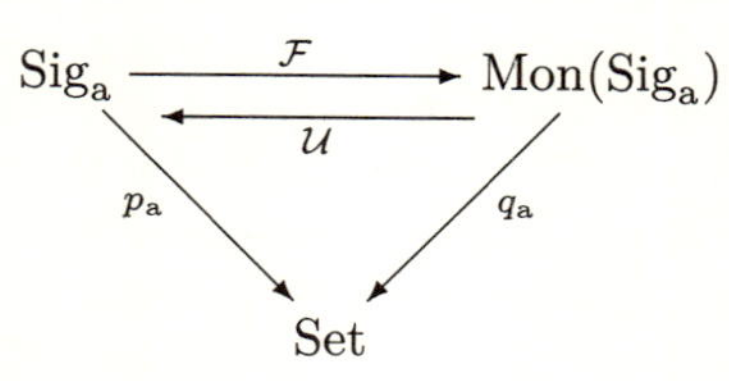

has a fibred left adjoint $\mathcal{F}$, the free monoid functor.

PROOF. Strictly speaking the multicategories with nonstandard amalgamations were defined [12] from the single tensor and additional property, called commutativity there. But, as it is well known, they can be equivalently defined using the total tensor, i.e., the one we defined above. For more, see also Section 6.6. □

The free functor $\mathcal{F}$ mentioned in the proposition above was described in [12].

6.2. The action of p_a on the basic fibration. The lax monoidal fibration p_a comes equipped naturally with an action on the basic fibration cod: $\mathrm{Set}^{\to} \to \mathrm{Set}$

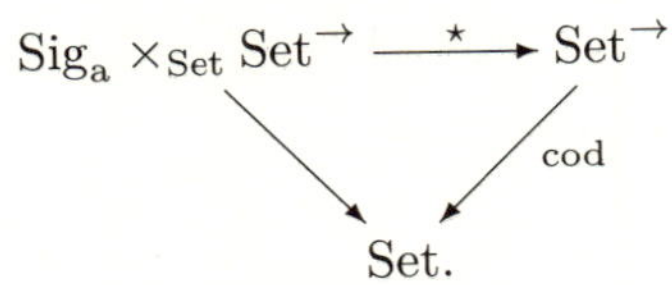

For (A, ∂^A, O) in Sig_a and (X, d, O) in $\mathrm{Set}^{\to}$, the set $A \star X$ is defined from the following diagram

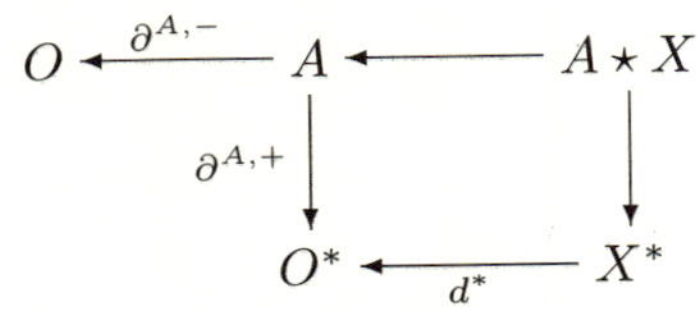

with the square being a pullback. $(-)^*$ is the free monoid functor, i.e.,

$$A \star X = \{(a, x_1, \ldots, x_{|a|}) : \partial_a^A(i) = d(x_i), i = 1, \ldots, |a|\}$$

and $\partial^\star\colon A \star X \to O$ is defined by

$$\partial^\star(a, x_1, \ldots, x_{|a|}) = \partial_a^A(0).$$

Thus it is the composition of the upper horizontal morphism in the above diagram. On morphisms the action $\star$ is defined in the obvious way. Thus we have an adjoint morphism of lax monoidal fibrations

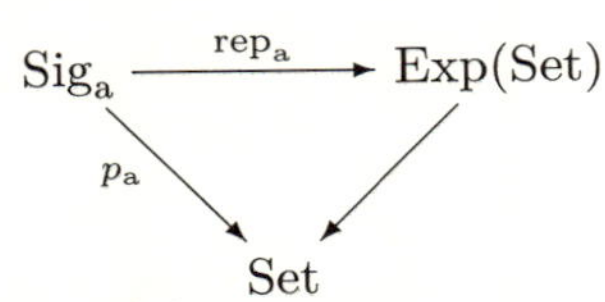

where, as usual, Exp(Set) is the exponential object in **Cat**/Set.

6.3. Polynomial diagrams and polynomial functors. In this subsection we collect the definitions and facts concerning polynomial diagrams and polynomial functors from the literature, that are needed in the following. For (much!) more, the reader should consult [8, 17] and bibliography there. We deal with polynomial functors based on an arbitrary locally cartesian closed category but with a special eye on Set and the presheaf category $\mathrm{Set}^{\mathcal{S}_*}$, where $\mathcal{S}_*$ is the coproduct in *Cat* of the (finite) symmetric groups. The later category will be important in Section 7. In this section, unless otherwise specified, $\mathcal{E}$ is an arbitrary locally cartesian closed category, and by this we mean that $\mathcal{E}$ has the terminal object, as well.

By a *polynomial diagram* (*over* O) in $\mathcal{E}$, we mean the following diagram in $\mathcal{E}$

$$O \overset{s}{\longleftarrow} E \overset{p}{\longrightarrow} A \overset{t}{\longrightarrow} O.$$

The object O is an object of types of the polynomial (t, p, s). We say that a polynomial diagram (t, p, s) in Set is *finitary* if and only if the function p has finite fibres. A *morphism of polynomial diagrams* (*over* $u\colon O \to Q$) in $\mathcal{E}$ is a triple (f, g, u)

of morphism making the diagram

$$\begin{array}{ccccccc} O & \xleftarrow{\ s\ } & E & \xrightarrow{\ p\ } & A & \xrightarrow{\ t\ } & O \\ {\scriptstyle u}\downarrow & & {\scriptstyle g}\downarrow & & \downarrow{\scriptstyle f} & & \downarrow{\scriptstyle u} \\ Q & \xleftarrow[s']{} & E' & \xrightarrow[p']{} & A' & \xrightarrow[t']{} & Q \end{array}$$

commute, and such that the square in the middle is a pullback. Morphisms of polynomial diagram compose in the obvious way, by putting one on top of the other. Let $\mathcal{PolyDiag}(\mathcal{E})$ denotes the category of the polynomial diagrams and morphisms between them.

Remark. If $\mathcal{E}$ is the category Set, we can think of A as the set of operations, and E as the set of arguments of all operations in A. Thus with this interpretation $p^{-1}(a)$ is the set of arguments (or arity) of a. Then $s(e)$, for $e \in p^{-1}(a)$, can be interpreted as the type of the argument e of the operation a, and $t(a)$ is the type of the values of the operation a.

We have an obvious projection functor

$$p_{\mathrm{pd},\mathcal{E}} \colon \mathcal{PolyDiag}(\mathcal{E}) \to \mathcal{E}$$

sending (f, g, u) to u, which is a lax monoidal fibration. The tensor in fibres is given by composition of diagrams, cf. [8, 1.11]. Let

$$p_{\mathrm{pd}} \colon \mathcal{PolyDiag} \to \mathrm{Set}$$

denote the finitary polynomial diagram fibration, the full subfibration of $p_{\mathrm{pd,Set}}$ consisting of finitary polynomial diagrams. By $p_{\mathrm{poly}} \colon \mathcal{Poly} \to \mathrm{Set}$ we denote the image of the fibration p_{pd} in $p_{\mathrm{exp,Set}}$. We shall see that p_{poly} is a lax monoidal subfibration of $p_{\mathrm{exp,Set}}$.

In our terminology, the connection between polynomial diagrams and polynomial functors can be expressed as the fact that the fibration of polynomial diagrams comes equipped with a representation morphism into $\mathrm{Exp}(\mathcal{E})$, i.e., a morphism of lax monoidal fibration, cf. Proposition 6.7 below,

$$\begin{array}{ccc} \mathcal{PolyDiag}(\mathcal{E}) & \xrightarrow{\ \mathrm{rep}_{\mathrm{pd},\mathcal{E}}\ } & \mathrm{Exp}(\mathcal{E}) \\ & {\scriptstyle p_{\mathrm{pd},\mathcal{E}}}\searrow \quad \swarrow{\scriptstyle p_{\mathrm{exp}}} & \\ & \mathcal{E} & \end{array}$$

whose essential image is, by definition, the (lax monoidal) fibration of (finitary) polynomial (endo)functors and polynomial transformations between them. We shall recall this now. Any morphism $u \colon O \to Q$ in a locally cartesian closed category $\mathcal{E}$ induces three functors

$$\mathcal{E}_{/O} \;\begin{array}{c} \xrightarrow{\ u_!\ } \\ \xleftarrow{\ u^*\ } \\ \xrightarrow[u_*]{} \end{array}\; \mathcal{E}_{/Q}$$

such that u^* is a pullback functor and $u_! \dashv u^* \dashv u_*$. The unit and counit of the adjunction $u_! \dashv u^*$ will be denoted by η^u and ε^u, respectively and the unit and counit of the adjunction $u^* \dashv u_*$ will be denoted by $\overline{\eta}^u$ and $\overline{\varepsilon}^u$, respectively.

For an object (t,p,s) over O, we define a functor

$$\mathrm{rep}_{\mathrm{pd},\mathcal{E}}(t,p,s) = t_! p_* s^* : \mathcal{E}_{/O} \to \mathcal{E}_{/O}.$$

For a morphism of polynomial diagrams $(f,g,u)\colon (t,p,s) \to (t',p',s')$, we define a morphism $\mathrm{rep}_{\mathrm{pd},\mathcal{E}}(f,g,u)\colon \mathrm{rep}_{\mathrm{pd},\mathcal{E}}(t,p,s) \to \mathrm{rep}_{\mathrm{pd},\mathcal{E}}(t',p',s')$ in $\mathrm{Exp}(\mathcal{E})$ over u, as follows. We have a diagram of categories, functors and natural transformations

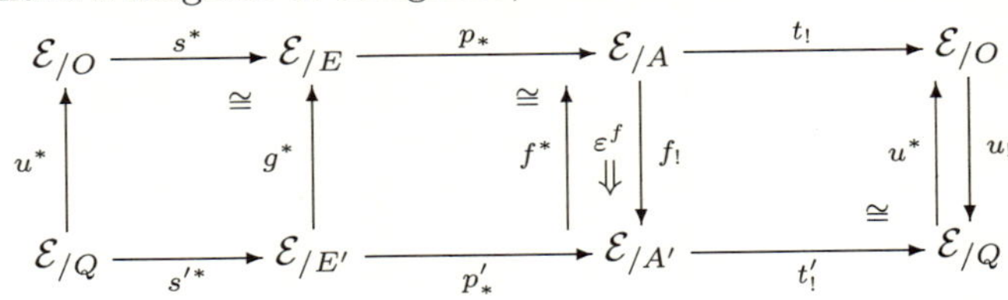

where $\varepsilon^f\colon f_! f^* \to 1_{\mathcal{E}/A'}$ is the counit of the adjunction $f_! \dashv f^*$. Thus we have a natural transformation

$$t'_!(\varepsilon^f)p'_* s'^* : t'_! f_! f^* p'_* s'^* \to t'_! p'_* s'^*$$

and passing through the natural isomorphisms indicated in the above diagram we get the corresponding a natural transformation $\mathrm{rep}_{\mathrm{pd},\mathcal{E}}(f,g,u)$ as follows:

$$u_! t_! f^* p'_* s'^* \to t'_! p'_* s'^*$$

via right square iso, and this

$$u_! t_! p_* g^* s'^* \to t'_! p'_* s'^*$$

via middle square iso, and this

$$u_! t_! p_* s^* u^* \to t'_! p'_* s'^*$$

via left square iso. Finally, taking the adjoint $(u_! \dashv u^*)$ of this morphism we get

$$\mathrm{rep}_{\mathrm{pd},\mathcal{E}}(f,g,u)\colon t_! p_* s^* u^* \to u^* t'_! p'_* s'^*$$

which is a morphism from $\mathrm{rep}_{\mathrm{pd},\mathcal{E}}(t,p,s)$ to $\mathrm{rep}_{\mathrm{pd},\mathcal{E}}(t',p',s')$ in $\mathrm{Exp}(\mathcal{E})$ over u. The essential image of the functor $\mathrm{rep}_{\mathrm{pd},\mathcal{E}}$ is, by definition, the fibration of polynomial (endo)functors and polynomial transformations $p_{\mathrm{poly},\mathcal{E}}\colon \mathcal{Poly}(\mathcal{E}) \to \mathcal{E}$.

By taking the exponential adjoint to $\mathrm{rep}_{\mathrm{pd},\mathcal{E}}$ we obtain an action of the fibration $p_{\mathrm{pd},\mathcal{E}}$ on the basic fibration $\mathrm{cod}\colon \mathcal{E}^{\to} \to \mathcal{E}$

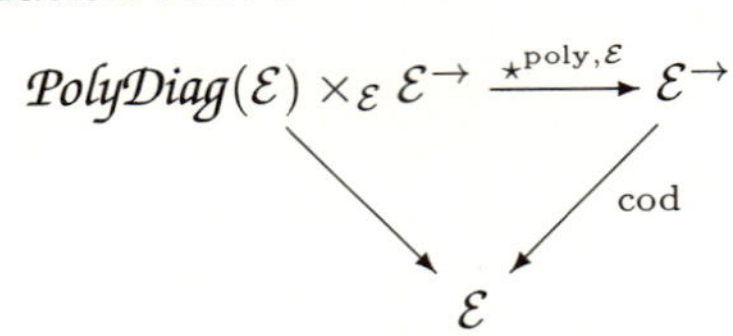

even if this point of view is less customary.

The case $\mathcal{E}$ = Set. We shall make here the above abstract definitions concrete in case $\mathcal{E} = \mathrm{Set}$. A functor $P\colon \mathrm{Set}_{/Q} \to \mathrm{Set}_{/Q}$ is a *polynomial functor*[11] if and only if it is isomorphic to one of form

$$\Pi_{(t,p,s)} = t_! p_* s^* : \mathrm{Set}_{/Q} \to \mathrm{Set}_{/Q}$$

[11] There is an obvious notion of a polynomial functor $P\colon \mathrm{Set}_{/O} \to \mathrm{Set}_{/Q}$ with O not assumed to be equal to Q, but this can be considered a special case of the above definition, as such functors are (some) polynomial functors $\mathrm{Set}_{/O+Q} \to \mathrm{Set}_{/O+Q}$. For details see [8].

for some polynomial diagram (t, p, s). Thus for $d\colon Y \to Q$ in $\mathrm{Set}_{/Q}$ we have

$$t_! p_* s^*(Y, d) = \{\langle b, \vec{y}\rangle : b \in B, \vec{y}\colon p^{-1}(b) \to Y, d \circ \vec{y} = s_{\lceil p^{-1}(b)}\}$$

where $s_{\lceil p^{-1}(b)}$ is the restriction of the function s to the fibre of the function p over the element $b \in B$. It is a routine to verify that a polynomial functor is finitary if and only if it comes from a finitary diagram.

A morphism of polynomial diagrams $(f, g, 1_Q)\colon (t, p, s) \to (t', p', s')$ defines a natural transformation $\Pi_{(f,g)}\colon \Pi_{(t,p,s)} \to \Pi_{(t',p',s')}$ so that for $d\colon Y \to Q$ in $\mathrm{Set}_{/Q}$, and $\langle b, \vec{y}\rangle \in t_! p_* s^*(Y, d)$, we have

$$\Pi_{(f,g)}(\langle b, \vec{y}\rangle) = \langle f(b), \vec{y} \circ (g_{\lceil p^{-1}(b)})^{-1}\rangle.$$

A natural transformation between polynomial functors is *polynomial* if and only if it is given by the morphism of polynomial diagrams defining them.

The following theorem is due to many authors. The precise account of this can be found in [8, 1.18 and 1.19]. However the proof, based on ideas of [2], seems to be new.

Theorem 6.4. *For any set Q, the functor*

$$\Pi_Q\colon (\mathcal{PolyDiag})_Q \to \mathrm{Nat}(\mathrm{Set}_{/Q}, \mathrm{Set}_{/Q})$$

defined above is faithful, full on isomorphisms and its essential image consists of finitary functors preserving wide pullbacks and cartesian natural transformations.

A functor $F\colon \mathrm{Set}_{/Q} \to \mathrm{Set}_{/Q}$ is *thin* if there is $q \in Q$ such that $F = \mathbf{i}_q \circ ev_q \circ F$ and $ev_q \circ F(1) = 1$. The functors $\mathbf{i}_q$ and ev_q are inclusion and evaluation functors, respectively, see Section 7.3.

Proof. First note that any functor $P\colon \mathrm{Set}_{/Q} \to \mathrm{Set}_{/Q}$ is a coproduct of thin functors and P preserves wide pullbacks and filtered colimits if all the thin factors in the coproduct do. As Π_Q preserves coproducts we can assume that P is thin say, $P = \mathbf{i}_q \circ \mathrm{ev}_q \circ P$ for some $q \in Q$. Thus $P_q = \mathrm{ev}_q \circ P\colon \mathrm{Set}_{/Q} \to \mathrm{Set}$ preserves all limits and is finitary. Hence, by the characterization of the representable functors cf. [20, p. 130], it is represented by an object $s\colon E \to Q$ in $\mathrm{Set}_{/Q}$ with E finite. Now it is a matter of a simple check, that with the diagram

$$Q \xleftarrow{\ s\ } E \xrightarrow{\ p\ } 1 \xrightarrow{\ t\ } Q$$

in $(\mathcal{PolyDiag})_Q$, where $t(*) = q$, the functor $\Pi(t, p, s)$ is isomorphic to P.

To show that the functor Π_Q is full and faithful it is enough to consider morphisms between diagrams with one operation and cartesian natural transformations between corresponding thin functors, as other cartesian natural transformations between other polynomial functors must come from those.

So suppose that we have a cartesian natural transformation between two such functors

$$\tau\colon P = t_! \circ p_* \circ s^* \to P' = t'_! \circ p'_* \circ s'^*$$

where

$$Q \xleftarrow{\ s'\ } E' \xrightarrow{\ p'\ } 1 \xrightarrow{\ t'\ } Q.$$

As P and P' are thin we must have $t = t'$, say $t(*) = q \in Q$. Thus $\tau_q = \mathrm{ev}_q(\tau)\colon P_q \to P'_q$ is a cartesian natural transformation between functors that preserves the terminal object. Hence, as any component of τ_q is a pullback of $(\tau_q)_{1_Q}$ which is a morphism from the terminal object to itself, τ_q is cartesian if and only if it is an isomorphism. By the Yoneda Lemma, the natural isomorphisms $\tau_q\colon P_q \to P'_q$ in $\mathrm{Cat}(\mathrm{Set}_{/Q}, \mathrm{Set})$ between the functors represented by $s\colon E \to Q$ and $s'\colon E' \to Q'$ correspond to isomorphisms

$$\begin{array}{ccc} E & \xleftarrow{\;g\;} & E' \\ & {\scriptstyle s}\searrow \quad \swarrow{\scriptstyle s'} & \\ & Q & \end{array}$$

in $\mathrm{Set}_{/Q}$. But those isomorphisms g are exactly the functions g making the diagram

$$\begin{array}{ccccccc} Q & \xleftarrow{\;s\;} & E & \xrightarrow{\;p\;} & 1 & \xrightarrow{\;t\;} & Q \\ {\scriptstyle 1_Q}\downarrow & & {\scriptstyle g^{-1}}\downarrow & & \downarrow & & \downarrow{\scriptstyle 1_Q} \\ Q & \xleftarrow[\;s'\;]{} & E' & \xrightarrow[\;p'\;]{} & 1 & \xrightarrow[\;t\;]{} & Q \end{array}$$

a morphism in $(\mathcal{PolyDiag})_Q$, i.e., making the left square commute and the middle square a pullback. The reader may verify that if we take a morphism of diagrams corresponding to τ_q, then Π_Q will send it back to τ_q. □

Remark. An analog of Theorem 6.4 does not hold in all locally cartesian closed categories. Even if $\mathcal{E}$ is a presheaf category, then the endofunctors on slices of $\mathcal{E}$ that are finitary and preserve wide pullbacks do not necessarily come from polynomial diagrams in $\mathcal{E}$. For $\mathcal{E} = \mathrm{Set}^{\to}$, the functor sending $(x\colon X_0 \to X_1)$ to $(\langle 1_{X_0}, x\rangle\colon X_0 \to X_0 \times X_1)$ is finitary and preserves all limits but is not polynomial.

6.4. Some properties of the representation $\mathrm{rep_a}$. The main objective of this section is to establish some properties of the representation $\mathrm{rep_a}$ and then show (Corollary 6.13) that the 1-level multicategories with nonstandard amalgamations are the same as the cartesian monads on slices of Set whose functor part is finitary and preserves wide pullbacks.

As some statements concerning polynomial diagrams and functors hold in greater generality, in arbitrary locally cartesian closed categories, we start in this more general context.

First, we describe supine and prone morphisms in $p_{\mathrm{pd},\mathcal{E}}\colon \mathcal{PolyDiag}(\mathcal{E}) \to \mathcal{E}$. Let $u\colon O \to Q$ be a morphism in $\mathcal{E}$. The supine morphism $\mathrm{su}_{u,(t,p,s)}\colon (t,p,s) \to (u \circ t, p, u \circ s)$ over u with the domain being the polynomial diagram (t,p,s) in $\mathcal{E}$ over O is defined by the diagram

$$\begin{array}{ccccccc} O & \xleftarrow{\;s\;} & E & \xrightarrow{\;p\;} & B & \xrightarrow{\;t\;} & O \\ {\scriptstyle u}\downarrow & & {\scriptstyle 1_E}\downarrow & & \downarrow{\scriptstyle 1_B} & & \downarrow{\scriptstyle u} \\ Q & \xleftarrow[\;u\circ s\;]{} & E & \xrightarrow[\;p\;]{} & B & \xrightarrow[\;u\circ t\;]{} & Q, \end{array}$$

i.e., $\mathrm{su}_{u,(t,p,s)} = (1_B, 1_E, u)$. The prone morphism $\mathrm{pr}_{u,(t,p,s)}\colon (\tilde{t}, \tilde{p}, \tilde{s}) \to (t, p, s)$ over u with the domain being the polynomial diagram (t,p,s) in $\mathcal{E}$ over Q is defined by

the diagram

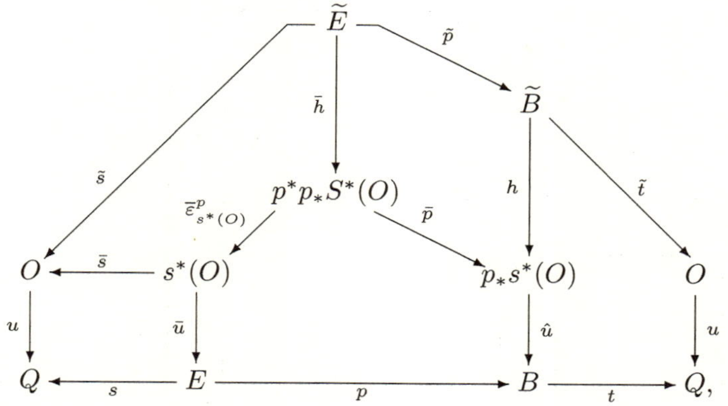

i.e., $\mathrm{pr}_{u,(t,p,s)} = (\hat{u} \circ h, \bar{u} \circ \bar{\varepsilon}^p_{s^*(O)} \circ \bar{h}, u)$. The above diagram is constructed in the following way. First we apply the functor $t_! p_* s^*$ to the (left most) morphism $u : O \to Q$ (in $\mathcal{E}_{/Q}$) to get $t \circ \hat{u} : p_*s^*(Q) \to Q$. $\bar{\varepsilon}^p$ is the counit of the adjunction $p^* \dashv p_*$. Then we pull it back along u to get $\tilde{t}$ and h. Finally, pulling back h along $\bar{p}$ we get $\tilde{p}$ and $\bar{h} = \bar{p}^*(h)$, the last part of the polynomial diagram $(\tilde{t}, \tilde{p}, \tilde{s})$. The morphism $\tilde{s}$ is defined as the composition $\bar{s} \circ \bar{\varepsilon}^p_{s^*(O)} \circ \bar{p}^*(h)$, and the objects $\widetilde{B}$ and $\widetilde{E}$ are $u^*t_!p_*s^*(O)$ and $\bar{p}_*(\widetilde{B})$, respectively.

We note for the record

Proposition 6.5. *The prone and supine morphisms in the bifibration $p_{\mathrm{pd},\mathcal{E}}$: $\mathcal{PolyDiag}(\mathcal{E}) \to \mathcal{E}$ are as described above.*

PROOF. Routine verification. □

The following lemma collects some known facts that will be used in Proposition 6.7.

Lemma 6.6. *Let u, u', s and p be morphisms in a locally cartesian closed category $\mathcal{E}$, such that* $\mathrm{cod}(u') = \mathrm{dom}(u)$, $\mathrm{cod}(u) = \mathrm{cod}(s) = \mathrm{dom}(p)$. *Then*

(1) $s^*(\varepsilon^u) \cong (\varepsilon^{s^*(u)})s^*$;
(2) $p_*(\varepsilon^u) \cong (\varepsilon^{p_*(u)})p_*$;
(3) $\varepsilon^{u \circ u'} \cong \varepsilon^u \circ (u_!(\varepsilon^{u'})u^*)$.

Proposition 6.7. *Let $\mathcal{E}$ be a locally cartesian closed category. The morphism*

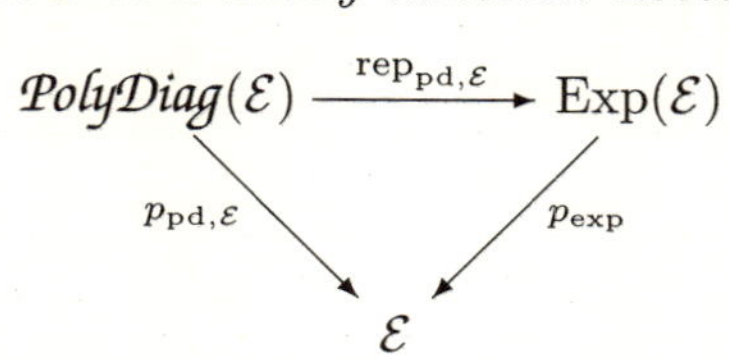

defined in the previous section is a morphism of bifibrations.

PROOF. We need to show that $\mathrm{rep}_{\mathrm{pd},\mathcal{E}}$ preserves prone and supine morphisms.

For the prone morphism we use the notation introduced at the beginning of the section. Let $\mathrm{pr}_{u,(t,p,s)} = (\hat{u} \circ h, \bar{u} \circ \bar{\varepsilon}^p_{s^*(O)} \circ \bar{p}^*(h), u) \colon (\tilde{t}, \tilde{p}, \tilde{s}) \to (t, p, s)$ be a prone morphism. We have a diagram of categories, functors and natural transformations

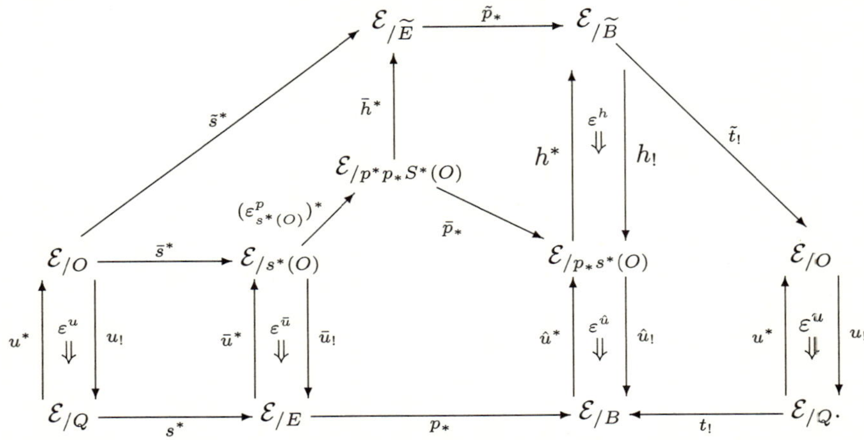

Roughly speaking, the adjoint ($u_! \dashv u^*$) natural transformation to $\mathrm{rep}_{\mathrm{pd},\mathcal{E}}(\mathrm{pr}_{u,(t,p,s)})$ is isomorphic to $t_!(\varepsilon^{\hat{u}h})p_*s^* = t_!\big(\varepsilon^{\hat{u}} \circ (\hat{u}_!(\varepsilon^h)\hat{u}^*)\big)p_*s^*$ i.e., it is defined with the help of the counits $\varepsilon^{\hat{u}}$ and ε^h. The adjoint to the natural transformation $\mathrm{pr}_{u,\mathrm{rep}_{\mathrm{pd},\mathcal{E}}(t,p,s)}$, being the prone morphism in Exp(Set) over u with the codomain $\mathrm{rep}_{\mathrm{pd},\mathcal{E}}(t,p,s)$, is $\big((\varepsilon^u)t_!p_*s^*\big) \circ \big(u_!u^*t_!p_*s^*(\varepsilon^u)\big)$. To show that these adjoints are isomorphic, we show, using the above Lemma 6.6, that the counit $\varepsilon^{\hat{u}}$ can be 'moved left' to the 'left' counit ε^u and the counit ε^h can be 'moved right' to the 'right' counit ε^u.

In the sequence of morphisms below, we mark on the right side of the line how we pass from a line to another. Numbers (1), (2), and (3) refer to Lemma 6.6, MEL is the middle exchange law. We have

$$\begin{array}{cl}
\mathrm{rep}_{\mathrm{pd},\mathcal{E}}(\tilde{t},\tilde{p},\tilde{s}) \xrightarrow{\mathrm{rep}_{\mathrm{pd},\mathcal{E}}(\mathrm{pr}_{u,(t,p,s)})} \mathrm{rep}_{\mathrm{pd},\mathcal{E}}(t,p,s) & \\ \hline
 & \text{def pr} \\
\tilde{t}_!\tilde{p}_*\tilde{s}^*u^* \xrightarrow{\mathrm{rep}_{\mathrm{pd},\mathcal{E}}(\hat{u}h,\bar{u}\varepsilon^p_{S^*(O)}\bar{h},u)} u^*t_!p_*s^* & \\ \hline
 & u_! \dashv u^* \\
u_!\tilde{t}_!\tilde{p}_*\tilde{s}^*u^* \longrightarrow t_!p_*s^* & \\ \hline
 & \text{def rep} \\
t_!(\hat{u}h)_!(\hat{u}h)^*p_*s^* \xrightarrow{t_!(\varepsilon^{\hat{u}h})p_*s^*} t_!p_*s^* & \\ \hline
 & (3) \\
t_!\hat{u}_!h_!h^*\hat{u}^*p_*s^* \xrightarrow{t_!(\varepsilon^{\hat{u}}\circ(\hat{u}_!(\varepsilon^h)\hat{u}^*))p_*s^*} t_!p_*s^* & \\ \hline
 & = \\
t_!\hat{u}_!h_!h^*\hat{u}^*p_*s^* \xrightarrow{(t_!(\varepsilon^{\hat{u}})p_*s^*)\circ(t_!\hat{u}_!(\varepsilon^h)\hat{u}^*p_*s^*)} t_!p_*s^* & \\ \hline
 & (1) \\
u_!u^*t_!\hat{u}_!\hat{u}^*p_*s^* \xrightarrow{(t_!(\varepsilon^{\hat{u}})p_*s^*)\circ((\varepsilon^u)t_!\hat{u}_!\hat{u}^*p_*s^*)} t_!p_*s^* & \\ \hline
 & \text{MEL} \\
u_!u^*t_!\hat{u}_!\hat{u}^*p_*s^* \xrightarrow{((\varepsilon^u)t_!p_*s^*)\circ(u_!u^*t_!(\varepsilon^{\hat{u}})p_*s^*)} t_!p_*s^* & \\ \hline
 & (2) \\
u_!u^*t_!p_*\bar{u}_!\bar{u}^*s^* \xrightarrow{((\varepsilon^u)t_!p_*s^*)\circ(u_!u^*t_!p_*(\varepsilon^{\bar{u}})s^*)} t_!p_*s^* & \\ \hline
 & (1) \\
u_!u^*t_!p_*s^*u_!u^* \xrightarrow{((\varepsilon^u)t_!p_*s^*)\circ(u_!u^*t_!p_*s^*(\varepsilon^u))} t_!p_*s^* & \\ \hline
 & u_! \dashv u^* \\
u^*t_!p_*s^*u^*u_! \xrightarrow{u^*t_!p_*s^*(\varepsilon^u)} u^*t_!p_*s^* & \\ \hline
 & \text{def rep} \\
u^*\,\mathrm{rep}_{\mathrm{pd},\mathcal{E}}(t,p,s)u_!u^* \xrightarrow{u^*\,\mathrm{rep}_{\mathrm{pd},\mathcal{E}}(t,p,s)(\varepsilon^u)} u^*\,\mathrm{rep}_{\mathrm{pd},\mathcal{E}}(t,p,s) & \\ \hline
 & \text{def pr} \\
u^*\,\mathrm{rep}_{\mathrm{pd},\mathcal{E}}(t,p,s)u_! \xrightarrow{\mathrm{pr}_{u,\mathrm{rep}_{\mathrm{pd},\mathcal{E}}(t,p,s)}} \mathrm{rep}_{\mathrm{pd},\mathcal{E}}(t,p,s). &
\end{array}$$

Thus $\mathrm{rep}_{\mathrm{pd},\mathcal{E}}$ preserves the prone morphisms.

Now we show preservation of supine morphisms. Let $\mathrm{su}_{u,(t,p,s)} = (1_B, 1_E, u)$: $(t,p,s) \to (u \circ t, p, u \circ s)$ be a supine morphism in $\mathcal{PolyDiag}(\mathcal{E})$. The representation of the supine morphism $\mathrm{rep}_{\mathrm{pd},\mathcal{E}}(\mathrm{su}_{u,(t,p,s)})$ is a natural transformation which is isomorphic to the adjoint $(u_! \dashv u^*)$ to

$$(ut)_!(1_B)_! 1_B^* p_*(us)^* \xrightarrow{(ut)_!(\varepsilon^{1_B})p_*(us)^*} (ut)_! p_*(us)^*.$$

The above morphism is isomorphic to the identity natural transformation

$$u_! t_! p_* s^* u^* \xrightarrow{1_{u_! t_! p_* s^* u^*}} u_! t_! p_* s^* u^*$$

and the supine morphism with the codomain $\mathrm{rep}_{\mathrm{pd},\mathcal{E}}(t,p,s)$ over u is a natural transformation

$$(t_! p_* s^*)u^* \xrightarrow{\eta^u_{(t_! p_* s^*)u^*}} u^* u_!(t_! p_* s^*)u^*$$

adjoint to the above identity morphism. Thus $\mathrm{rep}_{\mathrm{pd},\mathcal{E}}$ preserves the supine morphisms, as well. □

From now on till the end of this subsection we shall consider fibrations over Set only. The following appears in [8]. It is an immediate consequence of Theorem 6.4 and Proposition 6.7.

Proposition 6.8. *The representation morphism* $\mathrm{rep}_{\mathrm{pd},\mathrm{Set}}$ *is a morphism of lax monoidal fibrations which is faithful and full on isomorphisms.*

Corollary 6.9. *We have a sequence of morphisms of lax monoidal fibrations*

$$\mathcal{PolyDiag} \xrightarrow{\mathrm{rep}_{\mathrm{pd},\mathrm{Set}}} \mathcal{Poly} \to \mathrm{Cart}(\mathrm{Set}) \to \mathrm{Exp}(\mathrm{Set})$$

with the first being an equivalence of bifibrations and the following two being inclusions full on isomorphisms. The composition of these morphisms is (isomorphic to) $\mathrm{rep}_{\mathrm{pd},\mathrm{Set}}$.

PROOF. This is an immediate consequence of Propositions 6.7 and 6.8. □

We shall construct a morphism of lax monoidal fibrations

$$\begin{array}{ccc} \mathrm{Sig}_{\mathrm{a}} & \xrightarrow{\iota_{\mathrm{a}}} & \mathcal{PolyDiag} \\ & {\scriptstyle p_{\mathrm{a}}}\searrow \quad \swarrow{\scriptstyle p_{\mathrm{pd}}} & \\ & \mathrm{Set}. & \end{array}$$

Let (A, ∂, O) be a signature in $(\mathrm{Sig}_{\mathrm{a}})_O$. The functor ι_{a} sends this signature to a polynomial functor defined by the following polynomial diagram

$$O \xleftarrow{s^A} E^A \xrightarrow{p^A} A \xrightarrow{t^A} O$$

where

$$E^A = \coprod_{a \in A} (|a|] = \{\langle a, i\rangle : a \in A,\ i \in (|a|]\}$$

p^A is the first projection, $s^A(a,i) = \partial^A_a(i)$, and $t^A(a) = \partial^A_a(0)$, for $a \in A$ and $i \in (|a|]$.

Moreover, for the morphisms of signatures $(f, \sigma, u)\colon (A, \partial^A, O) \to (B, \partial^B, Q)$ in Sig_a over $u\colon O \to Q$ we define a commuting diagram

$$\begin{array}{ccccccc} O & \xleftarrow{s^A} & E^A & \xrightarrow{p^A} & A & \xrightarrow{t^A} & O \\ {\scriptstyle u}\downarrow & & {\scriptstyle g}\downarrow & & \downarrow{\scriptstyle f} & & \downarrow{\scriptstyle u} \\ Q & \xleftarrow[s^B]{} & E^B & \xrightarrow[p^B]{} & B & \xrightarrow[t^B]{} & Q \end{array}$$

where $g(a, i) = \langle f(a), \sigma_a^{-1}(i)\rangle$, for $\langle a, i\rangle \in E^A$. The square in the middle is easily seen to be a pullback. Thus the above diagram is a morphism of polynomial diagrams

$$(f, g, u)\colon (t^A, p^A, s^A) \to (t^B, p^B, s^B)$$

in $\mathcal{PolyDiag}$.

We have

Proposition 6.10. *The morphism ι_a defined above is a morphism of lax monoidal fibrations, an equivalence of bifibrations, and it makes the triangle of morphisms of lax monoidal fibrations over* Set

$$\begin{array}{ccc} \mathrm{Sig}_\mathrm{a} & \xrightarrow{\iota_\mathrm{a}} & \mathcal{PolyDiag} \\ & {\scriptstyle \mathrm{rep}_\mathrm{a}}\searrow \quad \swarrow{\scriptstyle \mathrm{rep}_\mathrm{pd}} & \\ & \mathrm{Exp}(\mathrm{Set}) & \end{array}$$

commute, up to a fibred natural isomorphism.

PROOF. ι_a is faithful from the construction. Let $(f, g, u)\colon (t^A, p^A, s^A) \to (t^B, p^B, s^B)$ be a morphism of polynomial diagrams over $u\colon O \to Q$. Then $(f, \sigma, u)\colon (A, \partial^A, O) \to (B, \partial^B, Q)$ such that $\sigma_a = (g_{\lceil p^{-1}(a)})^{-1}$, for $a \in A$, is a morphism in Sig_a over u. Moreover $\iota_\mathrm{a}(f, \sigma, u) = (f, g, u)$. Thus ι_a is full.

To see that ι_a is essentially surjective as well, fix a diagram

$$O \xrightarrow{s} E \xrightarrow{p} A \xrightarrow{t} O$$

in $\mathcal{PolyDiag}$. For $a \in A$, choose bijections $\tau_a\colon (n_a] \to p^{-1}(a)$, for some $n_a = |p^{-1}(a)| \in \omega$. Putting

$$\partial_a^A(i) \begin{cases} t(a) & \text{if } i = 0, \\ s\tau_a(i) & \text{otherwise,} \end{cases}$$

we have that $\iota(A, \partial^A, O)$ is isomorphic to (t, p, s), i.e., ι_a is essentially surjective as well.

The verification that the triangle commutes is also easy and we leave it to the reader. □

Corollary 6.11. *We have a sequence of morphisms of lax monoidal fibrations*

$$\mathrm{Sig}_\mathrm{a} \xrightarrow{\iota_\mathrm{a}} \mathcal{PolyDiag} \xrightarrow{\mathrm{rep}_\mathrm{pd}} \mathcal{Poly}^{c} \to \mathrm{Cart}(\mathrm{Set}) \to \mathrm{Exp}(\mathrm{Set})$$

with the first two being equivalences of bifibrations and the following two being inclusions of bifibrations full on isomorphisms. The composition of all four morphisms is (isomorphic to) rep_a*, and hence* rep_a *is a morphism lax monoidal fibrations, morphism of bifibrations, faithful, and full on isomorphisms.*

PROOF. This is an immediate consequence of Corollary 6.9 and Proposition 6.10. □

Theorem 6.12. *The essential image of the morphism of lax monoidal fibrations* $\mathrm{rep}_{\mathrm{a}}$ *in* Exp(Set) → Set *consists of the finitary endofunctors preserving wide pullbacks and cartesian natural transformations.*

PROOF. By Proposition 6.10 the image of $\mathrm{rep}_{\mathrm{a}}$ in $p_{\exp}$ is $p_{\mathrm{poly}}\colon \mathcal{Poly} \to \mathrm{Set}$, the image of rep_{pd}. By Proposition 6.7, $\mathrm{rep}_{\mathrm{pd}}$ preserves prone morphisms. By Corollary 6.9, the image of $\mathrm{rep}_{\mathrm{pd}}$ is contained in the lax monoidal subfibration Cart(Set) → Set. Thus, it is enough to verify the statement on fibres. But this is the content of Theorem 6.4. □

By Proposition 2.1, any morphism of lax monoidal fibration induces a morphism of the corresponding fibrations of monoids. We finish this section by spelling the most important instance of this fact, announced at the beginning of this subsection, that follows from Corollary 6.11.

Corollary 6.13. *The fibration of multicategories with nonstandard amalgamation is equivalent to the fibration that has as objects finitary cartesian monads on slices of* Set *whose functor part preserves wide pullbacks. A morphism in that fibration between monads over* $\mathrm{Set}_{/O}$ *and* $\mathrm{Set}_{/Q}$ *over a function* $u\colon O \to Q$ *is a cartesian morphism of monads whose functor part is the pullback functor* $u^*\colon \mathrm{Set}_{/Q} \to \mathrm{Set}_{/O}$.

Example. The following example shows that some polynomial diagrams can be equipped with a monoid structure in the fibration of polynomial diagrams $\mathcal{PolyDiag}$ but that this monoid structure (unique in this case) cannot be lifted to the fibration of monotone diagrams $\mathcal{MPolyDiag}$. This is to explain why there are fewer monotone monads than polynomial ones. The signature has three types $O = \{x_0, x_1, x_2\}$ and seven operations. Three operations that will serve as units in the monoid $1_{x_0}\colon x_0 \to x_0$, $1_{x_1}\colon x_1 \to x_1$, $1_{x_2}\colon x_2 \to x_2$ and four other with typing as displayed:

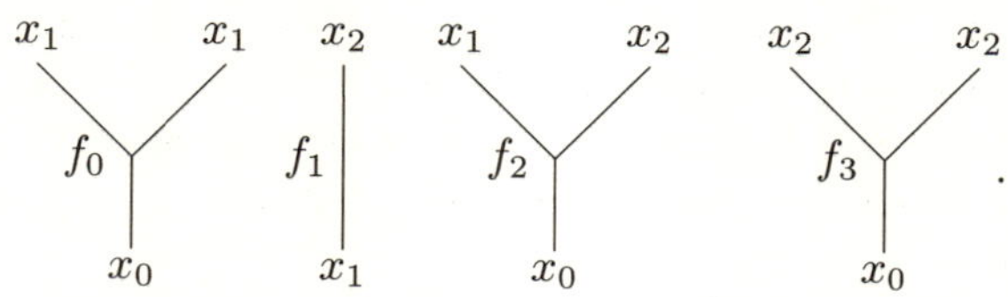

Then, no matter how we order entries in f_2: either $x_1 < x_2$ or $x_2 < x_1$, we won't be able to define one of the multiplications

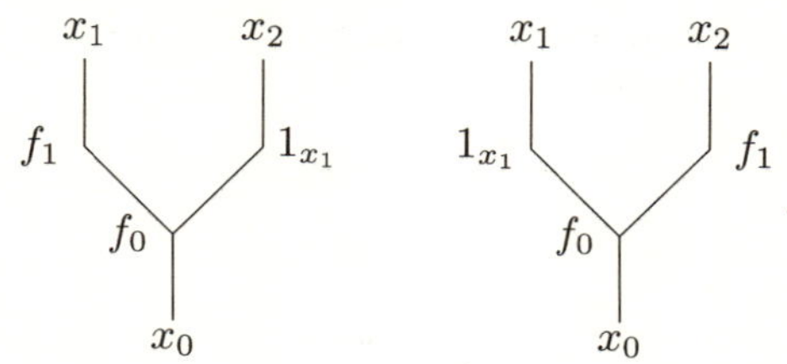

This problem disappears if we can switch entries in the result, which is possible in amalgamated signatures and polynomial diagrams.

6.5. The 2-level amalgamated signatures fibration $p_{2a}\colon \mathbf{Sig}_{2a} \to \mathbf{Set}^{\to}$. We describe below a lax monoidal fibration such that its category of monoids contains as a full subcategory the category of 2-level multicategories with nonstandard amalgamation, cf. [12]. It is fairly clear that this construction can be farther generalize by making the structure of objects even more involved but right now we don't see any real applications for such structures.

The fibration $p_{2a}\colon \mathrm{Sig}_{2a} \to \mathrm{Set}^{\to}$. The base category of our fibration is $\mathrm{Set}^{\to}$. A typical object of $\mathrm{Set}^{\to}$ is a function $(\dot{-})\colon O \to \ddot{O}$, denoted by $\overrightarrow{O}$. O is referred to as the set of objects of $\overrightarrow{O}$, $\ddot{O}$ is referred to as the set of types of $\overrightarrow{O}$, and $(\dot{-})$ is the typing of $\overrightarrow{O}$. A morphism in $\mathrm{Set}^{\to}$, denoted $\vec{u} = (u, \ddot{u})\colon \overrightarrow{O} \to \overrightarrow{Q}$, is a pair of morphism making the square

$$\begin{array}{ccc} O & \xrightarrow{\;u\;} & Q \\ {\scriptstyle (\dot{-})}\downarrow & & \downarrow{\scriptstyle (\dot{-})} \\ \ddot{O} & \xrightarrow[\;\ddot{u}\;]{} & \ddot{Q} \end{array}$$

commute. The notation $\overrightarrow{O}$ and $\vec{u}$ will be used exclusively in this subsection. In spite of the fact that we think of O and $\ddot{O}$ as disjoint sets, it is convenient to 'test' elements of those sets for equality in the sense that either they both belong to one set and are equal or otherwise we move one of the elements to $\ddot{O}$ and there they are equal. Formally, we define the 'graded equality' $\doteq$ so that if $x, y \in O + \ddot{O}$, then

$$x \doteq y \quad \text{iff} \quad \begin{cases} x = y & \text{if } x, y \in O \text{ or } x, y \in \ddot{O}, \\ x = \dot{y} & \text{if } y \in O \text{ and } x \in \ddot{O}, \\ \dot{x} = y & \text{if } x \in O \text{ and } y \in \ddot{O}. \end{cases}$$

By $\overrightarrow{O}^{\dagger}$ we denote the sum $\bigcup_{n\in\omega} \overrightarrow{O}^{\dagger}_{n}$, where

$$\overrightarrow{O}^{\dagger}_{n} = \{d\colon [n] \to O + \ddot{O} : d((n]) \subseteq O\}$$

is the set of functions from $[n]$ to the disjoint sum $O + \ddot{O}$ such that positive integers are sent to objects and 0 is sent either to an object or a type. Extending the previous conventions we will write If $(\dot{-})\colon O \to \ddot{O}$ is an identity, we write $O^{\dagger}$ for $\overrightarrow{O}^{\dagger}$. For $d\colon [n] \to O + \ddot{O}$ we have restrictions of d to $d^{+}\colon (n]^{+} \to O + \ddot{O}$ and to $d^{-}\colon \{0\} \to O + \ddot{O}$.

The total category of our fibration $p_{2a}\colon \mathrm{Sig}_{2a} \to \mathrm{Set}^{\to}$ has as objects triples $(A, \partial^{A}, \overrightarrow{O})$, such that A is a set, $\overrightarrow{O}$ is an object of $\mathrm{Set}^{\to}$ and $\partial^{A}\colon A \to \overrightarrow{O}^{\dagger}$ is a function. A morphism $(f, \sigma, \vec{u})\colon (A, \partial, \overrightarrow{O}) \to (B, \partial, \overrightarrow{O})$ is a triple such that $f\colon A \to B$ is a function, $\vec{u}\colon \overrightarrow{O} \to \overrightarrow{Q}$ is a morphism in $\mathrm{Set}^{\to}$ and for any $a \in A$ with $|a| = n$, $\sigma \in \mathrm{S}_n$ is a permutation such that

$$\begin{array}{ccc} [n] & \xleftarrow{\;\sigma_a\;} & [n] \\ {\scriptstyle \partial_a}\downarrow & & \downarrow{\scriptstyle \partial_{f(a)}} \\ O + \ddot{O} & \xrightarrow[\;u+\ddot{u}\;]{} & Q + \ddot{Q} \end{array}$$

commutes. The projection functor $p\colon \mathrm{Sig}_{2a} \to \mathrm{Set}^{\to}$ sends $(f, \sigma, \vec{u})\colon (A, \partial, \overrightarrow{O}) \to (B, \partial, \overrightarrow{Q})$ to $\vec{u}\colon \overrightarrow{O} \to \overrightarrow{Q}$.

For $\vec{u}\colon \vec{O} \to \vec{Q}$ in $\mathrm{Set}^{\to}$, we have a pullback operation

$$\begin{array}{ccc} u^*(B) & \xrightarrow{\ \pi\ } & B \\ \partial \big\downarrow & & \big\downarrow \partial \\ \vec{O}^{\dagger} & \xrightarrow[\ \vec{u}^{\dagger}\ ]{} & \vec{Q}^{\dagger} \end{array}$$

where $\vec{u}^{\dagger}(d) = (u + \ddot{u}) \circ d$, where $d\colon [n] \to O + \ddot{O}$. Then

$$\vec{u}^*(B) = \{\langle d, b\rangle : d\colon [|b|] \to O + \ddot{O},\, b \in B,\, \vec{u}^{\dagger}(d) = \partial_b\}.$$

The monoidal structure in p_{2a}. We have two lax morphisms of fibrations

$$\begin{array}{ccc} \mathrm{Sig}_{2a} \times_{\mathrm{Set}^{\to}} \mathrm{Sig}_{2a} & \xrightarrow{\ \otimes\ } & \mathrm{Sig}_{2a} \\ & {\scriptstyle p'} \searrow \quad \swarrow {\scriptstyle p_{2a}} & \\ & \mathrm{Set}^{\to} & \end{array} \qquad \begin{array}{ccc} \mathrm{Set}^{\to} & \xrightarrow{\ I\ } & \mathrm{Sig}_{2a} \\ & {\scriptstyle \mathrm{id}} \searrow \quad \swarrow {\scriptstyle p_{2a}} & \\ & \mathrm{Set}^{\to}. & \end{array}$$

The tensor $(A \otimes_{\vec{O}} B, \partial^{\otimes}, \vec{O})$ of two objects $(A, \partial, \vec{O})$ and $(B, \partial, \vec{O})$ in a fibre $(\mathrm{Sig}_{2a})_{\vec{O}}$ is defined as follows

$$A \otimes_{\vec{O}} B = \{\langle a, b_i\rangle_{i\in(|a|]} : a \in A,\, b_i \in B,\, \partial_{b_i}(0) \doteq \partial_a(i)\}$$

and for $\langle a, b_i\rangle_{i\in(|a|]} \in A \otimes_O B$,

$$\partial^{\otimes}(\langle a, b_i\rangle_i) = [\dot{\partial}_a^-, \partial_{b_i}^+]_i \colon \left[\sum_{i=1}^{|a|} |b_i|\right] = [0] + \coprod_i (|b_i|] \to O + \ddot{O}$$

where $\dot{\partial}_a^-(*) = \partial^{\otimes}(\langle a, b_i\rangle_i)(*)$ is '$\partial_a(*)$ as much as possible,' i.e.,

$$\dot{\partial}_a^+(*) = \begin{cases} (\dot{\partial_a(*)}) & \text{if } \partial_a(*) \in O \text{ and } \exists_{i\in(|a|]}\, \partial_{b_i}(*) \in \ddot{O}, \\ \partial_a(*) & \text{otherwise.} \end{cases}$$

Remark. This definition is slightly more complicated than possible to make sure that composition with identities (that have objects from O, rather than types from $\ddot{O}$, as codomains) is neutral.

The tensor on morphisms is defined as in Sig_a. For a pair of maps in Sig_{2a}

$$\bar{f} = (f, \sigma, \vec{u})\colon (A, \partial, \vec{O}) \to (A', \partial, \vec{Q}), \quad \bar{g} = (g, \tau, \vec{u})\colon (B, \partial, \vec{O}) \to (B', \partial, \vec{Q})$$

over the same map $\vec{u}\colon \vec{O} \to \vec{Q}$ in $\mathrm{Set}^{\to}$ we define the map

$$\bar{f} \otimes_{\vec{u}} \bar{g} = (f \otimes_{\vec{O}} g, \sigma \otimes_{\vec{u}} \tau, \vec{u})\colon (A \otimes_{\vec{O}} B, \partial^{\otimes}, \vec{O}) \to (A' \otimes_{\vec{Q}} B', \partial^{\otimes}, \vec{Q})$$

so that, for $\langle a, b_i\rangle_{i\in(|a|]} \in A \otimes_{\vec{O}} B$,

$$f \otimes_{\vec{u}} g(\langle a, b_i\rangle_{i\in(|a|]}) = (\langle f(a), g(b_{\sigma_a(j)})\rangle_{j\in(|f(a)|]}).$$

Clearly, $|a| = |f(a)|$ and

$$n = \left|\langle a, b_i\rangle_{i\in(|a|]}\right| = \sum_{i\in|a|} |b_i| = \sum_{i\in|f(a)|} |g(b_i)| = |f \otimes_{\vec{u}} g(\langle a, b_i\rangle_{i\in(|a|]})|.$$

Moreover, we put

$$(\sigma \otimes_{\vec{u}} \tau)_{\langle a, b_i\rangle_i} = \left[\sigma_a^-, \tau_{b_{\sigma_a(i)}}^+\right]_i$$

making the triangle

$$\begin{array}{ccc}
[n] & \xleftarrow{\;(\sigma\otimes_{\vec{u}}\tau)_{\langle a,b_i\rangle_i}\;} & [n] \\
{\scriptstyle \partial^{\otimes}_{\langle a,b_i\rangle_i}}\downarrow & & \downarrow{\scriptstyle \partial^{\otimes}_{\langle f(a),g(b_i)\rangle_i}} \\
O & \xrightarrow[\vec{u}]{} & Q
\end{array}$$

commute. This ends the definition of the tensor $\otimes$.

The unit $I_{\vec{\mathcal{O}}}$ in the fibre $(\mathrm{Sig}_{2a})_{\vec{\mathcal{O}}}$, is $(O, \partial^{I_O}, \vec{\mathcal{O}})$ where, for $x \in O$, the function $\partial^{I_{\vec{\mathcal{O}}}}_x : [1] \to O$ is constant equal x.

Remark. As we mentioned earlier, we would like to put $\partial^{\vec{\mathcal{O}}}_x(*) = \dot{x}$ to be sure that the codomains are always the types of $\vec{\mathcal{O}}$ but this would not work as the compatibility of the identities on the left in 2-level multicategories with nonstandard amalgamation must be on the objects of $\vec{\mathcal{O}}$.

Lemma 6.14. *The fibration $p_{2a}\colon \mathrm{Sig}_{2a} \to \mathrm{Set}^{\to}$ together with the above defined tensor $\otimes$ and unit I is a lax monoidal fibration whose fibres are strong monoidal categories. The fibred forgetful functor $\mathcal{U}_{2a}$*

$$\begin{array}{ccc}
\mathrm{Sig}_{2a} & \underset{\mathcal{U}_{2a}}{\overset{\mathcal{F}_{2a}}{\rightleftarrows}} & \mathrm{Mon}(\mathrm{Sig}_{2a}) \\
& {\scriptstyle p_{2a}}\searrow \quad \swarrow{\scriptstyle q_{2a}} & \\
& \mathrm{Set}^{\to} &
\end{array}$$

has a fibred left adjoint $\mathcal{F}_{2a}$, the free monoid functor.

PROOF. The fact that p_{2a} is a lax monoidal fibration is a simple check. The construction of a free monoid functor $\mathcal{F}_{2a}$ is essentially the same as the one given in [12], see also [1, 4, 16]. □

The full embeddings. The 2-level amalgamated signatures fibration p_{2a}: $\mathrm{Sig}_{2a} \to \mathrm{Set}^{\to}$ contains as a lax monoidal subfibration the amalgamated signatures fibration $p_a\colon \mathrm{Sig}_a \to \mathrm{Set}$, and as a consequence the category of monoids $\mathrm{Mon}(\mathrm{Sig}_a)$ with respect to p_a has a full fibred embedding into the category of monoids $\mathrm{Mon}(\mathrm{Sig}_{2a})$ with respect to p_{2a}. Moreover, the category of 2-level multicategories with nonstandard amalgamation of [12] is a full subcategory of $\mathrm{Mon}(\mathrm{Sig}_{2a})$, as well. Below we describe this in detail.

The first embedding

$$\begin{array}{ccc}
\mathrm{Sig}_a & \xrightarrow{\;\iota\;} & \mathrm{Sig}_{2a} \\
{\scriptstyle p_a}\downarrow & & \downarrow{\scriptstyle p_{2a}} \\
\mathrm{Set} & \xrightarrow[\delta]{} & \mathrm{Set}^{\to}
\end{array}$$

is given by the diagonal functor δ and an inclusion ι. The functor δ has both adjoints, say $l \dashv \delta \dashv r$, associating codomain and domain, respectively. The functor ι has also both fibred adjoints $L \dashv \iota \dashv R$. The left adjoint L is defined by composition. For ξ in Sig_a we have

$$L(\xi\colon Y \to \dot{O} \times O^*) = \dot{\xi}$$

where $\dot{\xi}$ is the composition

$$Y \xrightarrow{\xi} \dot{O} \times O^* \xrightarrow{1\times(\dot{-})^*} \dot{O} \times \dot{O}^* = \dot{O}^\dagger.$$

The right adjoint R is defined by the pullback. We have

$$R(\xi \colon Y \to \dot{O} \times O^*) = \tilde{\xi}$$

where $\tilde{\xi}$ is given by the following pullback

$$\begin{array}{ccc} \widetilde{Y} & \longrightarrow & Y \\ {\scriptstyle\tilde{\xi}}\downarrow & & \downarrow{\scriptstyle\xi} \\ O^\dagger & \xrightarrow[(\dot{-})\times 1_{O^*}]{} & \dot{O} \times O^*. \end{array}$$

The second embedding is a functor

$$\Phi \colon \text{Multicat} \to \text{Mon}(\text{Sig}_{2a})$$

from the category Multicat of 2-level multicategories with nonstandard amalgamation to the category of monoids in the lax monoidal fibration $p_{2a} \colon \text{Sig}_{2a} \to \text{Set}^{\to}$. Let $\mathbf{C}$ be an object of Multicat. Let $A = A(\mathbf{C})$, $O = O(\mathbf{C})$, $\ddot{O} = \ddot{O}(\mathbf{C})$ denote arrows, objects and lower level objects (i.e., types) in $\mathbf{C}$, respectively. The monoid $\Phi(\mathbf{C})$ is in the fibre over $(\dot{-}) \colon O \to \ddot{O}$, i.e., $\overrightarrow{\mathcal{O}}$. The universe of $\Phi(\mathbf{C})$ is A and the typing function $\partial^A \colon A \to \overrightarrow{\mathcal{O}}^\dagger$ is defined as follows. For $a \in A$ we have source of a, $s(a) \colon (|a|] \to O$ and target of a, $t(a) \in \ddot{O}$. The function $\partial_a \colon [|a|] \to O + \ddot{O}$ is equal $s(a)$ on $(|a|]$ and

$$\partial_a(0) = \begin{cases} x & \text{if } a = 1_x, \\ t(a) & \text{otherwise,} \end{cases}$$

i.e., it is the object x if a is the identity on x and it is the type $t(a)$ otherwise. The unit map $(e, \sigma, \mathrm{id}_{\overrightarrow{\mathcal{O}}}) \colon (O, \partial^{I_{\overrightarrow{\mathcal{O}}}}, \overrightarrow{\mathcal{O}}) \to (A, \partial^A, \overrightarrow{\mathcal{O}})$ is sending $x \in O$ to $1_x \in A$, and $\sigma_x = 1_{[1]}$. The multiplication

$$(\mu_A, \sigma, \mathrm{id}_{\overrightarrow{\mathcal{O}}}) \colon (A \otimes_{\overrightarrow{\mathcal{O}}} A, \partial^\otimes, \overrightarrow{\mathcal{O}}) \to (A, \partial, \overrightarrow{\mathcal{O}})$$

is defined with the help of simultaneous composition operation, see [12]. We first describe exactly the tensor $(A \otimes_{\overrightarrow{\mathcal{O}}} A, \partial^\otimes, \overrightarrow{\mathcal{O}})$. We have

$$A \otimes_{\overrightarrow{\mathcal{O}}} A = \{\langle a, a_i \rangle_{i \in (|a|]} : a_i, a \in A, \partial_{a_i}(0) \doteq \partial_a(i)\}$$

and for $\langle a, a_i \rangle_{i \in (|a|]} \in A \otimes_{\overrightarrow{\mathcal{O}}} B$,

$$\partial^{\otimes,+}_{\langle a, a_i \rangle_i} = \coprod_i \partial^+_{a_i} \colon (|\langle a_i, b \rangle_i|] = \coprod_i (|a_i|] \to O + \ddot{O}$$

and

$$\partial^{\otimes,+}_{\langle a, a_i \rangle_i}(0) = \begin{cases} (\dot{\partial_a(0)}) & \text{if } \partial_a(0) \in O \text{ and } \exists_{i \in (|a|]}\, \partial_{a_i}(0) \in \ddot{O} \\ \partial_a(0) & \text{otherwise.} \end{cases}$$

For $\langle a, a_i \rangle_{i \in (|a|]} \in A \otimes_{\overrightarrow{\mathcal{O}}} A$, we have the simultaneous composition $b = a(a_i/i : i \in (|a|])$ together with amalgamating maps $\phi_i \colon s(a_i) \to s(b)$ for $i \in (|a|]$ such that the function

$$[\phi_i]_i \colon \coprod_i s(a_i) \to s(b)$$

is a bijection. We put $\mu_A(\langle a, a_i\rangle_i)$ to be equal b, and σ_b is such that $\sigma_b^- = [\phi_i]_i$.

The remaining details, as well as the definition on morphisms, are easy. The category Sig_{2a} and the tensor $\otimes_{\vec{O}}$ are so defined that identities must have objects as codomains. As we mentioned at the beginning, this is so to mimic the behavior of identities in Multicat. However, all the other arrows in the monoids coming from Multicat have types as codomains. In fact, this characterizes the monoids coming from Multicat. We have

Proposition 6.15. *The functor*

$$\Phi\colon \mathrm{Multicat} \to \mathrm{Mon}(\mathrm{Sig}_{2a})$$

is full and faithful and its essential image consists of those monoids in which all arrows but identities have types as codomains (i.e., either $\partial_a(0) \in \ddot{O}$ or $a = 1_{\partial_a(0)}$, for $a \in A$).

The muticategories with the object of objects equal to $\dot{(-)}\colon O \to \ddot{O}$ are sent to the monoids in the fibre over $\vec{O}$.

Proof. Simple verification. □

6.6. Single tensor in the fibration p_a. In this subsection we describe another widely used tensor in the fibration p_a. To distinguish these two tensors, we shall call the one considered so far the *total tensor* and denoted it, in this subsection only, by $\otimes^t$. The tensor we are going to discuss here, the *single tensor*, will be denoted by $\otimes^s$. Both tensors have the same unit.

Let (A, ∂^A, O) and (B, ∂^B, O) be two object in the fibre $(\mathrm{Sig}_a)_O$. The *single tensor* $(A \otimes^s B, \partial^{\otimes^s}, O)$ is defined as follows

$$A \otimes^s B = \{\langle a, i, b\rangle : a \in A, b \in B, i \in (|a|], \partial^A_a(i) = \partial^B_b(0)\}$$

and for $\langle a, i, b\rangle \in A \otimes^s B$,

$$\partial^{\otimes^s}(\langle a, i, b\rangle) = \partial^A_a{}_{\lceil (|a|] - \{i\})} + \partial^{B,-}_b.$$

Note that contrary to the case of the total tensor $\otimes^t$, the coherence morphisms for the associativity α and the right unit ρ are not isomorphisms. This example together with the Burroni fibrations were the main motivation for choosing the directions of coherence morphisms in the definition of lax monoidal fibrations.

There is a long debate whether single or total tensor is more convenient. There are arguments for each. The fibration $p_a\colon \mathrm{Sig}_a \to \mathrm{Set}$ equipped with single tensor also acts on the basic fibration, but this action is not so much in use. The action $\star^s$

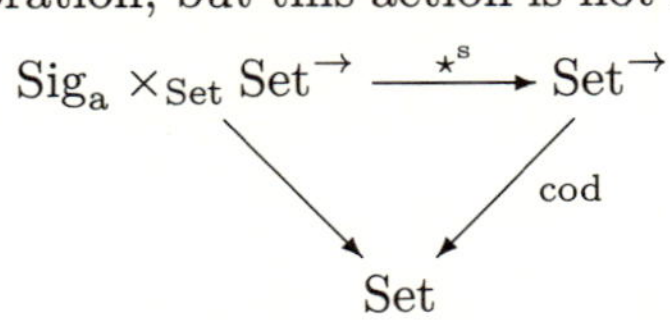

is defined as follows. For (A, ∂^A, O) in Sig_a and (X, d^X) in $(\mathrm{Set}^{\to})_O$ we put $(A, \partial^A, O) \star^s (X, d^X) = (A \star^s X, d^s)$, where

$$A \star^s X = \{\langle a, i, x\rangle : \partial^A_a(i) = d^X(x)\}$$

and $d^s(a, i, x) = \partial^A_a(0)$. One can easily verify that this extends to a definition of an action of the lax monoidal fibration $(\mathrm{Sig}_a, p_a, \otimes^s, I)$ on the basic fibration. So, by adjunction we get a morphism of lax monoidal fibrations

$$\begin{array}{ccc} \mathrm{Sig}_{\mathrm{a}} & \xrightarrow{\mathrm{rep}^{\mathrm{s}}_{\mathrm{a}}} & \mathrm{Exp}(\mathrm{Set}) \\ & {\scriptstyle p_{\mathrm{a}}}\searrow \quad \swarrow{\scriptstyle p_{\exp}} & \\ & \mathrm{Set}. & \end{array}$$

We can describe this representation equivalently as a representation of the fibration of polynomial diagrams. That is, if we consider in the fibration of polynomial diagrams the image under equivalence of categories ι_{a} of the tensor $\otimes^{\mathrm{s}}$ in p_{a}, we have a tensor, also denoted $\otimes^{\mathrm{s}}$, in $p_{\mathrm{pd}}\colon \mathcal{PolyDiag} \to \mathrm{Set}$. Then p_{pd} considered with this tensor has a representation $\mathrm{rep}^{\mathrm{s}}_{\mathrm{pd}}$ corresponding to the representation $\mathrm{rep}^{\mathrm{s}}_{\mathrm{a}}$, i.e., a morphism of lax monoidal fibrations

$$\begin{array}{ccc} \mathcal{PolyDiag} & \xrightarrow{\mathrm{rep}^{\mathrm{s}}_{\mathrm{pd}}} & \mathrm{Exp}(\mathrm{Set}) \\ & {\scriptstyle p_{\mathrm{pd}}}\searrow \quad \swarrow{\scriptstyle p_{\exp}} & \\ & \mathrm{Set} & \end{array}$$

that sends the diagram (t, p, s) to the functor $t_! p_! s^*$.

The representation $\mathrm{rep}^{\mathrm{s}}_{\mathrm{pd}}$ is faithful but not full even on isomorphisms, as the diagrams

$$O \xleftarrow{s} E \xrightarrow{p} A \xrightarrow{t} O \qquad O \xleftarrow{s} E \xrightarrow{p'} A \xrightarrow{t'} O$$

have isomorphic representations if and only if $t \circ p = t' \circ p'$. We want to describe the image of $\mathrm{rep}^{\mathrm{s}}_{\mathrm{pd}}$.

We shall call a polynomial diagram (t, p, s), a *linear diagram*[12] if and only if p is an isomorphism. We shall denote by $p_{\mathrm{ld}}\colon \mathcal{LinDiag} \to \mathrm{Set}$ the full subfibration of p_{pd} whose objects in the total category are linear diagrams. The image of $\mathrm{rep}^{\mathrm{s}}_{\mathrm{pd}}$ is the same as $\mathrm{rep}_{\mathrm{ld}}$, the image of $\mathrm{rep}^{\mathrm{s}}_{\mathrm{pd}}$ restricted to p_{ld}. The linear diagrams are closed under both tensors $\otimes^{\mathrm{t}}$ and $\otimes^{\mathrm{s}}$ and both tensors agree on them. Thus both representations $\mathrm{rep}^{\mathrm{t}}_{\mathrm{pd}}$ and $\mathrm{rep}^{\mathrm{s}}_{\mathrm{pd}}$ coincide on linear diagrams. This statement characterizes the linear diagrams. This is why the representation of linear diagrams is denoted by $\mathrm{rep}_{\mathrm{ld}}$, with no superscript.

Proposition 6.16. *The total category of the image of the morphism of lax monoidal fibrations*

$$\begin{array}{ccc} \mathcal{LinDiag} & \xrightarrow{\mathrm{rep}_{\mathrm{ld}}} & \mathrm{Exp}(\mathrm{Set}) \\ & {\scriptstyle p_{\mathrm{ld}}}\searrow \quad \swarrow{\scriptstyle p_{\exp}} & \\ & \mathrm{Set} & \end{array}$$

consists of endofunctors on slices of Set *that preserves colimits and wide pullbacks as objects and cartesian natural transformations as morphisms.*

PROOF. From the characterization of the image of p_{pd} in $p_{\exp}$ the necessity of the conditions is obvious. On the other hand, by Theorem 6.4, any endofunctor P

[12] This name and notion is taken from [17].

on a slices $\mathrm{Set}_{/O}$ that preserves colimits and wide pullbacks is of the form $t_! p_* s^*$ for some polynomial diagram

$$O \xleftarrow{s} E \xrightarrow{p} A \xrightarrow{t} O.$$

Recall that p has finite fibres. Since P preserves the initial object the fibres of p cannot be empty. Since p preserves binary coproduct the fibres of p cannot have more than one element. Thus p is an isomorphism, and the diagram representing P is linear. The characterization of natural transformations in the image of rep_{ld} follows directly from Theorem 6.4. □

The polynomial diagrams of form

$$O \xleftarrow{s} E \xrightarrow{p} O \xrightarrow{1_O} O$$

are also closed under the tensor $\otimes^t$ in p_{pd}, and form a monoidal full subfibration of p_{pd} considered with the total tensor $\otimes$. Such diagrams correspond to signatures that have exactly one operation of each (output) type. The representations of such diagrams are endofunctors P on $\mathrm{Set}_{/O}$ that send a function d to a function $P(d)$ that has as fibres finite products of fibres of d. For this reason, we call such polynomial diagrams *monomial diagrams* and the full subfibration of monomial diagrams will be denoted by $p_{\mathrm{md}}\colon \mathcal{MonoDiag} \to \mathrm{Set}$. The composition of $\mathrm{rep}_{\mathrm{pd}}$ with the inclusion gives a representation morphism $\mathrm{rep}_{\mathrm{md}}$

$$\begin{array}{ccc} \mathcal{MonoDiag} & \xrightarrow{\mathrm{rep}_{\mathrm{md}}} & \mathrm{Exp}(\mathrm{Set}) \\ & {\scriptstyle p_{\mathrm{md}}} \searrow \quad \swarrow {\scriptstyle \mathrm{rep}_{\mathrm{exp}}} & \\ & \mathrm{Set}. & \end{array}$$

We have

Proposition 6.17. *The total category of the image of the morphism* $\mathrm{rep}_{\mathrm{md}}$ *of lax monoidal fibrations*

$$\begin{array}{ccc} \mathcal{MonoDiag} & \xrightarrow{\mathrm{rep}_{\mathrm{md}}} & \mathrm{Exp}(\mathrm{Set}) \\ & {\scriptstyle p_{\mathrm{md}}} \searrow \quad \swarrow {\scriptstyle p_{\mathrm{exp}}} & \\ & \mathrm{Set} & \end{array}$$

consists of finitary endofunctors on slices of Set *that preserves limits and cartesian natural transformations.*

PROOF. This follows from Theorem 6.4 and the observation, that for any polynomial diagram (t,p,s) the polynomial functor $t_! p_* s^*$ preserves the terminal object 1 if and only if t is an isomorphism. □

Remarks. (1) We want to point out that the fibrations of linear diagrams p_{ld} and monomial diagrams p_{md} are certain fibrations of graphs. The category *Gph* has as objects parallel pairs of functions $s,t\colon A \to O$. A morphism of graphs $(f,u)\colon (s,t) \to (s',t')$ is a pair of functions $f\colon A \to A'$, $u\colon O \to Q$ making the diagram

$$\begin{array}{ccccc} O & \xleftarrow{s} & E & \xrightarrow{p} & O \\ {\scriptstyle u}\downarrow & & {\scriptstyle f}\downarrow & & \downarrow{\scriptstyle u} \\ Q & \xleftarrow[s']{} & E' & \xrightarrow[p']{} & Q \end{array}$$

commute. The category of cartesian graphs cGph is a subcategory containing the same objects as Gph and a morphism (f, u) in cGph if and only if the right square above is a pullback. These categories are fibred over Set and the projection functors $p_{\mathrm{G}}\colon \mathrm{Gph} \to \mathrm{Set}$, $p_{\mathrm{cG}}\colon \mathrm{Gph} \to \mathrm{Set}$ send a morphism (f, u) to u. Both fibrations have a lax monoidal structure with tensor given by the obvious pullbacks. Moreover, we have equivalences of lax monoidal fibrations

$$\begin{array}{ccc} \mathrm{Gph} & \longrightarrow & \mathcal{LinDiag} \\ & {\scriptstyle p_{\mathrm{G}}}\searrow \quad \swarrow{\scriptstyle p_{\mathrm{ld}}} & \\ & \mathrm{Set} & \end{array} \qquad \begin{array}{ccc} \mathrm{cGph} & \longrightarrow & \mathcal{MonoDiag} \\ & {\scriptstyle p_{\mathrm{cG}}}\searrow \quad \swarrow{\scriptstyle p_{\mathrm{md}}} & \\ & \mathrm{Set} & \end{array}$$

sending a graph $(t, s\colon A \to O)$ to a linear diagram $(t, 1_A, s)$ and a cartesian graph $(p, s\colon A \to O)$ to a monomial diagram $(t, p, 1_O)$.

(2) So far we haven't said anything about monoids in p_{a} with the single tensor $\otimes^{\mathrm{s}}$. We can pass from multiplication with respect to the total tensor to multiplication with respect to the single tensor by putting identities into all places but one. Thus we have an embedding of the monoids with respect to the total tensor into the the monoids with respect to the single tensor. To characterize the image of this embedding we shall use a certain natural isomorphism involving $\otimes^{\mathrm{s}}$ and the binary coproduct in fibres $+$. Note that for any signatures A, B, C in the same fibre over O of $\mathrm{Sig}_{\mathrm{a}}$, we have an isomorphism:

$$\big((A \otimes^{\mathrm{s}} B) \otimes^{\mathrm{s}} C\big) + \big(A \otimes^{\mathrm{s}} (C \otimes^{\mathrm{s}} B)\big) \cong \big((A \otimes^{\mathrm{s}} C) \otimes^{\mathrm{s}} B\big) + \big(A \otimes^{\mathrm{s}} (B \otimes^{\mathrm{s}} C)\big)$$

that 'repairs' the lack of strong associativity for the tensor $\otimes^{\mathrm{s}}$. Intuitively, both sides of the isomorphism contain the part

$$\big(A \otimes^{\mathrm{s}} (C \otimes^{\mathrm{s}} B)\big) + \big(A \otimes^{\mathrm{s}} (B \otimes^{\mathrm{s}} C)\big),$$

i.e., the A' into which we plug either a B with plugged in a C or a C with plugged in a B. The remaining part on both sides of the above isomorphism is A's into which we plug directly a B and a C. Clearly, this isomorphism, $\upsilon_{A,B,C}$ is natural in A, B and C. $\upsilon_{A,A,A}$ may look trivial but it is not! Then one can verify that a monoid (M, m, e) in $\mathrm{Sig}_{\mathrm{a}}$ with respect to $\otimes^{\mathrm{s}}$ comes from a monoid in $\mathrm{Sig}_{\mathrm{a}}$ with respect to $\otimes^{\mathrm{t}}$, if the following diagram

$$\begin{array}{ccccc} \big((M \otimes^{\mathrm{s}} M) \otimes^{\mathrm{s}} M\big) + \big(M \otimes^{\mathrm{s}} (M \otimes^{\mathrm{s}} M)\big) & \cong & & & \big((M \otimes^{\mathrm{s}} M) \otimes^{\mathrm{s}} M\big) + \big(M \otimes^{\mathrm{s}} (M \otimes^{\mathrm{s}} M)\big) \\ {\scriptstyle (m\otimes 1_M)+(1_M\otimes m)}\downarrow & & & & \downarrow{\scriptstyle (m\otimes 1_M)+(1_M\otimes m)} \\ (M \otimes^{\mathrm{s}} M) + (M \otimes^{\mathrm{s}} M) & \xrightarrow[{[m,m]}]{} & M & \xleftarrow[{[m,m]}]{} & (M \otimes^{\mathrm{s}} M) + (M \otimes^{\mathrm{s}} M) \end{array}$$

commutes, where the unnamed isomorphism is $\upsilon_{M,M,M}$. This condition corresponds to the commutativity condition in the multicategories with nonstandard amalgamations in [12].

7. Symmetric signatures vs analytic functors

7.1. The symmetric signature fibration $p_{\mathrm{s}}\colon \mathrm{Sig}_{\mathrm{s}} \to \mathrm{Set}$.

The category of symmetric sets. The category of symmetric sets is equivalent to the category of species, cf. [14], however the presentation is slightly different.

A symmetric set (A,α) is a graded set $\{A_n : n\in\omega\}$ with (right) actions of symmetric groups $\alpha_n\colon A_n\times \mathrm{S}_n \to A_n$, for $n\in\omega$. We write $a\in A$ to mean that $a\in \coprod_n A_n$ and if $a\in A$ then we write $|a|=n$ to mean that $a\in A_n$. Thus, for $a\in A$, we have $a\in A_{|a|}$. In case $\alpha(a,\sigma)$ is defined we usually write it as $a\cdot\sigma$ if it does not lead to a confusion. A morphism of symmetric sets $f\colon (A,\alpha)\to(B,\beta)$ is a family of morphisms of actions $f_n\colon (A_n,\alpha_n)\to(B_n,\beta_n)$ for $n\in\omega$, i.e., it is a function $f\colon A\to B$ commuting with the actions α and β, in short. We call such morphisms *equivariant*. The category of symmetric sets will be denoted by σSet. σSet is (equivalent to) of the presheaf category $\mathrm{Set}^{S_*^{\mathrm{op}}}$, where S_* the coproduct of (finite) symmetric groups in Cat. Clearly the groups S_* act on $O^\dagger$ on the right by composition, leaving 0 fixed. This symmetric set on $O^\dagger$ will be denoted by $O^\ddagger$. Any function $u\colon O\to Q$ induces an equivariant map $u^\ddagger\colon O^\ddagger\to Q^\ddagger$, so that $u^\ddagger(d) = u\circ d$. Thus we have a functor

$$(-)^\ddagger\colon \mathrm{Set}\to\sigma\mathrm{Set}.$$

The operad of symmetries S. Recall[13] that the universes of symmetric groups $\langle \mathrm{S}_n\rangle_{n\in\omega}$ form the underlying sets of an operad, called the operad of symmetries **S**. The compositions

$$\begin{aligned} *\colon S_k\times(S_{n_1}\times\cdots\times S_{n_k}) &\to S_{\sum_{i=1}^k n_i}\\ (\tau,\sigma_1,\dots,\sigma_k) &\mapsto \tau*(\sigma_1,\dots,\sigma_k)\end{aligned}$$

where, for $\tau\in S_k$, $\sigma_1\in S_{n_1},\dots,\sigma_k\in S_{n_k}$, $1\le m_0\le k$, $1\le m_1\le n_{k_{m_0}}$ are given by

$$\tau*(\sigma_1,\dots,\sigma_k)(k_1+\ldots+k_{m_0-1}+m_1) = k_{\tau^{-1}(1)}+\ldots+k_{\tau^{-1}(\tau(m_0)-1)}+\sigma_{m_0}(m_1).$$

The fibration p_{s}. Taking the pullback of the basic fibration on the category of symmetric sets $\mathrm{cod}\colon \sigma\mathrm{Set}^{\to}\to\sigma\mathrm{Set}$ along the functor $(-)^\ddagger$

$$\begin{array}{ccc} \mathrm{Sig}_{\mathrm{s}} & \longrightarrow & \sigma\mathrm{Set}^{\to}\\ {\scriptstyle p_{\mathrm{s}}}\downarrow & & \downarrow{\scriptstyle \mathrm{cod}}\\ \mathrm{Set} & \xrightarrow[(-)^\ddagger]{} & \sigma\mathrm{Set}\end{array}$$

we obtain the symmetric signature fibration p_{s}. We describe the category $\mathrm{Sig}_{\mathrm{s}}$ explicitly. An object of $\mathrm{Sig}_{\mathrm{s}}$ over the set O is a quadruple (A,α,∂^A,O) such that (A,α) is a symmetric set, $\partial^A\colon (A,\alpha)\to O^\ddagger$ is an equivariant map called the *typing* (or profile in [3]) map of the signature. We write $\partial^A_a\colon [n]\to O$ for the effect of ∂^A on $a\in A$, and n in this case can be referred to as $|a|$. The fact that ∂^A is equivariant means that we have $\partial^A_{a\cdot\sigma} = \partial^A_a\circ\sigma$, for $a\in A$ and $\sigma\in S_{|a|}$. A

[13] For example from [19, pp. 51–54].

morphism $(f, u)\colon (A, \alpha, \partial^A, O) \to (B, \beta, \partial^B, Q)$ in $\mathrm{Sig}_{\mathrm{s}}$ over a function $u\colon O \to Q$ is a commuting square of equivariant maps:

$$\begin{array}{ccc} (A,\alpha) & \xrightarrow{\ f\ } & (B,\beta) \\ {\scriptstyle \partial^A}\downarrow & & \downarrow{\scriptstyle \partial^B} \\ O^{\ddagger} & \xrightarrow[u^{\ddagger}]{} & Q^{\ddagger}. \end{array}$$

The monoidal structure in the fibres of p_{s}. We define two lax morphisms of fibrations

$$\begin{array}{ccc} \mathrm{Sig}_{\mathrm{s}} \times_{\mathrm{Set}} \mathrm{Sig}_{\mathrm{s}} & \xrightarrow{\ \otimes\ } & \mathrm{Sig}_{\mathrm{s}} \\ & {\scriptstyle p'_{\mathrm{s}}}\searrow \quad \swarrow{\scriptstyle p_{\mathrm{s}}} & \\ & \mathrm{Set} & \end{array} \qquad\qquad \begin{array}{ccc} \mathrm{Set} & \xrightarrow{\ I\ } & \mathrm{Sig}_{\mathrm{s}} \\ & {\scriptstyle \mathrm{id}}\searrow \quad \swarrow{\scriptstyle p_{\mathrm{s}}} & \\ & \mathrm{Set}. & \end{array}$$

Let $(A, \alpha, \partial^A, O)$, $(B, \beta, \partial^B, O)$ be two objects in the fibre over O in the fibration p_{s}. The *tensor product*

$$(A, \alpha, \partial^A, O) \otimes_O (B, \beta, \partial^B, O) = (A \otimes_O B, \alpha \otimes_O \beta, \partial^{\otimes}, O)$$

is defined as follows

$$(A \otimes_O B)_n = \left\{ \langle a, \langle b_i \rangle_{i \in (|a|]}, \sigma \rangle \colon \sum_i |b_i| = n, b_i \in B, a \in A, \right. \\ \left. \partial_{b_i}(0) = \partial_a(i), \text{ for } i \in (|a|],\ \sigma \in \mathrm{S}_n \right\}_{/\sim}$$

where the equivalence relation $\sim$ is defined as follows:

$$\langle a \cdot \tau; \langle b_{\tau(i)} \cdot \sigma_{\tau(i)} \rangle_i; \sigma \rangle \sim \langle a, \langle b_i \rangle_i, \tau * (\sigma_{\tau(1)}, \ldots, \sigma_{\tau(|a|)}) \circ \sigma \rangle$$

where $\tau \in S_{|a|}$, $\sigma_i \in S_{|b_i|}$, $\sigma \in S_{\sum_i |b_i|}$, $*$ is the composition in the operad of symmetries, and $\circ$ is the usual composition of permutations. The equivalence class of the element $\langle a, \langle b_i \rangle_{i \in (|a|]}, \sigma \rangle$ will be denoted by $[\langle a, \langle b_i \rangle_{i \in (|a|]}, \sigma \rangle]_{\sim}$.

Let $\kappa_i\colon (|b_i|] \to (\sum_i |b_i|]$ be the ith inclusion into the coproduct, for $i = 1, \ldots, |a|$. Clearly, there are many such inclusions that make $(\sum_i |b_i|]$ into a coproduct of $(|b_i|]$'s (in Set) but we will always mean the simplest, that is embedding blocks $(|b_i|]$ one after the other into $(\sum_i |b_i|]$ (i.e., $\kappa_i(j) = j + \sum_{k=1}^{i-1} |b_k|$ for $j \in (|b_i|]$).

We define

$$\partial^{\otimes}([\langle a, \langle b_i \rangle_{i \in (|a|]}, \sigma \rangle]_{\sim}) \colon \left[\sum_i |b_i| \right] \to O$$

so that

$$\partial^{\otimes}([\langle a, \langle b_i \rangle_{i \in (|a|]}, \sigma \rangle]_{\sim})(0) = \partial_a(0)$$

and the squares

$$\begin{array}{ccc} (\sum_i |b_i|] & \xrightarrow{\ \sigma^{-1}\ } & (\sum_i |b_i|] \\ {\scriptstyle \kappa_i}\uparrow & & \downarrow{\scriptstyle \partial^{\otimes}_{[\langle a, \langle b_i \rangle_i, \sigma \rangle]_{\sim}}} \\ (|b_i|] & \xrightarrow[\partial_{b_i}]{} & O \end{array}$$

commute, for all $i \in (|a|]$. The type of the codomain of the 'operation' $[\langle a, b_i, \sigma \rangle_{i \in (|a|]}]_{\sim}$ in $A \otimes_O B$ is the same as the type of the codomain of a in A and the types of the domain of the 'operation' $[\langle a, \langle b_i \rangle_{i\in(|a|]}, \sigma\rangle]_\sim$ in $A \otimes_O B$ are the types of the domains of b_i's in B put one next to the other and permuted by σ.

The action of S_n on $(A \otimes_O B)_n$ is defined so that

$$[\langle a, \langle b_i \rangle_{i\in(|a|]}, \sigma\rangle]_\sim \cdot \sigma' = [\langle a, \langle b_i \rangle_{i\in(|a|]}, \sigma \circ \sigma'\rangle]_\sim$$

for $\sigma' \in S_{\sum_i |b_i|}$.

For two morphisms

$$(f,u)\colon (A,\alpha,\partial,O) \to (B,\beta,\partial,Q), \quad (f',u)\colon (A',\alpha',\partial,O) \to (B',\beta',\partial,Q)$$

over u, we define their tensor to be

$$(f \otimes_u f', u)\colon (A \otimes_O A', \alpha \otimes_O \alpha', \partial^\otimes, O) \to (B \otimes_Q B', \beta \otimes_Q \beta', \partial^\otimes, Q)$$

in the following way. For $\langle a, \langle a'_i \rangle_{i\in(|a|]}, \sigma\rangle \in A \otimes_O A'$, we put

$$f \otimes_u f'([\langle a, \langle a'_i \rangle_{i\in(|a|]}, \sigma\rangle]_\sim) = [\langle f(a), \langle f'(a'_i)\rangle_{i\in(|a|]}, \sigma\rangle]_\sim.$$

Note that $|a| = |f(a)|$ and we have

$$\partial^B_{f(a)}(i) = u \circ \partial^A_a(i) = u \circ \partial^{A'}_{a'_i}(0) = \partial^{B'}_{f'(a'_i)}(0)$$

so $[\langle f(a), \langle f'(a'_i)\rangle_{i\in(|f(a)|]}, \sigma\rangle]_\sim$ belongs to $B \otimes_Q B'$ indeed. This ends the definition of the tensor product functor $\otimes$ in p_{s}.

The unit $I_O = (O, 1, \partial, O)$ for the tensor $\otimes_O$ in the fibre $(\mathrm{Sig_s})_O$ is defined as follows. For $x \in O$, $\partial_x\colon [1] \to O$ is a function such that $\partial_x(0) = \partial_x(1) = x$. So only the group S_1 acts on O and it acts trivially. The association $O \mapsto I_O$ is clearly the object part of a lax morphism of fibrations, as it should be.

Lemma 7.1. *The functors $\otimes$ and I together with obvious associativity, left unit, and right unit isomorphisms α, λ, ρ make the fibres of p_{s} into (strong) monoidal categories.*

Pulling back the monoidal structure in p_{s}. Any object $\langle B, \beta, \partial, Q\rangle$ in the fibre $(\mathrm{Sig_s})_Q$ can be pulled back along a function $u\colon O \to Q$

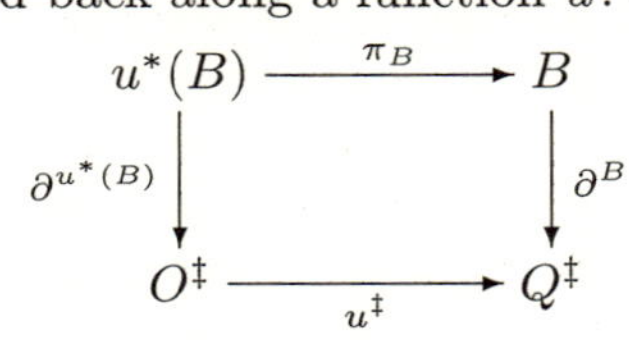

where $u^\ddagger(d) = u \circ d$. We have

$$u^*(B) = \{\langle b, d\rangle : b \in B,\ d\colon [|b|] \to O, \text{ such that } u \circ d = \partial_b\}$$

and

$$\partial^{u^*(B)}_{\langle b,d\rangle} = d.$$

The action in $u^*(B)$ applies the permutation to both arguments, i.e.

$$\langle b, d\rangle \cdot \sigma = \langle b \cdot \sigma, d \circ \sigma\rangle.$$

Let, for $x \in O$, $d_x[1] \to O$ be the function such that $d_x(0) = d_x(1) = x$. We have

$$u^*(I_Q) = \{\langle x, d_x\rangle : x \in O\}$$

and

$$\begin{aligned} \phi_0\colon I_O &\to f^*(I_{O'}) \\ x &\mapsto \langle u(x), d_x\rangle. \end{aligned}$$

Moreover, for $\langle A, \alpha\rangle$ and $\langle B, \beta\rangle$ in $(\mathrm{Sig_s})_Q$, we have

$$u^*(A \otimes B)$$
$$= \left\{ \left\langle \langle a, \langle b_i\rangle_{i\in(|a|]}, \sigma\rangle, d_a^+ + \coprod_{i\in(|a|]} d_{b_i}^- \right\rangle : u^\ddagger\left(d_a^+ + \coprod_{i\in(|a|]} d_{b_i}^-\right) = \partial_a^+ + \coprod_{i\in(|a|]} \partial_{b_i}^- \right\}$$

$(d_a^+ {:} [0]{:} \to O,\ d_{b_i}^- {:} (|b_i|]{:} \to O$, for $i \in (|a|])$ and

$$u^*(A) \otimes u^*(B) = \left\{ \left\langle \langle a, d\rangle, \left\langle \langle b_i, d_i\rangle \right\rangle_{i\in(|a|]}, \sigma \right\rangle : u^\ddagger(d) = \partial_a,\ u^\ddagger(d_i) = \partial_{a_i}, \text{ for } i \in (|a|] \right\}$$

$(d{:}[|a|]{:} \to O,\ d_i{:}[|b_i|]{:} \to O$, for $i \in (|a|])$. Thus we have a transformation

$$\phi_{2,A,B} \colon u^*(A) \otimes u^*(B) \to u^*(A \otimes B)$$

such that

$$\left\langle \langle a, d\rangle, \left\langle \langle b_i, d_i\rangle \right\rangle_{i\in(|a|]}, \sigma \right\rangle \mapsto \left\langle \langle a, \langle b_i\rangle_{i\in(|a|]}, \sigma\rangle, d_a^+ + \coprod_{i\in(|a|]} d_{b_i}^- \right\rangle.$$

Lemma 7.2. *The map* $(\pi_B, u) \colon u^*(B) \to B$ *is a prone arrow over* u. *The data* u^*, ϕ_0, ϕ_2 *above make the usual (three) diagrams of a (lax) monoidal functor commute, i.e.,* p_s *equipped with* $\otimes, I, \alpha, \lambda, \rho$ *is a lax monoidal fibration.*

Moreover, we have

Proposition 7.3. *The total category of the fibration of monoids* $q_s \colon \mathrm{Mon}(\mathrm{Sig_s}) \to \mathrm{Set}$ *is equivalent to the category of symmetric multicategories. The fibred forgetful functor from the fibration of monoids to the fibration of symmetric signatures* $\mathcal{U} \colon \mathrm{Mon}(\mathrm{Sig_s}) \to \mathrm{Sig_s}$

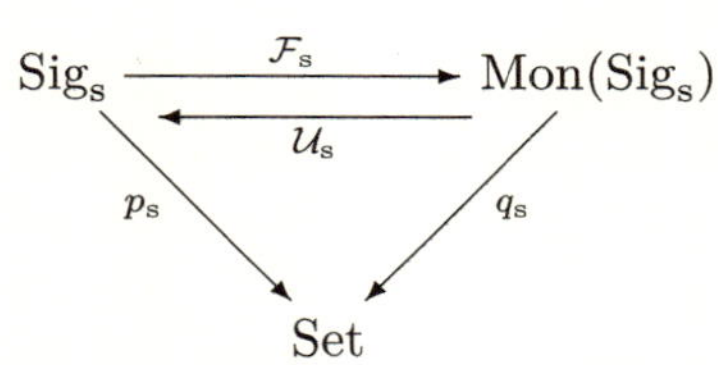

is a morphism of fibrations and has a left adjoint $\mathcal{F}_s$, *the free monoid functor, which is a lax morphism of fibrations.*

Remark. The p_s is a bifibration as, for an object $(A, \alpha, \partial^A, O)$ and a function $u \colon O \to Q$ the morphism

$$(1_A, u) \colon (A, \alpha, \partial^A, O) \to (A, \alpha, u^\ddagger \circ \partial^A, Q)$$

is a supine morphism.

7.2. The action of p_s on the basic fibration and analytic functors. The fibration p_s acts on the basic fibration $\mathrm{cod}\colon \mathrm{Set}^{\to} \to \mathrm{Set}$ as follows

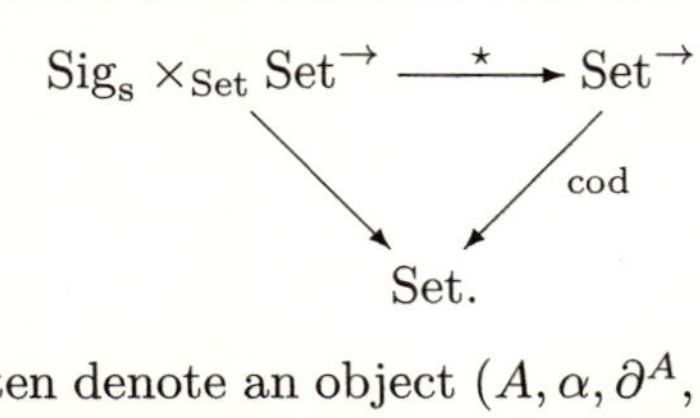

In the following, we often denote an object $(A, \alpha, \partial^A, O)$ in $\mathrm{Sig_s}$ as A and an object $d^X \colon X \to O$ in $\mathrm{Set}^{\to}$ as X, when it does not lead to confusion.

The object $d^\star \colon A \star X \to O$ is defined as the quotient of the set

$$\{(a, \vec{x}) : a \in A, \vec{x} \colon (|a|] \to X, \partial^{A,+}_a = d^X \circ \vec{x}\}$$

by an equivalence $\sim$ so that

$$(a, \vec{x}) \sim (a \cdot \sigma, \vec{x} \circ \sigma)$$

for $a \in A$, $\vec{x} \colon (|a|] \to X$, and $\sigma \in S_{|a|}$. The function $d^\star \colon A \star X \to O$ is defined as

$$d^\star([a, \vec{x}]_\sim) = \partial_a(0).$$

The action $\star$ is defined on morphisms as follows. For maps

$$(f, u) \colon (A, \alpha, \partial^A, O) \to (B, \beta, \partial^B, Q)$$

in $\mathrm{Sig_s}$,

$$(g, u) \colon (X, d^X) \to (Y, d^Y)$$

in $\mathrm{Set}^\to$ over $u : O \to Q$ and for an element $[a, \vec{x}]_\sim \in A \star X$ we put

$$f \star_u g([a, \vec{x}]_\sim) = [f(a), g \circ \vec{x}]_\sim$$

so that the square

$$\begin{array}{ccc} A \star X & \xrightarrow{f \star_u g} & B \star Y \\ {\scriptstyle d^\star}\downarrow & & \downarrow{\scriptstyle d^\star} \\ O & \xrightarrow[u]{} & Q \end{array}$$

commutes.

We have an adjoint morphism of lax monoidal fibrations

$$\begin{array}{ccc} \mathrm{Sig_s} & \xrightarrow{\mathrm{rep_s}} & \mathrm{Exp(Set)} \\ & {\scriptstyle p_s}\searrow \quad \swarrow & \\ & \mathrm{Set}. & \end{array}$$

For an object A in $(\mathrm{Sig_s})_O$, we have a functor

$$\mathrm{rep_s}(A) = A \star (-) \colon \mathrm{Set}_{/O} \to \mathrm{Set}_{/O}$$

and for a morphism $(f, u) \colon A \to B$ in $\mathrm{Sig_s}$ over $u \colon O \to Q$ we have a natural transformation

$$\mathrm{rep_s}(f, u) \colon A \star u^*(-) \to u^*\big(B \star (-)\big)$$

so that for a morphism $g : Y \to Y'$ in $(\mathrm{Set}^\to)_Q$ we have a commuting square

$$\begin{array}{ccc} A \star u^*(Y) & \xrightarrow{\mathrm{rep_s}(f,u)_Y} & u^*(B \star Y) \\ {\scriptstyle v1_A \star u^*(g)}\downarrow & & \downarrow{\scriptstyle u^*(1_B \star g)} \\ A \star u^*(Y') & \xrightarrow[\mathrm{rep_s}(f,u)_{Y'}]{} & u^*(B \star Y'). \end{array}$$

We note for the record, that for $[a, \vec{x}]_\sim \in A \star u^*(Y)$ so that $a \in A$ and $\vec{x} \colon (|a|] \to u^*(Y)$ we have

$$\mathrm{rep_s}(f, u)_Y([a, \vec{x}]_\sim) = \langle \partial^A_a(0), [f(a), u^Y \circ \vec{x}]_\sim \rangle$$

where

$$\begin{array}{ccc} u^*(Y) & \xrightarrow{u^Y} & Y \\ {\scriptstyle u^*(d^Y)}\downarrow & & \downarrow{\scriptstyle d^Y} \\ O & \xrightarrow[u]{} & Q \end{array}$$

is a pullback in Set. We have

Proposition 7.4. *The morphism of lax monoidal fibrations* rep_s *defined above is a morphism of bifibrations and it preserves coproducts in the fibres.*

PROOF. The proof of this proposition can be made more abstract but we prefer it to be concrete. Preservation of coproduct is trivial.

First, we shall show that rep_s is a morphism of fibrations. Let $B = (B, \beta, \partial^B, Q)$ be a symmetric signature. The prone morphism over $u\colon O \to Q$ in the fibration p_s is defined via pullback in the category of symmetric sets

$$\begin{array}{ccc} u^*(B,\beta) & \xrightarrow{u^B} & (B,\beta) \\ {\scriptstyle \partial^{u^*(B)}}\downarrow & & \downarrow{\scriptstyle \partial^B} \\ O^{\ddagger} & \xrightarrow[u^{\ddagger}]{} & Q^{\ddagger}. \end{array}$$

We write $u^*(B)$ for $(u^*(B,\beta), \partial^{u^*(B)}, O)$. Then the prone morphism is $\mathrm{pr}_{u,B} = (u^B, u)\colon u^*(B) \to B$. We will use the usual representation of this pullback in Set, i.e.,

$$u^*(B) = \{\langle b, d\rangle : b \in B, d\colon [|b|] \to O, u \circ d = \partial^B_b\}$$

and

$$\partial^{u^*(B)}(\langle b, d\rangle) = d, \quad u^B(\langle b, d\rangle) = b$$

for $\langle b, d\rangle \in u^*(B)$.

We will show that the following natural transformations

$$u^*(B) \star u^*(-) \xrightarrow{\mathrm{rep}_s(\mathrm{pr}_{u,B}) = \mathrm{rep}_s(u^B,u)} u^*\big(B \star (-)\big)$$

and

$$u^*\big(B \star u_! u^*(-)\big) \xrightarrow{\mathrm{pr}_{u,\mathrm{rep}_s(B)} = u^*\left(B \star \varepsilon^u_{(-)}\right)} u^*\big(B \star (-)\big)$$

in $\mathrm{Cat}(\mathrm{Set}_{/Q}, \mathrm{Set}_{/O})$ are isomorphic as objects of $\mathrm{Cat}(\mathrm{Set}_{/Q}, \mathrm{Set}_{/O})_{/u^*(B\star(-))}$. To this end we shall define a natural isomorphism

$$u^*(B) \star (-) \xrightarrow{\xi} u^*\big(B \star u_! u^*(-)\big)$$

(and its inverse) so that $\mathrm{rep}_s(u^B, u) = u^*(B \star \varepsilon^u_{(-)}) \circ \xi$, i.e., for any $d^Y : Y \to Q$ the triangle

(1)
$$\begin{array}{ccc} u^*(B) \star (d^Y) & \xrightarrow{\mathrm{rep}_s(u^B,u)_{d^Y}} & \\ {\scriptstyle \xi_{d^Y}}\downarrow & & u^*(B \star d^Y) \\ u^*\big(B \star u_! u^*(d^Y)\big) & \xrightarrow{u^*(B\star\varepsilon^u_{d^Y})} & \end{array}$$

commutes. We fix $d^Y : Y \to Q \in \mathrm{Set}_{/Q}$. The following diagram

$$
(2)\qquad
\begin{array}{ccccc}
(|b|] & \xrightarrow{\ \vec{x}\ } & u^*(Y) & \xrightarrow{\ u^Y\ } & Y \\
\Big\downarrow & & \Big\downarrow{\scriptstyle u^*(d^Y)} & & \Big\downarrow{\scriptstyle d^Y} \\
[|b|] & \xrightarrow[\ d\ ]{} & O & \xrightarrow[\ u\ ]{} & Q
\end{array}
\qquad \text{with } \vec{y}\colon (|b|] \to Y,\ \ \partial_b^B\colon [|b|] \to Q
$$

where the right hand square is a pullback is to fix the notation. We do not assume now that other morphisms exist, but if they do they have domains and codomains as displayed. Similarly, other figures in this diagram are not assumed to commute unless we explicitly say so. We will refer often to this diagram in the rest of the proof.

We note that, for $b \in B$,

$$
\begin{array}{rll}
\langle o, [b, \vec{y}]_\sim \rangle \in u^*(B \star d^Y) & \text{iff} & o \in O, u(o) = \partial_b^B(0), \partial_b^{B,+} = d^Y \circ \vec{y} \\
\langle o, [b, \vec{x}]_\sim \rangle \in u^*(B \star u_! u^*(d^Y)) & \text{iff} & o \in O, u(o) = \partial_b^B(0), \partial_b^{B,+} = u \circ u^*(d^Y) \circ \vec{x} \\
[\langle b, d \rangle, \vec{x}]_\sim \in u^*(B) \star u^*(d^Y) & \text{iff} & u \circ \partial_b^B = d, d^+ = u^*(d^Y) \circ \vec{x}.
\end{array}
$$

With the above notation we spell out the three functions occurring in (1):

$$
\begin{array}{rcl}
u^*(B) \star u^*(d^Y) & \xrightarrow{\mathrm{rep_s}(u^B, u)_{d^Y}} & u^*(B \star d^Y) \\
[\langle b, d \rangle, \vec{x}]_\sim & \longmapsto & \langle d(0), [b, u^Y \circ \vec{x}]_\sim \rangle \\
u^*(B \star u_! u^*(d^Y)) & \xrightarrow{u^*(B \star \varepsilon^u_{d^Y})} & u^*(B \star d^Y) \\
\langle o, [b, \vec{x}]_\sim \rangle & \longmapsto & \langle o, [b, u^Y \circ \vec{x}]_\sim \rangle
\end{array}
$$

and

$$
\begin{array}{rcl}
u^*(B) \star u^*(d^Y) & \xrightarrow{\ \xi_{d^Y}\ } & u^*(B \star u_! u^*(d^Y)) \\
[\langle b, d \rangle, \vec{x}]_\sim & \longmapsto & \langle d(0), [b, \vec{x}]_\sim \rangle \\
[\langle b, \bar{d} \rangle, \vec{x}]_\sim & \longleftarrow\!\shortmid & \langle o, [b, \vec{x}]_\sim \rangle
\end{array}
$$

where $\bar{d}\colon [|b|] \to O$ is so defined that $\bar{d}(0) = o$ and $\bar{d}^+ = u^*(d^Y) \circ \vec{x}$. A simple check shows that (1) commutes, i.e., $\mathrm{rep_s}$ preserves prone morphisms.

Now we shall show that $\mathrm{rep_s}$ preserves supine morphisms, i.e., it is a morphism of opfibrations. Let $(A, \alpha, \partial^A, O)$ be a symmetric signature. The supine morphism $\mathrm{su}_{u,A}$ in p_s over $u\colon O \to Q$ with domain A is defined from the square

$$
\begin{array}{ccc}
(A, \alpha) & \xrightarrow{\ 1_A\ } & (A, \alpha) \\
{\scriptstyle \partial^A}\Big\downarrow & & \Big\downarrow{\scriptstyle u^\ddagger \circ \partial^A} \\
O^\ddagger & \xrightarrow[\ u^\ddagger\ ]{} & Q^\ddagger
\end{array}
$$

we write $A_!$ for $(A, \alpha, u^\ddagger \circ \partial^A, Q)$ and $\mathrm{su}_{u,A} = (1_A, u)\colon A \to A_!$.

We shall show that the natural transformations

$$
A \star u^*(-) \xrightarrow{\mathrm{rep_s}(su_{u,A}) = \mathrm{rep_s}(1_A, u)} u^*(A_! \star (-))
$$

and

$$A \star u^*(-) \xrightarrow{\mathrm{su}_{u,\mathrm{rep_s}(A)} = \eta^u_{A\star u^*(-)}} u^* u_! A \star u^*(-)$$

are isomorphic in ${}_{A\star u^*(-)}\backslash \mathrm{Cat}(\mathrm{Set}_{/Q}, \mathrm{Set}_{/O})$.

We shall define a natural isomorphism

$$u^* u_!\big(A \star u^*(-)\big) \xrightarrow{\zeta} u^*\big(A \star (-)\big)$$

(and its inverse) so that $\mathrm{rep_s}(1_A, u) = \zeta \circ \eta^u_{A\star u^*(-)}$, i.e., for any $d^Y : Y \to Q$ the triangle

$$(3) \qquad \begin{array}{ccc} & & u^*(A_! \star d^Y) \\ & \nearrow^{\mathrm{rep_s}(1_A,u)_{d^Y}} & \uparrow \zeta_{d^Y} \\ A \star u^*(d^Y) & \searrow_{\eta^u_{A\star u^*(d^Y)}} & \\ & & u^* u_!\big(A \star u^*(d^Y)\big) \end{array}$$

commutes. Using the notation from diagram (1) we note that, for $a \in A$, we have

$$\begin{aligned} [a, \vec{x}]_\sim \in A \star u^*(d^Y) \quad &\text{iff} \quad \partial^{A,+}_a = u(d^Y) \circ \vec{x} \\ \langle o, [a, \vec{y}]_\sim \rangle \in u^*(A_! \star d^Y) \quad &\text{iff} \quad o \in O, o = \partial^A_a(0), u \circ \partial^{A,+}_a = d^Y \circ \vec{y} \\ \langle o, [a, \vec{x}]_\sim \rangle \in u^* u_!\big(A \star u^*(d^Y)\big) \quad &\text{iff} \quad o \in O, u(o) = u \circ \partial^A_a(0), u \circ \partial^{A,+}_a = d^Y \circ \vec{x}. \end{aligned}$$

Now we spell out explicitly the function occurring in (3).

$$\begin{array}{ccc} A \star u^*(d^Y)) & \xrightarrow{\mathrm{rep_s}(1_A,u)_{d^Y}} & u^*(A_! \star d^Y) \\ [a, \vec{x}] & \longmapsto & \langle \partial^A_a(0), [a, u^Y \circ \vec{x}]_\sim \rangle \\ A \star u^*(d^Y)) & \xrightarrow{\eta^u_{A\star u^*(d^Y)}} & u^* u_!(A \star u^*(d^Y)) \\ [a, \vec{x}] & \longmapsto & \langle \partial^A_a(0), [a, \vec{x}]_\sim \rangle \end{array}$$

and

$$\begin{array}{ccc} u^* u_!\big(A \star u^*(d^Y)\big) & \xrightarrow{\zeta_{d^Y}} & u^*(A_! \star d^Y) \\ \langle o, [a, \vec{x}]_\sim \rangle & \overset{\zeta_{d^Y}}{\longmapsto} & \langle o, [a, u^Y \circ \vec{x}]_\sim \rangle \\ [\langle o, [a, \vec{x}]_\sim \rangle & \longleftarrow\!\shortmid & \langle o, [a, \vec{y}]_\sim \rangle. \end{array}$$

In the last correspondence, $\vec{y} \mapsto \vec{x} \colon (|a|] \to u^*(Y)$ is defined using the fact that right square in (2) is a pullback and $d^Y \circ \vec{y} = u \circ \partial^{A,+}_a$.

Again, a simple check shows that (3) commutes, i.e., $\mathrm{rep_s}$ is a morphism of opfibrations, as well. □

Later we will show that $\mathrm{rep_s}$ is faithful and full on isomorphisms.

The fibration that is the essential image of the representation $\mathrm{rep_s}$ will be denoted by $p_{\mathrm{an}} \colon \mathcal{An} \to \mathrm{Set}$, we take it as the definition of the fibration of (multi-variable) analytic (endo)functors and analytic transformations between them. Thus by an *analytic functor* on $\mathrm{Set}_{/O}$, where O is a set, we understand a functor (isomorphic to one) of the form $A\star(-) \colon \mathrm{Set}_{/O} \to \mathrm{Set}_{/O}$ for a symmetric signature $A = (A, \alpha, \partial^A, O)$. Moreover, by an *analytic transformation* over a function $u \colon O \to Q$ between two analytic functors $A\star(-) \colon \mathrm{Set}_{/O} \to \mathrm{Set}_{/O}$ and $B\star(-) \colon \mathrm{Set}_{/Q} \to \mathrm{Set}_{/Q}$

we mean a natural transformation of the form $\mathrm{rep}_s(f,u)\colon A \star u^*(-) \to u^*(B \star (-))$ for a morphism symmetric signatures $(f,u)\colon (A,\alpha,\partial^A,O) \to (B,\beta,\partial^B,Q)$.

Remarks. (1) Note that for the one element set, say $[0]$, we have $\mathrm{Set}_{/[0]} \cong \mathrm{Set}$ and the fibre of p_{an} over $[0]$ is (isomorphic to) the category of usual (one-variable) analytic functors that was characterized in [15], see also [2], as the category of finitary endofunctors on Set that weakly preserve pullbacks and weakly cartesian natural transformations between them. We will see in the next section how this characterization extends from this fibre to the whole fibration.

(2) In [15] multivariable analytic functors were defined as certain functors $\mathrm{Set}^I \to \mathrm{Set}$, were I is a finite set. Ignoring the 'size problems,' this definition can be extended to infinite sets Q by saying that the class(!) of multivariable analytic functors $\mathrm{Set}_{/Q} \to \mathrm{Set}$ is a cofiltered 'limit' of the classes(!) of analytic functors $\mathrm{Set}^I \to \mathrm{Set}$ where $I \subseteq Q$ and I is finite. In these terms, what we call an analytic functor on $\mathrm{Set}_{/Q}$, is just a Q-tuple of multivariable analytic functors $\mathrm{Set}_{/Q} \to \mathrm{Set}$.

(3) The multivariable analytic functors $\mathrm{Set}_{/Q} \to \mathrm{Set}$ can be also described more explicitly avoiding all the 'size problems.' Let Q be a set and $q \in Q$. We have the evaluation functor and the inclusion of a fibre

$$\mathrm{Set}_{/Q} \xrightarrow{\mathrm{ev}_q} \mathrm{Set} \xrightarrow{\mathbf{i}_q} \mathrm{Set}_{/Q}$$

such that $\mathrm{ev}_q(X,d) = d^{-1}(z)$ and $\mathbf{i}_q(B)\colon B \to Q$ is the function defined by $\mathbf{i}_q(B)(b) = z$ for $b \in B$. A *multivariable analytic functor* from $\mathrm{Set}_{/Q}$ to Set is a functor of the form

$$\mathrm{Set}_{/Q} \xrightarrow{B\star(-)} \mathrm{Set}_{/Q} \xrightarrow{\mathrm{ev}_q} \mathrm{Set}$$

where (B,β,∂^B,Q) is a symmetric signature, $z \in Q$ and ev_z is the evaluation functor such that $\mathrm{ev}_z(d\colon X \to Q) = d^{-1}(z)$. If we restrict symmetric signatures to those for which $\partial^B_b(0) = z$ then such functors, as we shall see, determine the signatures up to an isomorphism.

7.3. A characterization of the fibration of analytic functors. The following extends the characterization of analytic functors and analytic transformations, cf. [15], from the fibre over $[0]$ to the whole fibration of analytic functors p_{an}.

Theorem 7.5. *The lax monoidal fibration $p_{\mathrm{an}}\colon \mathcal{A}n \to \mathrm{Set}$ has as its objects finitary endofunctors on categories $\mathrm{Set}_{/Q}$ that weakly preserve wide pullbacks and weakly cartesian natural transformations as morphisms between them.*

Before we prove a series of lemmas needed to establish the above theorem, we shall immediately present the following obvious corollary that is even more interesting.

Corollary 7.6. *The fibration of symmetric multicategories is equivalent to the fibration of weakly cartesian analytic monads and weakly cartesian morphisms of monads whose functor parts are pullback functors between them. Under this correspondence the free symmetric multicategories correspond to the free analytic monads.*

The following definition is an extension of a notion from [2]. Let Q be a set. The functor $F\colon \mathrm{Set}_{/Q} \to \mathrm{Set}$ is *superfinitary* if and only if there is an object $d\colon I \to Q$

in $\mathrm{Set}_{/Q}$ with I finite such that, for any $d^X : X \to Q$ in $\mathrm{Set}_{/Q}$

$$F(X, d^X) = \bigcup_{f:(I,d)\to(X,d^X)} F(f)\big(F(I,d)\big),$$

i.e., the elements of $F(I,d)$ generates the whole functor F. The following two Lemmas and their proofs are 'colored' versions of Theorem 2.6 and Corollary 2.7 and their proofs from [2].

Lemma 7.7. *Let $F\colon \mathrm{Set}_{/Q} \to \mathrm{Set}$ be a superfinitary functor. Then F is a multivariable analytic functor if and only if F weakly preserves pullbacks.*

PROOF. By observations of V. Trnková, cf. [26], any functor $F\colon \mathrm{Set}_{/Q} \to \mathrm{Set}$ is a coproduct (indexed by $F(1_Q)$) of functors that preserves the terminal object. If $F = \coprod_i F_i$, then F weakly preserves pullbacks (is analytic) if and only if all F_i's do (are). Thus it is enough to prove the lemma for a superfinitary functor $F\colon \mathrm{Set}_{/Q} \to \mathrm{Set}$ such that $F(1_Q) = 1$.

So suppose that F weakly preserves pullbacks. Fix a minimal object $d^F : (n] \to Q$ such that for any $d^X : X \to Q$

$$F(X, d^X) = \bigcup_{f:((n],d^F)\to(X,d^X)} F(f)\left(F\big((n], d^F\big)\right).$$

By 'minimal' we mean that there is no proper subobject of $\big((n], d^F\big)$ in $\mathrm{Set}_{/Q}$ with this property. Thus there is an element $g^F \in F(d^F)$ that it is not in the image of any proper inclusion into d^F. The pair (g^F, d^F) or just the element g^F if d^F is understood, will be called *generic*, cf. [2, 15]. Thus, if we have a morphism

$$\begin{array}{ccc} X & \xrightarrow{\quad f \quad} & (n] \\ & {\scriptstyle d^X}\searrow \quad \swarrow{\scriptstyle d^F} & \\ & Q & \end{array}$$

in $\mathrm{Set}_{/Q}$ such that $g^F \in F(f)(X, d^X)$ then f is onto. Therefore, any endomorphism of $\big((n], d^F\big)$ in $\mathrm{Set}_{/Q}$ leaving g^F fixed, is a bijection. We can define a subgroup of S_n as follows

$$\mathcal{G}^F = \{\sigma\colon d^F \to d^F \in \mathrm{Set}_{/Q} : F(\sigma)(g^F) = g^F\} \subseteq \mathrm{S}_n.$$

Let F^o be $\mathrm{S}_{n/\mathcal{G}^F}$, the set of left cosets of S_n over $\mathcal{G}^F$. The class of τ in $\mathrm{S}_{n/\mathcal{G}^F}$ will be denoted by $[\tau]_{\sim F}$. We have a right action, say ϕ, of S_n on F^o acting by composition on the right. We define $\partial^{F^o} : F^o \to Q^\ddagger$ so that, for $\tau \in \mathrm{S}_n$,

$$\partial^{F^o,+}_{[\tau]_{\sim F}} = d^F \circ \tau\colon (n] \to Q$$

and $\partial^{F^o}_{[\tau]_{\sim F}}(0) = z$ where z is any element of Q (if Q is empty there is nothing to prove).

The functor $\mathrm{ev}_z \circ (F^o, \phi, \partial^{F^o}, Q) \star (-)$ will be denoted from now on simply as $F^o \star (-)$. For $d^X : X \to Q$ in $\mathrm{Set}_{/Q}$ we put

$$\begin{array}{rcl} F^o \star (d^X) & \xrightarrow{\kappa^F_{(d^X)}} & F(d^X) \\ \big[[\mathrm{id}_{(n]}]_{\sim F}, u\colon d^F \to d^X\big] & \longmapsto & F(u)(g^F). \end{array}$$

We shall show that $\kappa^F : F^o \star (-) \to F$ is a natural isomorphism. To this aim it is enough to show

(a) for every $d^X \colon X \to Q$ and $x \in F(X, d^X)$ there is $u \colon d^F \to d^X$ in $\mathrm{Set}_{/Q}$ such that $x = F(u)(g^F)$;
(b) for $u, v \colon d^F \to d^X$ we have, $F(u)(g^F) = F(v)(g^F)$ if and only if there is $\sigma \in \mathcal{G}^F$ such that $u = v \circ \sigma$;
(c) κ^F is natural.

We establish (a) and (b). Then, as F is a functor, (c) will be obvious.

Ad (a). It is enough to show (a) for elements $x \in F\big((n], d^F\big)$. As F weakly preserves pullbacks and $F(1) = 1$, F weakly preserves binary products. We have a binary product

$$\begin{array}{ccccc} (n] & \xleftarrow{\pi_1} & (n] \times_Q (n] & \xrightarrow{\pi_2} & (n] \\ & {\scriptstyle d^F}\searrow & \downarrow {\scriptstyle d^\pi} & \swarrow{\scriptstyle d^F} & \\ & & Q & & \end{array}$$

of d^F with itself in $\mathrm{Set}_{/Q}$ and hence a weak product

$$F(d^F) \xleftarrow{F(\pi_1)} F(d^\pi) \xrightarrow{F(\pi_2)} F(d^F)$$

in Set. Hence there is $p \in F(d^\pi)$ such that

$$F(\pi_1)(p) = g^F, \quad F(\pi_2)(p) = x.$$

Since F is superfinitary and by assumption on d^F, there are a morphism $f \colon d^F \to d^\pi$ and $y \in F(d^F)$ such that $F(f)(y) = p$. Thus $g^F = F(\pi_1)(p) = F(\pi_1 \circ f)(y)$, and hence $\pi_1 \circ f \colon d^F \to d^F$ is epi and then iso. Putting $u = \pi_2 \circ f \circ (\pi_1 \circ f)^{-1} \colon d^F \to d^F$, we have

$$x = F(\pi_2)(p) = F(\pi_2 \circ f)(y) = F(\pi_2 \circ f \circ (\pi_1 \circ f)^{-1})(g^F) = F(u)(g^F)$$

as needed.

Ad (b). If for some $\sigma \in \mathcal{G}^F$ we have $u = v \circ \sigma$, then

$$F(u)(g^F) = F(v \circ \sigma)(g^F) = F(v)\big(F(\sigma)(g^F)\big) = F(v)(g^F).$$

Now assume that $F(u)(g^F) = F(v)(g^F)$. We form a pullback in $\mathrm{Set}_{/Q}$

$$\begin{array}{ccc} (P, d^P) & \xrightarrow{\bar{v}} & \big((n], d^F\big) \\ {\scriptstyle \bar{u}}\downarrow & & \downarrow{\scriptstyle u} \\ \big((n], d^F\big) & \xrightarrow[v]{} & (X, d^X). \end{array}$$

As F weakly preserves pullbacks, there is $p \in F(d^P)$ such that $F(\bar{u})(p) = g^F = F(\bar{v})(p)$. Using (a) we get $f \colon \big((n], d^F\big) \to (P, d^P)$ such that $F(f)(g^F) = p$. Thus $\bar{u} \circ f, \bar{v} \circ f \in \mathcal{G}^F$. We have

$$u = u \circ (\bar{v} \circ f) \circ (\bar{v} \circ f)^{-1} = v \circ (\bar{u} \circ f) \circ (\bar{v} \circ f)^{-1}$$

and $(\bar{u} \circ f) \circ (\bar{v} \circ f)^{-1} \in \mathcal{G}^F$.

Now suppose that F is a composition of functors

$$\mathrm{Set}_{/Q} \xrightarrow{B \star (-)} \mathrm{Set}_{/Q} \xrightarrow{\mathrm{ev}_z} \mathrm{Set}$$

for a symmetric signature $(B, \beta, \partial^B, Q)$ and $z \in Q$. Since $F(1_Q) = 1$, (B, β) has just one orbit. We can assume that $\partial^B_b(0) = z$ for $b \in B$. Thus by a slight abuse we shall identify $B \star (-)$ with F. Let

$$\begin{array}{ccc} (P, d^P) & \xrightarrow{\pi_Y} & (Y, d^Y) \\ {\scriptstyle \pi_X}\downarrow & & \downarrow{\scriptstyle g} \\ (X, d^X) & \xrightarrow[f]{} & (Z, d^Z) \end{array}$$

be a pullback in $\mathrm{Set}_{/Q}$. We need to show that

$$\begin{array}{ccc} B \star P & \xrightarrow{B\star\pi_Y} & B \star Y \\ {\scriptstyle B\star\pi_X}\downarrow & & \downarrow{\scriptstyle B\star g} \\ B \star X & \xrightarrow[B\star f]{} & B \star Z \end{array}$$

is a weak pullback. Fix $b \in B$ and let $n = |b|$. Suppose for

$$\begin{array}{ccc} (n] & \xrightarrow{u} & X \\ & {\scriptstyle \partial^{B,+}_b}\searrow \quad \swarrow{\scriptstyle d^X} & \\ & Q & \end{array} \qquad\qquad \begin{array}{ccc} (n] & \xrightarrow{v} & Y \\ & {\scriptstyle \partial^{B,+}_b}\searrow \quad \swarrow{\scriptstyle d^Y} & \\ & Q & \end{array}$$

we have

$$[b, f \circ u]_\sim = (B \star f)([b, u]_\sim) = (B \star g)([b, v]_\sim) = [b, g \circ v]_\sim,$$

i.e., there is $\sigma \in \mathrm{S}_n$ such that

$$b \cdot \sigma = b, \quad f \circ u \circ \sigma = g \circ v.$$

Using the property of the above pullback, we get $w\colon (n] \to P$ such that

$$\pi_X \circ w = u \circ \sigma, \qquad \pi_Y \circ w = v.$$

Then

$$B \star \pi_Y([b, w]_\sim) = [b, \pi_Y \circ w]_\sim = [b, v]_\sim$$

and

$$B \star \pi_X([b, w]_\sim) = [b, \pi_X \circ w]_\sim = [b \cdot \sigma, u \circ \sigma]_\sim = [b, u]_\sim. \qquad \square$$

Recall that the functor $F\colon \mathrm{Set}_{/Q} \to \mathrm{Set}_{/Q}$ is *thin* if there is $z \in Q$ such that $F = \mathbf{i}_q \circ \mathrm{ev}_q \circ F$ and $\mathrm{ev}_q \circ F(1) = 1$.

Clearly, every functor $F\colon \mathrm{Set}_{/Q} \to \mathrm{Set}_{/Q}$ is a coproduct of thin functors indexed by the domain of $F(1_Q)$ and every natural transformation between such functors is a coproduct of transformations between thin functors. We have

Lemma 7.8. *Let $F\colon \mathrm{Set}_{/Q} \to \mathrm{Set}_{/Q}$ be a finitary functor. The following are equivalent*

(1) *F is a multivariable analytic functor;*
(2) *F weakly preserves wide pullbacks of power $\leq \aleph_0 + |Q|$;*
(3) *F weakly preserves wide pullbacks.*

PROOF. As every functor $\mathrm{Set}_{/Q} \to \mathrm{Set}_{/Q}$ is a coproduct of thin functors, we can assume that F is thin. Thus we consider F as a functor $\mathrm{Set}_{/Q} \to \mathrm{Set}$ such that $F(1) = 1$.

(3) $\Rightarrow$ (2) is obvious and (1) $\Rightarrow$ (3) can be proved as in Lemma 7.7.

In order to show (2) $\Rightarrow$ (1) we shall show that F is superfinitary and use Lemma 7.7. Suppose on the contrary that for each $n \in \omega$ and $f\colon (n] \to Q$ there is $d^f : X^f \to Q$ and $x^f \in F(d^f)$ such that $x^f \notin \bigcup_{h:f\to d^f} F(h)(f)$. Since F weakly preserves pullbacks of power $\leq \aleph_0 + |Q|$, there is $p \in F\big(\prod_{n\in\omega, f:(n]\to Q}(X^f, d^f)\big)$ such that $F(\pi^f)(p) = x^f$ for $n \in \omega$ and $f\colon (n] \to Q$. Since F is finitary, there are

$$\begin{array}{ccc} (m] & \xrightarrow{\;g\;} & \prod_{n\in\omega, f:(n]\to Q}(X^f, d^f) \\ & {\scriptstyle f_0}\searrow \quad \swarrow & \\ & Q & \end{array}$$

and $y \in F\big((m], f_0\big)$ such that $F(g)(y) = p$. Then $\pi^{f_0} \circ g\colon \big((m], f_0\big) \to (X^{f_0}, d^{f_0})$ and

$$F(\pi^{f_0} \circ g)(y) = F\big(\pi^{f_0}(p)\big) = x^{f_0}$$

contrary to the assumption. □

The following fact is not needed for the proof of Theorem 7.5 but it follows easily from the proofs of the above lemmas and puts some light on the correspondence between orbits of (B, β) and elements of $B \star (1_Q)$.

Scholium 7.9. *Let $F\colon \mathrm{Set}_{/Q} \to \mathrm{Set}_{/Q}$ be a finitary functor that preserves wide pullbacks. Then F is isomorphic to a functor*

$$(B, \beta, \partial^B, Q) \star (-)\colon \mathrm{Set}_{/Q} \to \mathrm{Set}_{/Q}$$

for a symmetric signature $(B, \beta, \partial^B, Q)$ such that (B, β) contains as many orbits as the cardinality of the domain $F(1_Q)$ orbits. In particular, if F is thin iff (B, β) has exactly one orbit.

Remark. In the proof of the above Lemma 7.7 we introduced the notions of a minimal object, a generic element and a generic pair. The notion of a generic element is a variant of a notion introduced in [15], see also [2]. For the later use, we describe below such generic pairs for the functors of form $B \star (-)$, where (B, β) has one orbit.

Lemma 7.10. *Let $(B, \beta, \partial^B, Q)$ be a symmetric signature such that (B, β) has a single orbit, $b \in B$ and let $n = |b|$. Then the positive typing for b, i.e., $\partial^{B,+}_b\colon (n] \to Q$ is the minimal object for functor $B \star (-)\colon \mathrm{Set}_{/Q} \to \mathrm{Set}_{/Q}$ and $[b, 1_{(n]}]_\sim \in B \star \big((n], \partial^{B,+}_b\big)$ is a generic element for $B \star (-)$. More generally, let $d^Y : Y \to Q$, $\vec{y}\colon (n] \to Y$ and $b \in B$. Then $\langle b, \vec{y}\rangle$ represents a generic element $[b, \vec{y}]_\sim \in B \star d^Y$ for $B \star (-)$ if and only if $\vec{y}$ is a bijection and $\partial^{B,+}_b = d^Y \circ \vec{y}$.*

PROOF. Exercise. □

Lemma 7.11. *Let $(f, u)\colon (A, \alpha, \partial^A, O) \to (B, \beta, \partial^B, Q)$ be a morphism of symmetric signatures in $\mathrm{Sig_s}$ over a function $u\colon O \to Q$. Then the natural transformation in $\mathrm{Cat}(\mathrm{Set}_{/Q}, \mathrm{Set}_{/O})$ representing (f, u)*

$$\mathrm{rep_s}(f, u)\colon \mathrm{rep_s}(A) \circ u^* \to u^* \circ \mathrm{rep_s}(B)$$

is weakly cartesian.

PROOF. Thus we have a commuting square

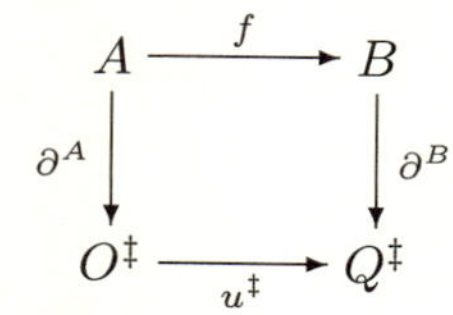

in the category of symmetric sets. Let

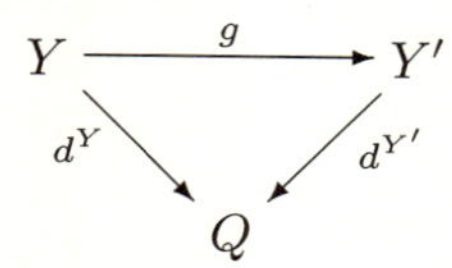

be a morphism in $\mathrm{Set}^{\to}_{/Q}$, i.e., in $\mathrm{Set}_{/Q}$. Then we pullback this morphism along u, and we get a diagram

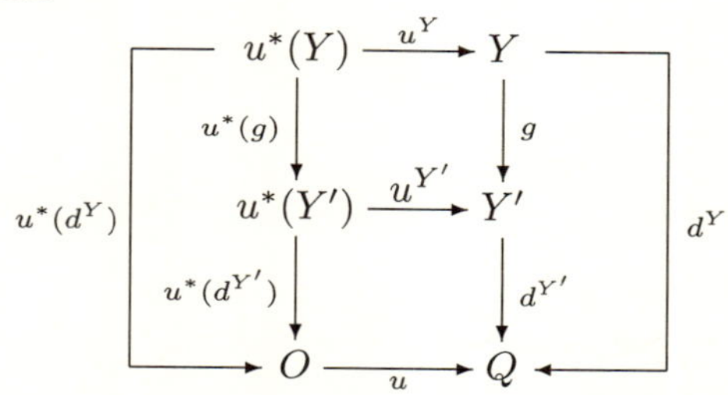

in which three squares are pullbacks. We need to show that the square

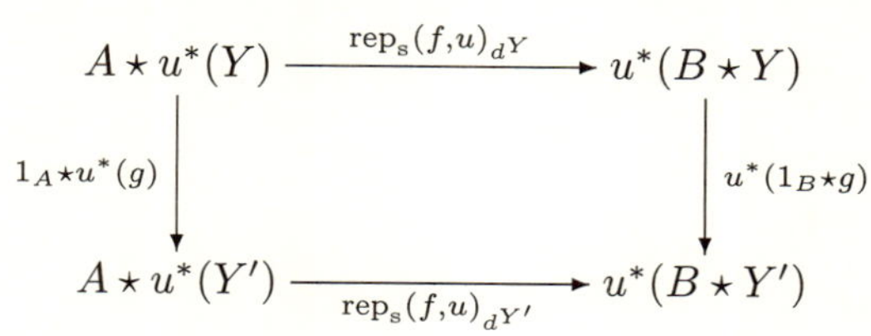

is a weak pullback. So let $[a, \vec{x}]_\sim \in A \star u^*(Y')$ i.e., $a \in A$, $\vec{x} : (|a|] \to u^*(Y)$, so that $\partial_a^{A,+} = u^*(d^Y) \circ \vec{x}$ and let $\langle o, [b, \vec{y}]_\sim\rangle \in u^*(B \star Y)$ i.e., $o \in O$, $b \in B$ and $\vec{y} : (|b|] \to Q$, so that $u(o) = \partial_b^B(0)$ and $\partial_b^{B,+} = d^Y \circ \vec{y}$. Moreover, assume that

$$\begin{aligned}\mathrm{rep}_{\mathrm{s}}(f,u)_{d^{Y'}}([a,\vec{x}]_\sim) &= \langle \partial_a^A(0), [f(a), u^{Y'} \circ \vec{x}]_\sim\rangle \\ &= \langle o, [b, g \circ \vec{y}]_\sim\rangle = u^*(1_B \star y')(\langle o, [b,\vec{y}]_\sim\rangle),\end{aligned}$$

i.e., $\partial_a^A(0) = o$, and there is $\sigma \in S_{|a|}$ such that $f(a) = b \cdot \sigma$ and $u^{Y'} \circ \vec{x} = g \circ \vec{y} \circ \sigma$. Thus, using the upper pullback above, we get a function $\vec{z}$: $(|a|] \to u^*(Y)$ such that

$$u^Y \circ \vec{z} = \vec{y} \circ \sigma, \quad u^*(g) \circ \vec{z} = \vec{x}.$$

Then

$$1_A \star u^*(g)([a, \vec{z}]_\sim) = [a, u^*(g) \circ \vec{z}]_\sim = [a, \vec{x}]_\sim.$$

Moreover

$$o = \partial_a^A(0) \quad \text{and} \quad \langle f(a), u^Y \circ \vec{z}\rangle = \langle b \cdot \sigma, \vec{y} \circ \sigma\rangle \sim \langle b, \vec{y}\rangle,$$

i.e., $\mathrm{rep_s}(f,u)_{d^Y}([a,\vec{z}]_\sim) = \langle o, [b,\vec{y}]_\sim \rangle$. Thus the above square is a weak pullback, as required. □

Lemma 7.12. *Let $(A, \alpha, \partial^A, O)$ and $(B, \beta, \partial^B, O)$ be two symmetric signatures in $(\mathrm{Sig_s})_O$. If $\xi\colon A\star(-) \to B\star(-)$ is a weakly cartesian natural transformation then there is a unique morphism of symmetric signatures $(f, 1_O)\colon (A,\alpha,\partial^A,O) \to (B,\beta,\partial^B,O)$ in $(\mathrm{Sig_s})_O$ such that $\mathrm{rep_s}(f,1_O) = \xi$.*

PROOF. Let $(A, \alpha, \partial^A, O)$, $(B, \beta, \partial^B, O)$ be two symmetric signatures in $(\mathrm{Sig_s})_O$ and let $\xi\colon A\star(-) \to B\star(-)$ be a weakly cartesian natural transformation. By remark after Scholium 7.9, we can assume that both (A,α) and (B,β) have one orbit. Then the existence of ξ as above implies that for some $z \in O$ we have $\partial^A_a(0) = \partial^B_b(0)$ for all $a \in A$ and $b \in B$. Hence we shall not consider these values any more. Fix $d^A\colon (n^A] \to O$ and $x^A \in A\star d^A$ so that (d^A, x^A) is a generic pair for the functor $A\star(-)$, and $d^B\colon (n^B] \to O$ and $x^B \in B\star d^B$ so that (d^B, x^B) is a generic pair for the functor $B\star(-)$. Thus there is a morphism

$$\begin{array}{ccc} (n^B] & \xrightarrow{\;u\;} & (n^A] \\ & {\scriptstyle d^B}\searrow \quad \swarrow{\scriptstyle d^A} & \\ & O & \end{array}$$

in $\mathrm{Set}_{/O}$, such that $B\star u(x^B) = \xi_{d^A}(x^A)$. Since ξ is weakly cartesian, the square

$$\begin{array}{ccc} A\star d^B & \xrightarrow{\;\xi_{d^B}\;} & B\star d^B \\ {\scriptstyle A\star u}\downarrow & & \downarrow{\scriptstyle B\star u} \\ A\star d^A & \xrightarrow[\;\xi_{d^A}\;]{} & B\star d^A \end{array}$$

is a weak pullback and there is $x \in A\star d^B$, such that

$$\xi_{d^B}(x) = x^B \quad \text{and} \quad A\star u(x) = x^A.$$

Since x^A is generic for $A\star(-)$ there is a morphism

$$\begin{array}{ccc} (n^A] & \xrightarrow{\;v\;} & (n^B] \\ & {\scriptstyle d^A}\searrow \quad \swarrow{\scriptstyle d^B} & \\ & O & \end{array}$$

in $\mathrm{Set}_{/O}$, such that $A\star v(x^A) = x$. Thus $A\star(u\circ v)(x^A) = x^A$ and $u\circ v$ is iso. By naturality of ξ

$$B\star v\big(\xi_{d^A}(x^A)\big) = x^B$$

and $B\star(v\circ u)(x^B) = x^B$ and $v\circ u$ is iso. Therefore, both u and v are bijections, and $n^A = n^B = n$. Thus we can assume that u is an identity, $d^A = d^B = d$ and $x^B = \xi_d(x^A)$. Moreover, by Lemma 7.10, we can assume that $x^A = [a, 1_{(n]}]_\sim$ for some $a \in A$, $n = |a|$ and $d = \partial^{A,+}_a$. Furthermore, again by Lemma 7.10, we can assume that $\xi_d(x^A) = x^B = [b,\vec{x}]_\sim$ for some $b \in B$ and a bijection $\vec{x}\colon (n] \to (n]$ such that $\partial^{B,+}_b = \partial^{A,+}_a \circ \vec{x}$.

Thus $\partial^{B,+}_{b\cdot(\vec{x})^{-1}} = \partial^{B,+}_b \circ (\vec{x})^{-1} = \partial^{A,+}_a$. Hence the association $a \mapsto b\cdot(\vec{x})^{-1}$ extends to a morphism of symmetric signatures $(f,1_O)\colon (A,\alpha,\partial^A,O) \to (B,\beta,\partial^B,O)$ such that $f(a\cdot\sigma) = (b\cdot(\vec{x})^{-1})\cdot\sigma$. We shall show that $\mathrm{rep_s}(f,1_O) = \xi$.

First note that

$$\begin{aligned} \mathrm{rep_s}(f, 1_O)_d(x^A) &= \mathrm{rep_s}(f, 1_O)([a, 1_{(n]}]_\sim) = [f(a), 1_{(n]}]_\sim \\ &= [b \cdot (\vec{x})^{-1}, 1_{(n]}]_\sim = [b, \vec{x}]_\sim = x^B = \xi_d(x^A), \end{aligned}$$

i.e., $\mathrm{rep_s}(f, 1_O)$ and ξ agree on x^A. Now let $d^X : X \to O$ and $x \in A{\star}d^X$ be arbitrary. Since x^A is generic we have a morphism $w\colon d \to d^X$ such that $A \star w(x^A) = x$. Now using the naturality of ξ and $\mathrm{rep_s}(f, 1_O)$ on w, i.e., serial commutativity of the diagram

$$\begin{array}{ccc} A \star d & \overset{\xi_d}{\underset{\mathrm{rep_s}(f,1_O)_d}{\rightrightarrows}} & B \star d \\ {\scriptstyle A\star w}\big\downarrow & & \big\downarrow{\scriptstyle B\star w} \\ A \star d^X & \overset{\xi_{d^X}}{\underset{\mathrm{rep_s}(f,1_O)_{d^X}}{\rightrightarrows}} & B \star d^X \end{array}$$

we get that $\mathrm{rep_s}(f, 1_O)_{d^X}(x) = \xi_{d^X}(x)$ and hence $\mathrm{rep_s}(f, 1_O) = \xi$.

If $(g, 1_O)\colon (A, \alpha, \partial^A, O) \to (B, \beta, \partial^B, O)$ is another morphism of symmetric signatures such that $\mathrm{rep_s}(g, 1_O) = \xi$, then in particular

$$[g(a), 1_{(n]}]_\sim = \mathrm{rep_s}(g, 1_O)_d(x^A) = \xi_d(x^A) = \mathrm{rep_s}(f, 1_O)_d(x^A) = [f(a), 1_{(n]}]_\sim.$$

This implies that $g(a) = f(a)$ and, since (A, α) has one orbit, $(g, 1_O) = (f, 1_O)$, as required. □

Proof of Theorem 7.5. From Lemma 7.8 we know that the objects in the essential image of the representations rep_s are finitary functors that weakly preserve wide pullbacks. From Lemma 7.11 we know that morphisms in the essential image are weakly cartesian natural transformations. Let $\xi\colon A{\star}(-) \to B{\star}(-)$ be a morphism in Exp(Set) over $u\colon O \to Q$ which is a weakly cartesian natural transformation. By Proposition 7.4 $\mathrm{rep_s}$ is a prone morphisms of fibrations. Hence ξ can be factored in Exp(Set), in an essentially unique way, via a prone morphism $\mathrm{rep_s}(\mathrm{pr}_{u,B})\colon u^*(B) \star (-) \to B \star (-)$ in $p_{\exp}$ and vertical morphisms $\xi'\colon A \star (-) \to u^*(B) \star (-)$ in the fibre over O, so that $\xi = \mathrm{rep_s}(\mathrm{pr}_{u,B}) \circ \xi'$. From Proposition 4.3 both morphisms $\mathrm{pr}_{u,B\star(-)}$, ξ' are weakly cartesian. As again by Lemma 7.12 rep_s is faithful and full on weakly cartesian arrows in fibres, we obtain that $\mathrm{rep_s}$ is faithful and full on weakly cartesian arrows in the whole fibration. □

7.4. The analytic diagrams vs. analytic functors. In Section 6, we have shown that the concepts of amalgamated signature, polynomial diagram and polynomial functor are equivalent when organized into lax monoidal fibrations. The signatures are the most explicit and the functors are the most abstract among these concepts. The diagrams constitute a useful and important link between them. In Section 7.2, we have described the direct connection between lax monoidal fibrations of symmetric signatures and of analytic functors. We provide here the missing link in this approach, the analytic diagrams. They correspond to analytic functors in much the same way as polynomial diagrams correspond to polynomial functors. In fact, these representing diagrams will constitute a full subcategory of the category of polynomial diagrams in the category of symmetric sets σSet. However the monoidal structure is *not* inherited from $p_{\mathrm{poly},\sigma\mathrm{Set}}\colon \mathcal{PolyDiag}(\sigma\mathrm{Set}) \to \sigma\mathrm{Set}$. In the remaining part of the paper we want to indicate the relevant definitions and the obvious facts leaving a more comprehensive study of the analytic diagrams to another paper.

Recall that the diagonal functor $\delta\colon \mathrm{Set} \to \sigma\mathrm{Set}$ induced by the unique functor $\mathcal{S}_* \to 1$, has both adjoints orb $\dashv \delta \dashv$ fix. This adjunction can be also sliced. For details concerning all such functors that we shall consider in the following see the Appendix.

By a *pseudo-analytic diagram* (*over a set* O) in Set we mean a diagram in σSet

$$\delta(O) \xleftarrow{s} (E,\varepsilon) \xrightarrow{p} (A,\alpha) \xrightarrow{t} \delta(O)$$

such that the fibres of p are finite.

The object O is an object of types of the polynomial (t,p,s). A *morphism of pseudo-analytic diagrams* (*over a function* $u\colon O \to Q$) is a triple (f,g,u), with f and g morphisms in σSet, making the diagram

$$\begin{array}{ccccccc}
\delta(O) & \xleftarrow{s} & (E,\varepsilon) & \xrightarrow{p} & (A,\alpha) & \xrightarrow{t} & \delta(O) \\
\downarrow{\scriptstyle \delta(u)} & & \downarrow{\scriptstyle g} & & \downarrow{\scriptstyle f} & & \downarrow{\scriptstyle \delta(u)} \\
\delta(Q) & \xleftarrow[s']{} & (E',\varepsilon') & \xrightarrow[p']{} & (A',\alpha') & \xrightarrow[t']{} & \delta(Q)
\end{array}$$

commute and the square in the middle is a pullback. Morphisms of pseudo-analytic diagrams compose in the obvious way, by putting one on top of the other.

Let $\mathcal{PAnDiag}$ denotes the category of pseudo-analytic diagrams and morphisms between them. The category of analytic diagrams $\mathcal{AnDiag}$ is the slice $\mathcal{PAnDiag}$ over the pseudo-analytic diagram $\mathcal{R}$

$$\delta(1) \leftarrow (R,\rho) \to \delta(1) \to \delta(1)$$

where $R_n = \{n\} \times (n]$ with action $\rho_n(\langle n,i\rangle,\tau) = \langle n, \tau^{-1}(i)\rangle$ for $\tau \in \mathrm{S}_n$. As $\delta(1)$ is the terminal symmetric set all the morphisms in the above diagram are uniquely determined. Thus an analytic diagram is a pseudo-analytic diagram

$$\begin{array}{ccccccc}
\delta(O) & \longleftarrow & (E,\varepsilon) & \xrightarrow{p} & (A,\alpha) & \longrightarrow & \delta(O) \\
 & & \downarrow & & \downarrow & & \\
 & & (R,\rho) & \longrightarrow & \delta(1) & &
\end{array}$$

so that p is a pullback of $(R,\rho) \to \delta(1)$ along $(A,\alpha) \to \delta(1)$. Thus the fibre of p over $a \in A$ can and will be identified as $\{a\} \times (|a|]$ with the action of $S_{|a|}$ so that $\langle a,i\rangle \cdot \tau = \langle a\cdot\tau, \tau^{-1}(i)\rangle$, for $\tau \in S_{|a|}$. With such an identification if $(f,g,u)\colon (t,p,s) \to (t',p',s')$ is a morphism of analytic diagrams then $g(a,i) = \langle f(a), i\rangle$, for $\langle a,i\rangle \in E$. Hence we shall not specify g in the morphism of analytic diagrams anymore and we shall denote it when necessary as $\bar{f}$.

We have an obvious projection functor

$$p_{\mathrm{and}}\colon \mathcal{AnDiag} \to \mathrm{Set},$$

sending (f,u) to u, which is a lax monoidal fibration. However the tensor in fibres is *not* the one induced by the tensor of those diagrams as if they were polynomial diagrams in $\mathcal{PolyDiag}(\sigma\mathrm{Set}) \to \sigma\mathrm{Set}$. It will be described below in an indirect way.

As in the case of the polynomial diagram fibration, the fibration of analytic diagrams comes equipped with a representation morphism into the exponential fibration $\mathrm{Exp(Set)} \to \mathrm{Set}$. The representation functor

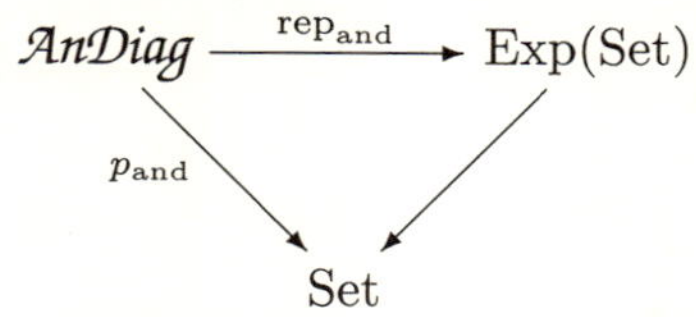

is defined as follows. For an analytic diagram (t, p, s) over O as displayed above, we define a functor $\mathrm{rep}_{\mathrm{and}}(t, p, s)$ from $\mathrm{Set}_{/O}$ to $\mathrm{Set}_{/O}$ as the composition of five functors

$$\mathrm{Set}_{/O} \xrightarrow{\delta_{/O}} \sigma\mathrm{Set}_{/\delta(O)} \xrightarrow{s^*} \sigma\mathrm{Set}_{/(E,\varepsilon)} \xrightarrow{p_*} \sigma\mathrm{Set}_{/(A,\alpha)} \xrightarrow{t_!} \sigma\mathrm{Set}_{/\delta(O)} \xrightarrow{\mathrm{orb}_{/O}} \mathrm{Set}_{/O},$$

i.e., we take a diagonal functor $\delta_{/O}$ to move from the 'set context' to the 'symmetric set context,' then we apply the usual polynomial functor over the category of symmetric sets and we come back again to the set context via $\mathrm{orb}_{/O}$, by taking orbits of whatever we collected on the way.

We describe it with more details in three steps. For an object $(X, d^X\colon X \to O)$ in $\mathrm{Set}_{/O}$ the domain of $p_*s^*(\delta_{/O})(X, d^X)$ in $\sigma\mathrm{Set}_{/(A,\alpha)}$ is the set

$$\{\langle a, \vec{x}\rangle : a \in A, \vec{x}\colon p^{-1}(a) \to X, d^X \circ \vec{x} = \pi_O \circ s_{\lceil p^{-1}(a)}\}$$

where $\pi_O\colon \delta(O) = \omega \times O \to O$ is the obvious projection, equipped with an action ξ acting by conjugation, i.e., for $\sigma \in S_{|a|}$

$$\xi(\langle a, \vec{x}\rangle, \sigma) = \langle a, \vec{x}\rangle \cdot \sigma = \langle a \cdot \sigma, \vec{x}\big((-) \cdot \sigma^{-1}\big) \cdot \sigma\rangle.$$

The typing function sends $\langle a, \vec{x}\rangle$ to $a \in A$. The functor $t_!$ changes only the typing, i.e., $t_!p_*s^*(\delta_{/O})(X, d^X)$ in $\sigma\mathrm{Set}_{/\delta(O)}$ has typing sending $\langle a, \vec{x}\rangle$ to $\langle |a|, t(a)\rangle \in \delta(O)$. Finally, $\mathrm{orb}_{/O}$ associates the orbits to what we've got so far, i.e., the domain of $(\mathrm{orb}_{/O})t_!p_*s^*(\delta_{/O})(X, d^X)$ in $\mathrm{Set}_{/O}$ is the set of equivalence classes $[a, \vec{x}]_\sim$ of pairs $\langle a, \vec{x}\rangle$ as above, divided by the action ξ. The value of this morphism on the class $[a, \vec{x}]_\sim$ is sent to $\pi_O t(a) \in O$.

For a morphism of analytic diagrams $(f, u)\colon (t, p, s) \to (t', p', s')$ over u as defined above, we define a morphism in $\mathrm{Exp(Set)}$ over u, i.e. a natural transformation

$$\mathrm{rep}_{\mathrm{and}}(f, u)\colon (\mathrm{orb}_{/O})t_!p_*s^*(\delta_{/O})u^* \to u^*(\mathrm{orb}_{/Q})t'_!p'_*s'^*(\delta_{/Q})$$

using the diagram

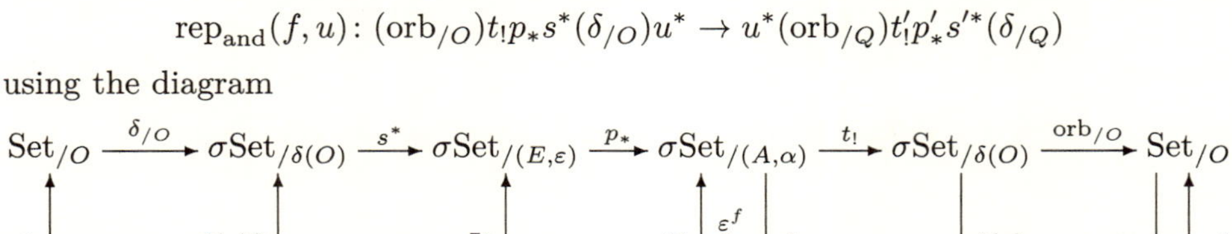

as follows. By adjunction $u_! \dashv u^*$ it is enough to define a natural transformation between these functors

$$u_!(\mathrm{orb}_{/O})t_!p_*s^*(\delta_{/O})u^* \to (\mathrm{orb}_{/Q})t'_!p'_*s'^*(\delta_{/Q})$$

and using the commutativity of some squares (including Beck – Chevalley condition) we define a natural transformation between functors isomorphic to those above

$$(\mathrm{orb}_{/Q})t'_!(\varepsilon^f)_{p'_*s'^*(\delta_{/Q})}\colon (\mathrm{orb}_{/Q})t'_!f_!f^*p'_*s'^*(\delta_{/Q}) \to (\mathrm{orb}_{/Q})t'_!p'_*s'^*(\delta_{/Q}).$$

Tracing this definition back through the adjunctions we find that the so defined natural transformation $\mathrm{rep}_{\mathrm{and}}(f,g,u)$ applied to an object $(Y, d^Y : Y \to Q)$ in $\mathrm{Set}_{/Q}$ sends the element

$$[a, \vec{y}\colon p^{-1}(a) \to u^*(Y)]_\sim$$

in the domain of $(\mathrm{orb}_{/O})t_!p_*s^*(\delta_{/O})u^*(Y, d^Y)$ to the element

$$\langle t(a), [f(a), u^Y \circ \vec{y}]_\sim\rangle$$

in the domain of $u^*(\mathrm{orb}_{/Q})t'_!p'_*s'^*(\delta_{/Q})(Y, d^Y)$, where the notation is as in the following diagram

$$\begin{array}{ccccccc}
Y & \xleftarrow{u^Y} & u^*(Y) & & \xleftarrow{\quad\vec{y}\quad} & & \\
{\scriptstyle d^Y}\downarrow & & \downarrow{\scriptstyle u^*(d^Y)} & & & & \\
Q & \xleftarrow[u]{} & O & \xleftarrow[\pi_O]{} & \omega \times O & \xleftarrow[s\lceil p^{-1}(a)]{} & p^{-1}(a) \subseteq E
\end{array}$$

We leave for the reader the verification that the so defined $\mathrm{rep}_{\mathrm{and}}$ is a lax morphism of fibrations.

In order to show that the essential image of rep_{and} is the fibration of analytic functors, we shall define first a morphism (in fact an equivalence) of fibrations

$$\begin{array}{ccc}
\mathrm{Sig}_s & \xrightarrow{\ \iota_s\ } & \mathcal{AnDiag} \\
 & {\scriptstyle p_s}\searrow \quad \swarrow{\scriptstyle p_{\mathrm{and}}} & \\
 & \mathrm{Set}. &
\end{array}$$

To an object $(A, \alpha, \partial^A, O)$ in Sig_s, ι_s assigns an analytic diagram as follows

$$\delta(O) \xleftarrow{\ s^A\ } (E^A, \bar{\alpha}) \xrightarrow{\ p^A\ } (A, \alpha) \xrightarrow{\ t^A\ } \delta(O)$$

where

$$E^A = \{\langle a, i\rangle : a \in A, i \in (|a|]\}$$

and

$$\bar{\alpha}(\langle a, i\rangle, \sigma) = \langle a, i\rangle \cdot \sigma = \langle a \cdot \sigma, \cdot\sigma^{-1}(i)\rangle.$$

Moreover

$$s^A(a,i) = \langle |a|, \partial^A_a(i)\rangle, \quad p^A(a,i) = a, \quad t^A(a) = \langle |a|, \partial^A_a(0)\rangle$$

for $\langle a, i\rangle \in E^A$, $a \in A$.

If $(f,u)\colon (A, \alpha, \partial^A, O) \to (A', \alpha', \partial^{A'}, Q)$ is a morphism in Sig_s over $u\colon O \to Q$, then ι_s assigns to it the following morphism of diagrams

$$\begin{array}{ccccccc}
\delta(O) & \xleftarrow{\ s^A\ } & (E^A, \bar{\alpha}) & \xrightarrow{\ p^A\ } & (A, \alpha) & \xrightarrow{\ t^A\ } & \delta(O) \\
{\scriptstyle \delta(u)}\downarrow & & {\scriptstyle \bar{f}}\downarrow & & \downarrow{\scriptstyle f} & & \downarrow{\scriptstyle \delta(u)} \\
\delta(Q) & \xleftarrow[s'^{A'}]{} & (E'^{A'}, \overline{\alpha'}) & \xrightarrow[p'^{A'}]{} & (A', \alpha') & \xrightarrow[t'^{A'}]{} & \delta(Q)
\end{array}$$

so that $\bar{f}(a,i) = \langle f(a), i\rangle$ for $\langle f(a), i\rangle \in E^A$, as before.

Proposition 7.13. *The association ι_s is an equivalence of fibrations.*

PROOF. ι_s is in fact an isomorphism if we restrict only to those analytic diagrams that

$$\delta(O) \xleftarrow{s} (E,\varepsilon) \xrightarrow{p} (A,\alpha) \xrightarrow{t} \delta(O)$$

for which (E,ε) is identified with $(A,\alpha) \times_{\delta(O)} (R,\rho)$. Thus it is an equivalence indeed. □

Proposition 7.14. *The following triangle of morphisms of fibrations*

$$\begin{array}{ccc} \mathrm{Sig}_s & \xrightarrow{\iota_s} & \mathcal{A}n\mathcal{D}iag \\ & {\scriptstyle \mathrm{rep}_s}\searrow \quad \swarrow{\scriptstyle \mathrm{rep}_{\mathrm{and}}} & \\ & \mathrm{Exp}(\mathrm{Set}) & \end{array}$$

commutes up to an isomorphism.

PROOF. All the necessary items were defined. We shall check that the values of both functors on objects (that are functors on slices of Set) agree on objects. The remaining details are left for the reader.

Let $A = (A, \alpha, \partial^A, O)$ be a symmetric signature and $X = (X, d^X)$ an object in $\mathrm{Set}_{/O}$. Both values $\mathrm{rep}_s(A)(X)$ and $\mathrm{rep}_{\mathrm{and}} \circ \iota_s(A)(X)$ are functions into the set O whose domains are (= can be identified) with the set

$$\{\langle a, \vec{x}\rangle : a \in A, \vec{x} \colon (|a|] \to X, d^X \circ \vec{x} = \partial_a^{A,+}\}$$

divided by an equivalence relation. In the former case the relation identifies the pair $\langle a, \vec{x}\rangle$ with $\langle a \cdot \sigma, \vec{x} \circ \sigma\rangle$ for $\sigma \in S_{|a|}$. In the second case the relation identifies the elements of the same orbit of the action such that for $\sigma \in S_{|a|}$

$$\langle a, \vec{x}\rangle \cdot \sigma = \langle a \cdot \sigma, \vec{x}\big((-) \cdot \sigma^{-1}\big) \cdot \sigma\rangle.$$

For $i \in (|a|]$, we have

$$\big(\vec{x}\big((-) \cdot \sigma^{-1}\big) \cdot \sigma\big)(i) = \vec{x}(i \cdot \sigma^{-1}) \cdot \sigma = \vec{x}\big(\sigma(i)\big) \cdot \sigma = \vec{x} \circ \sigma(i).$$

The last equality follows from the fact that the action in X (in fact $\delta(X)$) is constant. Thus these equivalence relations are the same and hence the whole morphisms into O sending the equivalence class of $[a, \vec{x}]_\sim$ to $\partial_a^{A,+}(0)$ are the same. □

As ι_s is an equivalence of fibrations by Proposition 7.13 and the essential image of rep_s is (by definition) the fibration of analytic functors, we get from the above Proposition 7.14

Corollary 7.15. *The essential image of the representation functor*

$$\begin{array}{ccc} \mathcal{A}n\mathcal{D}iag & \xrightarrow{\mathrm{rep}_{\mathrm{and}}} & \mathrm{Exp}(\mathrm{Set}) \\ & {\scriptstyle p_{\mathrm{and}}}\searrow \quad \swarrow & \\ & \mathrm{Set} & \end{array}$$

is the fibration of analytic functors.

7.5. Comparing the polynomial and the analytic approaches. In Section 6 we have shown that the lax monoidal fibrations of amalgamated signatures, polynomial diagrams and polynomial functors are equivalent. In the previous Section 7.4 we have introduced the notion of an analytic diagram and we have shown that the lax monoidal fibrations of symmetric signatures, analytic diagrams, and analytic functors are equivalent. Thus in each case, we have three different ways of presenting essentially the same notion. Below we compare these notion at all three levels, i.e., we shall define the missing functors $K_{\rm sig}$, $K_{\rm diag}$ and natural transformations Φ, Ψ in the following diagram

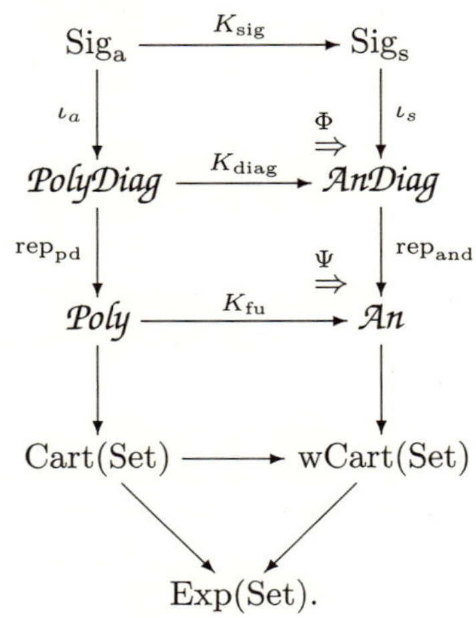

All the arrows are morphisms of lax monoidal fibrations and of bifibrations over Set. The four named vertical arrows are equivalence of lax monoidal fibrations. The five unnamed arrows are inclusions. The three named horizontal arrows are morphisms comparing signatures, diagrams, and functors, respectively.

So we begin by describing the functor

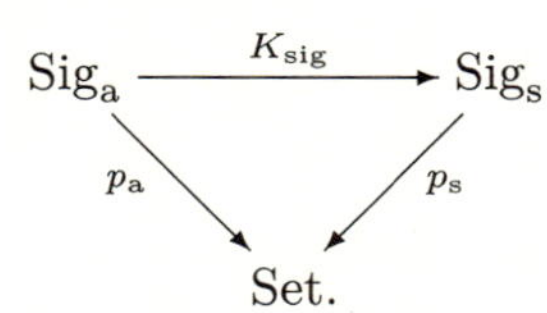

The morphism $(f,\sigma,u)\colon (A,\partial^A\colon A\to O^\dagger,Q)\to(B,\partial^B,Q)$ in $\mathrm{Sig_a}$ over u is sent to a morphism

$$(s(f,\sigma),u)\colon (s(A),\alpha,\partial^{s(A)}\colon s(A)\to O^\ddagger,O)\to(s(B),\beta,\partial^{s(B)},Q)$$

so that

$$s(A)=\{\langle a,\tau\rangle : n\in\omega, a\in A_n, \tau\in \mathrm{S}_n\}$$

and, for $\langle a,\tau\rangle\in s(A)$

$$\partial^{s(A)}(a,\tau)=\partial^A_a\circ\tau : [|a|]\to O,\quad s(f,\sigma)(a,\tau)=\langle f(a),\sigma_a^{-1}\circ\tau\rangle.$$

We have, for$\langle a,\tau\rangle\in s(A)$,

$$\begin{aligned}\partial^{s(B)}\circ s(f,\sigma)(a,\tau)&=\partial^{s(B)}(f(a),\sigma_a^{-1}\circ\tau)\\&=\partial^B_{f(b)}\circ\sigma_a^{-1}\circ\tau=u\circ\partial^A_a\circ\tau=u^\ddagger\circ\partial^A(a,\tau),\end{aligned}$$

i.e., the square

$$\begin{array}{ccc} (s(A),\alpha) & \xrightarrow{s(f,\sigma)} & (s(B),\beta) \\ {\scriptstyle \partial^{s(A)}}\downarrow & & \downarrow{\scriptstyle \partial^{s(B)}} \\ O^{\ddagger} & \xrightarrow[u^{\ddagger}]{} & Q^{\ddagger} \end{array}$$

commutes and K_{sig} is a well-defined functor. We note for the record

Proposition 7.16. *The functor K_{sig} is full, faithful, and its essential image consists of those symmetric signatures that have free actions.*

PROOF. Simple check. □

Next we define the functor

$$\begin{array}{ccc} \mathcal{PolyDiag} & \xrightarrow{K_{\mathrm{diag}}} & \mathcal{AnDiag} \\ & {\scriptstyle p_{\mathrm{pd}}}\searrow \quad \swarrow{\scriptstyle p_{\mathrm{and}}} & \\ & \mathrm{Set}. & \end{array}$$

To a polynomial diagram

$$O \xleftarrow{s} E \xrightarrow{p} A \xrightarrow{t} O$$

K_{diag} associates an analytic diagram

$$\delta(O) \xleftarrow{\tilde{s}} (\widetilde{E},\tilde{\varepsilon}) \xrightarrow{\tilde{p}} (\tilde{A},\tilde{\alpha}) \xrightarrow{\tilde{t}} \delta(O)$$

so that

$$\tilde{A} = \{\langle a,h\rangle : a \in A, h : (|a|] \xrightarrow{\cong} p^{-1}(a)\}$$
$$\widetilde{E} = \{\langle a,h,i\rangle : a \in A, h : (|a|] \xrightarrow{\cong} p^{-1}(a), i \in (|a|]\}$$

where $|a|$ is the number of elements of $p^{-1}(a)$. For $\langle a,h\rangle \in \tilde{A}$ we have

$$\tilde{\alpha}(\langle a,h\rangle,\tau) = \langle a, h\circ\tau\rangle, \qquad \tilde{t}(a,h) = \langle |a|, t(a)\rangle$$

and for $\langle a,h,i\rangle \in \widetilde{E}$,

$$\tilde{\varepsilon}(\langle a,h,i\rangle,\tau) = \langle a, h\circ\tau, \tau^{-1}(i)\rangle, \quad \tilde{p}(a,h,i) = \langle a,h\rangle, \quad \tilde{s}(a,h,i) = \langle |a|, s\circ h(i)\rangle.$$

To a morphism of polynomial diagrams

$$\begin{array}{ccccccc} O & \xleftarrow{s} & E & \xrightarrow{p} & A & \xrightarrow{t} & O \\ {\scriptstyle u}\downarrow & & {\scriptstyle g}\downarrow & & \downarrow{\scriptstyle f} & & \downarrow{\scriptstyle u} \\ Q & \xleftarrow[s']{} & E' & \xrightarrow[p']{} & A' & \xrightarrow[t']{} & Q \end{array}$$

K_{diag} associates a morphism of analytic diagrams

$$\begin{array}{ccccccc} \delta(O) & \xleftarrow{\tilde{s}} & (\widetilde{E},\tilde{\varepsilon}) & \xrightarrow{\tilde{p}} & (\tilde{A},\tilde{\alpha}) & \xrightarrow{\tilde{t}} & \delta(O) \\ {\scriptstyle \delta(u)}\downarrow & & {\scriptstyle \tilde{g}}\downarrow & & \downarrow{\scriptstyle \tilde{f}} & & \downarrow{\scriptstyle \delta(u)} \\ \delta(O) & \xleftarrow[\widetilde{s'}]{} & (\widetilde{E'},\widetilde{\varepsilon'}) & \xrightarrow[\widetilde{p'}]{} & (\widetilde{A'},\widetilde{\alpha'}) & \xrightarrow[\widetilde{t'}]{} & \delta(Q) \end{array}$$

so that, for $\langle a,h\rangle \in \tilde{A}$ and $\langle a,h,i\rangle \in \widetilde{E}$

$$\tilde{f}(a,h) = \langle f(a), g_{\lceil p^{-1}(a)} \circ h\rangle, \quad \tilde{g}(a,h,i) = \langle f(a), g_{\lceil p^{-1}(a)} \circ h, i\rangle.$$

This ends the definition of K_{diag}.

Now we shall define the natural isomorphism Φ. Fix (A, ∂^A, O) in $\mathrm{Sig}_{\mathrm{a}}$. We need to define a morphism

$$\Phi_{(A,\partial^A,O)} \colon K_{\mathrm{diag}} \circ \iota_{\mathrm{a}}(A, \partial^A, O) \to \iota_{\mathrm{s}} \circ K_{\mathrm{sig}}(A, \partial^A, O)$$

in $\mathcal{A}n\mathcal{D}iag$ in the fibre over O, i.e., a morphism of analytic diagrams

$$\begin{array}{ccccccccc}
K_{\mathrm{diag}} \circ \iota_{\mathrm{a}}(A,\partial^A,O) = & \delta(O) & \xleftarrow{\widetilde{s^A}} & (\widetilde{E^A}, \tilde{\varepsilon}) & \xrightarrow{\widetilde{p^A}} & (\tilde{A}, \tilde{\alpha}) & \xrightarrow{\widetilde{t^A}} & \delta(O) \\
\Big\downarrow {\scriptstyle \Phi_{(A,\partial^A,O)}} & \Big\downarrow {\scriptstyle \delta(1_O)} & & \Big\downarrow {\scriptstyle \Phi_1} & & \Big\downarrow {\scriptstyle \Phi_0} & & \Big\downarrow {\scriptstyle \delta(1_O)} \\
\iota_{\mathrm{s}} \circ K_{\mathrm{sig}}(A,\partial^A,O) = & \delta(O) & \xleftarrow[s^{s(A)}]{} & (E^{s(A)}, \bar{\alpha}) & \xrightarrow[p^{s(A)}]{} & (s(A), \alpha) & \xrightarrow[t^{s(A)}]{} & \delta(O)
\end{array}$$

An element of $\widetilde{A}$ is a pair $\langle a,h\rangle$ such that $a \in A$ and $h\colon (|a|] \to p^{-1}(a) = \{\langle a,i\rangle : i \in (|a|]\}$ is a bijection. An element of $s(A)$ is a pair $\langle a,\tau\rangle$ so that $a \in A$ and $\tau \in S_{|a|}$. Thus we can put

$$\Phi_0(a,h) = \langle a, \pi_2 \circ h\rangle$$

with $\pi_2(a,i) = i$, for $i \in (|a|]$. Clearly, Φ_0 is a bijection. An element of $\widetilde{E^A}$ is a triple $\langle a,h,i\rangle$ so that $\langle a,h\rangle \in \widetilde{A}$ and $i \in (|a|]$. An element of $E^{s(A)}$ is a triple $\langle a,\tau,i\rangle$ such that $\langle a,\tau\rangle \in s(A)$ and $i \in (|a|]$. Clearly, we put $\Phi_1(a,h) = \langle a, \pi_2 \circ h, i\rangle$, and Φ_1 is a bijection as well.

We have

Proposition 7.17. *The transformation $\Phi\colon K_{\mathrm{diag}} \circ \iota_{\mathrm{a}} \to \iota_{\mathrm{s}} \circ K_{\mathrm{sig}}$ defined above is a natural isomorphism.*

PROOF. We have already seen that the components of Φ are isomorphisms. The verification that Φ is natural is left for the reader. □

Finally, we define the natural isomorphism Ψ. We fix a polynomial diagram

$$O \xleftarrow{s} E \xrightarrow{p} A \xrightarrow{t} O$$

in $\mathcal{P}oly\mathcal{D}iag_O$. We need to define a morphism

$$\Psi_{(t,p,s)} \colon K_{\mathrm{fu}} \circ \mathrm{rep}_{\mathrm{pd}}(t,p,s) \to \mathrm{rep}_{\mathrm{and}} \circ K_{\mathrm{diag}}(t,p,s)$$

in $\mathcal{A}n$ in the fibre over O, i.e., a natural transformation

$$\Psi_{(t,p,s)} \colon t_! p_* s^* \to (orb_{/O}) \tilde{t}_! \tilde{p}_* \tilde{s}^* (\delta_{/O})$$

between endofunctors on $\mathrm{Set}_{/O}$. To this end, we need to define its components

$$(\Psi_{(t,p,s)})_{(X,d^X)} \colon t_! p_* s^*(X, d^X) \to (\mathrm{orb}_{/O}) \tilde{t}_! \tilde{p}_* \tilde{s}^* (\delta_{/O})(X, d^X)$$

for any object $(X, d^X \colon X \to O)$ in $\mathrm{Set}_{/O}$. Fix (X, d^X) in $\mathrm{Set}_{/O}$. An element of $t_! p_* s^*(X, d^X)$ is a pair $\langle a, \vec{x}\rangle$ so that $a \in A$, $\vec{x}\colon (|a|] \to X$ is a function such that $d^X \circ \vec{x} = s_{\lceil p^{-1}(a)}$. An element $[\langle a,h,\vec{x}\rangle]_\sim$ of $(\mathrm{orb}_{/O}) \tilde{t}_! \tilde{p}_* \tilde{s}^* (\delta_{/O})(X, d^X)$ is an equivalence class of triples $\langle a,h,\vec{x}\rangle$ so that $\langle a,\vec{x}\rangle$ is an element of $t_! p_* s^*(X, d^X)$ and $h\colon (|a|] \to p^{-1}(a)$ is a bijection. The action of $S_{|a|}$ is defined so that $\langle a,h,\vec{x}\rangle \cdot \tau =$

$\langle a, h \circ \tau, \vec{x}\rangle$. Thus any two triples $\langle a, h, \vec{x}\rangle$ and $\langle a', h', \vec{x'}\rangle$ are identified if and only if $a = a'$ and $\vec{x} = \vec{x'}$. Thus we can put

$$(\Psi_{(t,p,s)})_{(X,d^X)}(a, \vec{x}) = \{\langle a, h, \vec{x}\rangle : h\colon (|a|] \xrightarrow{\cong} p^{-1}(a)\},$$

i.e., we associate to $\langle a, \vec{x}\rangle$ the equivalence class of all triples whose second component is a bijection of $(|a|]$ and $p^{-1}(a)$. As these sets have, by definition, the same number of elements, Ψ is well defined.

We have

Proposition 7.18. *The transformation* $\Psi\colon K_{\mathrm{fu}} \circ \mathrm{rep}_{\mathrm{pd}} \to \mathrm{rep}_{\mathrm{and}} \circ K_{\mathrm{diag}}$ *defined above is a natural isomorphism.*

PROOF. As before, the verification that Ψ is natural is left for the reader. From the considerations above it should be clear that for any polynomial diagram (t, p, s) and (X, d^X) in $\mathrm{Set}_{/O}$ $(\Psi_{(t,p,s)})_{(X,d^X)}$ is a bijection. So $(\Psi_{(t,p,s)})$ is a natural isomorphism and hence Ψ is an isomorphism, as well. $\square$

In that way we have completed the description of the diagram of categories, functors and natural transformations from the beginning of this subsection. Thus, we know that the whole diagram commutes (at least up to an equivalence), moreover the named horizontal functors are equivalences of categories. As we note, Proposition 7.16, K_{sig} is full and faithful. Therefore both K_{diag} and K_{fu} are full and faithful, as well. Hence using the characterizations of fibrations of polynomial and analytic functors, Proposition 6.12, Theorem 7.5, we obtain a statement, a bit surprising at first sight.

Corollary 7.19. *Any weakly cartesian natural transformation between polynomial functors is cartesian.*

Appendix

We spell below in detail some well known definitions of various adjoint functors between slices of Set and σSet.

First, recall that the unique functor $\mathcal{S}_* \to 1$ induces by composition the diagonal functor δ that has both adjoints

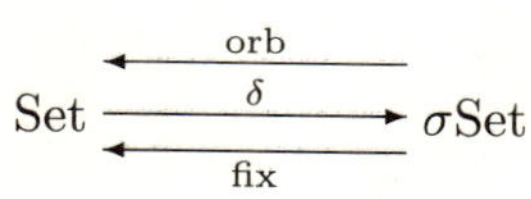

$\mathrm{orb} \dashv \delta \dashv \mathrm{fix}$. The functor δ sends set X to $\omega \times X$, i.e., to ω copies of X, with nth copy of X equipped with a trivial action of S_n. The functor orb sends a symmetric set (A, α) to the set of its orbits with respect to all actions $A_{/\alpha}$. The functor fix sends a symmetric set (A, α) to the product over ω of the sets of fix points with respect to each action S_n, i.e.

$$\mathrm{fix}(A, \alpha) = \prod_{n \in \omega} \mathrm{fix}_n(A_n, \alpha_n)$$

where $\mathrm{fix}_n(A_n, \alpha_n) = \{a \in A_n : a \cdot \sigma = a \text{ for } \sigma \in \mathrm{S}_n\}$. The functor fix is not used directly but its existence shows that δ preserves colimits.

For any set O the above adjunction can be sliced, i.e., we have functors

$$\mathrm{Set}_{/O} \underset{\mathrm{fix}_{/O}}{\overset{\mathrm{orb}_{/O}}{\overset{\longleftarrow}{\underset{\longleftarrow}{\xrightarrow{\delta_{/O}}}}}} \sigma\mathrm{Set}_{/\delta(O)}$$

such that $\mathrm{orb}_{/O} \dashv \delta_{/O} \dashv \mathrm{fix}_{/O}$. For $d^X : X \to O$ in $\mathrm{Set}_{/O}$

$$\delta_{/O}(X, d^X) = \delta(d^X) \colon \delta(X) \to \delta(O)$$

is the sliced diagonal functor. Moreover, for $d^Y : (Y, \zeta) \to \delta(O)$ in $\sigma\mathrm{Set}_{/\delta(O)}$, we have

$$\mathrm{orb}_{/O}\big((Y,\zeta), d^Y\big) \colon \mathrm{orb}(Y,\zeta) \to O$$

so that $orb_{/O}\big((Y,\zeta), d^Y\big)([y]_\sim) = o$ if $d^Y(y) = \langle n, o\rangle$ for some $n \in \omega$. Finally, for $\big((Y,\zeta), d^Y\big)$ as above

$$\mathrm{fix}_{/O}\big((Y,\zeta), d^Y\big) \colon \coprod_{o\in O} \prod_{n\in\omega} \mathrm{fix}_{n,o}\big((Y,\zeta), d^Y\big) \to O$$

is the obvious projection function, where

$$\mathrm{fix}_{n,o}\big((Y,\zeta), d^Y\big) = \{y \in Y_n : d^Y(y) = o,\ y \cdot \sigma = y \text{ for} \sigma \in \mathrm{S}_n\}.$$

Next we recall the pullback functor and its adjoint in the category of symmetric sets σSet. Any morphism $p\colon (E,\varepsilon) \to (A,\alpha)$ in σSet induces three functors

$$\sigma\mathrm{Set}_{/(E,\varepsilon)} \underset{p_*}{\overset{p_!}{\overset{\longrightarrow}{\underset{\longrightarrow}{\xleftarrow{p^*}}}}} \sigma\mathrm{Set}_{/(A,\alpha)}$$

so that $p_! \dashv p^* \dashv p_*$. p^* is defined by pulling back along p, $p_!$ is defined by composing with p. The actions are defined in the obvious way. For $d^X : (X,\xi) \to (E,\varepsilon)$ the universe of $p_*\big((X,\xi), d^X\big)$ is

$$\{\langle a, \vec{x}\colon E_a \to (X,\xi)\rangle : a \in A, d^X \circ \vec{x} = i_a\}$$

where E_a and i_a are defined from the following pullback in σSet

$$\begin{array}{ccc} (E,\varepsilon) & \xrightarrow{\ p\ } & (A,\alpha) \\ {\scriptstyle i_a}\uparrow & & \uparrow{\scriptstyle \bar{a}} \\ E_a & \longrightarrow & S_{|a|} \end{array}$$

and $\bar{a}\colon S_{|a|} \to (A,\alpha)$ is the morphism from the symmetric set $S_{|a|}$ (with action of $S_{|a|}$ on the right) sending identity on $(|a|]$ to a.

The action in $p_*\big((X,\xi), d^X\big)$ is defined by conjugation

$$\langle a, \vec{x}\rangle \cdot \sigma = \langle a \cdot \sigma, \vec{x}\big((-) \cdot \sigma^{-1}\big) \cdot \sigma\rangle$$

and the typing sends $\langle a, \vec{x}\rangle$ to a.

Thus we can draw a diagram of categories and functors

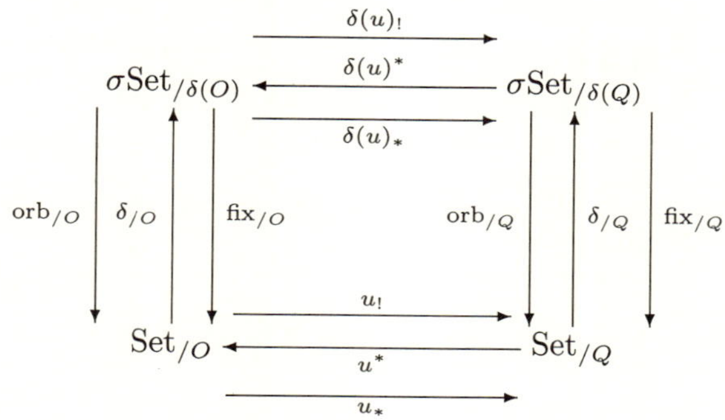

in which we have a natural isomorphism of functors

$$\delta(u)^* \circ \delta_{/Q} \cong \delta_{/O} \circ u^*$$

and hence of their left

$$\mathrm{orb}_{/Q} \circ \delta(u)_! \cong u_! \circ \mathrm{orb}_{/O}$$

and right adjoints

$$\mathrm{fix}_{/Q} \circ \delta(u)_! \cong u_! \circ \mathrm{fix}_{/O}.$$

References

1. J. Adámek, *Free algebras and automata realizations in the language of categories*, Comment. Math. Univ. Carolinae **15** (1974), 589–602.
2. J. Adámek and J. Velebil, *Analytic functors and weak pullbacks*, Theory Appl. Categ. **21** (2008), no. 11, 191–209.
3. J. C. Baez and J. Dolan, *Higher-dimensional algebra.* III: *n-categories and the algebra of opetopes*, Adv. Math. **135** (1998), no. 2, 145–206.
4. H.-J. Baues, M. Jibladze, and A. Tonks, *Cohomology of monoids in monoidal categories*, Operads: Proceedings of Renaissance Conferences (Hartford, CT – Luminy, 1995) (J.-L. Loday, J. D. Stasheff, and A. A. Voronov, eds.), Contemp. Math., vol. 202, Amer. Math. Soc., Providence, RI, 1997, pp. 137–165.
5. D. Bourn, *La tour de fibrations exactes des n-catégories*, Cahiers Topologie Géom. Différentielle Catég. **25** (1984), no. 4, 327–351.
6. A. Burroni, *T-catégories (catégories dans un triple)*, Cahiers Topologie Géom. Différentielle **12** (1971), 215–321.
7. A. Carboni and P. Johnstone, *Connected limits, familial representability and Artin glueing*, Math. Structures Comput. Sci. **5** (1995), no. 4, 441–459.
8. N. Gambino and J. Kock, *Polynomial functors and polynomial monads*, available at `arXiv:0906.4931`.
9. J. Giraud, *Méthode de la descente*, Bull. Soc. Math. France Mém. **2** (1964).
10. M.-F. Gouzou and R. Grunig, *Fibrations relatives*, Séminaire de théorie des catégories dirigé par J. Bénabou, 1976.
11. V. Harnik, M. Makkai, and M. Zawadowski, *Computads and multitopic sets*, available at `arXiv:0811.3215`.
12. C. Hermida, M. Makkai, and J. Power, *On weak higher-dimensional categories.* I. 1, J. Pure Appl. Algebra **154** (2000), no. 1-3, 221–246; 2, **157** (2001), no. 2-3, 247–277; 3, **166** (2002), no. 1-2, 83–104.
13. B. Jacobs, *Categorical logic and type theory*, Stud. Logic Found. Math., vol. 141, North-Holland, Amsterdam, 1999.
14. A. Joyal, *Une théorie combinatoire des séries formelles*, Adv. in Math. **42** (1981), no. 1, 1–82.
15. ———, *Foncteurs analytiques et espèces de structures*, Combinatoire énumérative (Montréal, QC – Québec, QC, 1985) (G. Labelle and P. Leroux, eds.), Lecture Notes in Math., vol. 1234, Springer, Berlin, 1986, pp. 126–159.

16. G. M. Kelly, *A unified treatment of transfinite constructions for free algebras, free monoids, colimits, associated sheaves, and so on*, Bull. Austral. Math. Soc. **22** (1980), no. 1, 1–83.
17. J. Kock, *Polynomial functors and trees*, Int. Math. Res. Not. IMRN **2011**, no. 3, 609–673.
18. J. Kock, A. Joyal, M. Batanin, and J.-F. Mascari, *Polynomial functors and opetopes*, Adv. Math. **224** (2010), no. 6, 2690–2737.
19. T. Leinster, *Higher operads, higher categories*, London Math. Soc. Lecture Note Ser., vol. 298, Cambridge Univ. Press, Cambridge, 2004.
20. S. Mac Lane, *Categories for the working mathematician*, 2nd ed., Grad. Texts in Math., vol. 5, Springer-Verlag, New York, 1998.
21. N. Saavedra Rivano, *Catégories tannakiennes*, Lecture Notes in Math., vol. 265, Springer, Berlin, 1972.
22. M. Shulman, *Framed bicategories and monoidal fibrations*, Theory Appl. Categ. **20** (2008), no. 18, 650–738.
23. T. Streicher, *Fibred categories à la Jean Bénabou*, available at `http://www.mathematik.tu-darmstadt.de/~streicher/FIBR/FibLec.pdf`.
24. S. Szawiel, *Poprawność definicji zbior"ow opetopowych w rozwl"oknieniach monoidalnych*, Master's Thesis, Warsaw University, 2009.
25. S. Szawiel and M. Zawadowski, *Comparing algebraic defintions of opetopic sets*, in preparation.
26. V. Trnková, *Some properties of set functors*, Comment. Math. Univ. Carolinae **10** (1969), 323–352.
27. M. Zawadowski, *On ordered face structures and many-to-one computads*, available at `arXiv:0708.2659`.

Instytut Matematyki, Uniwersytet Warszawski, ul. S. Banacha 2, 00-913 Warszawa, Poland

E-mail address: `zawado@mimuw.edu.pl`

Titles in This Series

53 **Bradd Hart, Thomas G. Kucera, Anand Pillay, Philip J. Scott, and Robert A. G. Seely, Editors,** Models, logics, and higher-dimensional categories, 2011

52 **Dmitry Jakobson, Stéphane Nonnemacher, and Iosif Polterovich, Editors,** Spectrum and dynamics, 2010

51 **Javad Mashreghi, Thomas Ransford, and Kristian Seip, Editors,** Hilbert spaces of analytic functions, 2010

50 **P. Robert Kotiuga, Editor,** A celebration of the mathematical legacy of Raoul Bott, 2010

49 **Miguel Abreu, François Lalonde, and Leonid Polterovich, Editors,** New perspectives and challenges in symplectic field theory, 2009

48 **David Avis, David Bremner, and Antoine Deza, Editors,** Polyhedral computation, 2009

47 **John Harnad and Pavel Winternitz, Editors,** Groups and symmetries: From Neolithic Scots to John McKay, 2009

46 **Jean-Marie De Koninck, Andrew Granville, and Florian Luca, Editors,** Anatomy of integers, 2008

45 **Panos M. Pardalos and Pierre Hansen, Editors,** Data mining and mathematical programming, 2008

44 **Stanley Alama, Lia Bronsard, and Peter J. Sternberg, Editors,** Singularities in PDE and the calculus of variationa, 2007

43 **Andrew Granville, Melvyn B. Nathanson, and Jozsef Solymosi, Editors,** Additive combinatorics, 2007

42 **Donald A. Dawson, Vojkan Jakšić, and Boris Vainberg, Editors,** Probability and mathematical physics: a volume in honor of Stanislav Molchanov, 2007

41 **André Bandrauk, Michel C. Delfour, and Claude Le Bris, Editors,** High-dimensional partial differential equations in science and engineering, 2007

40 **V. Apostolov, A. Dancer, N. Hitchin, and M. Wang, Editors,** Perspectives in Riemannian geometry, 2006

39 **P. Winternitz, D. Gomez-Ullate, A. Iserles, D. Levi, P. J. Olver, R. Quispel, and P. Tempesta, Editors,** Group theory and numerical analysis, 2005

38 **Jacques Hurtubise and Eyal Markman, Editors,** Algebraic structures and moduli spaces, 2004

37 **P. Tempesta, P. Winternitz, J. Harnad, W. Miller, Jr., G. Pogosyan, and M. Rodriguez, Editors,** Superintegrability in classical and quantum systems, 2004

36 **Hershy Kisilevsky and Eyal Z. Goren, Editors,** Number theory, 2004

35 **H. E. A. Eddy Campbell and David L. Wehlau, Editors,** Invariant theory in all characteristics, 2004

34 **P. Winternitz, J. Harnad, C. S. Lam, and J. Patera, Editors,** Symmetry in physics, 2004

33 **André D. Bandrauk, Michel C. Delfour, and Claude Le Bris, Editors,** Quantum control: Mathematical and numerical challenges, 2003

32 **Vadim B. Kuznetsov, Editor,** The Kowalevski property, 2002

31 **John Harnad and Alexander Its, Editors,** Isomonodromic deformations and applications in physics, 2002

30 **John McKay and Abdellah Sebbar, Editors,** Proceedings on moonshine and related topics, 2001

29 **Alan Coley, Decio Levi, Robert Milson, Colin Rogers, and Pavel Winternitz, Editors,** Bäcklund and Darboux transformations. The geometry of solitons, 2001

28 **J. C. Taylor, Editor,** Topics in probability and Lie groups: Boundary theory, 2001

27 **I. M. Sigal and C. Sulem, Editors,** Nonlinear dynamics and renormalization group, 2001

26 **J. Harnad, G. Sabidussi, and P. Winternitz, Editors,** Integrable systems: From classical to quantum, 2000

25 **Decio Levi and Orlando Ragnisco, Editors,** SIDE III—Symmetries and integrability of difference equations, 2000

24 **B. Brent Gordon, James D. Lewis, Stefan Müller-Stach, Shuji Saito, and Noriko Yui, Editors,** The arithmetic and geometry of algebraic cycles, 2000

23 **Pierre Hansen and Odile Marcotte, Editors,** Graph colouring and applications, 1999

22 **Jan Felipe van Diejen and Luc Vinet, Editors,** Algebraic methods and q-special functions, 1999

21 **Michel Fortin, Editor,** Plates and shells, 1999

20 **Katie Coughlin, Editor,** Semi-analytic methods for the Navier-Stokes equations, 1999

19 **Rajiv Gupta and Kenneth S. Williams, Editors,** Number theory, 1999

18 **Serge Dubuc and Gilles Deslauriers, Editors,** Spline functions and the theory of wavelets, 1999

17 **Olga Kharlampovich, Editor,** Summer school in group theory in Banff, 1996, 1998

16 **Alain Vincent, Editor,** Numerical methods in fluid mechanics, 1998

15 **François Lalonde, Editor,** Geometry, topology, and dynamics, 1998

14 **John Harnad and Alex Kasman, Editors,** The bispectral problem, 1998

13 **Michel C. Delfour, Editor,** Boundaries, interfaces, and transitions, 1998

12 **Peter C. Greiner, Victor Ivrii, Luis A. Seco, and Catherine Sulem, Editors,** Partial differential equations and their applications, 1997

11 **Luc Vinet, Editor,** Advances in mathematical sciences: CRM's 25 years, 1997

10 **Donald E. Knuth,** Stable marriage and its relation to other combinatorial problems: An introduction to the mathematical analysis of algorithms, 1997

9 **D. Levi, L. Vinet, and P. Winternitz, Editors,** Symmetries and integrability of difference equations, 1996

8 **J. Feldman, R. Froese, and L. M. Rosen, Editors,** Mathematical quantum theory II: Schrödinger operators, 1995

7 **J. Feldman, R. Froese, and L. M. Rosen, Editors,** Mathematical quantum theory I: Field theory and many-body theory, 1994

6 **Guido Mislin, Editor,** The Hilton Symposium 1993: Topics in topology and group theory, 1994

5 **D. A. Dawson, Editor,** Measure-valued processes, stochastic partial differential equations, and interacting systems, 1994

4 **Hershy Kisilevsky and M. Ram Murty, Editors,** Elliptic curves and related topics, 1994

3 **Rémi Vaillancourt and Andrei L. Smirnov, Editors,** Asymptotic methods in mechanics, 1993

2 **Philip D. Loewen,** Optimal control via nonsmooth analysis, 1993

1 **M. Ram Murty, Editor,** Theta functions: From the classical to the modern, 1993